MW00686398

Astronomy in Prehistoric Britain and Ireland

# Astronomy in Prehistoric Britain and Ireland

*Clive Ruggles*

Yale University Press
New Haven and London

*To*

*Sue, Emma, Nick, Alice and Patrick*

for all they have endured along the way



Designed by Kate Gallimore

Set in Baskerville MT and Syntax by Best-set Typesetter Ltd., Hong Kong

Printed in Hong Kong

Library of Congress Cataloging-in-Publication Data

Ruggles, C. L. N. (Clive L. N.)
Astronomy in prehistoric Britain and Ireland / Clive Ruggles.
p. cm.
Includes bibliographical references and index.
ISBN 0-300-07814-5
1. Astronomy, Prehistoric—Great Britain. 2. Astronomy, Prehistoric—Ireland. 3. Megalithic monuments—Great Britain. 4. Megalithic monuments—Ireland. 5. Great Britain—Antiquities. 6. Ireland—Antiquities. I. Title.
GN805.R84 1999
936.1—dc21 98-40922
CIP

# Contents

Preface and acknowledgements vii

Some conventions ix

Access to sites and monuments x

List of boxes xi

Introduction *Prehistoric Astronomy: Tales of Two Cultures* 1

public and academic faces • academic mainstream and fringe • different mainstreams • disciplines and interdisciplines

## PART I PAST DIRECTIONS

1 *Sun, Moon and Stones: Some 'Classic' Astronomical Sites* 12

Newgrange: symbolic orientation on the sun? • Ballochroy: precise solstitial foresights? • Kintraw: a solar observation platform? • Brainport Bay: a calendrical complex? • Le Grand Menhir Brisé: a universal lunar foresight? • Stonehenge: the astronomer's dream? • Classic sites: some lessons

2 *Backsights and Foresights: the Work of Alexander Thom and its Reassessment* 49

Thom's approach: the four levels • Level 1: solar calendar, moon and stars • Level 2: the limbs of the sun and moon • Level 3: high-precision lunar foresights • Level 4: the very highest precision • Concluding remarks

3 *Sitelines and Statistics: an Independent Statistical Study of 300 Western Scottish Sites* 68

An emphasis on rigour • The selection of sites and indications • The analysis and its results • The moon and the stone rows • Lessons learned • Strengths and limitations of the statistical approach • From statistical abstraction to social context

4 *Alignment and Artefact: Reconciling the Archaeological and Astronomical Evidence* 79

Megalithic science? • Astronomy and society: seeking a social context for prehistoric astronomy • Beyond alignments • Orientations revisited

## PART II CURRENT DIRECTIONS

5 *Orientation and Astronomy in Two Groups of Stone Circles* 91

An ideal group • Manifestations of astronomy in recumbent stone circles • Interpreting the Scottish recumbent stone circles • Similar but different? The axial stone circles of Cork and Kerry

6 *Orientation and Astronomy in Two Groups of Short Stone Rows* 102

Deceptively simple? Short stone rows from an archaeological perspective • A fresh start: astronomy and the Cork–Kerry rows • Reinterpreting the Scottish rows • Linearity and linkages: some general conclusions

7 *Astronomy and Sacred Geography: the North Mull Project* 112

From alignment to landscape: a different approach • Monuments, a mountain, and the moon • Why there? The locations of the north Mull rows • Overall conclusions: the sacred geography of Bronze-Age Mull

8 *Astronomy in Context: a Synthesis* 125

Early developments • Astronomy and ancestors: alignments in chambered tombs • In tune with the cosmos: circles, landscape and astronomy in the Late Neolithic • Stonehenge and its landscape • The Bronze Age and Beyond • Overview: continuity and change

PART III FUTURE DIRECTIONS

9 *Wider Issues* 144

The big question: so what? • Casting aside the baggage • Lateral thinking • Themes for the future

10 *Looking Further* 156

Seeking the evidence • Beyond the green and the brown • Drawing together the strands

Appendix *Horizon survey and data reduction techniques* 164

Preparations • Data capture in the field • Data reduction

Reference lists of monuments 172

Tables 200

Notes 224

Bibliographical abbreviations 268

Bibliography 269

Index 277

Figure acknowledgements 286

# Preface and acknowledgements

This book concerns a field of enquiry that has existed on the fringes of mainstream archaeology for many decades. Those working in this area, more often trained as physical scientists than archaeologists or historians, have often found it difficult to communicate with colleagues who focus upon wider issues in British and Irish prehistory. Straddling disciplinary boundaries, prehistoric astronomy is a topic rife with misunderstandings and prejudices, one that 'while less vituperatively debated, still seethes'.[1] This should not be the case. It is a field of enquiry which, as I shall argue, has a great deal of relevance in studies of prehistory today.

One of the frustrations of working in this area is the extent to which pertinent work of excellent quality by non-archaeologists is often ignored completely by archaeologists, and vice versa, while statements which to anyone with a basic knowledge in one of the relevant disciplines are evidently insupportable continue to be cited and recited year after year. This seems no less true as this book goes to press than it was thirty, or indeed ninety, years ago. At the end of the twentieth century archaeoastronomy is a thriving interdiscipline shedding light upon past societies throughout the world, yet even archaeoastronomers from outside Europe are often unfamiliar with recent research concerning prehistoric Britain and Ireland—the very area whose controversies sparked the development of the wider endeavour back in the early 1970s.

If this book has a single aim then it is to bring the wide range of people interested in prehistoric astronomy to a common starting point. One of the biggest challenges in trying to do this is that the subject matter straddles three very different academic disciplines—archaeology, astronomy, and statistics—as well as straying into a number of others. In attempting to address a cross-disciplinary audience, I am aware that I cannot expect every archaeologist or historian to wish to grasp the relevant mathematical details of positional astronomy or the physics of atmospheric refraction and extinction, any more than the average astronomer will wish to become embroiled in the fine details of absolute dating techniques or of different theoretical approaches to archaeological explanation. Yet I would contend that no expert in any one constituent discipline can reasonably expect to approach a full critical appreciation of the evidence presented here without acquiring at least a basic understanding of levels of discourse in the others.

For this reason I have chosen to write a self-sufficient account that attempts to introduce the basic concepts in each field in a minimal way, comprehensible to all yet painless to the specialist, while presenting optional background material in each of the three distinct disciplinary areas in boxes set apart from the main text. These boxes contain self-contained explanations and pointers to further reading for those who want them, but can equally well be passed over, either by those to whom their contents are already familiar or by those to whom they are of no interest.[2] If the approach has succeeded, then the subject matter should by implication be accessible to non-specialists in any of the main academic fields. In addition, I hope that I have managed to cater for the reader who is drawn to the topic simply by a general interest in people's perception and use of the sky in times long before literacy, and is perhaps intrigued or confused by the extraordinary disparity between relevant academic publications over the last thirty years, many of which treat the subject in great detail and complexity, while many others ignore it completely.

In exploring the different disciplinary perspectives that I believe to be the key to this inconsistency, I hope that I will be able introduce the reader to some of the difficulties, as well as to some of the rewards, in tackling a highly interdisciplinary problem within an increasingly specialised and compartmentalised academic system.

This book is a product of fieldwork spanning twenty-five years, undertaken mostly in western Scotland and more recently in south-west Ireland. A measure of how long an initial idea can take to come to fruition is that I first negotiated with Yale University Press about a possible book on prehistoric astronomy as far back as 1981. Since then, as my other work has led me into a variety of academic disciplines and finally into full-time archaeology, my own ideas on the subject have altered almost beyond recognition, something that could also fairly be said of theory and practice relating to British and Irish prehistory. The ways in which my own ideas and work have developed, and the context into which they can now be set, would both have been quite unthinkable two decades ago.

That the book has appeared at all owes a very great deal to the tolerance and support over many years of colleagues and friends who are far too numerous to mention. Special thanks are due to Anthony Aveni, Gordon Barclay, Richard Bradley, Douglas Heggie, Michael Hoskin, Stephen McCluskey, Graham Ritchie, and Nicholas Saunders, each of whom has read several chapters, and each of whom responded amazingly promptly with detailed comments and suggestions when my own timescale seemed so agonisingly slow. Many others offered comments on single chapters, specific passages, boxes, and points of detail: particular thanks are due to Patrick Ashmore, John Barnatt, David Batchelor, Caitlin Buck, Aubrey Burl, Von Del Chamberlain, Andrew Fitzpatrick, Peter Freeman, Julie Gardiner, Mark Gillings, Ann Lynch, Derek McNally, Frank Prendergast, and Marijke van der Veen. I am especially indebted to Paul Walsh for undertaking the thankless task of checking carefully through—and correcting—the Irish monument lists, as well as providing many other useful comments. The responsibility for all remaining shortcomings is of course mine alone.

As regards the photographs, I am especially grateful to Chris Jennings for permission to include several of his marvellous and evocative pictures. Virtually all of the maps and line drawings in this book are the work of Deborah Miles-Williams, whose efforts to keep within deadlines despite many other pressures are very much appreciated. Mike Middleton kindly drew Fig. 5.2.

Chapters five to seven contain a few passages reproduced verbatim, or almost so, from papers by this author that have already appeared in *Archaeoastronomy*, the supplement to *Journal for the History of Astronomy*. Thanks are due to Michael Hoskin and Science History Publications Ltd for their permission to do this.

# Some conventions

Because of the cross-disciplinary nature of this book, a number of conventions need some explanation. The first concerns dates. The archaeologist has many methods for date determination, such as radiocarbon (carbon-14, or $^{14}C$) dating, dendrochronology (tree-ring dating), and luminescence dating, each with its own inherent errors and uncertainties.[1] The radiocarbon technique produces 'radiocarbon dates' that have to be converted to calendar dates using a calibration curve derived using a different dating method. The earlier convention in archaeological publications was to use 'bc', 'ad' and 'bp' (years before 'present', defined as AD 1950) to denote uncalibrated radiocarbon dates and 'BC', 'AD' and 'BP' for calibrated ones, but this has now largely been superseded. The prevailing convention is to use 'BC', 'AD' and (most commonly) 'BP' for uncalibrated radiocarbon dates and 'cal BC', 'cal AD' or 'cal BP' for calibrated ones. In view of the possible confusion between this convention and the expectations of non-archaeologists, this book will reverse that convention, using 'uncal BC' and 'uncal AD' for uncalibrated dates and 'BC' and 'AD' for calibrated ones, thus retaining the meanings commonly associated with the latter terms.

The astronomer uses a numerical scale in which the year before +1 (AD 1) is expressed as 0, and the year before that as −1; thus the astronomer's year 0 is the archaeologist's year 1 BC and the astronomer's year −2800 is the archaeologist's year 2801 BC. In this book BC dates are used exclusively in preference to numeric ones, and explanations and arguments are couched in terms of calendar or calibrated radiocarbon dates, uncalibrated radiocarbon dates only being used to describe raw data. Laboratory references are given alongside uncalibrated dates as is the archaeological convention. Calendar dates are estimated using calibration data published in 1993,[2] although precise estimates of calendar dates are seldom essential to the development of ideas in this book and quoted dates are often estimated merely to the nearest 250 years.

The second convention concerns the terms 'accuracy' and 'precision'. These convey distinct meanings, and the difference between them is important, although the terminology may be confusing. Similar conventions operate within archaeology and civil engineering. 'Precision' is generally taken as referring to the degree of refinement of a measurement, that is the size of the units in which it is quoted (to the nearest half-degree, to the nearest minute of arc, etc.),[3] the implication being that there would be an appropriate level of agreement between several measurements of the same thing.[4] 'Accuracy', on the other hand, refers to how well the measurement of an attribute conforms to its true value.[5] To an engineer, a series of measurements may not be as accurate as their precision would suggest because of the possibility of systematic error in the measuring instrument. For the archaeologist, accuracy can also relate to a testable archaeological idea: for example, although it may be possible to obtain repeatable theodolite measurements of horizon features precise to 1′ or better, the accuracy of the measurements may be rather lower because of the arbitrariness of the theodolite position in relation to relevant archaeological features. The term 'precision' is also used here to refer to the activities of prehistoric people as opposed to our attempts to measure what we find in the archaeological record: so, for instance, we refer to the precision with which a structure was aligned upon an intended target.[6] Thus on the basis of declination measurements precise (in the first sense) to 1′ and considered accurate to within 0°·1, we might conclude that a solstitial alignment was set up to a precision (in the second sense) of around 0°·5.

Similarly, the terms 'altitude', 'elevation' and 'height' can have particular and quite distinct meanings. Here, different and often contradictory conventions are found in common use. In this book the term 'elevation' is used exclusively to mean the height of a location above sea level, while 'altitude' is used to mean the vertical angle between a viewed point and the horizontal plane through the observer. In other words, 'elevation' always refers to a distance and 'altitude' to an angle. The word 'height' is generally avoided in this context except to mean height above local ground level. Thus a row of three standing stones of heights up to 1·5 m may be found at an elevation of 150 m where the horizon altitude in one of the directions of alignment is 4°·5.

Following common usage, the terms 'midsummer' and 'midwinter' are employed to denote the summer (June) and winter (December) solstices respectively: thus 'midsummer sunrise' means, specifically, sunrise on the day of the June solstice.

A further issue concerns site names. In the literature these follow certain conventions—for example in Ireland townland names are generally used—but here and elsewhere individual sites are often known by different names, sometimes several. Every attempt has been made to make clear and consistent choices, and alternative names have been cited wherever possible in the text and/or lists. Site locations are generally given as the name of the relevant county (e.g. Co. Cork; Kincardine).[7] However, in certain areas of greater site density further subdivisions are used (e.g. mid-Argyll; Kintyre) and in the case of larger islands, the island name is generally used (e.g. Islay; Mull).

Grid references on the British or Irish National Grid are normally given in the text when a site is first mentioned. They are given in six- or eight-figure form, depending upon the precision to which the position is quoted, preceded by letters that identify the 100 km grid square, two in the case of a British grid reference and one in the case of an Irish one.[8] It is also possible to express grid references in all-figure form, so that for example the grid reference of the Dervaig North stone row might be expressed as either NM 4390 5202 or $^{1}$4390 $^{7}$5202 and that of the Gurranes stone row as either W 174315 or $^{1}$174 $^{0}$315. This all-figure form is used in many of the line drawings as well as in the data reduction programs mentioned in the Appendix. Information for converting between the two forms

| | | | | | |
|---|---|---|---|---|---|
| **4** | A | B | C | D | E |
| **3** | F | G | H | J | K |
| **2** | L | M | N | O | P |
| **1** | Q | R | S | T | U |
| **0** | V | W | X | Y | Z |
| | **0** | **1** | **2** | **3** | **4** |

| | | | | | | | |
|---|---|---|---|---|---|---|---|
| **12** | | | | | HP | | |
| **11** | | | | HT | HU | | |
| **10** | | | | HY | HZ | | |
| **9** | NA | NB | NC | ND | | | |
| **8** | NF | NG | NH | NJ | NK | | |
| **7** | NL | NM | NN | NO | | | |
| **6** | | NR | NS | NT | NU | | |
| **5** | | NW | NX | NY | NZ | | |
| **4** | | | SC | SD | SE | TA | |
| **3** | | | SH | SJ | SK | TF | TG |
| **2** | | SM | SN | SO | SP | TL | TM |
| **1** | | SR | SS | ST | SU | TQ | TR |
| **0** | SV | SW | SX | SY | SZ | TV | |
| | **0** | **1** | **2** | **3** | **4** | **5** | **6** |

is given in the above diagram for both Irish (left) and British (right) National Grid references. The column number of the grid letter(s) gives the 100 km digit for the easting (first coordinate) and the row number that of the northing (second coordinate).

# Access to sites and monuments

While several of the monuments described in this book are in the care of organisations such as Historic Scotland and English Heritage and are open to the public, many are situated on private land. Although not always a necessity, it is a politeness as well as a service to future visitors to seek the permission of the farmer or landowner before approaching them. It is also in the interests of both the landowner and the visitor for the latter to be shown the exact location and the best route of approach, as anyone will be quick to confirm who has searched long and hard for an overgrown monument, or fought across various obstacles to reach a remote location only to discover that a far easier route exists from a different starting point.

# List of boxes

## Archaeology

1 Megalithic monuments and prehistoric archaeology 4
2 Conspicuous prehistoric monuments in Britain and Ireland 14
3 Stonehenge and its archaeological context 44
4 Society in the fourth, third and second millennia BC: the evidence of the monuments 84
5 People and their environment in the fourth, third and second millennia BC: the wider evidence 86
6 People in the landscape 116
7 Sacred geographies 120
8 Modes of explanation 146

## Astronomy

1 The concept of declination 18
2 Determining the declination of a horizon point 22
3 The annual motions of the sun 24
4 The motions of the moon, 1: lunistices, standstills, and limiting declinations 36
5 Dividing the solar year 54
6 Variations in the longer-term 57
7 The motions of the moon, 2: high-precision complications 60
8 Conceptualising the equinoxes 150

## Statistics

1 Probability and odds 39
2 Probabilities of chance alignments upon astronomical targets 42
3 Probability distributions 50
4 Hypothesis testing 73
5 Megalithic mensuration and geometry 82
6 What is the probability of $n$ orientations falling within $\theta$ degrees? 95
7 Classical and Bayesian approaches 160

# Introduction

# Prehistoric Astronomy

*Tales of Two Cultures*

The separation between the two cultures has been getting deeper under our eyes; there is now precious little communication between them, little but different kinds of incomprehension and dislike.

C. P. Snow, 1956[1]

One of the curious features of megalithic science is the fact that different people can come to widely differing conclusions on the basis of the same body of evidence. In many cases, these differences seem to arise because of [their] widely differing backgrounds.

Douglas C. Heggie, 1981[2]

Nowhere do the 'two cultures' clash at such a head-on trajectory as here, where archaeological science is at its most densely numerate and social archaeology is at its most speculative.

Julian Thomas, 1989[3]

## Public and academic faces

Around the shortest day each year, just after dawn, a shaft of sunlight suddenly penetrates deep into the interior of the five-thousand year old passage grave at Newgrange, Co. Meath (Fig. 0.1). The phenomenon lasts only for a few minutes, and is dependent upon the weather, yet each year hundreds of people clamour to see it at first hand. Only a privileged few actually get to do so. The waiting list is several years long, and the queue can be jumped only by dignitaries such as government ministers. At the winter solstice in 1987, the skies were completely cloud-covered only twenty minutes before sunrise. The Taoiseach, Charles J. Haughey, arrived on site and was immediately ushered inside the tomb. No sooner had he turned to face the famous roof-box, than the skies cleared on the horizon, long enough to allow the legendary shaft of sunlight to illuminate the passage and chamber. This was widely reported in the national press[4] and was generally seen as a good omen for his future political career.[5]

Witnessing the spectacle of sunrise at Stonehenge on the longest day of the year, whether or not in search of similar portents, has tended to be a much less exclusive affair. Less constricted to public view than its Irish counterpart, the classic image of the sun rising behind the Heel Stone (Fig. 0.2) has been in the public eye since at least 1905, when the so-called 'Ancient Order of Druids' (actually founded in 1781) first performed a ceremony at the site.[6] During the First World War, a hut in the military camp at Larkhill, 3 km to the north-east of Stonehenge, was demolished in order to prevent it obscuring the all-important view of the midsummer sunrise.[7] As the years went on, more and more people gathered at Stonehenge each June, until by the early 1960s the interior had to be enclosed within a protective barbed wire barricade within which only police, 'Druids' and press photographers were allowed. By the mid-1970s the seasonal gathering had developed into an annual 'free festival' lasting for several days and attracting crowds several thousand strong.[8]

As these widely known examples confirm, the general idea that there is a connection between ancient stone monuments (see Archaeology Box 1)—or at least the best known ones—and astronomy is firmly engrained in the popular culture. Common belief held that the sun entered Newgrange long before this was demonstrated archaeologically.[9] Folklore has it that the Water Stone long barrow in Somerset dances on midsummer's day when there is a full moon.[10] Popular legend has long attributed healing properties to ancient standing stones in association with the sun.[11] The US National Air and Space Museum in Washington DC has for many years held a display on Stonehenge astronomy, and in 1990 the Royal Mail issued a set of four commemorative stamps 'tracing the story of astronomy from the old star-watchers . . . to the great institutions of today', with Stonehenge occupying pride of place on the 37 pence stamp. An article published in *The Independent on Sunday* in 1991 announced the discovery in Cambridgeshire of 'a sun temple that puts Stonehenge in the shade'.[12] Another, in *The Times Higher Education Supplement* in 1995, reported the suggestion that early structures at Stonehenge itself were built to 'observe terrifying meteor storms plaguing the skies'.[13] On 22 December 1997 the *Irish Times* published two separate articles on winter solstice alignments, at Newgrange and a stone row in Connemara.[14] The list could be extended almost indefinitely.

Yet if this is the public face of prehistoric astronomy, the academic literature appears to present a very different one. To take one example, in *The Oxford Illustrated Prehistory of Europe*, an authoritative general work on European prehistory published

in 1994,[15] astronomy is mentioned in only two contexts: the possibility that very early bone engravings, dating from Upper Palaeolithic times around 31,000 BC, represent a tally of days relating to the phases of the moon, and the possible calendrical significance of Dacian sanctuaries of the first millennium BC.[16] There is not the slightest mention of the possible astronomical significance of Neolithic and Bronze Age monuments of the fourth to second millennia BC in north-west Europe, monuments that include Newgrange and Stonehenge. This is not exceptional: the same is true of numerous other accounts of European prehistory. Yet it is not that there has been any lack of academic interest in prehistoric astronomy during the last few decades—far from it—and it is not that prehistorians are unaware of it. The stark fact is that astronomy is not at the top of most of their agendas.

To the non-archaeologist, this may seem quite remarkable. In the public perception, astronomy is amongst the most important characteristics of Newgrange and Stonehenge. How can there be such a disparity between this and the view of many leading prehistorians?

Fig. 0.1 The midwinter sun illuminates the interior of the passage tomb at Newgrange.

Fig. 0.2 The Heel Stone viewed from the interior of Stonehenge at dawn, in the general direction of midsummer sunrise. Many published photographs (e.g. Hoyle 1972, 23) show the midsummer sun rising exactly over the Heel Stone, but see page 38.

## Academic mainstream and fringe

Part of the answer becomes clear if we examine a period when the difference between academic and public perceptions of prehistory was at its most evident—the late 1960s and early 1970s—and in particular if we consider the effect of the publication in 1965 of Gerald Hawkins's *Stonehenge Decoded*.[17] This book was without question the single most influential publication in terms of moulding today's public perception of Stonehenge astronomy. It generated a wave of publicity in the form of newspaper articles and television documentaries, and achieved immense popularity in the years following its publication. Doubtless one of the reasons for this was its very title, a masterpiece of marketing which carried the clear implication that all the mystery of Stonehenge's purpose had been suddenly and completely solved (despite, in fact, statements to the contrary in the text). The text itself is an easy-to-read step-by-step account in which the reader is carried along in the excitement of discovery.[18]

Also critically important was the timeliness of the book's appearance. On the one hand, it made great play of the fact that the discoveries of various solar and lunar alignments at Stonehenge were made using (or, as it must have seemed to many readers, by) an IBM computer. In the 1960s the computer was not the commonplace object it is today: this was an age when the computer held great wonder and promise, the age that saw the appearance (in the Stanley Kubrick film *2001: A Space Odyssey*) of the archetypal thinking and talking computer 'HAL'.[19] It was also an age in which astronomy generated great public enthusiasm and excitement: the race was on to reach the moon and, to many, the universe seemed about to be opened up by space travel. The vision of a computer solving the enigmas of Britain's best known prehistoric monument and relating them to the phenomena of space certainly held great contemporary appeal.[20]

On the other hand, curiously enough, *Stonehenge Decoded* was also in tune with a very different group of people in the late sixties, the emerging 'alternative movement'. Deriving its momentum from fears about the Cold War and possible use of the atomic bomb, growing concerns about the environment, and disillusionment with the materialistic society, the movement openly questioned the value of scientific and economic progress and, at this early stage in its development, led many people to seek alternatives in pseudo-science. Within this general context, archaeology had also acquired a vigorous and vociferous 'alternative'.

Archaeology was particularly susceptible to an alternative approach. It was a discipline founded upon the enthusiasm of amateurs, yet their status had declined markedly as the practical work of excavation was increasingly controlled and directed by professionals. Long gone were the days of the amateur antiquarian of the nineteenth and early twentieth centuries, who could 'dig' a barrow on a Sunday afternoon (oblivious to the irreparable damage being done in the process) and later write up the results for a book or respectable local journal.[21] Learned journals in the 1960s were publishing the results of professional excavations, undertaken using a range of scientific techniques, with interpretations being based increasingly upon quantitative analyses rather than qualitative arguments.[22] In the process, the past had largely become dead and remote for the layman, accessible only in the sterility of libraries and museums. True, the amateur could still participate in archaeological research and discovery, as a volunteer on an excavation or through other activities such as field walking, but this was merely as a cog in an establishment wheel, feeling increasingly removed from the real frontiers of knowledge. The 'alternative archaeology' responded to this deprivation in a very direct way. It restored to the non-professional the excitement of personal discovery, as well as fulfilling a deeper, but evidently very real, need for direct personal interaction with the past.

The dominant theme was that of 'lost knowledge', the idea being that prehistoric times somehow represented a spiritual golden age when people lived in harmony with their environment, tapping currents of 'earth energy' (a power accessible to our ancient ancestors but now lost to narrow-minded twentieth-century scientific thought) in order to heal, restore fertility, control the weather, levitate, and so on. Ley lines—the most notorious manifestation of alternative archaeology—formed 'a striking network of lines of subtle force across Britain, and elsewhere on spaceship Earth, understood and marked in prehistoric times by men of wisdom and cosmic consciousness'.[23] 'Ley hunting' became a popular activity holding the obvious attraction that years of specialist training were immaterial; indeed, a scientific training could be seen as counter-productive, since an open mind was necessary in order to detect and interact with the earth currents.[24] To professional archaeologists, this escalation of the 'lunatic fringe' was tiresome in the extreme. Few of them could be bothered to waste their time on a topic which was to them clearly spurious, yet their silence (interspersed with occasional bursts of derision) was taken as a sure sign of academic arrogance and narrow-mindedness, and simply fed the flames of the alternative movement. In their isolation, ley-liners raised their study to the status of a fully developed alternative discipline, with its own sub-fields, experts, doctrinal disputes and journals, fighting for its very existence against the academic establishment.[25]

In this battle between the orthodox and the alternative, *Stonehenge Decoded* was enthusiastically drawn in on the latter side. The astronomical sophistication inherent in Hawkins's interpretation of Stonehenge fitted well into the framework of lost knowledge which is an integral part of the alternative picture.[26] Solar and lunar cycles duly began to appear in explanations of the supposed earth forces.[27] Prehistoric astronomy became an integral part of an alternative framework challenging the view of the archaeological establishment, 'revealing a prehistoric science so advanced that its achievements can be evaluated only in the light of our own',[28] and thereby also revealing a considerable irony, since the alternative movement was founded upon a questioning of those very achievements.

Its association with alternative archaeology provides one obvious reason why professional archaeologists tended to be wary of prehistoric astronomy in the years following the publication of *Stonehenge Decoded*, generally preferring to stand well clear than to get embroiled in arguments that might involve the lunatic fringe. But there were also purely academic reasons for their reticence. Hawkins's ideas had been subjected in 1966 to a detailed and damning critique by Richard Atkinson,[29] a leading authority on Stonehenge who had undertaken excavations there in the 1950s and had written a classic book on the

## ARCHAEOLOGY BOX 1

### MEGALITHIC MONUMENTS AND PREHISTORIC ARCHAEOLOGY

Monuments built of large stones, including chambered tombs, stone circles, stone rows and single standing stones, remain in their hundreds as a conspicuous legacy of a period spanning some three thousand years of prehistory in Britain and Ireland. In most of the lowlands of south and east Britain, where stone was not readily available or was not the preferred medium of construction, large burial and ceremonial structures were built instead of earth and timber. These edifices dominate the archaeological record during the fourth, third and second millennia BC, a period spanning the Neolithic and Early to Middle Bronze Age.[1] The principal types of monument are described in Archaeology Box 2. Some examples are shown here in Fig. 0.3.

#### INVENTORIES OF MONUMENTS

Regional inventories of ancient monuments have been published by the Royal Commission on the Historical Monuments of England, RCHME; the Royal Commission on the Ancient and Historical Monuments of Wales, RCAHMW; the Royal Commission on the Ancient and Historical Monuments of Scotland, RCAHMS; and in Ireland by the Stationery Office for the Office of Public Works.[2] Not all regions are covered by such inventories and where they exist they may be of greatly different dates: for example, Argyll is covered by a series of volumes published by RCAHMS between 1971 and 1992, while those for Caithness and Sutherland were published as far back as 1911, and other parts of Scotland have no inventories at all.[3] Modern inventories generally include plans and descriptions together with references to source material such as excavation reports. On the other hand, classifications and descriptions of monuments in older inventories may be misleading.

Useful information was once also held on record cards and large-scale maps by the Archaeological Branches of the British and Irish Ordnance Survey. The latter Branch still exists, but the responsibilities of the former have been transferred to the three Royal Commissions. The information that used to be held on record cards is now held in the form of computer databases and, increasingly, Geographical Information Systems (see Archaeology Box 6).[4] The same is true of the county Sites and Monuments Records (SMRs) held by County Council Planning Departments in Britain and by the Department of the Environment for Northern Ireland. In England, for example, RCHME has integrated the available SMR information into a National Heritage Information System,[5] while English Heritage maintains, and publishes, a complete list of scheduled ancient monuments. In the Republic of Ireland, SMRs have been compiled for all twenty-six counties by the Archaeological Survey, *Dúchas*, the Heritage Service of the Department of Arts, Heritage, Gaeltacht and the Islands.[6] An Archaeology Data Service has been established to foster on-line access to archaeological data in the UK.[7]

Nowadays, the appropriate medium for primary site data is increasingly being seen as the computer database rather than the published book. Recent RCAHMS publications, for example, present distillations of the raw data and thematic discussions rather than detailed descriptions of monuments.[8]

In addition to general inventories, a number of catalogues and descriptive accounts of particular classes of monument have been published in recent years. See Archaeology Box 2 for further details.

#### THE WIDER PICTURE

The primary concern of modern archaeology is not monuments but people. Archaeologists no longer concentrate first and foremost on describing what is in the archaeological record, but in trying to understand various aspects of prehistoric society (see Archaeology Box 4). In recent years there has been a necessary shift of emphasis away from the most conspicuous ancient monuments towards studying the wider activities of people, as reflected in settlement evidence, artefact scatters, field systems, and so on, together with a wide range of environmental data (see Archaeology Box 5). Considerable insights can be gained by looking at the spatial distribution of evidence of human activity in relation to the landscape and how this changed over time (see Archaeology Boxes 6 and 7). A topic of particular interest, and one to which this book is particularly relevant, is the question of what prehistoric people thought and believed (see Archaeology Box 8). Nonetheless, the monuments themselves are still of very great importance within this much wider picture.

a b

c d

e f

g

Fig. 0.3 Prehistoric stone monuments: some British examples.
a. Ruined chambered tomb at Unival, North Uist (UI28 in List 2).
b. Stone circle at Moel ty Uchaf, Gwynedd.
c. Double stone row at Merrivale, Devon.
d. Short stone row at Duachy, Lorn (LN22 in List 2; SSR28 in List 6).
e. Four-poster at Spittal of Glenshee, Perthshire.
f. Single standing stone at Gartacharra, Islay.
g. Standing stone and ring cairn at Strontoiller, Lorn (LN17 in List 2).

site.[30] Atkinson condemned Hawkins's book as 'tendentious, arrogant, slipshod and unconvincing',[31] and concluded his detailed report: 'It is a great pity that Professor Hawkins has allowed his undoubted enthusiasm for his subject to lead him beyond the bounds of logic and accuracy, imposed alike by the obligations of scholarship and by the parameters of the problem he has tried to solve.'[32]

Such criticism of *Stonehenge Decoded* from within the academic mainstream seemed to fall largely upon deaf ears amongst the public at large. The ground swell seemed equally oblivious to another line of argument emanating from the standpoint of cultural history. From this perspective, the idea of Stonehenge as an 'astronomical observatory', eclipse predictor, or Neolithic 'computer'—phrases that became synonymous with Stonehenge in the popular imagination during the 1970s—clearly represented naïve attempts to explain Stonehenge in terms of the cultural artefacts of the times, and so were by their very nature highly questionable: no better, perhaps, than Inigo Jones's attempts in the seventeenth century to portray Stonehenge in the architectural style of the late Renaissance.[33] As early as 1967 Jacquetta Hawkes had issued her famous aphorism 'Every age has the Stonehenge it deserves—or desires'.[34] Why did even serious-minded onlookers apparently pay little heed to this warning?

The answer is mainly that detailed critiques focusing on the scientific and archaeological issues were drowned out by the outpourings—mostly rhetorical—that followed the publication of Hawkins's ideas. Another significant factor was the appearance during the 1970s and into the 1980s of a series of popular books which, capitalising on the public interest in prehistoric astronomy generated by Hawkins, picked up on the theme in order to communicate it to an enthusiastic public.[35] These books tended to be sympathetic to Hawkins,[36] and where they did mention the serious voices raised against his ideas it was often merely to hit back in general terms, rather than to counter the detailed arguments.[37] While hostile to the claims of the fringe, these books often did little to represent the views of mainstream professional archaeologists, and generally had little intersection with the popular books being written at the same time by the archaeologists themselves. On the other hand, the archaeologists' books, rather than giving detailed refutations of the astronomical ideas, tended to do so only on a rhetorical level or else simply ignored them altogether.[38] And so, while the academic debates raged, the public simply became confused by three types of book—mainstream archaeological, serious astronomy-sympathetic, and fringe—saying very different things but usually to be found inextricably mixed up with one another on the archaeology shelves of the local academic bookshop.

Even if this goes some way to explaining how and why public and academic perceptions of prehistoric astronomy began to diverge in the years following *Stonehenge Decoded*, a major question still remains: why haven't the disparities been resolved in more recent years? Advances in academic thought, especially in popular subjects like astronomy and archaeology, are usually quick to perpetrate into non-specialist publications and thence to an eager public.[39] If, after due consideration amongst the professionals, the conclusion has been reached that there really is no genuine astronomical significance to Newgrange and Stonehenge, then why hasn't this news filtered through? On the other hand, if the opposite conclusion has been reached, how can archaeological publications possibly ignore it?

## Different mainstreams

To begin to answer this question we must look more closely at the nature of the academic debate about prehistoric astronomy that both preceded and followed the publication of *Stonehenge Decoded*. It was a debate quite unlike most specialist controversies within academia, in that it brought into direct contact two widely disparate areas of academic endeavour, with fundamentally different sets of principles, views and approaches and which seldom had cause to intersect. Ultimately far more intriguing than its role in conflicts between orthodoxy and the fringe, prehistoric astronomy was set to play a central role in a conflict entirely within that orthodoxy, an interdisciplinary battle that, while it had been smouldering on and off for many years, burst fully into life in 1965 and continued vigorously, and often truculently, for almost twenty years. Like professional archaeology and its fringe, it represented a clash of two cultures, but this time they were the two cultures of C. P. Snow.[40]

The debate arguably had its origins some sixty years earlier, following the work of Sir Norman Lockyer, a distinguished physicist and astronomer and founder editor of the scientific journal *Nature*. Lockyer himself followed in a tradition dating back a further two centuries, for ever since the earliest British antiquarians had roamed the countryside recording ancient remains, certain individuals had become fascinated by the astronomical possibilities of megalithic monuments, and particularly stone circles, and often dedicated large amounts of time and energy to their investigation. Alluring, spectacular, and poorly understood, these sites are found in large numbers in northern and western parts of Britain and in Ireland, as well as in north-west France, and have long provided highly tempting fodder for speculation.[41] Names such as Martin Martin,[42] William Stukeley,[43] John Wood,[44] John Smith,[45], H. du Cleuziou,[46] and A. L. Lewis[47] all have their place in a succession of ideas ranging from Druidical astronomical temples in the eighteenth century through to 'rude astronomical observatories'[48] in the early twentieth, with alignments upon horizon astronomical events forming an increasingly prominent theme.[49]

Lockyer, however, was the first investigator to back up astronomical speculations with accurate measurements, and the first to attempt to put the idea of prehistoric astronomy on some sort of scientific footing. Turning his interests from the temples of ancient Egypt,[50] he visited and investigated dozens of stone monuments from Cornwall to Aberdeenshire. His book *Stonehenge and Other British Stone Monuments Astronomically Considered* sold widely, running to a second edition three years after its first appearance in 1906. Concluding from his researches that there were very many significant alignments upon the sun, moon and stars at a variety of British megalithic monuments, Lockyer could confidently state: 'For my own part I consider that the view that our ancient monuments were built to observe and to mark the rising and setting places of the heavenly bodies is now fully established'.[51] Lockyer's book attracted an eager response from many academics, including astronomers, architects and engineers, but archaeologists' response was generally muted, even though some follow-up work did find its way into the archaeological literature, including two papers in the very first issue of the journal *Antiquity* in 1927.[52] According to Aubrey Burl 'it was the start of an unreasoned antagonism that even today has not entirely vanished.'[53] But what caused this reaction? Certainly, many valid criticisms can be made of Lockyer's work itself,[54] and a bigger problem may have been the association in the 1920s and 1930s of astronomical ideas with an early wave of activity on the archaeological fringe, including ley lines in their first manifestation.[55] A major problem, undoubtedly, was that most astronomical speculation was centred upon free-standing orthostatic monuments (settings of standing stones) and particularly upon stone circles. As opposed, say, to megalithic tombs, these sites were very poor in the sort of evidence traditionally provided by excavation and central to archaeological investigations: stratigraphy, human bone, charcoal, and pottery and other artefacts. As far as most archaeologists at the time were concerned, orthostatic monuments such as stone circles were a troublesome topic best avoided.[56]

Yet archaeologists who did try to interpret these sites were not necessarily unsympathetic to the idea of astronomy *per se*. Writing at the end of the 1930s, Gordon Childe had no trouble in accepting that the solstitial orientations of Stonehenge and Woodhenge were intentional,[57] and a decade later Stuart Piggott was happy to concede that 'some [monuments] may have been planned in relation to a celestial phenomenon such as sunrise at the summer solstice'.[58] In 1962, shortly before the appearance of *Stonehenge Decoded*, Jacquetta Hawkes would write of Stonehenge: 'The whole temple with its many openings upon the sky may well have been used for measuring the rising and setting of moon, planets, stars and constellations, but the orientation of the mighty trilithons must prove that here at Stonehenge, as elsewhere, the supreme deity was the Sun God himself'.[59] What went wrong?

Some vital clues to the fundamental causes of the rift between astronomers and archaeologists that was to follow can be gleaned from Childe's response to an early paper by the engineer Alexander Thom, whose name would feature so prominently later, presented to a mixed audience at the Royal Statistical Society in 1955.[60] A keen amateur astronomer inspired by a paper on the standing stones of Callanish (Isle of Lewis) published in 1912 by Boyle Somerville, one of those inspired in their turn by Lockyer,[61] Thom had already by 1955 visited some 250 megalithic monuments and accumulated a large quantity of survey data, which he had proceeded to analyse statistically. Thom believed he had uncovered evidence of precise geometrical and astronomical knowledge far exceeding anything that had hitherto been suspected. Childe expressed doubts on grounds such as the effect of stone movements, site degradation and destruction, and changes in the landscape since prehistoric times,[62] but undoubtedly the real problem was that he was uneasy with the social implications of Thom's interpretations. Put simply, astronomers and geometers did not fit easily within cultures dependent upon 'primitive subsistence-farming'[63] and whose chieftains 'were warriors armed with daggers and bows'.[64] In addition, Childe's opening remarks are particularly revealing:

> While archaeologists greatly appreciate the immense amount of work, learning and technical skill which Professor Thom has put into his paper, many of them, I must confess, when faced with mathematical symbols which they do not understand have aroused in them severe emotions which are somewhat overbalanced; some of us think that sigmas and symbols of that kind are 'words of power' and any result from their use must have a very high degree of truth. Others are liable to feel that they are deadly weapons which are being used against them to put over a 'fast one', and it is only fair to state that this is the attitude that archaeologists are likely to display at the start.[65]

This is a clear indication of C. P. Snow's cultural divide beginning to open up. The astronomer-engineer and the archaeologist are speaking different languages: Childe does not understand, and refuses to try to understand, the language of mathematics, while Thom is equally ignorant of the wider framework of evidence and reasoning that had led archaeologists to a social model quite different from his own. It is scarcely surprising, then, that while Thom's ideas began to gain favour amongst a small group of astronomers and others during the 1950s and early 1960s, his work in particular, and the idea of prehistoric astronomical practices in general, continued to generate little interest amongst professional archaeologists.

Indifference was not, however, an option in 1965. Following the intense wave of public interest unleashed by *Stonehenge Decoded*, Glyn Daniel, then editor of *Antiquity*, invited two colleagues to comment on Hawkins's ideas: the archaeologist Richard Atkinson, and a leading astronomer, Fred Hoyle. Atkinson, as we have already noted, produced a swingeing critique, but if Daniel had expected something similar from Hoyle, what he received and was obliged to publish was very different. Hoyle actually produced an extension of Hawkins's ideas, and it was expressed in mathematical terms that were undoubtedly completely baffling to the great majority of *Antiquity*'s regular readers.[66] This led to a situation in which leading astronomers and leading archaeologists were taking diametrically opposite views. Some major reputations were at stake, yet somebody had to be wrong.

Remarkably, little clarification was achieved during the years that followed. Indeed, as the debate developed and further key figures from various disciplines became involved, it became increasingly complex and fraught. A significant development was that, although the argument had been triggered by the Stonehenge controversy, its focus rapidly switched to the work of Thom. Since his retirement in 1961 Thom had devoted his main efforts to his studies of megalithic monuments, and during the 1960s articles by him had appeared in a bewildering mixture of academic journals.[67] The appearance in 1967 of his first book *Megalithic Sites in Britain*[68] was also timely. Thom's conclusions carried far more revolutionary implications than those of Hawkins: according to Thom 'megalithic man' laid out configurations of standing stones all over Britain[69] using precisely-defined units of measurement and particular geometrical constructions, and carried out meticulous observations of the sun, moon and stars. This widespread fascination with astronomy and geometry was unsuspected by

a

Fig. 0.4 The standing stone at Beacharr, Kintyre, one of Alexander Thom's 'megalithic lunar observatories'.

a. The standing stone viewed from the north-east.
b. The view from the south-east, showing the standing stone in the distance and the mountains of Jura in the background.
c. Moonset over the Paps of Jura, after Thom 1971, fig. 6.3.

b

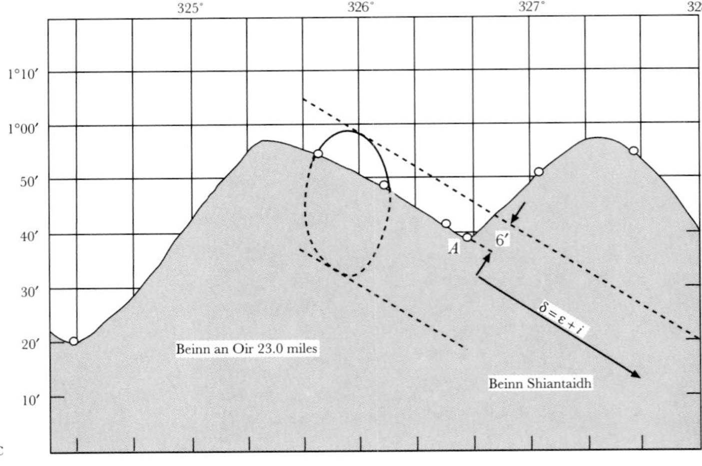


c

archaeologists on the basis of the other evidence available to them; yet Thom's conclusions were not formed on the basis of a single site like Stonehenge, but backed up by widespread data, gleaned from several hundred monuments, and supported by rigorous statistical analysis. The social implications of Thom's conclusions posed a direct challenge to the ideas generally accepted by archaeologists (see chapter four), yet the mathematical evidence was largely unfathomable. The distance between the archaeologists' understanding and that of the astronomers, engineers and statisticians who had become interested in this field, was greater than ever.

Despite this, some rapprochement was achieved at this time. Atkinson, who had been so scathing about Hawkins's work, was impressed by the quality of Thom's fieldwork and analysis and this led him to be sympathetic and open-minded towards Thom's ideas: 'It seldom happens that a single book, by an author who makes no claim to be an archaeologist, compels archaeologists themselves to re-examine their assumptions about a whole section of their past. This one does . . .'.[70] And in 1969, Glyn Daniel invited Thom to join him to study the large stone rows at Carnac in Brittany, where he subsequently worked for several years.[71]

Nonetheless, no sooner had paths to reconciliation begun to be identified when the stakes were raised. Thom's second book, *Megalithic Lunar Observatories*, appeared in 1971.[72] Here, Thom claimed that prehistoric people painstakingly observed the complex variations in the horizon rising and setting positions of the moon, standing stones being erected to mark spots where its long-term movements could be recorded against distant natural horizon features (Fig. 0.4). 'Megalithic man', it seemed, had developed an intimate knowledge of the moon's motions not to be rivalled until the seventeenth and eighteenth centuries.

Opinions became more deeply divided than ever. Statistician David Kendall, reviewing *Megalithic Lunar Observatories* for *Antiquity*, described it as 'a remarkable book by a remarkable man' that was 'compulsory reading for all archaeologists'.[73] Astronomer Douglas Heggie, also writing in *Antiquity*, concluded that 'Thom's evidence that megalithic man observed the moon is so strong that it may be accepted without hesitation.'[74] And Thom and his collaborators began to find a regular outlet for his publications in the recently founded *Journal for the History of Astronomy*.[75] Most archaeologists, on the other hand, were openly sceptical, and Thom's description of his 'megalithic man' as 'a competent engineer . . . [who] had an extensive knowledge of practical geometry' and studied the movements of the sun and moon in great detail[76] did nothing to diminish their strong suspicions that Thom was somehow, despite the mathematical façade, falling into the trap identified earlier by Hawkes[77] and simply seeing his own reflection in the past.

The nature of the gulf between the two sides was summed up well by Richard Atkinson, writing in the *Journal for the History of Astronomy* in 1975. He pointed out that many prehistorians, like himself, had been trained only in the humanities, and thus lacked the numeracy required to understand Thom's arguments in detail. On the other hand, non-archaeologists failed to appreciate 'how disturbing to archaeologists are the implications of Thom's work, because they do not fit the conceptual model of the prehistory of Europe which has been current during the whole of the present century, and even now is only beginning to crumble at the edges.'[78] Andrew Fleming, writing in *Nature* later the same year, also summed up the cautious attitude of most prehistorians of the time: 'The suggestion that Neolithic and Bronze Age peoples of north-west Europe may have been brilliant astronomers and geometers has been met with stupefaction by prehistorians, who faced the recent spate of papers on the subject in total disarray. Most have not felt competent to judge the statistical and astronomical significance of the work . . .' yet 'Until some reconciliation can be achieved between the sites as ceremonial monuments and as complicated solutions to astronomical puzzles, it will take more than clever statistical arguments to convince prehistorians that more than a handful of the present claims can be justified.'[79] As this remark clearly illustrates, ten years on from *Stonehenge Decoded* each side was still largely arguing past the other, starting from different tenets, addressing different aspects of the evidence, using different methods, and—hardly surprisingly—coming to totally different conclusions. There was a fundamental lack of communication between those with backgrounds in the numerate disciplines such as astronomy or engineering, who tended to provide the astronomical deductions based upon their own survey work, and those with backgrounds in archaeology and the social sciences who tried to place them in the context of prehistoric society. In ten years of argument, the academic community had come little nearer any sort of consensus.

## Disciplines and interdisciplines

Only in the mid-1970s did the situation begin to improve noticeably. While the gulf between the two sides in the prehistoric astronomy debate was as great as ever, a few signs were appearing that the demarcation was no longer a straightforward one along disciplinary lines. On the one hand, astronomical lines of enquiry were being pursued actively and seriously by a number of archaeologists. Michael O'Kelly, who had first witnessed the midwinter sunrise phenomenon at Newgrange in 1969,[80] asked surveyor Jon Patrick to confirm the details.[81] John Barber had complemented Thom's data by surveying stone circles in south-west Ireland.[82] In Scotland, Euan MacKie had attempted archaeological tests of Thom's astronomical hypotheses.[83] Although most archaeologists remained wary of Thom's ideas, a few of them were prepared to accept that the alignment of monuments upon the sun and moon had been demonstrated convincingly by Thom and that 'it can no longer be denied that a great deal of time and effort, both physical and intellectual, was expended at that time on observational astronomy'.[84] According to Colin Renfrew, writing in 1973, 'These findings . . . have created a great furore among British archaeologists, some of whom are reluctant to believe that the barbarian inhabitants of prehistoric Britain were capable of such ingenuity. But there are good ethnographic parallels . . .'.[85] Atkinson went even further:

> It is hardly surprising . . . that many prehistorians either ignore the implications of Thom's work, because they do not understand them, or resist them because it is more comfortable to do so. I have myself gone through the latter process; but I have come to the conclusion that to reject Thom's thesis because it does not conform to the model of prehistory on which I was brought up involves also the acceptance of improbabilities of an even higher order. I am prepared, in other words, to believe that my model of European prehistory is wrong, rather than that the results presented by Thom are due to nothing but chance.[86]

At the same time, some astronomers had begun to provide more cautious assessments of what, following Thom, they had come to know as 'megalithic astronomy'.[87] Heggie, writing in 1972, had in fact expressed a number of reservations about Thom's claims for lunar alignments of very high precision.[88] Sensing controversy in the air, astronomers flocked to a session on 'megalithic astronomy: fact or speculation?' at the International Astronomical Union in Grenoble in 1976. The session 'drew an audience so large that some were unable even to find standing room in the lecture hall' but concluded that 'even the broadest questions in megalithic astronomy remain unsettled'.[89]

The controversy over megalithic astronomy served as the catalyst for some wider developments during the early 1970s. In December 1972 the Royal Society and the British Academy agreed to co-sponsor a symposium entitled 'The place of astronomy in the ancient world', which brought together Hawkins, Thom, MacKie, and Atkinson amongst others and which also broadened the discussion to literate as well as preliterate societies.[90] In a review paper published in the journal *Current Anthropology* in 1973, Elizabeth Baity set the work of Hawkins and Thom in a world context, and considered evidence for astronomical practice not just from the archaeological record but from historical documents, written texts, iconography, and from myth and ritual persisting to the present day.[91] This wider approach was inherent in investigations of pre-Columbian and indigenous American astronomy that took off during the 1970s,[92] largely independently of the arguments going on back in Britain and Ireland. The term 'archaeoastronomy' was coined to describe this wider field of activity,[93] and by the end of the 1970s it had become fully established as a new 'interdiscipline' with two publications devoted to archaeoastronomy having appeared.[94] As one archaeologist has aptly described it, 'Archaeoastronomy, the study of the astronomy practiced in ancient times, is a hybrid science, part archaeology, part astronomy (often using formidable statistics), which requires from its practitioners the skills of both. It is a rare scholar who is fluent in the two disciplines . . .'.[95]

This difficulty partly accounts for the continuing polarisation of attitudes with regard to megalithic astronomy during the late 1970s, a polarisation which is perhaps revealed most starkly in the contrast between the words of astronomer Edwin Krupp, who could claim in 1977 that 'Now, in an era of wild, unsubstantiated claims about ancient astronauts by Erich von Däniken and his fellow travelers, their pseudoscientific misconceptions of these earlier peoples are in part dispelled by the reliable, scientific findings of archaeoastronomy'[96] and those of archaeologist Glyn Daniel, whose opinion in 1980 was that

> Many people, no doubt bored by the prosaic account of megaliths to be got from archaeological research, jumped on the Hawkins–Thom bandwagon, accepting the builders of megaliths not only as experts in Pythagorean geometry and possessors of accurate units of mensuration but also as skilled astronomers who studied eclipses, the movements of the moon and the positions of the stars. To me this is a kind of refined academic version of astronaut archaeology. The archaeoastronomy buffs, although they very properly eschew wise men from outer space, very improperly insist on the presence in ancient Europe of

> wise men with an apparently religious passion for astronomy.[97]

In fact, one or two archaeologists were prepared to consider Thom's ideas and to work actively in an attempt to integrate them with other more conventional archaeological evidence. For example Aubrey Burl, a specialist on stone circles, gave serious consideration to Thom's ideas in a series of publications from 1970 onwards.[98] The greatest notoriety was achieved by Euan MacKie, whose book *Science and Society in Prehistoric Britain*, which appeared in 1977, proposed that prehistoric societies included Mayan-style astronomer-priests.[99] Stuart Piggott, reviewing *Science and Society* for *Antiquity*, expressed the views of the majority of archaeologists when he declared that '[MacKie's] treatment of Thom's hypotheses is that of the passionate conviction of a disciple rather than that of a disinterested critic of a number of propositions which have not received by any means universal acceptance from either astronomers and mathematicians, or from archaeologists' and summed up his feelings in the dictum 'Great Thom is cast, and all else must follow'.[100]

The real turning point came at the end of the 1970s when the disciplinary lines began to be breached in earnest. Astronomers, mathematicians and surveyors began to look fully and critically at the ideas of other astronomers, mathematicians and surveyors.[101] A conference held in Newcastle in 1979 which brought together archaeologists, astronomers and others to debate the issues of prehistoric astronomy, saw many speakers presenting critiques of others in their own discipline, and though the exchanges were often fiery, helped to engender a spirit of interdisciplinary understanding and constructive criticism.[102] The year 1981 saw the appearance of Douglas Heggie's *Megalithic Science*, a book by an astronomer which considered the evidence for megalithic astronomy (as well as mensuration and geometry) comprehensively and in detail.[103] And later in the same year the first international symposium on archaeoastronomy took place in Oxford. This not only provided another landmark in the interaction between astronomers and archaeologists but also, significantly, exposed both sides in the megalithic astronomy debate to the views and experience of American archaeoastronomers, who had progressed a good way further towards the integration of astronomy and anthropology.[104]

It was well into the 1980s before the heat of the debate finally died down. Rhetorical outbursts were gradually replaced by careful critiques, collaboration gradually took the place of antagonism, and real attempts began to be made to understand the different disciplinary approaches and to integrate them. It had taken more than twenty years of debate to arrive at a viable starting point for more serious investigations of astronomical practice within the mainstream of north-west European prehistory.

Part One of this book takes a detailed, retrospective look at the arguments and counter-arguments presented between 1965 and 1985, together with the evidence upon which they were based, and focuses upon the lessons—both interpretative and methodological—that can be learned. The first chapter introduces basic concepts and issues, by way of a number of key individual sites. Background astronomical and archaeological material is introduced here in boxes for non-specialist readers. This is followed in chapter two by a detailed examination of Alexander Thom's data and interpretations, and in chapter three by a description of a large independent investigation that attempted to remedy a number of the methodological shortfalls identified by Thom's critics. Finally, chapter four concentrates on the social context of prehistoric astronomy, addressing such issues as its nature and purpose, who practised it, and its place within the wider social picture that archaeologists strive to understand.

Much has happened in recent years. Important new data have been obtained, especially from groups of monuments in western and north-eastern Scotland and south-western Ireland. New investigations have been undertaken in which prehistoric astronomy has been considered in its wider archaeological context through integrated research programmes. The second part of the book describes these recent developments and uses them, together with wider developments in prehistoric archaeology as a whole, to begin to build up a new, and hopefully more reliable, picture of the nature and meaning of astronomical practice at certain times and places in prehistoric Britain and Ireland.

But it will not have escaped the reader's notice that this leaves open the fundamental question of why astronomy is still, apparently, not at the top of most prehistorians' agendas, despite much recent activity in this area. A communication gap between astronomers and archaeologists most evidently still exists. In 1993 the Royal Astronomical Society set up a special Working Group to advise English Heritage on the astronomical aspects of the displays at the proposed new Stonehenge Heritage Centre. Yet in an interview that took place shortly after the report had been received by English Heritage, its chief archaeologist, in the context of admitting that 'all the collated [archaeological] research still did not explain the meaning and purpose of Stonehenge', was reported as saying dismissively: 'But at least we now have all the archaeological facts to go along with the astronomers, the Druids, the Flat Earthers and all the rest'.[105]

As recently as 1994, a new book on early astronomy could still turn first to Stonehenge,[106] while archaeologist Christopher Chippindale, in the revised 1994 edition of his *Stonehenge Complete*, could dismiss Stonehenge astronomy as 'no longer a current affair'.[107] Admittedly, in the same book Chippindale sympathetically remarks that

> Neither the particular schemes of Stonehenge astronomy nor the larger vision of megalithic science have found favour with archaeologists. I am not sure they were ever given a really fair consideration, or that they could have been. Although some astronomers tried to grasp the archaeology, and some archaeologists the astronomy, practically no one had or has sufficient grasp of *both* subjects fairly to explore them together. And the statistical issues, far beyond the mathematics of most archaeologists, taxed the methods of first-rate statisticians.[108]

Yet he also pointedly distinguishes 'basic facts and archaeological understandings' from the work of 'astronomers and earth-mystery researchers'.[109]

Then again, the year 1996 saw the appearance of a large book by a leading historian of science devoted to astronomical interpretations of Stonehenge and other Neolithic and Bronze Age monuments in Wessex.[110] It was followed in 1997 by a book dealing with interpretations of prehistoric monuments over the last 500 years,[111] treating astronomical ideas in a historical context, alongside ley hunting and earth-mystery research, as a 'non-intellectual notion of the past'.[112]

The continuing communications gap is due in part to the fact that much of the work described in this book has been reported in specialist archaeoastronomical publications rather than in the mainstream archaeological literature more usually consulted by prehistorians. But there are deeper reasons. These are explored in the final two chapters, which examine recent developments in a wider theoretical context, address a number of more general and fundamental issues, and attempt to set broader agendas for the future. A number of particular questions that remain unanswered are also identified here, and possible research strategies are suggested for tackling them.

Discussing astronomical alignments at Stonehenge, Newgrange and elsewhere, archaeologist Richard Bradley commented in 1984: 'It is customary to question the reality of such evidence, arguing that a knowledge of astronomy goes beyond the likely capacity of the contemporary population. This view is largely a measure of our ignorance, and betrays an ethnocentric attitude to the question.'[113] In other words, by dismissing prehistoric astronomy altogether we may be just as guilty of projecting our own prejudices into the past as by recreating it in our own image. This book results from a will to do neither.

# Part I PAST DIRECTIONS

# 1

# Sun, Moon and Stones

*Some 'Classic' Astronomical Sites*

From a country home of mine near Florence I plainly observed the Sun's arrival at, and departure from, the summer solstice, while one evening at the time of its setting it vanished behind the top of a rock on the mountains of Pietrapana, about 60 miles away, leaving uncovered a small streak of filament of itself towards the north, whose breadth was not the hundredth part of its diameter. And the following evening, at the similar setting, it showed another such part of it, but noticeably smaller, a necessary argument that it had begun to recede from the tropic.

Galileo Galilei, 1632[1]

Now when it is recalled that 24 hours before and 24 hours after the actual solstice the Sun's declination is only about $0'{\cdot}2$ less than its maximum, it seems wellnigh impossible to develop any observing technique which will differentiate the actual day of the solstice itself. If any arrangement will make this possible it is that at Ballochroy.

Alexander Thom, 1954[2]

It must be remembered that Lockyer's observations [of midsummer sunrise at Stonehenge] were made with instruments of the highest precision, whereas the instruments used by the original builders were confined to their own naked eyes and, at the most, a number of straight sticks cut from the nearest hazel-thicket.

Richard Atkinson, 1956[3]

## Newgrange: symbolic orientation on the sun?

There is no better place to begin a discussion of astronomy in prehistoric Britain and Ireland than the magnificent passage tomb at Newgrange (O 007727). It represents a relatively uncontentious example of an imposing prehistoric monument incorporating a simple, yet elegant and spectacular, solstitial alignment. Both archaeologists and their colleagues in other disciplines have played a part in the discovery and detailed investigation of the phenomenon of midwinter sunrise at the site.

Situated some 14 km west of Drogheda in Co. Meath, Ireland, Newgrange forms part of the Boyne Valley passage tomb cemetery, a rich complex of prehistoric monuments built in the fourth and third millennia BC amidst fertile agricultural land along the northern banks of the River Boyne (Fig. 1.1; see also Archaeology Box 2). The tomb itself is situated on a long, low ridge with a commanding view over the valley. Around it is a scattering of what appear to be 'satellite' structures: three smaller passage tombs along the ridge to the east, and various barrows, standing stones and enclosures down towards the river. Upstream and downstream respectively are Knowth and Dowth, companion great tombs each with their own cluster of satellites.[4]

Newgrange has been described as a *tour de force* in megalithic tomb architecture.[5] The mound, carefully constructed of layer after layer of pebbles and turf, is over 80 m across. The fine façade that confronts the visitor today, with its high walls gleaming with white quartz (Fig. 1.1d), is the product of restoration following excavations by Michael O'Kelly between 1962 and 1975.[6] This frames an entrance on the south-east side from which a 19 m-long passage, with walls of large upright stones and ample headroom for the modern visitor, leads to a large central chamber. The chamber itself is remarkable for its fine corbelled vault some 6 m high. Three recesses open out from it, two at the sides and one at the end, and each of these originally housed a stone basin in which cremated bone and grave goods seem to have been placed. Cunningly conceived drainage channels, carved into the tops of the roof slabs, divert rainwater away from the tomb's interior. Radiocarbon dates suggest a date of construction somewhat before 3000 BC.[7]

The stones in most chamber tombs are rough, with non-planar surfaces and edges far from straight. Inside Newgrange, the majority of the stones are carefully dressed. There is also a profusion of artwork decorating the walls and roof of the chamber, the recesses, the passage and the kerbstones on the outside: intricate and beautiful designs made up of spirals and cup marks, triangles and lozenges, zig-zags and lattices. Across the tomb entrance is the finest carved stone of all, its outward side completely covered by a pattern containing five interlocking spirals surrounded by curves and lozenges (Fig. 1.3b).[8]

The entire passage tomb is surrounded by a so-called 'Great Circle' of large standing stones a little over 100 m in diameter. These are large, rough blocks of stone, providing a considerable contrast to the dressed stones in the tomb. In fact, only

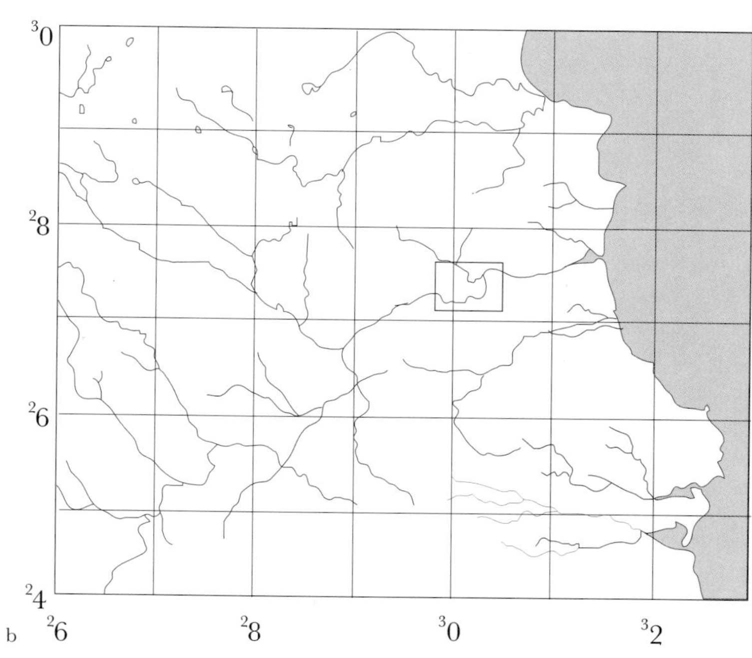


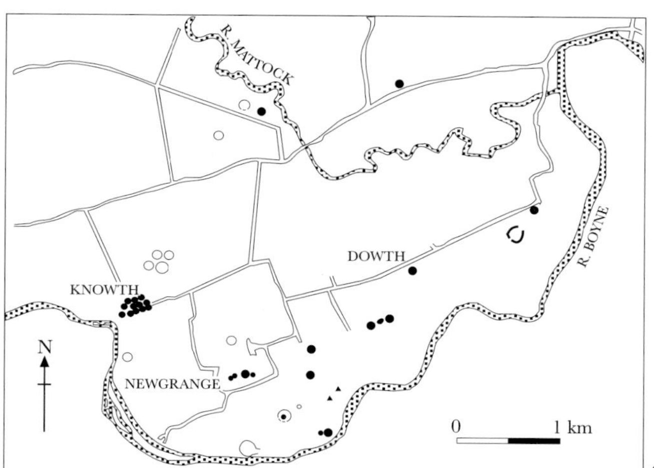


Fig. 1.1 Newgrange, Co. Meath.
a,b. Location of the Boyne valley monuments.
c. Situation of Newgrange with respect to the other Boyne Valley monuments. Filled circles mark tombs and cairns and open circles denote enclosures. After O'Kelly 1982, fig. 2.
d. Newgrange passage tomb, as reconstructed, viewed from the south-east.

## ARCHAEOLOGY BOX 2

### CONSPICUOUS PREHISTORIC MONUMENTS IN BRITAIN AND IRELAND

This box describes some of the main types of conspicuous Neolithic and Early Bronze Age monuments, often characterised as 'ritual' monuments, encountered in Britain and Ireland and gives an idea of their chronology.

#### Long barrows and chambered tombs

Chambered tombs are amongst the earliest stone monuments erected by agricultural communities in Britain and Ireland. They were mostly collective burial places, and were erected in considerable numbers in England, Wales, Scotland, and Ireland between about 4250 and 2750 BC:[1] over 1500 examples remain in Ireland alone.[2] There is a broad dichotomy in design between 'long tombs' whose central chambers open straight to the outside and 'passage tombs' (formerly known as passage graves) containing an entrance passage, although there are a great many variations on these themes and the boundary between the two main categories is far from clear-cut. The long tombs represent the earlier development, paralleling that of the long barrows of earth, turf, timber and chalk, constructed in lowland England and Scotland between about 4250 and 3250 BC.[3]

Major typological groupings of long tombs include the Cotswold–Severn tombs in central southern England and south Wales, the Clyde tombs of western Scotland, and the court tombs in Ireland. Passage tombs are found around Ireland but in particular concentrations at the famous cemeteries in the Boyne Valley and Loughcrew, Co. Meath, and Carrowkeel and Carrowmore, Co. Sligo. They are also found in the Hebridean islands, northern Scotland and the islands of Orkney.[4] The dating of all these monuments is hampered by the fact that they were often used and reused over a considerable period, some being built on top of smaller, simpler tombs from earlier times and many being used for later interments.[5]

Two variants of passage tomb, the Clava cairns in the area around Inverness, Scotland[6] and the wedge tombs of south-west Ireland, may represent the last vestiges of this tradition. The wedge tombs were still being erected late in the third millennium BC, and their use may have extended well into the second millennium.[7] Recent radiocarbon dates from four Clava cairns centre on about 2200–2000 BC.[8]

#### Round enclosures and circles of timber and stone

Round enclosures are found throughout Britain and Ireland and date from the early fourth millennium BC onwards. The earliest are the causewayed enclosures, large irregular earthwork rings with many entrances built in southern England up to about 3000 BC. These possibly served a range of functions, related to settlement, defence, ceremonial activity, and even the exposure of dead bodies prior to burial (cf. Archaeology Box 4). Amongst the best known examples are Hambledon Hill in Dorset and Windmill Hill in Wiltshire.[9]

Henge monuments are later than the causewayed enclosures, smaller, and more regular. They consist of a roughly circular ditch with outer bank, usually with a single entrance or two entrances on opposite sides.[10] Henges are found all over Britain and Ireland, but in greatest concentration in the east of Britain. Their construction appears to span the period from about 3250 to 2250 BC.[11] On the other hand, timber and stone circles (more correctly termed rings, since many are nowhere near exactly circular) are found mainly in the north and west of Britain and in Ireland. Around forty examples of timber circles—which do not remain as conspicuous monuments in the modern landscape—are now known.[12] They appear to span a period from about 3000 to 1500 BC.[13] Many hundreds of stone circles have been documented, of various sizes and forms,[14] and Fig. 1.2 gives an idea of their geographical distribution within Britain and Ireland. Stone circles are notoriously difficult to date, but it has been suggested that they began to appear not long after the first henges, and indeed that the great stone circles of western Britain, such as Castlerigg in Cumbria, may well have been 'the counterparts of henges in those highland areas where it was hard to dig but where building stone was plentiful'.[15]

Sometimes these monuments are found in association with each other. Settings of wood or stone, and particularly rings of upright wooden posts or standing stones, are found inside some henges,[16] and there is evidence that some timber circles were later replaced by henges, or reconstructed in stone.[17] Some of the best known henges with associated stone circles are the Ring of Brodgar and the Stones of Stenness in Orkney, Balfarg in Fife, Arbor Low in Derbyshire, and Avebury in Wiltshire.

Amongst the latest stone circles may be some of those in south-west Ireland, which appear to have been constructed well into the second millennium, and often consisted of only five stones.[18] Examples of a type of monument consisting of just four standing

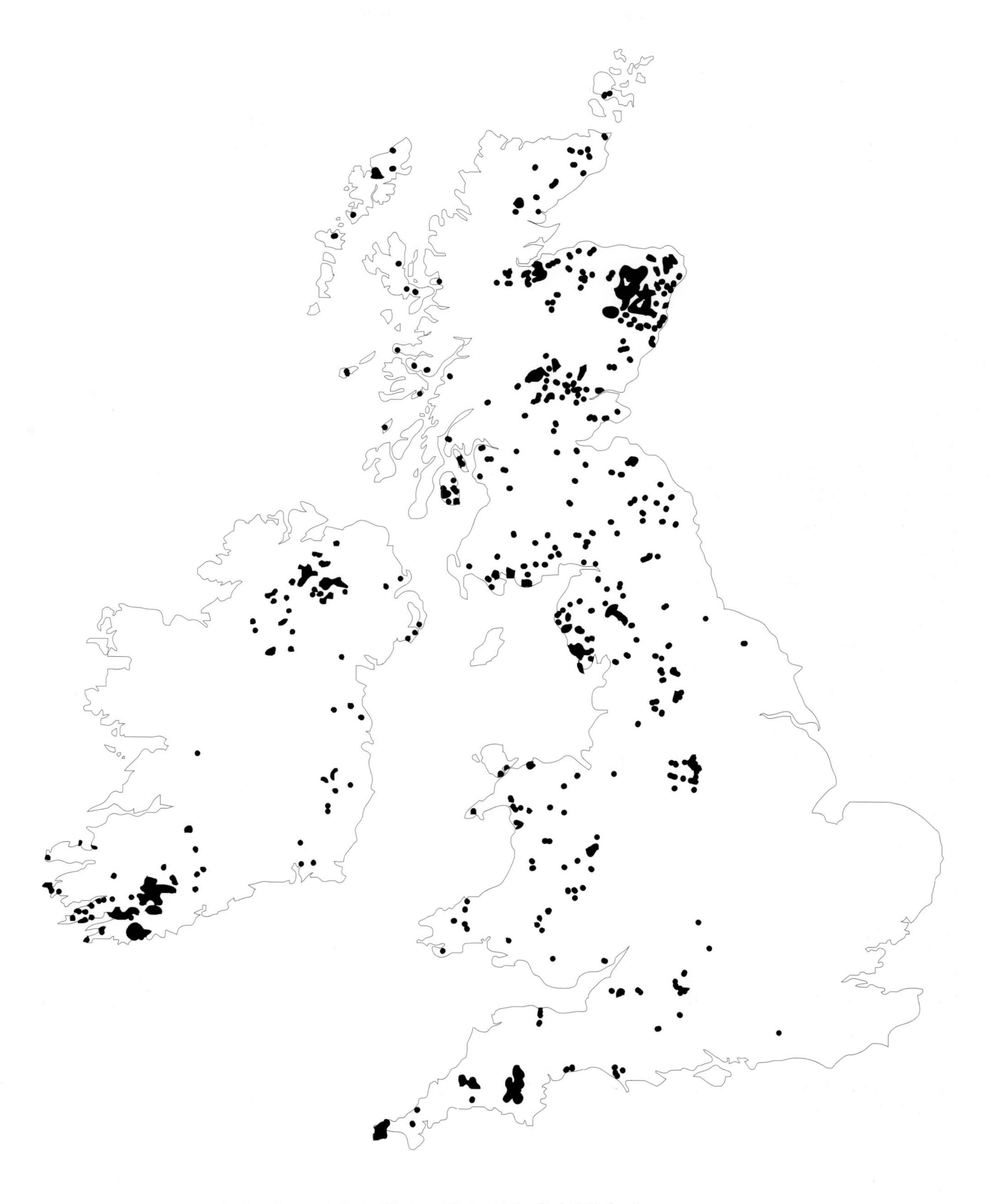

Fig. 1.2 The geographical distribution of stone circles in Britain and Ireland (after Burl 1976, fig. 1).

stones in a rectangular formation, known as a 'four-poster', are found throughout Britain and Ireland and may be a diminutive form of stone circle.[19]

### Linear monuments

Linear constructions appeared throughout much of lowland Britain in the mid-fourth millennium BC in the form of cursuses (or cursūs), linear earthworks comprising two widely separated ditches and banks running parallel. The longest of these, the Dorset cursus, is some 10 km long.[20] Bank barrows, over-long long mounds with ditches close to their edge, appear at about the same time and seem to be part of the same tradition.

Early in the third millennium avenues and rows of standing stones began to be attached to circles, famous examples being the 2 km-long West Kennet Avenue at Avebury and the radial avenue and rows at Callanish on Lewis in the Outer Hebrides. The existence of detached avenues, especially in northern Ireland, and double rows, especially in south-west England, suggests a transition towards the construction of linear stone monuments in their own right. Certainly, by around 2000 BC long, single rows of stones were being constructed in south-west England, south-west Wales and northern Ireland. Short rows of up to six stones also appeared at around this time, and were subsequently erected in their hundreds, with particular concentrations in western Scotland, northern Ireland, and south-west Ireland, possibly until late in the second millennium BC.[21]

### Round cairns and barrows

During the third millennium BC there was a gradual transition towards individual rather than collective burial, and towards small round cairns and barrows instead of large chambered tombs.[22] There was considerable regional variation in the type of burial (e.g. cremation or inhumation). In northern Britain and in Ireland especially, crouched inhumations were often placed in stone cists, accompanied by special pots called Beakers and by food vessels. Round cairns and barrows started to appear in a great variety of forms,[23] including the ostentatious Wessex bell barrows and disc barrows, and (by the mid-second millennium) the ring cairns and kerb cairns that are so widespread in Scotland. They were often constructed in groups (cemeteries). Some round barrows are complex in internal structure while others are simple mounds covering a single burial; superficially similar structures may turn out to be very different upon excavation. Round cairns and barrows continued to be erected until around the middle of the second millennium BC.[24]

### Standing stones

Large standing stones are often found in association with cairns (e.g. Strontoiller in Argyll), as well as with henge monuments (e.g. the Heel Stone at Stonehenge) and stone circles (e.g. Long Meg and her Daughters in Cumbria). Single standing stones are widespread in Britain and Ireland, but while many undoubtedly date to the later Neolithic and Early Bronze Age, they have also been erected throughout later prehistoric, historic and modern times as boundary markers, route markers, cattle rubbing posts, and for a variety of other purposes. On the other hand, a single standing stone may be the only visible remnant of a more complex prehistoric structure such as a stone row or stone circle. Of all types of prehistoric monument, single standing stones are perhaps the most difficult to date (often within extremely wide margins) and the most difficult to interpret.[25]

### Decorated stones

Many of the passage tombs of Ireland are highly decorated with abstract motifs such as concentric circles, spirals, zig-zags, and a variety of other abstract forms.[26] These may be related to the patterns on a type of Late Neolithic pottery known as Grooved Ware, often found in henges. While forms such as lozenges, chevrons and triangles are also found in Scotland, the dominant type of decoration consists of simple cup marks or cup-and-ring marks, sometimes in large groups. Such marks are found on standing stones, cist slabs, and natural outcrops.[27]

twelve circle stones survive, and they are concentrated on the southern side of the tomb and irregularly spaced, as can be seen in Fig. 1.3a. However, there is no evidence that the 'missing' stones ever existed and it seems probable that the circle, if intended as such, was never completed.[9] Other evidence indicates that the incomplete circle was erected many centuries after the tomb.[10]

Lockyer had noted in the 1900s that the passage at Newgrange was approximately aligned upon the rising sun at winter solstice.[11] However, the true nature of the interplay between the light from the rising solstitial sun and the architecture of the tomb (see Fig. 0.1) was only realised more than sixty years later, when it was witnessed at first hand by Michael O'Kelly on 21 December 1969 and again in 1970:

> At exactly 8.54 hours GMT the top edge of the ball of the sun appeared above the local horizon and at 8.58 hours, the first pencil of direct sunlight shone through the roof-box and along the passage to reach across the tomb chamber floor as far as the front edge of the basin stone in the end recess. As the thin line of light widened to a 17 cm-band and swung across the chamber floor, the tomb was dramatically illuminated and various details of the side and end recesses could be clearly seen in the light reflected from the floor. At 9.09 hours, the 17 cm-band of light began to narrow again and at exactly 9.15 hours, the direct beam was cut off from the tomb. For 17 minutes, therefore, at sunrise on the shortest day of the year, direct sunlight can enter Newgrange, not through the doorway, but through the specially contrived slit which lies under the roof-box at the outer end of the passage roof.[12]

The 'roof-box' referred to by O'Kelly is a feature uncovered in the excavations, which appeared at first to have no obvious function. It is situated above the long roof slab at the front of the passage, its floor being the top of the roof slab itself, while its own roof is formed by a lintel jutting clear at the front (only the tip of which was visible before excavation) and a corbel slab sloping down to meet the passage roof at the back (see Fig. 1.3c). The corbel slab was richly decorated with dot-in-circle and rayed dot and circle motifs.[13] It is this box, and not the entrance below, that admits the midwinter sun's light into the interior.

We immediately encounter a question that will recur time and time again throughout the investigation of prehistoric

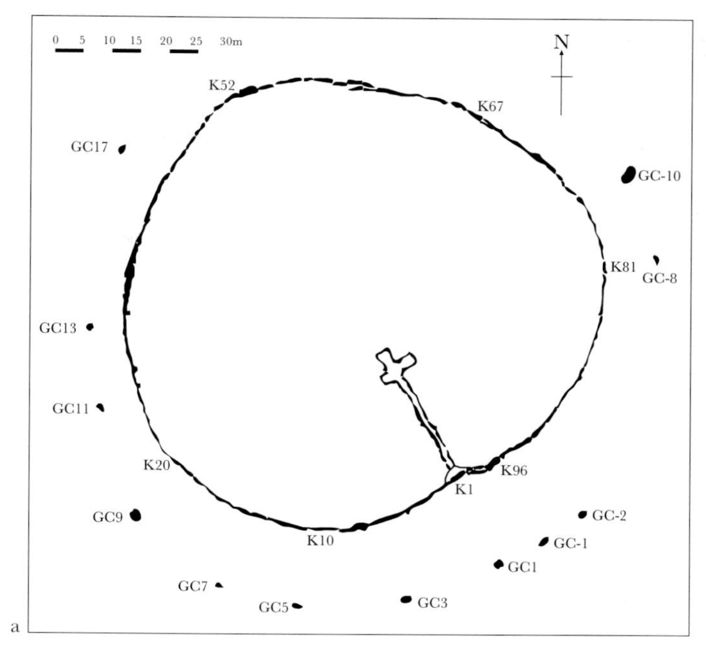

a

b

Fig. 1.3 The solstitial alignment at Newgrange.
a. Plan of Newgrange, showing the passage and surrounding stone circle. After O'Kelly 1982, fig. 3.
b. The entrance viewed from the exterior, showing the carved entrance stone (K1).
c. Cross-section of the passage at Newgrange, showing the path of the light from the rising midwinter sun. After Patrick 1974, fig. 1 and O'Kelly 1982, fig. 22.

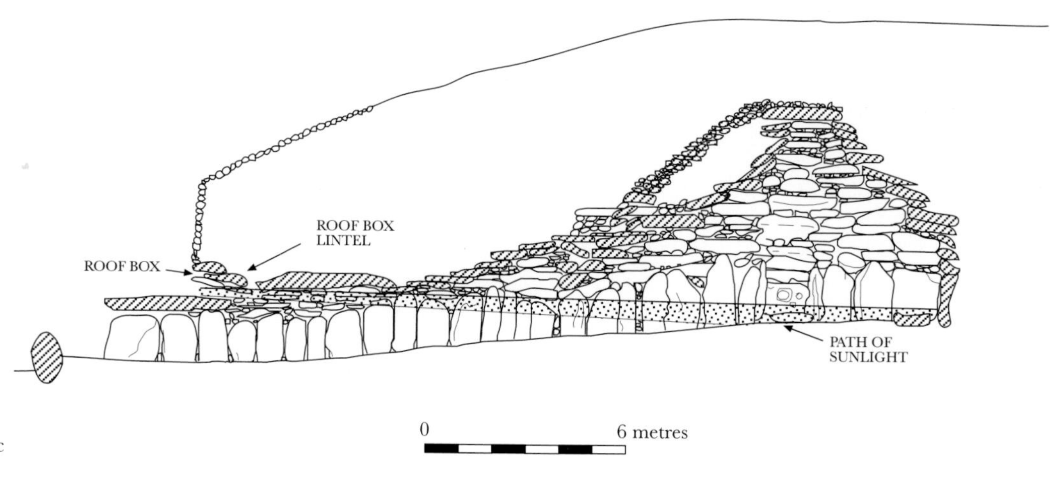

c

## ASTRONOMY BOX 1

### THE CONCEPT OF DECLINATION

In order to understand the arguments about which celestial phenomena may have concerned people in prehistoric Britain, it is necessary to have a basic understanding of the apparent movements of the heavenly bodies as seen from the earth's surface, in other words of what is known as positional astronomy. It is not generally necessary to know why these motions appear as they do, except insofar as this helps to understand the apparent motions themselves.

For our purposes, declination is the single most important concept in positional astronomy. It allows us to encapsulate, in a single number, certain key astronomical properties of any point in an observer's sky, and in particular of any point on their horizon. The main text will refer to the concept time and again.[1]

#### The celestial sphere

A cloudless sky, whether viewed on a clear evening, during a sunny day, or on a dark starlit night, appears to be a great hemispherical dome. Positioned upon it, moving around, are the discs of the sun and moon and the points of light forming the stars. This optical illusion is mimicked exactly by a planetarium, which projects images of the sun, moon, planets and stars onto a physical dome. If one were to stand for several hours at night in one spot, monitoring the slow collective movement of the stars, each rising behind the eastern horizon, progressing across the sky and setting behind the western horizon, one would soon gain the feeling of being at the centre of an entire celestial sphere that was slowly rotating.

This concept of a 'celestial sphere' in fact provides a remarkably convenient, as well as natural, way of describing the apparent motions of the heavenly bodies. All observers, at a particular position on the earth and at a particular time, see part of the sphere above their heads with the remainder being hidden below their horizon (exactly half and half, if the horizon is completely level and flat).[2]

From the point of view of an observer on the earth, the celestial sphere rotates once daily with all the heavenly bodies affixed to it. Thus we can identify its north and south poles (the celestial poles) as the points around which it pivots. From central parts of Scotland the celestial north pole is located at an altitude of about 57° above the north point of the horizon, whereas the celestial south pole is equally far below the south point and is never seen. Having defined the poles on the celestial sphere we can go on to identify its equator (the celestial equator) and its lines of latitude and longitude, just as on the earth.

Declination is simply a synonym for latitude on the celestial sphere.[3] The celestial equator, then, is the line where declination = 0°. By convention, declinations north of the celestial equator are positive and those to the south negative. The declination of the north celestial pole is +90° and that of the south celestial pole is −90° (Fig. 1.4).[4]

The significance of declination is that all the heavenly bodies move daily around lines of constant latitude on the celestial sphere, i.e. around lines of constant declination. Thus the Pleiades have a declination of about +24°, and Polaris, which is very near to the north celestial pole, has a declination of +89°. Hence, by determining the declination of a horizon point and referring to sources of astronomical information such as star lists or atlases, one knows at a stroke all the heavenly bodies that will rise or set there at some time during the daily cycle.

#### Longer timescales

On timescales longer than a single day, the simplification that each heavenly body can be associated with a single declination line begins to break down. The sun, moon and planets in fact move slowly about on the celestial sphere as it rotates, with the result that their declinations vary cyclically. The declination of the sun, for example, varies over an annual cycle between approximately +23°·4 at the summer solstice on 21 June and −23°·4 at the winter solstice on 21 December. The declinations of the stars vary over a much longer timescale owing to a phenomenon known as precession. These and other longer-term variations will be discussed in later boxes.

The important point for the moment is that all the relevant changes with time are well documented, so that the declination of a horizon point can tell us not only what will rise or set there at different times nowadays, but also what would have done so at any specified epoch in the past.[5]

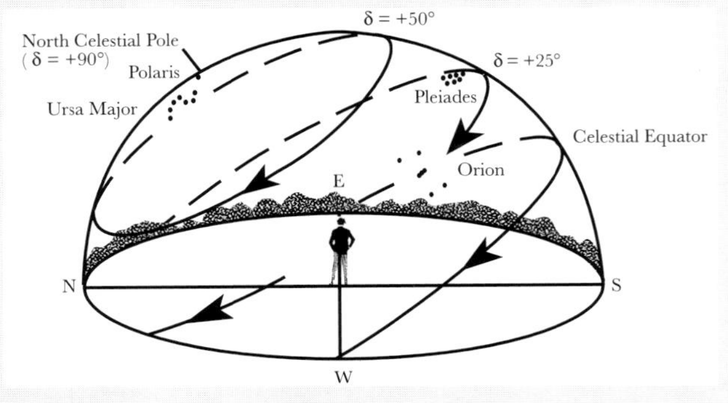


Fig. 1.4 The celestial sphere and lines of declination, as seen from the latitude of central Scotland.

astronomy. Was it deliberate? After all, any construction with a small opening will admit sunlight if and when the sun reaches the relevant part of the sky; is it possible that the Newgrange phenomenon could have arisen fortuitously rather than being an integral part of the design? The fortuitous occurrence of an alignment involving the sun at a special time of year (the solstice) and a special time of day (at sunrise) is more unlikely, but the question must still be asked.[14]

Claire O'Kelly herself was certainly convinced that the alignment was intentional:

> It is difficult . . . to remain sceptical once one has actually seen the thin thread of sunlight striking in along the passage at the winter solstice, this most dismal of all times of the year, until the dark of the chamber begins to disperse and more and more of it becomes visible as the sun rises and the light strengthens. Upon looking outward towards the entrance, one sees the ball of the sun dramatically framed in the slit of the roof-box and one realises that in the whole course of the year this brief spell is the only period when daylight has sway over the darkness of the tomb.[15]

Actually the last remark is a little misleading since 'direct sunlight penetrates to the chamber for about a week before and a week after the solstice but not as fully as on the few days centring on the 21st'.[16]

The sun's path through the sky is related to its declination, or latitude on the celestial sphere (see Astronomy Box 1). Over the year this varies between a maximum (most northerly path) of +23°·4 at the June solstice and −23°·4 at the December solstice. In prehistoric times the limits of the annual variation were somewhat greater: about ±24°·0 around 3000 BC. Following O'Kelly's discoveries, a survey of Newgrange was undertaken by Jon Patrick.[17] He calculated that the sun's rays would enter the roof-box, penetrate the entire length of the passage and illuminate the chamber just after sunrise whenever its declination lay between about −23°·0 and −25°·9. A resurvey by Tom Ray[18] has since raised the lower limit to about −25°·2. Even in 3000 BC the sun could never reach below −24°·0, but the range between −23°·0 and −24°·0 was sufficient to ensure that direct sunlight reached the centre of the tomb every morning for about two weeks on either side of the winter solstice. Nowadays the declination of the solstitial sun has increased to −23°·4 and the period of days during which dawn sunlight enters the tomb is shorter. At the time of construction, the rising solstitial sun would have cast a narrow beam down to the floor of the chamber immediately as it rose, rather than some four to five minutes later as at present.

But these are the details. More important at the outset is to avoid the purely subjective conclusion that a light-and-shadow phenomenon must have been deliberate simply because it is spectacular. The evidence at Newgrange does seem to weigh in favour of the deliberate rather than the fortuitous. First, the roof-box is an anomalous feature without any obvious function in utilitarian terms (as it seems to us). Second, if the gap in the roof-box were merely 20 cm lower or higher, or the passage a few metres shorter or longer, then sunlight would never have entered the chamber.[19] Third, at some time after its original construction, when the bones of a number of people had been placed within the tomb,[20] the entrance was permanently blocked with a large stone weighing about a tonne.[21] The roof-box, however, was only covered with two small quartz blocks which could be, and evidently were, moved to and fro to permit the roof-box to be opened and closed.[22] In other words, the design was such that although the living could no longer enter, by moving aside the quartz blocks at the relevant time the light of the midwinter sun could be allowed to continue to do so.

Nevertheless, the evidence is not conclusive. It could be that the roof-box had a function that seems obscure to us, yet was of great importance to people at the time: perhaps as an opening through which people could communicate with their ancestors,[23] or 'a soul-hole through which the spirits of the dead could come and go'.[24] We must also bear in mind that we are dealing with a reconstruction; the sides of the passage beneath the roof-box needed a good deal of rebuilding and some question must remain about the degree to which the stones and corbels of the reconstructed roof-box were replaced in their original positions, and to what extent this would effect the passage of sunlight.[25]

If astronomy really was involved we need to explore the possible reasons why it was important that the dawn sunlight around midwinter should light up the interior of the tomb. What *is* certain is that Newgrange was not an observatory, at least in any sense that would be meaningful to a modern astronomer. Its chief function was as a tomb for the dead (although see p. 89). Yet few people—archaeologists or astronomers—have doubted that a powerful astronomical symbolism was deliberately incorporated into the monument,[26] demonstrating a connection between astronomy and funerary ritual that, at the very least, merits further investigation.[27]

## Ballochroy: precise solstitial foresights?

No such consensus developed for many years with regard to the possible astronomical significance of Ballochroy (NR 731524) (see Fig. 1.5),[28] which for many people became the very embodiment of Alexander Thom's 'megalithic astronomy'. Unlike Newgrange, it is not a large, spectacular monument extensively visited by the general public, but merely a small group of megalithic structures situated on private ground, one of dozens of relatively unimposing sites to be found in the Argyll district of western Scotland alone. The Ballochroy site is located in the northern part of the Kintyre peninsula, close to the main A83 Tarbert–Campbeltown road, about 24 km beyond Tarbert. It can be accessed by walking up a steep, winding farm track which leaves the road about 400 m north of Ballochroy farmhouse.

From this vantage-point—a level area of high ground close to the west coast—one overlooks a broad stretch of the Sound of Jura including the small island of Gigha. Further away on a clear day can be seen the Isle of Jura itself, with the Paps of Jura, the group of peaks in its centre, being particularly impressive. Only some modern features (a nearby corrugated iron hay barn roof and some telegraph wires) serve to mar the view. The most conspicuous feature of the site itself is a 5 m-long row of three standing stones. The two southernmost stones (*a* and *b*) are a little over 3 m tall; the northernmost (*c*) is shorter but appears to have been broken off at the top. Stone *a* is roughly square at the base, tapering irregularly to a pointed top, but the

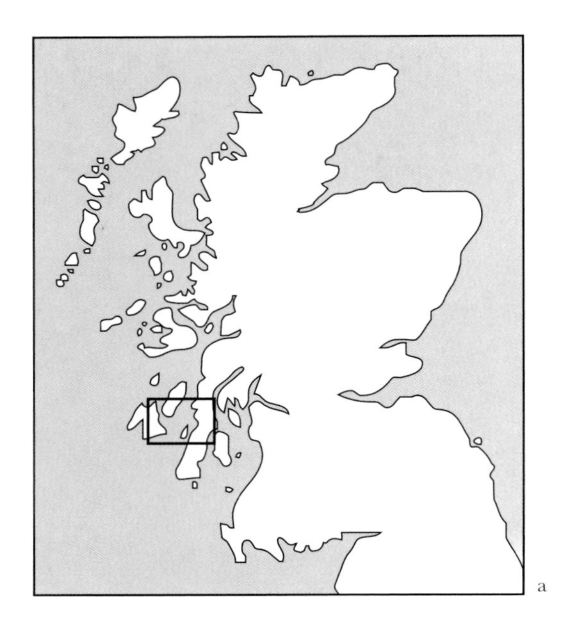

a

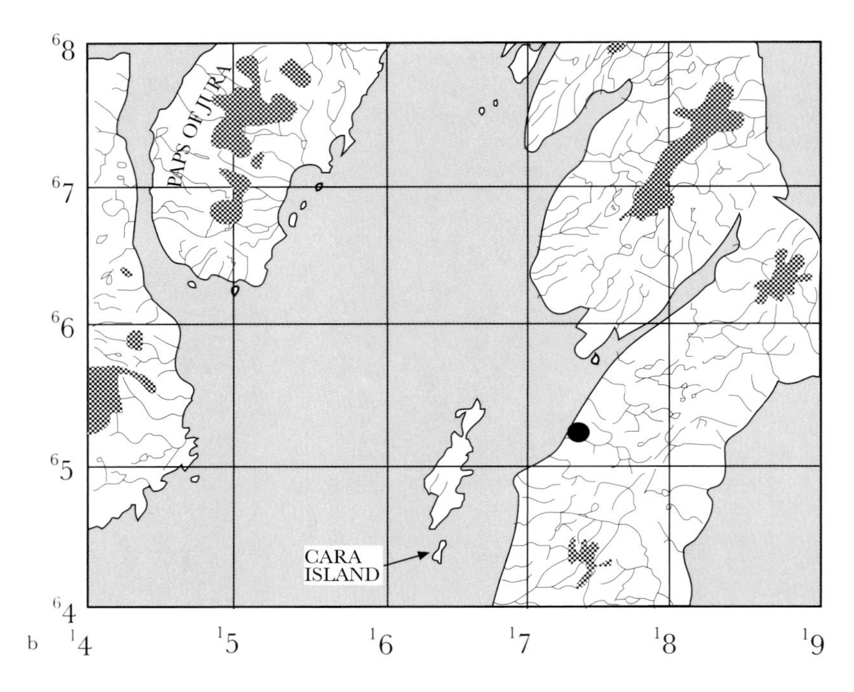


b

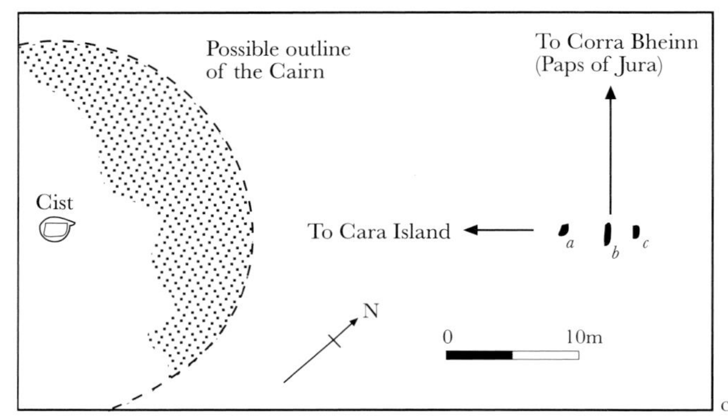


c

Fig. 1.5 The stone row at Ballochroy.
a,b. Location of the monument.
c. Plan of the site. After Thom 1954, fig. 3 and Burl 1983, fig. 1.
d. View from the south-east.

d

other two are thin slabs oriented across the alignment. Some 35 m away to the south-west, and protruding by about a metre above present ground level, is a rectangular burial cist with a large cap-stone. The cist is located in the same alignment as the stones, and its longer sides are also oriented in the direction of the alignment.[29] From the wider archaeological evidence, it is likely that megalithic constructions at Ballochroy commenced in the late third millennium BC at the earliest (see Archaeology Box 2), that is at least a thousand years after the construction of the tomb at Newgrange.

Thom, who had mentioned Ballochroy in his first published paper on megalithic astronomy,[30] came to regard it as one of the most important solar sites known to him.[31] Having carried out careful theodolite surveys to determine the declinations of conspicuous points on the horizon (see Astronomy Box 2), he suggested that the function of the monument was to pinpoint the longest and shortest days of the year by marking the exact setting position of the sun on both. This is by no means trivial to achieve. At Newgrange, as we have seen, sunlight entered the tomb not just on the solstice itself but for a period of several days before and after; this was inevitable because, close to the solstices, the daily change in the sun's path through the sky only alters by a very small amount (see Astronomy Box 3). At Ballochroy, however, the solstices could be discriminated by a subtle method using only natural features on the distant horizon.

The method is best appreciated by considering a hypothetical observer who stands at the same spot each evening in order to watch the sun setting. Suppose that from the chosen spot there is a clear view to a distant, hilly horizon in the west, containing plenty of fixed, identifiable landmarks such as hill summits and valleys. Given a run of clear evenings it would be easy to monitor the nightly progression of the sun's setting position along the horizon; at least for most of the year. It moves steadily northwards between late January and late May, and southwards between late July and late November, the daily change in the sun's sloping path when it is changing fastest—around March and September—being almost as great as its own diameter. By standing at a given spot, the daily change soon shows up against any distant fixed features such as hill-tops, valleys or trees.

However, within a week or two of the solstices the movement of the setting position becomes much smaller as it approaches its northern (June) or southern (December) limit. During a period of a week on either side of the solstices the sun's setting path changes by only about a third of its own diameter. In most circumstances it would be impossible to detect any difference in the sun's setting position throughout this period, even behind a mountainous horizon (Fig. 1.6a). How, then, could this have been achieved at Ballochroy?

The answer, according to Thom, was to use a distant, natural horizon feature as a foresight. Suppose that, as the sun sets on the day of the June solstice, there comes a moment where all that shows of it is the very tip of its upper limb, the point on the sun that traces out the top of its setting path. Because of the brightness of the sun even a tiny part of its surface can be clearly visible. The tip of the sun might be seen to twinkle down a hill slope parallel to its setting path, or else to gleam briefly in a horizon notch some while after the sun had set behind a hill to the left. In these special circumstances, even the very small lowering in the sun's setting path two or three days on either side of the solstice would be sufficient to ensure that the twinkle or reappearance would not take place (Fig. 1.6b, c). The December solstice could be marked in a similar way by noting the *non*-appearance of the tip of the sun behind a hill slope or notch, since on days adjacent to the solstice the sun's setting path would be higher up than on the solstice itself.

Fig. 1.6 Pinpointing the solstice, according to Thom. Generally, the minuscule difference between the sun's setting position on the summer solstice and two or three days earlier or later will be undetectable, even behind a mountainous horizon (*a*). However, by using a suitable hill slope (*b*) or notch (*c*), the last twinkle of the sun's upper limb will be seen on the solstice but not on the other days (after Ruggles, *New Scientist* 90 (1981), p. 751).

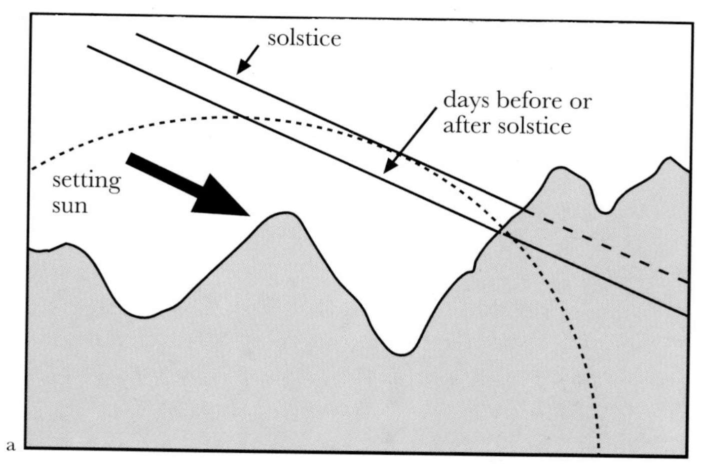


a

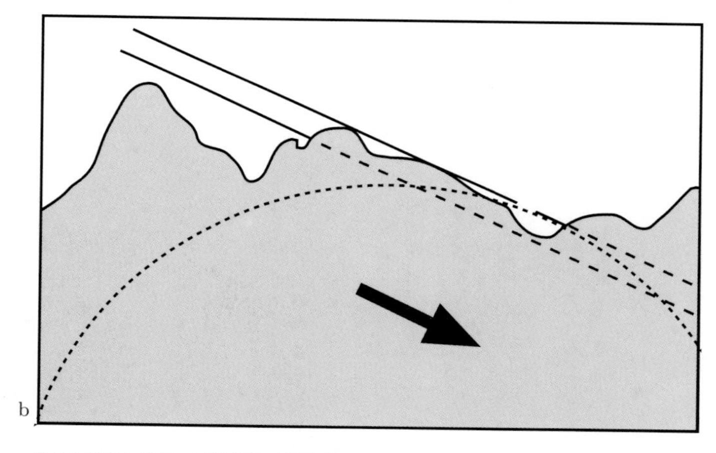

b

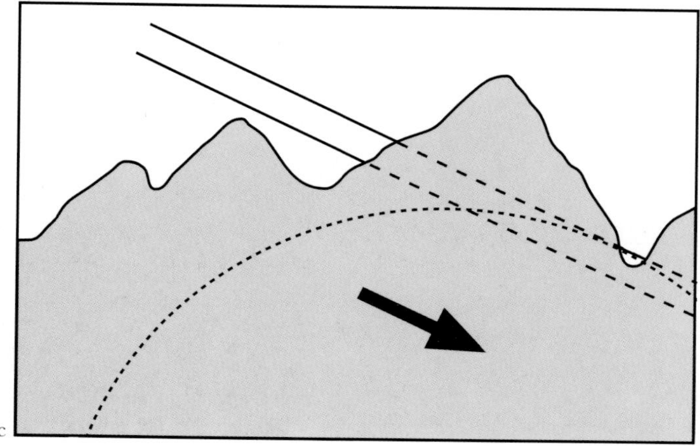

c

## ASTRONOMY BOX 2

### DETERMINING THE DECLINATION OF A HORIZON POINT

The *azimuth* of a horizon point is defined as its bearing from the observer measured clockwise round from true north, so that due east corresponds to an azimuth of 90°, south to 180°, west to 270°, and north to 0° and 360°. Its *altitude* is defined as the angle it subtends above the horizontal.

By surveying the azimuth $A$ and altitude $h$ of a point on the horizon from some fixed position (see Fig. 1.7),[1] and knowing the latitude $\lambda$ of that position, one can calculate the declination $\delta$ of the horizon point approximately (say, to the nearest degree) using the formula

$$\sin\delta = \sin\lambda \sin h + \cos\lambda \cos h \cos A \quad \ldots \text{(A2.1)}$$

The dependence on (terrestrial) latitude in this formula comes about because the whole celestial sphere tips over as the observer changes latitude on earth. For an observer at the terrestrial north pole, the north celestial pole is overhead and all the stars go round in horizontal circles. For an observer at the equator, the north celestial pole is on the horizon at the north point, the south celestial pole is on the horizon at the south point, and all the stars rise vertically in the east and set vertically in the west (see Fig. 1.8).

At the latitude of Britain and Ireland, heavenly bodies rising in the east and setting in the west do so at a shallow angle (about 30° in northern Scotland and 40° in southern England). Close to due north and south, however, they pass more or less horizontally above the horizon.

The dependence on horizon altitude is illustrated in Fig. 1.9. This shows lines of declination as viewed from a given observing position in central Scotland (latitude 57°). Given a low horizon (altitude 0°), an azimuth of 91° would yield a declination of about −1°, corresponding to the modern-day rising of ε Ori (the middle star of Orion's belt). On the other hand, if the horizon were higher, with an altitude of, say, 4°, then the declination corresponding to azimuth 91° would be around +3° and ε Ori would rise several degrees to the right (south).

The highest horizon declination attainable at a given location is around the co-latitude (i.e. 90° − 57° = +33° for our site in central Scotland), the exact figure depending upon the altitude of the horizon close to due north. Heavenly bodies at higher declinations than this will never set; instead they will circulate around the celestial pole in the sky. Similarly, the minimum possible declination is around minus the co-latitude; heavenly bodies at lower declinations than this will remain below the horizon and never be seen.

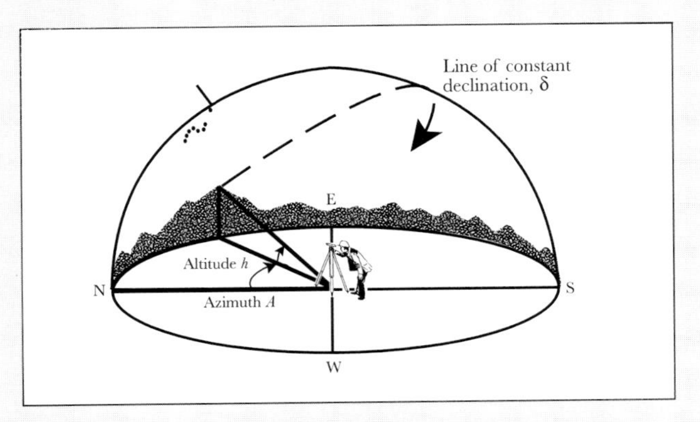


Fig. 1.7 Determining the declination of a horizon point.

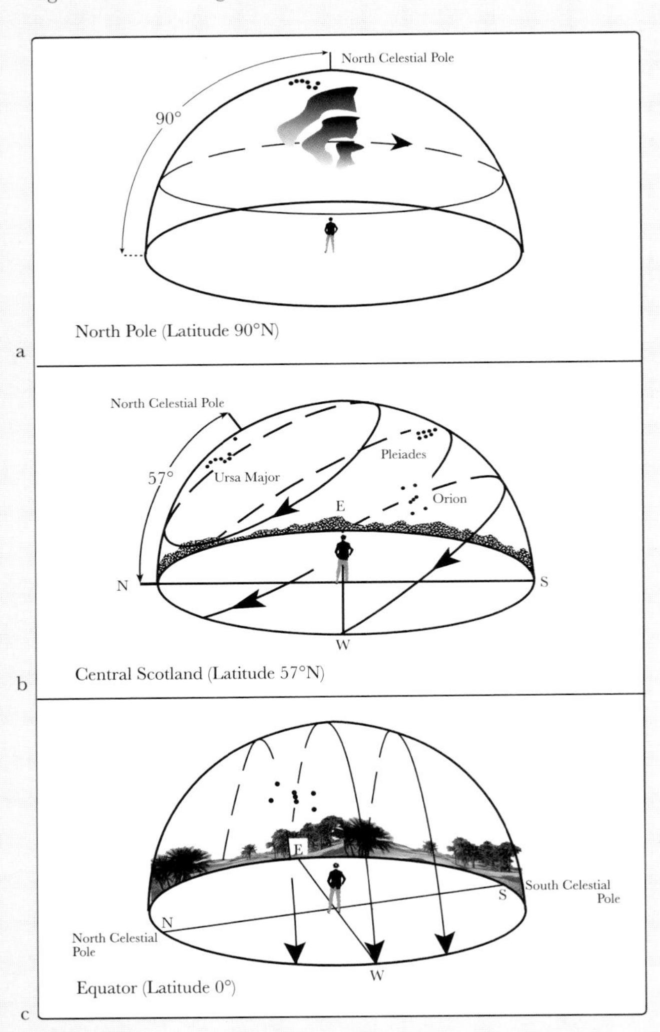


Fig. 1.8 The appearance of the celestial sphere at different latitudes. Inspired by Aveni 1980, fig. 20.

## Refraction and parallax

For more accurate work, other factors need to be taken into account before formula (A2.1) can be applied. The most serious of these is atmospheric refraction, which bends downwards rays of light reaching an observer from a distant object. This means, for example, that the rising or setting sun can be seen when it is in fact below the horizon. If it were not for refraction, the sun rising with declination 0° over a level horizon (altitude 0°) would appear exactly due east (i.e. at azimuth 90°).[2] In fact, as is evident in Fig. 1.9, the rising sun will be seen due east over a level horizon when its declination is somewhat less than 0°.

In order to take account of this effect, a correction must be applied to the measured ('observed') altitude, which is often denoted by $h_0$, in order to arrive at the appropriate value $h$ to be inserted in formula (A2.1). For many purposes it is adequate to apply a 'mean refraction' correction which is simply dependent upon $h_0$ itself, the correction being smaller for higher altitudes.[3] For more accurate work still, it may be necessary to take account of variations about the mean resulting from different atmospheric conditions (see also Astronomy Box 3).

A further effect, which is of concern (only) in the case of the moon, is parallax.[4] The problem arises because formula (A2.1) assumes that we stand at the earth's centre, whereas we actually stand on its surface. Allowance can be made for this in two different ways. For work of the highest accuracy it is necessary to apply an additional 'lunar parallax' correction to the measured altitude if the moon is of interest as a possible target. Formula (A2.1) then produces a 'geocentric lunar declination' for a given point on the horizon, distinct from the declination calculated for other purposes, and this is used exclusively in analyses concerning the moon.[5] This approach is necessary in discussing high-precision lunar phenomena (see Astronomy Box 7). Fortunately for all other needs it is sufficient to make a 'mean parallax' correction when calculating the expected declination of the moon at a given stage in its motions. This, which is the approach taken in Astronomy Box 4 and used throughout the book except in a small part of chapter two, is much simpler because it enables us to treat declination as a universal concept applying to all astronomical bodies.

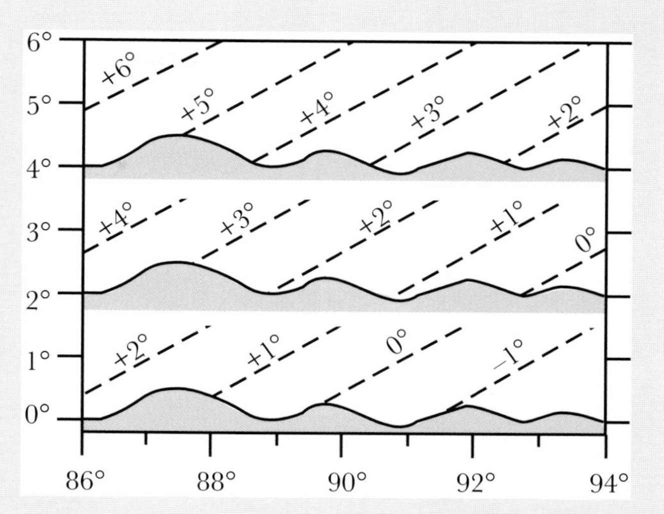


Fig. 1.9 The dependence of declination on horizon altitude. The figure shows fictitious horizon profiles for a location in the vicinity of Callanish, Isle of Lewis.

A widespread misconception is that Thom's method of using distant horizon foresights allows the solstice to be pinpointed to the very day. Thom himself certainly believed this, and it is inherent in his own scenario for how the sightlines were set up, involving setting out stakes at the position where the last gleam of the sun could be seen on successive nights.[32] However, the maximum possible difference between the sun's declination at the sunset nearest the solstice and that two days earlier or later is less than a minute of arc, or one thirtieth of the sun's own diameter.[33] Such a minuscule variation is almost certainly swamped by daily changes in atmospheric conditions (see Astronomy Box 3).[34] This rider notwithstanding, the use of distant natural foresights to observe and mark horizon astronomical events is unquestionably the most far-reaching single idea about preliterate astronomy propounded by Thom. This technique could have enabled the prehistoric inhabitants of Britain, with no instruments apart from the horizon available to them, to observe and mark particular astronomical events to a precision that is still remarkable.

Ballochroy attracted particular attention because of the presence of not just one but two high-precision solstitial foresights, one marking midsummer sunset and one marking midwinter sunset, both of which are 'indicated' by the standing stones themselves. The flat faces of the wide central slab *b* are oriented upon the right-hand slope of Corra Bheinn on Jura (NR 526755) at a distance of 31 km (Fig. 1.12a). According to Thom, in prehistoric times the tip of the setting midsummer sun would have twinkled down this slope (Fig. 1.12b). The alignment of the three stones points in the south-west to Cara Island (NR 638438), a small island some 12 km distant (Fig. 1.12c). The tip of the setting midwinter sun just clipped the right-hand end of the island, which could have acted as a foresight (Fig. 1.12d).

At first sight, the simultaneous presence of these two indicated foresights seems difficult to put down to chance. The implication that the site of Ballochroy was specially and carefully chosen for this reason seems reinforced by the fact that the declinations obtained correspond to a similar date, within

## ASTRONOMY BOX 3

### THE ANNUAL MOTIONS OF THE SUN

The declination of the centre of the sun varies annually between limits of $\pm\varepsilon$, where $\varepsilon$ is the obliquity of the ecliptic.[1] This has a value of about 23°·4 in the present day but was slightly greater in prehistoric times: about 24°·0 around 3000 BC.

#### THE SUN'S DECLINATION OVER THE YEAR

To a first approximation, the sun's declination $\delta$ at any time in the year is given by

$$\sin\delta = \sin\varepsilon\cos(0{\cdot}9856n) \qquad \ldots \text{(A3.1)}$$

where $n$ is the number of days that have elapsed since the June solstice[2] and all angles are expressed in degrees. The annual variation in the sun's declination calculated from this formula is shown in Fig. 1.10: it corresponds quite closely to a sine wave.[3] The magnitude of the daily change in the sun's declination is given by

$$\Delta\delta = 0{\cdot}9856\sin\varepsilon|\sin(0\cdot9856n)|/\cos\delta \qquad \ldots \text{(A3.2)}$$

where once again all angles are expressed in degrees.

Around the equinoxes ($n = 91$ and $n = 274$) the daily change reaches 24′, so that the sun's setting path is displaced from that on the previous day by three-quarters of the solar diameter.[4] However, near to the solstices the value is much smaller. By seven days from the solstice the daily change reaches just 3′, or one tenth of the solar diameter (Fig. 1.11).

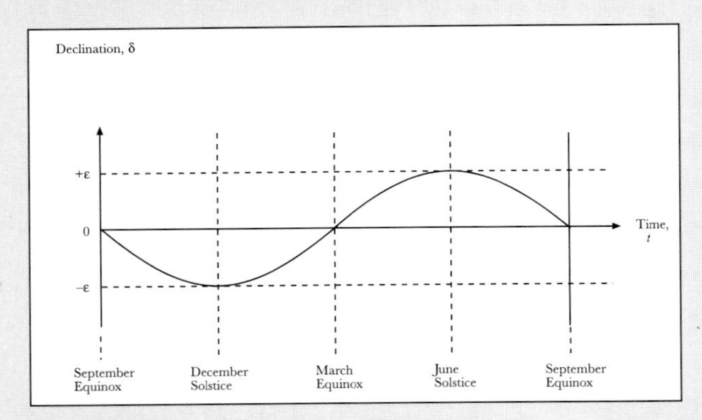


Fig. 1.10 The sun's annual variation in declination.

#### THE SUN'S SEMIDIAMETER

In any observations of high precision, it is not the centre of the solar disc that is likely to be important but its edges. If the declination of the centre of the sun at some moment is $\delta$, then the declination of its 'upper limb'—that part of the sun that traces the upper (northerly) limit of its path through the sky—will be $(\delta + s)$, where $s$ is the sun's semidiameter, approximately 16′. Similarly, the declination of its lower (southerly) limb will be $(\delta - s)$.[5]

In particular, the declinations of the sun's limbs at the solstices will be as given in the following table:

| | Upper limb | Centre | Lower limb |
|---|---|---|---|
| Summer Solstice | $+(\varepsilon + s)$ | $+\varepsilon$ | $+(\varepsilon - s)$ |
| Winter Solstice | $-(\varepsilon - s)$ | $-\varepsilon$ | $-(\varepsilon + s)$ |

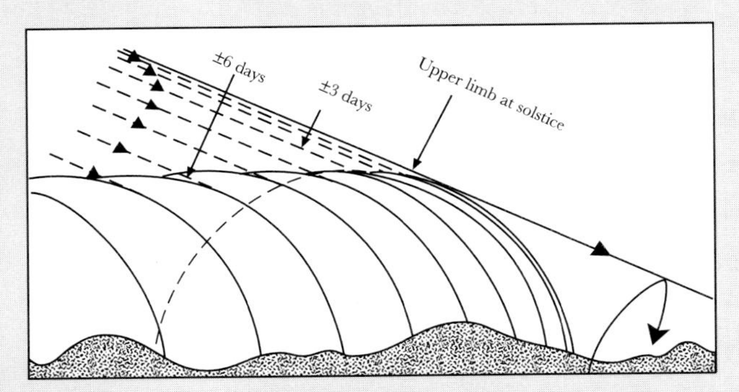


Fig. 1.11 A pictorial representation of the variation in the sun's declination on days close to the solstices, based on MacKie 1974, 172. Note that while the sun is shown each day with its centre on the horizon, sunset actually occurs when the upper limb disappears. The diagram shows the theoretical situation. In practice, the precise daily variation will depend upon factors such as the time of day at which the moment of solstice occurs. The apparent setting position may also vary from day to day by several arc minutes owing to variations in atmospheric conditions.

It is the upper limb that is involved when the tip of the sun just gleams behind the distant horizon, as is suggested by Thom at Ballochroy and Kintraw. Thus, for example, the bottom of the col between Beinn Shiantaidh and Beinn a'Chaolais as seen from the platform at Kintraw has a declination of −23°38′, corresponding to the upper limb of the setting sun when the declination of its centre is 16′ lower, i.e. −23°54′ (Fig. 1.14c).

#### THE SUN'S MOTIONS CLOSE TO A SOLSTICE AND THEIR DETECTABILITY

The difference between the sun's limiting declination at the exact solstice and at times $t_1$ and $t_2$, twenty-four hours before and after, is minuscule: only about 0′·2. If the sun sets just as it reaches its limiting declination then this value will represent the difference between the declination at sunset on the solstice and that on the preceding and following days. In general, however, two sunsets will occur between $t_1$ and $t_2$. In this case the declination difference between the sunset nearest the solstice and the other that occurs within this period will be even less.

The maximum possible difference in declination between the sunset nearest the solstice and that two days earlier or later is still less than a single arc minute. Three days before and after the solstice the sun's declination still differs from its solstitial limit by at most 2′; by seven days this has risen to 11′. Only by twelve days before and after the solstice does the difference approach a whole solar diameter, although by five further days before and after the solstice the sun is two full diameters away from its solstitial position.

The extent of atmospheric refraction is dependent upon air temperature and pressure. From his own measurements Thom calculated that daily and seasonal variations in atmospheric conditions would lead to an uncertainty in apparent declination of at most one minute of arc.[6] Others, however, warned that the uncertainty could be rather greater.[7] More recent work indicates that variations in atmospheric conditions can alter the apparent declination of an observed low-altitude object in Britain or Ireland by several arc minutes, and possibly by as much as half a degree.[8]

The evidence available today suggests strongly that the effects of variable refraction make it impracticable to detect the small differences in the sun's motions for at least two or three days on either side of the solstice, and possibly for a considerably longer period.

about one hundred years of 1600 BC,[35] a date comfortably within the bounds of archaeological possibility for the row of standing stones.[36] But there are a number of problems.

One difficulty is that the line along the stone row is not well defined, because two of the stones are slabs oriented across it. The alignment might be taken to point anywhere within an azimuth range of some 10° or more, depending upon whether one lines up the left- or right-hand sides of the stones, the centroids, the tops, and so on. Certainly the mean alignment is well to the left of Thom's foresight,[37] much closer in fact to the prominent bump on the left-hand end of Cara Island. The right-hand end of this bump would form a suitable foresight—indeed, arguably a better one than that proposed by Thom, since it is at higher altitude and is more prominent—except for the fact that it is not close to the setting path of the solstitial sun. We have no *a priori* reason for supposing the stone row to have been indicating one potential foresight rather than the other; only the *a posteriori* argument that one is astronomically significant whereas the other is not.

Similar criticisms apply to the Corra Bheinn foresight. While stone *b* indicates the hill slope in question, stone *c*, whose faces are equally flat, indicates a different hill slope that is without solar significance.[38] The wide face of stone *b* is not perfectly flat, and the stone may well have shifted somewhat since prehistoric times, so that it is impossible to say exactly which part of the slope of Corra Bheinn (if any) was intended; the slope is not quite parallel to the setting sun's path, so that the exact declination obtained depends upon which part is selected.[39]

There are also difficulties on archaeological grounds. It was pointed out in 1974 that the cist, which is on the line of sight to midwinter sunset, would almost certainly have been covered by a large cairn, and this would have obscured the view of Cara Island.[40] While we have no direct evidence on the relative chronology of the cairn and standing stones at Ballochroy, we can rule out archaeologically the possibility that such a cairn would have been constructed much later than the mid-second millennium BC; if anything, it is more likely that it predated by several centuries the astronomical date obtained for the standing stones on the assumption that the high-precision foresights were intentional.[41] Perhaps inevitably, the suggestion was then made that the cairn might have been destroyed in prehistoric times in order to permit the observations.[42] However, this idea is ruled out by a sketch of the site made by Edward Lhuyd around 1699[43] which clearly shows an alignment consisting not only of the three stones and the cairn (still erect) but also of two further, smaller cairns and a fourth standing stone. In summary, the Cara Island foresight appears to have been obscured throughout the assumed period of use, and furthermore the original monument was more extensive than what remains for us to see today, which casts further doubt upon interpretations based upon currently obvious features alone.

Taken as a whole, then, the evidence weighs overwhelmingly against the idea of there being intentional solstitial alignments of the high precision envisaged by Thom. Not that such a debate is ever finally settled: an astronomical enthusiast could, for example, always suggest that an observer in prehistoric times might have stood to one side of, or even on top of, the cairn to make midwinter sunset observations.[44] However, it is more plausible to suggest that the solstitial orientation, if deliberate, was of a much lower level of precision: an alignment of symbolic significance associated with the rituals of death and burial,[45] similar, in that respect, to Newgrange. The approximately solstitial orientation of one of the slabs set roughly at right angles to the alignment might well have arisen simply because the azimuths of midsummer and midwinter sunset happen to be roughly 90° apart at this latitude.[46]

Nonetheless the example of Ballochroy is a salutary one, serving to expose some of the dangers of embracing astronomical interpretations of orthostatic monuments too enthusiastically and uncritically.

## KINTRAW: A SOLAR OBSERVATION PLATFORM?

An obvious response to the dangers just mentioned is to try to derive ways of testing astronomical hypotheses concerning prehistoric monuments. And an obvious way of attempting this is by archaeological means. The classic site in this context is

+ 1°0′
+ 0°30′
316°
317°
a

Fig. 1.12 The solar foresights at Ballochroy.
a. Corra Bheinn, Jura. The mountain is seen to the right of the two highest Paps in Fig. 1.5d. The drawing here shows midsummer sunset at around 1600 BC, after Bailey *et al.* 1975, fig. 2.
b–d. Cara Island. In photograph (b), note the remains of the cairn Carn Mor in the alignment. (c) shows midwinter sunset at around 1600 BC, after Bailey *et al.* 1975, fig. 1.

b

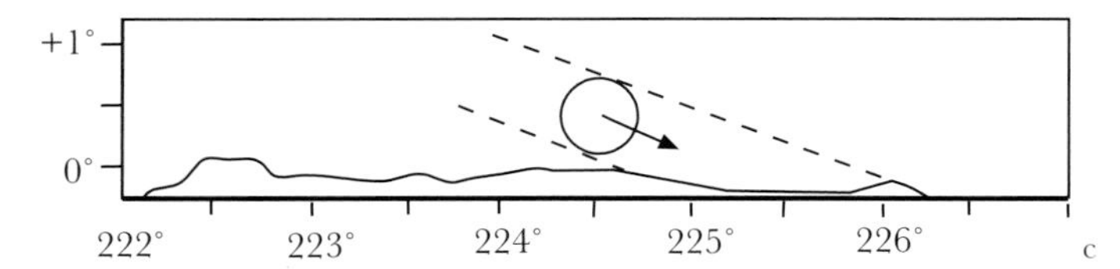

c

d

Kintraw (NM 831050).[47] Situated on a level terrace overlooking the head of Loch Craignish on the west coast of Argyll (Fig. 1.13), the most conspicuous feature here is a single, 4 m-high standing stone, which can be clearly seen from the main A816 road as it winds steeply inland on its way south from Oban to Lochgilphead. Until 1978 this stone leaned at a noticeable angle; then, during the following spring, it finally fell over, and was subsequently re-erected in its original position as determined by excavation of the socket hole.[48] Flanking the standing stone are the remains of two cairns, excavated in 1959 and 1960.[49] To the north-west is an enclosure, possibly also prehistoric.[50]

The story of Kintraw is one of the best-known in the entire debate about 'megalithic astronomy'. The flat faces of the standing stone are roughly oriented upon the deep col between Beinn Shiantaidh and Beinn a'Chaolais on Jura, some 45 km distant, which from this direction appears as a prominent V-shaped notch. The mountains form part of the Paps of Jura, the cluster of high mountains in the centre of the island that were also visible from Ballochroy, some 50 km further south along the coast.

Unlike at Ballochroy, there is no question that this distant notch is the most prominent horizon feature in the direction concerned. The south-westerly orientation of the local topography draws the eye to the left along the Craignish peninsula on the opposite side of the Loch, out to where the Paps of Jura suddenly appear. Nor is there much doubt that the stone was closely aligned upon this notch, whether intentionally or not, since its current position and orientation are as close to the original as can be determined from the available archaeological evidence.

Thom obtained a declination of $-23°54'$ for the setting sun whose upper limb would briefly reappear in the col.[51] This value is exactly the same in magnitude as those obtained at Ballochroy, and corresponds to winter solstice within the first half of the second millennium BC.

There is just one problem. As viewed from the standing stone, the bottom of the col is obscured by an intervening ridge less than 2 km away. In order to see it an observer would have to be raised off the ground by about 2 m. Thom initially suggested that observers stood on one of the nearby cairns, which he assumed to have been level-topped.[52] But this left the problem of how they knew where to place the cairn, when they could not see the col from ground level.[53] Clearly it was more satisfactory to postulate a different observing position. If the standing stone was acting not as an indicator but as a 'backsight', to be lined up with the distant foresight, then possible observing positions might be found to the north-east. Unfortunately, the ground to the north-east of the standing stone runs level for about 80 m and then falls steeply into a deep gorge.

The far side can best be reached by crossing the valley by the road bridge and then approaching from the north-west. This side of the valley is generally steep but it rises well above the level of the standing stone and cairns. When Thom first explored it he (actually, his granddaughter)[54] discovered a narrow ledge with a boulder at its edge at exactly the point where the standing stone and foresight were in line. Further up the hillside was another small stone that could have marked a position from which someone might have given an 'early warning' of the sun's reappearance in the col.[55]

The ledge, Thom now suggested,[56] was the intended observing position (Fig. 1.14), and he surmised that it represented the

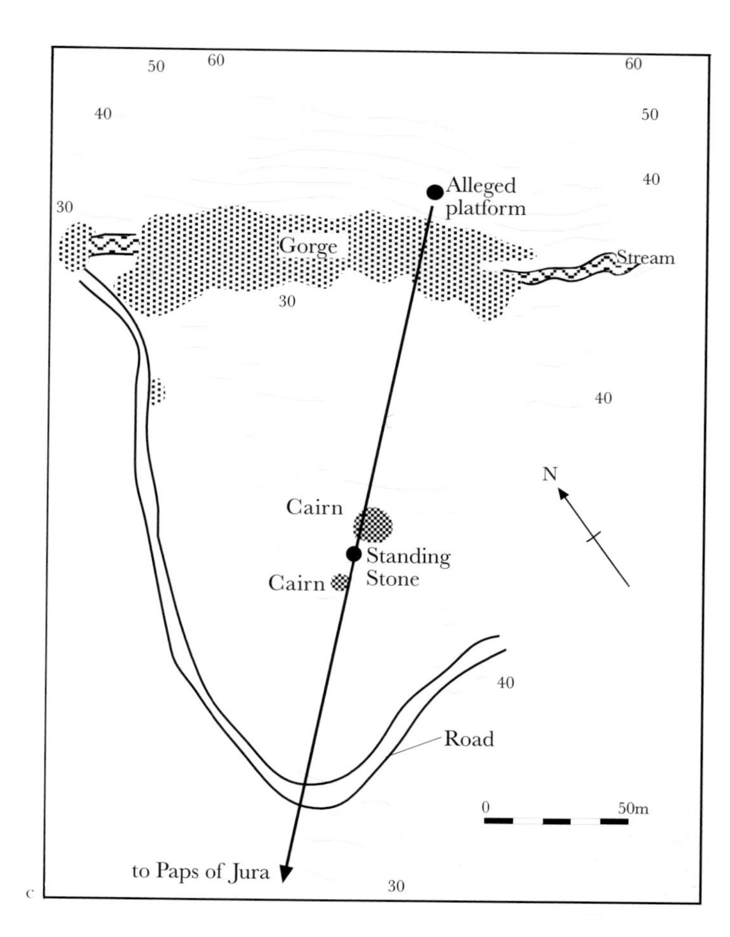


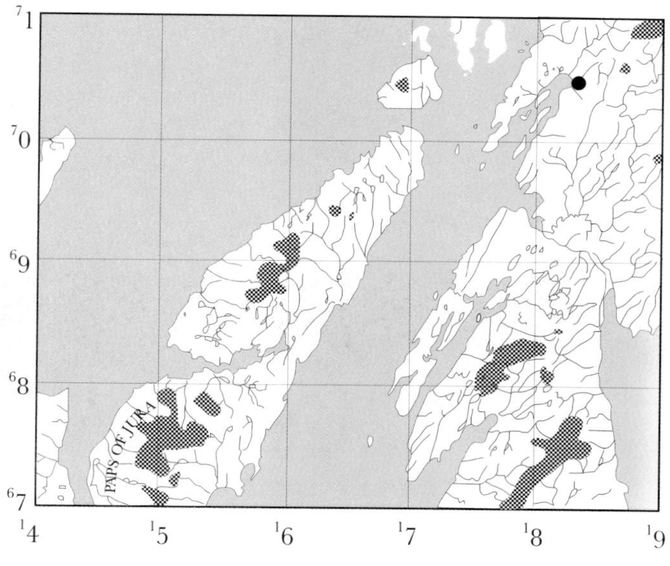


Fig. 1.13 Kintraw, Argyll.
a,b. Location of the monument.
c. Plan of the features near to the standing stone, based on MacKie 1974, 179 and RCAHMS 1988, 64. The road has subsequently been widened.
d. The standing stone and cairns from the east.

d

a

b

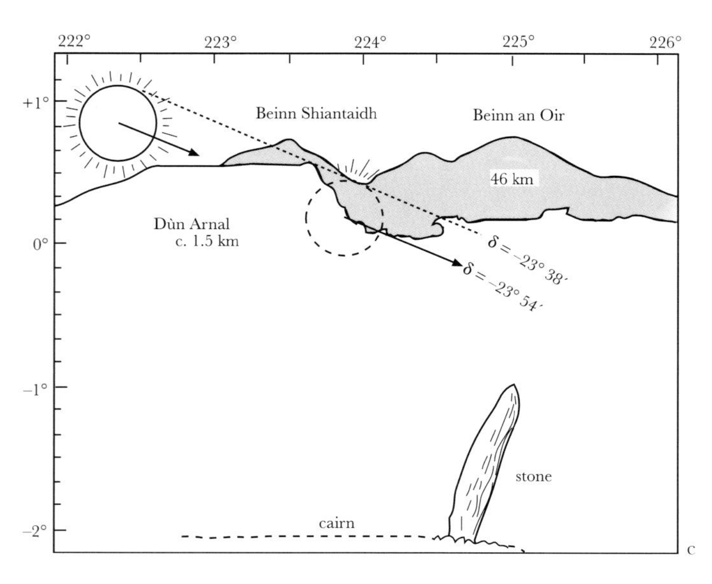


c

Fig. 1.14 The solstitial alignment at Kintraw.
a. The boulders and platform at Kintraw, during excavation.
b. The view from the platform at Kintraw (taken before 1978, when the standing stone was leaning).
c. Midsummer sunset at around 1800 BC as viewed from platform. The standing stone in the foreground is shown leaning, as it did prior to 1978. After MacKie 1974, 180.

remains of an artificially levelled platform. Euan MacKie recognised this as a unique opportunity to test, using standard techniques of archaeological excavation, a prediction following directly from the general idea of astronomical alignments upon distant natural foresights.[57] He duly excavated the platform in 1970 and 1971.[58]

Evidence that the platform was artificial would have given strong support to the astronomical interpretation of Kintraw.[59] Unfortunately, no direct trace of human activity was found there, whether in the form of flints, potsherds, postholes or charcoal. The boulder at the leading edge of the platform, however, was found to comprise two rocks lying end to end, their pointed ends just touching each other and forming a notch on the inner side. While it was possible that they had rolled down the hill and chanced to come to rest in contact, MacKie felt it more likely that they had been placed there deliberately.[60]

Covering the ledge behind these rocks was a compact and fairly level layer of small stones several centimetres thick. Such a layer could well have arisen naturally, so a technique was required to test whether it could be artificial. A suitable technique, known as petrofabric analysis, was long established within geomorphology.[61] The idea is that the overall distribution of orientation and dip of the stones would be different if such a platform were man-made from what would occur if the layer of stones had arisen naturally. By comparing the Kintraw distribution with various 'controls', it should be possible to decide whether the platform is artificial or not. An analysis of the orientation and dip of the long axes of the stones in the Kintraw platform was duly carried out.[62] On the basis of a visual examination of the results obtained, MacKie[63] argued that the distribution pattern of stones in the Kintraw stone layer appeared to resemble those found in other known artificial layers and not those found in natural ones. He concluded that this left 'little room for doubt that the stone layer behind the "boulder-notch" at Kintraw was made by man' and that Thom's interpretation was 'decisively vindicated'.[64]

Others disagreed. A major difficulty is the complete lack of human debris found during the excavation, which is very surprising if the platform was used as the astronomical theory supposes. A number of criticisms were also made of the way in which the petrofabric analysis had been carried out.[65] While MacKie responded forcefully to these criticisms,[66] he later conceded that, since no material was found to allow the platform to be dated, the evidence from Kintraw was inconclusive.[67]

Another doubt was cast upon the interpretation of the Kintraw ledge by the assertion[68] that the bottom of the distant col could not, in fact, quite be seen from the platform because the intervening ridge just obscures it. MacKie flatly denied this,[69] and indeed the present author has seen the notch from the platform; however a movement of less than 0·5 m from the centre of the platform made a crucial difference.[70] The disagreement may be accounted for by differences in eye height of different investigators, by differences in vegetation levels on the intervening ridge at different times of year, and perhaps most importantly by differences in weather conditions altering atmospheric refraction.[71]

This discussion leads one to ask why the solstice observations were not made from further back up the hillside in the first place,[72] either from the vicinity of Thom's 'early warning' stone or, even better, from higher up still, where 'the steep slope flattens out to a much more gentle crest where it is possible to walk or even run freely . . . without the risk of falling into the gorge, unlike the perilous situation on the "platform". More significantly there is a very scenic panoramic view of Loch Craignish and its islands with a *totally* unobscured view of the Paps of Jura.'[73] The answer is that as the observer's elevation rises, so the observed altitude of the distant notch decreases, necessitating a movement to the south-east in order to compensate. The hill-crest drops away sharply in this direction, and it is simply not possible to gain enough height to bring the foresight back into view beyond the intervening ridge.[74]

There would also have been severe difficulties in setting up the sightline in the first place. One method suggested by Thom is that an observer 'would seek by rapid movement across the line of sight to reduce the brilliant light to a point while the limb slid past the bottom of the notch',[75] but this would involve stepping rapidly sideways along a narrow path above a steep gorge. Another suggested method uses a row of observers:

> To those at the left the Sun would not reappear at all. The first man along the line to see the twinkle of light would be in the correct position, which would immediately be marked by a stake. If this process were repeated on several evenings around the time of the solstice the stake position would move first towards the right and then towards the left. The extreme right position would mark the day of the solstice.[76]

The problem here is that moving away from the platform along the ridge to the south-east (that is, to the left) causes the bottom of the col to disappear from view completely behind the intervening ridge.[77] As we have already noted, this difficulty can not be resolved by placing the row of observers further back up the hillside.

It is clear, then, that there are severe difficulties for the theory that Kintraw represented a high-precision solstitial sightline using the deep col in the Paps of Jura as a foresight. It is also clear that what promised to be an elegant and satisfying procedure—the archaeological verification of an astronomical hypothesis—has turned out in practice to be messy and inconclusive. This is not to say that the idea of such a verification is necessarily a bad thing, but rather that it has to be handled carefully within the context of many related questions such as how the astronomical hypothesis was arrived at, whether it was necessarily the best one to suggest on the evidence available before the test,[78] what other evidence bears upon the interpretation, and the extent to which that evidence is supportive or contradictory. In short, the integration of astronomical and archaeological evidence is a more complex process than it may seem at first.

## Brainport Bay: a calendrical complex?

It is instructive to elaborate upon this theme in the context of another site that has attracted a good deal of astronomical attention: Brainport Bay, near Minard in Argyll (NR 976951).[79] Here, on the western shores of Loch Fyne, are man-made platforms, standing stones, cup-marked stones and several other features, most of which were discovered and investigated between the mid-1970s and mid-1980s (see Fig. 1.15).

What is most striking about Brainport Bay is that a number of stone structures occur in a single NE–SW alignment. These were explored from 1976 onwards by Colonel P. Fane Gladwin and members of the mid-Argyll Archaeological Society.[80] At the south-west end, at an elevation of some 15 m, is the so-called *back platform and projection* (*A*), an area of natural outcrops that have apparently been converted into a platform by shaping and by infilling using small boulders and slabs of schist. Surface scatters of quartz chippings were found here. Some 60 m downhill to the north-east are two large boulders (*B*) oriented across the alignment, separated by a flat rubble surface about 0·5 m wide. The so-called *main platform* (*C*) starts some 10 m further again to the north-east, and extends for some 30 m to the north-east. It is a rocky outcrop up to 12 m in width (NW–SE), that has been modified and added to with paving and revetted platforms.[81]

At first, these features were thought to represent the remains of some kind of ancient settlement. However, there was a troublesome lack of any clear evidence of dwellings or midden deposits. Furthermore, a number of other features seemed without explanation. These included two socket holes—one ($C_2$) at the north-east end of the central, highest area of the platform, and the other ($C_3$) in a V-shaped cleft between two boulders at its south-west end—together with three small stone slabs between 1·1 m and 1·3 m long lying nearby on the surface, two of which appeared to fit well into the sockets.[82]

The idea that they were related to the distant horizon seems to provide a promising explanation of a number of otherwise anomalous features. The alignment (see *a* in Fig. 1.15e) is oriented to the north-east towards the only distant horizon visible from the site, at the far end of Loch Fyne. The rubble surface between the two large boulders at *B* provides an ideal position from which to view along the alignment to the north-east, the front stone being at a convenient height (1·6 m) to look over. From this vantage-point the rock outcrops of the main platform are silhouetted beneath the distant skyline, and the two small standing stones (as re-erected) appear, one behind the other, in the V-shaped cleft formed by the rocks on either side of the nearer stone (Fig. 1.16). Above the standing stones is a horizon notch formed by the junction between Beinn Dubhchraig and Beinn Oss, some 45 km distant. The dramatic effect of the two stones lining up upon the distant notch through a nearby cleft has been likened by Euan MacKie to the sights of a rifle barrel.[83]

MacKie, whose attention had been drawn to the site in 1976,[84] viewed midsummer sunrise from Brainport Bay in the following year, and the horizon was subsequently surveyed by Alexander Thom's son Archie. He determined that the upper limb of the sun rising with a declination of $+23° 6'$ would just

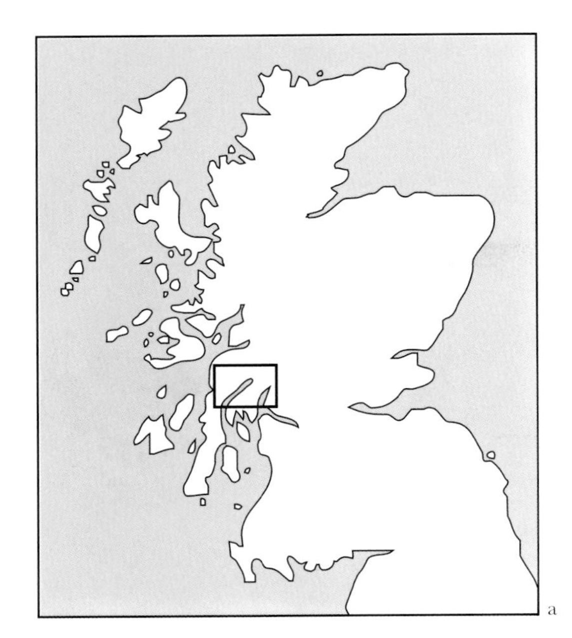


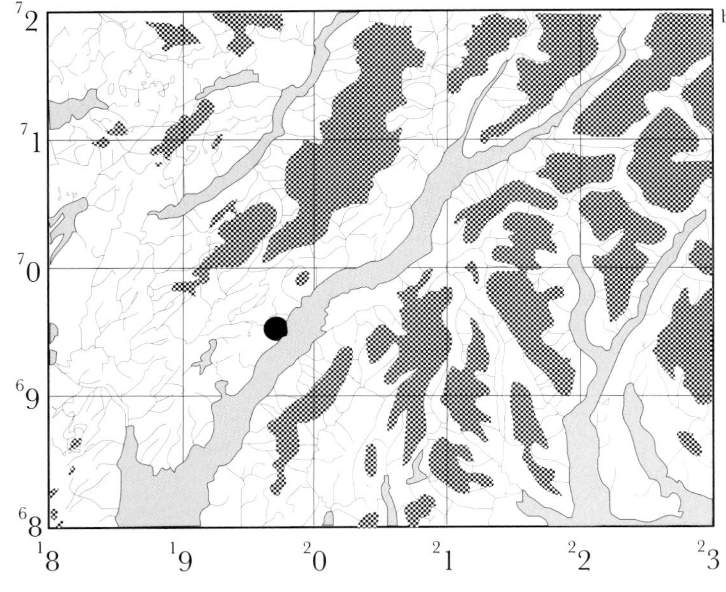


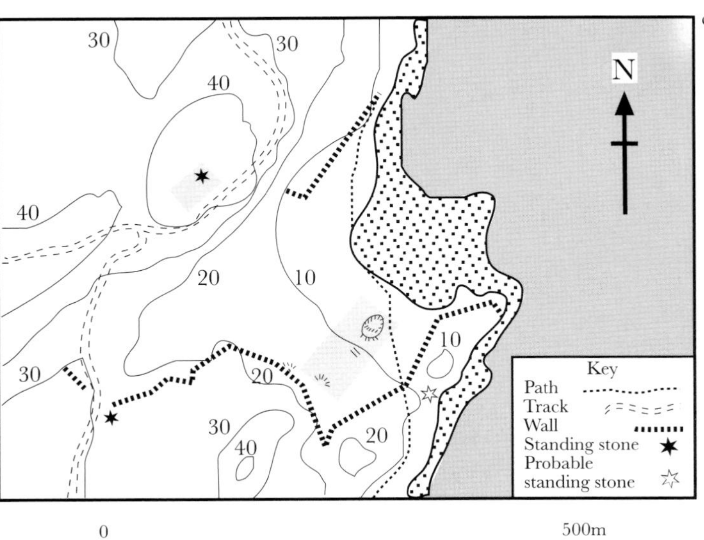

Fig. 1.15 Brainport Bay, Minard, Argyll.

a,b. Location of the site.

c. Overall plan of the site. After Fane Gladwin 1985, fig. 1. The shaded rectangles indicate the areas shown in more detail in Figs 1.15e and 1.17a.

d. The main alignment at Brainport Bay, viewed from the 'back platform'.

e. The main alignment at Brainport Bay, and features in the vicinity. Based on Fane Gladwin 1985, fig. 2; cf. RCAHMS 1988, 209. Scales and orientation are approximate.

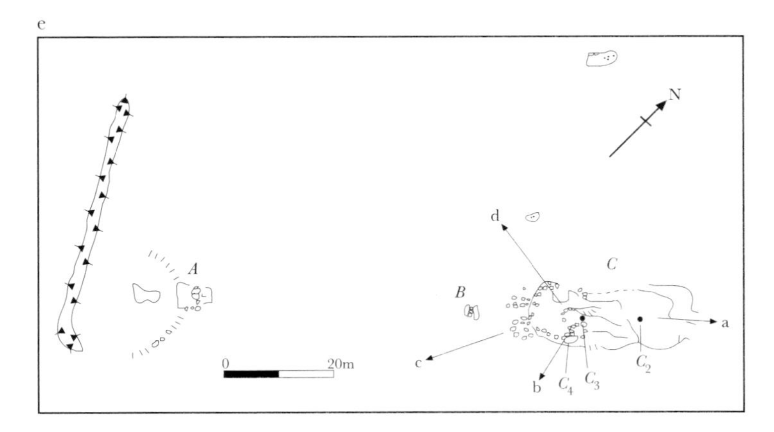

a b

c

Fig. 1.16 The 'main platform' and solstitial alignment at Brainport Bay.

a. The 'observation boulders' viewed across the alignment from the south-east.
b. The 'main platform', with the two pointer stones as reconstructed, and the alignment to the north-east, as viewed from the 'observation boulders'.
c. Midsummer sunrise between Beinn Dubhchraig and Beinn Oss, as viewed today.
d. Sunrise at Brainport Bay in *c.* 1800 BC, at the solstice (left sun) and approximately fifteen days before and after (right sun). Based on Fane Gladwin 1985, fig. 4, cf. MacKie 1981, fig. 3.8.

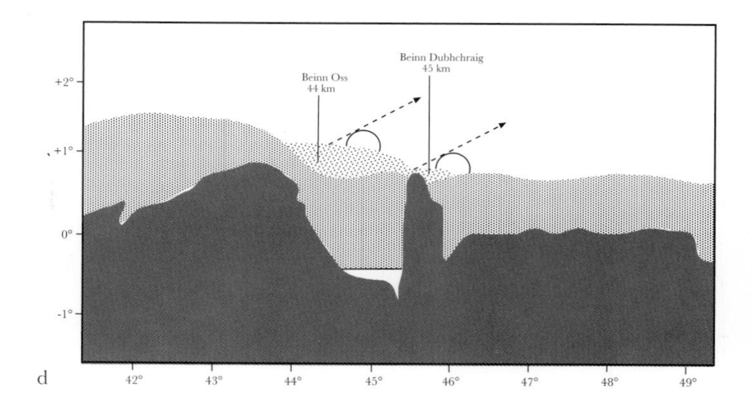


d

appear in the notch. Unlike at Ballochroy and Kintraw, this does not correspond to the exact time of solstice, but some days before and after (about fifteen days in 1800 BC).[85]

As a candidate for an astronomical alignment, the linear features at Brainport Bay seem to compare very favourably with the ledge and standing stone at Kintraw. Unlike at Kintraw, the platforms here (although basically formed by natural outcrops) were indisputably paved, terraced and modified by man; in addition, flint flakes and artefacts and shattered quartz fragments were found on various parts of the site. Excavations in 1982 established a $^{14}$C date from the back platform corresponding to the fourteenth century BC.[86] Furthermore the linear structures have no obvious function apart from indicating an alignment. Unlike any of the three sites so far discussed, there does not seem to be a funerary association: there are no cairns and no evidence of cremated bones. There are no artefacts or structures indicating domestic use, and no evidence of defensive structures.

However, the astronomical interpretation is not free from difficulties. If the 'rifle-barrel' effect is obtained from the two 'observation boulders' (*B*), what is the purpose of the back platform? While one can obtain a good general view of sunrise over the distant hills from here, it is no longer framed by the rock cleft in the main platform which, together with the standing stones, is a long way below the horizon. MacKie surmised at first that the back platform had a ceremonial function, where large numbers of people could have observed sunrise, with the precise indication of the horizon notch obtained from the observation boulders 'allowing the apparatus to function as an accurate calendrical instrument if needed'.[87] However a shock was in store. Later excavations showed that, contrary to earlier expectation, the two boulders had actually not been moved into place but were entirely natural.[88] It therefore had to be assumed that people in the second millennium BC made a chance discovery of a set of features already roughly oriented towards midsummer sunrise, and improved them. Furthermore, four radiocarbon assays obtained from the main platform yielded dates corresponding to the first millennium AD rather than the second millennium BC.[89]

There are two ways of interpreting the astronomical significance of the horizon foresight itself. The first is that the solstitial alignment was never used to obtain any great precision, so that the distant notch was irrelevant. The second is that the site was used as a precise solstitial marker, the notch being used to determine the solstice by the method of 'halving the difference'. The idea here is that, instead of attempting to pinpoint the exact solstice, people might have used a distant horizon notch to indicate the sun's position a few days earlier and later. This avoids the severe practical difficulties in pinpointing the exact solstice that were discussed in the context of Ballochroy and Kintraw: it is a good deal easier to indicate a day somewhat away from the solstice because the sun's daily movement is much greater (see Astronomy Box 3).[90] The exact date of the solstice could then be determined (retrospectively in the first instance) by counting the days between the pinpointed ones and halving the answer. This presupposes, of course, that it was important to determine the essentially unobservable event of the solstice itself;[91] and also has the pragmatic implication that people must have had some means of recording or remembering the required number of days from one year to the next.

A serious difficulty with the 'halving the difference' argument is the large number of distant horizon notches that can be seen from virtually any location in mountainous country. Picking such a location at random, there is only a relatively small chance that a fortuitous notch will be found exactly at a solstitial rising or setting position of the sun. However, the chances are appreciable of finding an acceptable notch within a degree or so in the appropriate direction from one of the extreme positions. Any such notch is susceptible to the 'halving the difference' argument, so that the mere presence of such a notch in the general direction of the alignment of features at a monument such as Brainport Bay does not constitute any sort of proof that it was used in this way by the people who occupied the site.

Such arguments, if they are to be properly followed through, need to be made more quantitative, and we shall encounter arguments about numerical probabilities shortly, in the context of Stonehenge. But a useful point can be made here. It is possible to formulate a convincing argument on practical grounds that if prehistoric people did precisely pinpoint the solstice at all, they almost certainly did it by using a precise horizon marker not at the solstice itself but close to it, and then 'halving the difference'. Not only was this easier than trying to distinguish tiny changes from one day to the next; it would also have given them a good deal more flexibility in choosing a suitable place from which to make the observations. But on the other hand, the abundance of suitable horizon features contrives to make it almost impossible for us to convince ourselves that any particular one was indeed used for the purpose of solar observations close to the solstice. Neither can we ever prove that it *wasn't* so used; the point is merely that on this subject, as in so many other matters relating to the activities of prehistoric people, the archaeological record (here taken in the broader sense to include the natural environment as well as material remains) is unable to speak to us clearly.

Perhaps the most striking thing about Brainport Bay (as opposed to Kintraw) is that we have platforms that were indisputably levelled by man, apparently in the mid-second millennium BC, although the picture is complicated by a range of activities on the site apparently stretching through to early historic times. The archaeological discoveries seem to indicate that this was not a domestic, defensive, industrial or funerary site, and hence perhaps one intended almost exclusively for ritual use. Furthermore, its situation where a striking distant horizon is seen to the north-east, roughly in the direction of midsummer sunrise, and the fact that several features are aligned in this direction, suggest that the solar connection did not go unnoticed and may well have been of considerable significance to the people who came here.

In trying to take the astronomical interpretation any further than this, however, things rapidly become more complex and fraught with problems. MacKie himself, acknowledging many of the difficulties in interpreting the main alignment, proceeded to place great emphasis upon subsequent discoveries in his efforts to demonstrate that Brainport Bay was a 'calendrical' site where the horizon sun was observed and marked with great precision. There are several further alignments. Fane Gladwin,[92] for example, lists three in the vicinity of the main alignment: one upon midwinter sunrise, one upon midwinter sunset and one upon sunset at the equinox (*b*, *c* and *d* in Fig. 1.15e). The first of these involves viewing the so-called 'pyramid stone' ($C_4$), a pointed-topped boulder 1·4 m high set in position on the south-east edge of the main platform, from the opposite side of the platform to the north-west. This stone, it is claimed, is sited at the only possible spot where an observer on the north-west side could look upwards from lower ground and align its tip with the high south-east skyline. From here it indicates midwinter sunrise.[93] However, the tip of the pyramid stone is aligned with the skyline from a range of positions to the north-west of the upper platform, not just a single spot. Choosing the optimum observing position to fit the astronomical theory is unconvincing without independent evidence since there is much room for manoeuvre. Similar problems attach to the other proposed alignments.

Upon a ridge some 240 m to the north-west of the main alignment is the so-called Oak Bank stone (*L*), a recumbent stone 3·4 m long which appears to have fallen from its south-east end (Fig. 1.17a).[94] If it did once stand it would have been plainly visible against the sky from the main outcrop. At first, it was thought to be a foresight for midsummer sunset, and a small platform was discovered at the required observing position, a few metres north of the main platform. However, to the disappointment of the excavator, excavation showed that activities at the platform were relatively modern.[95]

But all was not lost. Some 20 m south-east of the Oak Bank stone two outcrops were discovered with unusual carvings on their flat, upper surface in the form of a straight groove *c.* 0·5 m long with a cup mark at the centre (Fig. 1.17b). One of the grooves (*J* in Fig. 1.17a) is oriented at about 80°/260° and the other (*K*) at about 130°/310°. Some 35 m south-west of *L* is a further natural outcrop with a single cup mark (*N*). The first of the cup-and-groove marks was found to be oriented in the west (*e* in Fig. 1.17a) upon a small horizon notch in Siaradh Druim, about 1 km distant, yielding a declination very close to 0° and corresponding to sunset at the equinox.[96] The line from the Oak Bank stone to the cup-marked stone (*g*) indicates a horizon notch about 1·5 km distant. When its upper limb just appeared in this notch, the declination of the centre of the setting sun would have been −23° 51′,[97] very close to the figure obtained for the midwinter sunset alignments at Ballochroy and Kintraw.

These alignments seem impressive but the sceptic may still raise many doubts and objections. First, the notches in question are a mere 1 km and 1·5 km away from the observer. At such small distances changes in ground level and vegetation since prehistoric times may well be significant. Second, despite the astronomical importance attached to one of the two cup-and-groove marks, there does not seem to be any astronomical explanation for the other (*f*), which points to a featureless stretch of horizon well to the right of the midsummer sunset position.[98]

The significance of the cup-marked rock marking the solstitial alignment must be questioned on a number of grounds. We know that the notch in the approximate direction of midwinter sunset from the Oak Bank stone was identified during general forays to explore its calendrical potential, and that the cup

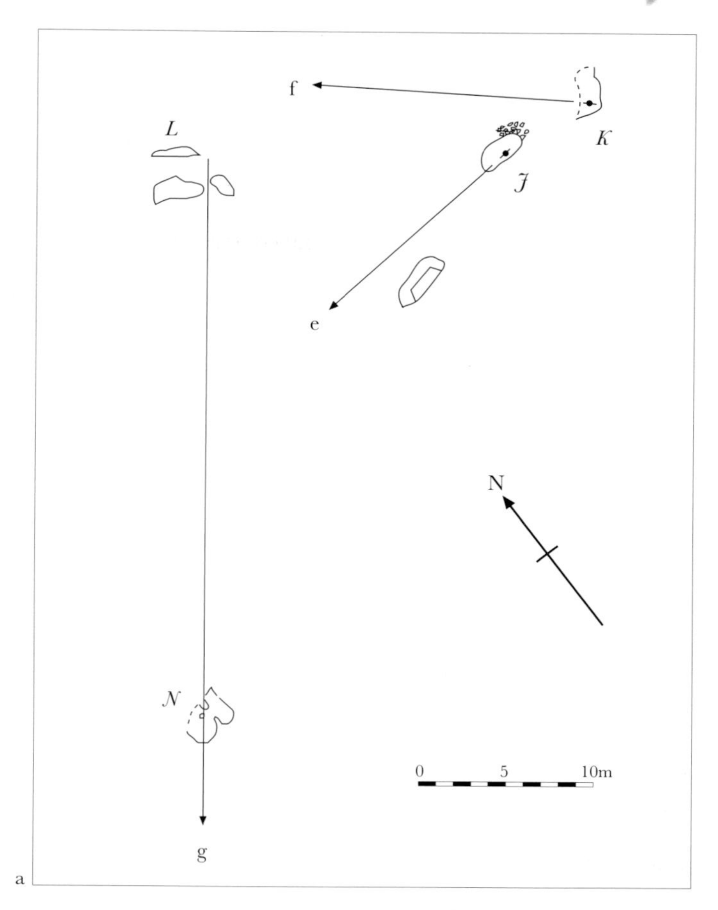

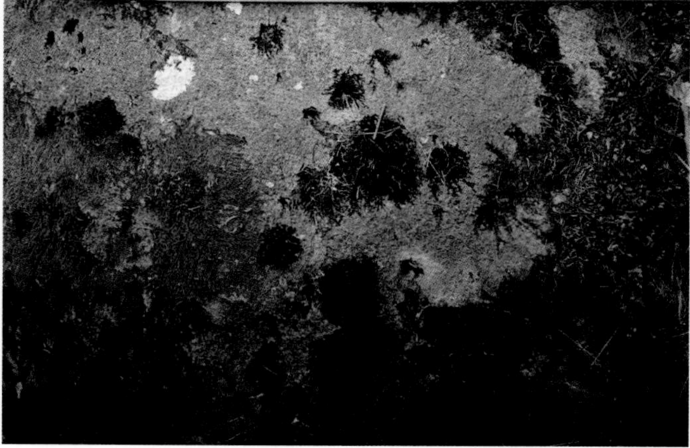


Fig. 1.17 The Oak Bank stone at Brainport Bay and associated solar alignments.
a. Features in the vicinity of the Oak Bank stone. Scales and orientation are approximate.
b. The cup-and-groove mark on stone *K*, emphasised using chalk, viewed from the north.

mark on stone *N* was discovered during a subsequent search along the alignment from the standing stone to the notch.[99] We have no idea how many equally prominent notches were ignored because they were in astronomically uninteresting directions, and how many other cup marks might be lurking unnoticed in the vicinity of the standing stone because they do not lie on astronomical sightlines and were not searched for.

A further problem is that there is no uniformity in the form of proposed horizon indicators: one is provided by the alignment from a standing stone to a cup mark on an outcrop, and the other by a cup-and-groove mark; furthermore, both are entirely different from the platforms, boulders and aligned pair of standing stones of the main alignment. While it would be unreasonable to expect that Bronze Age solar observers used exactly the same sort of device to indicate each alignment, it must nonetheless be acknowledged that the diversity of proposed methods of indication gives the present-day investigator a great deal of scope for assigning spurious astronomical explanations to chance alignments of archaeological features.

A final consideration about the possible astronomical significance of Brainport Bay concerns the nature and precision of the equinoctial alignment. Depending how it is determined, the declination of what is taken to be the equinoctial setting sun might be anywhere from $+0°8'$ to $+0°31'$ (for an explanation of this, see ahead to Astronomy Box 5). The declination of the sun setting with its upper limb in the Siaradh Druim notch at Brainport Bay is about $-0°7'$,[100] which is below the possible range. The Siaradh Druim notch can only be considered a precise indication of the equinox if 'the whole sun was observed in the notch instead of only the upper edge',[101] a proposal that seems to contradict the idea of using the upper limb to obtain great precision in the first place. Finally, there is a deeper problem in relation to supposed alignments upon the equinox, which is a concept not necessarily meaningful outside the Western scientific tradition (see chapter nine and Astronomy Box 8).

In short, it is difficult to agree that the evidence from Brainport Bay 'demonstrates to a high degree of probability that very long, potentially accurate [calendrical] alignments had . . . been devised'.[102] However, the example does serve to demonstrate some of the serious methodological pitfalls that can arise, even where, as in this case, an effort has been made to define a clear philosophical starting point.[103] In practice, the archaeological verification of an astronomical hypothesis has been turned into a cyclical process in which astronomical predictions are tested by directed fieldwork and excavation, then modified and re-tested, and so on. The general principle is fair enough; the problem is that while the archaeological evidence is allowed to modify the specific predictions, for example by adding more potential alignments, it is never allowed to influence the more fundamental hypothesis that Brainport Bay was a high-precision 'calendrical' site.[104] Thus, as contradictory data confront each suggested alignment, more are suggested in an attempt to bolster the calendrical idea, and the structure of 'supporting' evidence becomes steadily more cumbersome. Yet the increasingly attractive alternative, that the astronomy of the main alignment was of lower precision and all other alignments were fortuitous, is never considered. Instead, the idea of archaeological verification has been turned into mere *post hoc* justification, which in this case becomes less and less viable as the weight of evidence builds up against the hypothesis being proposed.

## Le Grand Menhir Brisé: a universal lunar foresight?

We shall stray briefly outside Britain and Ireland in order to provide a further illustration of the dangers of *post hoc* justification. Although the Brittany megaliths fall outside the remit of this book they form part of a related tradition that has been extensively studied.[105] Undoubtedly its most spectacular manifestations are the magnificent rows of Carnac, the longest of which, Kermario, consists of seven rows of stones ranging from half a metre to over four metres in height and runs for over a kilometre.[106]

Thom, together with his son Archie and a team of willing helpers, spent many years examining the megalithic monuments of Carnac in detail. A monument that began to attract his special attention was a massive fallen stone known simply as Le Grand Menhir Brisé, the great broken menhir (Fig. 1.18a). Now broken into four pieces, it appears originally to have stood to a height of more than 20 m, and would thus have been the tallest standing stone in Brittany. It is not of local stone and the task of hauling it over 4 km to its present position must have been almost unimaginable.[107]

Over time, the Thoms developed the idea that this huge standing stone was a foresight against which the rising or setting moon could be viewed from a number of different directions.[108] Their fieldwork established that suitable 'backsights' existed at observing positions from which the great menhir could be used to mark each of the eight principal limiting rising or setting positions of the moon (Fig. 1.18b).

The eight horizon positions arise because the motions of the moon are more complex than those of the sun. The moon's declination, like the sun's, varies cyclically between northerly and southerly limits, but on a timescale of a month rather than a year. Added to this, over a longer timescale—on a cycle of roughly 18·6 years—the limits themselves vary. The result is that during every nineteenth year the rising or setting position of the moon sweeps between monthly limits several degrees wider than the annual (solstitial) limits of the sun—these we shall refer to as the 'major limits' or 'major standstill limits'—whereas between nine and ten years later it only moves between narrow limits several degrees inside the annual sweep of the sun—these we shall refer to as the 'minor limits' or 'minor standstill limits'. For a more detailed explanation see Astronomy Box 4. The major limits represent the furthest positions to the north and south ever reachable by the moon. Fig. 1.20 shows schematically the directions of the solstices and the major and minor lunar limits from a location with a level horizon somewhere within Britain or Ireland; the exact azimuths are dependent upon the latitude and horizon altitude.

The eight rising and setting positions of the moon at its major and minor limits were considered to be significant by Thom. At first sight, the existence of suitable backsights in all eight directions from Le Grand Menhir Brisé seems very impressive indeed. But we must ask a crucial question: how confident can we be that the putative backsights identified by Thom and his team were actually intended as such? We know how the team went about locating candidates for backsights:

a

Fig. 1.18 Le Grand Menhir Brisé in Brittany.
a. View of the great menhir from the south-east.
b. The great menhir as universal lunar foresight with eight proposed backsights. After Thom and Thom 1971, fig. 2.

they identified the eight lines radiating out from the great menhir upon which such backsights would have to lie and examined what could be found along those lines.[109] Such an approach has an evident danger. The area around the great menhir contains the densest concentration of dolmens and standing stones in the whole of Europe,[110] so one would surely encounter features at least as promising in a great many different directions.[111] It would strengthen the argument in favour of the astronomical interpretation if the putative backsights formed a coherent set of structures archaeologically. They do not. On the contrary: the backsights turn out to be a diverse collection of stones, many of which are small and of doubtful prehistoric provenance: they include several boulders that are certainly natural and one cairn probably built a millennium later in the Iron Age.[112] Finally, there is considerable doubt as to whether the great menhir itself ever actually stood at all: it is possible that it broke during the process of erection.[113]

The lesson to be learned from this is that having developed a particular idea—in this case the notion that the great menhir might have functioned as a universal lunar foresight—a major aim of the original research design should have been to scrutinise that idea as thoroughly, and as pitilessly, as possible in the light of data that could be obtained in the field. An approach which involves walking out along alignments looking for potential backsights, failing to examine other evidence concerning the nature of the features encountered, and ignoring the likelihood of picking up similar features elsewhere, singularly fails to do this. If we are to have any hope of providing meaningful tests of our ideas, and hence of progressing those ideas while keeping them fully in tune with the evidence, then at the very least we must strive to collect and examine our evidence fairly, facing what is contradictory as well as what is confirmatory, coping with the negative as well as the positive.

An important aspect of this process, and one that we have already encountered in passing,[114] can be to estimate probabilities (see Statistics Box 1). The questions may be specific—what is the probability that the Newgrange solstice phenomenon could have arisen through factors quite unrelated to astronomy?; what is the probability that eight putative lunar

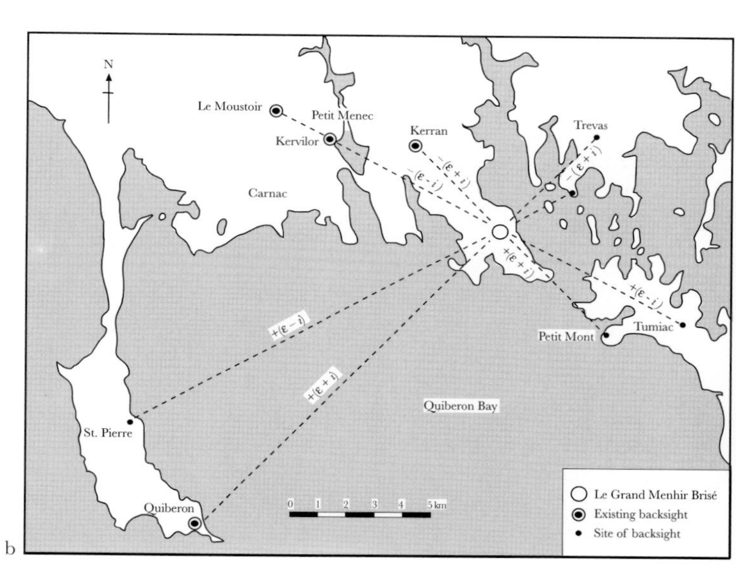


b

backsights for Le Grand Menhir Brisé could have arisen fortuitously?—but the general aim is always the same: to assess the degree to which the available data lend support to a given idea. The formulation of probability estimates has arisen in discussions of a number of 'classic' astronomical sites, but nowhere more prominently than at Stonehenge itself, to which we shall now turn.

## Stonehenge: the astronomer's dream?

Stonehenge (SU 122422) is a monument of considerable complexity. What we see there now (Fig. 1.21) is the cumulative result of a series of constructions and modifications on the same spot spanning as much as fifteen hundred years,[115] as they have survived after a further three and a half millennia. A basic (relative) chronology of the different features at the site was established by Richard Atkinson following his excavations during the 1950s,[116] although more recent investigations and reinterpretations have influenced the picture and the radiocarbon revolution has shifted the absolute dates.[117]

## ASTRONOMY BOX 4

### THE MOTIONS OF THE MOON, 1: LUNISTICES, STANDSTILLS, AND LIMITING DECLINATIONS

Just as the declination of the centre of the sun varies annually between limits of $\pm\varepsilon$, which it reaches at the solstices (Astronomy Box 3), so that of the moon varies also. In the case of the moon, however, the cycle takes only a month,[1] not a year. Furthermore, there is a complication: the most northerly and southerly declinations reached at the monthly 'lunistices'[2] are not constant from month to month, but themselves vary, over a period of some 18·61 years. The general way in which the moon's declination varies over this longer cycle, which we shall refer to as the lunar node cycle,[3] is shown in Fig. 1.19.

At one stage in each node cycle, the moon travels significantly further north and south each month than the sun does each year. Over a period lasting for about a year, the moon's declination at each lunistice comes close to theoretical outer limits given by $\pm(\varepsilon + i) - P$, where $i$ has the value 5°·15.[4] The amplitude of the monthly oscillation then begins to decrease until, some nine years later, its limits are well within the annual limits of the sun, so that the moon never ventures outside the declination limits $\pm(\varepsilon - i) - P$.

$P$ is a factor that has to be subtracted because of lunar parallax (see Astronomy Box 2) and has a value of around 0°·85 in Britain and Ireland.[5] Thus, for these latitudes in around 3000 BC, the widest monthly declination limits were approximately +28°·3 and −30°·1 while the narrowest were about +18°·0 and −19°·7.

#### TERMINOLOGY, AND SOME MISUNDERSTANDINGS

The time when the monthly swing of the moon's motions is at its widest is often, following Thom,[6] referred to as the 'major standstill'. Similarly, the time when the swing is at its narrowest is known as the 'minor standstill'. The term is convenient but misleading because the moon in no sense ever stands still. What reaches a maximum or minimum is a theoretical curve: the curve marked by a dashed line in Fig. 1.19b. The moon itself never ceases to hurry backwards and forwards between its monthly declination limits in the north and south.

Often the term 'standstill' is used more loosely, to refer to the limiting declinations themselves, which can cause confusion.[7] Other authors refer to the 'major moon' and 'minor moon'[8] or the 'maximum moon' and 'minimum moon'.[9] The declinations $\pm(\varepsilon + i) - P$ and $\pm(\varepsilon - i) - P$ are sometimes referred to respectively as the 'outer extremes' and 'inner extremes'.[10] This is misleading, at least in the case of the inner extremes, since they are not extreme in any sense; away from the minor standstill the moon passes each of these declinations twice every month.

In this book we attempt to avoid some of the confusion and overtones by using the term 'major standstill limits', or just 'major limits', to refer to the declinations $\pm(\varepsilon + i) - P$ and 'minor standstill limits', or just 'minor limits', for $\pm(\varepsilon - i) - P$. We can then refer, for example, to the moon setting at its northern major limit $[+(\varepsilon + i) - P]$, although when, if ever, it does so within a given lunar node cycle is another matter. It will only set exactly at its northern major limit if the northern lunistice coincides with the major standstill (which can occur at any time in the month), *and* if the time of setting coincides with the lunistice (which can occur at any time of day). There are also practical difficulties and, for higher-precision

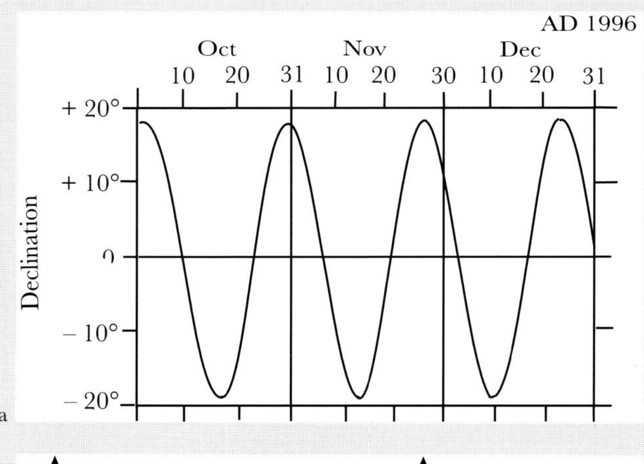


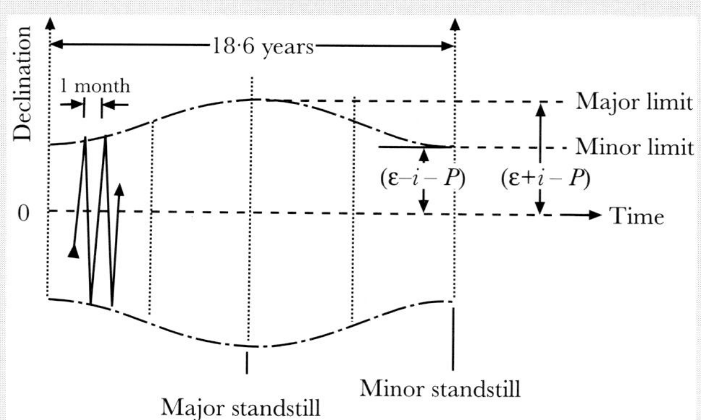


Fig. 1.19 The variations in the moon's declination.
a. The moon's monthly variation in declination (around the time of minor standstill). Adapted from Thom 1967, fig. 3.5b.
b. The form of the variation over a lunar node cycle. The solid line represents the actual path of the moon. The major and minor limits are *declinations*; the major and minor standstills are *times* (see text). Adapted from Thom 1971, fig. 2.2 and Ruggles 1997, fig. 1b.

observations, further complicating factors, which will be discussed in Astronomy Box 7.

### Directions of possible lunar significance

From any location in Britain or Ireland or at lower latitudes, the points where the major and minor limiting declinations intersect the horizon define eight directions which have often been considered to be of possible significance in the interpretation of alignments of prehistoric monuments in Britain and Ireland. They are illustrated schematically, along with the directions of solstitial sunrise and sunset, in Fig. 1.20.

Exceptions to this rule are locations, particularly in northern Britain, where the southern horizon is so high that the moon will stay below it when close to its southern major limit. An example is Ardnacross on Mull, discussed in chapter seven. At latitudes much above 61° (northern Shetland) the moon stays above the horizon when close to its northern major limit unless there is a high horizon in the north.

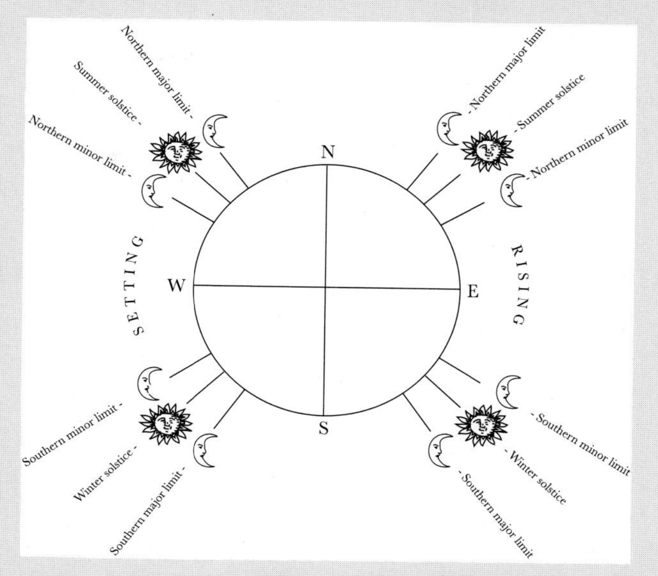


Fig. 1.20 Schematic representation of the directions of solstitial sunrise and sunset together with the major and minor lunar limits for a level horizon at a location in Britain or Ireland. Over the year, the sun rises and sets respectively within the eastern and western horizon arcs delimited by the solstices. Around the time of major standstill, which occurs every 18·6 years, the moon rises and sets within wider arcs delimited by the major limits, moving between these limits and back again once every month. Around minor standstill, mid-way between these times, the rising and setting positions are confined to the arcs between the minor limits. First published as Ruggles 1997, fig. 1a.

### Observations of the lunar limbs

The position of the upper or lower limb of the moon as it intersects the horizon can be directly observed, whereas the position of its centre can only be judged. Thus, if we are interested in the idea that lunar observations were made to reasonably high precision, we should consider the possibility that the declinations of the upper or lower limb, rather than the centre, were marked. The various configurations may be summarised as follows, where $s$ is the moon's semidiameter:[11]

| | Upper limb | Centre | Lower limb |
|---|---|---|---|
| Northern major limit | $+(\varepsilon + i + s) - P$ | $+(\varepsilon + i) - P$ | $+(\varepsilon + i - s) - P$ |
| Northern minor limit | $+(\varepsilon - i + s) - P$ | $+(\varepsilon - i) - P$ | $+(\varepsilon - i - s) - P$ |
| Southern minor limit | $-(\varepsilon - i - s) - P$ | $-(\varepsilon - i) - P$ | $-(\varepsilon - i + s) - P$ |
| Southern major limit | $-(\varepsilon + i - s) - P$ | $-(\varepsilon + i) - P$ | $-(\varepsilon + i + s) - P$ |

For the actual declination values see Astronomy Box 6.

Broadly speaking, Stonehenge 1 was a circular ditch and bank erected in an already well-used landscape around or a little after 3000 BC. The circle of 56 Aubrey Holes was probably (but not certainly) dug inside it at about this time and probably (but not certainly) held a ring of timber posts. Stonehenge 2, around 2900 to 2550 BC, marks a change in emphasis towards timber structures: complex patterns of postholes are found in the interior, as is an enigmatic grid-like formation in the north-east entrance, while the ditch and Aubrey Holes (now without posts) were generally left alone. Later, cremation burials were placed in many of the Aubrey Holes, as well as in the ditch, on the bank, and around the outside of the monument, and parts of the ditch were deliberately backfilled. In broad terms, Stonehenge 1 and 2 replace Atkinson's 'Stonehenge I'.

Stonehenge 3, which broadly replaces Atkinson's 'Stonehenge II' and 'Stonehenge III' and starts around 2550 BC, marks fresh activity on a much grander scale: the arrival of the bluestones and, probably at much the same time, the erection of the Heel Stone and a companion outside the entrance, and

Fig. 1.21 Aerial view of Stonehenge from the north-east, taken in 1963.

the construction of the Station Stone rectangle. About a century later, the giant uprights and lintels of sarsen stone were hauled to the site, carefully dressed, and erected to form the famous circle and horseshoe. Three large unlintelled sarsens were placed across the entrance, possibly at this time also. During the centuries that followed, some of the bluestones were dressed and set first into one position and then another, and a 2·5 km-long avenue was constructed linking the site to the River Avon. The site finally fell into disuse around 1600 BC. For further details see Archaeology Box 3.[118]

The various astronomical ideas associated with Stonehenge are extensive and it would serve no useful purpose to go through them all in great detail here.[119] However, a number of useful general principles can be illustrated by examining some of the arguments and counter-arguments relating to astronomy at Stonehenge.

A recurring theme in these arguments is the position of the Heel Stone. Despite persistent popular belief, it did not provide an exact marker of sunrise on the longest day of the year. It is not in the correct position: as viewed from the vicinity of the geometrical centre of the sarsen circle, the first gleam of the solstitial sun appears well to its left, only moving into line as the sun rises further and moves to the right. In the third millennium BC the sun would have risen even further to the left and been well clear of the horizon before it aligned with the Heel Stone.[120] In any case, an alignment of this nature could never achieve the required precision. This is because the distant horizon is flat, devoid of any natural reference point of the kind discussed in relation to Ballochroy and Kintraw. Instead, the Heel Stone must itself be used as foresight; but since it is only some 75 m distant from the observer, the observing position to be used from one day to the next would need to be specified to within a mere one or two centimetres, less than the distance from one eye to the other.[121] While it has been suggested that the Heel Stone might have served to determine the solstice exactly by a process of halving the difference, even this would have required an observing position precise to 10 cm or so.[122] In general, astronomical observations of the high precision needed to pinpoint a solstice, whether by direct observation or by halving the difference, require a suitably distant foresight. The closer the foresight, the smaller the room for manoeuvre at the observing position.[123]

Hawkins famously claimed that the phenomenon of midsummer sunrise over the Heel Stone was not isolated but merely one of several dozen astronomical alignments incorporated in the architectural design of Stonehenge at different stages in its construction. Working with his interpretation of Atkinson's chronology, Hawkins claimed that Stonehenge I contained no fewer than twenty-four putative alignments upon horizon solar and lunar targets such as the solstices and major and minor limits, marked by pairs of stones and stoneholes and by the directions of stones and stoneholes from the centre of the site (cf. Fig. 1.22a).[124] The probability of this happening fortuitously, Hawkins estimated, was only about six in a million.[125]

## STATISTICS BOX 1

### PROBABILITY AND ODDS

Terms such as 'probability', 'chance', and 'likelihood', although widely used informally in everyday language—usage that is not avoided in this book—have particular meanings when applied in a formal (mathematical) context. This and subsequent boxes will attempt to elaborate these meanings as far is necessary for a full understanding of the wider arguments being discussed.

Fundamental to many of these arguments is the notion of probability. This is a more complex concept than may appear at first, and one which can be approached in a number of ways and from different philosophical standpoints.[1] The explanation below is intended to be sufficient for most of the discussions in this book. For a different view see Statistics Box 7.

#### THE CONCEPT OF PROBABILITY

The concept of probability is often conceived through the analogy of a repeated trial or experiment. Suppose that a fair, six-sided die is thrown over and over again a very large number of times, perhaps thousands of times or more, and the results noted. In what proportion of the throws was the result a six? It will be found that the answer is almost exactly one sixth of the throws. If the number of throws is increased to millions the result will be incredibly close to—to most intents and purposes, exactly—$^1/_6$.

The question 'What is the probability that a single throw of the die will yield a "6"?' can be conceived as equivalent to the question 'If I throw the die a very large number of times indeed, in what proportion of throws will the result be a six?'. The answer, then, is $^1/_6$, or 0·1667. Similarly, the probability of throwing a four is also $^1/_6$; the probability of throwing a number greater than three is $^1/_2$. If something is impossible, it will occur with probability 0.[2] If something must happen, it will occur with probability 1. The probability of throwing a number between one and six inclusive is 1; the probability of throwing a seven is 0.

The notation P($A$) is generally used to denote the probability of event $A$. Thus if $A$ is the 'event' that the next throw of a die will yield a six, $P(A) = {}^1/_6$.[3]

The probability of an event such as throwing a six with a die is independent of what has gone before. Even if a six has already been thrown six times in a row, the probability of it happening again is still $^1/_6$ (always assuming the die is not a loaded one). The probability of two independent events both occurring is obtained by multiplying together their individual probabilities: thus, the probability of throwing a 6 twice in a row is $^1/_6 \times {}^1/_6$, or $^1/_{36}$. The general rule is

$$P(A \,\&\, B) = P(A) \times P(B) \qquad \ldots \text{(S1.1)}$$

where $A$ and $B$ are independent. It extends to any number of events $A$, $B$, $C$, etc. provided they are all independent of one another.

#### ODDS

Some people prefer to think in terms of odds rather than probabilities. Suppose that the odds of 'Astronomer's Apprentice' winning the 4.30 race at Ascot are 10 to 1 against. This means that if I place a bet of £1 on the horse, I will win £10 (plus getting my stake of £1 back) if it wins the race. If the bookmaker's odds are a fair reflection of the probability of the horse winning the race, then if it were possible to run the race over and over again very many times, I would eventually break even. In what proportion of races does the horse need to win so that this is the case?

The answer is not one race in every ten, as might be expected, but one race in every eleven. In this case, for every expenditure of 11 × £1, I will receive one bounty of £11 (£10 winnings + £1 stake). Thus, odds of 10 to 1 against correspond to a probability of $^1/_{11}$. Similarly, odds of 100 to 1 against correspond to a probability of $^1/_{101}$.

In general, odds of $n$ to 1 against correspond to a probability of $1/(n + 1)$ and vice versa. 'Evens' (odds of 1 to 1) correspond to a probability of $^1/_2$ or 0·5. Odds of '$m$ to 1 on' are equivalent to 1 to $m$ against, and hence correspond to a probability of $1/((1/m) + 1) = m/(1 + m)$. Thus odds of 6 to 4 on (= 1·5 to 1 on) correspond to a probability of $^{1\cdot5}/_{2\cdot5} = 0\cdot6$.

#### 'THE CHANCES ARE . . .'

When people say that the chances of something happening are 68%, they tend to mean that the probability of that something happening is 0·68. However, to say 'there is one chance in three' of something happening is more ambiguous: it may be interpreted as 'the probability of that thing happening is $^1/_3$' or as 'the odds against that thing happening are 3 to 1' (i.e. the probability of it happening is $^1/_4$). The difference between these two possibilities for the statement 'there is one chance in $n$' may be negligible if $n$ is large, but clearly for small $n$ it can be very significant. It is best simply to avoid this turn of phrase where a quantitative argument depends on it.

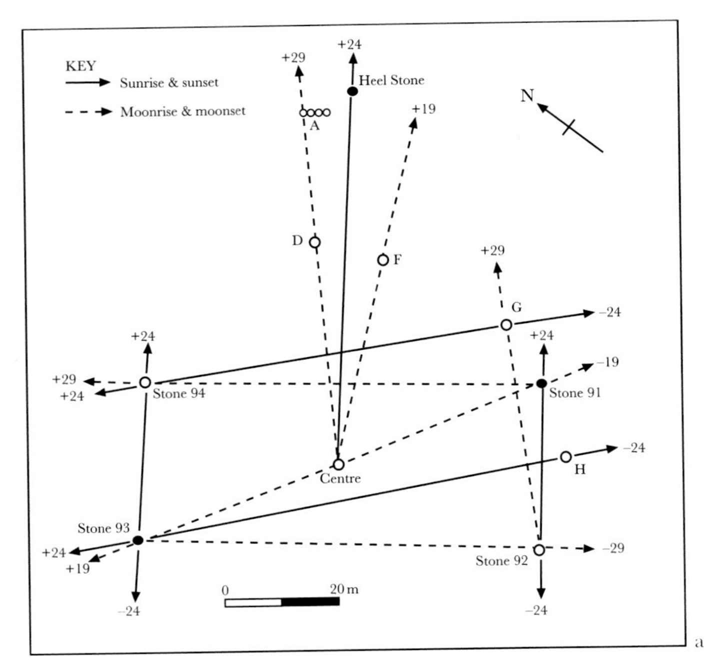


a

Fig. 1.22 Hawkins's and Hoyle's interpretations of what they took to be 'Stonehenge I'.

a. Some of Hawkins's alignments between pairs of stones and stoneholes. After Hawkins and White 1970, fig. 11. The full set of alignments discussed in the text is shown in *ibid.*, fig. 14.

b. Hoyle's scheme for eclipse prediction using the Aubrey Holes. After Hoyle 1977, fig. 5.1.

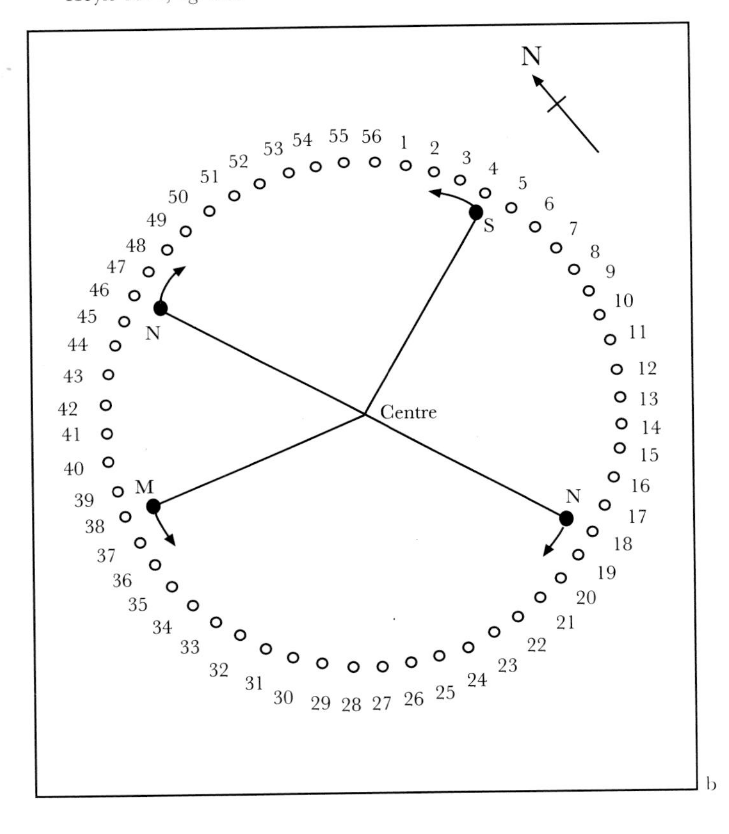


b

There were, however, three fundamental flaws in Hawkins's calculations, and when these are corrected, the probability of chance occurrence increases to better than evens (Statistics Box 2). The probability calculation is further undermined by the fact that the data are non-independent,[126] by the lack of *a priori* justification for the points chosen in the first place,[127] and by archaeological doubts about some of those that were.[128] To cap it all, some of the chosen targets make no sense astronomically.[129]

Another notorious idea worthy of mention is that the 56 Aubrey Holes could have been used to predict eclipses, by moving stakes or stones around the holes according to given rules and noting when they reach certain significant configurations (see Fig. 1.22b).[130] This raises questions of how such elaborate devices could have been set up without extensive record-keeping, for which there is no independent evidence.[131] There is also the fact that as eclipse predictors (as opposed to mere predictors of periods when eclipses definitely could *not* occur) they would have been highly unreliable,[132] and that simpler methods for doing the same thing are possible anyway.[133] Finally, other explanations of the 56 Aubrey Holes abound,[134] demonstrating just how easy it is to fit an explanation to this particular number (and perhaps to any other number) of holes.

The main problem with all these explanations is that they pay little or no attention to the archaeological evidence. Atkinson believed that the Aubrey Holes had been dug and almost immediately refilled many centuries before the station stones were put in place, so that the two aspects of Hawkins's astronomical interpretation of Stonehenge I could not have functioned simultaneously.[135] It is now thought likely that the Aubrey Holes first held a ring of timber posts, many of them being used later for cremations and other ritual—or at least formal—deposits (see Archaeology Box 3). Different as these two scenarios are, it is difficult to square either of them with the idea of the holes being containers for tally-markers. Perhaps most importantly of all, excavations elsewhere have shown that in its early phases Stonehenge was merely one of a number of broadly contemporary circular enclosures and henge monuments around Britain and Ireland, inside many of which were placed rings of timber posts and/or pits. For example, Aubrey Burl has listed ten sites containing pit circles where the number of pits in the ring varies from seven to forty-four or forty-five.[136] This must cast the severest doubt upon arguments that the actual number of Aubrey Holes at Stonehenge was of particular significance.[137]

The generation of new astronomical interpretations of Stonehenge gradually lost momentum during the 1970s, although in 1976 and 1977 some outlandish theories still continued to be published in *Nature*.[138] In 1975 the Thoms published a theory of Stonehenge as a universal lunar backsight, paralleling their earlier theory of Le Grand Menhir Brisé as a universal lunar foresight,[139] but their proposed artificial foresights did not stand up to archaeological reappraisal.[140] A decade passed before Stonehenge astronomy began to make a reappearance in the context of wider archaeological studies of Stonehenge,[141] which will be examined in detail later in chapter eight.

A number of points concerning procedure can nonetheless be made at this stage. The first is that attempting probability estimates from multiple alignments at a single site, although attractive in theory, can be highly misleading for a number of reasons, even when the calculations are done correctly. In particular it is very difficult to ensure, and hence to demonstrate to others, that the selection of data for such an analysis has been done in such a way that it is not influenced by the astronomical possibilities in the first place; and it is almost impossible to isolate questions of possible astronomical alignment from other design criteria that would directly relate one structural orientation to another, and hence to arrive at a set of data that can reasonably be argued to be independent of one another. Yet both criteria must be satisfied if the probability estimates derived are to have any useful meaning at all.

A second point is that it can be dangerous to rely on excavated features, even if the currently available evidence is fully taken into account and fairly interpreted. Not all of the site has been excavated and new discoveries can completely dislocate current theories, the prime example being the stone adjacent to the Heel Stone whose existence was discovered during a small salvage excavation in 1979.[142] On the other hand attempts to use archaeological evidence to refute astronomical theories can backfire. Thus Atkinson's argument against Hawkins's and Hoyle's interpretations of the Aubrey Holes as an eclipse predictor, on the grounds that the Aubrey Holes were refilled very soon after being dug,[143] is undermined by more recently published evidence suggesting that the holes may well, in fact, have contained timber posts at first and been subsequently left open after the posts were removed.[144]

Archaeologists have been quick to criticise Stonehenge astronomy where it has been subsequently invalidated by archaeological discoveries, as in the case of Peter Newham's interpretation of alignments from the Heel Stone and station stones to three large postholes some 250 m to the north-west, discovered when the present car-park was being built in 1966.[145] They subsequently turned out to be of Mesolithic date, holding great timber poles (very possibly more akin to totem poles than to posts, and not necessarily standing erect at the same time) predating Stonehenge 1 by over four thousand years.[146] Yet it is surely unreasonable to expect any astronomical interpretation to be any less reliant upon the current state of knowledge than wider archaeological interpretations, or any less susceptible to change in the light of new evidence. On the other hand, it is equally unreasonable for anyone putting forward an astronomical interpretation not to take the available archaeological evidence fully into account.

A final point concerns differing interpretations. At Stonehenge, as at Ballochroy, many debates are never finally settled. Interpretations may and likely will continue to differ, even where the archaeological evidence seems unequivocal. Thus Rodney Castleden postulates that a cult-house might have existed at the site of Stonehenge in Mesolithic times, from which the timber poles in what is now the car park might indeed have marked midsummer sunset.[147] Then again, Christopher Chippindale uses environmental evidence that indicates a discontinuity of occupation at Stonehenge to argue that there could not have been a continuously developing astronomical tradition,[148] whereas Castleden uses evidence from a wider archaeological context to argue that there was social continuity and continuity of tradition in the general vicinity of Stonehenge, so that even if the site itself was temporarily abandoned, the astronomical traditions that were reflected there when significant activity resumed were directly developed from those that had gone before.[149] Archaeological evidence can often cut both ways.

## Classic sites: some lessons

Undoubtedly, great personal satisfaction may be derived in exploring a stone monument, investigating the potential alignments, identifying horizon features, taking measurements, calculating declinations, and formulating an astronomical explanation for the site's construction and layout. In doing so, one may feel somehow 'in tune' with the mysterious people who erected the stones. Such a feeling doubtless motivated Sir Norman Lockyer as he surveyed numerous stone rings in the early years of this century, and it certainly motivated Alexander Thom in later decades.[150] There are, however, a great many pitfalls awaiting the aspiring archaeoastronomer.

In examining past interpretations of some of the 'classic' sites of megalithic astronomy a number of dangers will have become evident to the reader, such as *post hoc* justification, circular argument, and perhaps most of all the tendency to emphasize those data that confirm a preconceived set of ideas while ignoring those that do not. How then can we prevent this sort of thing continuing to happen?

One possible answer, as we have seen, is to try to estimate the probability that observed astronomical alignments could have arisen fortuitously, that is through a combination of factors quite unrelated to astronomy. But probability estimates of this sort at a single site are fraught with difficulties, not just because of the need to select the basic data fairly but also because of the whole question of non-independence.

If it is dangerous to rely on excavated features, as we concluded at Stonehenge, how much more dangerous it must be to rely on those features that remain conspicuous above the surface, in the absence of excavation. Standing stones may well have shifted in the thousands of years since their erection, or even been re-erected in relatively recent times, unbeknown to the modern investigator. Features just as important as the large standing stones that we can pick out so easily now may well be inconspicuous in the absence of excavation, or may have disappeared completely, as at Ballochroy. Even what is apparently a simple monument may have been used and modified over a considerable period, and excavations can reveal unexpected complexity beneath the ground.[151]

When all these procedural dangers and restrictions are recognised, the potential for saying anything with any reasonable degree of confidence at a single site may seem severely limited, certainly frustratingly so for the would-be archaeoastronomer hoping to derive the astronomical significance of a particular monument. There is a further problem. Evidence on astronomical alignments is one step further removed from the interpretative process than most material evidence studied by archaeologists: no-one would deny that a piece of Bronze Age pottery was intentionally made, recognised as a piece of pottery and used by Bronze Age people, although we can argue over its function and meaning; but was an astronomical alignment of Bronze Age structures that may be obvious to us necessarily intended, recognised or used by the people who built those structures? Some archaeologists may expect the archaeoastronomer to prove beyond all doubt that a given astronomical alignment was intentional before they consider it further:

## STATISTICS BOX 2

### PROBABILITIES OF CHANCE ALIGNMENTS UPON ASTRONOMICAL TARGETS

Any alignment of man-made structures upon a horizon point of apparent astronomical significance, such as a particular rising or setting point of the sun or moon, is susceptible to the question 'Did it arise by intention, or as a result of factors quite unrelated to astronomy'? Where several alignments can be considered together, we can pose the question 'What are the chances that this many alignments of apparent astronomical significance could have arisen fortuitously (that is, as a result of factors quite unrelated to astronomy)?' An answer may be obtained by the use of a straightforward mathematical formula. How meaningful this answer is will depend on what initial assumptions are made and upon a number of other factors.

#### Shooting at a target: the formula

Suppose that we have a number of alignments pointing at places on the horizon which have apparent astronomical significance, and wish to estimate the probability that this many astronomical alignments could have arisen fortuitously.

Borrowing from Hawkins and White,[1] we can use the analogy of a blindfold marksman, who is repeatedly spun round before firing a shot which hits the horizon in a random place.[2] Scattered around the horizon are a number of astronomical 'targets': ranges of horizon that could be construed as astronomically significant. We can calculate the probability that the marksman will hit exactly $r$ targets in $n$ shots. The answer, known as Bernoulli's formula, is

$$\frac{n!}{r!(n-r)!}p^r(1-p)^{n-r} \qquad \text{(S2.1)}$$

where $p$ is the proportion of the horizon occupied by astronomical targets or, equivalently, the probability that the marksman will hit an astronomical target in a single shot.[3] $r!$ indicates the factorial of $r$, namely $r \times (r-1) \times (r-2) \times \ldots \times 1$ (note that $0! = 1$).

It follows that the probability that the marksman will hit at least $r$ targets in $n$ shots is

$$P = 1 - \sum_{s=0}^{r-1} \frac{n!}{s!(n-s)!}p^s(1-p)^{n-s} \qquad \text{(S2.2)}$$

where the sigma symbol indicates that the formula must be evaluated with $s = 0, 1, 2$, etc. up to $(r-1)$ and the results summed.

For example, if $p = 0{\cdot}1$ and the marksman has 100 shots, then the probability of him hitting at least ten targets is 0·55, or somewhat over ½; but the probability of him hitting sixteen or more is 0·04 or only 1/25, and the probability of him hitting at least twenty-eight is less than 0·000001, or one in a million.

The smaller the probability that the astronomical alignments could have arisen by chance, the more likely it is that they were in fact intentional. There is a tendency to adopt a certain probability level, perhaps 0·05 or one in twenty, as indicative that the astronomical hypothesis should be accepted in favour of non-astronomical alternatives. However, to do this is to introduce an artificial barrier in what is in fact a continuous scale. It is clear, though, that we would be unwise to take a set of astronomical alignments as intentional if they could have arisen by chance with a probability as great as, say, one in five; whereas we would be unwise to ignore them if the probability of chance occurrence is as low as, say, 0·001 or one in a thousand.

#### Applying the formula in practice: determining the parameters

Before formula (S2.2) can be applied, it is vital to follow certain procedures in order to arrive at suitable values for the proportion of the horizon occupied by astronomical targets, $p$, the number of shots, $n$, and the number of astronomical 'hits' $r$. If unsuitable values are used, the result obtained may be at best misleading and at worst completely meaningless.

First, there is the question of how the astronomical targets are defined (and hence the determination of $p$). This will involve assumptions about perceived margins of error: how far away from a theoretical astronomical target must an alignment be before we would cease to accept it as indicating that target? We have no *a priori* justification for making any particular choice; and worse, if our choice is based upon the data themselves, then it may be tempting to choose that value that gives the smallest value of $P$ and hence makes the alignments appear least likely to have arisen by chance. Suppose, for example, that we notice that several alignments fall between 1°·0 and 1°·1 away from theoretical astronomical targets; if we then assume that the margin of error is 1°·1 rather than 1°·0, so as to be sure to include all these extra alignments as 'hits' without expanding the target area very much, and $P$ is reduced as a result, then we are weighing the odds in favour of the alignments appearing deliberate. The least we can do in this circumstance is to consider both margins of error in turn, and compare the results.

Related to this is the issue of which theoretical targets might be construed as potentially significant in the first place. There are many possibilities, including particular rising and setting positions of the sun and moon as well as those of other celestial bodies. If the choice is made on the basis of the data (for example, if lunar targets are included but solar ones excluded on the basis that many lines appear to point at lunar targets in the first place) then, once again, a circularity will have entered and the result will be weighed unfairly in favour of the alignments appearing deliberate.

A set of issues attaches to the selection of data, i.e. what constitute the alignments that are chosen for consideration in the test. If $n$ is to be a true reflection of the number of shots made by the marksman, we need to include not only all the alignments of possible astronomical significance but also all those alignments that might equally well have been identified as putative astronomical indications had they pointed at an astronomical target. It is all too easy to spot the former, but requires much greater discipline and determination to make a fair estimate of the latter. An ideal procedure from the statistical point of view might be to select all alignments for consideration before any oriented site plans are examined or before any surveys are carried out, but this may not be feasible or desirable in practice. Many more specific issues relating to data selection are addressed in Part One of this book.

### The example of Stonehenge I

Many of these issues are best illustrated by an example, and it is here that Hawkins's interpretation of Stonehenge I is useful. We shall examine the claim that twenty-four out of fifty possible alignments are of astronomical significance (see chapter one, note 124), and that the probability of this happening fortuitously is only 0·000006.[4] In fact, three errors have been made.[5]

First, one should calculate the probability of at least, rather than exactly, twenty-four alignments having arisen fortuitously, using formula (S2.2) rather than (S2.1). This increases the probability slightly, to 0·000008.

Second, there are many more than fifty possible alignments between pairs of those points considered by Hawkins at Stonehenge I. There are fourteen such points, comprising stones and stoneholes and also the geometrical centre of the monument. Within fourteen points there are ninety-one possible pairs. But Hawkins is prepared to consider alignments in either (or, in many cases, both) directions between the points in a pair, so the total is actually twice this, or 182. The number of possibilities may be reduced by eliminating alignments which can be argued to be inherently unlikely, such as those marked by two points very close together and those towards, rather than away from, the centre (which to our knowledge was unmarked); but the total is still at least 111.[6] The probability of at least twenty-four fortuitous hits out of 111 shots is 0·37, more than one in three.

Third, there is the question of margins of error in the astronomical targets. Hawkins and White considered eighteen astronomical targets, allowing a possible error of up to 2° in each. Eighteen targets each 4° wide cover 72°, or a fifth of the horizon, so Hawkins assumed $p = 0{\cdot}2$. In fact, three of the twenty-four alignments are over 2° from their target, so shouldn't have been included as hits at all.[7] The probability of at least twenty-one fortuitous hits out of 111 shots is 0·65, considerably better than evens. Alternatively, we could increase the width of the target ranges to 5°, enough to include two of the remaining three alignments, but now the probability that what is observed could have arisen by chance increases to 0·88.[8]

Other difficulties with Hawkins's interpretation of Stonehenge I are dealt with in the main text.

### A cautionary note: the assumption of independence

The alignments being considered must effectively be independent of each other, otherwise the analogy of the blindfold marksman breaks down. Take, for example, a row of five standing stones, oriented in one direction to an acceptable degree of precision upon an astronomical target such as midsummer sunrise. If we were to consider the alignments formed by all pairs of stones at that monument, we would arrive at the conclusion that ten out of twenty possible alignments were astronomical, since they all point in the same direction. Really we only have one hit out of two shots,[9] not ten out of twenty, because since the stones form a row, all ten orientations in each direction must of necessity be the same. If, say, $p = 0{\cdot}1$, the probability of at least one hit out of two shots arising fortuitously (from formula S2.2) is 0·19, or roughly one in five, whereas that of ten hits out of twenty is 0·000007, or seven in a million. It is clear, then, that making the assumption of independence falsely can lead to grossly misleading results.

## ARCHAEOLOGY BOX 3

### STONEHENGE AND ITS ARCHAEOLOGICAL CONTEXT

The earliest constructions at Stonehenge of which we are aware were built around or a little after 3000 BC[1] in an area where there had been considerable human activity for at least a thousand years (Fig. 1.23a, b). During the previous millennium the site of the future Stonehenge appears to have been within the domain of influence of a causewayed enclosure known as Robin Hood's Ball, one of five major causewayed enclosures in the Wessex chalk uplands. Each of these was an important focal point and was surrounded by a cluster of long barrows built during the earlier part of the fourth millennium BC[2] (see also Archaeology Boxes 2 and 4). Two cursuses, the longer of which runs for almost 3 km across undulating landscape, were probably constructed shortly before 3000 BC, a little to the north of the later site of Stonehenge.[3] The presence of such monuments indicates that the process of forest clearance was by this time well advanced, and environmental evidence (Archaeology Box 5) confirms that earlier patches of woodland had now largely disappeared and the landscape consisted predominantly of open grassland.[4]

#### STONEHENGE 1: *c.* 3000–2900 BC

Stonehenge 1 (Fig. 1.24a) was built at a time when causewayed enclosures were falling into disuse. It was a circular earthwork enclosure; a ditch was excavated using antler picks and the chalk fill was used to build a bank on the inside, forming a ring about 100 m across. There was a 15 m-wide entrance gap to the north-east and two narrower entrances to the south, one of which was later blocked. The ditch was built over a short space of time, possibly fifty years at most; the date can be established quite accurately from radiocarbon dating of antler picks found in its base.[5]

A circular ring of fifty-six holes known as the Aubrey Holes—spaced quite regularly at intervals of between 4·5 m and 4·8 m centre to centre[6]—is encountered just inside the bank. Exactly when the Aubrey Holes were dug is more difficult to determine. The fact that the geometrical centre of the Aubrey ring coincides with that of the ditch and bank suggests that these features were broadly contemporary, but there is no conclusive evidence.[7] The Aubrey Holes probably held a ring of timber posts but, again, this is not certain.[8]

A number of other features—postholes in the interior of the monument (see below), over fifty smaller stakeholes in the outer part of the north-east entrance (also see below), and three postholes discovered under the bank on the south-east side—may also belong to this phase, and one or more of them may even predate the construction of the ditch and bank, although there are arguments why this seems unlikely.[9]

#### STONEHENGE 2: *c.* 2900–2550 BC

Within no more than about a century of its construction, the ditch and bank monument began to be modified, apparently in line with the tradition of 'henge' monuments (see Archaeology Box 2) that were now starting to be built in many parts of Britain, including three—Durrington Walls, Woodhenge, and Coneybury—within just 4 km of Stonehenge (Fig. 1.23c).[10] Land in the region was increasingly coming under cultivation, and there is evidence of settlement within 1 km of Stonehenge.[11]

The emphasis at Stonehenge itself seems to have switched. A number of timber structures were built in the interior: this is clear from the presence of a large number of postholes within 20 m of the centre, variously interpreted as a roundhouse or one or more circles of wooden posts,[12] together with what appears to have been a 'passageway' leading towards the southern entrance (Fig. 1.24b).[13] Some fifty postholes in a grid-like formation across the north-east entrance were probably (but not certainly) also dug at this time,[14] as were a line of (at least) four holes, labelled the 'A' holes, placed across the axis some 20 m outside the north-east entrance.[15]

Meanwhile, much of the ditch was allowed to fill naturally, as were the Aubrey Holes, now devoid of posts.[16] Later, cremation burials—sometimes along with other items such as bone skewer pins, pottery fragments, and objects of flint and chalk—were placed in several of these; cremation deposits were also placed in the upper ditch, on or just inside the bank, and around the outside of the monument.[17]

#### STONEHENGE 3: *c.* 2550–1600 BC

During the next thousand years the landscape in the vicinity became intensively farmed, with the appearance of permanent field systems, farmsteads and settlements.[18] When it was not being used for food production, organised labour was clearly available on a vast scale, not only at Stonehenge but also at the great henge of Durrington at Woodhenge.[19] Stonehenge itself underwent one modification after another. It is not possible, at least on current evidence,

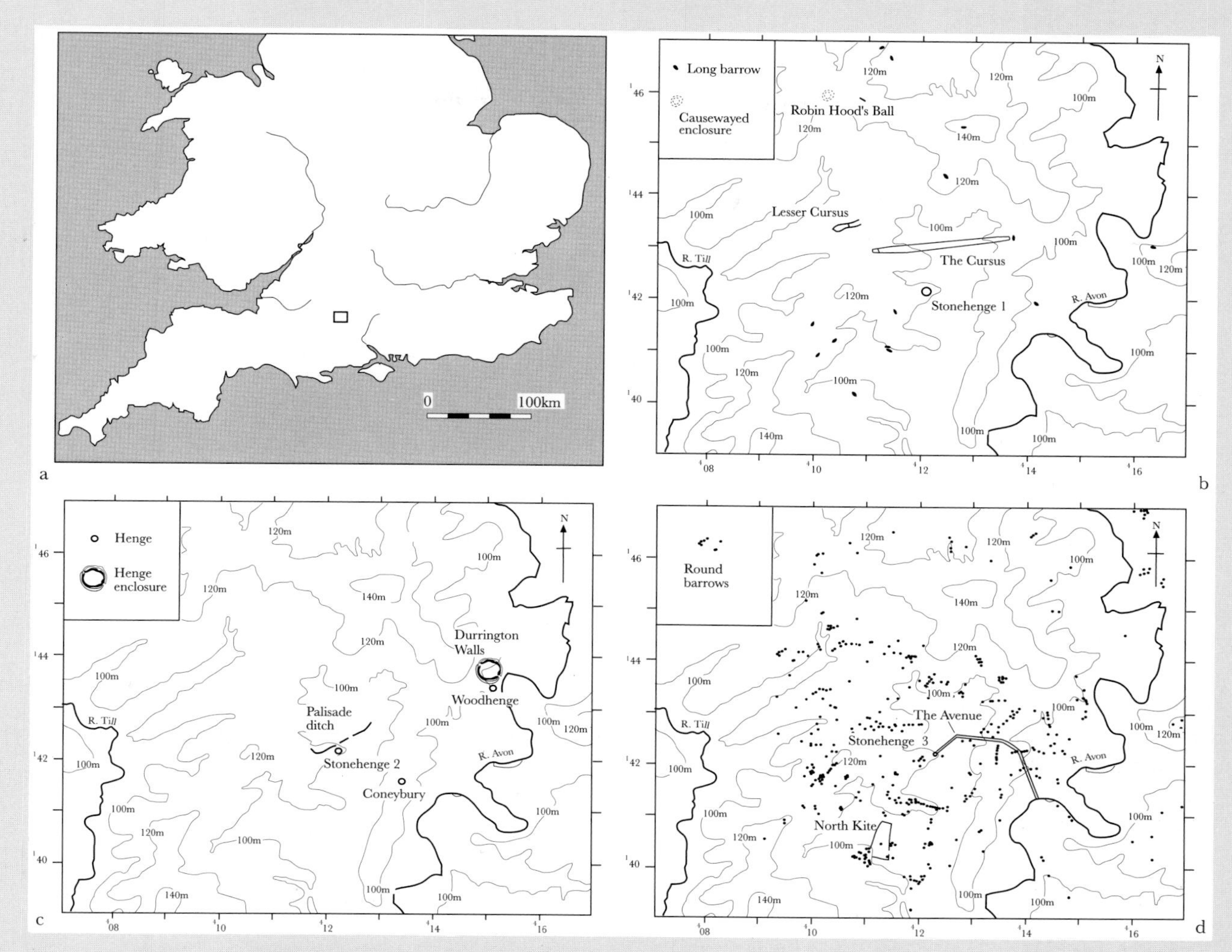


Fig. 1.23 Landscape and monuments in the vicinity of Stonehenge.
a. Location of the Stonehenge area.
b. *c.* 4000 BC to 3000 BC (after Cleal *et al.* 1995, figs 33 and 35).
c. *c.* 3000 BC to 2500 BC (after Cleal *et al.* 1995, fig. 57).
d. *c.* 2500 BC to 1500 BC (after Cleal *et al.* 1995, fig. 78).

to provide a definitive sequence of events but a tentative succession of construction episodes can be identified within the interior of the monument and, separately, on the periphery by the entrance and outwards to the north-east.

## *The interior*

At around 2550 BC the decision was made to construct circular settings in stone rather than wood. The bluestones were brought to the site then, on an extraordinary journey of several hundred kilometres, mainly by river and sea, from the Preseli Mountains of south-west Wales.[20] They were erected, possibly already dressed, in a setting that may have formed the north-eastern arc of a double circle, or a more irregular figure (Fig. 1.24c).[21] In any case, the focus of attention may well have been the Altar Stone, a slab of green sandstone acquired from a place on the bluestone route from the Preseli Mountains and erected on the south-west side of the double circle, opposite the entrance.[22] This subphase is now identified as Stonehenge 3i.

About a century later, the giant sarsen stones weighing 25 tonnes or more were hauled from the Marlborough Downs some 29 km to the north of Stonehenge, and erected to form the famous ring of thirty uprights joined by lintels together with an inner horseshoe of five trilithons (Stonehenge 3ii) (Fig. 1.24d).[23] It is likely that the smaller bluestones, now dressed, were also erected at this time, but exactly where and in what formation is not clear (Stonehenge 3iii).[24] Subsequently, they were set out as an oval and circle respectively within the sarsen trilithon horseshoe and circle (Stonehenge 3iv). The northern

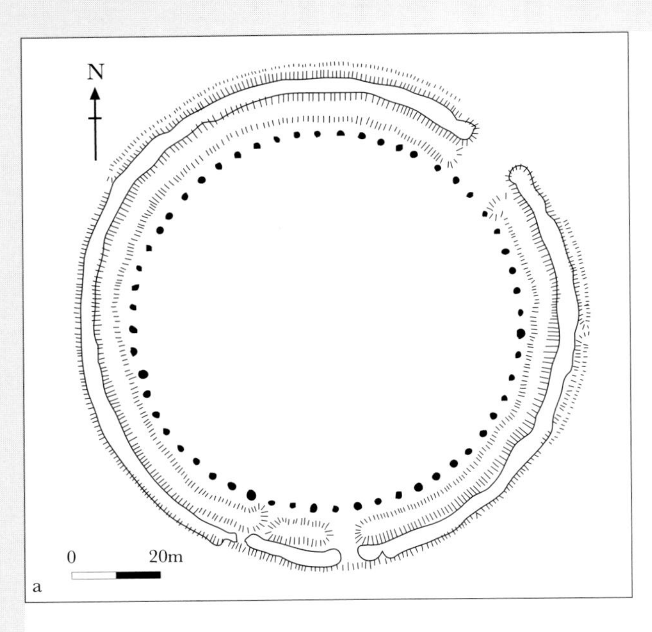


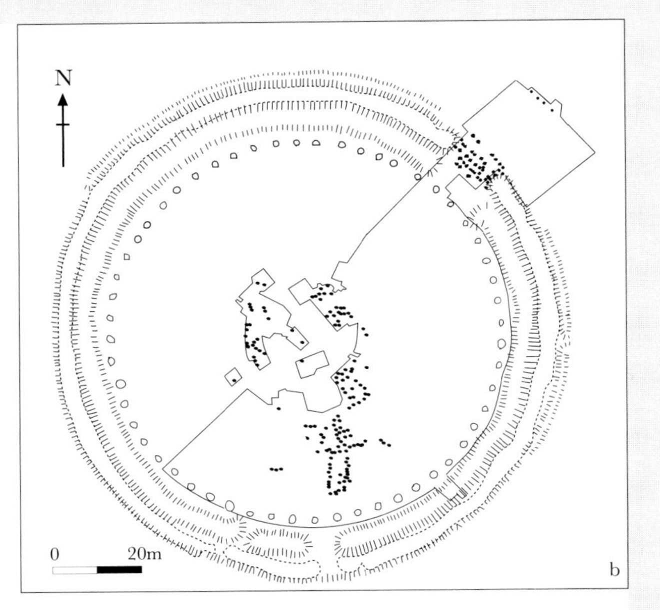


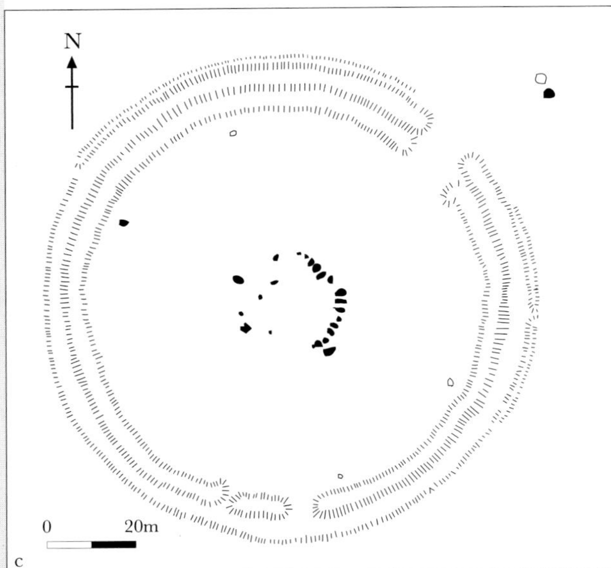


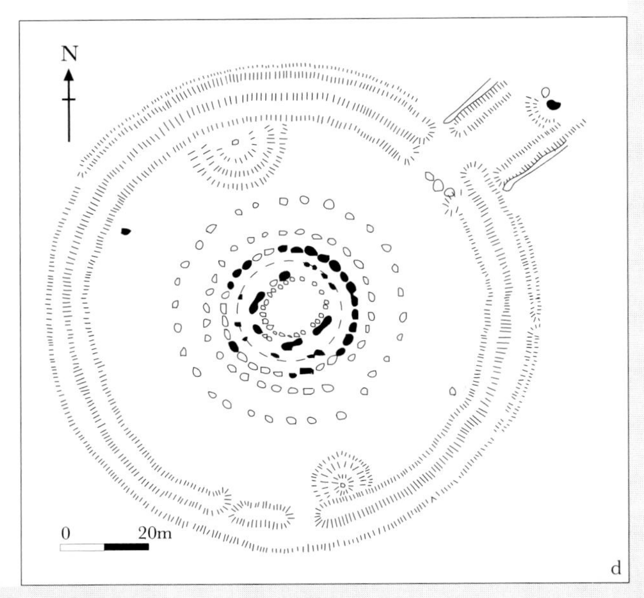


Fig. 1.24 Phases in the development of Stonehenge.

a. Stonehenge 1 (*c.* 2950 BC) (after Cleal *et al.* 1995, fig. 256).

b. Stonehenge 2 (*c.* 2900 to 2550 BC). The limits of excavated areas are shown (based on Cleal *et al.* 1995, fig. 66).

c. Stonehenge 3i and 3a (*c.* 2500 BC) (after Cleal *et al.* 1995, fig. 256).

d. Stonehenge 3ii and 3b onwards (*c.* 2400 BC to 1600 BC), simplifying the different subphases (after Cleal *et al.*, 167).

arc of the oval was later removed to produce an inner bluestone horseshoe (Stonehenge 3v). Finally, two concentric circles of holes were dug outside the sarsen ring (Stonehenge 3vi), apparently as part of a remodelling plan that was subsequently abandoned.[25]

### *The periphery*

It is unclear precisely how these events related to the sequence of events at the periphery. The Heel Stone,[26] a large, unworked sarsen block, was probably erected at an early stage, together with a companion stone a couple of metres to its north-west,[27] forming a pair some 20 m away outside the north-east entrance.[28] It seems most likely, but is by no means certain, that the four Station Stones were also erected at about this time,[29] forming a subphase (Stonehenge 3a) oriented on a different axis from the earlier phases and which is broadly contemporary with Stonehenge 3i (Fig. 1.24c).[30]

Later, a narrow ditch was dug around the Heel Stone, which probably (but not certainly) implies that the companion had been removed by this time, and two of the Station Stones were replaced by mounds (the 'North Barrow' and 'South Barrow') surrounded by ditches. These events may well have been contemporaneous with each other (Stonehenge 3b) and roughly contemporaneous with Stonehenge 3ii (Fig. 1.24d). The Heel Stone ditch was soon refilled.[31] The avenue, a pair of earthen banks roughly 20 m apart running for 2·5 km from the north-east entrance of Stonehenge to the River Avon (Fig. 1.23d), was constructed between about 2250 BC and 1900 BC (Stonehenge 3c), and on current evidence seems most likely to have been broadly contemporaneous with Stonehenge 3iv. It is not known whether it was constructed as one (long) operation or in a number of separate stages.[32]

At some time during Phase 3 three sarsen uprights were placed close together, but without a lintel, across and just inside the entrance. One of these survives as the so-called 'Slaughter Stone', which stood until the seventeenth century.[33] There are two possible stoneholes situated on the axis, midway between the two sides of the avenue, between the entrance and the position of the Heel Stone and its companion, but it is not certain that they actually held stones and whether, if they did, the stones that they held were placed elsewhere before or after.[34]

While some of the details are unclear, it is indisputable that during the early stages of Stonehenge 3 there was a marked change in direction of the axis of symmetry of the monument, which was shifted clockwise by several degrees.[35] However, the evidence does not support Atkinson's idea that the earthwork entrance was widened so as to bring it into line with the avenue and the new orientation, by backfilling the first few metres of the ditch on the south-east side.[36] Instead, it seems that backfilling took place at various points around the ditch, including both sides of the entrance, during the earlier Stonehenge 2.[37] When the avenue was built, the bank to the south-east of the old entrance remained, extending across about a third of the end of the avenue and perhaps continuing to restrict access to the monument.[38]

### *The environs*

During this time, round barrows were built in concentrations in the vicinity of Stonehenge (Fig. 1.23d). There are a number in the immediate vicinity (within 500 m),[39] but the main concentrations are along ridges about 1 km away to the north-west (Monarch of the Plain and others), north-east (Cursus group), east (King Barrow group) and south (Normanton Down group) where they were prominently visible from Stonehenge.[40]

> There is a profound difference between the data collected by an archaeoastronomer and an archaeologist. Data for the latter is something that can be grasped, measured, conserved or visited; it has tactile qualities. The archaeoastronomer has first to prove the existence of the data that is used, before any analysis from which valid conclusions are sought, can be attempted.[152]

While it is unrealistic to expect 'proof', it does not seem unreasonable to demand a fair degree of confidence that a given astronomical alignment was intentional before proceeding towards interpretation. But how, then, can such confidence be established? We have seen that probability arguments are highly problematic at single sites. A clue is provided by the example of the Aubrey Holes at Stonehenge. In attempting to answer his critics, Newham asked the question 'Why the necessity for 56 holes? Why go to the trouble of dividing a circle into a relatively complex number?'[153] The probable answer is that no-one did, very possibly no-one even counted them; and that the number 56 arose through a combination of factors quite unrelated to astronomy. As we have seen, this becomes evident as soon as a comparison is undertaken with similar sites[154] and no repetition of the number 56 is found.

A further example of the value of looking beyond a single site is provided by the example of Carn Ban (NR 991262), a Clyde tomb (see Archaeology Box 2) on the Isle of Arran.[155] The passage of this tomb is oriented upon a horizon point with a declination of +23°·9, close to the position of the midsummer rising sun.[156] Taken by itself it may be tempting to surmise that this astronomical orientation was intentional, but there are twenty-one other Clyde tombs on Arran with measurable orientations, and these face all around the compass, with a wide spread in declinations.[157]

Just as looking at groups of monuments can help to spot those 'one-off' astronomical alignments at individual sites that are most probably fortuitous,[158] so repeated trends amongst groups can begin to provide evidence of a statistical nature that astronomical alignments really were intentional. The bulk of Thom's work on 'megalithic astronomy' was aimed at providing just such evidence. It consisted of the analysis of measurements from large numbers of sites taken together, and it is to this work that we turn in chapter two.

# 2

# Backsights and Foresights

*The Work of Alexander Thom and its Reassessment*

> He made an instrument to know
> If the moon shine at full or no.
>
> Samuel Butler, *Hudibras*, 2:3 (1663), 261.

> We estimate that the Ballinaby site was used at the spring equinox about 4am and at the summer solstice about 10pm when the temperatures are about 40° and 50°F.
>
> Archibald S. Thom, 1981[1]

> In 1977 I visited [Callanish, Kintraw, Ballochroy, Temple Wood (Kilmartin) and Brodgar]. These sites proved psychologically devastating to my tentative acceptance of precision astronomy in ancient Britain. . . . By focusing his attention on the specific astronomical sightlines, Thom neglected to inform his readers of the richer archaeological context of many of the megaliths.
>
> Owen Gingerich, 1981[2]

## Thom's approach: the four levels

Alexander Thom began to survey megalithic monuments in the 1930s and continued to do so, whenever time permitted, until almost fifty years later. He did this with considerable vigour and enthusiasm:

> From [his] notebooks it is possible to follow Thom's travels around the country. To take 1955 as an example: in early April Thom was in Perthshire and Angus, having travelled north from Oxford on 28 March (Easter Monday). On his return journey he visited eight sites in the Lake District on the weekend of 15–17 April. In mid-July we find him in Devon and Cornwall, and then in Aberdeenshire, Perthshire and Inverness in August, where at least 19 sites, scattered from Perth to Culloden, were visited in the space of just six days. On Wednesday 13 September he was at Stainton Dale and Fylingdales in Yorkshire, and on the weekend of 24–25 September he visited three sites in Derbyshire. In total, notes were taken at 60 sites; no mean feat when it is remembered that this was the age before motorway travel.[3]

As a result he accumulated survey data from several hundred 'megalithic sites', and it is through analyses of data from many of these sites taken together, rather than from discussions of individual monuments, that by far the most important evidence in favour of 'megalithic astronomy' derives. This evidence is cumulative in nature, and is most conveniently divided into four stages, or 'levels'. Each stage involves analyses that test for astronomical alignments of greater precision than the previous stages, and at each stage evidence emerges of greater observational exactitude than before.[4]

*Level 1.* The earliest such analysis, published in 1955, involved the declinations 'indicated' by seventy-two structures at thirty-nine megalithic sites.[5] This was extended in 1967 to 261 indications at 145 sites.[6] On the basis of these data Thom suggested the existence of deliberate solar, lunar and stellar alignments set up to a precision of (at least) about half a degree, roughly equal to the diameter of the sun or moon. The solar alignment targets include the solstices, equinoxes and intermediate declinations representing equal divisions of the year into eight and possibly sixteen parts. The lunar alignments are upon the major and minor limits.

*Level 2.* In 1967, Thom published further analyses of those Level 1 indications falling near the solar solstitial declinations and the major and minor standstill limits, about thirty of the former and forty of the latter.[7] This suggested that the upper and lower limbs of the sun and moon were preferentially observed, and increases the inferred precision to at least about ten minutes of arc, or roughly a third of the solar or lunar diameter.

*Level 3.* In work first published in 1969,[8] Thom concentrated exclusively on the idea that natural foresights on the distant horizon were used to mark the motions of the moon with great precision, with megalithic structures serving merely to identify the observing position and the relevant foresight. His analysis[9] suggested the use of distant foresights for observations precise to at least 3′, or about a tenth of the diameter of the moon. Just setting up suitable sightlines must have involved co-ordinated observing programmes spanning one or more 18·6-year cycles.

*Level 4.* The analysis at Level 3 took no account of a number of small corrections that vary from one site to another and one indication to another. In three papers published by Alexander Thom and his son Archie in the late 1970s and early 1980s,[10] each sightline was considered on its own merits, taking into account the time of year and the time of day of presumed use. They concluded that the horizon markers studied were precise

## STATISTICS BOX 3

### PROBABILITY DISTRIBUTIONS

Statistics Box 1 was concerned with the probability of particular events occurring. Statistics Box 2 introduced general formulae (S2.1 and S2.2) for calculating probabilities that depended on the values of three parameters $r$, $n$, and $p$. Given $n$ and $p$, for example, the probability that a blindfold marksman will hit exactly $r$ targets (S2.1) could then be calculated and plotted for different values of $r$, resulting in what is known as a 'probability distribution' over $r$. $r$ is known in this context as a 'random variable'. The distribution is illustrated in Fig. 2.1 for $n = 111$ and $p = 0{\cdot}2$.

Probability distributions may be discrete (as in this example) or continuous. They are theoretical constructs generated using a particular model (in this case the blind marksman model) which permit us to make precise statements about the anticipated outcome of an experiment—or, more rigorously, the anticipated value of a random variable (in this case the number of hits, $r$). From such abstractions, we can make more specific inferences about actual data, such as whether the alleged solar and lunar alignments at Stonehenge I are likely to have been intentional.

In order to answer particular questions it is convenient to be able to summarise a probability distribution in various ways. Two well known and particularly useful 'summary statistics' are the mean and standard deviation. The mean of the distribution, $\mu$,[1] yields the 'expected value' of the random variable, the average value that would be expected if a large number of similar experiments were performed. In the example shown in Fig. 2.1, the mean is 22·2. The standard deviation, $\sigma$, is a measure of how spread out we would expect the different values obtained in repeated experiments to be, higher standard deviations indicating greater spread.[2] The value for the distribution in Fig. 2.1 is 4·2.[3]

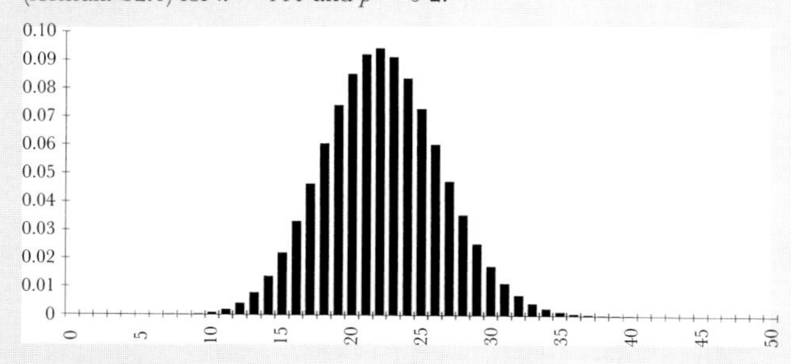


Fig. 2.1 The binomial distribution generated by Bernoulli's law (formula S2.1) for $n = 111$ and $p = 0{\cdot}2$.

#### THE NORMAL DISTRIBUTION

While there is no limit to the number of different probability distributions that can be generated to model the expected result of various processes, a relatively small number arise sufficiently often to be widely useful, and their properties have been studied extensively by statisticians. The one that has been by far the most important in the development of modern statistics, and remains the most ubiquitous, is the normal distribution. For each mean and standard deviation there is a single corresponding normal distribution $N(\mu, \sigma)$ whose general form is always as shown in Fig. 2.2, although the numerical values on the axes refer to particular values ($\mu = 22{\cdot}2$, $\sigma = 4{\cdot}2$).

The normal distribution[4] arises as a result of many different models. For instance, the probability distribution described above and illustrated in Fig. 2.1 (known as the binomial distribution) approximates to a normal distribution; the larger the value of $n$, the better the approximation. This is evident from a comparison of the two figures, where the mean and standard deviation have been deliberately chosen to be the same. But certain general properties of the normal distribution apply whatever the values of $\mu$ and $\sigma$. It is always true that 68·3% of the area under the graph lies between $\mu - \sigma$ and $\mu + \sigma$. This implies that, in an actual experiment, the probability that the value of the random variable will turn out to lie within one standard deviation of the mean is 0·683; the probability that it will lie outside this range is only 0·317, or under one third. The probability that it will turn out to be outside the wider range $\mu - 2\sigma$ to $\mu + 2\sigma$ is a mere 0·045, or under one in twenty.[5]

When Thom uses gaussian humps in his curvigrams, he is assuming implicitly that the processes of degradation which convert an intended declination $\delta$ into a measured declination $\delta_m$ will result in a probability distribution for $\delta$ given $\delta_m$ that is

Fig. 2.2 The normal distribution $N(\mu, \sigma)$. The numerical values on the axes refer to particular values of $\mu$ and $\sigma$ that are the same as the mean and standard deviation of the distribution in Fig. 2.1.

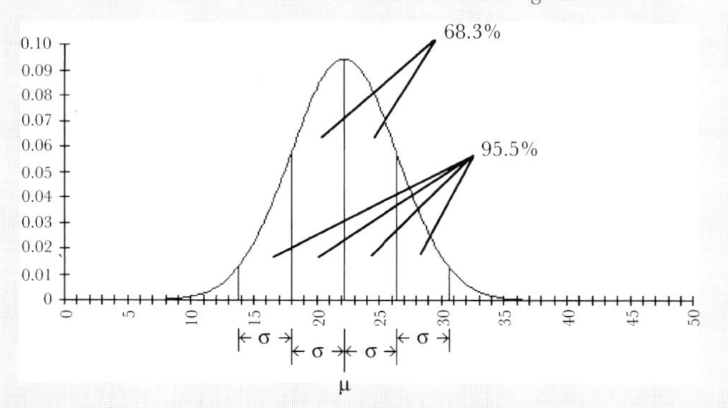

adequately modelled by a normal distribution. Similar assumptions are commonly made and in this case it does not seem unreasonable. In recent years, however, it has become possible for this powerful but simple and elegant analytical tool to be superseded by complex computer simulations for modelling distributions, including those encountered within archaeology.[6]

### THE EXAMPLE OF RADIOCARBON DATES

The notion of normal distributions is most familiar to archaeologists in the context of radiocarbon dating. Dates are obtained either by counting the number of spontaneous decays of $^{14}C$ isotopes to $^{12}C$ during a measured time period in a given sample of organic material, or by directly measuring the ratio of $^{14}C$ to $^{12}C$ isotopes using accelerator mass spectrometry. When an animal or plant dies and ceases to absorb carbon from its surroundings, the carbon within it contains a proportion of unstable $^{14}C$ isotopes similar to that prevailing in the ecosystem of the time. Subsequently, half of the remaining $^{14}C$ isotopes will spontaneously decay to $^{12}C$ during any given period of 5730 years. Using this information the age of a given sample can be estimated.[7]

The results are modelled by a normal distribution. Uncalibrated radiocarbon dates are always quoted within a margin of error, representing one standard deviation from the mean: thus '4530 ± 60 BP' implies that the date estimate is normally distributed with mean $\mu = 4{,}530$ and standard deviation $\sigma = 60$. It is not always appreciated that in over 30 per cent of radiocarbon dates the actual date can be expected to lie outside the quoted range. In this example, there is a probability of just 0·68 that the date lies between 4590 BP and 4470 BP. By doubling the margin of error the probability is increased to 0·95, but this means that there is still a one in twenty chance that the date falls outside the range 4650–4410 BP.

Because $^{14}C$ dating relies on the assumption that the proportion of $^{14}C$ in the atmosphere does not vary over time—an assumption known to be false—it is necessary to calibrate radiocarbon dates by some independent means, such as dendrochronology (tree-ring dating). The resulting 'calibration curves' are far from regular,[8] so that while the normal approximation may be adequate before calibration, much more complex probability distributions result on the calibrated timescale.[9]

to better than a single minute of arc; indeed, so precise that it seems they could only have been set up at the end of an averaging process lasting some 180 years.[11]

A number of publications during the late 1970s and early 1980s challenged Thom's results at the different levels from different standpoints. Amongst these were three comprehensive critiques by the present author based upon archaeological reappraisals and first-hand measurements in the field.[12] In this chapter we examine in some detail the data and their interpretation at each of the four levels, and draw general conclusions that will help us to take the discussion forward.

## LEVEL 1: SOLAR CALENDAR, MOON AND STARS

The central question at each of the four levels is whether each given set of putative astronomical alignments can quite adequately be explained away as a fortuitous occurrence. The first step in answering it, though, is not to enter into statistical arguments. Assessing the formal statistical significance of any set of results is relatively easy,[13] but it is only worth doing once we have satisfied ourselves that the results will be valid and meaningful; and this will only be true if the sightlines chosen for analysis have been selected fairly in the first place, that is in a manner totally uninfluenced by the astronomical possibilities. Thus it is the question of data selection that will be the first, and main, point needing to be tackled at each of the levels.

Thom summarised the evidence at Level 1 graphically, by plotting a cumulative probability histogram, or 'curvigram',[14] of the indicated declinations. The motivation for this is as follows. The declination that we measure today may not accurately reflect the intentions of the builders: for example the stones may well have shifted in the millennia since a monument was erected. While the measured declination may be the most likely one, it is also possible that the actual intended declination was a little to either side of it. For this reason Thom plotted each indication on the graph not as a point but in the form of a 'gaussian hump' (i.e. a normal curve), representing the spread of probability over declinations close to the measured one. (For more on normal distributions see Statistics Box 3). The humps were then plotted cumulatively in order to reveal whether any declinations were particularly preferred.

Displaying the data in this way has two great advantages. First, the process will tend to even out errors such as those due to deterioration since prehistoric times, and thus stands the best possible chance of showing if there are any obvious trends over and above the 'background noise' of alignments that have no consistent astronomical significance. Second, it makes no presuppositions about the nature of the astronomical targets (if any) that were aligned upon: if particular celestial bodies and events were of interest, then this should result in accumulations of alignments around particular declinations. If we find peaks at certain declinations, we can then seek to interpret the

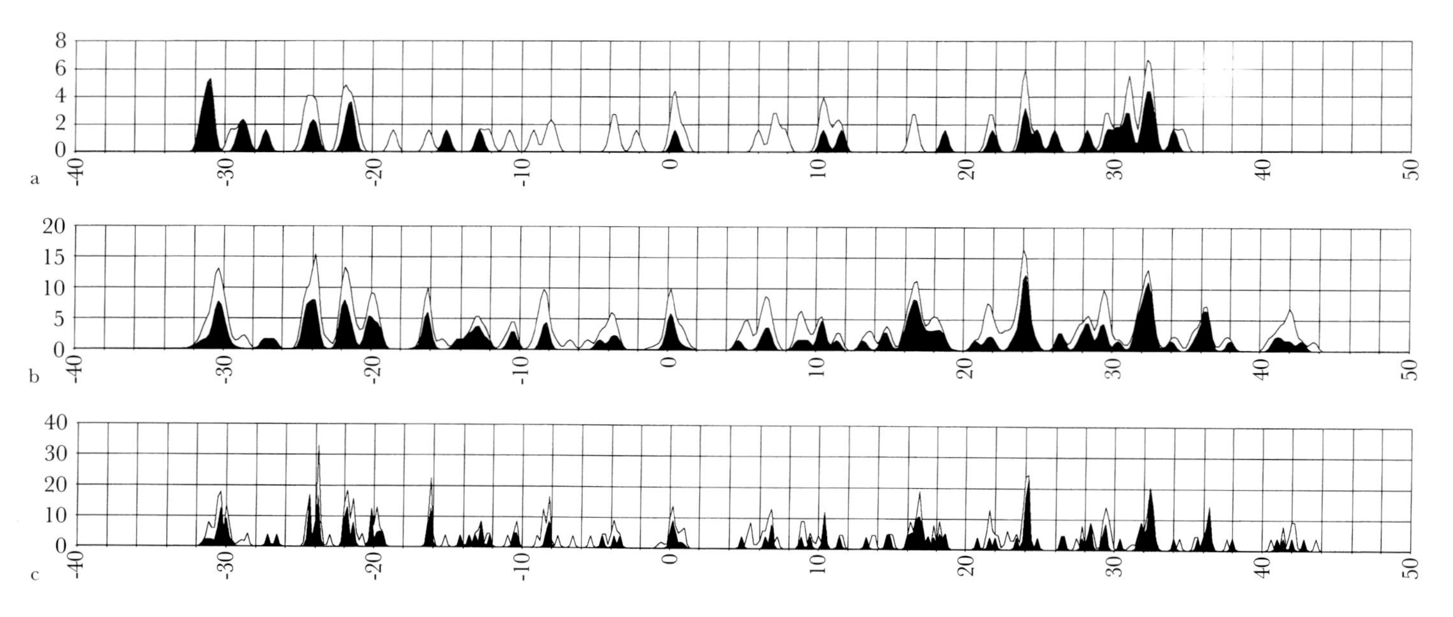


Fig. 2.3 Graphical summary of the indicated declinations in Thom's 'Level 1' data, in the form of 'curvigrams' (cumulative probability histograms).

a. Declinations indicated by seventy-two structures at thirty-nine sites, from the data in Thom 1955, tables 5 and 6. The shaded area represents data from short stone rows and aligned pairs. The unshaded area represents indications defined by a ring centre to an outlier. The area under each constituent gaussian curve is 1.0 and the standard deviation σ is 0·25 in every case.

b. Declinations indicated by 261 structures as 145 sites, from the data in Thom 1967, table 8.1. The shaded area represents 'Class A' alignments considered by Thom to be objective. The unshaded area represents 'Class B' alignments whose selection, by Thom's own admission, contains a subjective element. The area under each constituent gaussian curve is 1·0. σ is taken as 0°·25 except for those lines marked by Thom as less accurate, where it is taken as 0°·5.

c. As (b), except that σ is taken as 0°·1 (and 0°·2 for less accurate lines).

particular astronomical body or bodies to which they might correspond.

Fig. 2.3 is derived from the declination data produced by Thom,[15] and shows both the 1955 and 1967 datasets.[16] The area under each constituent hump is the same, so that equal weight is given to each indication.[17] In fact, the visual effect is dependent on the standard deviation (σ) chosen. For the 1955 data we have followed Thom and taken σ to be 0°·25 in every case, so that each constituent hump is of the same height and width (Fig. 2.3a). In the 1967 dataset Thom distinguished between ordinary data and a few cases where the indicated declination was determined less accurately. In Fig. 2.3b we continue to use σ = 0°·25 for the ordinary data and use 0°·5 for the 'less accurate' data. In Fig. 2.3c we use 0°·1 and 0°·2 respectively, a choice that assumes that the measured declinations are much more accurate reflections of the intended ones, and seems to reflect the values used by Thom in plotting his own graph.[18]

It is clear that there are considerable accumulations of probability at certain declinations and complete avoidance of others. This is quite different from the relatively smooth curve that would be expected if the indications measured had nothing to do with astronomy.[19] Peaks are evident around the solar solstices (±24°) and also to some extent around the four lunar limits (+28°, +18°, −20° and −30°), especially in the south. Thom, however, was struck by the fact that accumulations occur at six other positions representing the sun's declination at eight equal intervals throughout the year, counted from the solstices. This led him to postulate the existence of a calendar of eight and possibly sixteen equal divisions, with indicators of sunrise or sunset on important calendrical 'epoch' dates being erected throughout much of Britain. For the technical details of Thom's 'megalithic calendar' see Astronomy Box 5.

There are also peaks in the curvigrams that do not obviously correspond to significant positions of the sun or moon, including a sharp peak at about +32°, well outside the range of both. Thom interpreted these declinations in terms of bright stars. The problem here is that the declinations of stars change a good deal over a timescale of centuries (see Astronomy Box 6). This means that it is relatively easy to fit a stellar explanation to an orientation by choosing an appropriate date: thus, for example, over a period of 500 years the fifteen brightest stars cover about one third of the horizon.[20] Although Thom's stellar explanation was backed up with statistical arguments,[21] recent evidence from the physicist Bradley Schaefer indicates that atmospheric scattering and haze near to the horizon is a more severe problem than Thom realised,[22] and that all but the brightest stars under the most exceptional conditions will cease to be visible well before they reach the horizon, or only become visible after they have risen considerably above it.[23] Furthermore, according to Schaefer, changes in atmospheric conditions will vary the altitude (and hence the azimuth) of appearance or disappearance significantly from day to day.[24] If this is the case, then orientation upon stars simply could not have resulted in peaks as sharp as those observed in Fig. 2.3, which leaves us with the problem of finding an explanation for some of the most prominent peaks on the chart.

First, however, we must be satisfied that the data have been selected fairly. A number of questions arise here that are not immediately answerable from Thom's publications.[25] First, how were sites chosen for consideration? The 1955 reference list contained 'all the available data'[26] but the general reference list of sites for the 1967 book[27] comprised (only) those which were surveyed and 'contribute to the material of this book'.[28] As is clear from the sequence of site reference numbers used by Thom—A1/1, A1/2 etc.—for each site included about two others, which were present in an unpublished site list, were missed out.[29] Furthermore, not all the sites appearing in the general reference list are carried through to the source list which is used for astronomical indications.[30] Amongst those omitted are types of indication considered important by Thom such as alignments (rows) of standing stones;[31] more are included in the unpublished list but are not carried through to the general reference list.[32]

We must ask why such sites were found unsuitable for inclusion. 'Legitimate' reasons, that is reasons that would not introduce any overall bias, are unrelated to the possible astronomical function of a monument; its being in too bad a state of repair, for example, or weather conditions being too bad for a survey when the site was visited. It is all too easy, however, to conceive of ways in which large-scale bias could have been introduced. Suppose, for example, that monuments were first examined roughly, possibly using a magnetic compass, and only those with indications in astronomically 'interesting' directions were subsequently surveyed.[33] Unfortunately, it is not possible to explore procedural issues such as these on the evidence in Thom's publications.[34]

A second, more tricky question concerns the selection of potential indications at each site. In the 1955 analysis rigid selection criteria were adhered to. Thom considered only indications defined by the line from the centre of a megalithic ring to an outlier,[35] by two slabs in line, or by a row of three or more stones. In the 1967 analysis, however, there are no such clear-cut selection criteria, and subjective judgements are involved.[36] For example, there are a number of cases where the indication in one direction along a stone row has been included, but not the other.[37] If the decision to include or exclude a potential indication was influenced by the astronomical possibilities, wittingly or otherwise, then the data will mislead us.[38]

A further complication arises from the fact that two types of indication are included in the 1967 list. Most of the declinations quoted there are those of a point on the horizon, generally otherwise indistinguishable, in line with the mean orientation of some man-made structure such as a row of standing stones. Some 20 per cent of the quoted declinations, however, are those of candidates for natural foresights—prominent points on a distant horizon such as mountain peaks or notches between hills. The assumption here is that the structure on the ground does no more than roughly point out (preferably uniquely) which foresight is to be used. The problem with this is that at many sites there are a large number of equally prominent horizon features, and the selection of any particular one may well be influenced by the astronomical possibilities.[39]

Furthermore, and finally, we must ask on what grounds a given indication has been taken to be an indicated foresight. This very choice can be influenced by the astronomical possibilities,[40] and there is clear evidence that it has been.[41] For example, the declination (−21°·3) recorded for the indication to the SSE along the three-stone row at Duachy (Loch Seil) in Lorn (part of Thom's A1/4; LN22 in List 2 in the Reference Lists of Monuments on pp. 172–99) is that of the bottom of a depression towards the left of the range of horizon indicated by the actual alignment. This is listed as an indicated foresight and corresponds to sunrise on one of Thom's calendrical epoch dates. However, at Ballymeanoch (Duncracaig), mid-Argyll (Thom's A2/12; AR15 in List 2), the declination given for the four-stone row to the south-east is towards the top of a hill slope. It corresponds to solstitial sunrise (−23°·7). The foot of the slope is ignored as a plausible candidate for an indicated foresight, apparently because its declination (about −26°·3) does not fit an obvious lunar or solar explanation in Thom's scheme. Another example is Comrie (Tullybannocher) in Perthshire (Thom's P1/8; L22 in List 1), where the declination recorded is that of a horizon foresight some 4° to the left of the alignment of the two stones; Thom notes elsewhere that 'This might be lunar to the west where there is a little peak, but it is not a convincing site. The stones do not lie along the line'.[42] Thom himself believed that some monuments, such as the Castlerigg stone circle in Cumbria, incorporated only low-precision alignments while others were intended for far more refined observations.[43] But we must now question whether such a belief was really justified by the evidence. At this stage, the argument looks dangerously circular.

There are other worries. Some archaeological misinterpretations are included amongst the data. For example, Clachan Sands (Clach an't Sagairt), North Uist (Thom's H3/2), a backsight for a calendrical indication,[44] is simply a large natural block with a Latin cross incised near one corner.[45] There is no evidence that the site had any significance in prehistoric times. A supposed 'stone circle' at Upper Fernoch (Tayvallich), Knapdale (part of Thom's A3/4), is a natural formation of large boulders.[46] Another, Auldgirth, Dumfriesshire (Thom's G6/2), is an imitation built (probably) in the early nineteenth century.[47] The dataset also includes a circle-to-outlier line at Castlerigg (Thom's L1/1), in fact 'one of the lines which convinced the author of the necessity to examine the calendar hypothesis in detail';[48] yet the outlier was in fact moved to its present position at the edge of a field in recent times.[49]

But perhaps most worrying of all is the general lack of coherence in the data. While some of this admittedly arises from misidentifications,[50] the main problem is that there is no obvious evidence of consistency in the initial choice of sites and/or geographical areas. The north-eastern Scottish recumbent stone circles represent a group of over ninety similar monuments in north-eastern Scotland, yet only five of them are represented in the 1967 source list for astronomical indications. On the other hand, the dataset includes a variety of types of megalithic monument from all over Britain: stone circles, short rows of standing stones, pairs of standing stones, single standing stones, and longer rows of small stones, as well as various types of cairn, from northern Scotland to Cornwall. The types of indication are equally varied: along the stones of a row, along the flat face of a single stone, from the centre of a circle to an outlier, between the centres of two circles, between two stones on the opposite side of a circle (Fig. 2.5 shows one of three different indications across the ring at Castlerigg included in the dataset), along the passage of a cairn, and several more.[51] This wide variety of astronomical indicating devices seems odd if there really was uniform astronomical, and particularly calendrical, practice throughout Britain.[52] The diversity may well simply reflect how easy it is to fit theories to a site rather than revealing a function that the monuments actually served.[53]

Nonetheless, it would be unwise simply to dismiss the whole of Thom's Level 1 evidence because of these doubts. Instead,

## ASTRONOMY BOX 5

### DIVIDING THE SOLAR YEAR

#### THOM'S SOLAR CALENDAR

Thom suggested that prehistoric people divided the time interval from one solstice to the next into eight, or perhaps sixteen, periods of equal length.[1] (The word 'month' should be avoided since these time periods are not directly related to the cycles of the moon.) The start- and end-points of these divisions formed what Thom called 'epochs', dates upon which the sunrise or sunset position would have been of special significance, including the solstices themselves and what we might refer to as the 'megalithic equinoxes'[2] (but see Astronomy Box 8).

In considering these ideas it is necessary to take into account the fact that the sun's distance from the earth is not constant, and for this reason the earth does not travel around the sun at a constant rate. The day-to-day variation in the sun's declination is slightly less than expected from formula (A3.1) during the half of the year when it is farther away from the earth, and slightly greater in the other half, when it is nearer. A formula that expresses this additional effect is

$$\sin\delta = \sin\varepsilon\cos[0{\cdot}9856n + 2{\cdot}07\sin(0{\cdot}9856(n - n_p))] \quad \ldots \text{(A5.1)}$$

where $n_p$ is the time of perihelion (the point when the earth is closest to the sun) measured in days forward from the June solstice.[3] As in previous boxes, all angles are expressed in degrees. In AD 2000, perihelion occurs on about 3 January ($n_p = 196$). In 2000 BC it occurred on about 24 October ($n_p = 125$). In 4000 BC it occurred around the autumnal equinox ($n_p = 91$).[4]

Owing to this effect, the actual declination δ 'runs ahead of' that of the 'mean sun' given by formula (A3.1), which we now denote by $\delta_m$, during the half-year following perihelion, and 'lags behind it' for the half-year approaching perihelion. The meaning of 'running ahead of' and 'lagging behind' itself depends on the time of year: in the half-year from winter solstice to summer solstice, when the sun's declination is increasing, δ 'runs ahead of' $\delta_m$ means $\delta > \delta_m$, but in the other half of the year the opposite is true. This principle is illustrated in Fig. 2.4.

The outcome is to alter the declination of the sun at certain times of year by up to about half a degree. For $n_p = 125$, corresponding to about 2000 BC, the difference between δ and $\delta_m$ is greatest when $n$ is between about 47 and 76 and about 227 and 260 (early August to early September and early February to early March), when it exceeds 0°·6. Around the equinoxes, δ exceeds $\delta_m$ by a little over 0°·5.

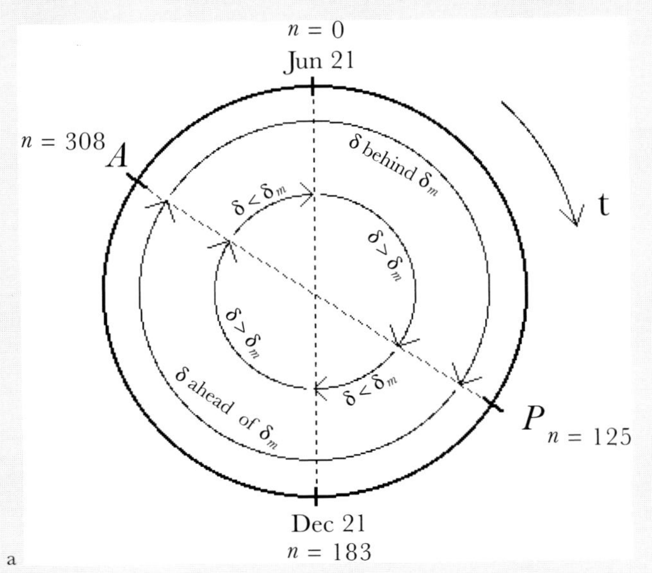


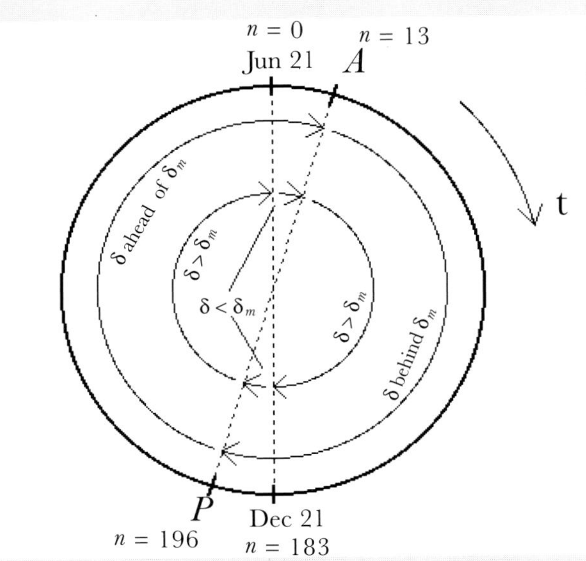


Fig. 2.4 The effect of the ellipticity of the earth's orbit on the sun's declination during the year. The summer solstice ($n = 0$) is shown at the top, and the winter solstice ($n = 183$) at the bottom. Time proceeds clockwise. Perihelion (the point when the earth is closest to the sun) is marked by $P$; aphelion (the point when it is furthest from the sun) is marked by $A$.
a. Perihelion at $n = 125$, corresponding to about 2000 BC.
b. Perihelion at $n = 193$, corresponding to the present day.

#### THE SUN'S DECLINATION AT THE 'EPOCHS'

The observable declination of the sun at the time of an 'epoch' varies slightly from year to year over the four-year leap-year cycle, owing to the fact that the length of the year is not an exact number of days, and a horizon marker can only be set up at the time of sunrise or sunset. In addition, the mean length of a sixteenth-part of the year is 365·25/16 = 22·83 days, which means that three of the sixteen divisions must

| Thom's Epoch no. | Mean no. of days from solstice | Approximate date | Mean declination (°) | Minimum declination (Thom) (°) | Maximum declination (Thom) (°) |
|---|---|---|---|---|---|
| 4 | 0·00 | Jun 21 | +23·9 | +23·9 | +23·9 |
| 5 | 22·83 | Jul 14 | +22·3 | +21·9 | +22·2 |
| 6 | 45·66 | Aug 6 | +17·2 | +16·5 | +16·9 |
| 7 | 68·48 | Aug 28 | +9·6 | +8·9 | +9·5 |
| 8 | 91·31 | Sep 20 | +0·5 | +0·1 | +0·7 |
| 9 | 114·14 | Oct 14 | −8·8 | −8·8 | −8·2 |
| 10 | 136·97 | Nov 6 | −16·8 | −16·5 | −16·1 |
| 11 | 159·80 | Nov 29 | −22·2 | −22·1 | −21·8 |
| 12 | 182·63 | Dec 22 | −23·9 | −23·9 | −23·9 |
| 13 | 205·45 | Jan 13 | −21·6 | −21·9 | −21·6 |
| 14 | 228·28 | Feb 5 | −16·0 | −16·5 | −16·0 |
| 15 | 251·11 | Feb 28 | −8·3 | −8·7 | −8·2 |
| 16/0 | 273·94 | Mar 23 | +0·5 | +0·2 | +0·8 |
| 1 | 296·77 | Apr 15 | +9·1 | +8·9 | +9·4 |
| 2 | 319·59 | May 7 | +16·5 | +16·4 | +16·9 |
| 3 | 342·42 | May 29 | +21·8 | +22·0 | +22·2 |
| 4 | 0·00 | Jun 21 | +23·9 | +23·9 | +23·9 |

have only 22 days, while the rest have 23. Unfortunately, we can not know which three, and this gives greater room for manoeuvre in fitting a theory to the available evidence. In the table below, we show the theoretical mean declination for each epoch in 2000 BC assuming a mean division length of 22·83 days, derived using formula (A5.1). Thom speculated that the number of days in each division was chosen so that non-solstitial horizon markers would fit two epochs at different times of the year as closely as possible. He then calculated the sunrise and sunset declinations that would be obtained as a result. The minimum and maximum declinations in Thom's scheme are given alongside for comparison.[5]

Declinations in the vicinity of the following, then, might be construed as fitting the eight-division calendar suggested by Thom: −23°·9; −16°·8 to −16°·0; +0°·1 to +0°·8; +16°·4 to +17°·2; and +23°·9. These, together with the following, might be construed as fitting the sixteen-division calendar: −22°·2 to −21°·6; −8°·8 to −8°·2; +8°·9 to +9°·6; and +21°·8 to +22°·3.

it is important to see whether, when the selection criteria are clarified and the other criticisms satisfied, statistical evidence still remains to support any of the categories of astronomical alignment claimed by Thom. For example, the archaeological misinterpretations included amongst the alleged indications only represent a small minority of the lines, and simply require identification and removal.[54] Yet, as we have seen, it is not possible to answer these questions merely by re-examining Thom's data as published, because of all the uncertainties about prior data selection. For this reason, an extensive independent survey of megalithic monuments in western Scotland was conducted between 1975 and 1981 under severe methodological constraints. This will be described in chapter three.

Long before this, however, Thom had moved his focus of interest onto higher-precision hypotheses. Fortunately, it is possible to give an adequate re-examination of these ideas, and in particular of the data selection that has given rise to them, without going beyond the sites considered by Thom himself.

## Level 2: the limbs of the sun and moon

Having produced the overall histograms published in 1967, Thom proceeded to examine more closely those lines with a possible solar or lunar explanation, in order to see whether there was any evidence of greater precision. In the lunar case, he did this by superimposing the four declination intervals centred upon the major and minor standstill limits, in order to examine more closely how each line deviates from the mean

Fig. 2.5 One of three indications across the ring at Castlerigg included in the Level 1 dataset. It is along a diameter towards a horizon notch and indicates the 'candlemas rising sun' with declination −16°·0 (see Thom 1967, fig. 12.10).

limit to which it appears to be related. In Fig. 2.6a we have regenerated the resulting curvigram from Thom's data.[55]

If alignments upon the major and minor lunar limits were deliberate but a precision of about half a degree was the best that was achieved, then we would expect a concentration of humps building up to an overall peak at about the origin. Instead, Fig. 2.6a shows two definite peaks, one at about +0°·25 and one at about −0°·3. As the middle of the graph represents the centre of the moon's disc when the moon is at any one of the lunar limits, and as the moon's semidiameter is about 0°·25, the peaks are suggestive that the upper and lower limbs of the moon were preferentially indicated, increasing the inferred precision to about ten minutes of arc.

The graphs produced by superimposing the solar solstitial alignments are similarly bimodal in form. In the case of the sun, however, the very presence of apparent alignments upon the lower limb seems problematic in itself, since observations of the lower limb of the sun are rendered difficult if not impossible by the glare of the solar disc.[56]

Only the double-peaked shape of the build-up of humps, rather than the actual number amassed, is relevant to the conclusions of Level 2. A formal statistical test could be devised to assess whether the data really do fit a bimodal distribution with peaks at ±0°·25, as would be predicted by the hypothesis that the solar or lunar limbs were preferentially observed, or whether a unimodal distribution centred upon zero would fit the data just as well;[57] but as at Level 1 we must first satisfy ourselves that the data have been selected fairly. Fortunately, doing so is easier than at the lower Level. The

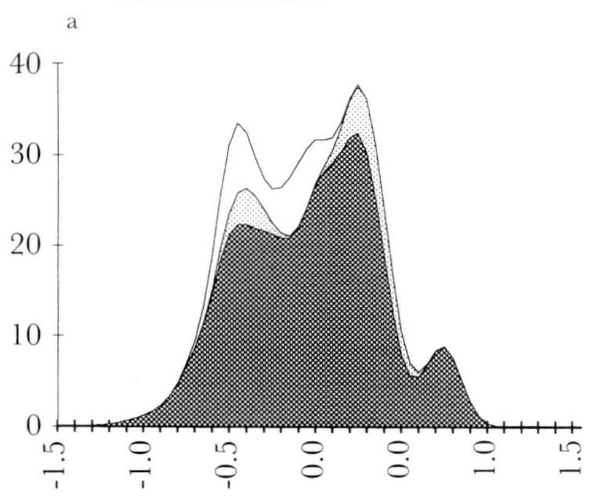


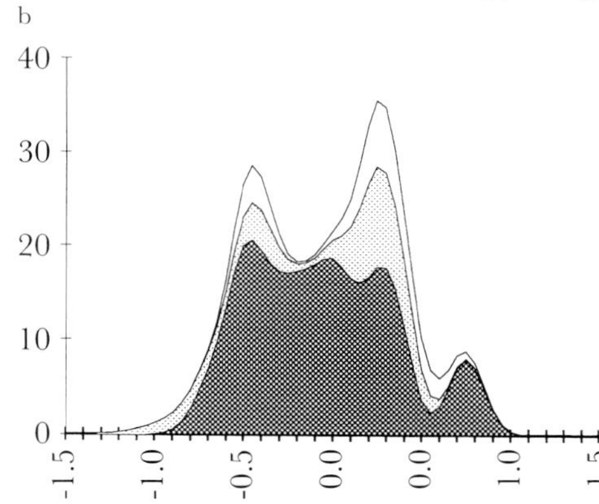


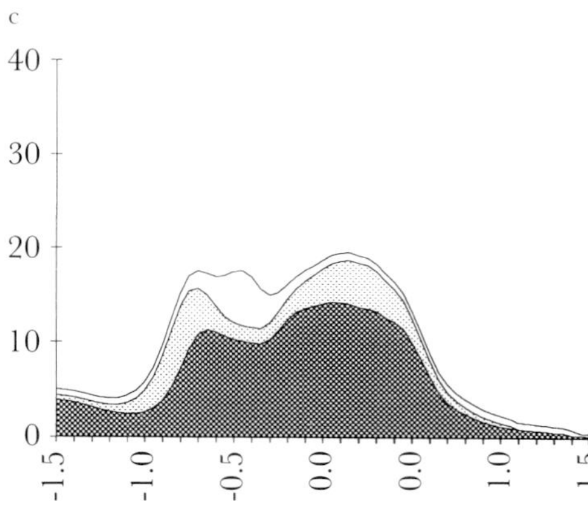


Fig. 2.6 Graphical summary of the indicated declinations in Thom's 'Level 2' data, in the form of curvigrams.

a. Declinations, plotted relative to the nearest major or minor limit, indicated by thirty-eight structures at thirty-four sites, from the data in Thom 1967, table 10.1. The unshaded area represents data of doubtful archaeological status, light shading is used where there is some doubt, and dark shading where there is no doubt. The area under each constituent gaussian curve is 1·0. The standard deviation σ is 0°·1 except for those lines marked by Thom as less accurate, where it is taken as 0°·2.

b. Remaining indications after reappraisal. The darkly shaded area represents data considered reasonable, light shading represents lines considered somewhat dubious, and lines considered very dubious are unshaded. Lines ruled out altogether have been omitted completely.

c. The picture following re-examination and resurvey. Shading has the same significance as in (b). The constituent humps now have a range of standard deviations.

## ASTRONOMY BOX 6

### VARIATIONS IN THE LONGER TERM

#### THE STARS

On a timescale of centuries the declinations of the stars gradually change. This is not because the individual stars move slowly about on the celestial sphere as it rotates: this does happen, but generally on an even longer timescale.[1] It is because, relative to the distant stars, the earth slowly pivots on its axis like a spinning top over a period of some 26,000 years. If we regard the earth as fixed, this means that over the centuries the entire network of stars on the celestial sphere gradually shifts position, so that, for example, different stars are now located near to the celestial poles and different ones now fall on the celestial equator.[2]

Reverting to a 'real' view rather than an earth-centred one for the moment, the solstices occur at the points on the earth's orbit around the sun where one of the earth's poles leans towards the sun. The effect of the earth's pivoting is that the position of the solstices and equinoxes gradually shift around the earth's orbit, each of them completing a circuit in about 26,000 years. Because of this shifting, the phenomenon is known as the 'precession of the equinoxes'.

#### THE SUN AND MOON

The limiting annual and monthly declinations of the sun and moon are not affected by the precession of the equinoxes, but they have changed noticeably over the past few millennia. This is because of the gradual decrease in the obliquity of the ecliptic $\varepsilon$ already noted in Astronomy Box 3. Using the analogy of the spinning top, it is as if the amount by which the top is tilted out of the vertical is gradually decreasing. Since 2000 BC each of the limiting declinations has changed by about 0°·5, an amount roughly equal to the width of the solar or lunar disc.

The table shows the declinations of the centre of the solstitial sun $\pm\varepsilon$ and of the centre of the moon at the major and minor standstill limits (see Astronomy Box 4) at 500-year intervals from 4500 to 1000 BC. The present-day declinations are also shown for comparison. All values are quoted to the nearest 0°·05, greater precision being unjustified for a variety of reasons.[3] To obtain the declination of the upper limb of the sun or moon, add 0°·25. For the lower limb, subtract 0°·25.

#### EVEN LONGER CYCLES (AND CLIMATE CHANGE)

When publications on positional astronomy, history of astronomy, and archaeoastronomy—this one included—speak of the precession of the equinoxes[4] they mean precession relative to the background stars. This affects the position of the stars in the sky, but in itself has no direct effect on global climate. Palaeoecologists, on the other hand, are concerned with long-term periodicities that *can* have such an effect. One of these is the precession of the equinoxes relative to the earth's perihelion,[5] which itself drifts round the earth's orbit relative to the background stars. For example, when perihelion occurs in June and the magnitude of the obliquity of the ecliptic (which, on very long timescales, oscillates periodically) is close to its maximum, so that the sun is both closest and highest in the sky in northern hemisphere summers, these will be at their warmest, but winters will be at their most cold. This 'climatic' precession is easily confused with what we might call the 'absolute' (or 'axial') precession spoken of in the astronomical and archaeoastronomical literature, but it has a different periodicity.[6]

| Date | $+(\varepsilon + i) - P$ | $+\varepsilon$ | $+(\varepsilon - i) - P$ | $-(\varepsilon - i) - P$ | $-\varepsilon$ | $-(\varepsilon + i) - P$ |
|---|---|---|---|---|---|---|
| 4500 BC | +28·4 | +24·15 | +18·15 | −19·85 | −24·15 | −30·2 |
| 4000 BC | +28·35 | +24·1 | +18·15 | −19·8 | −24·1 | −30·15 |
| 3500 BC | +28·35 | +24·05 | +18·1 | −19·75 | −24·05 | −30·1 |
| 3000 BC | +28·3 | +24·05 | +18·05 | −19·7 | −24·05 | −30·05 |
| 2500 BC | +28·25 | +24·0 | +18·0 | −19·65 | −24·0 | −30·0 |
| 2000 BC | +28·2 | +23·95 | +17·95 | −19·6 | −23·95 | −29·95 |
| 1500 BC | +28·15 | +23·85 | +17·9 | −19·55 | −23·85 | −29·9 |
| 1000 BC | +28·1 | +23·8 | +17·85 | −19·5 | −23·8 | −29·85 |
| AD 2000 | +27·7 | +23·45 | +17·45 | −19·1 | −23·45 | −29·5 |

double-peaked shape could not have been prejudiced by subjective data selection prior to an accurate survey, since rough compass measurements could not have determined the exact values of any declination within one of the four general 'target' areas. Wilfully biased selection, after the careful reduction of survey measurements, could admittedly have influenced the result; but as Thom twice refers to the unexpectedness of the double peak[58] we may assume that no such bias was present. This means that we can adequately reassess Level 2 purely on the basis of the sample of lines provided by Thom. The question is simply whether the characteristic bimodal shape can survive such a reassessment.

A detailed re-examination of the lunar data was undertaken by this author in 1979.[59] This began by selecting from Thom's 1967 dataset those lines with a listed declination within 0°·8 of a mean major or minor limit. The figure of 0°·8 was chosen as the tolerance so as to include the double peak but exclude lines that could be equally well interpreted as calendrical.[60] This leaves thirty-eight lines at thirty-four sites. The sites concerned are included in Table 2.1, where the number of Level 2 lines is shown in column 3. Archaeological information on the monuments is available by cross-reference to List 1.

An archaeological reappraisal was attempted first. This identified one monument (L49) that is probably attributable to the Early Christian period, another (L25) where the proposed alignment is part of an enclosure wall, and a third (L58) where the proposed alignment is part of a row of stones marking a modern parish boundary. The authenticity of another three monuments (L11, L12, L29) is in some doubt.[61] None of these lines was omitted from further consideration at this stage but the doubtful archaeological status of these six sites has been indicated by differential shading in Fig. 2.6a.[62] A worrying feature emerges even amongst the remaining twenty-eight sites whose status as prehistoric monuments is not in question. It is evident from an examination of Table 2.1 and List 1 that the nature of the sites and indications is every bit as diverse here as amongst the Level 1 data as a whole. If these sites really did have something special in common—their lunar significance—it seems likely that they would have other things in common as well. Instead, they very much resemble a random selection from the Level 1 data.

The next stage of the reappraisal was to examine the 'intrinsic' status of the putative lunar alignments, that is, their inherent likelihood as potential astronomical indicators. This revealed that two of the thirty-eight indications simply could not work in the manner claimed: in one case (L9) the claimed horizon can not be seen from the structure postulated to be indicating it, and in the other (L17) the foresight can not be seen from the backsight. A further four foresights listed simply as 'stones' (L14, L25 and two at L49) could not be located at all,[63] making a total of six lines that had to be dismissed from further consideration. Six more cases were considered doubtful or highly doubtful: four (L22, L50, L55, L57) because indications that seem impressive now were once part of a more complex structure, and two (L8, L18) because they involve 'outliers' of doubtful authenticity.[64]

The outcome of these reappraisals is that only twenty-three of the original thirty-eight lines seem wholly reasonable both archaeologically and intrinsically;[65] six are somewhat dubious for one reason or another, three are very dubious and six can be ruled out altogether. The resulting effect on the lunar histogram is shown in Fig. 2.6b. It is already clear that when the dubious lines are omitted (only the darkly shaded area remains), virtually all trace of a double-peaked structure disappears. In other words, the Level 2 evidence does not seem to be withstanding reassessment. This conclusion is reinforced by examining the remaining indications which, like the sites themselves, represent a wide variety of types and show no obvious sign of coherence.[66]

The last aspect of the Level 2 reassessment was to identify and attempt to eliminate any subjective bias that would influence the results as a whole. This involved independent site examinations and resurveys. The biggest problem uncovered by the fieldwork was that the accuracy of an indication, given its nature and present state of repair, was often considerably less than would justify the standard deviation (hump width) used by Thom, and there seemed to be no *a priori* reason for selecting the particular mean declination value quoted by Thom. For this reason, an independent estimate was made of the accuracy of each line, with different standard deviations being assigned to different lines. Horizon profiles were completely resurveyed wherever possible.[67] Putative horizon foresights were ignored, on the grounds that including them where they appear to fit an astronomical explanation but ignoring them otherwise biases the overall result, as we have already discussed in the context of Level 1;[68] thus the reassessed declinations are arrived at on the basis of the alignments of archaeological structures only.

The results are summarised in Table 2.2[69] and plotted in Fig. 2.6c. There is clearly no convincing evidence of a double peak amongst these data, especially when lines of reasonable status are considered alone. From this we conclude that no overall evidence remains on the basis of the Level 2 data for the preferential observation of the lunar limbs.

## Level 3: high-precision lunar foresights

Nonetheless, by 1967, Thom himself had become convinced that a great many megalithic monuments deliberately incorporated high-precision lunar alignments. He was also convinced that distant horizon foresights were used in some cases to mark particular rising or setting positions of the moon to high precision. In the work that we identify as Level 3, he developed this idea in detail.

The fullest data set is published in Thom's second book, *Megalithic Lunar Observatories*, and consists of forty horizon foresights at twenty-three sites.[70] The book also contains horizon profile diagrams for all but one of the lines concerned,[71] and site plans and descriptions for a few.[72] A distant horizon feature such as a pointed hilltop or notch defines a position in the sky much more precisely than a structure on the ground, and in his table Thom goes so far as to quote the relevant declinations to the nearest 0′·1, though the nearest 1′ would be more justifiable.[73] At this level of precision there is an added complication in that the parallax correction that must be applied because we do not observe from the centre of the earth (see Astronomy Box 2) is slightly different for different lines. For this reason here (and *only* here and this chapter) it is necessary to use a declination already corrected for this effect. We shall refer to this as the geocentric lunar declination.[74]

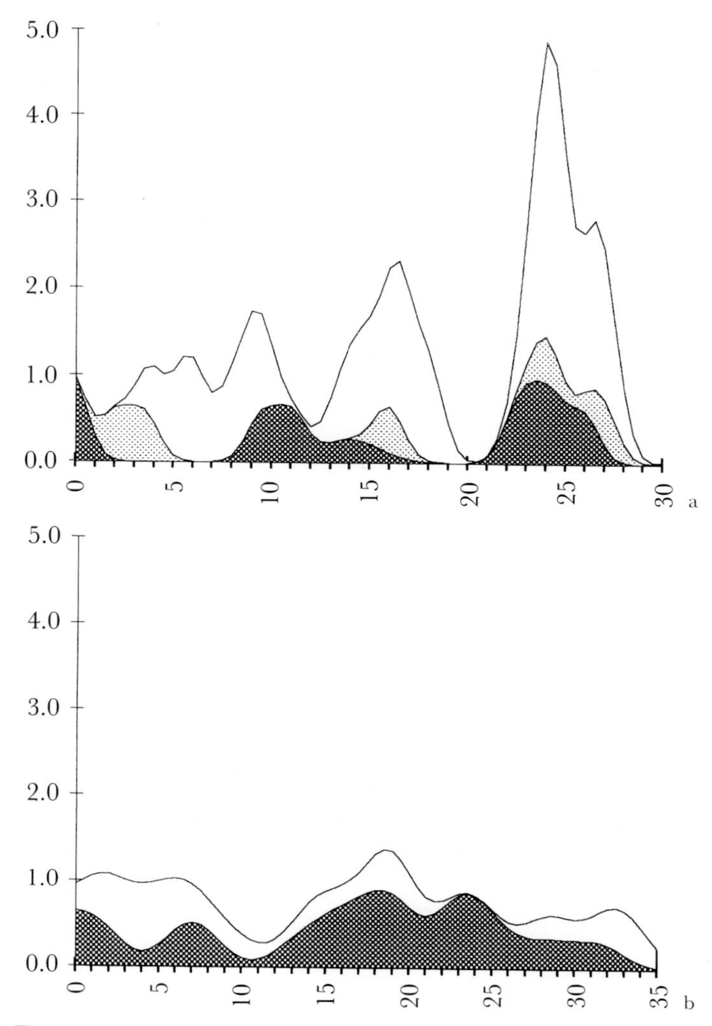


Fig. 2.7 Graphical summary of the indicated declinations in Thom's 'Level 3' data, in the form of curvigrams.

a. Declinations of forty putative foresights at thirty-four sites, from the data in Thom 1971, table 7.1. The graph shows the difference (positive or negative) from the nearest major or minor limit. The unshaded area represents unindicated foresights or lines dismissed out of hand (classified 'Y' or 'Z' in Table 2.3, col. 8). Light shading denotes data of doubtful status (classified 'B' or 'C' in Table 2.3, col. 7 or 8). Dark shading is used for the remainder. The area under each constituent gaussian curve is 1·0. The standard deviation $\sigma$ assumed is 0′·75 except for two lines where the declination is quoted to 1′ by Thom, and $\sigma$ is taken as 1′·5.

b. An independent assessment of indicated horizon notches and dips. The standard deviation assumed here is 1′·5 throughout.

Thom's data are plotted in Fig. 2.7a.[75] This curvigram differs from those in Fig. 2.6 in that the two halves are folded together, so as to plot the difference of the measured declination from the mean major or minor limit, whether negative or positive. The results are surprising. There are probability accumulations at around 6′ to 9′ and 15′ to 17′, with an especially large peak at around 24′ to 25′. The semidiameter of the moon is around 16′, which might explain the central peak, but why should so many putative horizon foresights mark positions almost exactly 25′ away from the mean lunar limits?

The answer, Thom suggested, lay in a tiny perturbation in the moon's motions that causes an additional wobble in its declination ($\Delta$) of amplitude 9′·4 and with a period of 173 days (Astronomy Box 7). The moon's semidiameter ($s$) is about 15′·9,[76] so that if a foresight marked the rising or setting position of one of the moon's limbs at a major or minor limit at the maximum, mean or minimum of the additional wobble, the resulting declination would differ from the mean limit $\pm(\varepsilon \pm i)$ by about 25′·3 ($s + \Delta$), 15′·9 ($s$ alone), or 6′·5 ($s - \Delta$). In addition, marking the centre of the moon at a maximum or minimum of the wobble would give 9′·4 ($\Delta$ alone). The correspondence with the observed peaks is manifest.

At the Nether Largie standing stones, mid-Argyll (part of Thom's 'Temple Wood' A2/8; L31 in List 1), according to Thom, the same foresight was used from several different observing positions to mark the moon setting at the major limit (declination $+(\varepsilon + i)$) at different positions of the wobble (see Fig. 2.9).[77] At many other sites only one configuration appears to have been marked. The data used to generate Fig. 2.7a are listed in Table 2.3.

What is being proposed at Level 3 is formally quite distinct from that at the earlier levels, and has to be reappraised in a different way. The focus of interest is now the precise foresight formed by a natural feature on a distant horizon, with man-made structures on the ground doing no more than marking the observing position and pointing out which horizon feature is to be used. An item of evidence in support of this idea, then, consists of three elements: a backsight marker, an indicator of the foresight, and the foresight itself.[78]

The idea immediately runs into serious trouble because, as emerged in the reassessment by this author in 1981, no fewer than twenty-one of the forty horizon features in the dataset are not actually indicated at all (or else the supposed indication is not genuine or is some degrees off line).[79] A further five foresights can not in fact be seen from the backsight because of the intervention of local ground, and one is non-existent. Thus only thirteen of the forty actually represent indicated horizon features in the first place.[80]

Of course, it can be argued that indicators may have disappeared since prehistoric times, or even that they were never necessary anyway, since if people were using horizon features for important astronomical observations they are likely to have known where to look, and only needed the observing position to be marked. The problem is that if we simply speculate that this was the case wherever we find a promising potential foresight, then we are going far beyond what the archaeological record actually tells us.[81] The chances are not inconsiderable that if we went to any arbitrary point in hilly country, even where there is no monument and no evidence of prehistoric activity, we could find at least one or two horizon features interpretable as lunar foresights.[82] In order to test the idea that horizon foresights were used for which no indication (now) exists, we would need a different and much more careful methodology capable of extracting something believable from the data and avoiding circular argument.[83]

The effect of restricting our attention to the indicated foresights in the Level 3 dataset can be seen in Fig. 2.7a, where only they are shaded. Of the thirteen cases, five were considered somewhat doubtful for one reason or another (see Table 2.3), and are shaded more lightly. Amongst these data, little more than a hint remains of the peaks at 25, 16 and 9 minutes from the mean. Nonetheless it is important, as well as informative, to complete the reassessment by examining the question of the fair selection of foresights.

## ASTRONOMY BOX 7

## THE MOTIONS OF THE MOON, 2 HIGH-PRECISION COMPLICATIONS

This box concerns high-precision phenomena that are only of concern in relation to the discussion of Level 3 and Level 4 phenomena in chapter two. In this context, it is necessary to undertake analyses and frame overall conclusions in terms of geocentric lunar declinations (see Astronomy Box 2), which is why the parallax factor *P* (cf. Astronomy Box 4) no longer appears explicitly.

### The lunar bands

In Astronomy Box 4 we presented a table of possible lunar 'target' declinations, corresponding to the four standstill limits and, in each case, observations of the centre of the lunar disc or of one or other limb. At a level of precision greater than about 0°·1, this picture has to be modified. The main reason is an additional 'wobble', or perturbation, which shifts the declination of the moon from what we would otherwise expect by up to 9′ in each direction over a period of 173 days.[1] Because of this, we must now consider eight 'lunar bands' on the horizon,[2] within which a variety of specific targets might have been of interest (see Fig. 2.8). The various possible configurations are shown in the table:

### Marking lunar targets to high precision: some practicalities

In practice, it is extremely difficult to determine any of these targets to high precision simply from a series of observations of the moon rising or setting. There are several reasons for this. First, the moon only approaches either lunistice once a month, and will only be at all close to its monthly maximum or minimum declination for two or three days around this time. Unless a rise or set happens to occur very close to the precise hour of the lunistice, the moon will not be directly observable rising or setting at the declination limit for the month, but will always be some way south (for northern limits) or north (for southern ones). In the worst case, where the lunistice falls mid-way between two risings or settings, the horizon moon will never be seen closer than about 10′ to the monthly maximum or minimum declination.[3]

Second, how close the monthly declination limit is to the actual major or minor limit will depend upon how much earlier or later the lunistice in question is than the standstill. Third, there will be an additional displacement owing to the 173-day wobble, and this will change significantly from month to month. In short, the moon will only be directly observable, say, setting at $+(\varepsilon + i + \Delta)$ to within two or three arc minutes if the time of setting coincides with (i) the lunistice to within about five hours, (ii) the major standstill to within about twenty weeks, and (iii) the maximum of the 183-day wobble to within about twenty days.

| | Upper limb | Centre | Lower limb |
|---|---|---|---|
| Northern major limit, wobble north | $+(\varepsilon + i + s + \Delta)$ | $+(\varepsilon + i + \Delta)$ | $+(\varepsilon + i - s + \Delta)$ |
| Northern major limit, mean wobble | $+(\varepsilon + i + s)$ | $+(\varepsilon + i)$ | $+(\varepsilon + i - s)$ |
| Northern major limit, wobble south | $+(\varepsilon + i + s - \Delta)$ | $+(\varepsilon + i - \Delta)$ | $+(\varepsilon + i - s - \Delta)$ |
| Northern minor limit, wobble north | $+(\varepsilon - i + s + \Delta)$ | $+(\varepsilon - i + \Delta)$ | $+(\varepsilon - i - s + \Delta)$ |
| Northern minor limit, mean wobble | $+(\varepsilon - i + s)$ | $+(\varepsilon - i)$ | $+(\varepsilon - i - s)$ |
| Northern minor limit, wobble south | $+(\varepsilon - i + s - \Delta)$ | $+(\varepsilon - i - \Delta)$ | $+(\varepsilon - i - s - \Delta)$ |
| Southern minor limit, wobble north | $-(\varepsilon - i - s - \Delta)$ | $-(\varepsilon - i - \Delta)$ | $-(\varepsilon - i + s - \Delta)$ |
| Southern minor limit, mean wobble | $-(\varepsilon - i - s)$ | $-(\varepsilon - i)$ | $-(\varepsilon - i + s)$ |
| Southern minor limit, wobble south | $-(\varepsilon - i - s + \Delta)$ | $-(\varepsilon - i + \Delta)$ | $-(\varepsilon - i + s + \Delta)$ |
| Southern major limit, wobble north | $-(\varepsilon + i - s - \Delta)$ | $-(\varepsilon + i - \Delta)$ | $-(\varepsilon + i + s - \Delta)$ |
| Southern minor limit, mean wobble | $-(\varepsilon + i - s)$ | $-(\varepsilon + i)$ | $-(\varepsilon + i + s)$ |
| Southern major limit, wobble south | $-(\varepsilon + i - s + \Delta)$ | $-(\varepsilon + i + \Delta)$ | $-(\varepsilon + i + s + \Delta)$ |

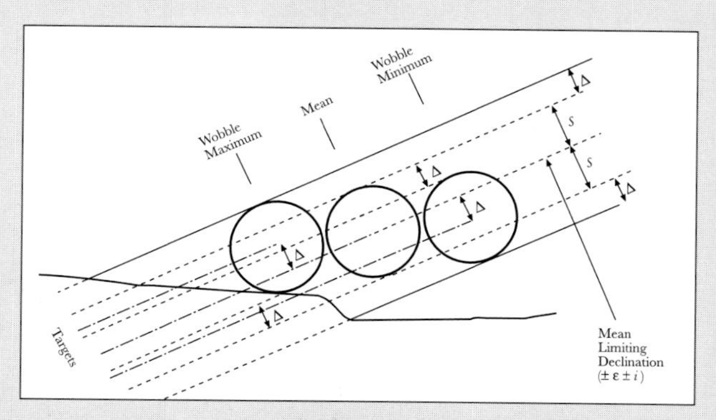


Fig. 2.8 A 'lunar band' and the targets within it. Adapted from Thom 1981, fig. 1.7.

An additional problem, hidden when we talk in terms of geocentric lunar declinations but important in practice, is the monthly variation in the lunar parallax,[4] which serves in practice to impose a variation in declination of $\pm 3'\cdot 5$ over a period of a month, on top of all the cycles already mentioned. This would further reduce the attainable precision in locating any particular target.

Even this is not all. Whether the critical risings or settings are directly observable also depends upon the phase of the moon (the limb being observed needs to be illuminated) and whether the event in question occurs during daylight hours. For many horizon markers, these together will eliminate three possibilities out of four.[5] Finally, and certainly not least, there is the possibility of bad weather.

A number of authors have tackled these issues in detail[6] and they agree that prehistoric observers probably could not have established the exact period of the 18·6-year lunar node cycle, or the existence of the 173-day wobble, without programmes of observations lasting at least several node cycles, i.e. several scores of years, and some means of recording their results. They certainly could not have done so without some means of extrapolating between nightly risings or settings close to the lunistice in a given month.

### Further complications at the highest precision

At higher precision still, as is postulated at Level 4, several other effects need to be taken into account. These include the gradual decrease in $\varepsilon$ (see Astronomy Box 6) of about 3′ every 500 years; a sinusoidal variation in the lunar parallax which alters the apparent declination of the moon by up to 3′ over a 179-year cycle;[7] differences in mean refraction corrections owing to the fact that particular events can only be observed at particular times of year and day;[8] and the fact that the magnitude of day-to-day variations in refraction owing to the daily changes in weather conditions may be much greater than supposed by Thom (see Astronomy Box 3).

There is in fact strong evidence that Thom did not select horizon features fairly, that is without regard to the astronomical possibilities. This can be seen amongst the profile diagrams presented by Thom himself, both where a horizon indication exists and where it does not. For some examples see Fig. 2.10.[84] In the case of the indicated foresights, eliminating such preselection means trying to identify for each 'line' the entire range of horizon that might have been indicated, and then to identify every feature that could be construed as a potential foresight without regard for its declination. This second step is particularly problematic, since it involves a hypothetical judgement: what features might Thom have been prepared to consider as putative foresights if their declinations had been 'interesting'? A relatively straightforward choice, and one that reflects a form of foresight frequently proposed by Thom, is to include just notches, hill junctions and the bottoms of dips. The result of selecting all such features, and only all such features, falling within the indicated horizons from Level 3, is shown in Fig. 2.7b (for the data see column 13 of Table 2.3).[85] These data show no trace of Thom's peaks.

A number of other criticisms have been made of the Level 3 data and analysis.[86] As at the previous levels, there is little consistency in the types of backsights, indicators and foresights (see Table 2.3). At some sites, much more plausible indicators exist than the lunar ones proposed, yet seem to have no astronomical significance at all.[87] Some foresights are too near to the observing position, so that vegetation poses considerable uncertainties and the precision to which the observing position must have needed to be specified is a problem.[88] Others are so distant and at such a low altitude that atmospheric conditions would render them invisible virtually all of the time.[89]

As if all this were not enough, there are considerable practical difficulties in actually observing the moon rising or setting at a major or minor limit at any of its wobble configurations. The main problem is that the moon only rises or sets roughly once a day, during which time it has moved considerably, so that in a given month the lunistice (see Astronomy Box 4) will generally fall between two consecutive risings or settings, and the moon will never actually be seen at its monthly maximum (northerly) or minimum (southerly) declination. In the worst case, it may get no nearer than about 10′ (Astronomy Box 7). Fully acknowledging this, Thom proposed that prehistoric observers could set up a backsight for a significant lunar event without actually observing it, but instead by extrapolating between the observed risings and settings. This could be accomplished by marking on the ground an 'extrapolation length', which depends upon the site and sightline, but is fixed in each case. By relating observations on two or three consecutive nights nearest a lunistice to this extrapolation length, the

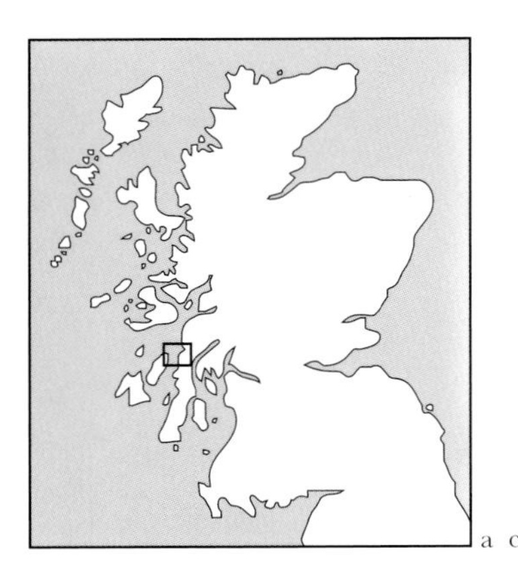


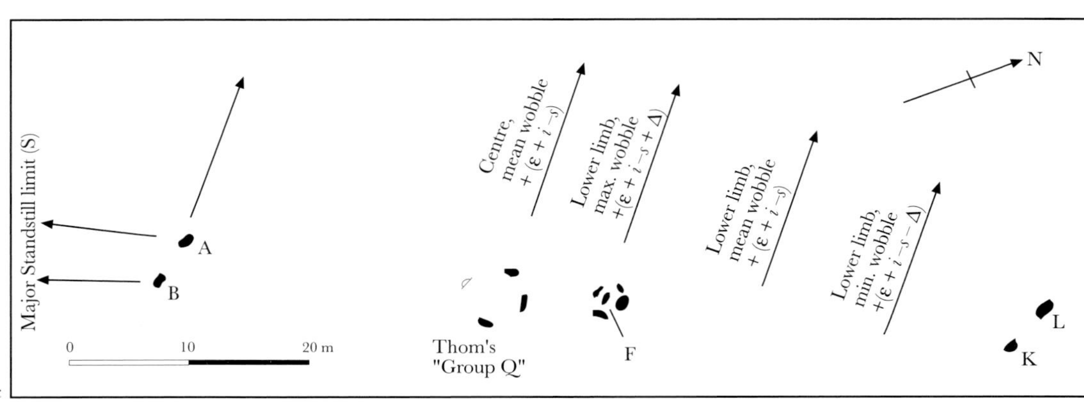


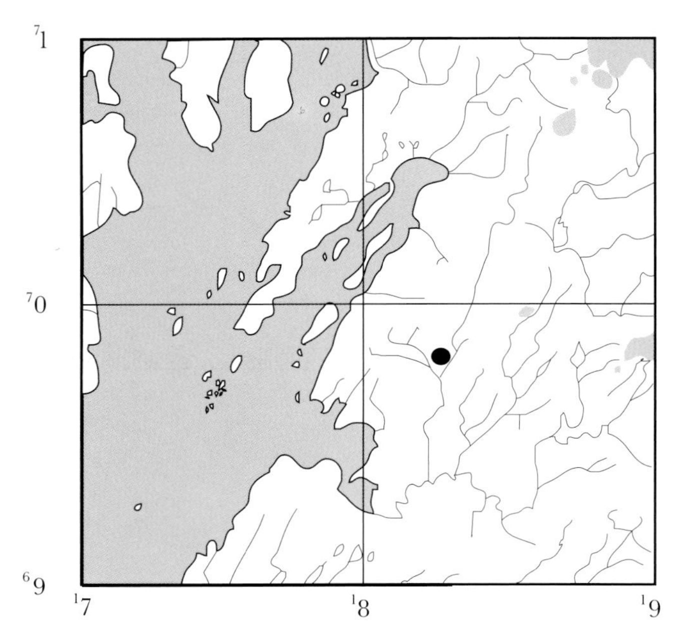


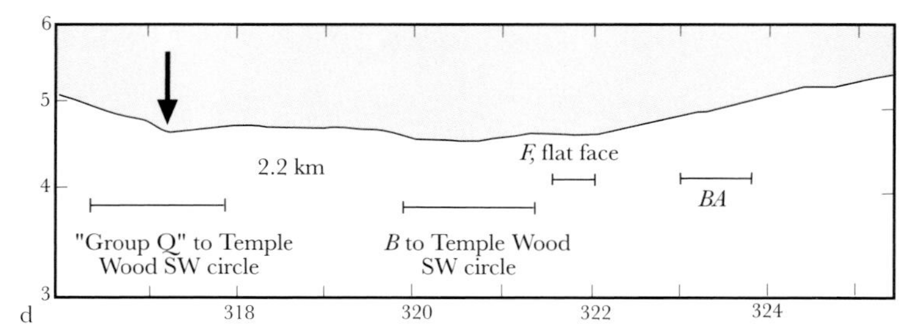


Fig. 2.9 Nether Largie, mid-Argyll.

a,b. Location of the Nether Largie standing stones.

c. Plan of the stones, showing the putative positions for observing Thom's notch in the north-west and the declination obtained in each case. Based on RCAHMS 1988, 136 and Thom 1971, fig. 5.1.

d. The horizon profile to the north-west. Thom's foresight, a small notch, is marked with an arrow. The horizontal bars show the directions 'indicated' by various alignments of standing stones, as determined during independent fieldwork published in 1981. The scales show azimuths and altitudes in degrees. Based on Ruggles 1981, fig. 4.3c.

e. The Nether Largie stones viewed from the north-east.

e

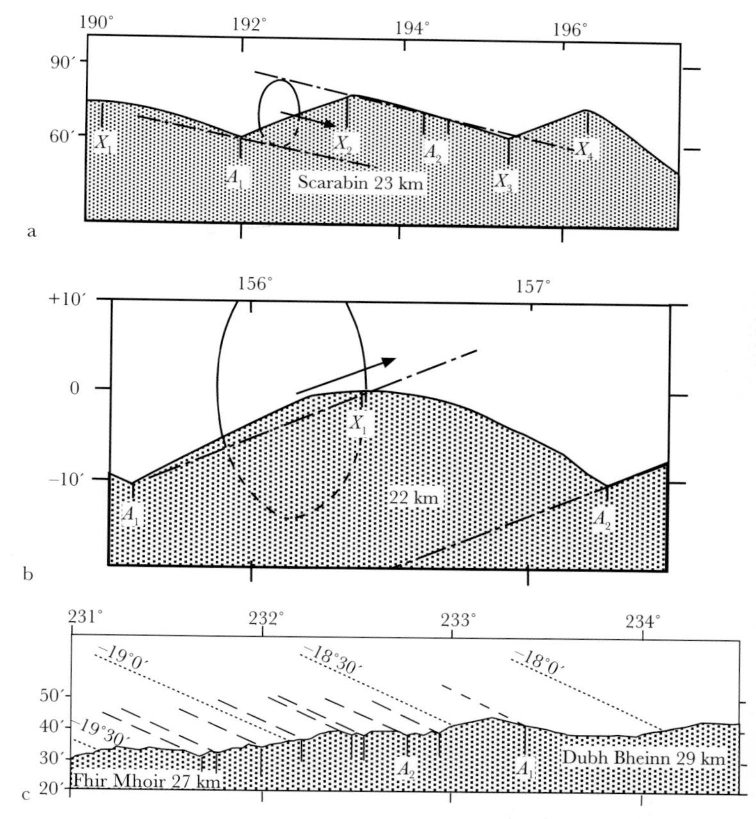


Fig. 2.10 Evidence for the preselection of foresights by Thom according to the astronomical possibilities.

a. Dirlot (L4) (unindicated). Profile diagram after Thom 1971, fig. 9.5. The moon is drawn, following Thom, setting with its centre at declination $-(\varepsilon + i)$. The bottom of a dip ($A_1$) and a small section of hill slope parallel to the moon's setting path ($A_2$) were included as potential foresights in Thom's analysis. Three hilltops ($X_1$, $X_2$, $X_4$), the bottoms of two further dips (one off to the left, $X_3$) and one prominent hill junction (off to the right) lying in the moon's path close to its major limit have all been missed out. $A_1$ and $A_2$ respectively yield declinations $-17'$ and $+14'$ away from the mean lunar limit, close to $-s$ and $+s$ respectively. According to a survey by the author in 1979 the eight potential foresights, from left to right, yield declinations $-29'$, $-16'$, $-18'$, $+12'$, $+12'$, $+11'$, $+31'$, and $+19'$ away from the mean lunar limit, a selection that includes many values that fall between the peaks in Fig. 2.7a.

b. Unival (L16) (indicated). Profile diagram after Thom 1971, fig. 6.15. The moon is drawn, following Thom, rising with its centre at declination $-(\varepsilon + i)$. The bottoms of two dips ($A_1$ and $A_2$) were included as potential foresights in Thom's analysis. The hilltop between them ($X_1$) and the bottoms of two further dips off to the left, lying close to the indication as well as in the moon's path close to its major limit, have both been missed out. $A_1$ and $A_2$ respectively yield declinations $-2'$ and $-27'$ away from the mean lunar limit, the latter close to $-(\Delta + s)$. According to a survey by the author in 1979 the five potential foresights, from left to right, yield declinations $+15'$, $+7'$, $-5'$, $-5'$, and $-34'$ away from the mean lunar limit. For an independent surveyed profile diagram, showing the additional dips and without the vertical exaggeration used by Thom, see Ruggles 1981, fig. 4.3a.

c. Kintraw (L30) (indicated). Independent resurveyed profile diagram after Ruggles 1983, fig. 7. This shows eight hill junctions, notches or bottoms of dips, which are given equal weight in the reappraisals by this author. Only points $A_1$ and $A_2$ were included as potential foresights in Thom's analysis. For the measured declinations see Table 2.3.

observing point for a theoretical limit could be determined and marked on the ground.[90]

According to Thom,[91] an extrapolation length associated with a lunar foresight was marked at nine sites. These are listed in Table 2.1. However, reassessment of the evidence showed that either the alleged foresight or the marker of the extrapolation length are of doubtful status in all but two cases,[92] and even these two remaining cases have little in common: at one site (L31), the marked extrapolation length is roughly perpendicular to the sightline and at the other (L38) they are parallel.[93] In addition, it transpires that Thom was wrong to assume that the extrapolation length was fixed for a given sightline; in fact, the use of a fixed extrapolation length would result in errors at least as great as the amplitude of the wobble itself.[94] Finally, *both* monthly maxima on either side of a standstill must be successfully determined in order for a particular sightline to be set up. Failure will result in a delay of nineteen years. Because six observations on particular days must be made, and because of the uncertainties of lunar phase, daylight and bad weather (see Astronomy Box 7), it is unlikely that a new sightline could successfully be erected more than once every seventy-five years on average.[95]

In short, the evidence against lunar observations of the high precision envisaged at Level 3 is quite overwhelming. But perhaps the last words on the subject should come from modern astronomy. First, Thom failed to take into account yet another cyclical variation in the moon's declination, due to variable parallax, amounting to about 6′ over a period of 180 years. Even if the 173-day wobble was observed, this effect should blur out any peaks such as those apparent in Fig. 2.7a.[96] Second, as we have mentioned in chapter one, evidence has emerged more recently that day-to-day variations in atmospheric refraction may be much greater than Thom assumed, rendering observations to a precision greater than about 6′ infeasible anyway (see Astronomy Box 3).

## Level 4: the very highest precision

While the reassessments described above were being undertaken, Thom (now working together with his son Archie) had gone on to examine a sample of putative lunar foresights even more painstakingly than before. The dataset at this level consists of forty-four examples, which overlap substantially with those at Level 3, but also include seventeen lines from eight new sites (see Table 2.1).[97] By taking into account the particular astronomical and atmospheric corrections for each foresight at the time of year and day when it would necessarily have been used, rather than assuming mean parallax and refraction corrections as at Level 3,[98] the Thoms concluded that horizon markers were precise to within as little as a single minute of arc.[99] These claims prompted further, detailed reassessment by this author.[100]

The most notable new site to enter at Level 4 is the Ring of Brodgar in Orkney (see L2 in List 1). This impressive henge and stone circle is arguably one of the most evocative prehistoric monuments in Britain, standing on a neck of land between two freshwater lochs, and in sight of the equally impressive Stones of Stenness as well as of Maes Howe chambered tomb.[101] For the Thoms, who first described the site in 1973, it also formed 'the most complete megalithic observatory remaining in Britain', with the outlying 'Comet Stone' and various large and small mounds in the vicinity forming

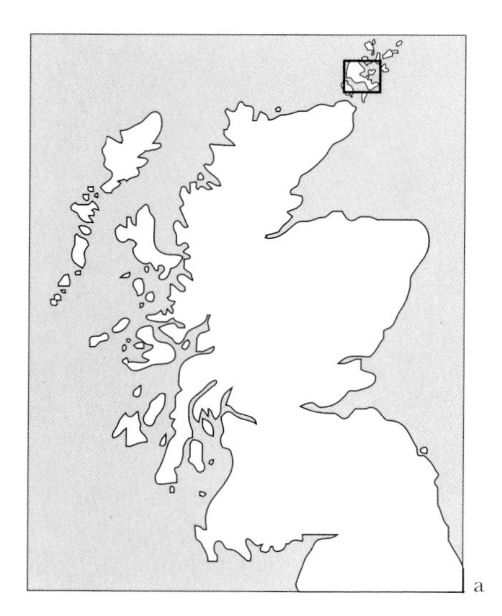


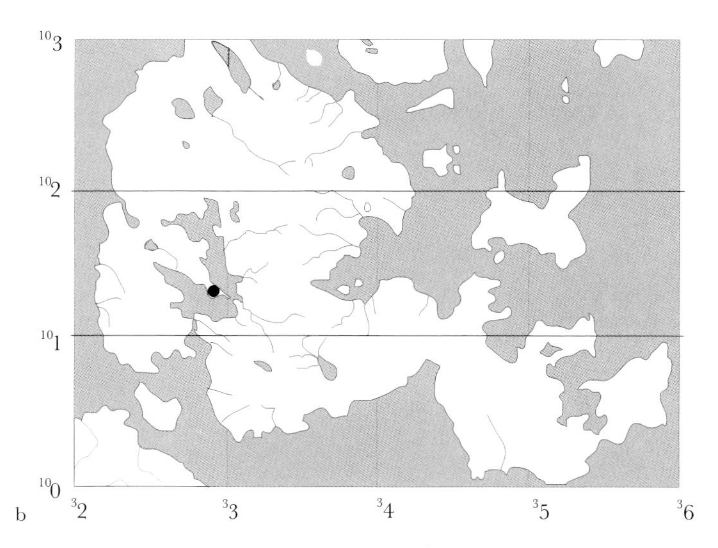


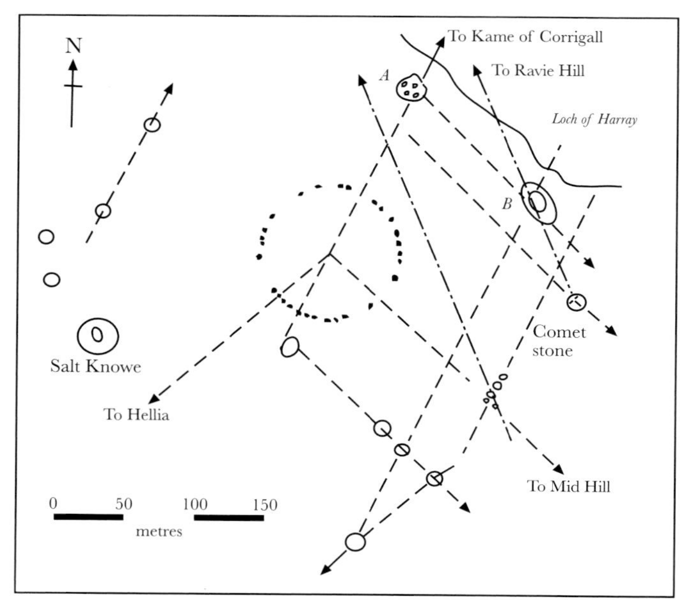


Fig. 2.11 Brodgar, Orkney.
a,b. Location of the Ring of Brodgar.
c. Plan of the Ring and mounds at Brodgar, showing the putative alignments to the four lunar foresights. Based on Thom and Thom 1975, fig. 3.
d. The Ring viewed from the south-west.

Fig. 2.12 (*facing page*) The Thoms' lunar sightlines at Brodgar. Profile diagrams (after Thom and Thom 1978a, fig. 10.2; 1973, fig. 2; 1975, fig. 2; and 1977, fig. 1 respectively) are shown on the left, and photographs with the alleged foresight marked are shown for comparison on the right. Note that the vertical scale is exaggerated in the profile diagrams.
a Hellia.
b. Mid Hill.
c. Kame of Corrigall.
d. Ravie Hill.

d

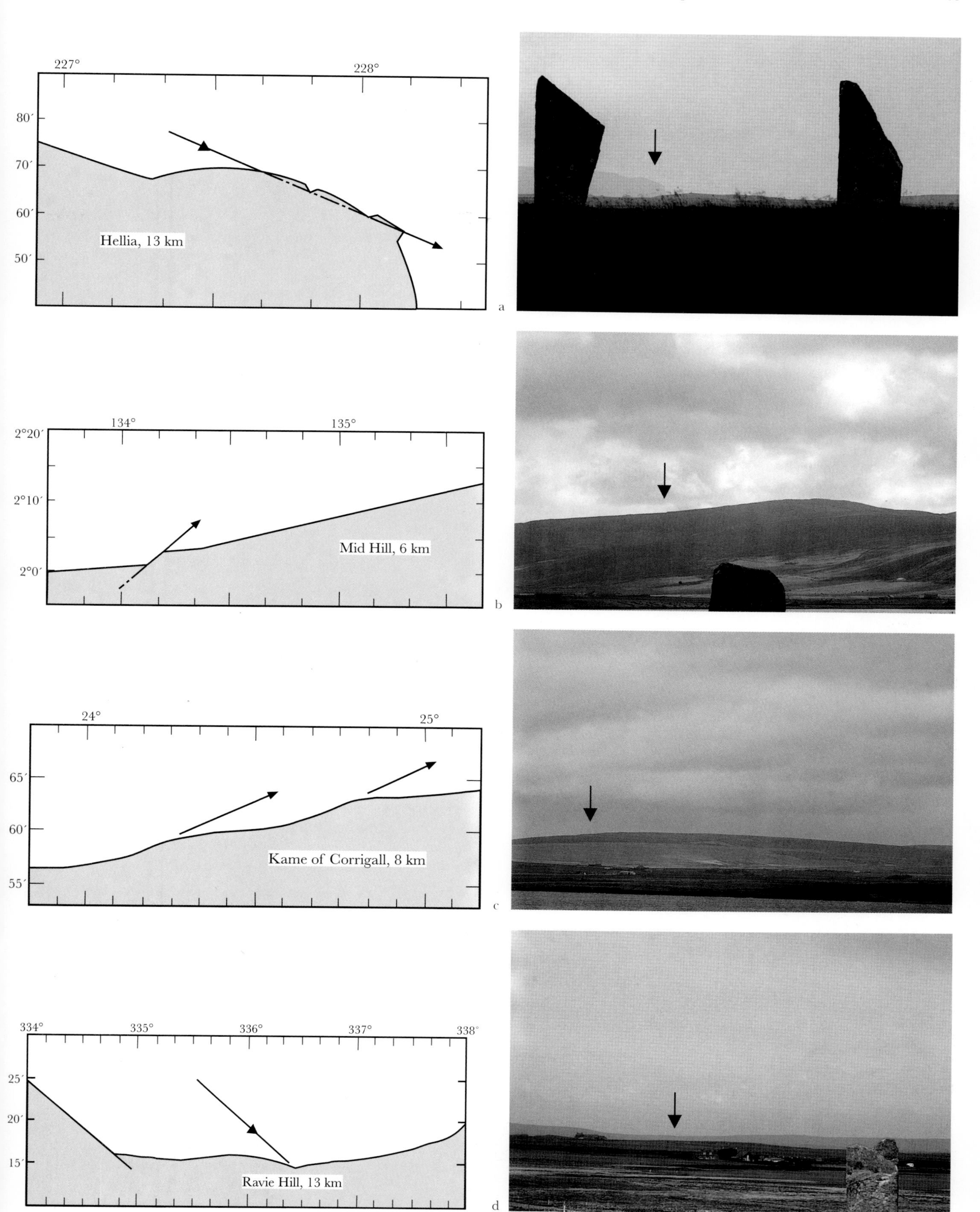
227°
228°
80′
70′
60′
50′
Hellia, 13 km
a
134°
135°
2°20′
2°10′
2°0′
Mid Hill, 6 km
b
24°
25°
65′
60′
55′
Kame of Corrigall, 8 km
c
334°
335°
336°
337°
338°
25′
20′
15′
Ravie Hill, 13 km
d

use: this turns out to be about 1700 BC, or possibly about 1500 BC.[104] The construction of the henge and stone circle, which according to the Thoms was carefully placed here in order to take advantage of the astronomical possibilities, predates even the earlier of these dates by several centuries.[105]

A now-familiar story is repeated when we examine the status of the remaining sightlines at Level 4. Three can be ruled out as intentional: the one at Stenness (L3), where the backsight was constructed some 1500 years before its supposed astronomical use;[106] one at Skipness, Kintyre (L40), where the backsight is merely a natural boulder (Fig. 2.13a); and one at Callanish, Lewis (L9), carried forward from Level 3, where the foresight cannot be seen from the backsight. Only fourteen of the remaining Level 4 sightlines, it transpires, actually represent cases where structures remaining today accurately indicate the proposed horizon foresight.[107] Amongst these are some of doubtful archaeological status, such as Dunskeig, Kintyre (L41) (Fig. 2.13b), where the two stones are probably the grounders of a modern field wall.[108] The remainder are scattered geographically, represent a diverse collection of monuments and indicating structures,[109] and show strong evidence that the selection of putative foresights within indicated horizons was strongly influenced by the astronomical possibilities.[110]

Fig. 2.13 Dubious backsights and indicators for alleged lunar sightlines of the highest precision.
a Skipness, Kintyre, which is a natural boulder rather than a genuine prehistoric monument. The proposed foresight is a deep notch amongst the hills of Arran in the background (A. Thom and A. S. Thom, 'Another lunar site in Kintyre', *AA* no. 1 (*JHA* 10) (1979), S97–8).
b. Dunskeig, Kintyre. A doubtful site; probably the remains of a relatively modern field wall. The alignment of the two stones does however indicate the hills of Arran, which include the lunar foresight.

backsights for four separate horizon lunar foresights (Figs. 2.11 and 2.12).[102] No fewer than nine lines from Brodgar are included in the Level 4 reassessment, together with one from nearby Stenness.

The presence of as many as four distinct lunar foresights at the one site seems difficult to explain away by chance until the foresights are examined on the ground. It is then discovered that only the cliffs of Hellia are at all imposing; Mid Hill, albeit noticeable with the naked eye, is merely a small step in an otherwise straight hill slope; and Kame and Ravie Hill are utterly unimpressive and almost impossible to spot without the benefit of a theodolite. Certainly, all but Hellia are outweighed in prominence by scores of other visible horizon features. The proposed indicators, mainly involving despoiled mounds and hence ill-defined, are unconvincing. A detailed critique of the proposed sightlines is given elsewhere,[103] but for the archaeologist there is a simpler objection to the Thoms' interpretation. At this level of precision, the small change in ε over the centuries (see Astronomy Box 6) becomes significant and one can deduce a rough date (to within a few centuries) of presumed

Turning to the practicalities of the observations, it is clear that many of the criticisms made at Level 3, to do with the complexity of the moon's motions and the practical difficulties in observing and marking them, still stand at Level 4. Indeed, the precision is now so great that observing programmes lasting 180 years are now needed, by the Thoms' own admission.[111] At this level their technical arguments are complex,[112] and they are not considered further here. They require a technical response, and the interested reader is referred elsewhere.[113] But two things are worthy of brief mention. First, the fact that ε gradually changes means that the date of presumed use is deduced from the data for each of the sightlines at Level 4, as at Brodgar. Additionally, at this level the Thoms introduce a 'graze effect' which purports to make our measurement of sightlines even more precise by taking into account the bending of light rays passing close above intervening ground.[114] In practice, however, the extent of this effect for different sightlines is also deduced from the data.[115] In both cases, the values of the relevant parameters are adjusted so as to provide the best fit to the data, but the argument is circular: in reality, the more different parameters that can be adjusted in order to provide a good fit to the measured data, the easier it is to fit something very close to whatever we measure. Surely this, rather than anything actually achieved in Neolithic or Bronze Age Britain, is the reason for the staggeringly small statistical residuals obtained by the Thoms.

## Concluding remarks

To sum up, we have seen that apparent trends in the crucial data at Levels 2, 3 and 4 can quite adequately be explained away by selection effects and the large number of free parameters that can be adjusted to provide a close fit between the high-precision lunar theory and the measured data. In any case, once we reach Level 3 there are enormous—almost certainly insurmountable—practical difficulties involved in observing and marking the moon's motions to the precision claimed. Taken together, these factors lead us to the unavoidable conclusion that lunar motions were *not* in fact observed and recorded to high precision in prehistoric times. For these reasons the idea of lunar observations and markers precise to a few minutes of arc will concern us no further in this book.

It should be pointed out that what we have just said does not conflict with the Thoms' statement that 'at no stage have we made any attempt to pull the values this way or that way to produce a better fit'.[116] We are certainly not suggesting that the Thoms were deliberately misleading people by carefully choosing only those lines which best fitted the theories they were trying to prove. Rather, the problem is one of implicit methodology: the values used in the Thoms' analyses are ones favourable to the lunar hypotheses that have already been singled out from less favourable data by their prior selection. This serves to emphasize that in collecting data on possible astronomical alignments, methodology is a critically important consideration.

The extent to which astronomical orientations and indications of a rougher nature were incorporated into megalithic structures was a question that, in the early 1980s, had to await the outcome of reassessments at Level 1.

# 3

# Sightlines and Statistics

*An Independent Statistical Study of 300 Western Scottish Sites*

I can put two and two together, you know. Putting two and two together is my *subject.* I do not leap to hasty conclusions. I do not deal in suspicion and wild surmise. I examine the data; I look for logical inferences.

Tom Stoppard, *Jumpers*, 1972

Selection criteria are essential and, provided that they are decided upon in advance of an investigation and rigidly adhered to during it, they should be quite safe.

Douglas Heggie, 1981[1]

Once we have accepted the reality of even the simplest observations, like the orientation of Stonehenge on the midsummer sunrise, the question is no longer one of acceptance or rejection, but simply of degree.

Richard Bradley, 1984[2]

## An emphasis on rigour

Being keen to examine Thom's conclusions at first hand, the present author and three colleagues, John Cooke, Roger Few and Guy Morgan, embarked in 1975 upon a long-term project which attempted to examine a large number of megalithic monuments just as Thom had done, but which strove to avoid, as far as possible, the pitfalls of unwitting selective bias. As the project progressed, its principal aim shifted away from merely reassessing Thom's work and towards attempting to lay a new methodological framework for assessing alignments of possible astronomical significance. The project and its methodology evolved over several years, and was brought to completion by the present author during a total of eight months of fieldwork in 1979 and 1981. The results were published in 1984.[3]

The project put great stress on methodological rigour. Its objective was to survey and analyse, independently of Thom, a large number of putative astronomical alignments at prehistoric sites, paying paramount attention to the unbiased selection of data for analysis so that the results would be statistically meaningful. In order that selection decisions should be open to full discussion and criticism, great care was taken to document them in detail. It was considered particularly important to reveal, at each stage in the selection process, information about the data being rejected from further consideration, and the reasons for doing so.[4] Furthermore, no survey data were reduced until the entire sample had been collected, in order to avoid the possibility of the selection strategy being influenced by the results obtained along the way.

As the bulk of the monuments considered by Thom at Level 1 had been settings of standing stones (i.e. stone circles, short stone rows, and so on, as opposed to megalithic tombs), such sites were the focus of the new project. The decision was made to investigate, in the first instance, all documented monuments of this type in well-demarcated geographical areas; not just the ones listed by Thom. This still left the problem of the variety of designs encountered: circles, circles with outliers, rows, aligned and non-aligned pairs of stones, and single slab-like, rectangular and irregular standing stones, together with many variations. How should one decide at each type of stone setting what constitutes an orientation worthy of consideration? Indeed, what sets of structures should be taken to constitute an individual monument—a 'site'—in the first place?[5] It was considered crucial to avoid making a series of unpremeditated decisions in the field, decisions that might vary from place to place and could easily be influenced by modern predilections and prejudices regarding astronomy.

In order to address these issues, a very strict procedure was followed. Prior to any fieldwork, an explicit code of practice was laid down governing all stages in the data selection process. This code of practice consisted of arbitrary but fixed rules, chosen so that in the judgement of the investigators it would give as much chance as possible of isolating astronomical trends in the data if any were actually present.[6] It was then strictly adhered to throughout the subsequent process of selecting monuments for consideration, during the fieldwork itself and while analysing the site data afterwards.[7] The merits and inadequacies of particular rules might of course be argued at great length,[8] but to do so would be to miss the entire point of the exercise, which was simply to determine whether there were significantly more astronomical alignments than would be expected by chance amongst the dataset as a whole, and if so, to give an idea of which celestial bodies and events they might relate to and the nature of the sites and structures concerned.[9] Minor adjustments in the selection rules, it was argued, were unlikely to alter in any substantial way the overall results of the statistical tests. More important were questions of the whole methodological approach and major strategies in the data selection.[10]

## The selection of sites and indications

The first stage in the long selection process was to define geographical areas within which sites would be considered. The Hebridean Islands of western Scotland were felt to be particularly suitable, first because they form a region rich in megalithic monuments and one which had been of particular interest to Thom, and second because individual islands and groups of islands form areas with well-defined boundaries. Virtually all the Hebridean Islands were included in the analysis, the exceptions being passed over for practical, rather than archaeological, reasons. For example, Colonsay was excluded because camping and caravanning were not allowed on the island and project funds were insufficient to cover higher-grade accommodation. Certain archaeologically interesting and geographically well-defined areas of the western Scottish mainland, such as mid-Argyll and the Kintyre peninsula, were also included.

The areas were coded, for the purposes of analysis and discussion, as follows: Lewis and Harris, and associated smaller islands (LH); North and South Uist, Benbecula, Barra, etc. (UI); North Argyll, i.e. Ardnamurchan, Sunart and Morvern (NA); Coll and Tiree (CT); Mull (ML); Lorn, including Appin, Benderloch and Lismore (LN); Mid-Argyll, including northern Knapdale (AR); Jura (JU); Islay (IS); and Kintyre, including southern Knapdale (KT). These are shown in Fig. 3.1.

The next stage was to compile a source list of orthostatic monuments within these geographical areas for initial consideration. The most comprehensive and reliable sources available were considered to be the inventories of the Royal Commission on the Ancient and Historical Monuments of Scotland (RCAHMS)[11] and the National Monuments Record of Scotland.[12] In order to address the question of Thom's own selection of sites, all monuments within these regions included in Thom's full, unpublished site list (see pp. 52–3) were also considered.[13] In addition, Aubrey Burl's *Stone Circles of the British Isles*, published in 1976, listed thirty-four reported stone circles and rings in the relevant areas.[14] Finally, new discoveries reported in the annual publication *Discovery and Excavation in Scotland* between 1970 and 1980, too recent to be included in the other lists, were also considered.

Comparing the various lists, it soon becomes clear that the definition of a 'site', made in most cases in the total absence of any evidence from modern excavations, is extremely subjective. Yet for the purposes of the project, clear criteria were needed for actually defining a site, in order to demonstrate that the decision in any particular case had not been influenced by astronomical considerations. After some consideration distance was used as the main criterion: any collection of stone circles, standing or prostrate stones and sites of stones (where accurately known) was considered to constitute a single site if each feature is within 300 m of at least one of the others. If, however, two features are separated by a sea channel, or natural rises in intervening ground level prevent them being intervisible, then the features were counted as separate sites.[15]

Using these criteria an initial site list was compiled, com-

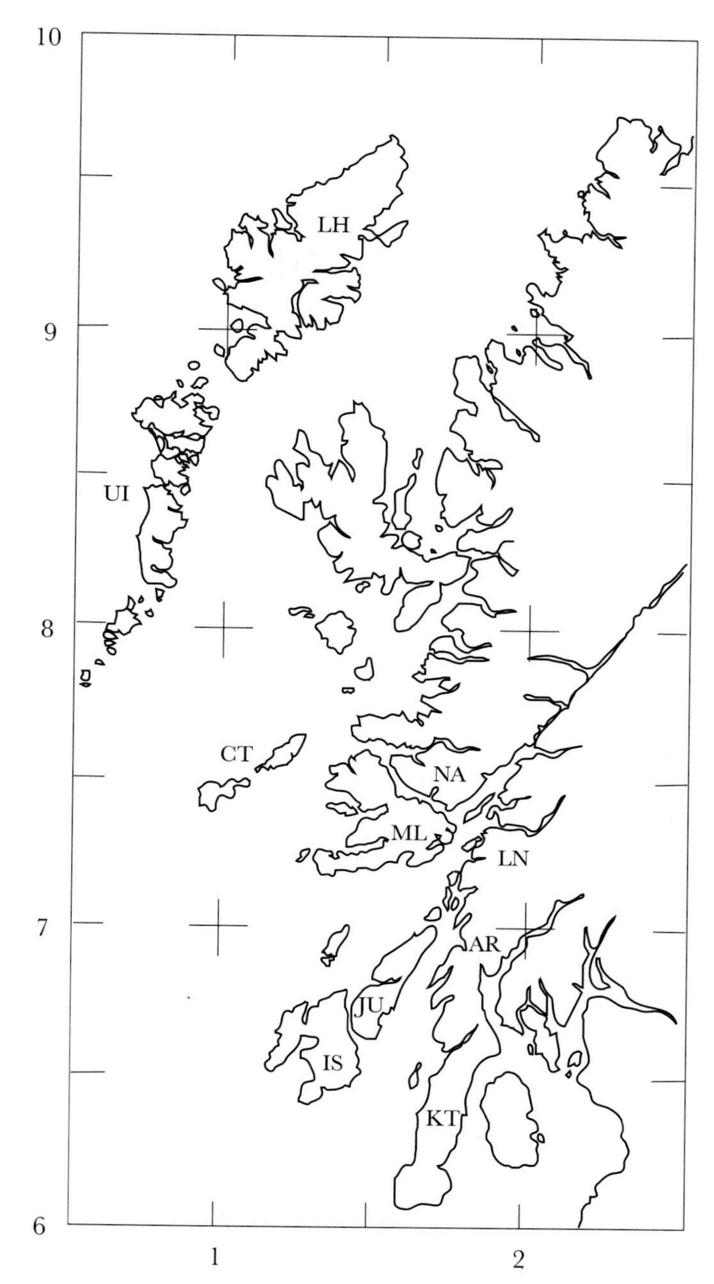


Fig. 3.1 Areas considered in the statistical analysis of 300 western Scottish sites.

prising 322 sites reported by at least one of the five sources.[16] No fewer than 133 of them, however, were then excluded from further consideration for various reasons unrelated to their astronomical potential. The authenticity of forty-seven sites was considered to be in serious doubt, since they comprised what were probably natural stones and boulders, more modern constructions such as enclosures and shielings, and the like; twenty-eight sites appeared not, after all, to be settings of standing stones but the remains of chambered tombs or cup-marked rocks; the exact (original) position of another thirty-one could not be determined, either because they had been completely destroyed or (in the case of a number of single standing stones) removed intact from their original positions; twelve could not be located during fieldwork; seven, in remote locations, could not be reached; six became known only after fieldwork on the relevant island (Islay) had been completed;[17] and two (Kintraw and Brainport Bay in mid-Argyll, described in chapter one) had already been excavated with a view to testing an astronomical hypothesis, rendering the statistical approach irrelevant.[18]

By these means the project finally arrived at 189 sites worthy of further consideration. For a reference list including brief descriptions and sources of further information, see List 2. Having arrived at a set of suitable sites, a pre-defined code of practice was needed which would specify unambiguously in each case, when it was later surveyed, which 'indications' should be measured and included in the dataset for statistical analysis. Devising such a code of practice is the trickiest part of the whole process. We are dealing with a variety of types of site, and the selection criteria must be flexible enough to cope with each different configuration of stones encountered. They must be tuned to give the best chance of isolating astronomical trends in the data if there are any, without allowing these valuable data to be submerged and lost amidst a welter of data of no consequence or pattern.[19]

At the heart of the code of practice adopted was a scale classifying likely structures for deliberate astronomical orientation into an order of preference. The rule at each site was: 'consider as potential indications only those structures with the highest classification that exists at this site'.[20] While the details are quite complex, in order to deal (for instance) with prostrate stones and stones of uncertain status,[21] the top six classes were essentially as follows:[22]

1. A row of three or more standing stones or of at least two aligned slabs.
2. One or two standing stones together with a number of prostrate stones that could have formed a row of at least three stones.
3. A pair of standing stones (not aligned), a single slab together with a prostrate stone that could have stood in line with its orientation, or three or more prostrate stones that could have formed a row.
4. Two stones, not both standing.
5. The flat faces of a single slab.
6. The flat faces of a single slab of uncertain status.

The site itself was then classified 1–6 according to the highest classification of structures occurring there. Some sites containing none of these, or more than six indications of the highest type, were given lower classifications.[23]

In his analyses at Level 1 Thom included a number of 'inter-site' indications, that is, indications formed by standing at one monument and using another as a foresight. Accordingly the project also considered all intervisible pairs amongst the 189 sites.[24] These were also selected and classified (1–3) according to a strict code of practice, taking into account the distance and visibility of the remote monument.[25]

The sites given a classification from 1 to 6, and the number of structures of the given classification found there, are listed in Table 3.1.[26]

The width of each possible indication was determined, in the field, from consideration of the possible ways in which an indication might have been provided and possible changes in the direction owing to more recent movement of individual standing stones. Each end of the resulting 'indicated azimuth range' (IAR) was determined to the nearest 0°·2 as viewed from a notional observing position 2 m directly behind the indicating structure.[27] The procedures for determining the limits of the IAR were subjective, but they should not have biased the overall result because no data were reduced, and hence declinations and astronomical potential were not determined, until after the fieldwork had been completed.[28] A wider 'adjacent azimuth range' (AAR), used in some of the statistical analyses, was defined as the range of horizon extending either side of the IAR by an amount equal to its own width, down to a minimum of 1° and up to a maximum of 2°.[29]

Since the project was not aiming to test ideas of high-precision astronomy at Levels 2 and beyond, measurements and calculations were carried out only to the nearest 0°·1. Where the entire horizon within an IAR was closer than 1 km, uncertainties about the exact observing position, original ground levels, and vegetation on the horizon could begin to have a significant effect, even at this level of precision. For this reason such cases, listed in the 'L' column of Table 3.1, were excluded from further consideration.[30] For IARs greater than 5° in width, the statistical weight attached to any declination within the range would be so small that its effect on the overall analysis would be negligible. Thus such lines, listed in the 'W' column of Table 3.1, were also excluded.[31] The two exclusion criteria were also applied to inter-site lines.

The remaining indicated horizons were surveyed wherever possible. In most cases horizon profile diagrams were constructed by combining theodolite measurements and photographs, usually taken using a 200 mm or longer lens. These are listed under 'A' in Table 3.1. Sometimes practical difficulties reduced the reliability, as where nearby buildings or trees obscured the horizon profile and necessitated a large adjustment from the theodolite station to the notional observing position. The lines concerned are listed under 'B'.[32] In some cases profile measurements could not be obtained directly and had to be calculated from large-scale Ordnance Survey maps. Such lines are listed under 'C'. The complete set of profile diagrams was duly published.[33] A typical diagram showing four profiles is reproduced in Fig. 3.2.

## The analysis and its results

The data obtained are summarised in Table 3.2. Taken as a whole, the azimuth and declination data seem unenlightening, showing no obvious trends and, indeed, looking much as would be expected by chance. Certainly the strong clustering around certain declinations that was so obvious in Thom's level 1 data is not evident here.[34] This is illustrated by the declination curvigram shown in Fig. 3.3a, whose construction differs significantly from those produced by Thom only in that the standard deviation for each constituent gaussian hump is a function of the width of the IAR, in preference to using a single value (or at most two values, for more or less reliable data) throughout.[35]

It is tempting to move on immediately and examine the data more closely in order to pick out interesting patterns from particular types of site or particular regions. However, it is important not to lose sight of the original aim of the whole exercise, which was simply to ascertain whether there was

Fig. 3.2 Four horizon profile diagrams from the western Scottish project, as published in Ruggles 1984a (fig. 7.11). The horizontal scales show azimuth, the vertical scales show altitude, and lines of declination at half-degree intervals are shown above the profiles. The horizontal lines below the profiles indicate the extent of the 'indicated azimuth range' (IAR). The arrows mark the limits of the wider 'adjacent azimuth range' (AAR).

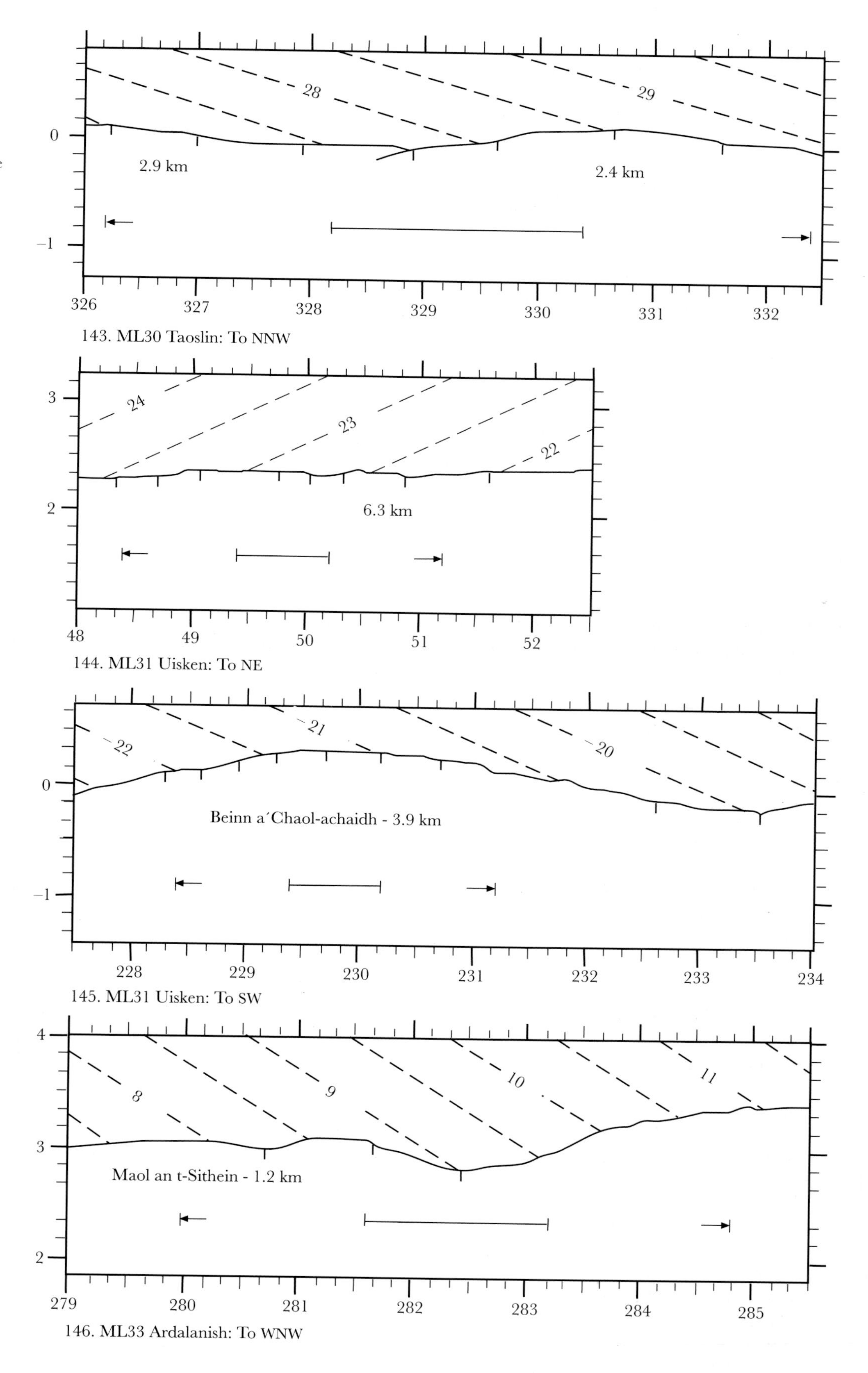

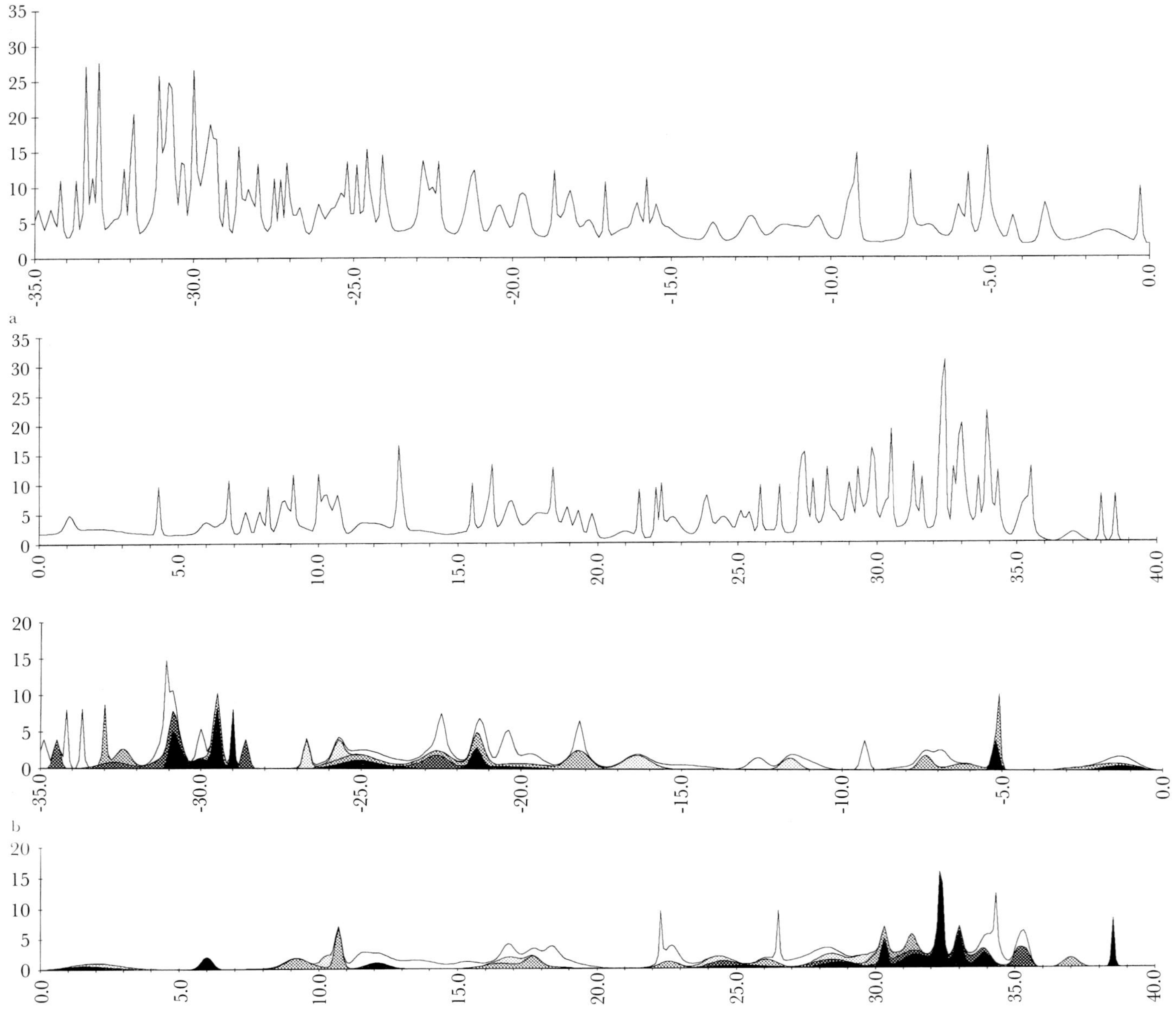


Fig. 3.3 Indicated declinations: the cumulative data from the western Scottish project. The curvigrams represent the indications in Table 3.2. Each constituent gaussian hump is centred upon the declination mid-way between the maximum and minimum values obtained within the IAR and has a standard deviation equal to half the difference between these values. Where the two values are equal the standard deviation has been taken to be 0·05 degrees. The area under each constituent hump is 1 unit.

a. All 276 indications.
b. The 130 on-site indications. Lines of classes 1, 2, 3, 4, and 5/6 are indicated by differential shading, class 1 lines being the darkest.

statistical evidence overall for significantly more astronomical alignments than would have been expected by chance.[36] For this reason, the next step taken was to devise a way of testing the hypothesis that there was significant evidence of deliberate astronomical orientation against the alternative (null) hypothesis that the structure orientations were effectively random.[37] Because the expected distribution under the null (random) hypothesis is non-uniform, no simple statistical test was available and Monte Carlo simulations were used in order to assess significance levels (see Statistics Box 4).

The principle of Monte Carlo testing is to run a series of simulations in which sets of pseudo-random data[38] are generated. We can then estimate how likely it is that a pattern observed in the 'real' data could have arisen fortuitously by determining what proportion of the 'random' sets of data manifest a similar pattern.[39] If the proportion is very low, then it is highly unlikely that the observed pattern could have arisen fortuitously. In this case, we wish to generate sets of sites with pseudo-random orientations and to examine the resulting set of indicated declinations.

The problem is that the indicated declination depends not only upon the azimuth, but also upon the latitude of the site and the altitude of the indicated horizon. Furthermore we are actually interested in an indication of finite width which will include a range of declinations. Thus the width of the indicated azimuth range (IAR) is relevant, and also the shape of the horizon. Each of these other properties needs to be distributed in exactly the same way amongst the simulated data as amongst the observed data.

## STATISTICS BOX 4

### HYPOTHESIS TESTING

We can spot patterns in sets of archaeological data and describe them; but we can not necessarily trust our judgement to decide if a pattern we observe is 'significant', i.e. if it is reasonable to postulate that the pattern has arisen for a certain set of reasons or whether it is quite likely to have arisen through the interaction of other factors entirely. One way forward is to make the assumption that the 'other factors' of the second alternative act in a similar way to a set of random processes. An attempt can then be made to estimate the probability that the observed patterns might have arisen as a result of such processes—i.e., fortuitously. If this probability is very small, then we have some justification for supporting the favoured explanation.[1]

Conventional statistical tests have their roots in the inductive paradigm, which stipulates the following. First, a hypothesis $H_0$ is formulated. Second, a set of data is collected (a number of experiments are performed) in order to test $H_0$. A test is then undertaken to determine whether or not to reject $H_0$ on the basis of the new data. Note that the hypothesis being tested in this way is the 'null hypothesis'—the uninteresting one that the data observed have arisen fortuitously. Only if this hypothesis is rejected should the 'alternative hypothesis' (a particular explanation) be taken seriously.

These two approaches are not obviously compatible. For one thing, the latter seeks a clear answer (accept or reject), while the former yields probability estimates—numbers on a continuous scale.[2] This is circumvented by adopting certain conventions regarding the 'significance level' at which $H_0$ will be rejected. Essentially, a set of data can be thought of as a collection of points in some 'sample space'. One first calculates the expected distribution of such sets of data within the sample space, i.e. the average over the various formations that hypothesis $H_0$ could produce. The significance level is an estimate of the percentage of cases in which sets of data produced under $H_0$ would have resulted in a formation deviating from the expected one by at least as much as the 'real data' actually did. The most common convention is to reject $H_0$ if the significance level is below 5 per cent ('at the 5 per cent significance level'), meaning that the probability that the observed data could have arisen fortuitously is estimated to be less than 0·05.

Where the null hypothesis only concerns the value of a single random variable, it is possible to speak of a 'confidence interval' (a band of values with the property that if the measured value falls outside it, the null hypothesis may be rejected—with confidence—at a given significance level) and 'confidence limits' (the edges of this band). For a normal distribution the 95 per cent confidence limits are $\mu \pm 1{\cdot}96\sigma$.

#### MONTE CARLO METHODS

In relation to a number of commonly encountered types of probability distribution, generic procedures have been developed for the purposes of hypothesis testing.[3] In other cases, for example where it is not possible to develop a simple analytical model of the probability distribution under the null hypothesis, Monte Carlo simulation may be useful.

Each Monte Carlo simulation produces a set of pseudo-data generated by assuming the null hypothesis to be true (generally these are attempts to simulate 'random' data in some sense). If we wish to test whether a particular property of the real data (more strictly, an observed deviation from what might be expected under $H_0$) is adequately explicable under $H_0$, we simply have to calculate the proportion of simulations in which that characteristic is also found (i.e. the data deviate by at least as much). It is not necessary to estimate the expected distribution; just to be able to determine rigorously what constitutes 'deviating by at least as much'. In most cases this is simpler than it sounds. For example, if the observed property of the real data was that '48 out of 100 declinations fall between −30° and −19°', then if we suspected that the expected number of declinations falling within this range under the null hypothesis should be far smaller, we would seek to determine the proportion of simulations in which *at least* forty-eight declinations fell in this interval. On the other hand, if we suspected that the expected number should be far larger, we would seek to determine the proportion with *at most* forty-eight declinations in this interval.

Using Monte Carlo simulation in particular, and statistical methods for hypothesis testing in the context of interpreting archaeological data in general, raises a number of broader questions and issues that are brought up elsewhere in the book.[4] In the context of understanding what we are doing and why, Monte Carlo tests do have the advantage that they make explicit some of the assumptions that tend to lurk implicit and unquestioned where hypothesis-testing is achieved using generic, 'off-the-shelf' statistical tests.

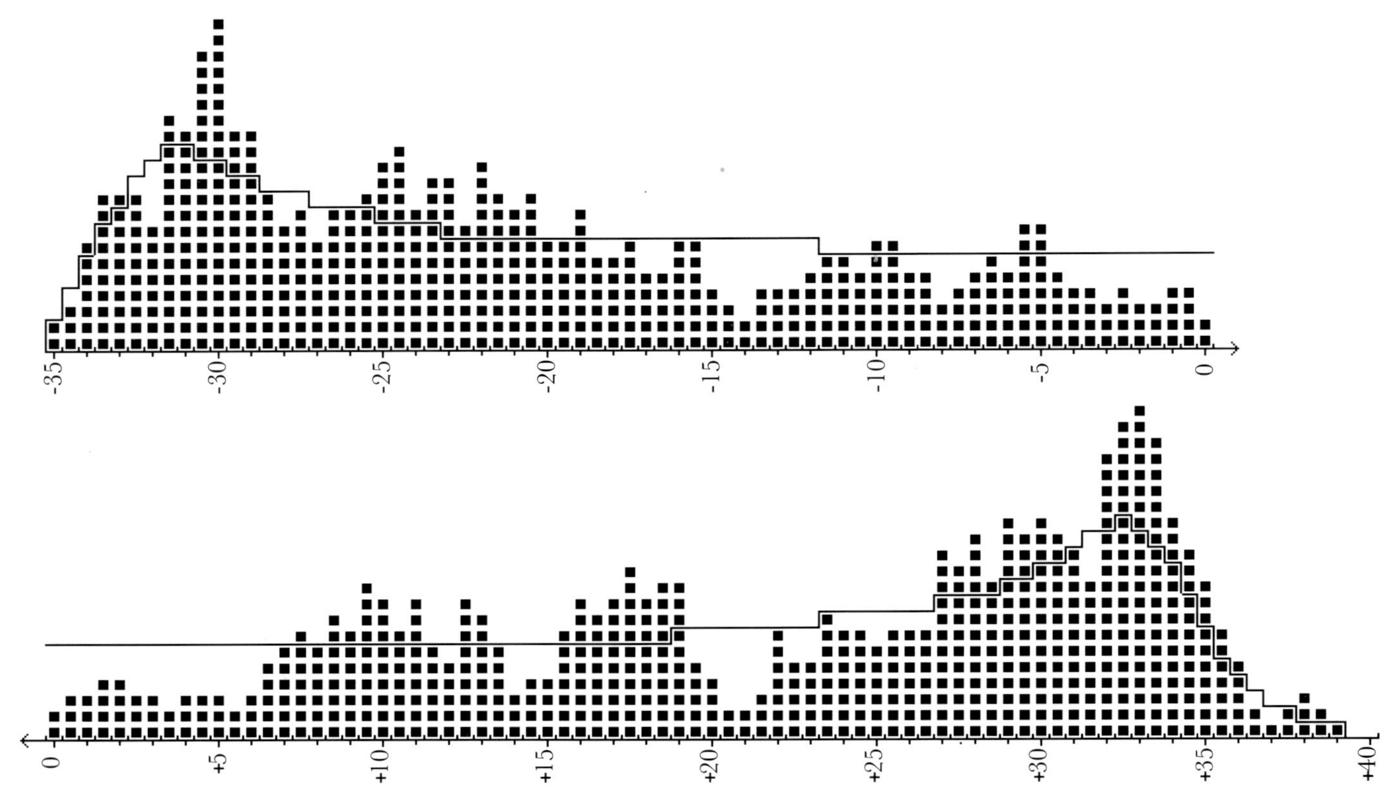


Fig. 3.4 The number of declinations falling within 1°·6 of a given value, compared with the most likely number that would have been expected fortuitously, as determined by Monte Carlo simulation. Values shown represent the centre of the bin, so that (e.g.) the bin marked '−30' represents the declination interval from −30°·8 to −29°·2. Each square indicates that one IAR in Table 3.2 'hits' the target in the sense that part of it falls within the target interval. The single line indicates the number of hits that would have been expected by chance.

The problem was solved by generating exactly 276 indications in each Monte Carlo simulation run, the same number as in the real data. The central azimuths were generated pseudo-randomly, but the width of the IAR, the horizon altitude and the site latitude were obtained by randomly permuting the actual IAR widths, and separately the altitudes and the latitudes, from the real data. In order to facilitate the process one simplification was made: it was assumed that all horizon profiles were flat, with their altitude taken as the mean of the actual altitude range within the IAR. One hundred simulations were produced in this way.[40]

For any 'target window' of declinations, it was then possible to calculate the number of 'hits' scored by the observed data upon that target and to ask in how many of the 100 simulations there were at least as many hits upon this target. For example, in the observed data seven lines hit the target interval from −24°·6 to −24°·4, but this many hits were only obtained in one of the hundred simulations, implying that the preference for declinations within 0°·1 of −24°·5 was significant at the 1 per cent level (see Statistics Box 4). However, in order not to prejudge the evidence for an interest in particular celestial bodies or events, and to avoid prejudging the precision to which astronomical alignments might have been set up, a comprehensive range of declination windows of varying centres and widths was examined. The question of whether certain windows were avoided was given as much attention as the question of whether they were preferred.[41]

The results at a fairly high level of precision can be visualised in Fig. 3.4, which shows the number of indicated declinations falling within 1°·6 of a given value, compared with the most likely number that would have arisen fortuitously, as determined by Monte Carlo simulation. This diagram does not indicate whether particular accumulations or areas of avoidance are statistically significant, which must be left to the analysis. This showed overall trends at three levels of precision:[42]

1. At the lowest level, declinations between about −15° and +15° were found to be strongly avoided.
2. At the second level, there was found to be a marked preference for southern declinations between −31° and −19°, and for northern declinations above +27°.
3. At the highest level of precision, there was found to be marginal evidence of a preference for six particular declination values to within one or two degrees: −30°, −25°, −22°·5, +18°, +27° and +33°.

How should these conclusions be interpreted? The first is the most straightforward: it could simply have arisen as the result of a strong preference for structures to be oriented roughly N–S, NW–SE or NE–SW rather than E–W. That such a preference should exist is of course interesting in itself, but beyond the rough determination of azimuthal direction it does not necessarily reflect anything astronomical at all.

Since the majority of the data represent opposite pairs of indications along the same structures, one of the second-level trends is almost certainly a simple consequence of the other.[43] The declination range above +27° corresponds to that part of the horizon farther to the north than the sun or moon ever rose or set. On the other hand the interval from −31° to −19° represents, to within a degree or so, the range of possible values of the southerly limit of the moon's monthly motions at different points in the 18·6-year cycle. A distribution over this range would be expected, for example, if structures were roughly oriented upon the most southerly rising or setting position of the moon over a short period of time (perhaps at most a year or two) without any recognition of—or at least without any interest in marking—the gradual change of this limit over the longer cycle. Thus, if the southern trend represents the cause and the northern one the effect, there may be some evidence here of preferential orientation upon the rising and setting positions of the moon.[44]

Three of the highest-level targets (−30°, +18° and +27°) may indicate a specific interest in the lunar standstill limits, and could be taken to imply that the 18·6-year cycle was recognised and deliberately marked, but it seems anomalous that there is no evidence of any interest in the fourth limit at around −19°·5. Another preferred declination (−25°) may indicate an interest in the winter solstice. The others have no obvious solar or lunar significance and one (+33°) falls well outside the solar or lunar range.

Amongst these data there is no evidence whatsoever for an interest in the summer solstice and the sunrise and sunset on the calendrical epoch dates suggested by Thom. As over 40% of Thom's own data came from this same region,[45] the general idea of a 'megalithic calendar' dividing the year into eight or sixteen equal parts is dealt a severe blow by these conclusions. Nor is there any evidence amongst these data of astronomical orientations of a precision greater than about one degree. Even the preferences for the other declinations at a precision of one or two degrees are only marginal, and an independent statistical appraisal of the data has concluded that they are insignificant.[46] In short, the data give no support at all to Thom's conclusions at Level 1,[47] and we are forced to conclude that the idea of prehistoric orthostatic monuments in Britain incorporating astronomical alignments precise to anything much greater than about a degree is completely unproven by the sort of approach taken by Thom.

## The moon and the stone rows

Something more positive does, however, emerge from the data when we examine the types of monument and indication that give rise to the overall trends. This reveals that the on-site data, and particularly the sites of classes 1 and 2, feature predominantly amongst the indications falling in particular 'preferred' declination intervals.[48] Taking just the class 1 and 2 sites, i.e. the rows of three or more stones and pairs of aligned slabs, it is evident that virtually all are oriented in the northern and southern quarters of the compass (Fig. 3.5a), and that these sites are largely responsible for the low-level trend identified overall.

In fact, the anomalous east–west indications come from the only two class 1 or 2 sites in the Outer Hebrides: Callanish (LH16), whose radial rows form part of a complex monument apparently very different from the isolated stone rows and pairs, and Blashaval, North Uist (UI19), where three stones are placed in a 50 m-long line, unlike the other short stone rows in the sample whose overall lengths never exceed 20 m.[49] If the data are restricted to the mainland and Inner Hebrides, the north–south tendency becomes total (Fig. 3.5b).[50] The trend can not be explained by orientations merely following the local topography, as the sites are scattered over a wide geographical area and there are several clear counter-examples.[51]

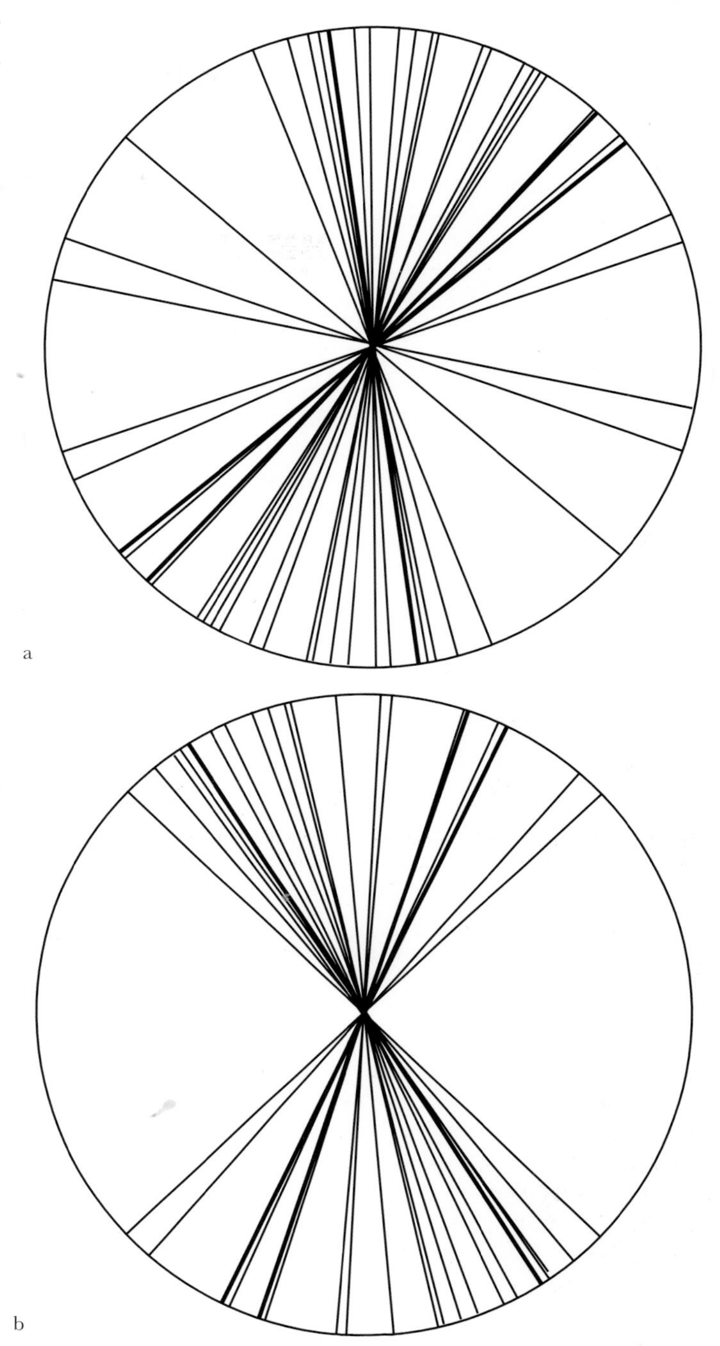


Fig. 3.5
a. Central azimuths of all class 1 and class 2 indications in Table 3.2.
b. The same, excluding the data from the Outer Hebrides, i.e. Callanish (LH16) and Blashaval (UI19). Cf. Ruggles 1984a, fig. 12.4.

Might this obvious trend in azimuths have arisen because the orientation of the rows in question was influenced by more particular astronomical considerations? Certainly, when the declination curvigram is restricted to on-site data only (Fig. 3.3b) some clustering is evident, especially amongst the higher classes of data. Restricting the data to the stone rows and aligned pairs in Mull and mainland Argyll, it transpires that these monuments are all, without exception, oriented in the south upon a declination between −31° and −19°, and that they are largely, if not entirely, responsible for the second-level trend,[52] the remaining indications in the interval forming part of the general 'background noise'.[53] Even amongst the non-aligned pairs and single flat slabs in this region, the data show quite a strong pattern of lunar orientation in the south.[54] Seen in this light, even the southern indication at Ballochroy seems more probably part of a wider pattern of lunar orientation than an isolated instance of solar orientation.[55]

Unwittingly, but fortunately, the project left open the opportunity of testing some of these conclusions in the future by collecting further data. This was because of the selection criterion that led indications to be omitted from further consideration when the indicated horizon was closer than 1 km. Since the conclusions now pointed to declination preferences precise to one or two degrees at most, there was no longer any *a priori* reason to exclude any but the very closest horizons (say, within 100 m) on the grounds that changes in ground level would significantly effect the measured declination.[56] A project to collect the new data was undertaken in 1985[57] and the results are described on pp. 109–10.

## Lessons learned

It is clear, then, that although it lent no support to Thom's conclusions at Level 1, the western Scottish project did begin to reveal patterns of low-precision lunar alignment amongst a concentration of architecturally similar monuments included amongst the data—the short stone rows of Argyll and Mull. This suggests that it would be more productive to concentrate on smaller groups of more similar sites in smaller regions. A similar conclusion had already been reached by Burl, who since 1969 had been urging archaeologists and archaeo-astronomers to turn from the study of single sites (and high-precision alignments) and to look instead for 'a group of similar monuments . . . in a restricted locality'.[58]

The outcome of the project re-emphasizes the dangers of concentrating on single sites and isolated alignments considered out of their archaeological context, even though examples of this practice continue to appear. Thus a recent paper extrapolates from the fact that sunset behind a single prominent peak as viewed from an 'ancient inscribed rock pile' in Co. Mayo, occurred on around 21 April and 21 August, to postulate that a calendar was in use that divided the year into three or six equal parts.[59] Yet the alignment is unindicated, no supporting evidence is provided from other monuments, and no evidence for such a calendar has been reported from elsewhere. The chances of being able to fit one of any number of 'one-off' explanations to fortuitous occurrences are very high indeed, and unsupported evidence such as this can be highly misleading.

The rigorous approach suggests some basic procedural principles that should be followed in all circumstances. These can be summarised as 'observe everything' and 'report all you observe'.[60] The point is to avoid, and be able to demonstrate that one has avoided, selecting data that fit an idea and ignoring the rest. It is an obvious principle but one which is flouted time and time again, for example in every site plan where astronomical lines have been superimposed but many other lines, equally impressive apart from their lack of astronomical potential, have been omitted.[61]

Ignoring the 'negative' evidence is often apparent in the selection of putative indications at a particular monument, but also arises when great emphasis is placed upon a particular site or sites while other superficially similar ones are ignored. As we saw in chapter one, speculative interpretations of the ring of Aubrey Holes at Stonehenge, placing great emphasis upon the importance of the number 56, ignore posthole rings and pit circles at many broadly contemporary henges and henge-like monuments. Likewise, Thom's high-precision lunar interpretation of the Nether Largie stones at Kilmartin (his 'Temple Wood') is weakened when the monument is compared with the superficially similar site of Barbreck, less than 10 km to the north (Fig. 3.6).[62] Similarly, Le Grand Menhir Brisé (see chapter one) lies at the end of the destroyed tomb of Er Grah, which suggests that it might have functioned as a tomb marker—albeit a particularly magnificent and prominent one—similar to several others in the vicinity;[63] at the same time, only one of the many other large standing stones in the area has been proposed as a lunar foresight.[64]

Perhaps the most important principle of all is to keep an open mind and constantly be prepared to change one's ideas as new data are revealed. Statements are often made to the effect that 'we know that . . . [people made certain observations]'. This is simply a self-delusion. We can never know: all we can ever have is a degree of belief in a certain idea according to the evidence available. Once this is admitted, it is easier to modify that belief in the light of new evidence. The present author first visited Ballochroy in 1973, but became cautious about Thom's solstitial interpretation of the site following a visit to another Scottish three-stone row—the one at Duachy, which was not solstitially oriented. Initial enthusiasm for Thom's ideas was followed by profound disillusionment, but led eventually to a more reasoned set of ideas which were then modified over the years as more three-stone rows were examined and wider evidence and experience were gained. The fact that theories and data develop in parallel, helping each other in some sort of loop, is a general principle applying throughout archaeology (see later chapters) and one that we cannot ignore.

## Strengths and limitations of the statistical approach

What, then, can we say generally about the sort of rigorous approach, accompanied by formal statistical analysis, epitomised by the western Scottish project? It clearly has some important strengths. Chief amongst these is the ability to discern whether astronomy was probably a major factor influencing monument orientations or whether those that appear astronomical might well have arisen through a combination of factors quite unrelated to astronomy. Provided that due attention is paid to questions of the fair selection of data, such an approach can handle a diversity of types of monument and possible indicating devices, drawing attention to any overall astronomical trends without prejudging the astronomical targets involved.[65] It also uses the evidence that remains above ground today, without the need for excavation; unavoidable

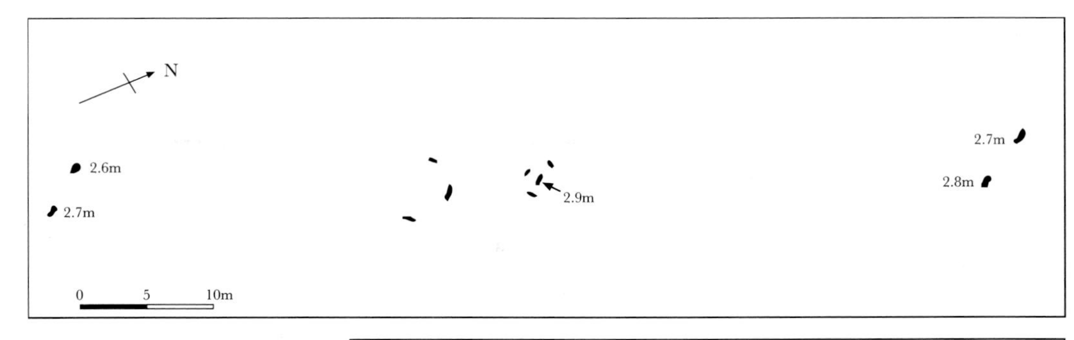


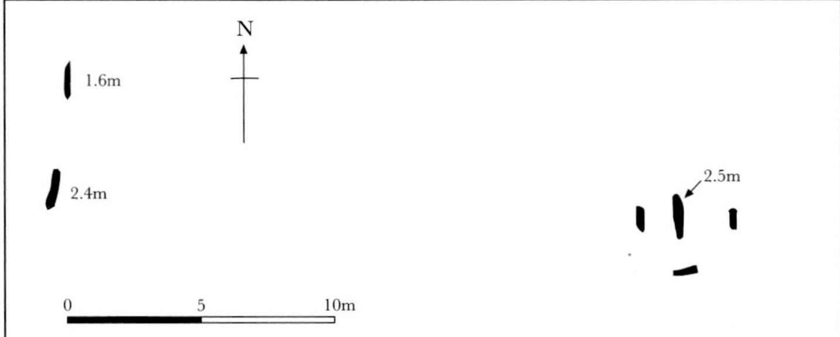


Fig. 3.6 A comparison of two superficially similar monuments. The heights of tall standing stones are marked. After Patrick 1979, figs 1 and 2.
a. Nether Largie (Thom's 'Temple Wood') (AR13(*b*)).
b. Barbreck (AR3).

errors and uncertainties, such as the difference between the present disposition of the material remains and that intended by the builders, can be overcome since they will tend merely to add to the general background 'random noise'.[66] Finally, as we have seen, it can help identify, within a larger dataset, particular groups of monuments worthy of further investigation.

But there are also limitations. Such an approach can only allow us to isolate the most general astronomical trends, widely and consistently adhered to in prehistoric times. It is inevitably limited by the quality of the evidence that ultimately survives above the ground surface today,[67] and tends to isolate large monuments built in stone from other constructions that might leave no obvious trace above the ground but could be detected and investigated using other archaeological techniques. It takes no account of the way practices might have differed from one region to another, and even from one monument to another. Neither does it make any attempt to identify how practices might have changed with time, either within a region or at a single location; yet excavation can reveal change through time far more complex than could have been expected from the surface record alone.[68] In terms of interpreting what is in the archaeological record, we quickly find ourselves wanting to focus upon particular groups of monuments or particular geographical regions, yet the available datasets rapidly become too small to carry the rigorous approach any further.[69]

Beyond these limitations, the rigorous approach also has some serious shortcomings: more subtle perhaps, but absolutely fundamental. The first concerns the choice of selection criteria. Even Douglas Heggie, a leading proponent of rigorous selection criteria, recognised that while they might be 'rigorous and systematic in one sense, [they might be] rather arbitrarily restrictive in another'.[70] As is clear from the western Scottish project, the diversity of types of monument and the states in which their remains are found today result in a complex set of criteria, arguably full of the influences of 'modern' prejudices about what might have been important in prehistoric times. There is an uncomfortable feeling that, in spite of striving for general principles to ensure objectivity, we are in fact imposing our own prejudices just as surely, if not as blatantly, as if we were simply selecting the lines that fit one theory and ignoring others that do not.[71] It is quite possible that we may be failing altogether to recognise signs of astronomical practice because we are 'selecting them out' at an early stage as a result of looking for the wrong things.

In view of this it is comforting to think that statistical rigour can at least prevent us from interpreting chance occurrences as significant. Yet even the extent to which this is true becomes questionable when we look more deeply still. The problem is that conventional ('classical') statistical methodology is predicated on the assumption that a hypothesis is formed before data are acquired in order to test it.[72] In practice, however, recognition of some pattern in the archaeological data always precedes the formation of a hypothesis, and the hypothesis chosen may in fact be one of several—indeed, many thousands—of possibilities.[73] To choose a 'likely' hypothesis and then test it back on the data that helped to suggest it is clearly to move dangerously in the direction of circular argument,[74] yet further datasets may not be available. One approach is to try to recognise 'families' of equally tenable hypotheses and to examine the apparent significance level at which we would be prepared to reject the null hypothesis in favour of each one of them.[75] The difficulty here is that we can not say how low the apparent significance level should be before we should reject the null hypothesis in favour of one of many alternatives.

We have returned, of course, to the point that ideas and data necessarily and inevitably develop in parallel, and we have run into the problem that classical statistics lacks the theoretical framework to deal with this procedure in a rigorous way.[76] The only other option seems to be to look for other methodologies for building and testing theoretical ideas in parallel with expanding the body of available data, methodologies that ensure that we will move forward rather than getting caught in circular argument. This is a theme of major importance, and one to which we shall return later.

In the meantime, it is clear that the western Scottish project occupied an important place in the development of ideas on astronomy in prehistoric Britain and Ireland. It also represented an important step in the development of methodology and procedure within archaeoastronomy, demonstrating that rigorous approaches have an important role in testing complex ideas. It is unlikely, though, that another project of the same nature or scale will ever be undertaken.

## From statistical abstraction to social context

It will be evident from what has gone before—indeed, from the entire discussion up to this point—that it has taken place almost without any reference to the people who practised the supposed astronomy, their social organisation, methods of subsistence, trade and exchange, technology, or even their ideology; indeed, to all other aspects of the social context within which any prehistoric astronomy would have been practised and within which it would have had meaning. This reflects the nature of the great majority of discussions on the topic of prehistoric astronomy up until the early 1980s. In these early days of archaeoastronomy it was symptomatic that it was conducted in almost complete isolation from the sorts of questions of interest to mainstream archaeologists.

From the archaeoastronomical point of view, there was good reason for this. Most archaeoastronomers—as well as a number of archaeologists—held the view, noted already in chapter one, that archaeoastronomical evidence is 'one step removed' from most other archaeological evidence.[77] This reinforced the notion that archaeoastronomers and archaeologists had separate roles. That of the archaeoastronomer was to assess, by statistical means, the nature and extent of astronomical influences on the design and placement of archaeological structures, accumulating data which could only then be considered alongside a variety of other archaeological data in order to interpret astronomical practice in its social context.

Thus Douglas Heggie, towards the end of his 1981 book containing a detailed critique of ideas on British prehistoric astronomy, shied away from interpreting the 'interesting core of facts' that remained from the statistical analyses.[78] Similarly, the western Scottish project aimed to establish general methodological guidelines and for this reason deliberately stopped short of any detailed interpretation of the evidence. 'Apt methods of analysis of putative astronomical orientations and alignments need to be formulated by the archaeoastronomers and presented clearly to the archaeological community at large for constructive criticism. Only when a reliable basic approach has been accepted can meaningful astronomical evidence be incorporated into theories of prehistoric society.'[79] A two-stage process was assumed, in which a statistical investigation was undertaken by the archaeoastronomer, whereupon any 'significant' results would be handed over to the archaeologist for interpretation. Unfortunately, this very assumption, and archaeoastronomy's evident prepossession with statistical methodology, left it open to the charge in the early 1980s that it was a study 'conducted . . . up to now, separately from consideration of the society in which it operated',[80] and as a result archaeologists tended to feel it was irrelevant and paid it little attention.

We cannot hope to interpret what we find—we cannot hope to understand anything about the nature of astronomical practice in prehistoric times and its meaning to prehistoric people—without beginning to think more seriously about the people themselves. Numerous examples from a variety of non-Western cultures, historial and contemporary, inform us that astronomy is not practised in isolation: astronomical beliefs, observations and rituals invariably form part of a world-view which derives from, and affects, many different aspects of life.[81] Thus evidence relating to a range of cultural phenomena could be relevant to studies of astronomical practice, and should not be dismissed out of hand because it is difficult to quantify.

Moreover, whether we acknowledge it or not, such 'subjective' evidence is implicit in the ideas that we choose to test in the first place. Everyone working in this field—and this was true of Thom himself—inevitably starts from the context of their own social model, their own set of ideas concerning the nature of people and society in prehistoric times. It is this that motivates them to collect data that they feel to be relevant to their theory and it is this that underpins their interpretation of those data.

It is to these social models, explicit and implicit, to the extent to which they were well grounded in the wider evidence available from the archaeological record, and to the wider social and archaeological context of megalithic astronomical alignments, that we now turn in order to complete our picture of people's understanding of prehistoric astronomy in Britain and Ireland by about the mid-1980s.

# 4

# Alignment and Artefact

*Reconciling the Archaeological and Astronomical Evidence*

These were the most primitive people I had ever come across, and their customs and methods were those of the Stone Age. I suppose I had subconsciously been expecting them to be little better than hominoids, crude and brutal figures in man-shapes. And yet they were as intelligent and sociable and swift in understanding as any other group of human beings, and gentler than most.

John Simpson, 1993, describing his first encounter with the Ashaninca [Campa] of western Brazil[1]

Whatever we do we must avoid . . . the idea that Megalithic man was our inferior in ability to think.

Alexander Thom, 1967[2]

Theodolites wink towards every skyline notch where the sun once set or moon rose or where Arcturus for a brief year or two shimmered dimly down into the mists of a prehistoric evening. . . . But any single-minded preoccupation with astronomy . . . must limit an investigation of the past. Prehistory should encompass all of man's activities, not just one or two aspects of our ancestors' lives, and too often . . . the people who built the stone circles have been ignored.

Aubrey Burl, 1979[3]

## Megalithic science?

Alexander Thom was an engineer, and his interests were technological. In Thom's view, prehistoric Britain was populated by 'megalithic man' who had

> . . . a solid background of technological knowledge. Here I am thinking not only of his knowledge of ceramics, textiles, tanning, carpentry, husbandry, metallurgy, and the like, but of his knowledge of levers, fulcrums, foundations, sheerlegs, slings, and ropes. . . . There was also his ability to use boats: he travelled freely as far as Shetland, crossing the wide stretch of open water north of Orkney, as well as the exceedingly dangerous Pentland Firth and the North Channel between Kintyre and Ireland. This involved a knowledge of the tides and tidal currents that rule those waters.[4]

Certainly, the very existence of large stone monuments bears witness to the engineering capabilities of people in the Neolithic and Bronze Age. It is also clear that people at these times—and even Mesolithic hunter-gatherers in earlier millennia[5]—did travel between the Scottish islands; hence, while no direct evidence remains, it is evident that they must have built boats and must have had a practical knowledge of the tides. The question is whether, as Thom went on to assume, 'the relation between the phase of the moon and the time, direction, and violence of the tide must [have been] clearly understood'[6]—at least, in terms familiar to us—and whether, as Thom believed, this was likely to have led in turn to the development of a precise solar calendar and a well-orchestrated and long-term programme of careful lunar observation.

The danger in any discussion of this issue is that it gets sidelined into emotive arguments about the intellectual capabilities of our ancestors, epitomised by the fiery debate between Hawkins and Atkinson over Atkinson's depiction of Stonehenge's constructors as 'howling barbarians'.[7] People in Neolithic and Bronze Age times did in fact have a general level of intellect equal to ours[8]—they are our near neighbours in evolutionary terms—and Atkinson, as a professional archaeologist, surely never intended to imply otherwise.[9] As Aubrey Burl has remarked:

> The people who put up the stone circles . . . were men and women physically like us, with the same limitations of mind and body. Their beliefs were different because their lives were different. They had an acute and sensitive awareness of animals, plants, clouds and weather because these things were their intimacies, the framework of the world in which they lived.[10]

The question, then, is not one of intellect. It is to some extent about ability, as constrained by existing technological achievement, experience and knowledge, but above all else it is one of motivation. People *could* have used the ring of 56 Aubrey Holes at Stonehenge, devoid of a permanent ring of posts, as an eclipse predicting device; the use of distant natural foresights *could* have enabled Neolithic and Bronze Age inhabitants of Britain, with no tools apart from large unworked stones and the horizon available to them, to observe and mark particular astronomical events to remarkable precision.[11] The crucial question, however, is *did* they?

In tackling the question of motivation the social context is of critical importance. A modern example conveniently illustrates the point. An anthropologist working with the Mursi, a small group of cultivators and cattle herders living in south-western Ethiopia, during the 1960s and 1970s, witnessed how

> on one occasion a man, who had been wearing a knotted cord round his ankle for several weeks, announced to a group of bystanders that 72 days had elapsed between the planting and first harvesting of his sorghum crop. He had been keeping track of this interval by successively knotting a piece of cord for every day that passed. The other members of the group treated this information as a curiosity without relevance to their daily lives, not as a 'discovery' to be added to their total stock of knowledge about the world. Their main reaction, in fact, was one of mild surprise that anyone should have taken the trouble to record such a trivial fact, and it was, without doubt, quickly forgotten.[12]

By assuming, explicitly or implicitly, that types of measurement that seem highly significant to us in the modern Western world were significant in another human society, or indeed that their whole framework for understanding the world around them was in any way similar to our own, we are plainly guilty of ethnocentrism—the tendency to create a privileged view of our own culture in relation to all others and hence to project our own ways of comprehending things onto the group of people we are interested in. As a result, our conclusions will be questionable at best. For this reason it is crucial to refer to the social context and to consider a wider range of possibilities.

The danger of ethnocentrism is evident throughout the terminology used by the early archaeoastronomers. By talking of 'megalithic man' and thinking of him [sic] as an engineer, Thom had already isolated himself from the social framework built up by archaeologists and distracted attention from his own scientific conclusions.[13] Archaeologist Graham Ritchie, speaking at the 1981 Oxford Archaeoastronomy Symposium, attempted to redress the balance by giving the archaeological perspective on the megalithic tradition, pointing out that there was no 'megalithic culture' as such, but instead that megalithic monuments of various types span a period of about four millennia, being built at various stages in a long and complex period of social change, and are related to many independent traditions.[14]

Another term that contributed to the suspicion felt towards Thom and his followers by many archaeologists during the 1970s and 1980s was 'megalithic astronomy', which—with its 'megalithic observatories'—was often discussed together with megalithic mensuration and geometry (see Statistics Box 5) as part of 'megalithic science'.[15] This vocabulary alone doubtless led many archaeologists to conclude that twentieth-century scientists and engineers were simply projecting images of themselves into the past and, as a result, to pay no further attention to their conclusions. Others, however, drew a careful distinction between the new evidence for technological achievement that had apparently been uncovered by some archaeoastronomers and its interpretation in manifestly ethnocentric terms.[16]

It is one thing to suggest, for example, that precise measuring rods of standard length were produced for century after century over an area from Orkney to Carnac,[17] or that high-precision observations of the moon were made over generations and recorded in stone. If such ideas really are justified by the evidence available, then archaeologists might well be interested in their implications for social organisation and the development of ideas.[18]

It is quite another matter to suggest (or worse, simply to assume implicitly) that prehistoric people perceived the world in anything resembling the way that we do. Here lies the real danger of ethnocentrism and it is much more difficult to avoid. Those of us raised in the Western scientific tradition[19] cannot help using a set of concepts formed in the context of that tradition as our basis for trying to understand things, and that includes trying to understand the very different traditions of other peoples; and it is often difficult to see where we are making unwarranted assumptions that other 'world-views' are similar to our own. In the context of megalithic geometry, one such assumption is that a particular geometrical shape meaningful to us (such as a circle or ellipse) was perceived as such by prehistoric people;[20] another is that what to us seems an obvious method of constructing a particular shape was actually seen as obvious, and hence used, by the builders.[21] For an astronomical example, consider the following statement by Thom relating to his megalithic calendar (see Astronomy Box 5): 'We imagine megalithic man experimenting for years with foresights for the rising and setting sun . . . [so] that he obtained declinations very close to those we have obtained as the ideal.'[22] The crux of the problem is the highly questionable assumption that their 'ideal' must have been the same as ours.[23]

One of the consequences of people starting to come to terms with these issues in the early 1980s was a glut of papers on whether prehistoric astronomy was 'scientific' or 'ceremonial' in nature.[24] The tacit assumption was that high-precision observations represented 'scientific astronomy' (as epitomised by Thom's megalithic observatories) whereas low-precision alignments must have been 'purely symbolic' (as epitomised by the solar alignment involving the roof-box at Newgrange). In fact, this is far from self-evident. The Greek poet Hesiod, for instance, writing in about the eighth century BC, describes ways in which farmers could use the heliacal rising and setting of various constellations[25] to regulate agricultural activities.[26] This is predictive, and hence would be described by some as scientific, but it is not very precise. On the other hand, observations matching the precision achieved several millennia later by Galileo[27] could have been recorded adequately using backsights consisting simply of stakes or poles inserted in the ground; if people really did go to the trouble of erecting stone monuments aligned upon horizon astronomical phenomena, these must surely also have served another and presumably ritualistic purpose.[28]

The most troublesome question is exactly what one might mean by 'scientific' in the first place. It is obviously central in this sort of argument but is not clearly defined; in fact, a number of different answers might be given.[29] The question is only relevant, though, if we seek to trace the origins of our own science or to compare other ways of conceptualising the world with the way in which we do so ourselves.[30] It is totally unhelpful if we are trying to understand the development of thought in another culture independently of the achievements of our own.[31]

It seems, then, that the idea of a rigid dichotomy between low-level ritual on the one hand and high-level science on the other is potentially misleading, and that seeking to clarify what we mean by science merely serves to distract us from more interesting issues.[32] In retrospect, the 'science or symbolism' debate was oversimplistic and ultimately fairly unproductive, serving only to polarise views towards extremes that have been remarkably persistent.

Nowhere is this more evident than in interpretations of the very term 'astronomy', which for some carries the inevitable overtone of Western analytical science,[33] a curious point of view since the term 'cosmology' is successfully used in two different ways by anthropologists and astrophysicists.[34] In order to assign a useful meaning to the term 'astronomy', it helps to focus on the distinction between the observation of celestial objects and phenomena, and their perception and use.[35] Different groups or individuals may 'see' the same objects in the sky, but the significance and meaning they attach to them, and the use they make of them, will be influenced by the whole way in which they strive to make sense of the world around them and of their own place within it. The term 'astronomy' will be used in this book to describe the process of observation, perception, and use of celestial objects and phenomena in the context of frameworks of thought that may bear little resemblance to our own.[36]

The 'science or symbolism' debate did, however, succeed in focusing attention on two vital issues that need to be addressed if we are to begin to understand the function that astronomical observations might really have served, and the meaning that astronomical alignments might really have had, to people in prehistoric Britain and Ireland. The first is the need to consider other ways of conceptualising the world. It is clear that in order to do this, we must start by identifying and making explicit those astronomical concepts that are tied to a modern, Western way of thinking, so that we can try to minimise the tendency to impose concepts that were not necessarily meaningful to our distant ancestors. The second is the need to consider the different social contexts within which prehistoric astronomy might have operated.

## Astronomy and society: seeking a social context for prehistoric astronomy

The social implications of Thom's claims had not in fact been entirely ignored, even back in the 1970s. If fully accepted, Thom's ideas seemed to imply a very high level of social organisation, not just providing enough labour to build huge public monuments in places such as Orkney and Wessex, but also sustaining astronomical research programmes lasting many generations (see chapter two), and presumably supporting the specialists capable of organising them.[37] Thom's results also implied that there existed a communications network supporting a remarkable uniformity of practice over a wide geographical area. The alleged use of precise 'standard' megalithic yardsticks has already been mentioned; his megalithic calendar was manifested not in one or two all-encompassing 'observatories' but in dozens of simpler calendrical markers, varying greatly in form, mostly marking sunrise or sunset at only a single 'epoch' date, scattered all over Britain. For the majority of archaeologists, these implications alone provided sufficient grounds to conclude that Thom's ideas did not merit further serious consideration.[38]

One archaeologist, Euan MacKie, controversially argued the other way, accepting Thom's evidence for the presence of high-precision measurement units and astronomical observing instruments and using it to construct a 'two-tier' social model. According to MacKie, there existed in Late Neolithic times an élite class of 'wise men, magicians, astronomers, priests, poets, jurists and engineers with all their families, retainers and attendant craftsmen and technicians', supported by the peasant population at large, living apart in places such as the large henge at Durrington Walls, 3 km from Stonehenge, and the Neolithic village of Skara Brae in Orkney.[39] Similar ideas could possibly be extended to the whole of Atlantic Europe in Neolithic times.[40] Unfortunately, the main tenets of MacKie's argument came in for heavy criticism: the analogy with the Maya, upon which MacKie's interpretation rested heavily;[41] evidence from the distribution of Grooved Ware pottery that certain types of special item were available only for the use of the privileged élite;[42] evidence that particular sites were the domestic centres for members of this élite;[43] and of course Thom's conclusions themselves, critiques of which have already been discussed in detail.[44]

Just one or two archaeologists were prepared to invoke a cautious appraisal of Thom's data in support of social models developed in the context of a much broader archaeological picture. According to Colin Renfrew:

> Religious specialization is now hardly to be doubted at the stone circles of Stonehenge and Avebury, and Alexander Thom . . . has shown how the observations of the sun and moon at such sites was part of a calendrical interest seen (perhaps later) over much of Britain, especially in the highland zone. Even if the megalithic unit of measure was related to the pace or span rather than to a fixed universal standard . . . there can be no doubting the precision and geometrical skill with which they were laid out. . . . Specialist observers or 'seers', in effect a priesthood, were a feature of this society . . .[45]

The wider picture is derived not only from studying the large stone monuments themselves (Archaeology Box 4) but also from a range of archaeological evidence relating to people and their environment (Archaeology Box 5). Summarised as briefly as possible, in the Early Neolithic period (late fifth to late fourth millennium BC) Britain and Ireland were populated by early farmers, rearing livestock and clearing areas of woodland for cultivation. These people built large communal tombs of earth and stone and, later, began to build public monuments such as the causewayed enclosures and cursuses. The following millennium, the Late Neolithic, saw organised activity apparently reach its peak in areas such as Orkney, the Boyne Valley and Wessex, with the construction of great public monuments such as the Rings of Stenness and Brodgar, Maes Howe, Newgrange, Knowth and Dowth, Avebury, Silbury Hill, Durrington Walls, and the sarsen monument at Stonehenge. Despite its obvious achievements this was still 'a population with a fairly low level of technology, lacking the use of metal, without writing, and without urban centres'.[46]

## STATISTICS BOX 5

### MEGALITHIC MENSURATION AND GEOMETRY

In addition to the evidence he accumulated in favour of 'megalithic astronomy', Alexander Thom also analysed data from many hundreds of megalithic monuments to support two ideas developed in parallel: mensuration (the possible use of 'standard' units of measurement for setting out megalithic rings and rows) and geometry (the methods used to set out the rings, many of which are markedly non-circular). The ideas put forward by Thom, and subsequent critiques of them, are described briefly here.

#### GEOMETRY: ELLIPSES AND FLATTENED CIRCLES

From the analysis of surveys of over 250 megalithic rings in England and Scotland, Thom concluded that several different geometric shapes were intended by their builders, and that the rings were laid out with considerable precision using techniques of Euclidean geometry. In addition to circles, he identified ellipses, two forms of flattened circle, ovals ('egg shapes'), and several more complex constructions (see Fig. 4.1).[1]

The idea that the shapes of megalithic rings fall predominantly into one of a few simple categories found some archaeological favour in the 1970s[2] but critics pointed out that Thom had done no more than demonstrate that his range of 'standard' shapes provided reasonable fits to the measured data; alternative sets of shapes could be found that seemed to fit the data equally well.[3] This leads on to the question of how one tests which of two different shapes provides a better explanation of a particular ring or set of rings. In general this is a difficult problem, since different shapes may require different numbers of parameters for their definition,[4] although some techniques have been suggested for its solution.[5] A thorough statistical critique of Thom's geometrical theories has been given by Heggie.[6]

Archaeological concerns include the fact that many rings have associated cairns, kerbs, banks and ditches,[7] so that the selection of standing stones is not always convincing as the crucial diagnostic of the shape of a monument. Furthermore, the positions of the stones themselves will generally have been affected by natural, animal, and human interference.[8] Those stone circles that have been excavated often reveal such a number of alterations, both in prehistoric periods and in modern times, that one must question the wisdom of using any site plan of extant stones to investigate the intended shape.[9]

Perhaps the least contentious geometrical interpretation of non-circular megalithic rings is that some were deliberately constructed as ellipses. A number of excavated stone rings have been found to be close to elliptical in shape, including Cultoon in Islay (IS28 in List 2)[10] and two north-eastern Scottish recumbent stone 'circles': Strichen (RSC7 in List 3), where the

Fig. 4.1 Five geometrical constructions proposed by Thom to explain the shapes of non-circular megalithic rings: (a) Merry Maidens, Cornwall; (b) Barbrook I, Derbyshire; (c) Esslie The Greater, Kincardine; (d) Burgh Hill, Roxburgh; and (e) Kerry Hill, Powys. After Burl 1976, 42.

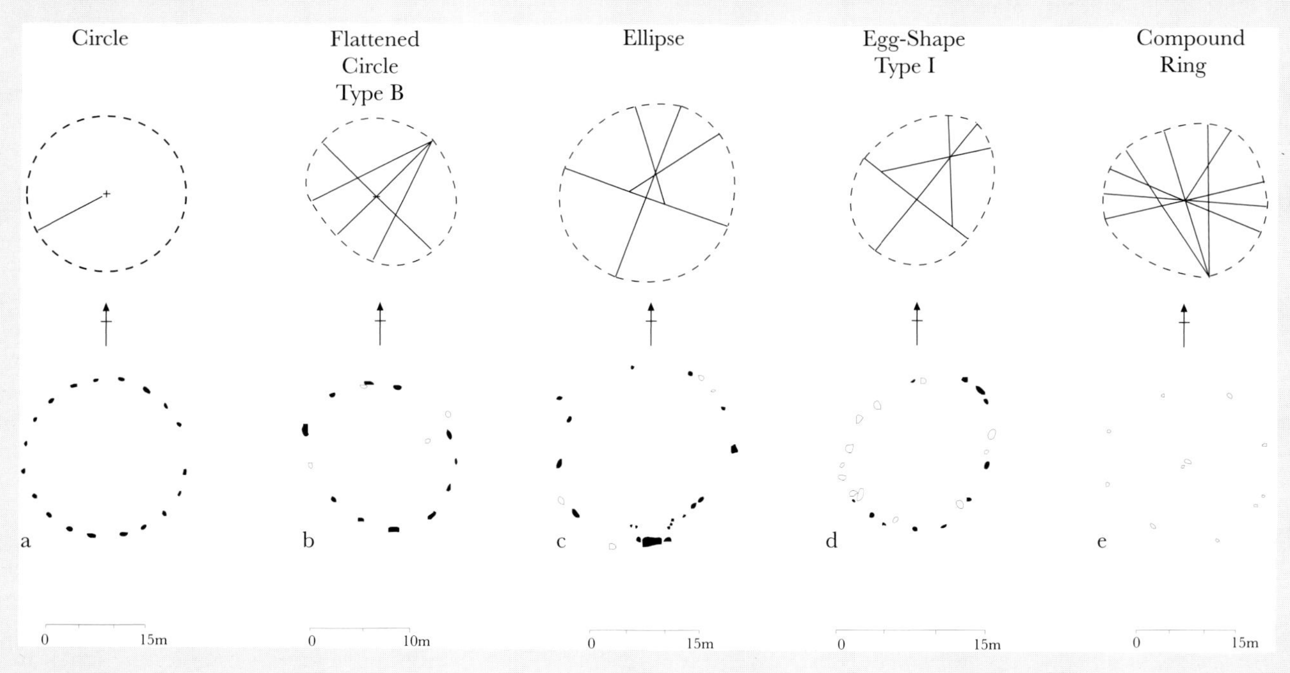

recumbent stone lies on the major axis of the ellipse,[11] and Berrybrae (RSC5 in List 3) where it lies on the minor axis.[12] On the other hand, the Thoms' claim that the great trilithons at Stonehenge originally lay on an ellipse[13] does not fit the archaeological evidence.[14]

### Mensuration and the 'megalithic yard'

In 1955, Thom analysed the diameters of forty-six circular stone rings and concluded that they were laid out as multiples of a standard unit of measurement which was used throughout Britain.[15] Following subsequent analysis of 112 circular rings together with thirty-three non-circular ones, where particular geometrical constructions were assumed, Thom concluded that a 'megalithic yard' (MY) of 0·829 m ± 0·001 m 'was in use from one end of Britain to the other'.[16] Thom and Thom later analysed the layout of such complex monuments as Avebury and the multiple stone rows in Caithness and Brittany, and obtained to their own satisfaction further confirmation of the existence, precision and universality of the MY.[17] Other publications also postulated the existence of a 'megalithic rod' equal to 2·5 MY, and a 'megalithic inch' equal to one fortieth of a megalithic yard, evidence for its existence being derived from the dimensions of cup-and-ring marks.[18]

For a thorough statistical critique of these ideas the best source, once again, is Heggie.[19] In brief, statistical reassessments of Thom's data both from classical and Bayesian (see Statistics Box 7) viewpoints reached the conclusion that the evidence in favour of the MY is at best marginal, and that even if it does exist the uncertainty in our knowledge of its value is only of the order of centimetres, far poorer than the 1 mm precision claimed by Thom.[20] In other words, the evidence presented by Thom could be adequately explained by, say, monuments being set out by pacing, with the 'unit' reflecting an average length of pace.

The Early Bronze Age, from the late third millennium onwards, was a time when, in many areas, the focus gradually shifted from communal activity towards one where power and control was in the hands of a privileged few. Nowhere is this more evident than in Wessex, where the skyline around the sarsen monument at Stonehenge gradually became filled with barrows containing single prestigious burials. Yet far away in north-eastern Scotland, western Scotland, and south-western Ireland small groups of people apparently continued to live a less centrally controlled existence in scattered homesteads.[47]

We can make a number of general remarks about the nature and purpose of astronomical observations during this time. The fact that such observations, even of quite considerable sophistication, might have been made at all is not actually surprising in itself: the movements of the heavenly bodies are of almost universal concern, even amongst the most technologically primitive of hunter-gatherer societies.[48] Many instances are known of careful and precise astronomical observations amongst small indigenous groups, a much-quoted example being that of the Hopi Indians, who (traditionally) are essentially egalitarian with no centralised political organisation, yet villages such as Walpi have an elaborate ceremonial calendar regulated by carefully tracking the horizon rising position of the sun from fixed points on the ground.[49]

The difficulty is in extracting direct evidence for astronomical practice from the material record, and in its interpretation. Many aspects of mythology and beliefs surrounding astronomical bodies, well known from indigenous groups world-wide, would simply leave no direct trace in the material record.[50] Some activities might give rise to particular forms of archaeological evidence (for example, the counting off of months by observing the phases of the moon, a very widespread activity, might leave its trace in the form of tally marks on recoverable artefacts)[51] but have no direct influence on ceremonial architecture since the observation can be made from any position with a clear view of the sky.

Even where astronomical influences on monumental architecture are demonstrable, interpretations are likely to be complex. Perhaps with the Hopi example in mind, calendar regulation is often quoted as a motivation for monumental orientations in the direction of sunrise or sunset at particular times of the year. However, even where sunrise observations are made from particular spots for this purpose, it does not follow that there is a necessity to erect markers of any great prominence (or permanence) at these positions or to orient them towards the horizon in question.[52] It is also wrong to think that a reasonably precise agricultural or ceremonial calendar can only be regulated by keeping track of the horizon rising or setting position of the sun: it could be done perfectly adequately, say, by observing the heliacal rising or setting of stars—observations that do not have to made from particular spots on the ground—as well as by indirect and composite methods, for example combining observations of a variety of astronomical events with observations of a range of other natural phenomena.[53] Even if a precise determination of the solstices was desired, this could have been better achieved by 'halving the difference' than by direct observation of the sun rising or setting at the solstices,[54] or by other methods completely.[55]

The obvious question then arises: why, when the time of year could be determined quite adequately for agricultural purposes by horizon observations without the necessity for prominent markers, or indeed in other ways entirely, would Neolithic and Bronze Age communities have gone to all the trouble of 'enshrining' astronomical observations in stone?

## ARCHAEOLOGY BOX 4

### SOCIETY IN THE FOURTH, THIRD AND SECOND MILLENNIA BC: THE EVIDENCE OF THE MONUMENTS

Spectacular burial and ceremonial monuments are only one manifestation—albeit a conspicuous one in the archaeological record that comes down to us today—of certain aspects of (certain) people's lives in prehistoric Britain and Ireland.[1] Nonetheless, a broad picture of the changing nature of social organisation and power relations may be gleaned by extrapolating from the evidence that the monuments present to us. For this purpose the period from the third quarter of the fifth millennium BC to the third quarter of the second millennium BC has been split into three subdivisions each of roughly one thousand years in duration. These periods correspond roughly to the Early Neolithic, Late Neolithic and Early to Middle Bronze Age periods respectively, and that terminology is retained here for convenience.[2]

#### THE EARLY NEOLITHIC: *c.* 4250–3250 BC

That some of the earliest farming communities in Britain and Ireland put a great deal of effort into burial and ceremonial activities is evident from the scale of the long tombs: for example, at Tinkinswood, Glamorgan and Brenanstown, Co. Dublin capstones in excess of forty tonnes were raised.[3] This implies organised labour, but whether this was achieved by co-operation within an egalitarian and classless society, or whether social ranking was beginning to appear even at this time, continues to be a matter for debate. Within the tombs, the bones of the dead were carefully arranged. They were generally disarticulated, often organised with similar bones grouped together, and sometimes graded according to age or sex. Possibly, bodies were exposed to birds or other animals prior to final burial. New bones were added, and existing ones cleared out or rearranged, over long periods—many hundreds of years in some cases. The social implications of this are unclear: on the one hand, it seems that only the bones of certain people were placed in the tombs; yet once they had arrived they were mixed up with all the rest. In other words only certain individuals seem to have been privileged with formal burial, but amongst that group all were treated equally.[4]

The later part of the fourth millennium tombs saw a decline in the construction and use of communal long tombs and in southern Britain the rise of public monuments such as the large causewayed enclosures Estimates of the labour involved in their construction vary considerably and they seem to have served a range of functions, but they most probably acted as regional centres of a ritual and/or practical nature. Very possibly, they were focal points for interaction between local groups or tribes in a 'segmentary' society without any overall chief or controlling group. On the other hand, their existence may indicate that differences of power and prestige, already present, were now being brought into the open instead of being veiled by the rituals associated with death and the celebration of ancestors. Indeed, defensive structures have been found at some enclosures. The enigmatic cursus monuments raise similar social questions. Turning to funerary practices, there is evidence that these began subtly to change while funerary monuments continued (at least at first) to retain the same general outward appearance. There was an increasing tendency for burials to remain articulated and some single individual burials began to appear, presumably of people with special status. In sum, the evidence for social ranking is now stronger, although it is still possible to argue that society remained essentially egalitarian.[5]

#### THE LATE NEOLITHIC: *c.* 3250–2250 BC

This period apparently marks a culmination of public religious activity, with burial, ritual and ceremonial practices being both extensive and highly organised. Obvious concentrations of activity, indicating significant social change, occurred in the Boyne Valley of Ireland, the Orkney islands of Scotland, and the Wessex chalklands of England. There is also evidence of privacy and control of access,[6] which suggests that sacred knowledge was no longer in the public domain but controlled by powerful élites.[7]

However, what has been called a 'fragmentation of the social landscape' also took place around this time, with widely different developments starting to take place in different regions.[8] It is often hard to be specific, because of the many factors affecting whether a monument will have survived or has yet been discovered.[9] In addition, there is still insufficient chronological evidence to generate a reliable general picture of developments through time. Nonetheless some things do seem reasonably clear.

In Ireland and northern Britain generally, existing practices of communal burial survived well into the third millennium BC, but while disarticulated bones continued to be mixed together unburnt in the Orkney tombs, the practice of cremation became more common elsewhere. In the Boyne valley and Orkney, burial monuments grew progressively grander, culminating in the great passage tombs of Newgrange and Maes Howe, which it has been estimated took something like 100,000 person-hours to construct, the equivalent of roughly 300 people working full-time for at least a month or thirty for at least a year.[10] In Orkney, ritual enclosures involving similar amounts of labour, in the form of large henges such as the Stones of Stenness and the Ring of Brodgar, were also constructed around the beginning of the third millennium. The geographical distribution of Orkney chamber tombs suggests that each served a territory of a few square kilometres, possibly being located in the domain of a group of a few dozen or hundreds of people, with these segmented groups interacting on an essentially equal-to-equal basis.[11] However, whether the largest monuments indicate that a centralised organisation subsequently developed—for example, a social hierarchy in which a central élite exercised power and control over smaller local groups—is still a matter of some dispute.[12]

Wessex is distinguished by the appearance during this period of massive public monuments such as Avebury, Silbury Hill and Durrington Walls,[13] as well as the construction of the bluestone and sarsen monuments at Stonehenge (see Archaeology Box 3). These constructions seem to represent millions of person-hours of labour, each million being the equivalent of some 300 people working full-time for about a year. There seems little doubt that organised labour on this scale requires social ranking, including a central chief or élite group with the power (either through force or persuasion) to mobilise—and feed and otherwise support—a workforce drawn from a considerable area.[14]

Despite substantial differences in social organisation from area to area, public monuments, in the form of henges and stone circles, were constructed all over Britain and Ireland during the third millennium BC, their scale perhaps reflecting the labour resources that could be mobilised for the project, and hence the nature of social organisation, in a particular place at a particular time. During the same period, smaller public monuments were constructed in areas where there is no reason to suppose that a ranked society appeared at all. An example is the north-eastern Scottish recumbent stone circles, which were very possibly built by small families working in co-operation.[15]

The diverse regional communities of the third millennium BC were certainly in contact through trade and exchange (see Archaeology Box 5), and elements of ritual practice often spread into new areas, or were mimicked elsewhere at a later time. Meanwhile existing practices were constantly evolving, both as a result of local developments and because of new contacts and influences.

## Early to middle bronze age: *c.* 2250–1250 BC

Towards the end of the third millennium BC the framework of public ritual tradition that had continued to develop over two millennia began to break down.[16] Existing public monuments were abandoned. The emphasis shifted inexorably away from communal activity towards one where power and control was in the hands of a privileged few. The increasing expression of this power—for example through individual burials, use of personal ornamentation, possession of prestige objects, and restricted access to sacred places—accompanied a range of technological innovations and may have been the result of increased competition for resources, for these changes were accompanied by an expansion of settlement into new and marginal areas (see Archaeology Box 5). In short, a 'social and spiritual upheaval' set in across Britain and Ireland.[17]

At Stonehenge, however, the bluestone and sarsen monument continued in use, with various rearrangements of the bluestones being attempted. Amongst the round barrows in the Wessex region appeared elaborate burials with rich grave goods. These developments may well reflect the actions of people struggling to maintain existing traditions and values (and hence their own power) in the face of increasing pressure for change, with the use of existing large public monuments, rooted in the authority of the past, together with increasingly ostentatious burials, serving to reaffirm their own status. By about 1500 BC, the struggle may have been getting desperate. Stonehenge was abandoned soon after this time, and a few bell barrows, built without burials, may represent the last vestiges of the 'Wessex culture' before it collapsed completely.[18]

Elsewhere by this time there is little evidence for the survival of traditional religious activities or for the social structures that had once supported them. In some parts of western Scotland and south-western Ireland it does seem that small ritual monuments were still being constructed and used until as late as around 1250 BC, possibly even 1000 BC, but the areas that had been chief centres of activity a millennium earlier had become marginalised. In Scotland, many existing ceremonial monuments (or their vicinity) were still being used for burials, but this more likely reflects a veneration for the conspicuous works of ancestors than a continuity of tradition.[19] Certainly, by the end of the second millennium BC nothing or virtually nothing seems to have remained of a tradition of communal monument construction that had existed for some three thousand years.

## ARCHAEOLOGY BOX 5

### PEOPLE AND THEIR ENVIRONMENT IN THE FOURTH, THIRD AND SECOND MILLENNIA BC: THE WIDER EVIDENCE

A wide range of data, much of it fragmentary, gives us vital clues about the activities of people in prehistoric times and how they exploited the environment around them. This includes evidence on climate and vegetation from pollen and other plant remains; evidence of agricultural activity, for example from plough- or ard-marks in the subsoil, or from field boundaries in the form of walls or banks, or showing up only as cropmarks visible from the air; evidence of subsistence and diet, from the waste products of food preparation (both plant and animal) and from human remains themselves; evidence on technology, from tools and other objects of stone, bone, wood, clay, metal, etc.; and evidence on trade and exchange, acquired by identifying the places of production and examining the final whereabouts of artefacts.[1]

The broad outline that follows is based mostly upon the interpretations of archaeologists, which are based in turn upon the work of palaeoenvironmentalists. It is split chronologically, as in Archaeology Box 4.

#### The early neolithic: *c.* 4250–3250 BC

When people first began to cultivate cereal crops in Britain and Ireland in the fifth millennium BC, the land was covered by almost unbroken forest. The transition from foraging to farming, and from mobility to permanent settlement, was neither sudden nor simple. In some places herding, hunting, fishing and gathering remained the main means of subsistence for many centuries, with only scattered areas of woodland being cleared for small-scale cultivation and quickly regenerating. In other places, though, the hoe gave way to the ard and the plough, larger areas were cleared, and the population increased. In Wessex, for example, large tracts of chalkland were open by about 3500 BC. By the end of this period, farming communities were established all over Britain and Ireland. It is likely that by this time people generally lived in small scattered farmsteads, perhaps in extended family groups, a pattern that persisted as the norm in many regions not only during this period but also throughout most of the subsequent two millennia.[2] Flint mining and stone quarrying for axe manufacture were established, with major production sites in southern Britain and northern Ireland.[3]

#### The late neolithic: *c.* 3250–2250 BC

Around the end of the fourth millennium BC there were significant changes in the pattern of farming. One interpretation of the evidence is that early agricultural practices did little to replenish the soil and much of the land that had been opened up had become infertile and was reverting to forest and scrub.[4] Certainly, agricultural development spread into new regions and elsewhere there were permanent changes such as the enclosure of large areas using field systems, as at Céide, Co. Mayo. These changes may have been triggered by increased competition for resources and even by resulting conflict. Such competition may also have been a major factor in prompting the social changes which took place in many areas (see also Archaeology Box 4).[5] Another suggestion is that major climatic aberrations just after 3200 BC, caused perhaps by large volcanic eruptions, had a noticeable effect.[6]

While the social characteristics of different regions diverged markedly at this time, interaction was nonetheless maintained (or soon redeveloped) between them. Trade and exchange networks carried goods around Britain and Ireland. For example, early in this period the main focus of axe production shifted from southern to northern England.[7] It is also possible that there was direct interaction between some widely separated areas, such as the Boyne valley and the Orkney islands.[8]

Evidence for a ranked society, not only in areas such as Orkney and Wessex but possibly much more widely, is reinforced by the appearance in the third millennium of prestige goods which were used exclusively (for an initial period at least) by a privileged few. A number of distinctive styles of decorated ceramics appeared in different places at different times, differing starkly from the existing plain pottery. Amongst these were Peterborough Wares in the south-east of England; Grooved Ware in Orkney and much of lowland Scotland, Yorkshire, Wessex and East Anglia; and Beakers, which are found widely distributed around Britain, Ireland and western Europe in general well into the Bronze Age.[9] Each of these styles may have represented high-status items at their first appearance, retaining their prestige as long as an élite group could continue to control supply or restrict exchange, but gradually becoming commonplace as they became more widely available.[10]

#### Early to middle bronze age: *c.* 2250–1250 BC

By around 2000 BC, the demands of population growth were beginning to exhaust the agricultural potential of the 'core' areas of greatest population. Early in the second millennium there was a significant expansion in settlement out into more marginal regions such as upland Scotland, Dartmoor and the North York Moors, into which existing subsistence patterns were extended. In Wessex, attempts to maintain the social *status quo* through large public projects (see Archaeology Box 3) must have depended increasingly upon the efficient movement of surplus food produce from peripheral areas. In all probability, they also relied heavily upon certain individuals' ability to reaffirm their own status by the possession of prestige objects such as bronze daggers, faience beads and goldwork, many of which were brought in from Europe through widespread trade and exchange networks.[11]

Two problems conspired to cause rapid change after about 1500 BC. First, the marginal regions could not support sustained agricultural activity. This, it seems, was exacerbated by a significant climatic change towards cooler, wetter conditions. There was a rapid reduction in the food supply from these regions, which were eventually abandoned again, whence they descended into moorland and peat bog. It has been suggested that competition for land led to the development of defended sites and the increased amount of weaponry visible in the archaeological record at the time. In addition, upheavals on the European mainland were affecting the wider exchange networks and the supply of prestige goods. Social élites, such as that in Wessex, were unable to maintain their level of supply either of food or of prestige objects, and thus became increasingly insupportable and rapidly collapsed.[12]

Elsewhere, much more efficient agricultural techniques emerged from the necessity to increase production in those areas that did remain fertile, in particular lowland areas well away from old centres of population and their barren peripheries. By 1250 BC the monumental landscape of three millennia had been almost completely replaced by a landscape of intensive agriculture, of extensive field systems and larger settlements, one that would eventually see the emergence of defensive sites and structures such as hill forts and brochs.[13]

There can only be one answer: that the architectural alignments had a symbolic function, rather than being intended solely for any use that would seem to us 'practical'.[56] This suggests that if we find convincing evidence that astronomical alignments were deliberately incorporated in public monuments, then a different question should be asked: why was it important for prehistoric people to encapsulate astronomical symbolism in those monuments? This, as we shall see, provides a much more productive framework for approaching the whole issue of the nature of prehistoric astronomy.

Our starting point, then, is that astronomical alignments in monumental architecture have to be understood in terms of prehistoric ritual and ideology. 'Astronomical alignments' may reflect the part played by celestial symbolism in whatever rites accompanied death and burial,[57] and may well have contributed to the importance of existing ancestral monuments in the landscape in later times, a possibility to which we shall return later. The obvious example is Newgrange, and many settings of standing stones that we have already discussed appear to be associated with tombs, cairns and cists. Wider questions are raised where astronomical symbolism seems to be incorporated in other public monuments such as stone circles and rows, and these lead on, as we shall see, to further, challenging issues relating to prehistoric cognition and thought.

## Beyond alignments

First, however, if we are to begin seriously to explore why and how what was seen in the sky was important to people in prehistory, then we must begin by considering a wider range of types of evidence that might be relevant.

Orientations and alignments at public monuments are certainly not the only conceivable way in which a concern with events in the sky might have left its mark in the material record. Astronomical alignments in domestic structures are an obvious possibility; we are imposing our own values in separating 'ritual' practices from the 'mundane' activities of everyday life,[58] and astronomy may have been just as influential in the latter as the former. Examples of astronomical alignments in domestic structures are known from a variety of modern indigenous groups, for whom a rich cosmological symbolism influences many different aspects of ceremonial and everyday life.[59] Surprisingly perhaps, the possibility of astronomical alignments being incorporated in buildings within prehistoric settlements had received little or no attention by the early 1980s, despite the discovery and excavation of several Neolithic and Bronze Age settlements around Britain and Ireland by this time.[60] Neither had the possibility that astronomical factors might have influenced the location of monuments within the landscape as a whole, something that has only begun to receive attention more recently in the context of broader studies of the human activity in the prehistoric landscape.

Astronomical concerns might also leave their mark in rather different ways, and by the early 1980s a good deal of interest had been generated around the world in, for example, possible calendrical tally marks on natural rocks or small portable objects, possible astronomical symbolism in rock art, and patterns of artefact deposition at and around monuments. Yet such claims from Neolithic and Bronze Age Britain and Ireland were still few and far between and often unconvincing. Aubrey Burl had, however, drawn attention to the possibility that cup markings and quartz assemblages might be related to lunar rituals at the eastern Scottish recumbent stone circles.[61] Apart from this, there had been some elaborate astronomical interpretations of Irish megalithic art[62] and suggestions of astronomical interpretations for cup-and-ring marks in southern Scotland.[63]

The sum total is not impressive. Of course, one can argue that more convincing evidence may not have been forthcoming because it was never sought. Nonetheless, these scattered examples suggest that seeking direct evidence of astronomy in art or artefact is unlikely to be a productive approach. Worse, it brings the very real danger of seeking meaning on entirely the wrong level, in effect by asking the wrong questions. The dangers of seeking to interpret rock art symbols as representations of astronomical bodies or events are best illustrated by a modern example. The following is an anthropologist's account of his first attempts to ascertain at first hand the meaning of geometrical patterns carved on wooden parts of their houses by a modern group of hunter-gatherers, the Zafimaniry of Madagascar:

> To the question 'what are those pictures of?' I was answered with great certainty that they were pictures of nothing. When I asked for a cause or the point of the carvings I triggered the ready-made phrase that there was no point, and when I asked what people were doing I was told 'carving'. There was actually one answer I was given very often, but I felt it was so bland and therefore frustrating that I payed [sic] no attention to it and did not even put it down in my field notes. It was that 'it made the wood beautiful'.[64]

Another sobering example concerns the famous paintings on a rock overhang at Chaco Canyon, New Mexico, one of over twenty examples of rock art in the US south-west which have been said to depict the Crab nebula supernova in AD 1054.[65] New dating evidence from two of these now shows that they were painted too recently for this to have been a possibility,[66] which must cast serious doubt upon the idea as a whole.

During the 1970s and early 1980s there were several discussions concerning environmental evidence in the context of prehistoric astronomy, and its potential relevance to astronomical interpretations is undeniable. This is most obvious because of two general objections that were often raised to the whole idea of horizon astronomical observations as well as to particular astronomical sightlines: first, that there was much more extensive tree cover in prehistoric times; and second, that the climate then (as now) was variable and bad weather conditions might well have interfered with critical observations of the rising or setting sun or moon.[67] The climate was certainly a serious difficulty in relation to Thom's high-precision lunar sightlines,[68] even considering the possibility that there has been a significant general deterioration since Neolithic and Early Bronze Age times, when conditions may have been somewhat warmer and drier than today.[69] However, for lower-precision observations, which could be postponed in case of bad weather or perhaps approached on a more casual basis, this would not generally be a critical problem.

More serious is the problem of vegetation. It is certainly true generally that during Neolithic and Bronze Age times changing subsistence practices gradually transformed the landscape of Britain and Ireland from one that was extensively afforested with trees such as birch, hazel, oak and elm, to one containing a great deal of open grassland, pasture and cultivated fields; but this sweeping statement hides a wealth of complex detail.[70] In many areas, for example, woodland regenerated after cultivation ceased.[71] Some Hebridean islands never were thickly afforested in the first place.[72] Clearly it is helpful, if there is a serious suggestion of astronomical alignments being set up in a given area at a given time, to examine pollen evidence wherever possible in order to see if the nature of the environment seems consistent with the idea.[73]

The material record from the Neolithic and Bronze Age is not necessarily the only source of relevant information on astronomical practices in Britain and Ireland at that time, although other sources are generally more controversial. Indirect evidence was used by a number of authors in support of their ideas. One argument is that certain practices and concepts might have continued, more or less unaltered, into Iron Age times and even later. Thus MacKie's astronomer-priests resembled in many ways the Iron-Age Druids, whom he suggested might have descended directly from a Neolithic theocracy.[74] Similarly, there have been several suggestions that the Celtic calendar, with its festivals at the solstices and equinoxes and the 'mid-quarter' (or 'cross-quarter' or 'half-quarter') days halfway between them, might have been descended from Thom's 'megalithic calendar'.[75] Unfortunately, this idea is critically weakened by the reassessments we have already described of the direct evidence in favour of Thom's calendar, and in addition there are problems regarding the Celtic calendar itself.[76]

There is historical evidence relating to indigenous practices at these later times, namely the writings of Classical authors, and this was often invoked in such arguments. A frequently quoted passage is one attributed to Hecateus of Abdera (*c.* 330 BC) which survives in the writings of the Greek historian Diodorus of Sicily in the first century BC:

> They say also that the moon, as viewed from this island, appears to be but a little distance from the earth and to have upon it prominences, like those of the earth, which are visible to the eye. The account is also given that the god visits the island every nineteen years, the period in which the return of the stars to the same place in the heavens is accomplished . . .[77]

The account refers to island of the Hyperboreans, which may well be (but is not certainly) Britain. An earlier passage describes a 'magnificent sacred precinct of Apollo' [the sun],[78] which has been widely interpreted as Stonehenge,[79] although others interpret it as Callanish.[80] 'The god' in the passage has been taken by some to be the sun and others to be the moon,[81] and the passage as a whole has been interpreted as referring to the 18·6-year lunar node cycle, from which it is deduced that the people described attached special importance to the extreme positions of the rising and setting moon at lunar standstills.[82] However, if the end of the last sentence in the quotation is included, then it becomes quite clear that the passage is describing the slightly longer, 19·0-year Metonic cycle marking the time after which the moon returns to the same phase on any given day in the year.[83] In any case some maintain that Diodorus is an unreliable historian and this passage is of doubtful value.[84]

A lesser-known passage from the Greek geographer and historian Strabo, writing at about the time of Christ, does actually give rather more direct evidence of the possible importance of the moon, at least amongst the Celtiberians, who inhabited north-eastern parts of the Iberian peninsula: 'The Celtiberians and their neighbours in the north offer sacrifice to a nameless god at the seasons of the full moon, by night, in front of the doors of their houses, and whole households dance in chorus and keep it up all night.'[85] Nonetheless, the very fact that such a variety of possible interpretations of the Diodorus quotation has been put forward adequately demonstrates the dangers in using passages such as these to support specific ideas. And even if, when interpreted with caution, they might really tell us something about practices in Britain and Ireland (or at least in Atlantic Europe) in the pre-Roman Iron Age, it is quite another question as to whether there is any evidence for a continuity of tradition extending back into the Bronze Age and before.

Suggestions of continuity of tradition have sometimes been carried a good deal further even than this. Modern place-names have sometimes been held to reflect earlier astronomical practices at megalithic monuments,[86] as have modern folk traditions.[87] There are even suggestions that practices continuing until modern times in the Basque country, where the use of what is classified as a non-Indo-European language may indicate relatively little 'cultural rupture' in the distant past, may bear some direct relation to the 'megalithic period' of the Bronze Age and before.[88] Generally, such ideas have drawn a highly cautious, if not totally dismissive, response.[89]

On the other hand, evidence from modern indigenous peoples can be used to highlight assumptions that we might tend to make rather too unquestioningly.[90] For example, in modern Indonesia we know something of the complex funerary ceremonies of which standing stones are the only visible remains; no astronomy was involved at all.[91] Various examples from the Americas (see chapter nine) serve to warn us against separating astronomical practice as part of ritual activity from the mundane. In south-west Ireland, we find that rural farmers in modern times are conscious of certain consequences of the moon's 18·6-year cycle, countering the assumption that painstaking horizon observations of the moon were needed in order to 'discover' it.[92] Finally, it continues to be tacitly assumed that if people observed the changing rising and setting position of the sun, then they must have understood that this regulated seasonal events.[93] The Mursi, however, have a lunar-regulated calendar and note rough correlations between the time of year and the rising position of the sun, but view the latter as no more reliable a seasonal indicator than the behaviour of various birds, animals or plants.[94] These diverse examples illustrate the value of cautionary evidence available from the ethnographic record.[95]

The fact that reassessments of Thom's work had already produced hints of systematic, low-precision astronomical alignments, and that other sorts of evidence were generally scattered and unimpressive, led this author in the 1980s to the conclusion that it was well worth exploring further the primary evidence from coherent groups of orthostatic monuments.[96] Nonetheless, even then it was clear that such work should not be undertaken in isolation from the social context, should take into account a wider range of evidence wherever possible, and should be driven by broader questions. Since astronomical alignments seemed to be a factor influencing monument orientations, it was necessary to proceed from a wider consideration of the possible influences upon, and meanings of, those orientations.

## Orientations revisited

'Megalithic monuments' had many purposes. They encapsulated symbolism on many levels; they conveyed different meanings to different people; and their significance may have changed radically through time.[97] Single standing stones or small stone settings may have served to mark places of special significance, such as ceremonial, meeting or trading places; as symbols of power; or more pragmatically as routeway markers for travellers by land or navigational aids for travellers by sea. They may have functioned as boundary markers, as cattle rubbing stones—the list of possible explanations is almost endless.[98] The various possibilities are not mutually exclusive; nor were the purposes for which they were originally erected necessarily reflected in the uses to which they were later put. Many later structural modifications were often made.[99] Large monuments, too, may well have been centres where social and religious activity was mixed with economic activity.[100] Even the great tombs, which were clearly used for burial, evidently did not have this as their chief significance, in the sense that there was no need to go to all the trouble of building them simply to solve the problem of disposing of the dead.[101] It is important, then, to dismiss at the outset any notion that astronomy, or for that matter any other single factor (including burial), is likely to have been the sole or overriding consideration influencing the design of a prehistoric stone monument, be it an elaborate tomb or large henge in Orkney or Wessex or a simple stone circle or row located in a marginal area.

From this perspective, the question of orientation is potentially of considerable importance in relation to our wider understanding of public monuments. There are, for example, a variety of possible influences upon the orientations of inhumations and tombs: the direction of significant places (mythical or holy places, areas of origin of ancestors, former burial places, the dead person's house, or natural features such as mountains, clefts, hills, rivers, and rocks); existing practice (the orientation of one grave may determine that of others that follow); the desire to make an impression (e.g. orientation upon a gate, pathway or road so that a grave or cemetery will impress, or at least be seen by, those passing by it); significant times of day (e.g. orientation generally towards the west, to the sunset, because evening was a significant time of day for that person in their daily activities); and particular celestial bodies that might be significant.[102] In addition, we should not ignore pragmatic considerations: the orientation of tombs, for example, may well have been influenced by the local lie of the land, with long tombs in some areas tending to be built along ridges[103] or with their entrances being oriented downhill.[104] Another possible influence is the direction of the prevailing wind.[105] Isolating any influences such as these, whether to our eyes subtle or crude, could give us insights into aspects of the symbolic significance of certain monuments.

In view of this it is perhaps surprising that, in Britain and

Ireland at least, the question of the orientation of prehistoric monuments and its interpretation had received relatively little attention in the mainstream archaeological literature, even well into the 1980s.[106] Where it was addressed it was often based on orientation measurements only in crude divisions of the compass (e.g. N, NNE, NE, etc.)[107] and interpretations tended to be considered 'purely speculative and incapable of proof'.[108] Quantitative analyses of monument orientations were virtually non-existent.[109] One reason for this is that, as is evident to the prehistorian, orientation is only one of a variety of ways in which symbolic meaning could be incorporated in the location and design of a public monument: choice of location, size and shape, choice of materials, texture, visual impact, the position of ritual offerings, and many other considerations might all have been important.[110] Ideas about monument orientation should not be developed and tested in isolation from wider interpretative theories concerning the sites.

Yet we should not ignore what orientation patterns tell us. There are many clear trends, with a strong preference for easterly orientation being evident amongst many groups of long barrows and chambered tombs from northern Scotland down to southern England,[111] while orientation towards the SSW was the norm amongst the Clava cairns and recumbent stone circles of north-east Scotland,[112] and an interest in the west and south-west is apparent amongst the axial stone circles and wedge tombs of south-west Ireland, reflecting a wider tradition of NE–SW orientation amongst the stone rows of the region.[113] These trends invite some sort of explanation; and this should inform wider interpretations. It would be perverse indeed to ignore a feature of monumental symbolism that is so conspicuous in the archaeological record.

For the purposes of analysis it is possible to divide the possible factors influencing monument orientation into three broad categories: terrestrial features, azimuthal (compass) directions, and astronomical targets.[114] The first category is very broad, covering orientations upon particular types of terrestrial feature as well as orientations influenced by the local topography, and for this reason the categorisation may seem imbalanced. It is useful, though, in that it helps to distinguish types of hypothesis testable in different ways.

Orientation influenced by, say, the direction of the prevailing wind or the direction whence ancestors came would tend to produce a non-random distribution of azimuths. Orientation upon the rising or setting of certain celestial bodies would tend to produce preferences for, or avoidances of, particular ranges of declination.[115] In these two cases it is possible to undertake a 'context-free' statistical test by simply considering the azimuths and horizon declinations respectively. If significantly more instances of a particular azimuth or declination are observed than would be expected if orientations were randomly distributed, then the reasons for this can be explored.[116]

In fact, the context-free approach hides a number of problems, some of which we have already encountered in relation to formal reassessments of Thom's ideas. First, there is no such thing in practice as a 'random' set of orientations. Each orientation will actually have arisen through a combination of factors, many of which may be particular to a single monument, or small number of monuments, even in the light of a wider tradition.[117] These may serve to 'dilute' even a fairly strong tradition and make it unrecognisable.[118] Second, a tradition that has nothing to do with astronomy, or with directional targets *per se*, might not actually lead to pseudo-random azimuths or declinations.[119] Third, classical tests to determine whether significantly more instances of a particular azimuth or declination are observed than would be expected by chance run into the 'multiple-hypothesis' problem discussed in chapter three.[120] Finally, the fact that the approach is context free would be seen by most archaeologists as a disadvantage rather than an advantage.[121]

Another approach is to try to identify 'significant' directions under any particular theory. Astronomical and directional influences may result in overall preferences for certain declinations and azimuths, but terrestrial 'targets' must be laboriously defined for each site. Such an approach can take into account ideas relating to the broader social context but is potentially far more complex and introduces a whole set of questions about the basis upon which certain targets are selected as significant. For example, if we suspect that certain monuments were preferentially oriented upon prominent hills, and wish to test the idea, we would need to find a way of identifying all the prominent hills visible from a given site. But we would need to avoid falling into one of two dual traps, either by being too subjective and possibly introducing selective bias, so that our conclusions are meaningless, or on the other hand by insisting upon rigorous 'objective' criteria defined in context-free terms that may have had little to do with the perceptions of prehistoric people and hence turn out to be just as unenlightening.[122] This theme of testing more context-rich theories is a complex one to which we shall return more than once in later chapters.

The way forward identified by this author and others in the early 1980s was a simpler one. Given that strong orientation trends were evident amongst groups of similar monuments and had been little studied, it made sense to investigate these further. Given a variety of reasons for believing that astronomical factors, albeit at a low level of precision, were likely to have influenced monument orientation,[123] it also made sense to measure declinations as well as azimuths. At the same time, it was clear that the work should no longer proceed in isolation from the wider context, and should permit the generation and testing of related ideas. The first two chapters of Part Two describe some of the results obtained by following this approach.

As ideas developed during the 1980s, it soon became clear that demonstrable astronomical symbolism might not just relate to the primary orientation of monuments, but also to their position within the landscape. Chapter seven describes a project which involved a regional examination of the landscape on the island of Mull, taking into account factors that are symbolic (including the potential for astronomical observations) as well as ecological.

A very different approach from considering groups of orientations, one which emphasizes specific instances of astronomical symbolism in a wider interpretative context, makes its appearance in chapter eight, where recent ideas relating to prehistoric astronomy are described that have emerged as part of broader archaeological investigations.

# PART II CURRENT DIRECTIONS

# 5

# Orientation and Astronomy in Two Groups of Stone Circles

To see the pair of milk-white quartz boulders newly exposed in the black trench of a turf bank on a hilltop in Crocknaraw, north of Clifden, and to realise that at least half-a-dozen other standing stones and several other megaliths are or were visible from that point, is to be given a glimpse of a cultural landscape the meaning of which has been lost beneath the bogs.

Tim Robinson, 1990[1]

The observation of this fact led me to notice that, in mountainous districts, hill tops appeared to take the place of outlying stones, and, if the outlying stones had anything to do with the relative position of the sun and the circles, prominent hill tops would certainly be in every respect superior to them, if the circles were so placed that they could be utilised.

A. L. Lewis, 1892[2]

The larger henges, the great stone circles, such as Callanish, and the monumental recumbent stone circles of the north-east are the cathedrals of prehistoric Scotland, spanning many centuries of building or reconstruction. Massive pillars of stone make such an important contribution to the Scottish landscape that it is easy to forget we know so little about them.

Graham and Anna Ritchie, 1981[3]

## AN IDEAL GROUP

The recumbent stone circles (RSCs) of north-eastern Scotland form an isolated regional grouping of small ceremonial monuments with obvious common features in their design and a remarkable consistency in orientation. These stone rings are distinguished by the presence of a single large stone—often by far the most massive stone present—placed on its side between two tall uprights (Fig. 5.1).[4] This 'recumbent stone' and its 'flankers' are placed with remarkable consistency on the SSW side of the ring. As viewed from the centre, they are oriented without exception between WSW and SSE, i.e. within a quarter of the available horizon.[5] The sites were not defensive, domestic or industrial; neither were they cemeteries: small pockets of human cremated bone were quite often placed within them, but these seem to have been formal deposits or offerings.[6] The obvious inference is that the RSCs were ceremonial monuments. The conformance of major elements of their design to what is clearly a strong ritual tradition raises important questions of the nature of that tradition, of the various purposes the rings served, and of the social context within which they were built and used. None seems significantly more grandiose than its fellows, and they seem to occupy small territories of no more than about 10 km$^2$, which suggests that they served small groups of subsistence farmers.[7] The absence of exotic artefacts in the RSCs seems to support the conclusion that these groups were largely independent and egalitarian, not in competition or subject to higher-level political control.[8] This is in distinct contrast to the area immediately to the south, where henges are very common.[9]

Any study of the RSCs is hampered by a lack of reliable excavated data. The only modern excavation, at Berrybrae (RSC5 in List 3), remains unpublished after over twenty years.[10] While Aubrey Burl sees the RSCs as dating from around 2500 to 1750 BC,[11] the dates he relied on (*c.* 1400 uncal BC)[12] were from the final phase of use of the monument.[13] Ian Shepherd has argued for an earlier date within the late Neolithic, perhaps nearer 3000 BC.[14] Burl's argument that the RSCs were descended from the Clava cairns, a cairn group to the north-west near Inverness (see chapter eight),[15] has been considerably weakened by recent radiocarbon dates from four Clava cairns which seem to place them firmly in the Early Bronze Age (see Archaeology Box 2).

Work by Burl in the 1970s had already led to the suggestion that the monuments had a ritual significance connected with the moon,[16] and this argument was consolidated in a paper published in *Antiquity* in 1980.[17] The path forged by the archaeoastronomical critiques of the late 1970s had also begun to approach the RSCs. As the project described in chapter three progressed, it had become clear that it would be valuable to seek a group of demonstrably similar monuments so as to avoid most of the problems about data selection.

Ideally, the group would be well-defined and confined to a given geographical area, yet sufficient in number to provide a reasonable set of data. In addition, we might seek sites with a design such that (say) one direction at each site is clearly of special importance. In that case we

Fig. 5.1 The recumbent stone circle at Easter Aquorthies (RSC63), viewed from the east. The recumbent stone and flankers are visible on the left.

> would avoid most of the problems about selecting possible 'indications' at each site.[18]

The RSCs came close to this ideal. About one hundred certain or possible RSCs exist, about half of which are in a reasonable state of preservation, confined to an area some 80 × 50 km (Fig. 5.2).[19] In each case the line joining the centre of the ring and the centre of the recumbent stone provides an obvious principal axis.

Burl's existing work showed an apparently strong correlation between the axial orientations of the RSCs and the rising and setting moon at its major and minor standstill limits. A diagram published in the *Antiquity* paper showed all the orientations falling neatly between the rising and setting moon at the southern major limit, with the exception of a small group falling close to moonset at the minor limit.[20] However, his conclusions were based mainly upon azimuths measured from published site plans rather than determined in the field, and took no account of altitude. For this reason, a field project was undertaken by this author in 1981, working initially with Burl, in order to discover whether new surveys would confirm the patterns suspected from the earlier work, and to study further the common features of this intriguing group of monuments.

As in the western Scottish project, careful criteria were developed for data selection and strict procedures were adhered to on site, all in an attempt to avoid prejudicing the results.[21] An initial list of ninety-nine monuments (see List 3) was obtained by comparing existing lists. The first step was to classify each monument as a certain, probable, possible or unlikely RSC on the basis of first-hand inspection and reference to other sources.[22] Of the ninety-nine, twenty were documented as destroyed or unrecognisable,[23] but during the summer of 1981 an attempt was made to visit each of the remainder, prioritising those in best condition. In the event, five sites were not reached,[24] two could not be located,[25] and a further eight were in fields under crop and could not be approached.[26] This left sixty-four that were successfully examined.

Burl's earlier conclusions had been based on the orientation of the principal axis of the monuments. A wider range of measurements was now taken, including horizon altitude data and the azimuths of the two ends of the recumbent stone and the inner edges of the flanking stones as viewed from various key points. Survey data were also collected to test related hypotheses of interest, for example that the top of the recumbent stone preferentially stood below the skyline as viewed from certain positions, and that RSCs were preferentially located so as to have a distant view over the recumbent stone.

## MANIFESTATIONS OF ASTRONOMY IN RECUMBENT STONE CIRCLES

As the 1981 fieldwork soon revealed, there are in fact two quite different ways of defining the principal axis of an RSC, and

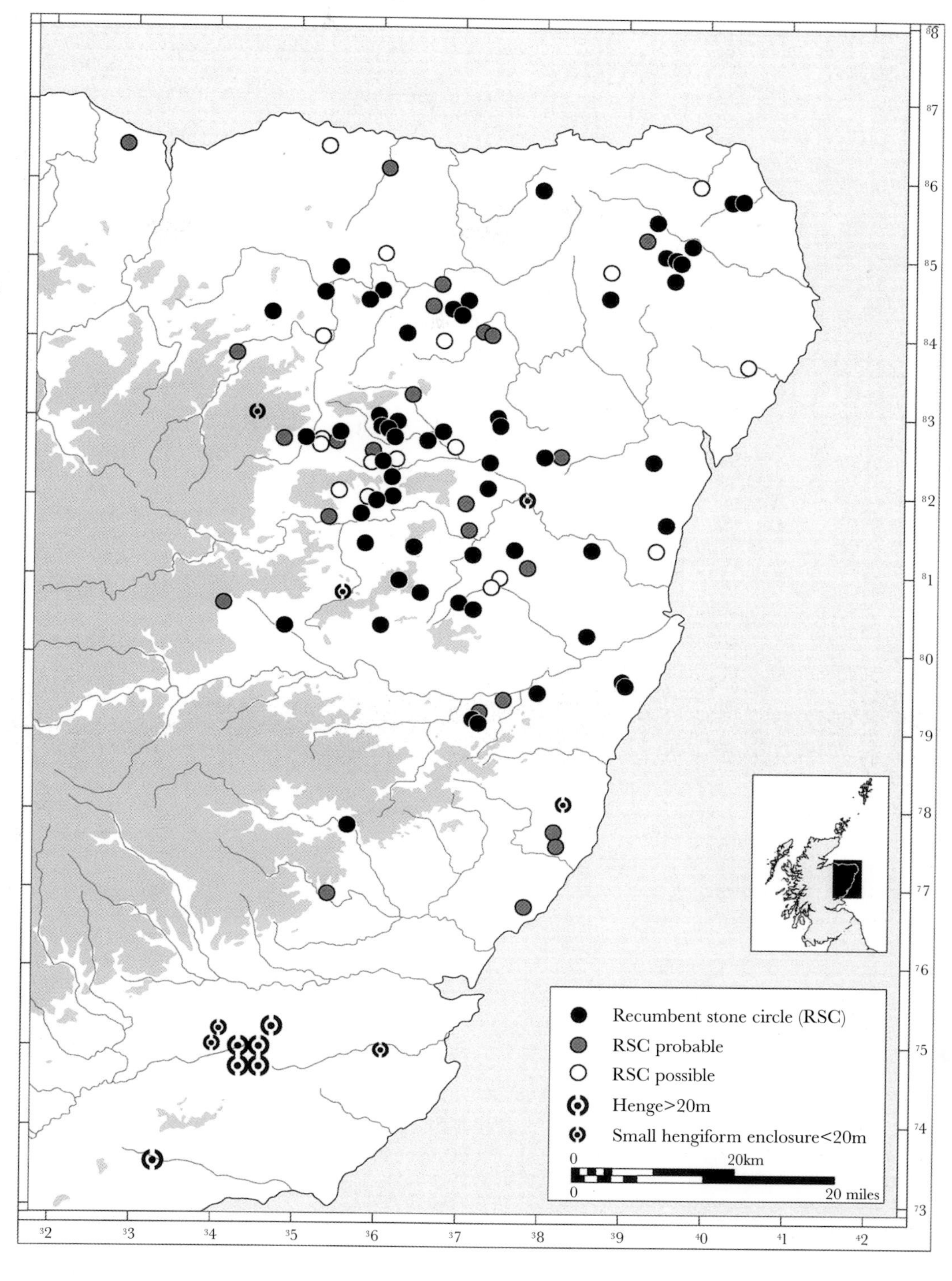


Fig. 5.2 Map of the Grampian region showing the locations of certain, probable and possible recumbent stone circles as well as large and small henges (see Archaeology Box 2). Land above an elevation of 300 m is shaded. Unlikely RSCs are not shown. Designations as certain, probable or possible follow List 3 except that Blue Cairn (RSC78) and The Cloch (Supp. List A) are included as probable RSCs and Ley, Bellman's Wood, and Leslie Parish (Supp. List B) are included as possible RSCs.

earlier analyses had confused them. One is the line joining the centre of the ring and the centre of the recumbent stone (the 'Centre Line') and the other is the line perpendicular to the long axis of the recumbent stone (the 'Perpendicular Line'). Depending on the present state of a particular monument, it may be possible to define either line, or both: and where only the latter exists, this had often been presented as the former.[27] However, one only needs to examine the site plans of a few better-preserved examples to see that the long axis of the recumbent stone was not always set exactly tangentially to the ring (see Fig. 5.3);[28] indeed, the azimuths of the principal axes defined in these two ways may differ by almost 20°.[29] Table 5.1 summarises the available information.[30]

It is possible to estimate the probability, given a set of $n$ azimuths selected at random, that all of them will fall within a range of $\theta°$ (see Statistics Box 6). In the case of the Centre Line data, thirty-eight azimuths fall between 157° and 236°, i.e. within a range of 79°. The probability of this happening fortuitously, according to formula (S6.1) of Statistics Box 6 ($n = 38$; $\theta = 79$), is of the order of $10^{-23}$. The Perpendicular Line data yield forty-seven azimuths between 147° and 237° and a probability of fortuitous occurrence less than 1 part in $10^{26}$. In other

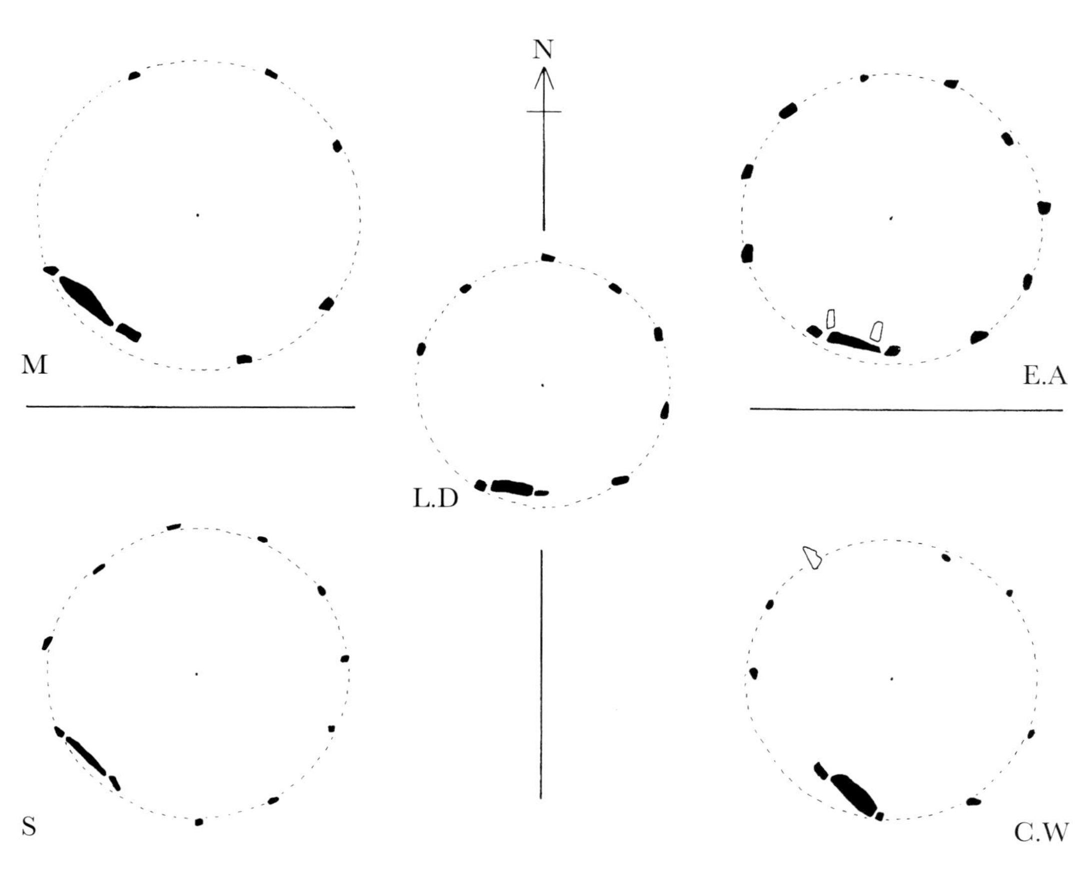


Fig. 5.3 Plans of five recumbent stone circles, showing the difference between the Centre Line and Perpendicular Line orientations. M = Midmar Kirk (RSC 71); EA = Easter Aquorthies (RSC 63); LD = Loanhead of Daviot (RSC 59); S = Sunhoney (RSC 72); and CW = Cothiemuir Wood (RSC 48). First published as Ruggles and Burl 1985, fig. 8.

words, whatever the details, it is quite clear that the remarkable consistency in orientation is wholly unattributable to chance. There is absolutely no doubt that the orientation of an RSC was of the utmost importance to its builders. Furthermore, the fact that this was achieved in a group of monuments scattered over such a wide area seems to rule out explanations related to prevailing wind direction, local topography,[31] or sightings upon any particular terrestrial feature.[32]

At first sight, it might seem that the overall orientation pattern between WSW and SSE could be explained most simply as a pattern of orientation in the general direction of (the quarter of the horizon centred upon) a skewed cardinal direction.[33] But what if the axial orientations of the RSCs were determined not simply in relation to the daily east-to-west swing of all objects in the sky but by longer-term observations of particular celestial bodies on or close to the horizon? In this case, we might expect that there would be a preference for distant, or at least reasonably distant, horizons over the recumbent stone, unobscured by local vegetation and perhaps providing fixed points of reference. This idea was tested by dividing horizon distances into four categories—A (up to 1 km), B (1–3 km); C (3–5 km) and D (over 5 km)—and then estimating the distance category of each stretch of horizon, noting the azimuths of junctions between different categories to the nearest degree. Such a procedure enables us to divide the horizontal circle into 1° intervals and then, for each interval, to examine the percentage of horizons falling into each category.[34] When this 'horizon scan' is produced for azimuths relative to the orientation of each RSC, it emerges that there is a complete avoidance of nearby (category A) horizons in the general direction of the primary axis. This is evident in Fig. 5.4, where the results are shown for the Perpendicular Line.[35] The fact that the zone of avoidance is almost 40° in width shows that, without exception, when the recumbent stone is viewed from within (or beyond) the circle, from a position some 5 m or more away along a line perpendicular to it, there is never any nearby horizon above any part of the recumbent stone.[36] In other words, it seems to have been vital that the entire stretch of horizon above the recumbent stone was at least moderately distant.[37] An obvious conclusion is that this might have been so that observations of celestial bodies on or directly above this stretch of horizon could be made from the interior of the ring. This is supported by evidence that the recumbent would invariably have appeared below the horizon as seen by an observer in the circle, while in the great majority of cases the flankers cut the horizon, appearing to partition off the section above the recumbent.[38]

Once we move beyond this general conclusion, it is not possible to give any clear and simple astronomical interpretation of the RSCs. The altitude data generally confuse rather than clarify or confirm the neat groupings bounded by lunar standstill limits originally identified by Burl, and the two possible ways of defining the principal axis of an RSC add further confusion.[39] Yet there is still the hint of a correlation between the axial orientations of the monuments and the motions of the moon. Figs 5.5a and 5.5b show the declinations of the horizon point over the centre of the recumbent stone, as viewed along the Centre Line and the Perpendicular Line respectively.[40] In the case of the Perpendicular line data there is a cluster of five declinations in the vicinity of moonset at the minor limit,[41] clearly separated from the main group of declinations below −25°. The difficulty with this interpretation is that the main

## STATISTICS BOX 6

### WHAT IS THE PROBABILITY OF *n* ORIENTATIONS FALLING WITHIN θ DEGREES?

If we find a group of monuments with similar orientations, the first question that suggests itself is whether this is likely to have arisen by intention. Specifically, if all the orientations fall within a relatively restricted range of azimuths, what is the likelihood that this could have arisen fortuitously (i.e. through a combination of factors quite unrelated to the azimuthal direction)?

The means of arriving at a probabilistic answer seems straightforward enough but the answer itself may be misleading. For example, suppose that we find that six out of six orientations fall between azimuth 30° and 150° (i.e. within 60° of east). The obvious question is: 'What is the probability that six orientations, each placed randomly, would all fall within 60° of east?' The probability of any one falling in this range is $^{120}/_{360} = {}^{1}/_{3}$, so the probability that all of them will do so must be $(^{1}/_{3})^{6}$, i.e. about 0·0013, or a little over one in a thousand.[1] Consequently we might conclude that the observed trend is very unlikely to have arisen fortuitously.

The problem is that we only asked the question that we did, as opposed to hundreds of others we might have asked, because of what we observed in the data in the first place. We did not ask, for example, what was the probability that six out of six orientations would fall between 130° and 250° because not all of our observed data fell in that range.[2]

In order to avoid falling into this trap, we need to provide a better estimate of the probability of fortuitous occurrence, without prejudging the issue by choosing the 'target range' on the basis of the data. This suggests the more general question 'What is the probability that *n* orientations will fall within an azimuth band of only θ degrees?'

This is not quite as straightforward to calculate, but the procedure is as follows. Suppose for simplicity that $\theta < 180$. If the $n$ azimuths are going to be clustered within a range of θ degrees, then there must be one—call it $A_m$—which marks the anticlockwise extent of the cluster of azimuths; there will be no other azimuth within 180° of $A_m$ going anticlockwise. The probability that each of the other $n$—1 azimuths fall between $A_m$ and $A_m + \theta^3$ is $(\theta/360)^{n-1}$. But any of the $n$ azimuths could turn out to be $A_m$, so the overall probability required is

$$n \times (\theta/360)^{n-1} \qquad \ldots \text{(S6.1)}$$

In the example above, $n = 6$ and $\theta = 120$, so formula (S6.1) gives $6 \times (^{1}/_{3})^{5} = 0{\cdot}025$, or only 1 in 40; a very different result which gives much weaker evidence for intentional orientation in a particular range of azimuths.

#### PROBLEMS WITH OUTLIERS

Choosing the right question is also critical if a handful of the sample fall outside the range being tested. For example, suppose that six out of seven orientations fall between 30° and 150° but there is an outlier with azimuth 180°.

The probability of fortuitously obtaining six orientations within 60° of east together with one outside this range is $(^{1}/_{3})^{6} \times {}^{2}/_{3} \times 7$. The probability of an azimuth falling outside the range is $^{2}/_{3}$, and the factor seven occurs because any of the seven orientations may be the outlier. The result is 0·0064, still well under one in a hundred.

On the other hand, the probability of obtaining six orientations within a 120° range together with one outside this range is approximately $6 \times (^{1}/_{3})^{5} \times {}^{2}/_{3} \times 7$,[4] i.e. about 0·12, more than one in ten, and we would conclude that there is no convincing evidence for intentional orientation in a particular range of azimuths.

group covers a range of declinations down to around −33° (where sites face due south), and so it provides no evidence of a particular interest in the rising or setting moon at the (more readily observable) major limit. Indeed, the most promising overall explanation of these data as they stand might be in terms of orientation upon the winter sun low in the sky.

In order to progress we need to move away from theoretical abstractions concerning the geometrical axes of symmetry towards considering the observations that might actually have taken place in practice. We have no evidence that the exact centre of the recumbent stone was marked in any way, and it is difficult to justify concentrating particularly upon it. It makes much more sense to focus attention upon the whole visible range above the recumbent stone. Neither can we assume that there was just one observer, who carefully placed themselves at a position that can still be identified, such as the exact geometric centre of the circle. Perhaps a group of people merely stood somewhere in the circle interior to make observations at auspicious times. All this makes the orientation data potentially much more complex. The width of horizon framed by the

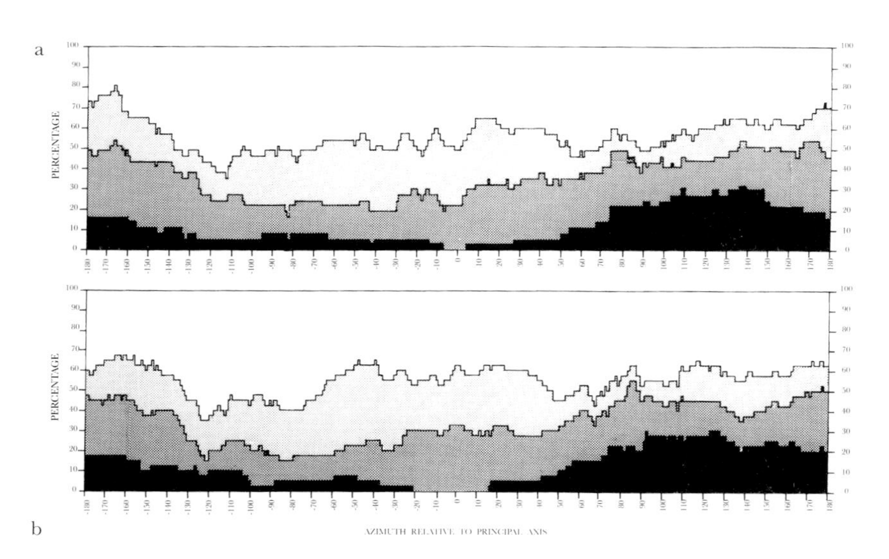


Fig. 5.4 Orientation of visibility. The frequency of occurrence of different horizon distances in various directions relative to the 'Perpendicular Line' axis, combining data from RSCs where the latter can be determined. For each 1° interval in azimuth relative to this axis (taken in the south-westerly direction) the plot shows the percentage of horizons in category 'A' (up to 1 km, black), 'B' (1–3 km, dark shading), 'C' (3–5 km, light shading), and 'D' (over 5 km, white). First published as Ruggles 1984c, fig. 4(b); also Ruggles 1984b, fig. 11.

recumbent and flankers depends upon the distance of the observer away from them, and its central azimuth depends upon the distance of the observer away from the axis of symmetry. Nonetheless, it is still possible to reach some important conclusions.

For the purposes of the analyses undertaken as part of the 1981 project, a consistent dataset was obtained by assuming the notional observing position to be at the geometrical centre of the ring for Centre Line data, while a position 10 m behind the recumbent stone was used for the Perpendicular Line data.[42] The relevant azimuth, altitude and declination measurements for 'indicated horizons' defined in this way are listed in Tables 5.2 and 5.3.[43] The most important thing these data reveal is that even including the whole visible range of horizon above the recumbent stone, there are still several RSCs where the entire indicated range falls entirely below −30° in declina-

Fig. 5.5 Declinations of the horizon point over the centre of the recumbent stone at Scottish RSCs, as viewed along (a) the Centre Line, (b) the Perpendicular Line, plotted to the nearest degree. Unshaded squares represent less reliable data (lines marked 'o' or 'u' in col. 13 of Table 5.2 or 5.3). After Ruggles and Burl 1985, figs 10c and 11c respectively. (c) shows, for comparison, the declinations of the horizon point over the centre of the axial stone at the south-west Irish ASCs.

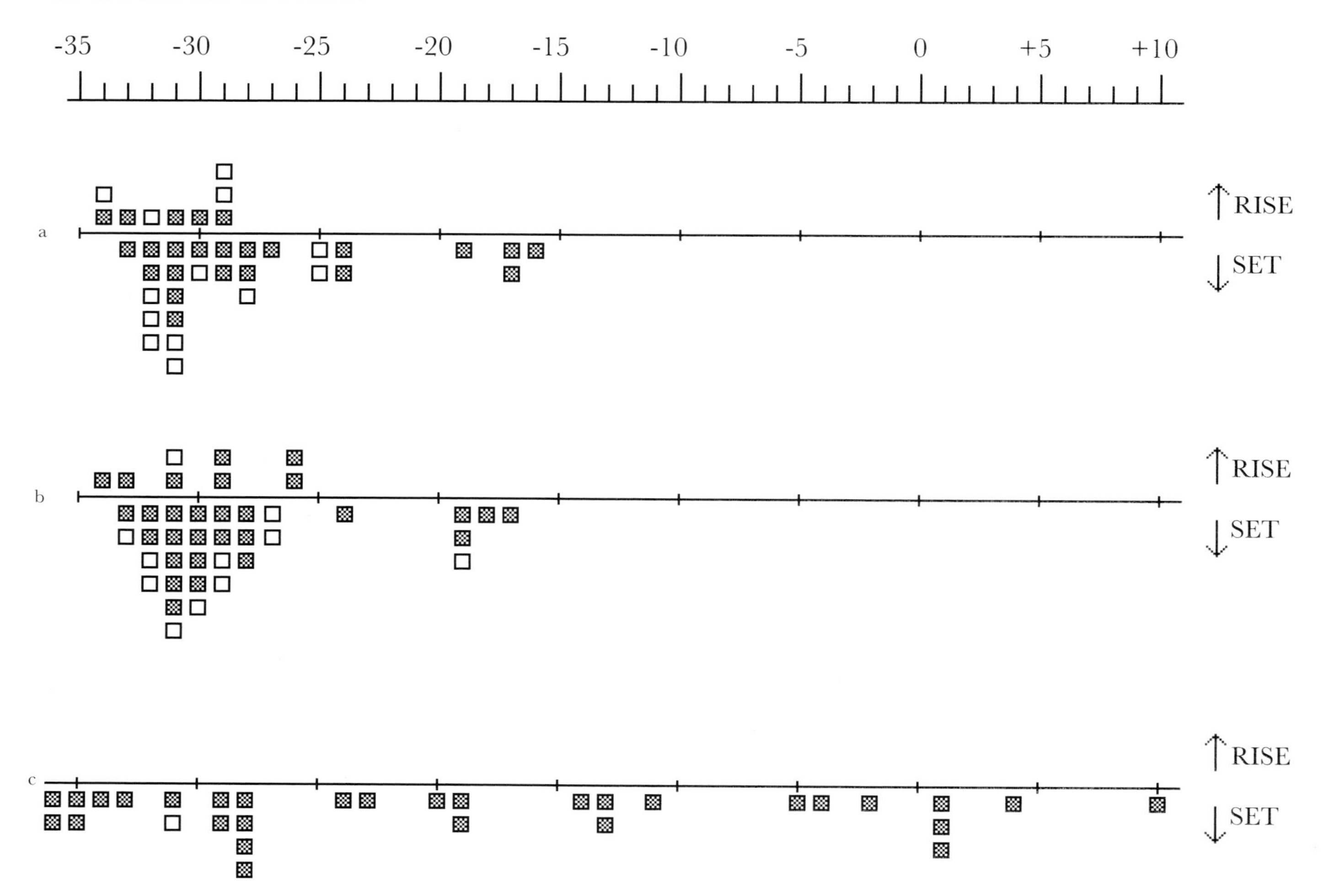

tion, and in some cases well below.[44] This is true regardless of whether Centre Line or Perpendicular Line data are chosen, and regardless of the exact observing position, unless this was very much closer to the recumbent and flankers than the circle centre. It is therefore impossible to conclude that the builders were concerned specifically with the rising or setting positions of the moon (or, of course, of the sun).

On the other hand, every indicated range without exception includes values below $-19°$, a conclusion that holds for all observing positions in the general vicinity of the circle centre.[45] While this could be explained by reference to the sun, an interpretation that fits the data better is that the aim was simply to orient the recumbent stone so that, as viewed from the circle interior, the full moon would pass low over it around midsummer each year. Given that construction took place at an arbitrary point in the lunar node cycle, this would result in a pattern of declinations very similar to that found. Perhaps it was important that the moon should pass into the space above the recumbent rather than rising there, which would explain the bias towards westerly rather than easterly orientations. There is, then, good reason for continuing to suspect that observations of the moon were of particular importance at these circles.

If this was the case, though, then the builders of some monuments—those who happened to choose an orientation closer to the moon's setting position and who happened to construct their monuments in a year close to a minor standstill—might have been surprised to discover in subsequent years (closer to the major standstill) that the midsummer moon, approaching the recumbent and flankers from the left, failed to pass over the recumbent, setting instead to the left of the left flanker. Indeed, at Midmar Kirk (RSC71), the horizon to the south is so high that the midsummer full moon would not have been visible at all around the time of a major standstill. If it was important to hold a ceremony at the time of the midsummer moon's passage every year, then we can only speculate as to how people reacted to this.

At this stage it may be tempting to explore the lunar possibilities in more detail, perhaps seeking further lunar associations in monumental alignments or looking for other evidence of an association between RSCs and the moon. Before rushing headlong in this direction, though, it is wiser to take stock and to consider the wider interpretative context.

## Interpreting the Scottish recumbent stone circles

Looking at the wider evidence draws attention to a variety of things, without obvious practical value in our terms, that seem to have been important to those who built and used the RSCs. The type of stone used was evidently a concern, particularly in the case of the recumbent itself, which was often transported from some distance, while readily-available local stone was used in the circle.[46] In many cases the heavy recumbent was positioned with great care and carefully chocked so that the top was flat and no more than two or three degrees off horizontal.[47] This can be seen clearly at Midmar Kirk (RSC71), where the tapering base of the recumbent stone and the chockstones are visible above present ground level (Fig. 5.6). Several things reinforce the idea that the recumbent and flankers were of central importance to the ritual function of the monuments. At many of the rings the heights of the circle stones are graded, rising towards the flankers, which are the tallest uprights.[48] Decoration in the form of cup marks (Fig. 5.7) is only ever found on the recumbent stone, flankers, or adjacent circle stones.[49] Quartz scatters are frequently found near the recumbent, as at the excavated site of Berrybrae (RSC5), but rarely elsewhere.[50]

There is also good reason to believe that where the circles were placed in the landscape was not just a matter of economic common sense or convenience. While they do seem often to be placed near to patches of deep, fertile and well-drained soil[51]—unsurprising if their planners were agriculturalists—

Fig. 5.6 The recumbent stone at Midmar Kirk (RSC71), viewed from the south-west.

Fig. 5.7 The cup-marked recumbent at Sunhoney (RSC72), viewed from the east. The stone has fallen into the ring, and the cup marks are visible on what is now its top surface.

they were also invariably placed in conspicuous settings, often on artificially levelled sites,[52] generally on hilltops or south- or south-east-facing slopes.[53] They tend to cluster in areas of low hills, away from more mountainous regions and from low-lying river valleys.

Finally, they were often oriented upon prominent landmarks. Very often the horizon above the recumbent contains a single conspicuous hilltop. According to the published results of the 1981 project, this is the case at no fewer than thirty-two of the forty-nine RSCs listed in Table 5.2 and/or Table 5.3.[54]

In short, the type of stone of which the RSCs were constructed, their form, their location, and their relationship to certain points in the landscape around them were all of importance to those who built and used these modest monuments. The aim, perhaps, was to ensure that the circles were at one with the world; that their place in the cosmos was symbolised and confirmed in all the right ways according to the prevailing traditions of sacred knowledge.[55]

Within this context, we can return to the astronomy. The consistency in orientation of the RSCs, together with the form of the flat recumbent stone and its upright flankers, clearly demarcating a stretch of horizon, invites the speculation that they were a focus for astronomical observations from the interior of the ring. Even Gordon Childe was tempted to speculate in 1940 that 'the Recumbent Stone with its straight upper edge would form an admirable artificial horizon for observing heliacal risings of stars for calendrical purposes'.[56] As we have seen, the correlation of horizon distance with azimuth relative to the recumbent stone gives strong quantitative support to the idea.[57] But, as we have also seen, the declination data suggest that the astronomical body involved could have been the midsummer full moon. In this case we should probably think not in terms of the moon being observed but rather of it shining over the recumbent and casting light upon the circle during a favourable midsummer night. It is possible that this annual interplay of nature and the monument may have been of the utmost symbolic significance to the builders, setting the stone circle in tune with the cosmos at an auspicious time of year, confirming its place in nature's rhythms, and providing the backdrop for sacred ceremonial.[58] Perhaps the white pieces of quartz were deposited around the recumbent stone at this time, as part of these ceremonies, because they symbolised and reflected the light of the moon.[59]

There is no evidence that the conspicuous hilltops appearing in the horizon above the recumbent at a high proportion of the RSCs are particularly correlated with specific lunar, or other astronomical, events. The azimuths and altitudes of single conspicuous hilltops appearing in the horizon above the recumbent were measured as part of the 1981 project, in order to investigate this question, and the results are listed in Table 5.4. The declinations obtained are shown graphically in Fig. 5.8a. Their spread does suggest a correlation with the moon, but it follows the same general pattern as the Centre Line data as a whole (compare Fig. 5.5a), making it unreasonable to conclude anything other than that this desirable attribute of orientation upon a prominent landmark was simply achieved, wherever possible, within the constraints of the general custom of lunar orientation.[60]

On the other hand, an analysis of the orientations of cup marks from the circle centre, at those RSCs where they have been found, has revealed a very intriguing pattern.[61] Cup marks occur at a dozen RSCs, eight of which were studied as part of the 1981 project.[62] The results are reproduced in Table 5.5, and the declinations obtained can be seen in Fig. 5.8b. Remarkably, seven of the eleven declination values are clustered between −31° and −28°, within 2° of the rising or setting moon at the major standstill limit, and three of the remainder fall close to −19°, within 1° of the setting moon at the minor limit. At Balquhain there appear to be cup marks lying beneath the setting position of the moon at both standstill limits. All of the cup marks found on (or below) the recumbent stone itself relate closely to the setting moon at one or other standstill limit. Only at Loanhead of Daviot does an anomalous declination value occur, and it may be significant that it is one of the few to the east of south and on an adjacent stone rather than the recumbent or flankers. (It is tempting to relate this to the solstitial sun.) If these apparently precise lunar alignments are not fortuitous,[63] what is the explanation? Were people really carrying out programmes of observation to determine the lunar standstill limits?

Perhaps, in view of what we have already said both here and in earlier chapters, a more likely explanation is to be found in the repeated ceremonials that, if our speculations so far are at all close to the truth, might have accompanied the annual passage of the midsummer full moon over the recumbent stone. The gradual widening or narrowing of the moon's path in the sky from one year to the next would surely have become obvious before very long, and perhaps it was as a result of this that the varying setting positions of the moon were noted with some interest. Perhaps in time the limiting directions themselves became sacred and this is the reason why cup marks were placed to mark those directions.[64]

Can we consolidate or extend these ideas by looking more closely at the form and structure of the RSCs, and particularly by examining individual or regional variations? Certainly there are morphological differences. For instance, some RSCs are associated with internal cairns or ring cairns; some are embanked; and some have the recumbent and flankers placed carefully as part of a large and near-perfect circle while others have

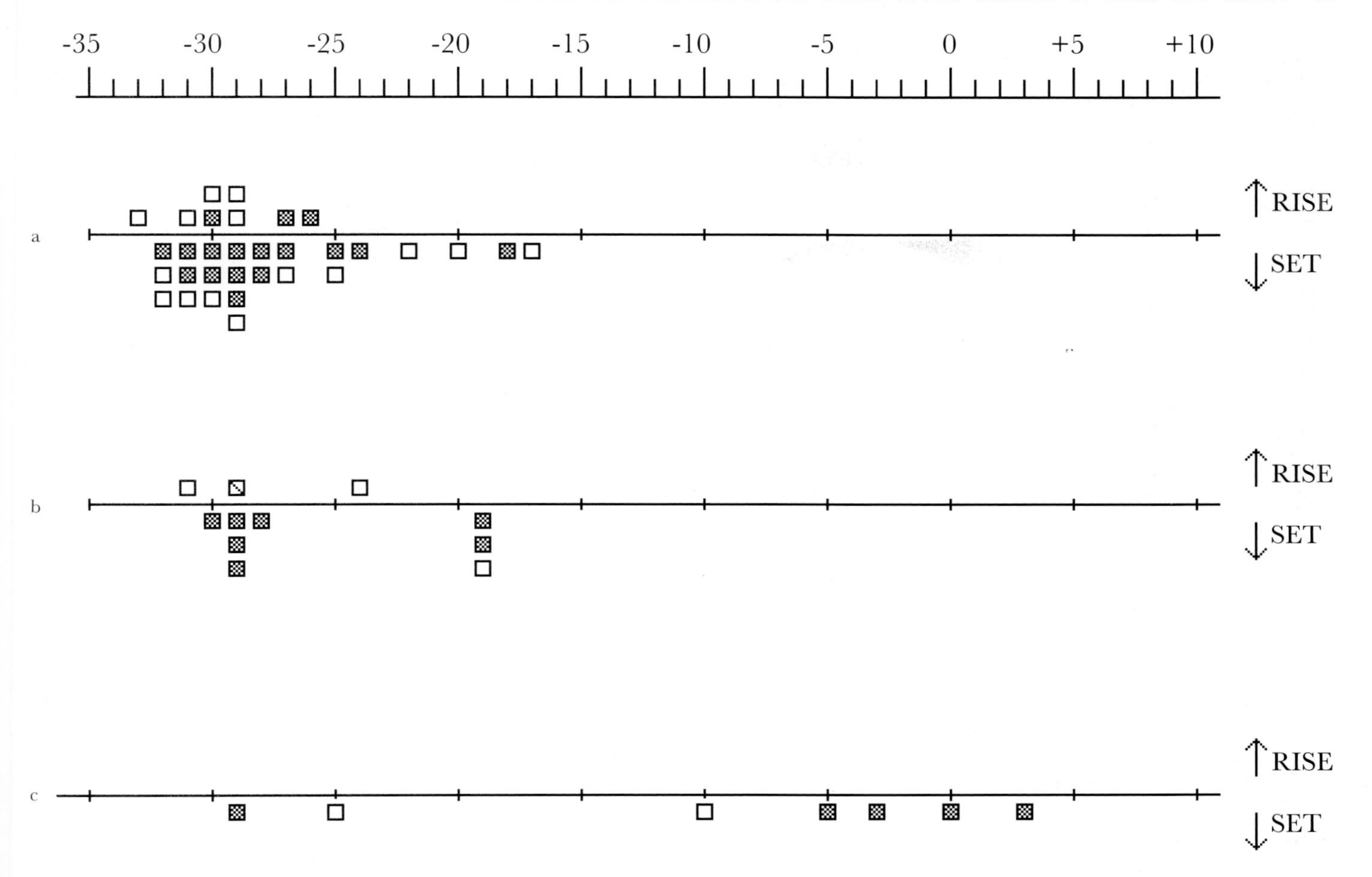


Fig. 5.8 (a) Declinations of single conspicuous hilltops within the indicated horizons at Scottish RSCs, plotted to the nearest degree. Shaded squares represent circles where the azimuth of the hilltop is within 3° of the Perpendicular Line azimuth. (b) Declinations of the horizon directly above cup marks at Scottish RSCs, as viewed from the centre of the ring, plotted to the nearest degree. Shaded squares indicate that the cup marks are on the recumbent stone; squares with a diagonal line that they are on a flanker; and unshaded squares that they are on an adjacent stone. (c) shows, for comparison, the declinations of single conspicuous hilltops within the indicated horizons at Irish ASCs, in the portal-to-axial direction, shaded where the azimuth of the hilltop is within 3° of the ASC axis.

them placed well inside a smaller and more distorted ring.[65] Burl has proposed an evolutionary model, interpreting the centrally placed examples as representing a core tradition with later outliers showing greater variation.[66] There are however, dangers in relying upon typological classifications of the present form of a monument to deduce its place in chronology.[67] Attempts to examine regional variations in orientation have also failed to reach any clear conclusion.[68] Perhaps only the sorts of questions that can be answered by archaeological excavation, such as the detailed sequence of events at individual sites, will ultimately prove more enlightening.

In the meantime, it is interesting to note that despite all the care apparently put into some features of RSC construction, other aspects were seemingly less important. An intriguing possibility is that once the three major stones had been carefully set in place, perhaps with the co-operation of neighbouring groups,[69] the circle stones were often left to be added in a more casual fashion later. Certainly this might explain the large difference between the Perpendicular Line and Centre Line orientations in some cases.[70] Indeed, at a few sites only the recumbent and flankers exist in various states of repair, and while it is generally assumed that the smaller circle stones have been removed in modern times, as happened at Strichen (RSC7)[71], it is possible that at some of these no-one ever got round to adding the circle stones at all.[72] Excavation might some day establish if this is the case.[73]

The fact that people never became centrally organised and controlled in this area at this time, has bequeathed to us a set of small, similar monuments, and enables us to catch a rare and valuable glimpse of some aspects of shared ritual tradition. The Scottish RSCs clearly demonstrate how detailed surveys of a group of monuments, taken in context with other archaeological evidence, can reveal much more than could be gleaned simply from examining the spread of orientations.

## Similar but different? The axial stone circles of Cork and Kerry

South-western Ireland represents an area where this line of research has been extended in recent years, with intriguing results. Counties Cork and Kerry are extremely rich in free-standing megalithic monuments, including over a hundred 'axial stone circles'[74] (ASCs) that are so similar in form to the Scottish RSCs that some authors, notably Burl, have actually referred to them as recumbent stone circles.[75] The most obvious similarity is that, like the Scottish RSCs, the Irish ASCs contain a single axial (recumbent) stone placed on its side facing the south-west or west. But there are important

Fig. 5.9 The axial stone circle at Gortanimill (ASC11), viewed from the east.

differences. Instead of being flanked by two uprights, the axial stone stands alone, and there are two 'portals' on the opposite side of the ring. The monuments are generally symmetrical about an axis through the axial stone, the rings consisting of paired stones with a tendency for height gradation upwards towards the north-easterly portals, rather than towards the south-westerly flankers as is the case for the RSCs.[76] In addition, the axial stone itself—and the rings as a whole—are generally smaller than in Scotland, with many monuments having only five stones in total.[77] Radiocarbon dates obtained at Cashelkeelty (ASC30 in List 4)[78] suggest that the ring there may have been constructed in the Late Bronze Age between about 1250 BC and 800 BC, implying a surprisingly late construction date for some, and perhaps all, of the Irish ASCs.[79]

The range of orientations of the Irish ASCs is no less concentrated than that of their Scottish counterparts, although it is slightly different, with the azimuths of the axial stones lying roughly between due south and due west.[80] As with the RSCs, such consistent orientations over such a wide area must have been astronomically determined in the broadest sense.

It is puzzling indeed that no stone rings of similar form are found anywhere apart from these two small concentrations in opposite corners of Britain and Ireland. The morphological similarities do suggest a direct link between the two groups,[81] but on the other hand their geographical and temporal separation seem to argue otherwise. Certainly it is possible that the ASC tradition somehow derived from the RSC tradition, but if this was the case then the interesting questions are how and why. If we can identify which elements of the older ritual tradition were incorporated into the new one, then this might help us to suggest some answers.

With such questions in mind, the Irish ASCs were examined in 1994 as part of an ongoing programme of archaeoastronomical surveys in south-west Ireland. The starting point was a list of forty-eight ASCs of between seven and nineteen stones compiled by Ó Nualláin (see List 4).[82] Visits were made to all of those where the axial stone and/or both the portals were still present,[83] thus allowing a reasonable estimate of the axial direction to be made, and a total of thirty-one were surveyed.[84]

The results were perhaps surprising. For a start, the data on the distribution of horizon distance with azimuth show no preference for distant horizons either behind the axial stone or behind the portals:[85] Gortanimill (ASC11), for example, has local ground rising immediately behind the axial stone and obscuring the more distant horizon (Fig. 5.9). Next, when we examine the declinations of the horizon point above the centre of the axial stone (these are listed in Table 5.6 and plotted in Fig. 5.5c)[86] it is clear that the distribution is completely unlike that of the RSCs. The horizons above the axial stone as a whole—even as viewed from some 5 m beyond the portals, where the portals appear to 'flank' the axial stone and the range is relatively restricted—still span declinations from the minimum possible around due south all the way up to +14°, and there is no readily apparent pattern to them. Finally, there is no evidence for preferential orientation upon conspicuous hill summits. The small number of indicated horizons that actually contain single conspicuous summits are listed in Table 5.7 and plotted in Fig. 5.8c.[87] There is no astronomical pattern to them.[88]

In short, the ASCs bear no consistent relationship with any specific astronomical body or event. Of course, it is always possible to relate particular monuments in the group to particular events, such as the setting sun at the solstice, equinox or mid-quarter day, or the setting moon at one or other lunar standstill limit;[89] but there are severe dangers in applying an 'archaeoastronomical toolkit' of targets like this, a topic to which we shall return in chapter nine. Claims of alignment upon Venus,[90] invoked to account for some of the orientations not otherwise explained by the setting sun or moon, can also be disregarded.[91] On the other hand it is hard to ignore the solstitial alignment of Drombeg (ASC52), one of the best-preserved and best-known ASCs. The axis of this monument is aligned upon a conspicuous horizon notch as well as upon winter solstice sunset (see Fig. 5.10a). While much commented upon,[92] though, this phenomenon is not precise;[93] nor, more importantly, is it repeated from one ASC to another, which means that whether or not it was intentional is unprovable statistically. Nearby Bohonagh, for example, is aligned upon a low flat hill and virtually due west (Fig. 5.10b).

Viewing the ASCs as a group, then, it seems that we can say little with conviction about their orientations beyond the basic fact that they aim for the quarter of the horizon between due south and due west.[94] They exhibit none of the indicators that led us to consider more specific astronomical interpretations for the Scottish RSCs. The temptation is to suggest, following Burl,[95] that the Irish ASC tradition somehow derives from the earlier RSC tradition in north-east Scotland, but that in the process various subtleties of symbolic association were lost. This position may certainly be supported by the fact that the Irish circles are generally smaller, and include the degenerate examples with only five stones, perhaps suggesting a dying tradition. But on the other hand the symmetrical ASCs give the impression on the whole of being more carefully planned than some of their counterparts in north-east Scotland, where the recumbents and flankers could often be set considerably askew to the rest of the circle. Perhaps there were new symbolic

a b

c d

Fig. 5.10 Portal-to-axial alignments at four axial stone circles.
a. Drombeg (ASC52).
b. Bohonagh (ASC51).
c. Derreenataggart West (ASC40).
d. Reanascreena South (ASC48).

associations here; ones of more localised significance. Certainly there is no simple interpretation.

It is possible to take this discussion further by moving away from stone circles. This is because, unlike the Scottish RSCs, the Irish ASCs are just one form of conspicuous monument found intermingled within a single geographical region, sharing a common, or at least very similar, pattern of orientation. In Cork and Kerry, as well as the ASCs, there are over ninety rows comprising three to six standing stones,[96] over a hundred stone pairs[97] and numerous single standing stones,[98] and around a hundred wedge tombs.[99] The orientations of the rows and pairs are highly clustered around NE–SW,[100] and even the single standing stones are often slabs and generally follow the same orientation pattern,[101] while the wedge tomb entrances all face the western arc of the horizon.[102] Sometimes different types of monument are found in close association, as at Cashelkeelty, where it seems that an ASC was constructed close by, and no more than two or three hundred years after, a four-stone row;[103] and overall there is little evidence to separate the rings and rows chronologically,[104] although the wedge tombs do seem to be earlier (see Archaeology Box 2).

The stone rows of Cork and Kerry were the focus of a programme of archaeoastronomical fieldwork initiated in the early 1990s, and it is to these that we now turn.

# 6

# Orientation and Astronomy in Two Groups of Short Stone Rows

Merry stared at the lines of marching stones: they were worn and black; some were leaning, some were fallen, some cracked or broken; they looked like rows of old and hungry teeth. He wondered what they could be . . .

J. R. R. Tolkien, *The Lord of the Rings*, 1955

For instance, one might entertain the virtually untestable idea that [megalithic monuments] were orientated on sacred places.

Douglas Heggie, 1981[1]

More research, coupled with scepticism but unimpeded by prejudice, will bring us closer to an understanding of what such alignments . . . meant to the communities that laid them out. It is, after all, the pursuit of the whole past, not just the comfortably preferred elements of it, that should be the preoccupation of all who profess an interest in antiquity. No less than architecture and artefacts, astronomy and its alignments are a legitimate part of that pursuit.

Aubrey Burl, 1988[2]

## Deceptively simple? Short stone rows from an archaeological perspective

Short rows of standing stones were erected in considerable numbers in north-west Europe during the Bronze Age. Several hundred rows of between three and six stones, mostly under 10 m in length and rarely longer than about 25 m,[3] are found in various parts of Ireland, Britain, and north-western France,[4] but in particular concentrations in Argyll and the Inner Hebrides in western Scotland, and Counties Cork and Kerry in the Irish south-west.[5] Fig. 6.1 shows the five-stone row at Dervaig N, Mull (SSR22 in List 6). These monuments seem relatively modest in terms of construction effort, superficially straightforward in conception, and apparently simple in purpose. However, just what that purpose was is completely unclear. There seems little question that they were not domestic sites, and it is quite clear that they were not defensive. They do quite often have funerary associations, but generally people resort to categorising them as ritual or ceremonial monuments, which is to admit to our lack of any real understanding of their meaning and use.

Aubrey Burl has proposed that the short rows represent the tail end of a long tradition that began early in the third millennium BC with avenues attached to circles.[6] As time went on the rows got ever shorter (degenerating eventually to only a pair) and were erected in ever more remote areas towards the west coasts. But the geographical distributions of rows of particular numbers of stones may be misleading, since the number of stones now existing in a row or pair may not be an accurate reflection of the number placed there in antiquity.[7] In any case, the few available radiocarbon dates suggest that three- and four-stone rows were being constructed through to the Late Bronze Age—even later than Burl envisaged—both in south-west Ireland and western Scotland.[8]

The short stone rows could quite easily have been erected by relatively small groups of people. It is uncommon to find a stone weighing more than three or four tonnes.[9] However, even fundamental questions relating to how they were used remain unanswered and debated. Are we right to assume tacitly that each monument was used over and over again by a single group of people, as a centre for ritual or ceremonial within their own territory? The fact that a row does not enclose an area of land—does not obviously mark off a piece of sacred space—may suggest other possibilities more readily than in the case of the stone circles. Perhaps the rows did not serve any consistent purpose at all after their construction—the act of building them being sufficient in itself.[10] Perhaps, especially in more marginal areas such as the Inner Hebrides, they served to mark points or routes in the landscape and were used by a broader community of roving animal herders? Perhaps they marked territorial boundaries or sacred divisions of the landscape? These general possibilities are discussed in chapter nine, but whichever is closest to the truth, one thing seems clear enough: that these ostensibly uncomplicated monuments had a strong material presence and symbolic importance, perhaps operating at many levels. By investigating repeated trends in their location and design, it is possible that we will start to identify some of the common elements of the ritual or symbolic tradition that the enigmatic stone rows reflect. The hope is that this will cast some light on why they were constructed and the nature of the use that was made of them.

Fig. 6.1 The five-stone row at Dervaig N, Mull.

That the rows were 'pointing at' something is an obvious suggestion. While we should not close our minds to a range of other possibilities, this one should obviously be investigated to the full. The archaeoastronomical work discussed in earlier chapters is clearly very relevant to this issue, and we have already seen how projects in the late 1970s and early 1980s began to draw particular attention to some of the stone rows in western Scotland. We shall return to the Scottish rows shortly, but first we will consider the group of short rows in south-west Ireland. The reasons for this are twofold. First, it is difficult to draw back completely from the baggage of previous interpretation that attaches to the Scottish monuments. By concentrating to begin with on a geographically separate group that has been virtually ignored from an archaeoastronomical point of view,[11] we avoid most of this baggage and can make a fresh start. Second, they have been studied more recently, in a programme of fieldwork undertaken by the author between 1991 and 1993.[12]

## A fresh start: astronomy and the Cork–Kerry rows

One of the main aims of the project that commenced in 1991 was to investigate the oriented monuments of south-west Ireland in a systematic way in the light of new ideas and approaches within archaeoastronomy. The work programme represented a return to a fieldwork methodology that had not been applied since the mid-1980s, in the sense that it concentrated on horizon survey in the general direction of alignment at a group of similar monuments. However, some of the principles underlying this procedure were entirely different.[13] One of these was that high-precision alignments were not being considered.[14] Another was that attention was not being closely restricted to the horizon in the actual direction of alignment, since it was felt that a prominent landscape feature (such as a conspicuous hill) located a few degrees off the apparent present-day mean alignment of the monument might be an important element in its intended symbolic function. One thing had not changed, though: there would be no attempt to 'fit' the data to predetermined astronomical targets. Instead, any accumulations of declinations would be allowed to speak for themselves.

Of the various types of monument in south-west Ireland with a common orientation pattern, the stone rows were selected as those where orientation evidence is likely to be best preserved. During the first season of fieldwork, attention was focused upon the longer, four- to six-stone rows, where the orientation we measure today is most likely to be an accurate reflection of that originally intended. Three-stone rows with all three stones still standing were investigated in the second stage of the project, in the two subsequent years.[15]

The starting point was a list of seventy-three rows of between three and six stones in Counties Cork and Kerry published by Seán Ó Nualláin in 1988,[16] supplemented by a corpus published by Burl[17] in advance of a major publication on stone rows.[18] Fig. 6.2 shows a distribution map based upon information known to this author at the time of his field project. A number of regional inventories have since been published,[19] which include details of further examples, so that the corpus provided here in List 5 is more up to date.[20] Of the seventy-nine rows in the initial list,[21] twenty-one were immediately excluded from further consideration and not visited,[22] four were not located,[23] and at one—Knocknanagh East (CKR6)—all of the uprights except one were found to have been cast down and moved, so even an approximate orientation could not be determined at first hand. Horizon surveys were undertaken at forty-nine of the remaining fifty-three rows, although in one case indicated azimuth data could not be obtained.[24] Site data (including approximate orientation and stone heights) were confirmed at the remainder but no survey was undertaken.[25] The data obtained are summarised in Tables 6.1 and 6.2.[26]

The strong clustering of orientations around NE–SW is evident from columns 2 and 3 of Table 6.1. Only seven examples fall outside the range 20°/200° to 70°/250°: two within 5° of due north–south[27] and five within 12° of due east–west.[28] For this reason it is convenient to refer to the 'NE' and the 'SW' directions, even in the limiting cases, and this we shall do in what follows, going so far as to drop the inverted commas for convenience.

Fig. 6.2 Map of south-west Ireland showing the locations of the Cork–Kerry three- to six-stone rows considered in the 1991–3 project. The six-stone row at Beal Middle (CKR1) is off the map to the north and the three-stone row at Garryduff (CKR16) is off the map to the east. First published as Ruggles 1996, fig. 1. Based on a map supplied by the Archaeological Branch, Ordnance Survey Ireland.

An obvious difference between these stone rows and the circles discussed in the previous chapter is that they point both ways: there is no innate feature in the architecture that draws attention to one direction rather than the other. If we wish to test the idea that the rows really were aligned upon something, there is no *a priori* reason for choosing one direction in preference, and to do so on the basis of the astronomical or other possibilities could well result in a circular argument. On the other hand, the gradation in the heights of the stones at a substantial number of the monuments does suggest an apparent direction of indication,[29] and this can be taken into account when analysing the data.[30] The surprise is that—at first, at least—this information seems to generate confusion rather than enlightenment. Given that stone height gradation seems to have been symbolically important, and consistently applied, at both the RSCs and ASCs, it is perhaps surprising to discover that amongst the Cork–Kerry rows the apparent direction of indication, where this can be determined, is almost as often to the NE as to the SW.[31]

If the rows really were pointing somewhere, then one of the obvious questions—especially in view of the contrasting results from the RSCs and ASCs—is whether we can detect any unusual variation in the proportion of non-local horizons in the direction of orientation. Of course, the situation with the stone rows is slightly more complicated because of their bidirectionality; but if the direction of interest was consistently to the SW, or else to the NE, then a horizon scan for azimuths relative to site orientation, in which (say) the direction of the north-easterly orientation is always taken as zero, might reveal something. In fact, such plots reveal nothing out of the

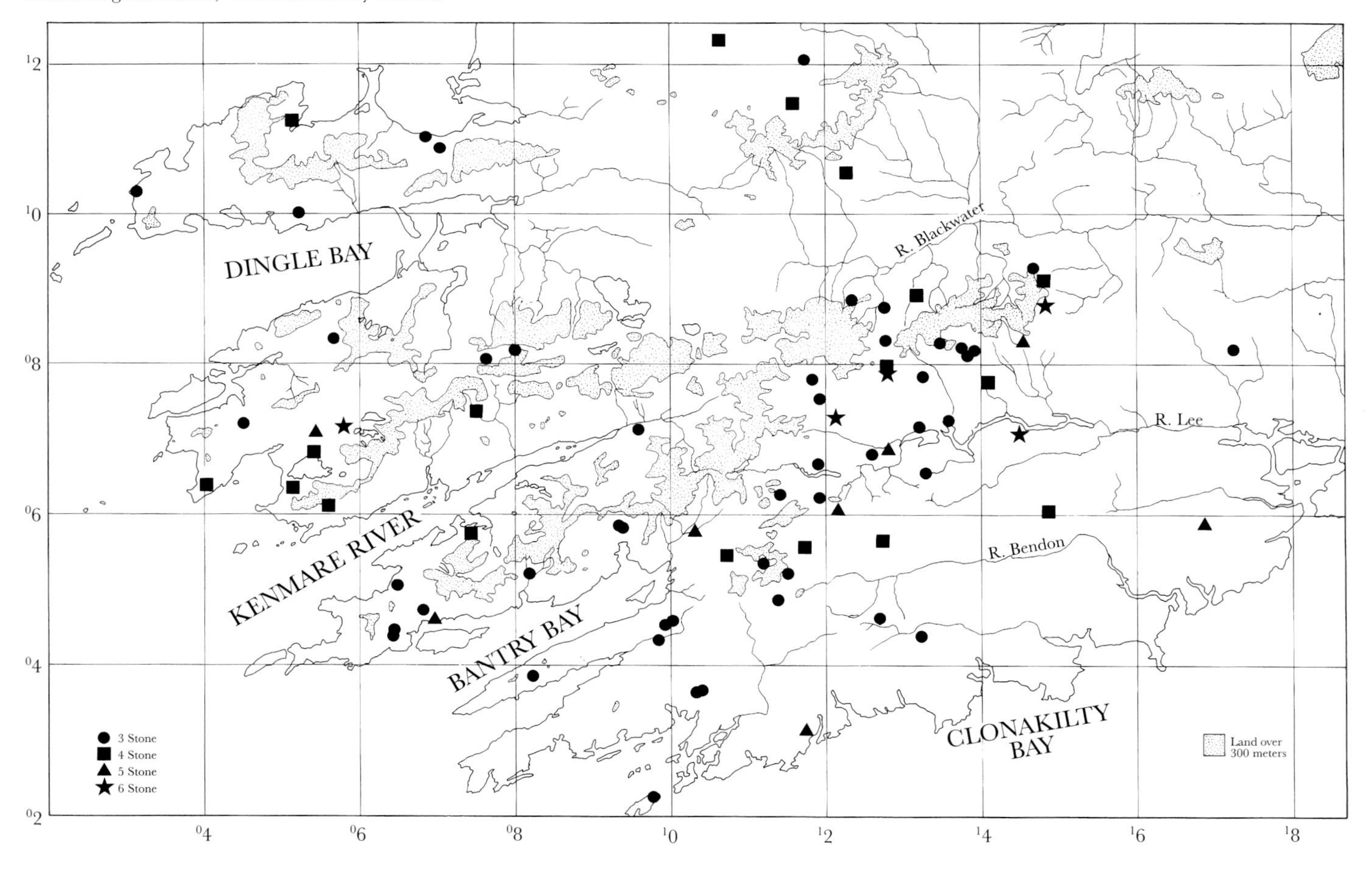

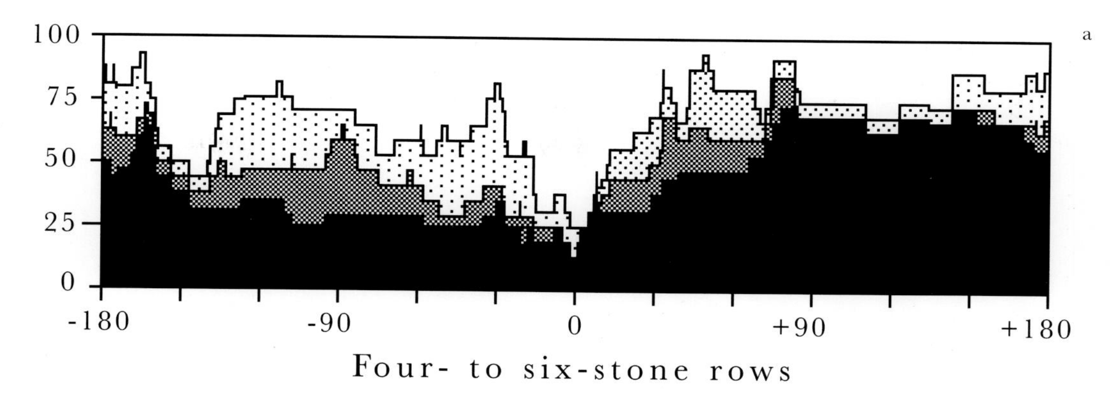


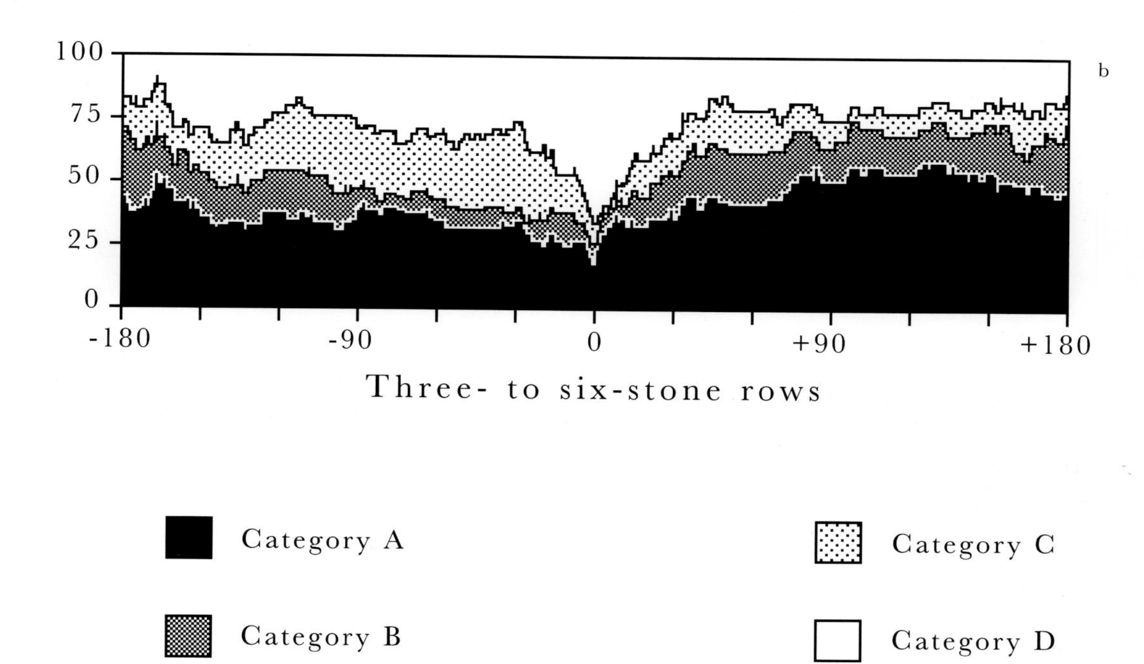


Fig. 6.3 Orientation of visibility. The frequency of occurrence of different horizon distances in various directions relative to the apparent direction of indication in Cork–Kerry stone rows. For the significance of the horizon categories and shading, see Fig. 5.4. (a) shows the data from four- to six-stone rows, after Ruggles 1994a, fig. 2b. (b) shows the data including three-stone rows (after Ruggles 1996, fig. 2b).

ordinary; there are no large and suggestive fluctuations in the proportions of category A, B, C, and D horizons at different relative azimuths.[32]

But surely the assumption that either the NE or the SW was consistently the direction of interest is questionable, in view of the data from the stone height gradations? Fig. 6.3 shows what results if we plot the proportions of the different horizon distances for azimuths relative to the apparent direction of indication—i.e. south-westwards in some cases and north-eastwards in others—where this information is available. The plots, both for the longer rows and for all the data together, reveal a sharp increase in the proportion of distant horizons, and a corresponding decrease in nearby ones, in a relatively narrow band centred upon the apparent direction of indication.[33] There are two possible explanations why the avoidance of nearby horizons in the apparent direction of indication is not total. The first is that what is being revealed is a trend rather than a universal custom; the second is that we have made some mistakes in identifying the apparent direction of indication. However, unless we can think of convincing independent indicators of the direction of indication in order to provide corroboration, then the latter is unprovable on this evidence alone. To suggest otherwise would certainly be to descend into a circular argument.[34]

While the Cork–Kerry rows had inevitably attracted some astronomical attention in the past,[35] the only substantial investigation of their astronomical potential prior to the work described here was undertaken by Ann Lynch in the 1970s.[36] Her conclusions are still quoted in the archaeological literature: a 1986 inventory of the Dingle peninsula mentions in its introduction that 'the alignments at Ardamore [CKR8] and Cloonsharragh [CKR4] were oriented respectively on the setting sun at the winter solstice and the rising sun at the summer solstice'[37] and an inventory of South Kerry published in 1996 notes the solstitial alignment of Dromteewakeen (CKR28) and Eightercua (CKR54) as well as the alignment of Doory (CKR46) and Kildreelig (CKR55) upon the major lunar limit.[38] Lynch's approach involved defining mean row orientations with great care and quoting declinations in both directions along those orientations to a precision of 0°·1.[39] From a probability analysis of the data obtained from thirty-seven rows she concluded that significantly many astronomical targets were indicated to a precision of somewhat under 2°. However, the targets were a more-or-less equal mix of solar and lunar events, and she expressed concern about 'the diverse events indicated within such a homogenous group of sites'.[40]

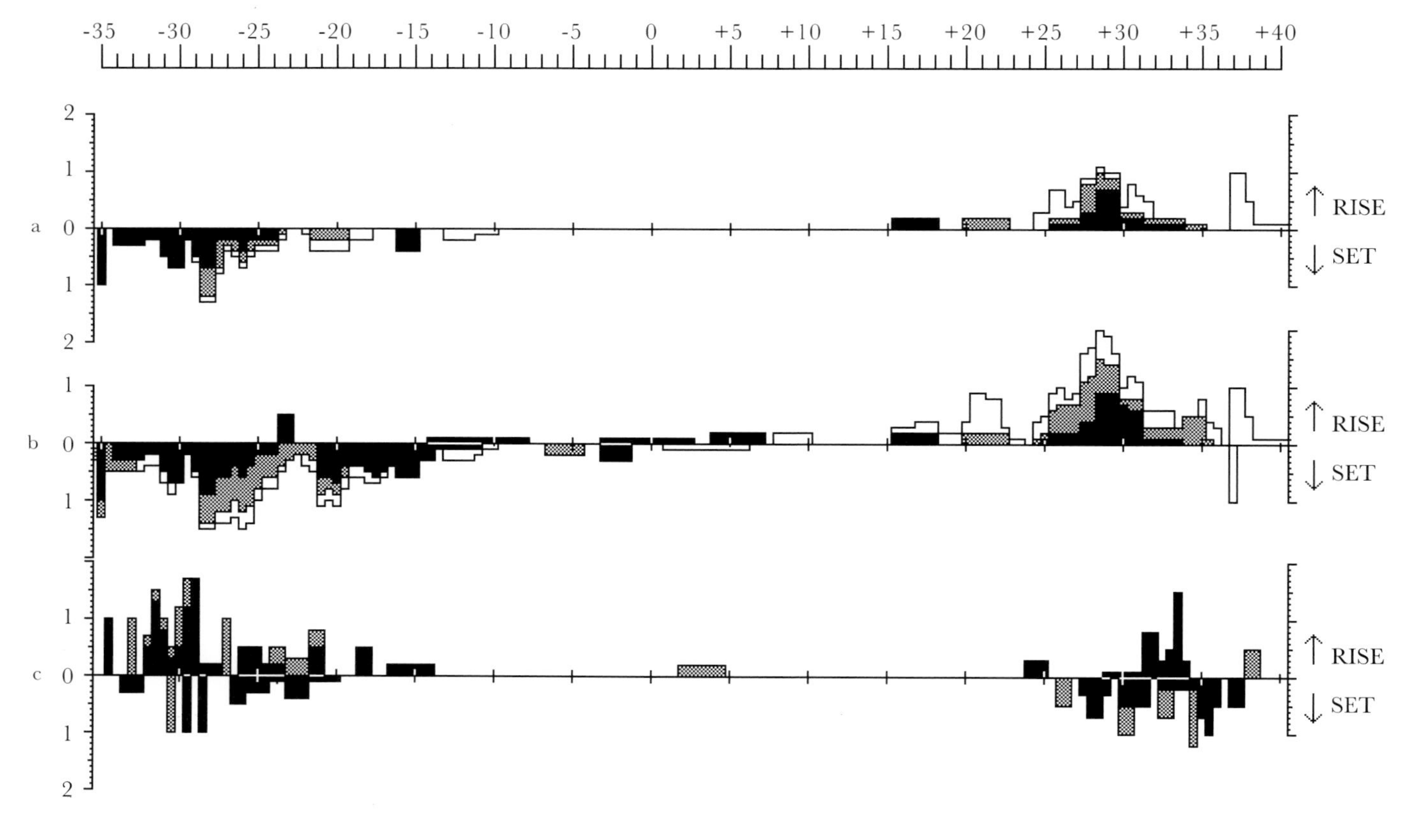


Fig. 6.4 Indicated declinations at (a) Cork–Kerry rows with four or more stones; (b) all Cork–Kerry rows surveyed; (c) western Scottish rows. Each range is plotted with equal weighting assigned to all declinations between the limits given in Table 6.1 or Table 6.3, the area under each being 1 unit. In order to display the data in a uniform way, the Cork–Kerry data have been rounded to the nearest 0°·5. In (a) and (b), dark shading denotes declinations in the apparent direction of indication and no shading denotes those in the opposite direction. Lighter shading is used where the apparent direction of indication was unidentified. In (c), the direction of indication is undifferentiated but data from aligned pairs are included, shown by lighter shading. Not shown in (b) is one indication between +48° and +50° (apparent direction of indication unidentified).

One of the drawbacks of Lynch's approach was that it was limited to quoting the precise declinations of alignments in the exact direction of orientation as defined by some predetermined means from the present-day placement of the stones.[41] Lynch's data were ignored prior to the fieldwork described here, but a careful comparison was undertaken afterwards. In the new approach, which followed that of the western Scottish project described in chapter three, an 'indicated horizon range' was defined, centred upon the mean orientation of the row, but stretching for a number of degrees on either side to reflect both the sinuous disposition of the stones[42] and uncertainties in how closely their present positions might reflect the original, intended orientation.[43] A total of seventy-nine horizon profiles were surveyed at forty-eight rows, the remaining seventeen being omitted owing to their proximity or because of visibility problems.[44] Some quite large discrepancies were found between the two sets of results where they overlap, and these are detailed in Table 6.1 along with the new data.[45]

The indicated declinations are plotted in Fig. 6.4. For each indication, equal weighting is assigned to all declinations between the limiting ones given in Table 6.1, the total weighting being the same in each case.[46] Darker shading is used for declinations in the apparent direction of indication. Fig. 6.4a shows the data from the four- to six-stone rows only; Fig. 6.4b includes the three-stone rows. Amongst the more restricted data, all the southern declinations represent setting objects and all northern declinations rising ones, reflecting the NE–SW orientation of the monuments. The three-stone row data introduce some exceptions, including the wholly anomalous row at Kippagh (CKR11), whose mean orientation is 356°/176° and which is aligned in the south upon the rising sun close to the winter solstice as it just scrapes above a horizon whose altitude exceeds 14°. In the main, though, the south-westerly alignments seem to fall within two broad clusters falling around but largely above the lunar standstill limits at −30° and −20°, suggesting some sort of relationship to the southern moon. The north-easterly data show quite a strong grouping around the northern major limit at +28°, although it is centred rather uncomfortably high, beyond where the moon can ever reach; and there is only a relatively small and much less convincing grouping around the minor limit at +18°.[47] Certainly no correlation is evident with solar horizon events.

Before taking the astronomical investigation any further we must give some attention to conspicuous hills, for there seems little doubt that these were of importance to the builders of many of the Cork–Kerry rows. A number point directly at the highest or most distant summit visible in a wide stretch of flatter or closer horizon, some of the best examples being Eightercua (CKR54) at Reeneragh, Reananerree (CKR40) at Douce Mountain, Gortnagulla (CKR19) at Benlee, Cloonshear Beg (CKR50) at the hill at W 151601, Cabragh North (CKR30) at Musherabeg, Tullig (CKR14) at Mulliganish, and Farrannahineeny (CKR61) at Nowen Hill. Some of these are shown in Fig. 6.5. The row at Garrough

(CKR59) is aligned upon Two-Headed Island, an isolated islet in Darrynane Bay which appears just below the sea horizon. For the purposes of the project, conspicuous hilltops were identified by applying criteria related to horizon altitude variations within the indicated range.[48] Under these criteria—and including the sw profile at Garrough, where the point on the sea horizon directly above the highest point on Two-Headed Island is very clearly defined—no fewer than thirty-nine of the seventy-nine horizon profiles surveyed contain conspicuous hilltops.[49] For the data in full see Table 6.2.

There seems to be a strong correlation between the presence of conspicuous hills and the directionality of a row as determined from the stone height gradation. In all, at twenty-three out of the forty rows for which an apparent direction of indication could be determined, a hilltop occurs in that direction but not in the opposite one; the reverse is true in only two cases.[50] Of the thirty-nine surveyed profiles containing hilltops, twenty-six of them are in the apparent direction of indication. This seems to provide confirmatory evidence that the alignments upon hills were intentional, and that height gradation was often used to emphasize the direction concerned, although once again it appears to be a case of a trend rather than any sort of universal custom.

Fig. 6.5 A stone row aligned upon a conspicuous hill summit: Gortnagulla (CKR19), aligned to the south-west upon Benlee, 11 km away.

On the premise that where prominent hills do occur they may give a better idea of the intended direction of indication than the present positions of the stones, it is of interest to examine the declinations of the hilltops themselves. The relevant data are listed in Table 6.2 and presented in Fig. 6.6, with darker shading indicating hilltops in the apparent direction of indication.[51] Amongst these, all but three of the south-westerly declinations fall between $-30°$ and $-19°$, while the north-easterly ones concentrate in a tight group between $+26°$ and $+30°$. The pattern is even more pronounced amongst the longer rows alone, with the larger sample introducing such conspicuous outliers as Curragh More (CKR25) and Cashelkeelty (CKR66),[52] whose apparent directions of indication to the ENE are aligned upon hilltops with declinations of $+7°$ and $+6°$, quite at odds with the general pattern.

On the whole, though, these data do seem to strengthen the general idea of a relationship with the moon, and they certainly confirm the lack of any apparent correlation with the sun; but there does not seem to be any simple interpretation. If, say, we argue that the southern indications are in some way related to the setting full moon close to midsummer, then we are left trying to explain why some rows were apparently aligned instead upon the rising full moon close to midwinter, while the setting midwinter moon and the rising midsummer moon were apparently ignored. It would be easier if the evidence pointed to solar alignments in one direction and lunar in the other; then we could suggest that there was a relationship with both, the monument being associated perhaps with a ceremony enacted on the occasion of the rising or setting sun close to midsummer or midwinter and the simultaneously setting or rising full moon.[53] But it does not.

To summarise: two things suggest strongly that the direction along the alignment was of importance, at least at the majority of the Cork–Kerry stone rows. First, there is a high proportion of distant horizons in the apparent direction of indication, as inferred from the height gradation of the stones; and second, most of the conspicuous hilltops that are found on the horizon in one or other direction along a row are also in the apparent direction of indication. However, while these three characteristics—stone height gradation, distant horizon, and the presence of a conspicuous hilltop—are clearly correlated with one another, they are not correlated at all with either compass direction, NE or SW. There is a fairly even split between rows where all these factors seem to indicate an interest in the alignment to the SW and those where the interest appears to have been in the NE. And as far as the astronomy is concerned, we can tacitly conclude that the moon was involved, but the details remain obscure.

## Reinterpreting the Scottish rows

It is of great interest, then, to see if the Scottish rows share similar properties. In doing so we shall try to extract the relevant data from fieldwork and analyses that have typically been undertaken with rather different objectives. List 6 is a list of Scottish short stone rows compiled from various published lists backed up by data from the National Monuments Record of

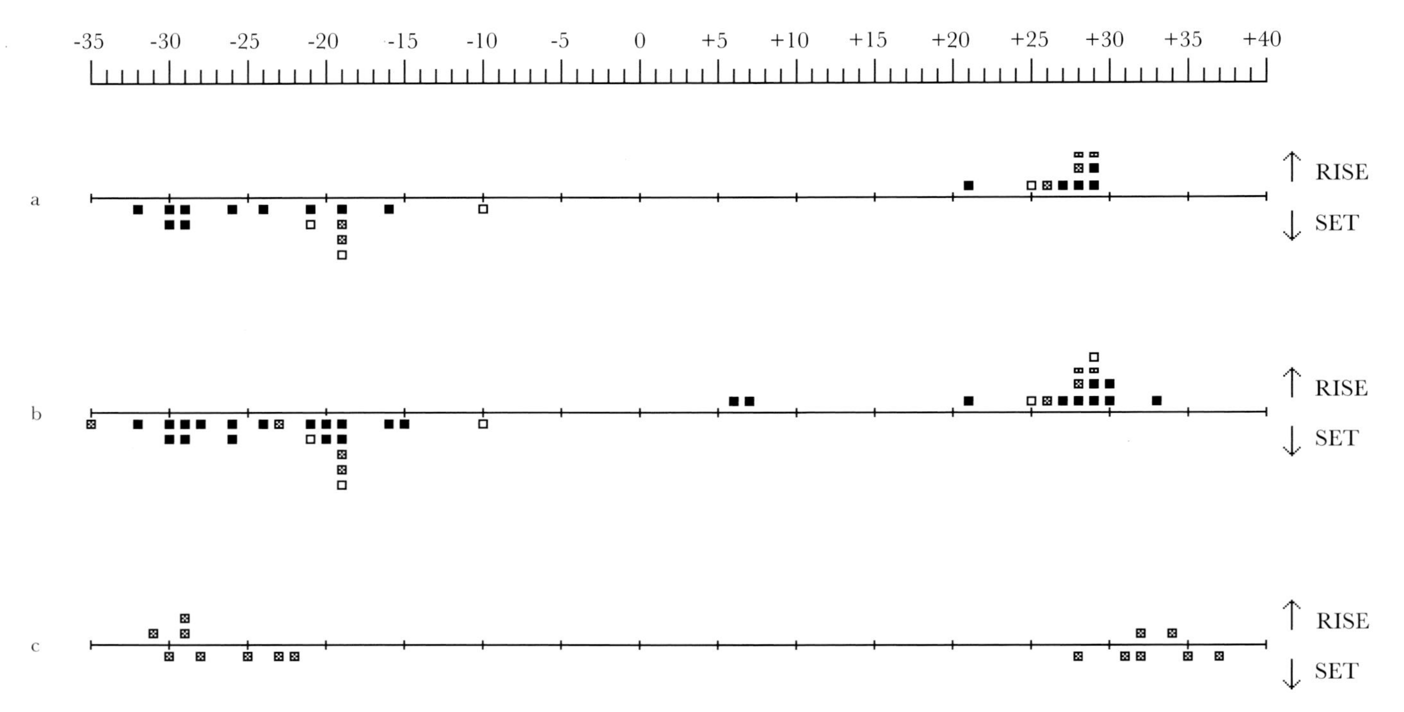


Fig. 6.6 Declinations of hilltops within the indicated horizons at (a) Cork–Kerry rows with four or more stones; (b) all Cork–Kerry rows surveyed; (c) western Scottish rows. Each square represents a single hilltop, plotted to the nearest degree. Where two hilltops of equal altitude occur in the same range, half squares are used. In (a) and (b), dark shading denotes declinations in the apparent direction of indication and no shading denotes those in the opposite direction. Lighter shading is used where the apparent direction of indication was unidentified. In (c), the direction of indication is undifferentiated but data from aligned pairs are included, shown by squares with a diagonal line. Not shown in (b) is one hilltop at declination +50° (apparent direction of indication unidentified). Part (a) follows Ruggles 1994a, fig. 4 and part (b) Ruggles 1996, fig. 4.

Scotland (NMRS). The majority of the monuments fall within the areas covered by the 'megalithic astronomy' project published in 1984,[54] and so there is a considerable overlap with List 2, but the essential data on length and number of stones are repeated here so that they can be examined in a form compatible with the Cork–Kerry rows in List 5. As in south-west Ireland, a length limit of 25 m has been applied. This has resulted in the exclusion of several monuments which appear in other lists of stone rows, including Burl's stone rows gazetteer.[55] Also excluded are some stone settings familiar from the earlier debates about megalithic astronomy, most notably the Nether Largie stones close to Temple Wood (AR13(b) in List 2). Finally, the five radial rows at Callanish (LH16 in List 2) have been excluded because they clearly form part of a more complex setting of standing stones rather than a simple linear monument.[56]

The 1984 project identified a number of pairs of broad slabs in alignment.[57] These 'aligned pairs' have been included as supplementary data (see Supp. A to List 6) but such data are only available for areas considered in the 1984 project. Outside those areas and without visiting the monuments concerned, it is not possible reliably to distinguish aligned pairs from non-aligned ones according to the same criteria, and no attempt has been made to do so.

Although List 6 covers all of Scotland, the short stone rows for which appropriate data are readily available are concentrated in mainland Argyll and the Inner Hebrides,[58] and for this reason we shall restrict attention to these 'western Scottish' examples in what follows. There is certainly an obvious difference in orientation pattern between the Cork–Kerry and the western Scottish rows. The latter are clustered around north-south rather than NE–SW (cf. Fig. 3.5b). It can be seen from columns 2 and 3 of Table 6.3 that all twenty measured examples fall between 302°/122° and 48°/228°, with all but six falling within 30° of true north-south (but, interestingly, only two examples within 10°). This trend extends to the eight aligned pairs, with the single exception of Lagavulin N (AP7) which is aligned almost east-west. The trend is even quite clear amongst single standing stones.[59] Without looking more deeply we might well conclude that a concern with dividing the horizon into four parts centred upon the cardinal points, and a desire to ensure that the rows' orientations fell within the north-south sectors, provide a simple and perfectly satisfactory explanation of the orientations.

Since most of the western Scottish rows were surveyed in the course of earlier projects with different goals, data on the variation of horizon distance with azimuth all round the horizon were not collected, and no attempt was made to identify a preferred direction of indication from stone height gradation. We can, however, examine two factors that might suggest whether there was an overall preference for either the northerly or the southerly indication. One is the relative distance of the horizon in the two directions along each row; the other is the presence of conspicuous hilltops in the indicated direction. In fact, the more distant horizons are more or less evenly split between the two possibilities,[60] as are the conspicuous hilltops.[61] The tentative conclusion so far must be, as with the Cork-Kerry data, that if the direction along the alignment was of significance at all to the builders, then there was no general preference for either the north or the south. It remains to be seen whether stone height gradation data will clarify the issue.

As far as the astronomy is concerned, the indicated declinations are listed in Table 6.3 and plotted in Fig. 6.4c.[62] The results as they stand are not highly impressive. As regards the southern declinations, the most that can be said is that (with a couple of exceptions) there is a cut-off above about −20°, which could reflect a desire to orient the rows within the range of the midsummer full moon low in the southern sky, as with the RSCs to the east. Despite the fact that the highest peaks in the graph occur at around −30°, there is no apparent correlation with the actual rising and setting points of the midsummer full moon at arbitrary points in the lunar node cycle, which should result in a distribution peaking around −30° and −20° and falling off in between. In any case, the near-equal spread of indications to the east and west of south seems to weaken any argument that the builders were concerned with specific rising or setting events of a particular astronomical body. The data from conspicuous hilltops (Table 6.4, Fig. 6.6c)[63] seem to do little to clarify the picture.

When we look to the north, however, we see a cut-off below +24°, which immediately suggests a solar, rather than a lunar, explanation. Could it be that the builders were doing no more than ensuring that the rows were oriented in the north towards the part of the sky between summer solstice sunrise and sunset—the part of the horizon where the sun never passes?[64] Perhaps they conceptually quartered the sky, the solstitial sunrise and sunset directions demarcating the four regions, something that is well known amongst indigenous peoples around the world (see chapter nine), and it was important to have the rows aligned upon the northern quarter. Another and rather different possible reason is that this would ensure that the sun always rose on one side of the row and set on the other, something that could conceivably have been of enormous symbolic significance. The problem with either of these explanations is that they do not fit the southern orientations as well, which in some cases fall well within the solar solstitial limit.[65] Once again, a simple solar explanation seems to fail.

The southern indications were investigated quite extensively in the mid-1980s in order to build upon the initial results from the 1984 project (see chapter three). The study area was confined to mainland Argyll and Mull, and then more intensively to the Kilmartin Valley area and northern Mull, which clarifies the southern declination pattern a good deal by removing several of the outlying values, including the three below −32° and the two above −19°.[66] On the other hand the two longer, NNE–SSW alignments at Nether Largie ('Kilmartin'), were included.

The main conclusions are best illustrated by plotting the southern indications of the rows and aligned pairs in the way it was originally done (see Fig. 6.7).[67] Where a row or aligned pair is found in isolation, it yields a declination very close to (say, within two degrees of) −30°. This is true in eleven cases, although the two declinations one or two degrees below −30° are from monuments in poor condition and might be disregarded. If this is done, then all the alignments are extremely close to moonrise or set at the major standstill limit, and where they are not they are slightly higher, as would be expected if the moon had not in fact been quite at the limit when the alignment was set up. Those rows and pairs that do not yield declinations around −30°, but fall instead within a wider spread between about −26° and −21°, are invariably found close to others that do.[68]

This 'primary–secondary' pattern of orientation does suggest that there was a real interest in those extreme directions where the rising or setting moon was only seen very rarely, perhaps just on two or three occasions during a few consecutive summers in any one generation.[69] Should we conclude that so much significance was attached to them that people watched, perhaps for several years, until the annual sweep of the midsummer full moon reached its maximum? Perhaps the purpose of the monument related to ceremonials to prevent it going further and disappearing completely? Such fanciful speculations are unproductive, but what is significant about this conclusion is that it seems to imply, unlike the cup marks at the RSCs, that people really were aware of the annual changes in the moon's motions, and took account of them, perhaps waiting for the best part of a lifetime before erecting one of these monuments.

The idea of preplanning over several years is an uncomfortable one, though. An alternative possibility that does not carry this implication is that the rows were simply built to align upon the midsummer full moon, but where the moon was seen to rise or set significantly further to the south in subsequent years, then a second monument was erected nearby. This would certainly imply that the monuments, once built, continued to be used, rather than the act of building them being sufficient in itself; otherwise the changes in the moon's motions would presumably not be noticed. The idea might also explain why at sites such as Ballymeanoch (SSR30/AP5) it is the longer row

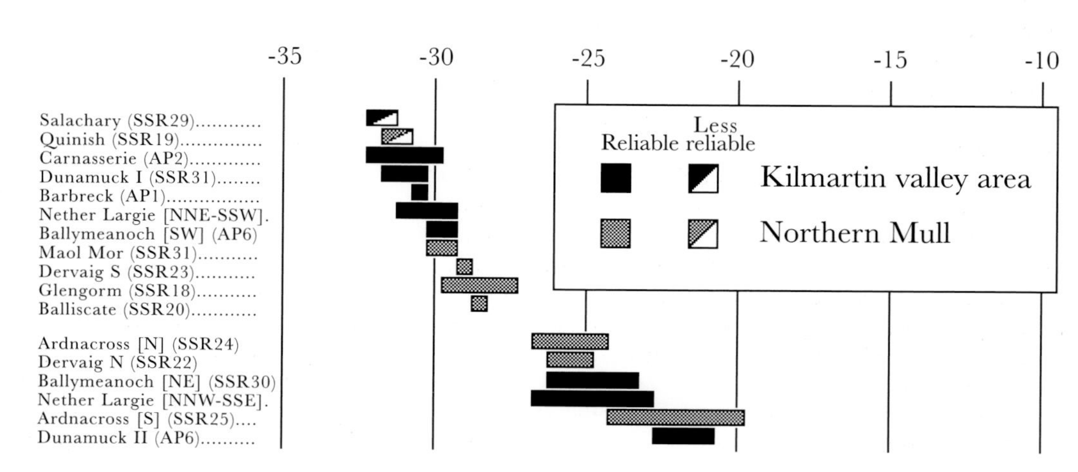


Fig. 6.7 A plot of the southern indications at stone rows and aligned pairs in Argyll and Mull, in the style of Ruggles 1985, fig. 3 (lower) and Ruggles 1988b, fig. 9.2, but with updated data. 'Reliability' refers here to the state of repair of the monument rather than the status of the survey (see Ruggles 1985, fig. 3 caption).

that is in the 'secondary' direction, while the 'primary', major standstill limit, direction is marked only by an aligned pair.[70]

But there are problems. First, the five large stones at Nether Largie (Fig. 2.11) only fit the 'primary–secondary' pattern under the assumption that the longer NNE–SSW alignment marked the primary direction while the NNW–SSE orientation of the individual aligned pairs and isolated slab *F* marked the secondary one.[71] While this may be a very elegant explanation of the layout of the site it is difficult to see how it could have come about in practice unless the entire complex was preplanned. Second, unlike the RSCs, there is no independent evidence that the southerly direction was really the direction of interest. Third, the mixture of rising and setting indications makes one a little uneasy; if ritual or custom made such clear demands about the orientation of these monuments upon the extreme southerly moon, why was it apparently left open to the builders to decide which? And fourth, we can never know whether a row that now stands alone originally stood in the vicinity of another, since destroyed; if so, there is no reason why the primary rather than the secondary orientation should be the one that remains.

The double three-stone row at Ardnacross (SSR24/25) seemed to provide a good way of testing the 'primary–secondary' idea, as the orientations here both seemed to be in the secondary direction, providing an isolated, but very clear, exception to the rule. However, only one stone stood at the site and the row orientations estimated from the positions of the five fallen stones were in considerable doubt. The 'primary–secondary' theory demanded that one of them must originally have been in the primary orientation, something that could be tested by excavation. When the excavation was carried out some years later as part of the North Mull project (see chapter seven), it merely confirmed, more or less, the orientations that had already been proposed.[72] Interestingly, while this deals a further blow to the 'primary–secondary' theory it is not actually inconsistent with the idea that if the moon was seen to rise or set significantly further to the south in subsequent years another row was built. Here, because of the high southern horizon, the moon completely fails to appear when it reaches its major standstill limit. The ridge 2 km away is so broad that it is not possible to overcome this difficulty by building the monument anywhere in the immediate proximity. Perhaps, then, even after two attempts, it simply proved impossible to build a third alignment to mark southernmost moonrise and set.

The picture, then, is a mixed one. The orientation trend amongst the western Scottish rows is very clear and—unlike the skewed distribution of the Cork–Kerry rows—invites the simple explanation that the orientations were merely confined within the bounds of a conceptual quartering of the horizon centred upon the cardinal directions. There is no clear evidence of a preference for distant horizons or for conspicuous hilltops in either direction along the row, certainly not consistently to the north or south. Amongst the two main concentrations of rows in the Kilmartin valley area and northern Mull, there is evidence, albeit rather equivocal, of southerly lunar orientations of some subtlety. There is no necessity, however, to postulate long periods of preplanning. A more straightforward explanation, that fits the data equally well, is that rows were constructed at various times within the 19-year node cycle, but, in those cases where the moon continued to appear further to the south, 'corrected' by the later construction of further rows.

## Linearity and linkages: some general conclusions

Having concluded the previous section with such detailed interpretations of the stone rows in mid-Argyll and Mull it is necessary to draw back to look at the wider picture. We have already noted that removing similar monuments located slightly further afield—Jura, Islay and even southern Mull—from the original dataset achieved a major clarification, because these rows had outlying declination values. There is every reason to suspect that the more widely scattered rows that are found, for example, elsewhere in Scotland will provide very much less in the way of evidence of common practice with regard to the more esoteric elements of material tradition such as orientation. Perhaps we should scarcely be surprised, for example, that there appears to be little uniformity in the orientation of the four surviving three-stone rows in south-east Perth.[73]

In this context, the question of degrees of relationship between the Cork–Kerry rows and the western Scottish ones, over and above their obvious similarity in form, seems a good deal more remote than it did in the case of the ASCs and RSCs. In addition, as with the ASCs and RSCs, the Cork–Kerry and western Scottish stone rows present us with two groups of superficially similar monuments that have rather different properties when looked at in more detail. Not only are their overall orientation patterns different (if equally well defined); the evidence of orientation upon conspicuous hilltops, so clear in Cork and Kerry, is rather more equivocal in the case of the western Scottish rows, and there is no evidence of a preference for more distant horizons either way along the alignment, although taking into account the directionality implied by stone height gradations may reveal some correlations. If this happens, this will fly in the face of the astronomical evidence which seems to draw attention exclusively to the southward indications.

Yet there is evident continuity of material tradition between southern Ireland and western and central Scotland. Unlike the RSCs and ASCs, the main groupings of short stone rows are not isolated, and indeed there is a significant grouping right between them in mid-Ulster.[74] Burl has argued that all of them shared a common ancestry in longer linear monuments.[75] The concentrations of stone pairs[76] and the presence of four-posters[77] in both Perthshire and in Cork and Kerry reinforces the idea of some form of linkage, direct or indirect, between the two areas. A difference between Scotland and south-west Ireland is that in the latter the short stone rows and stone circles are found in close association and evidently are closely related elements of a single tradition.[78]

It is perhaps reassuring, then, that there is some evidence to suggest that, whatever their other differences, both the Cork–Kerry and the Argyll–Mull short stone rows may have been linked into the cosmos through recurrent performance related to the moon. But where do we go from here? There remain the archaeoastronomically unstudied northern Irish rows, but there are also other ways to proceed. In order to progress our understanding of the symbolism incorporated in these monuments it may be crucial to consider their properties away from the direction of orientation. An idea that immediately suggests itself when the visitor sees monuments such as Cloghboula More (CKR13) and Rossnakilla (CKR53)—where the horizon in both directions along the row is close but there are wide views with prominent hills across the alignment—is that the directions perpendicular to the row might have been important. Such an interpretation also seems eminently plausible at Gneeves (CKR9), Kippagh (CKR11) and a number of other cases, including some in Scotland. The challenge is to derive from such speculations controlled sets of ideas susceptible to testing by looking at the material record.

This and other goals need to be met as part of the shift towards more contextual studies of the possible patterns of thought and symbolism that helped to define the location and orientation of these intriguing monuments within the Bronze Age sacred landscape and skyscape. The next chapter describes a project that was specifically designed in an attempt to rise to that challenge.

# 7

# Astronomy and Sacred Geography

*The North Mull Project*

The Isle of Mugg . . . lies among other monosyllabic protuberances. There is seldom clear weather in those waters, but on certain rare occasions Mugg has been descried from the island of Rum in the form of two cones. The crofters of Muck know it as a single misty lump on their horizon. It has never been seen from Eigg.

Evelyn Waugh, *Officers and Gentlemen*, 1955

The real problem of British archaeoastronomy [is] one of methodology. The nature of the problem is such that different methodologies give different results. It is unlikely that we have yet found the optimum methodology which, although free from subjective bias, yet cannot accidentally discriminate against a type of alignment favoured by the megalith builders.

Ray Norris, 1988[1]

An archaeology of place can never provide the guarantees of verification which processual approaches seem to demand if it interprets beyond the evidence, but it cannot provide a satisfactory understanding of the past if it does not.

Julian Thomas, 1996[2]

## From alignment to landscape: a different approach

In the last chapter we attempted to identify some common elements of the ritual traditions that dictated why, where and how the Scottish and Irish stone rows were constructed, focusing specifically upon the possibility that they were aligned upon something: astronomical events, perhaps, or prominent features in the landscape, or both. However, it would be unwise to concentrate exclusively on surveys of the horizon along the alignment undertaken at the individual monuments, even though this approach does seem to give some positive results. As with stone circles, there are many other ways in which stone rows may have served to express and reaffirm symbolically certain things about the world perceived as important by the people who built and used them. Investigating a wider range of possibilities requires a rather different approach.

For a start, we need to consider whether the standing stones that remain so conspicuous today really did represent the single most important architectural structures at the site as it would have been seen and understood at the time of its use. Perhaps there were additional wooden structures, platforms or paths, or stones laid flat that are now hidden beneath the ground?[3] Were the monuments modified over time? Did whatever ceremonies or other activities that took place there leave any trace in the form of deposits—ritual offerings, perhaps, or cremated remains—and if so are there obvious patterns to their spatial distribution, implying that different symbolic significance was attached, perhaps consistently, to different places within and around the monument? The only way to make real headway with such questions is through excavation.[4] This can also yield absolute dates and a variety of evidence on related human subsistence activities as well as the natural environment (cf. Archaeology Box 5). Of particular interest in the context of suspected relationships between monuments and distant features in the landscape, or celestial bodies close to the horizon, is the nature of vegetation in the vicinity; what was the likelihood that tree cover prevented a particular distant mountain or horizon profile being directly visible? This is a question that pollen cores may well help to answer.

We should also look more broadly at aspects of the rows' design (including their orientation) and their location. The monuments did not exist in isolation but were built and used in a landscape that was lived in and exploited. What environmental, ecological or topographic factors influenced where a stone row was placed? And in what social context did the monuments operate? Did each belong to a small group of people living in a fixed settlement, occupying their own small territory, or were the rows built within communal lands and accessed, perhaps seasonally, by numbers of wandering animal herders? We can only start to answer many of these questions by moving away from purely site-based investigations towards thinking about prehistoric landscapes as a whole, a shift in emphasis that is very much in tune with wider developments of thought in archaeology (Archaeology Box 6).

Traditionally, fieldwork used to be seen almost exclusively in terms of the discovery and excavation of sites. Today, however, while sites and their excavation remain of paramount importance, the focus has broadened to take in

> whole landscapes, and surface survey at sites in addition to—or instead of—excavation. Archaeologists have become aware that there is a great range of 'off-site' or 'non-site' evidence, from scatters of artifacts to features such as plowmarks and field boundaries, that provides important information about human exploitation of the environment. The study of entire landscapes by regional survey is thus now a major part of archaeological fieldwork.[5]

Then again, if a monument's relationship to the surrounding landscape was symbolically important, we might learn a good deal not just by studying where it was placed and how it was oriented, but also where it was not located and what potential alignments evidently went unnoticed or else were deliberately eschewed. In particular, it might be valuable to examine the whole of the landscape and horizon surrounding a monument for prominent natural features and celestial events of potential importance, since many different relationships perceived as significant may have helped to define the place even though they were not subsequently reflected in the orientation. We should also scrutinise the potential (defined in similar terms) of places where (as far as we know) monuments were *not* erected.

The problem, of course, is that this shift in emphasis brings severe challenges from the methodological point of view. Once we come to consider the prospect that significant associations existed not just in the direction of orientation, we encounter once again the problem identified in chapter two of the sheer number of possibilities.[6] In order to suggest that particular relationships might have been significant we must select them from a potentially very large set of options, and if we do this on purely *post hoc* grounds we sink immediately into a circular argument that relegates our conclusions to the realm of pure speculation. How, then, should we proceed?

In order to tackle some of these issues, a project was conceived in the mid-1980s that would concentrate upon a relatively small area and apply a variety of approaches to studying monuments in the landscape in an integrated way. Northern Mull was chosen as the study area,[7] with its short stone rows the main focus of attention. A major motivation was that these monuments seemed to represent one of two concentrations of short stone rows with a clear pattern of lunar orientation, as described in the previous chapter, and this was felt to be a result that needed to be examined in a broader context.[8] The outcome was the North Mull Project, a detailed investigation of the short stone rows of northern Mull in their fuller archaeological context through an integrated programme of excavation, landscape survey, archaeoastronomical investigation, and computer analysis.[9] While the project's aims and objectives and the research design were specified in advance,[10] the workplan was deliberately kept flexible so that it could be modified in the light of new discoveries and interpretations, and in order to permit methodological developments. Fieldwork on the project took place in five consecutive summers, starting in 1987.[11]

The Isle of Mull as a whole was not well suited for prehistoric agriculture, being generally rugged with poor soils. Yet there was certainly human activity there from Mesolithic times onwards. Neolithic axes and flint artefacts have been recovered, and there is one chambered cairn, although there is no evidence of settlements.[12] While the inland hills might have been suitable for animal herding, it was probably not until after about 2000 BC that occupation spread into the coastal lowlands and inland valleys.[13] It appears that northern Mull was never heavily afforested, and even in the Early Neolithic consisted predominantly of scrubland with grasses and ferns, some tall-herb communities and only scattered stands of hazel, birch and alder.[14] One of the earliest results from the North Mull Project was to confirm this in the vicinity of Glengorm (see below).[15] Vegetation, it seems, is unlikely to have been a major problem if it was desirable to see the wider landscape and distant horizon from the stone rows.

When the project commenced, the remains of seven short stone rows were known in North Mull: Quinish, Maol Mor, Dervaig N, Dervaig S, Balliscate, and the double row with associated cairns at Ardnacross (SSR19–25 in List 6).[16] In addition there was an anomalous triangular setting of three standing stones at Glengorm (SSR18), which it was suspected might once have been a stone row also.[17] These monuments are confined to the northern and north-eastern part of northern Mull, forming a restricted distribution among the more general scatter of stone settings and single standing stones (Fig. 7.1a).[18] There was no obvious consistency in the stone row locations, their elevations varying from under 40 m at Glengorm and Quinish, both close to the sea, to above 130 m at Maol Mor, situated on the broad crest of a ridge.[19] While the majority of the row orientations followed the local topography, Balliscate and Ardnacross stood out as conspicuous exceptions.[20]

The two main components of the project fieldwork were the excavation of two rows (together with related environmental and survey work), and all-round horizon surveys and locational analyses, which took place at, and in the vicinity of, all of the North Mull rows.

## Monuments, a mountain, and the moon

Excavations took place at Glengorm in 1987–8 and at Ardnacross in the three subsequent years. The main aim at Glengorm was to determine whether the standing stones there were in fact originally placed in a line, and if so, to establish whether their orientation fitted the lunar pattern observed at the other linear stone settings in the area (see chapter six).[21] The excavation was able to locate the shallow stonehole marking the original position of the southernmost stone (*C*), and showed that it had been dragged some 6 m to its current position in order to be wedged into a convenient crack in the bedrock. The central stone (*B*) was undisturbed, and the northernmost stone (*A*) had been re-erected approximately in its original position (Fig. 7.2).[22] The three stones did indeed originally form a row and, interestingly, the excavation also revealed an isolated small hole, suitable for holding a wooden post, carved in the bedrock in the alignment to the south, some 4·2 m from the original position of stone *C*. While it cannot be linked stratigraphically to the stones themselves, it is tempting to interpret it as part of an earlier timber alignment, or perhaps as a reference marker used in setting up the row.[23]

That two parallel rows of at least three stones had originally stood at Ardnacross was not seriously in question, but only one stone remained standing out of an apparent total of six. A major objective here, as at Glengorm, was to determine

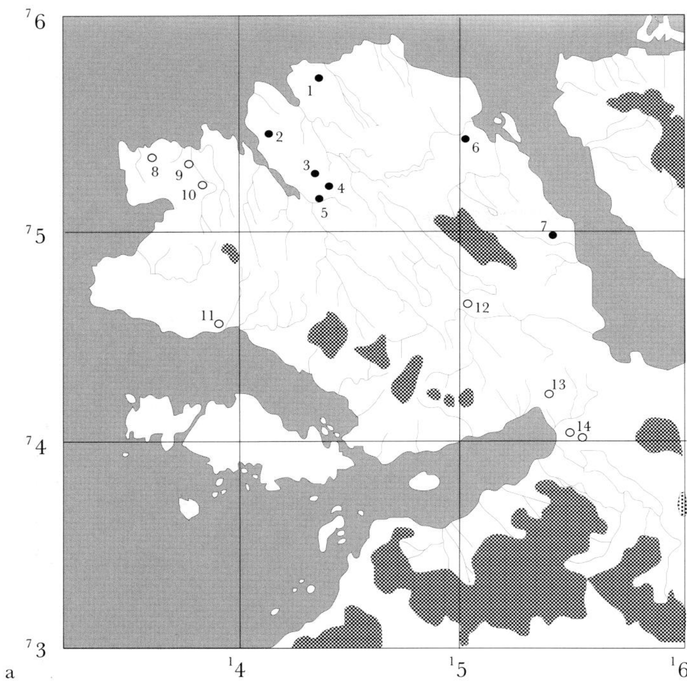


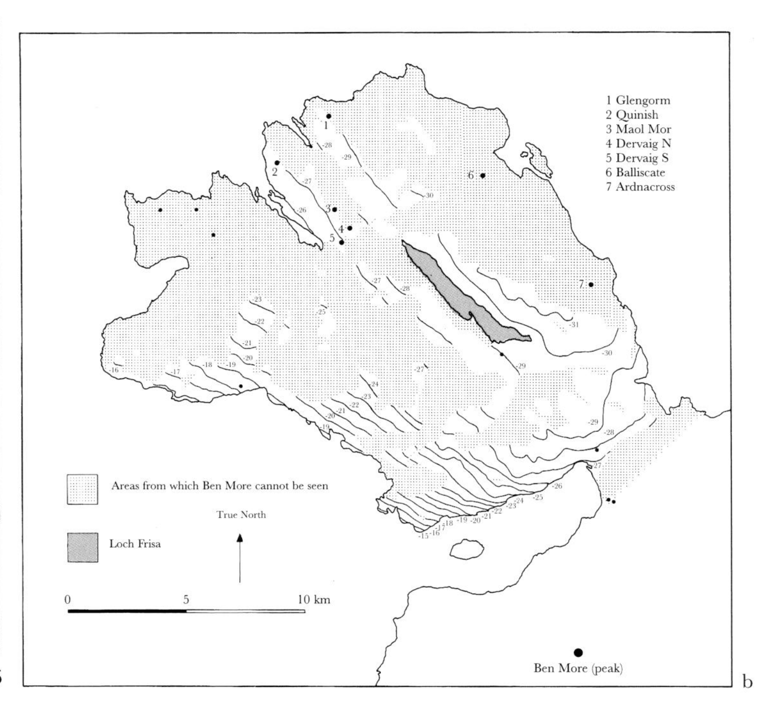


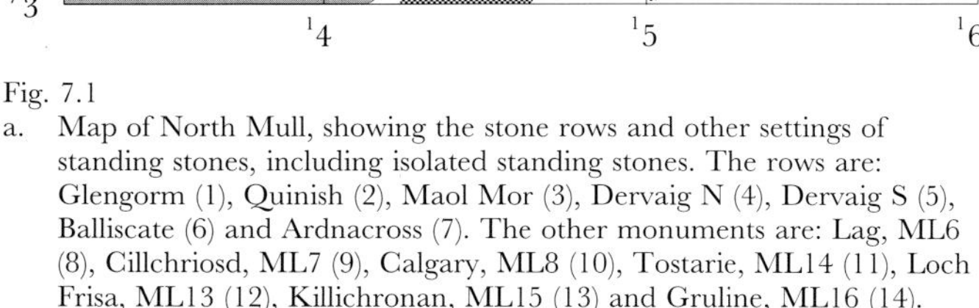

Fig. 7.1

a. Map of North Mull, showing the stone rows and other settings of standing stones, including isolated standing stones. The rows are: Glengorm (1), Quinish (2), Maol Mor (3), Dervaig N (4), Dervaig S (5), Balliscate (6) and Ardnacross (7). The other monuments are: Lag, ML6 (8), Cillchriosd, ML7 (9), Calgary, ML8 (10), Tostarie, ML14 (11), Loch Frisa, ML13 (12), Killichronan, ML15 (13) and Gruline, ML16 (14).

b. Declination contour map of the same area, with the same monuments marked. The shaded areas are those from which Ben More cannot be seen. The contours in the unshaded areas join locations from which the declination of Ben More is a whole number of degrees. First published as Ruggles and Martlew 1992, fig. 14.

the original orientations. Unexpectedly, the Ardnacross excavation revealed that the end stones of both three-stone rows had been deliberately cast down and partially buried in large pits. It seems quite possible that this represented an attempt to despoil the monument and obliterate the original symbolism,[24] although on present evidence it is impossible to be at all specific about when this happened.[25] Unfortunately, this has resulted in the near-total obliteration of the original stoneholes. The central stone of the northern row, on the other hand, appears to have fallen *in situ* and its stonehole was largely intact.[26] The upshot is that the original orientations of the rows appear to have been only a little different from those already deduced from surface features alone, although the margins of error are still quite wide.

At Ardnacross, unlike at Glengorm,[27] it proved possible to establish a reasonable stratigraphic sequence and hence to identify various phases of construction and associated activities. The subsoil into which the stoneholes were cut was found to be extensively scarred by ard-marks (Fig. 7.3),[28] implying that the area had previously been cultivated.[29] The stones themselves were erected between about 1250 and 900 BC,[30] and prior to this the area was cleared by burning off the vegetation, a process that was subsequently repeated, presumably to keep it clear.[31] The largest of three kerb-cairns was built between the stone rows at some time after the burning activity had ceased, perhaps marking a radical change in how the site was being used.[32] In addition, at some time between about 1150 and 800 BC, an enigmatic setting of four small stones was constructed in a pit near the southernmost stone of the northern row.[33]

Fig. 7.2 Excavated features at Glengorm. After Ruggles and Martlew 1989, fig. 3; first published as Martlew and Ruggles 1996, fig. 3.

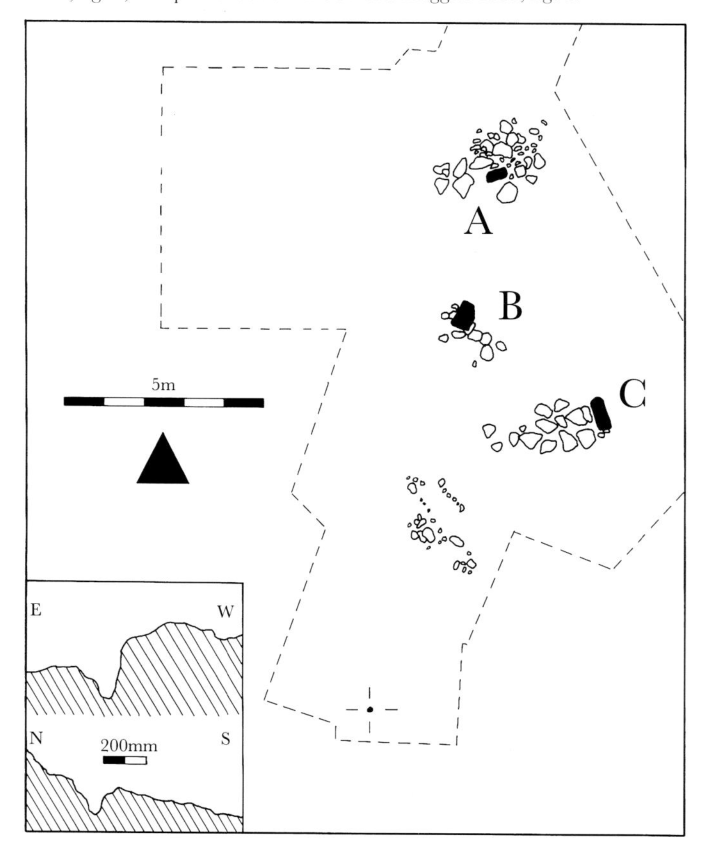

Cultivation continued around the stones into relatively recent times.[34]

At Glengorm the original orientation of the row (azimuth *c.* 156°) was several degrees farther to the east of south than had hitherto been suspected from the orientation of stone *B* alone. As viewed from the site, the southern indication is upon a nearby spur on the east side of Glen Gorm, no more than 2 km distant, yielding a declination of approximately −28°·5.[35] A more interesting discovery is revealed, however, when one walks up rising ground to the north of the stone row. Away on the far southern horizon, exactly in the original alignment, is the isolated and prominent peak of Ben More, the highest mountain on Mull, some 25 km distant (Fig. 7.5).[36] Owing to the forestry plantation that presently covers the intervening spur, it is unclear whether the peak could have been seen from the stones themselves if the spur were clear of vegetation. It is certainly marginal. On the other hand, one only has to move some 10 m westwards, to one side of the row, for the mountain to become clearly visible, and the same distance in the opposite direction for it to be completely hidden.[37]

Intriguingly, it gradually emerged in the course of the associated survey work that each of the other four rows concentrated towards the northern tip of the island—Quinish, Maol Mor, Dervaig N and Dervaig S—also seemed to be placed on the very limit of visibility for Ben More, that is in a position where the peak was clearly visible from some points within a few metres of the stones and totally obscured from others.[38] Up to this point, the row orientations had not provided any clue of a relationship with the mountain. In retrospect, though, it was clear enough that the direction of alignment and the azimuth of Ben More did indeed coincide to within 20° (see Table 7.1); it was just that this was a lower-precision correlation than had hitherto been considered.[39] Furthermore, the fact that in each case the peak of Ben More yields a declination between −29°·0 and −26°·5 was highly suggestive of an association with the southern moon.[40]

Of course it is exciting to make discoveries such as this, and to proceed immediately with possible interpretations, but it is also dangerous because, as we have pointed out earlier in the book, apparently significant patterns arise surprisingly often in random data. We need to be satisfied that the placing of these five rows on the very limits of visibility for Ben More is unlikely to have occurred fortuitously. We also need to assess the probability, if there was a relationship with Ben More, that the locations were deliberately selected with regard to the mountain's astronomical potential.[41]

There are different ways of tackling these questions, but an unavoidable necessity is to derive suitable control data. One approach is to consider the other possible factors that might have influenced where a monument was placed, asking whether favourable areas tend to coincide with those places from which Ben More happens to be visible anyway, and from which it has a declination within a certain range. We might find that we do not have to invoke Ben More or astronomy at all in order to provide a perfectly adequate explanation of the spatial distribution of the monuments.

An alternative approach—which avoids having to consider other possible factors influencing a monument's location but may still provide a useful indication of whether a certain property common to a set of locations is likely to be

Fig. 7.3 Ard-marks in the sub-soil around the central stone of the southern row at Ardnacross.

significant—is to estimate the proportion of places in the landscape as a whole that would share that property. In the present case, we need first to work out from which parts of northern Mull Ben More is visible. We can then estimate the chances that it will turn out to be on the limit of visibility from a location chosen for other reasons entirely (effectively, at random). Second, we need to know its declination from those locations from which it is visible. We can then estimate the probability that the observed declination trend would come about fortuitously if stone rows were sited without regard for the astronomical possibilities.

Following the latter approach for the moment, a good indication is provided by the 'declination contour map' shown in Fig. 7.1b. This was produced in 1991 by a team of Earthwatch volunteers working from Ordnance Survey maps and simple computer programs.[42] It confirms that each of the rows is located on the edge of visibility for Ben More and shows clearly that if five monuments were placed at random in the northern part of Mull, the chances of every one of them being on the limit of visibility (i.e. on the border between the shaded and unshaded areas) is very small indeed.[43] On the other hand, there is little to suggest that, in addition to seeking spots where Ben More was on the limit of visibility, any particular attention was paid to its astronomical potential. The range of available declinations amongst such spots in this northernmost part of

## ARCHAEOLOGY BOX 6

### PEOPLE IN THE LANDSCAPE

#### FROM SITES TO LANDSCAPE

The concept of an archaeological 'site', although central to the discipline as traditionally perceived, is questionable because it is operationally identified by the modern archaeologist. Stone monuments may happen to be readily identifiable and categorised as 'sites' but human activity took place in the landscape as a whole and, while much of the evidence of this will not have survived the test of time, we should strive to understand as much of the wider picture as possible.[1] Many archaeologists now argue that broad analyses of the landscape are likely to be more enlightening than piecemeal investigations of individual 'types' of site.

In recent years 'landscape archaeology' has acquired a distinct identity as the examination of human activity in the wider landscape.[2] This involves a set of field techniques that are complementary to site-based excavation and survey, such as field walking,[3] aerial photography, and geophysical survey. In fact, the main aim of using such techniques is often to identify new 'sites'[4] but they also provide valuable data on the spatial distribution of human activity in relation to the landscape, provided that it is felt possible to make reasonable assumptions about the processes of decay and destruction in the intervening millennia that have served to transform the material remains left at various stages of prehistory into the archaeological record we see today.[5]

Such data inform studies of how the prehistoric landscape was used and exploited, and how and why this changed over time. In the 1970s, a variety of quantitative techniques for recognising and analysing spatial patterning in archaeological data were developed and applied to archaeological problems.[6] An example is site catchment analysis, which can be used to explore the territories (and hence subsistence and economic resources) controlled by neighbouring settlements or the areas of influence surrounding different public monuments, the boundaries between different such areas being modelled, for example, by constructs known as 'Thiessen polygons'.[7]

Processual archaeologists (see Archaeology Box 8) believe that the distribution of archaeological artefacts and features relative to elements of the landscape provides valuable insights into social and economic organisation in the past, insights that are gained by modelling the natural and cultural formation processes that have brought about the distributions we see.[8] But more contextualised, and often more qualitative and subjective, arguments have also gained favour, where human processes in the landscape are considered in a much more interpretative and empirical way, and an attempt is made to take into account prehistoric ideology and cognition rather than relying on the premise that broad patterns of human action are solely determined by ecological factors (see also Archaeology Box 7). At the same time computers have hugely increased the size and complexity of analytical problems that can be tackled. Of great importance in this context are Geographical Information Systems.

#### GEOGRAPHICAL INFORMATION SYSTEMS (GIS)

Geographical Information Systems (GIS) are computer systems for managing databases of spatially-referenced information. They allow the user to organise data of different types relating to the same geographical area (e.g. topography, soil type, roads and settlements, ritual monuments) into separate layers or coverages that can be displayed and manipulated separately or together. By doing so they facilitate the input, storage and retrieval, visualisation and analysis of various types of spatial data.[9]

Fig. 7.4 The GIS viewshed function (after Renfrew and Bahn 1996, 192).

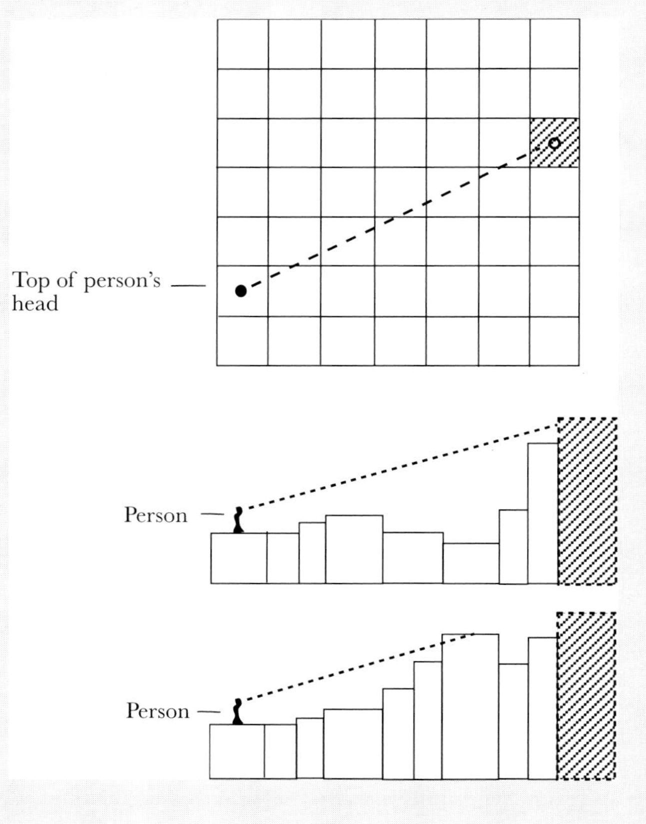

In the 1990s GIS have found a variety of applications in archaeology: as the appropriate database technology for sites and monuments information and archaeological inventories; as tools for visualising and exploring spatial data; and as analytical tools.[10] Initially, GIS were widely used for identifying the probable locations of new sites using predictive models relating human activity to ecological resources.[11] They have also, for example, permitted simple site catchment analysis to be replaced by more subtle techniques taking into account the form of the terrain: techniques such as 'cost benefit analysis' and 'optimum path [or corridor] analysis'.[12] Early uses of GIS were almost exclusively conducted within an implicitly or explicitly systematic paradigm,[13] which is ironic because GIS contain hitherto unavailable tools which, used in appropriate ways, can help inform questions of landscape perception and ideology.[14] An example is the viewshed (or 'line-of-sight') function which calculates the area visible from a given point (see Fig. 7.4) and can help determine whether monuments were situated with reference to (in this case, within sight of) things such as particular natural features in the landscape (perhaps considered sacred?) or existing older monuments.[15] Possible alignments upon horizon astronomical phenomena can also be taken into account as a part of such investigations.[16]

Is a particular mountain really visible from a given place on the ground? The answer given by a GIS viewshed function might be critically dependent upon the accuracy of the digital elevation data used to model the landscape topography and the assumed eye height of the observer.[17] It may be necessary actually to visit the place—to perform 'ground truthing'—in order to provide a definitive answer. In the future it should be possible, given sufficiently accurate topographic data, to generate the sort of horizon profile information used extensively in this book, together with a good deal of other data about the properties of certain places in relation to the surrounding topographic landscape and skyscape, providing the means to visualise a place without the need to visit it with a theodolite or camera. It may also be possible to model other aspects of the personal perception of a place[18] such as surrounding sounds and smells. Of course, the accuracy of any such reconstructions is always limited by the available evidence: the extent of topographic changes since prehistoric times may be uncertain;[19] environmental data only give a broad and uncertain picture of vegetation cover at different times; and evidence of human settlement and activity is patchy (see Archaeology Box 5).

The future should also bring new GIS functionality of direct use within archaeology. For example, probabilistic ('fuzzy') viewsheds could be used to model both the degree to which a distant object, although theoretically visible in perfect conditions, might actually be discernible in average circumstances;[20] or (given relevant information on the past environment) to estimate the likelihood of a sightline being obscured by intervening vegetation. The technological future also holds the prospect of temporal and three-dimensional GIS.[21] But the success or otherwise of GIS as a tool for archaeologists will ultimately be judged by their success in helping to give useful answers to archaeological questions.

Mull is not much greater than the range actually found at the sites chosen.

Interestingly, including on the declination contour map the other six standing stone sites known in northern Mull[44] does draw attention once again to the moon. Three of these are in the Mornish peninsula, from which Ben More is completely invisible, but the others are all within sight of the peak.[45] Tenga and Killichronan yield declinations for Ben More of −29°·1 and −27°·9 respectively, continuing the trend found at the five stone rows in the north. Yet Tenga is situated in the centre of northern Mull, overlooking Loch Frisa, and Killichronan is in the south, near to the narrow neck joining the two halves of the island. This means that the declination trend appears to extend to the other stone settings even though they are separated geographically from the northern group. Only Tostarie, on the west coast, yields a completely different declination: −18°·3. It is, of course, notable that while the other values fall one to three degrees within the major standstill limit of the moon, this value is one degree within the minor limit.[46]

On the other hand, widening the argument to include the two possible stone rows in the remainder of Mull leads to a rather different conclusion. One of these, at Uluvalt (SSR27), is also oriented upon Ben More. However, it is situated a mere 4 km away from the mountain, to its south-east, and the peak yields a declination of more than +37°.[47] This tends to suggest that Ben More itself was the focus of interest rather than any particular astronomical event with which it might happen to align.[48]

While the nature of the astronomy (if any) remains unclear at this stage, the idea that the visibility of Ben More was important is reinforced by the use of Geographical Information Systems (GIS). GIS have found many uses in archaeology because of their potential for visualising and analysing spatial data (see Archaeology Box 6). Of particular interest here is the viewshed (or 'line-of-sight') function, which enables maps showing 'visible areas', and even the equivalent of the declination contour map, to be produced far less labour intensively.[49] Fig. 7.6 shows the area immediately around Dervaig, including

Fig. 7.5 The Glengorm stones and the southern horizon, viewed from higher ground to the north, showing Ben More on the skyline.

the rows at Quinish, Maol Mor, Dervaig N and Dervaig S (marked by crosses). For comparison, the plot also includes the locations of Bronze Age cairns[50] (marked by diamonds) together with that of an isolated burial cist (marked by a square) which contained a Bronze Age food vessel and is thus supposed to be broadly contemporary with the stone rows.[51] Remarkably, the plot revealed that this cist is also at a limit of visibility for Ben More, something hitherto unsuspected because it had not been visited and was situated close to the coast at an elevation little above sea level. On the other hand, the cairns appear to have no association with Ben More, the peak being visible from none of them despite an even wider range of topographic positions than the rows: one is close to the coast, two are in the broad Bellart valley, and one possible cairn is on the summit of a hill (Carn Mor).[52] This implies that rather different factors determined the locations of the cairns.

If the arguments up to this point seem rather unfocused it is because they have developed rather haphazardly, in an exploratory fashion. It is not advisable to proceed indefinitely in this way. We may succeed in collecting control data in various ways as the interpretation is developed, but it is clearly possible to 'bend' the arguments to suit what is found. For example, if the cairns had been in sight of Ben More but the isolated cist had not, we might have argued that the symbolism of location that applied to the stone rows applied to the cairns also. We need to proceed more systematically.

## Why there? The locations of the north Mull rows

One of the main objectives of the survey element of the North Mull project was precisely this: to explore systematically the principles of ritual tradition that might have influenced why the rows were built in particular places. One way to do this is to compare the symbolic potential of the site locations actually chosen with those of alternative locations, either selected at random[53] or because they seem equally plausible on other grounds. The project concentrated upon two now-familiar indicators of symbolic potential: the variation of horizon distance with azimuth, and the visibility of prominent hills and their relationship to astronomical phenomena.

The two methods of defining sets of control points characterise two rather different approaches to interpreting the data. The 'random control points' approach involves comparing the properties of the locations of actual monuments with points in the landscape as a whole over a wide area. One problem with generating sets of control points entirely at random is that it may be impracticable to obtain survey data at sufficiently many of them for a meaningful comparison to be made with the actual monuments, because they are so scattered. To overcome this to some extent, strings of control points were generated along 'sampling paths', paths that could be walked in the landscape from one point to another.[54]

Fig. 7.6

a. Map of the Dervaig area, showing the stone rows (marked with crosses), and other Bronze Age monuments. Shading represents contours, lighter colours denoting higher elevations. The distance from north (top) to south (bottom) is approximately 12 km; that from east to west approximately 9 km.

b. The Ben More viewshed for the same area, with the same monuments marked. Note that, in contrast to Fig. 7.1b, the shaded areas are those from which Ben More can be seen.

Plots obtained from the same GIS files have already appeared as Ruggles and Medyckyj-Scott 1996, figs 6–2 and 6–3.

The second approach takes into account other factors influencing site locations. It is inherently more contextual, and is well suited to addressing more local questions. The general issue here is whether sacred or symbolic principles influenced where monuments were placed in the landscape or whether their locations can be adequately explained by purely pragmatic, ecological, and economic factors influencing human settlement and activity such as slope and aspect, topographic position, soil cover and land-use potential, drainage, ease of approach, proximity to fresh water and proximity to the sea. The very idea that sacred and symbolic principles may have played an important role in how people perceived and used the landscape has begun to be seriously considered by archaeologists in recent years (Archaeology Box 7).

More conventional modes of enquiry reveal certain common aspects of the locations of the North Mull stone rows: for example, they are all set on level or gently sloping ground, well drained and easily approachable.[55] With this in mind, teams of volunteers were asked to identify places in the vicinity of (within about 1 km of) a given row that appeared to satisfy all of the locational prerequisites just as well as the location actually chosen by the builders.[56] The result was a set of 'alternative locations' in the vicinity of the real ones: eight at Glengorm, five at Quinish, three at Ardnacross, four at Balliscate and eight at Dervaig (around the two separate rows).[57] Those around Glengorm are shown in Fig. 7.8.[58]

Is there a significant difference between the way in which horizon distance varies with azimuth at the actual stone rows and at the control points? The answer seems obvious when we examine a plot of the frequency of occurrence of the different horizon distance categories[59] at the stone rows themselves and at twenty-two control points sampled using the first of our two approaches (see Fig. 7.9), and this is confirmed by a simple statistical test.[60] One of the most striking differences is that from the random control points there is a preponderance of distant horizons around north-west, along the line of the ridges and valleys, while most horizons between south and west are restricted by the adjacent ridge. In contrast, the rows themselves are generally placed so as to give reasonably distant horizons between SSE and WSW, and appear to eschew distant horizons to the north-west. It seems clear that there was a conscious effort on the part of the builders to locate the North Mull rows according to horizon visibility criteria that were not easy to achieve given the general topographical constraints in the area.[61]

Interestingly, a similar exercise carried out in the Kilmartin

## ARCHAEOLOGY BOX 7

### SACRED GEOGRAPHIES

In recent years, archaeologists have become increasingly aware that patterns of human activity within the prehistoric landscape are likely to have been influenced by more than just ecological and economic factors. Ancient landscapes might also have been structured according to symbolic or cosmological principles, forming what have become known as 'sacred geographies'.[1] Specific places, and indeed whole landscapes, are 'contexts for human experience, constructed in movement, memory, encounter and association'.[2] Structuring human activity within a landscape charged with meaning in this way might have been vitally important as a way of reinforcing people's understanding of the structure of the world and their own place within it.[3] Being able to demonstrate that a group's activities were fully in tune with the cosmos may also have served as a mechanism for legitimising tenure of certain lands and of reinforcing political control.[4] The importance of context-rich perceptions of the landscape is now recognised amongst a wide variety of historical and indigenous non-Western societies, from hunter-gatherer groups to urbanised states. They contrast sharply with the Western view of the land as commodity.

Aspects of sacred geographies have been studied amongst a diversity of historical and indigenous non-Western societies. They repeatedly show the importance of sacred places—places imbued with special or sacred significance or supernatural power—which may be marked by natural features such as caves, boulders, springs, mountain tops, or trees, or by rock paintings, shrines or monuments. They often show the importance of seasonal patterns of movement in the landscape which take place along well-defined paths, particularly amongst hunter-gatherer groups but also amongst cultivators and pastoralists and even urbanised states. Much quoted examples include the 'dreamtime' (ancestral) landscape of Australian aboriginal groups,[5] the Hopi and the Lakota (see chapter nine) together with other US indigenous groups, and Incaic Cuzco;[6] but examples are found world-wide.[7]

Cosmological symbolism is frequently of central importance in the design of sacred or secular buildings (see chapter nine); in making sacred certain locations in the landscape;[8] and even in determining the layout of cities.[9] Folk traditions may be widespread even in the context of powerful prevailing ideologies,[10] and although the latter may be dominant in textual evidence, folk practices may lead to local sacred geographies whose nature may best, and perhaps only, be explored by studying the archaeological record and oral tradition. A good example of this comes from late medieval India.[11]

Bearing these ideas in mind, archaeologists have begun to consider how cosmological principles might have helped structure human activity in the Neolithic and early Bronze Age landscapes of Britain and Ireland, and how, conversely, monuments and landscapes that came to form the 'centre of the world'[12] influenced cosmological beliefs. Interpreting the material record is made all the more difficult because landscapes and monuments were continually created and transformed, with enduring monuments, whatever their original purposes, remaining as focuses for ritual and ceremonial, and continuing as conspicuous sacred (ancestral) places in the landscape long after their original purpose was lost to living memory.[13] The new insights that can nonetheless be gained from broad interpretations are well illustrated by Bradley's discussion of the changing perceptions of stone and earthwork alignments in the Neolithic landscapes of western Europe.[14]

Amongst the specific ideas that have emerged is the suggestion that henge monuments such as the Rings of Brodgar and Stenness in Orkney reflected

Fig. 7.7 The sacred geography of Carnac. The famous stone rows are located within a landscape containing monuments spanning the whole of the Neolithic period, some of which predate and some of which postdate them. After Patton 1993, fig. 5.10.

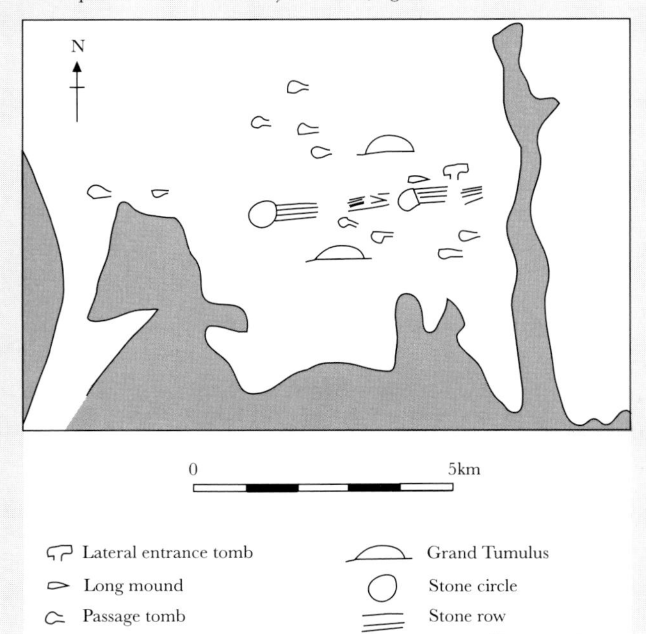

and represented the local topography, their 'rings of standing stones together with enclosing circles of water with external banks referenc[ing] the lochs and encircling hills';[15] similar suggestions have been made at Stonehenge.[16] And a variety of more regionally focused papers have appeared discussing sacred geographies in particular parts of Neolithic Britain and Ireland[17] as well as Brittany (Fig. 7.7).[18] This is an area where both ideas and their application in regional case studies are developing at a fast rate.

valley, mid-Argyll,[62] showed that the group of short stone rows there (the other concentration of short stone rows with a clear pattern of lunar orientation) shared this preference for non-local horizons between SSE and WSW, but clear views were generally available in this direction anyway throughout the valley. The stone row builders of northern Mull, it seems, had to go to some lengths to achieve a desired pattern of horizon visibility that their counterparts in mid-Argyll were generally able to take for granted.[63]

One of the main conclusions that emerges from an examination of the local alternative locations is that the amount of distant horizon visible from a row was not important in itself. Higher situations with wider views were often available within easy reach without compromising other considerations such as levelness of the site, good drainage and ease of access.[64] At Glengorm, for example, wide views to the north-east and north-west, in addition to those to the south and SSW, could have been achieved by placing the stones 150 m further to the north-east at Site *G* (see Fig. 7.8).[65] It seems that the important thing was to have a reasonably distant horizon in a particular direction, namely around SSW; wide views in other directions may even have been consciously avoided.[66]

Since it is necessary to have a good view in order to see prominent distant peaks, this leads naturally on to the question of whether prominent hills were important in themselves, possibly placing even tighter constraints upon where the stone rows were placed. Before we start to consider questions concerning their visibility and astronomical potential we need to recognise that there is not necessarily any correlation between what we consider a prominent hill, and what groups of prehistoric people may have considered a significant feature in the landscape.[67] Because of this, we cannot expect that any attempt on our part to define objective measures of prominence will isolate features consistently of interest to the builders or users of a particular group of monuments. Nonetheless if we do find recognisable patterns that cannot easily be explained away as fortuitous occurrences, then we do have reason to believe that we have isolated something 'real'.

In the North Mull project, attention was focused upon a small number of the most prominent hills.[68] No attempt was made to identify these at the control points in the wider landscape, but some interesting conclusions were reached both by comparing the locations of the stone rows themselves and by looking at the local alternatives. In order to shed light on the question of whether the rows might have been preferentially placed so as to associate prominent peaks with particular astronomical phenomena, regardless of their orientation, an effort was made to identify the most prominent hill summits visible from each monument (up to a limit of about ten).[69] A plot of the declinations obtained (Fig. 7.10) brings a surprise, for they are highly clustered, with concentrations between about −31° and −26° (rising and setting lines); around −20° (rising lines only); around +24° (rising and setting lines); and between about +31° and +34°.

It is important to realise that these data, unlike those that we have considered in the previous two chapters, are not confined (even roughly) to the direction of orientation; the whole horizon around each monument is being taken into account. That this reveals any obvious trends is itself remarkable, and it seems to imply that it was not just the orientation of the rows, but their very locations, that were chosen with regard to—perhaps even to emphasize—particular hills and certain astronomical events correlated with them. Furthermore, the first three of the four groups have a very obvious interpretation in terms of the midsummer full moon in the south and the midsummer sun in the north; was it at this time of year that the rows came into full harmony with their surroundings, both terrestrial and celestial?

The cluster between −31° and −26° in Fig. 7.10 contains at least one peak from each site. While some of the rows in the northern group contribute several peaks, Balliscate and Ardnacross each contribute only one. At Balliscate this is

Fig. 7.8 'Alternative' sites in the vicinity of Glengorm. First published as Ruggles, Martlew and Hinge 1991, fig. 5.

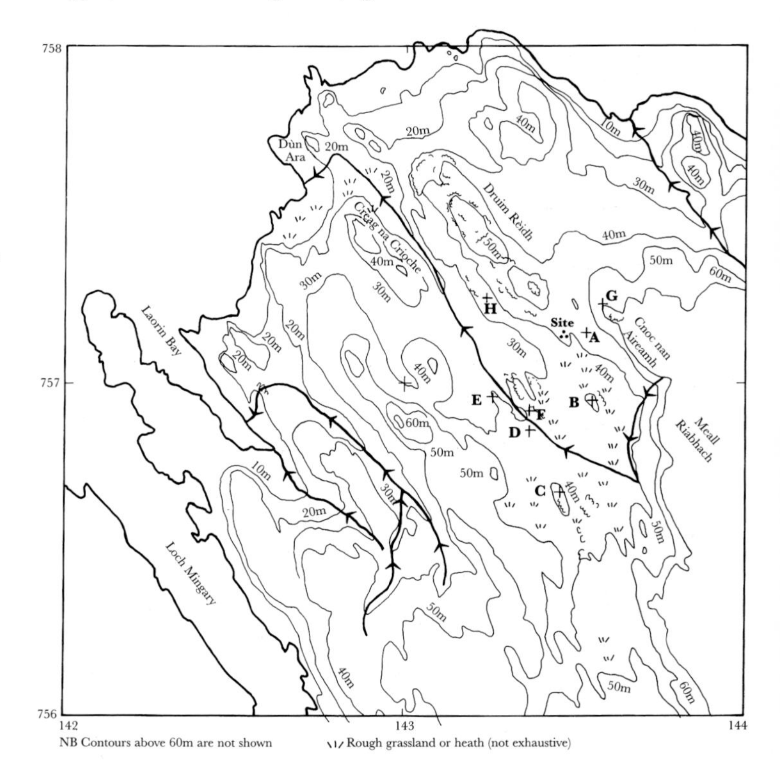

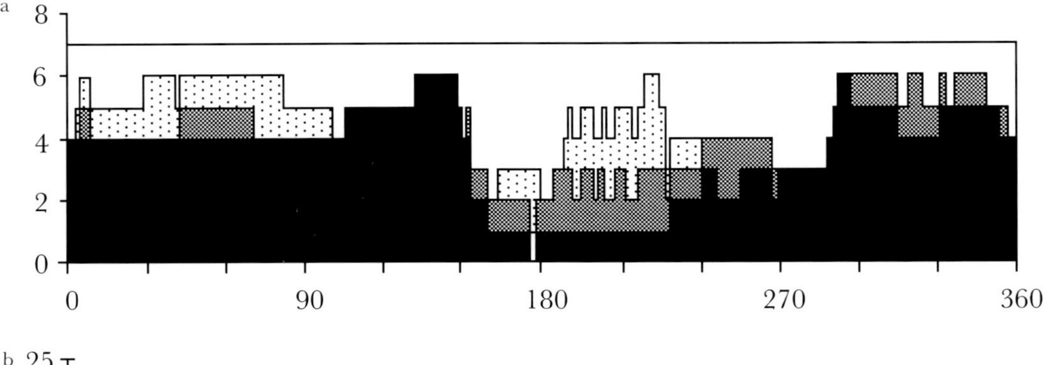


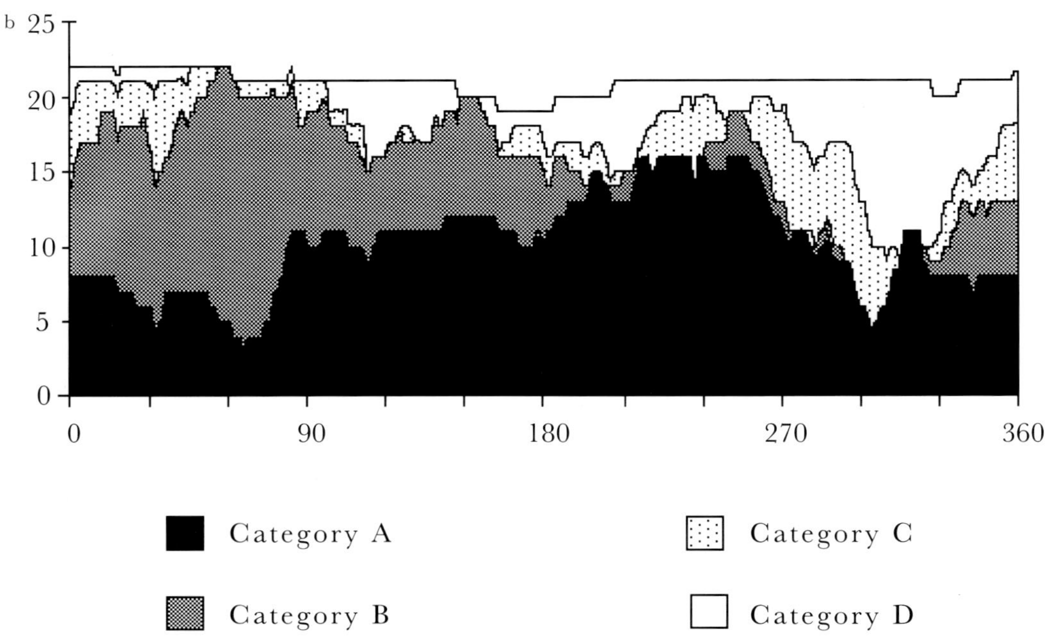


Fig. 7.9 The frequency of occurrence of different horizon distances in various directions (a) at the north Mull stone rows and (b) at twenty-two control locations in northern Mull. For the significance of the horizon categories see Fig. 5.4. After Ruggles, Martlew and Hinge 1991, fig. 9.

Speinne Mór, poking up through a small gap in local ground just east of south at a declination of −28°·9. At Ardnacross it is the distinctive wedge-shaped peak of Beinn Talaidh, just visible at the southern end of the wide eastern vista, yielding a declination of −25°·5. It is interesting that both these peaks marked a rising point (although Speinne Mór is very close to due south) suggesting perhaps that they might have performed the same function at the east coast sites as Ben More did at the rows in the northern group. At Balliscate, as at the Ben More sites, the orientation of the stone row is within 10° of the peak.[70] Furthermore, both Balliscate and Ardnacross are situated critically for sighting upon their respective peaks, in the same way as the northern rows for Ben More.[71]

The implication is that it was important to have a prominent peak associated with the rising of the midsummer full moon, although it does not seem to have been critical for the peak to mark its southernmost limit in the node cycle, at least more precisely than to within three or four degrees. Fig. 7.10 also provides evidence that at some rows the builders managed to make use of second peaks or groups of peaks, associating them with midsummer full moonrise at different times in the 19-year cycle (nearer the minor standstill limit) or with the rising or setting of the solstitial sun. It may not be coincidental, then, that from the vicinity of Quinish, uniquely amongst the North Mull rows, the rising full moon could have been observed behind Ben More shortly after the midsummer sun had set in the opposite direction behind a prominent hill—actually, the triple peaks of Beinn Mhor, Ben Corodale and Hecla in South Uist, prominent on the distant horizon across the waters of the Hebridean Sea on a clear evening.[72] Quinish was excellently placed to pick out the sun setting between the left-hand and central peaks as it approached midsummer and between the central and right-hand peaks around the solstice itself.

In moving on from the overall result to look at the details, we have left the relative safety of the systematic analysis and started to introduce rather more speculative interpretations once again. These are taken further elsewhere,[73] but the local analyses do generally confirm the conclusion that there was a desire to have a prominent peak associated with the rising of the midsummer full moon. For example, only Ben More is sufficiently far east of south to perform this function in the vicinity of Glengorm, and only Site *F* provides an alternative location from which it is visible.[74] If it was crucial that Ben More was visible from the immediate vicinity of the standing stones, then this would explain why Site *G* was eschewed, despite its wider all-round views, and also why Site *A* was not chosen, despite its being similar to the chosen site in every other respect.

To summarise: the more systematic analyses indicate that an important criterion in choosing where to position a stone row

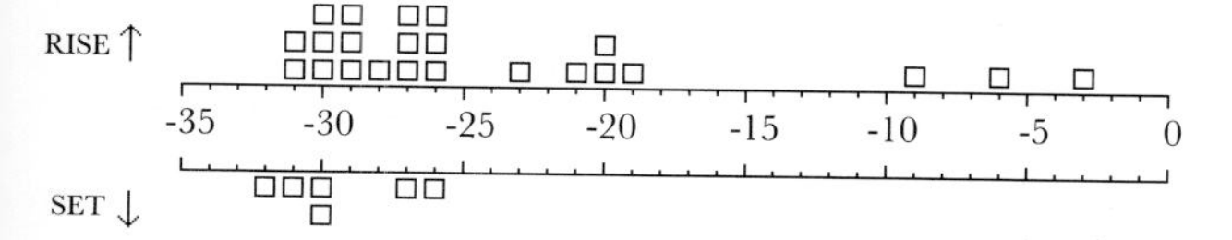


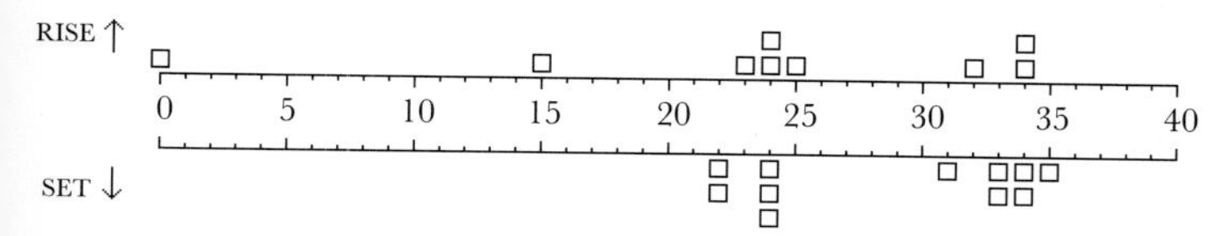


Fig. 7.10 Declinations of prominent hill summits at the North Mull stone rows. Easterly (rising) lines are plotted upwards; westerly (setting) ones downwards. After Ruggles and Martlew 1992, fig. 13.

was to have a non-local horizon in the south and to the west of south, and that this was achieved in northern Mull despite the general run of the local topography. An additional requirement, it seems, was to have a prominent peak on the skyline in the general direction of the rising midsummer full moon. Taken together, these suggest that it may have been important that, having risen, the moon should be clearly visible on its low path not very far above the horizon until it set again.[75] Great care seems to have been taken to position the monument so that the peak in question was not freely visible from all around. Finally, it appears that the stone rows' locations were chosen to emphasize the southerly direction while hiding from view most, and in some cases all, other distant horizons and prominent peaks.

## Overall conclusions: the sacred geography of Bronze-Age Mull

The North Mull project undertook a regional examination of the landscape, taking into account factors—such as astronomical potential—that might have been of symbolic rather than simply of economic or ecological significance. In doing so, it attempted to develop a more integrated way of tackling the issue of monuments and their relationship to the landscape. It also represented an attempt to find new ways in which the statistical evidence for lunar symbolism could be extended, reappraised, and set in a wider cultural context.

Factors unrelated to astronomy may well explain the general concentration of five rows towards the northern tip of the island. The area in which they are situated is one of the few parts of northern Mull that was reasonably suitable for settlement during prehistoric times. It is virtually the only part of the coastline which slopes gently to the sea[76] and which combined agricultural potential[77] with ease of access.[78] On the eastern coast, at Ardnacross, ard-marks found in the subsoil provide direct evidence that cultivation took place there prior to the erection of the stones.[79] The stone rows of northern Mull, then, seem to be associated with settlement and agriculture. It is only when we look at the more specific locations within these general areas that patterns emerge indicating that they were placed in very particular positions with regard to landscape and horizon visibility and astronomical potential.

This must certainly affect how we interpret the alignments. For example, we have said that it does not seem to have been critical for the monuments to be located so that the prominent peaks marked the southern limit of the moon in the lunar node cycle. We could argue from this that the major standstill limit was of interest but only at a very low level of precision. A rather more convincing conclusion, however, is that the general association of Ben More and the midsummer full moon, which would have been observable from anywhere in the northernmost part of the island with clear views in the relevant direction, and would clearly have been spectacular, itself led to the association between the two being perceived as important within this general area. In most years, the relevant moonrise would not have occurred right behind the mountain.[80] More often than not it would have risen a short distance to the left of the peak and then passed low over it.[81]

A question of considerable interest is what might have been the reason for constructing a monument in a place from which a prominent peak—a sacred hill or mountain, perhaps—or a special astronomical event, again possibly of great sacred significance, was on the very limit of visibility. Why should it not be easily visible from all around the monument? An obvious possibility is that this restricted access to the phenomenon in some sense, perhaps to a particular group of privileged people. But these monuments were quite common, and apparently served only small groups of people in total in an area, and at a time, when there is little evidence of developing social hierarchy or competition for power and resources. In this context, privileged status seems to make little sense, and in any case the phenomenon was by no means exclusively visible from the monument itself. Another possibility is that 'hiding' the phenomenon, in the sense of making it only just visible from a stone row directly associated with it, helped to bestow special meaning upon the monument, or to reinforce the sacredness of the place where it stood.[82] This, however, seems incompatible with the linear form of the monument, which clearly does emphasize what we take to be the significant direction.

A third possibility is that, just as a stone circle may have served to separate its interior space from the rest of the world, imbuing it with special meaning,[83] so these stone rows may have marked the boundary between two parts of the world charged with very different ideological significance: those from which a sacred mountain could, and could not, be seen.[84] While this interpretation is just as speculative as the others, it is perhaps more in tune with the sorts of ideas emerging from studies of sacred geographies world-wide (cf. Archaeology Box 7; see also chapter nine), and further relationships found in the archaeological record from Bronze Age Scotland might clarify whether the conjecture stands up to closer scrutiny.

One of the few direct clues to the sort of ritual or ceremonial activity that might have taken place at the stone rows is that hundreds of angular fragments of white or transparent quartz appear to have been scattered around the standing stones at both Glengorm and Ardnacross.[85] Of course, types of stone were important in a variety of contexts, technological as well as ritual: specific stone and mineral resources were exploited for tools and for jewellery, and particular types of rock were used in the construction of burial cairns, stone rows and stone circles. Quartz fragments have been found associated with a variety of monuments spanning a long period of prehistory, and no single symbolic meaning is likely to lie behind all the observed occurrences.[86] Nonetheless, scatters of quartz fragments have also been found at Irish short rows and circles such as Knocknakilla, Co. Cork (W 297842);[87] and amongst the mountains of Connemara several short rows were built entirely of white quartz blocks standing vividly in the landscape.[88] While there are no consistent patterns in the spatial distribution of the scatters that would suggest a specific interest in any particular part of the row, in contrast to the situation at the Scottish RSCs, the very fact that quartz is found once again at monuments apparently associated with the moon,[89] together with its white colour, itself reminiscent of moonlight, give further credence to the idea that this substance may have formed an essential element in ceremonies intimately associated with the movements of the moon.[90]

In addition to generating and refining various ideas relating to the northern Mull rows themselves, the North Mull project has drawn attention to a number of wider methodological issues. To begin with, it certainly seems that by considering the broader properties of place that may have determined where a monument was built and into which its orientation might merely have fitted or harmonised, rather than concentrating on the orientation *per se*, we may gain rather clearer insights into the builders' actual intentions. Faced, say, with the isolated fact that the East Cult alignment in Perthshire (SSR7) is placed along the crest of a broad ridge,[91] one could argue just as convincingly that a location was sought to suit the desired orientation as that the orientation was chosen to suit the topography at the chosen location, and neither option says anything about why the orientation or location was desirable in the first place. On North Mull, however, we can now say with some confidence that symbolism relating to the horizon and the sky had a strong influence upon where the rows were placed, and their orientation reflected rather than dictated this choice.

Certain patterns of behaviour only become evident when the strict methodological criteria employed in looking along the alignments are relaxed. The more rigid criteria of chapter six (cf. Table 6.4) completely failed to detect the significance of Ben More.[92] The horizon profiles in the precise directions of alignment are unenlightening, with little to suggest that the rows were consistently and precisely aligned upon prominent hill peaks.[93] Yet at a lower precision, the correlation between the orientations of the five northernmost rows and the direction of Ben More is quite clear. We must move on from just looking along the alignments at individual monuments.

A major problem is that, once we become interested in the wider landscape, tackling relevant questions may require an amount of time, or a level of resources, that is simply not viable. This is why the North Mull project produced no data from the 'random' control points relating to prominent hills and their astronomical potential. In some cases, GIS may provide a useful way forward. For instance, consider the following two questions:

(1) Were the stone rows placed with regard to the visibility (not necessarily on the horizon) of natural features of apparent significance in the landscape, such as prominent mountains?
(2) If so, were they preferentially placed so that these prominent horizon features would align with the rising or setting points of certain celestial bodies?

There are many issues here, but it is clear at the very least that we need to construct the equivalent of a declination contour map for each of several prominent peaks or other features in the landscape. A general strategy has been developed for tackling these questions quantitatively using 'multiple viewsheds' in a GIS which would allow us, for example, to identify the natural features that best 'explain' the placing of a set of monuments.[94] Of course, we might well wish to take into account the influence of ecological or economic factors upon monument locations. Again, GIS can be adapted to the purpose, since this information can be held in other map layers, and combined with the multiple viewshed in appropriate ways.[95] The resolution of the digital elevation data used to model the landscape topography is a critical factor.[96]

When trying to interpret the symbolism of stone monuments in the rich context of the local landscape it has not proved possible—and may never prove possible—to predefine meaningful general criteria for selecting and interpreting data; instead, specific procedural decisions have to be made along the way in the light of discoveries made so far and of the questions currently of interest as a result of those discoveries. This being the case, even if we successfully avoid the most obvious traps of circular argument, how can we ever have reasonable confidence in our conclusions? One answer is to try to develop quantitative methods suited to this way of working, and this theme is developed further in chapter ten. In the meantime, the most important thing is to be able subsequently to defend any conclusions reached. To this end it is absolutely crucial that all of the procedural decisions made, and all of the data collected (whether or not they support the general conclusions reached), are documented in full.[97]

Perhaps more than anything else, the North Mull project demonstrates the value of a multi-faceted, regionally based investigation in tackling issues relating to prehistoric astronomy. Studies of the properties of different locations within the landscape, taking into account esoteric factors such as visibility and astronomical potential, integrated with excavation and environmental studies, enable us to identify and investigate complex relationships between places in the landscape and prominent natural features visible from those places, or the motions of celestial bodies visible from them. We may even begin to think about how the significance of certain relationships changed over time. Much of this evidence would simply not be revealed by archaeoastronomical horizon surveys alone.

Yet the archaeoastronomy is important. The motions of celestial bodies such as the sun and moon, even in prehistoric times, are known to us; and where land forms are relatively unchanged since prehistoric times they are still available for analysis.[98] Constraints within social and political space may be difficult to recover archaeologically, but preferential location with respect to topography or astronomy are aspects of sacred geographies that are still available for analysis. Studies of astronomical alignments may tell us certain things, but an integrated approach can start to reveal more intricate patterns of behaviour. As a result, we may be able to gain some rather deeper insights into how the builders or users of a particular group of monuments envisaged, and through their own actions helped to bring order into, the landscape in which they lived. This conclusion is borne out well in northern Mull.

# 8

# Astronomy in Context

*A Synthesis*

At Swineshead . . . is a druidical temple, which the country people call Sunken Kirk . . . [The] entrance is nearly south-east . . . This monument of antiquity, when viewed within the circle, strikes you with astonishment, how the massy stones could be placed in such regular order, either by human strength or mechanical power.

Richard Gough, 1789[1]

L1/3. Sunkenkirk. SD 171882. Description = C. Diameter = 93·7 ft. $m_1$ = 34. $\varepsilon_1$ = +1·22. $m_2$ = 17. $\varepsilon_2$ = +1·22. Class = B. Az = 128°·8. $h$ = +0°·5. $\delta$ = −21°·5. Sun. 'Entrance'.

Alexander Thom, 1967[2]

This may have been what a stone circle was to its people, a place where axes and gifts were exchanged, a place where annual gatherings were held, a place to which the bodies of the dead were brought before burial, but, above all, a place that was the symbol of the cosmos, the living world made everlasting in stone, its circle the shape of the skyline, its North point the token of the unchangingness of life, a microcosm of the world in stone, the most sacred of places to its men and women.

Aubrey Burl, 1988[3]

## Early developments

In recent years, evidence relating to prehistoric astronomy has begun to emerge in the context of several wider archaeological investigations, as well as being the main focus of archaeoastronomical projects. A number of intriguing results also continue to emerge from the work of non-specialists, of which some at least deserve serious consideration and further investigation. In this chapter we bring together a range of evidence in an attempt to construct a broad picture of the nature and purpose of astronomical observations in the context of broader developments in British and Irish prehistory.

This synthesis will focus on the Neolithic and the earlier part of the Bronze Age, i.e. the period from about 4250 to 1250 BC, when the evidence that has been interpreted as relating to astronomical practice is most extensive. Yet simply because no conspicuous monuments remain from the preceding Mesolithic period, when, for several thousand years following the disappearance of the last ice sheets, Britain and Ireland were inhabited by groups of roaming forest dwellers living by fishing, foraging, and hunting,[4] we should by no means conclude that what Mesolithic people saw in the sky was unimportant to them. This would single them out from many hunter-gatherers in historical and modern times.[5] It is merely that this has left no trace in the archaeological record that has been identified to date. Going even further back, Alexander Marshack has suggested, in a series of publications, that markings found on fragments of bone and stone found in France and Russia and dating from Upper Palaeolithic times, as early as 30,000 BC, are tally marks related to the phases of the moon.[6] The idea is exciting and the evidence is highly intriguing, but it is unfortunately far from conclusive.[7]

Starting, then, in the Early Neolithic, we encounter a landscape, mostly thickly afforested, in which scattered groups of people herded sheep and other livestock and cultivated cereals (Archaeology Box 5). These people 'were intent on stamping their mark on the land, in contrast with the gatherer-hunters who, it seems, lived less intrusively in their environment'.[8] It has been suggested that the long barrows, large collective monuments for the dead constructed at this time (Archaeology Box 4), were used by the early farmers to lay claim to territory previously occupied by relatively high numbers of hunter-gatherers.[9] They 'would stand for ever to represent the permanent link between a community, the ancestral dead and the land which they farmed'.[10] Was astronomical symbolism incorporated in these early monuments? If so, was its purpose to reaffirm such links symbolically, forming one way in which the monument was demonstrably attuned to the workings of the cosmos?

Orientation was certainly important. There are strong general patterns amongst several regional groups of earthen long barrows and long houses in Britain and Ireland, the Netherlands, northern Germany, Denmark, and Poland,[11] with the spread of orientations generally being centred upon the east or south-east (Fig. 8.1). These broad trends bear witness to one way in which the design of tombs—'houses for the dead'—reflected the design of domestic structures[12] and the extent to which similar practices were widespread across northern

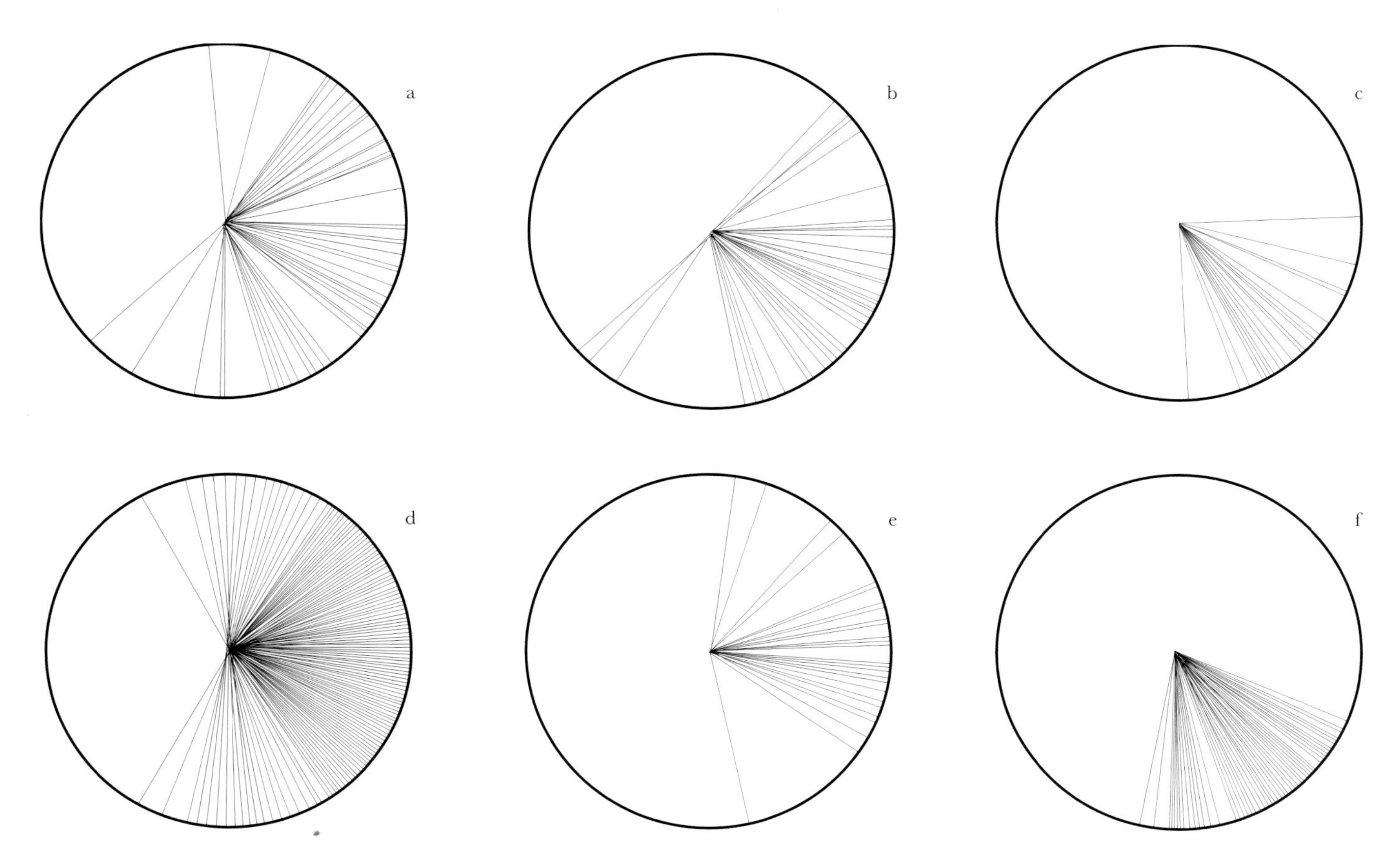


Fig. 8.1 Examples of orientations of regional groups of long tombs and long houses, after Hodder 1984, fig. 4.
a. 'Gallery graves' in Brittany.
b. Breton passage tombs with shorter passages.
c. Breton passage tombs with longer passages.
d. British earthen long barrows.
e. Megalithic tombs in the Netherlands.
f. Rectangular and trapezoidal Neolithic long houses.

Europe.[13] The conclusions are based on orientation data generally measured and quoted only to the nearest one-sixteenth division of the compass (i.e. to within a band of width 22°·5), but even at this level of precision such consistency over wide areas could only have been achieved by reference to the daily motions of the heavenly bodies. Was there more to it?

In possibly the only systematic regional survey of its kind, Aubrey Burl has examined the orientations, determined where possible to within ±2°, of sixty-five long barrows on Salisbury Plain.[14] Sixteen typical examples lie within 5 km of the future site of Stonehenge (Fig. 8.2).[15] Unfortunately Burl does not provide a complete list of the field data, but he does tell us that 'no fewer than 13 of the 65 barrows face either to the north or south of [the solar] arc', concluding that it is unlikely the monuments were aligned upon the sun.[16] On the other hand, only six examples fall outside the wider lunar arc, from which he concludes that 'it seems quite probable that they were intentionally aligned on the rising moon'.[17]

Despite Burl's confidence, it seems premature to distinguish with any great assurance between solar and lunar hypotheses on the basis of the data as presented,[18] especially as they contain some inconsistencies[19] and no account has been taken of horizon altitude, which means that the azimuths corresponding to given astronomical events at any given location are themselves uncertain within several degrees.[20] Yet it may well be significant that the bulk of the outlying orientations are to the south rather than to the north of the solar and lunar arcs. This suggests the possibility, which has also been raised in relation to certain groups of Neolithic tombs in southern

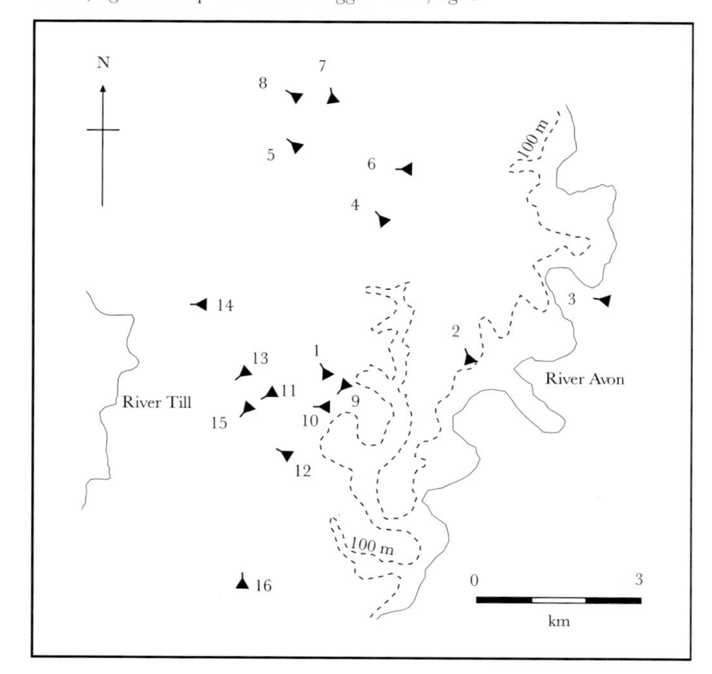


Fig. 8.2 Long barrows on Salisbury Plain: the locations and orientations of sixteen barrows within 5 km of the future site of Stonehenge. After Burl 1987a, fig. 3. First published as Ruggles 1997, fig. 3.

Fig. 8.3 The Dorset Cursus.
a. Plan of the south-west part of the monument and the associated long barrows. Based on Barrett *et al.* 1991, figs 2.16 and 2.14.
b. Midwinter sunset over the Gussage Cow Down barrow as viewed from the Bottlebush Down terminal. After Barrett *et al.* 1991, fig. 2.16 (inset), with approximate azimuths and altitudes added from a comparison with Penny and Wood 1973, fig. 14.
c. A visualisation of (b). After Bradley 1993, fig. 27.

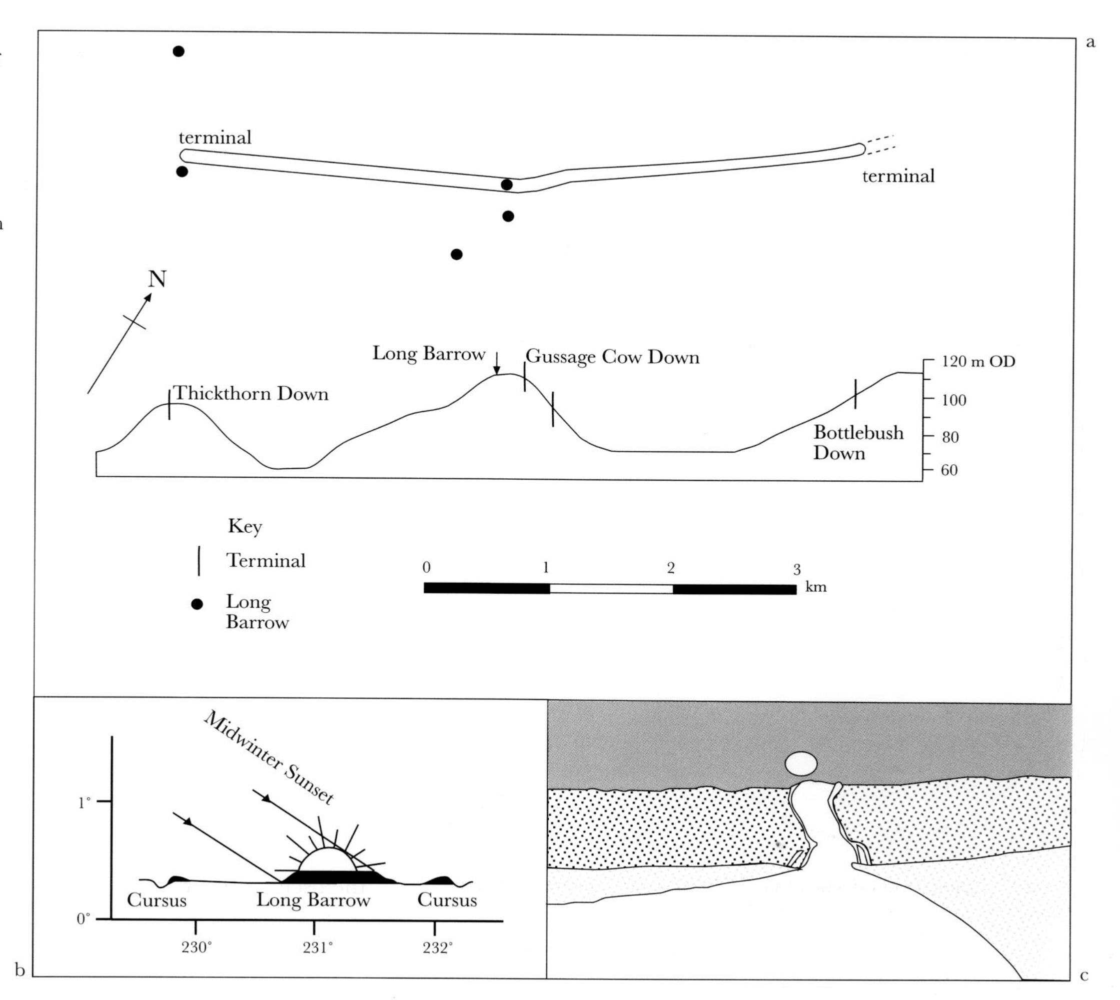


Europe,[21] that it was important to orient the entrance upon a direction where the rising sun or moon would sometimes be seen to pass, possibly climbing in the sky rather than necessarily at its actual point of rising.

Julian Thomas has suggested that while many of the earliest monuments in Britain and Ireland are relatively unsophisticated tombs serving to place the ancestors in conspicuous locations, the architecture of monuments increasingly served to constrain how people could approach them and move within them, and to restrict what would be open to view at any given time. By the middle of the fourth millennium BC there was a growing interest in the construction of linear monuments which he sees as not merely setting spaces apart from the world, but dictating particular patterns of movement through it. 'These include tombs with long passages leading to the burial chamber, often aligned on other monuments, landscape features or the movements of celestial bodies.'[22] Alignments with astronomical bodies, Thomas suggests, could have served to constrain movement through and within symbolically charged spaces to happen only at certain opportune times.

What is the evidence for an increased level of sophistication in astronomical symbolism amongst these monuments? The Dorset Cursus (ST 969125 – SU 040192) dating to between about 3400 and 3100 BC,[23] received attention in the 1970s as a possible solar and lunar observatory.[24] Consisting of two parallel banks and ditches nearly 100 m apart, this cursus runs for nearly 10 km across Cranborne Chase and it appears that its earliest section, whose north-eastern terminal was at Bottlebush Down, was aligned so as to 'enclose' an existing long barrow at Gussage Cow Down 3 km to the south-west (Fig. 8.3a). The Bottlebush terminal was located below the crest of a hill, from where the barrow appeared on the skyline framed by the parallel ditches and banks.[25] It is certainly true that, when viewed from this terminal, the sun around winter solstice would have set behind the Gussage Cow Down barrow (Fig. 8.3b,c),[26] and that this was intentional is accepted by a number of archaeologists. 'By linking [this] monument to the movements of the heavenly bodies [it is likely that] its builders were making the Cursus appear part of nature itself and freeing its operation from any challenge.'[27] Similarly, the western section of the Dorchester Cursus in Oxfordshire was oriented upon the midsummer sunset.[28]

Not all cursuses are aligned upon the solstitial sun. The Stonehenge Cursus (SU 1095 4290 – SU 1370 4320),[29] like its Dorset counterpart, may have been aligned in relation to a long barrow, which in this case lay across its axis on the skyline just beyond its eastern terminal.[30] But this cursus runs roughly east–west and from the far terminal the sun would have been seen to rise behind the barrow somewhat after the spring and before the autumn equinoxes.[31] Nor did astronomical alignments, if intentional, exist in isolation. Each of these monuments was also carefully placed with respect to existing

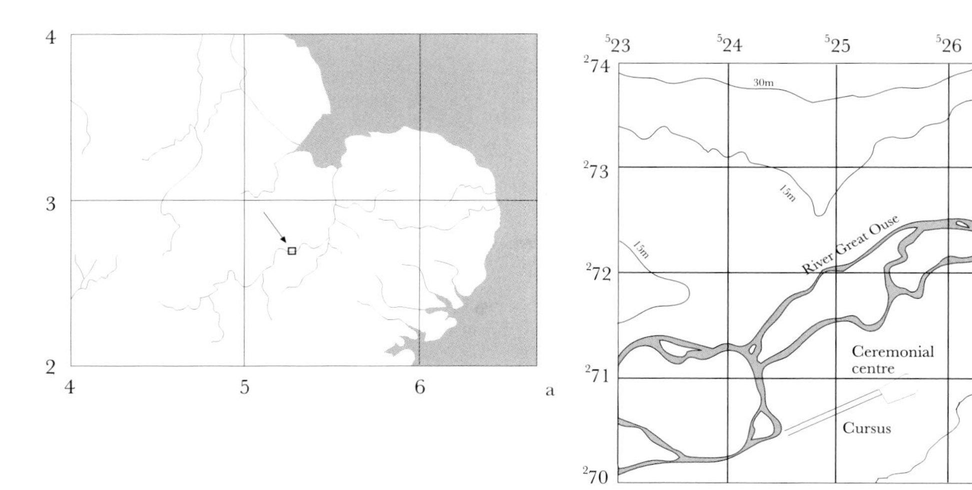


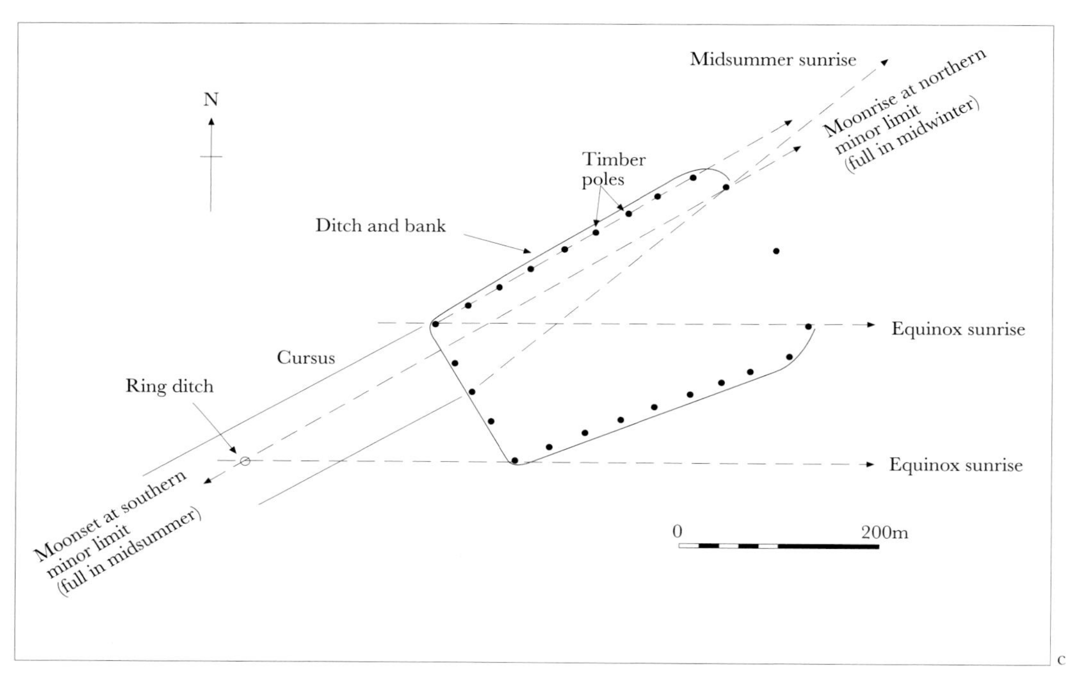


Fig. 8.4 The Godmanchester Neolithic enclosure and cursus.
a,b. Location of the monument.
c. Plan of the site showing proposed solar and lunar alignments. From a previously unpublished plan kindly supplied by Fachtna McAvoy, English Heritage.

monuments such as long barrows. In each case 'The key to understanding this particular monument might be the way in which it created a connection between the dead, the monuments that commemorated them, and the unchanging workings of the heavens. What better way could there be to demonstrate the timeless power of the ancestors?'[32] It has also been suggested that the solar alignment of the Dorset Cursus represented a fundamental change from lunar symbolism inherent in the earlier long barrows. Claims of precise lunar alignments involving the barrows around the Dorset cursus itself are susceptible to the sorts of criticism raised in Part One of this book,[33] but nonetheless there is a possibility that the construction of the cursus represented 'the imposition of a massive solar alignment upon a series of separate monuments which were orientated towards the rising moon'.[34] The argument is repeated at Dorchester,[35] and it is suggested that a similar, drastic change may have been repeated elsewhere in the middle Neolithic.[36] In view of the importance of this concept and uncertainties concerning the wider barrow orientation data that might underpin it, further field data are clearly needed.

Mention must be made at this point of the Early Neolithic complex near Godmanchester, Cambridgeshire (TL 256710). Excavations by English Heritage have revealed the site of a roughly rectangular ditch-and-bank enclosure, *c.* 325 × 200 m, open-ended to the north-east, constructed in about 3800 BC (Fig. 8.4). Flanking the interior was a 'U'-shaped configuration of twenty-three large timber poles, with a single pole marking the centre of the north-eastern 'entrance'. After a relatively short period of use the timber poles were destroyed by burning and a cursus was subsequently constructed extending to the south-west.[37] The monument has been the subject of ill-advised speculations based on solar and lunar alignments measured between various pairs of posts in the configuration, which are open to similar criticisms as those levelled at Hawkins's work at Stonehenge (see chapter one).[38] The enclosure's axis of symmetry, marked by posts at both ends, is oriented towards sunrise on about 1 May and 1 August, and one might speculate that these were times of year which may have been significant in the contemporary calendar, although it is dangerous indeed to extrapolate from isolated instances of alignments upon sunrise close to mid-quarter days in order to suggest that there was a continuity of tradition extending first to Thom's megalithic calendar, then to the Celtic calendar more than three thousand years later.[39] The equinoctial alignment between the post at the western corner and the opposite post at the eastern end of the entrance has also attracted particular attention, but its selection from numerous possibilities renders it unconvincing, as does recent work questioning whether the equinox was a meaningful concept in prehistoric times.[40] Interestingly, in view of the apparently solar connotations of other cursuses, it has been proposed that the Godmanchester example (which follows the line of the north-western edge of the enclosure) is aligned upon moonrise and set at the minor standstill limit.[41] Of course, its orientation also falls within the rising and setting range of the sun, and perhaps this simply re-emphasizes the need for cursus monuments to be examined as a group.

## Astronomy and ancestors: alignments in chambered tombs

More persuasive evidence of solar symbolism is found in certain chambered tombs. In particular, the idea that the solar alignment of the roof-box arrangement at Newgrange was intentional, as discussed in chapter one, seems very convincing. It would perhaps be surprising if solar symbolism were evident at Newgrange while being completely absent from other chambered tombs, particularly its two large companions in the Boyne Valley cemetery, Dowth and Knowth. Dowth is also quoted as being aligned upon winter solstice, albeit at sunset rather than sunrise, although closer examination of the relevant accounts[42] reveals there that it is not the long main passage of the largest tomb under this mound that is involved—the alignment here is upon sunset in early November and early February—but a much shorter passage leading to a smaller tomb further to the south.[43] No solstitial explanation has been offered at Knowth, whose two passages face roughly east and west. Here it has been claimed that sunlight would penetrate the tomb on certain days close to the equinox, around sunrise and sunset respectively,[44] and that a number of other effects of sunlight and shadow upon rock carvings were intentional and meaningful.[45] Perhaps the most intriguing suggestion is that a sequence of twenty-nine symbols on kerbstone SW22 on the south-west side of the tomb may represent the phases of the moon (Fig. 8.5b).[46] Recent work has uncovered other solar alignments and sunlight-and-shadow phenomena at Newgrange itself,[47] and it has also been suggested that hidden carvings within the roof-box contain symbolic representations of the sunrise.[48] While some of these interpretations may appear attractive, their diversity must urge caution, leading us to wonder to what extent data are being selected to fit theories, rather than the other way round.[49] We must also beware of basing interpretations of elaborate decorations simply and solely upon numerical correlations with certain astronomical or calendrical cycles.[50]

An important implication of the astronomical function of the roof-box at Newgrange is that the tomb was apparently built with the eventual blocking of its entrance in mind (chapter one), conflicting with the usual explanation that the practice of blocking off large communal tombs came as a result of social or ideological upheaval (see Archaeology Box 2). There are other cases where the sun's light could apparently still enter the tomb after blocking, most notably Maes Howe, the greatest of the chambered tombs in the Orkney islands, where monumental construction reached considerable heights in the Late Neolithic.[51] Maes Howe may well be broadly contemporary with Newgrange and Knowth.[52] Its entrance is oriented south-west, and although it has been stated a number of times that the setting sun's rays around the midwinter solstice illuminate, or at least once illuminated, the rear wall of the central chamber,[53] the mean axis of the inner part of the passage (azimuth 221°) was more in line with sunsets some three weeks earlier or later.[54] A stone that was used to block the passage was not quite tall enough to fit to the top of the entrance, and it has been suggested that the shortfall was to allow, as at Newgrange, the sun's light to enter the tomb after it had been sealed.[55] It has also been suggested that short blocking stones were used at Bryn Celli Ddu in Anglesey[56] and two smaller tombs close to Newgrange.[57] While Bryn Celli Ddu is said to be aligned upon the rising sun[58] the others are oriented towards the south, and one must resort to the speculation that this was so that 'the light of the low midwinter sun could shine down their passages at noon'.[59]

To what extent might we be uncovering meaningful solar symbolism, and to what extent could many of these isolated instances be better explained as fortuitous occurrences? In the past there have been numerous attempts to catalogue particular instances of solar (and lunar) alignment, but these raise

a

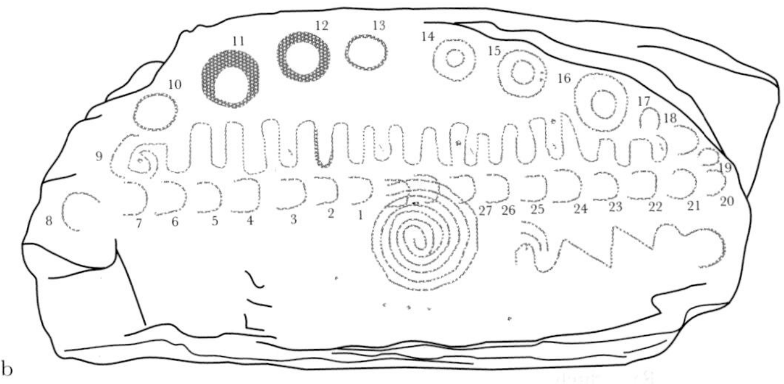


b

Fig. 8.5 'Calendar stones' from Knowth.
a. Decorated kerbstone SE4. The pattern is reminiscent of a sundial.
b. The interpretation of a cyclic arrangement of twenty-nine symbols on kerbstone SW22 as a lunar phase cycle. On days 1–9 the moon is represented as a crescent, on days 10–13 it is represented as an oval rising in the evening sky, on days 14–16 it is full, and then it returns to a crescent, disappearing on days 28 and 29 when the crescents are drawn 'hidden' behind the large spiral. After Brennan 1980, 98. See also Heggie 1982b, 20; Brennan 1983, 144. The stone is approximately 2 m wide.

the now-familiar question of data selection.[60] To obtain more persuasive evidence it is best to turn to systematic studies of patterns of orientation amongst regional groups, and here once again it is Aubrey Burl who is responsible for the most important pioneering work.[61] The orientations of the stone-lined passages of chambered tombs are generally easier to determine than those of earthen barrows,[62] and Burl's syntheses certainly reveal some strong regional trends. The orientations of the forecourts of the Clyde–Solway tombs of south-western Scotland are clustered within a narrow azimuth band around north-east.[63] Those of the Camster tombs in Caithness are strongly clustered about due east.[64] The heel-shaped cairns of the Shetlands are all oriented close to south-east. This is to name but a few Scottish examples.[65] Such overwhelmingly non-random trends cry out for explanation, and it is unfortunate that careful studies of groups of chambered tomb orientations, working from the material evidence at a suitable level of precision and taking into account horizon altitudes and astronomical potential—and interpreted in the context of the wider archaeological picture—are still virtually non-existent.[66]

It is possible that the builders of the Clyde–Solway, Camster, and heel-shaped tombs followed local traditions relating them to the sun rising around the longest, mid-year and shortest days respectively. It is also possible that the focus of attention was the moon, with the orientations perhaps relating to full moonrise around midwinter, the equinoxes and midsummer respectively. Given the current state of the evidence, it is impossible to distinguish with any confidence between the solar and lunar possibilities. An alternative explanation might be that it was desirable simply to keep within the range of horizon where the sun rises at any time of the year,[67] or during some part of it.

Yet lest we return to considering astronomy to the exclusion of everything else, it is important to remind ourselves that not all local groups of chambered tombs show distinct orientation trends and, even where they do, the simplest explanation may not always be an astronomical one. The orientations of twenty-two Clyde tombs on the Isle of Arran are scattered all round the compass,[68] and it has been suggested that some of the tombs in the Carrowkeel cemetery, Co. Sligo, may be aligned upon the passage tomb of Maeve's Cairn (G 626346) 27 km away on Knocknarea mountain.[69] It has been claimed that a number of Clava cairns are oriented upon mountain peaks,[70] as is Maes Howe itself upon conspicuous hills on the island of Hoy.[71] Even where an astronomical motivation can be postulated it is clear that no single, specific astronomical explanation fits the clearly-defined and often narrow—but nonetheless different—azimuth ranges into which the orientations of certain groups of chambered tombs fall.

Nonetheless, some conclusions are inescapable at the broadest level. Throughout Britain and Ireland, there is a strong preference against orienting tomb entrances upon the part of the horizon close to due north. In most regions it is avoided completely.[72] This does suggest that, whatever the local tradition, it was important to avoid that range of directions where the sun or moon never appeared, either on or above the horizon. It is also clear that the predominant overall trend in chambered tomb orientation throughout Britain and Ireland is eastwards, towards the part of the horizon where celestial objects rise. The Clava cairns of Inverness-shire and the Irish wedge tombs, two groups which face predominantly south-westwards or westwards, stand out as conspicuous exceptions.[73]

The Clava cairns have been studied in some detail by Burl. These monuments consist of two types, passage tombs and ring cairns, and each example without exception is oriented within the quarter of the compass centred upon SSW. Burl has calculated the horizon declinations in the direction of orientation on the basis of azimuths deduced from existing plans and considered accurate to ±2°, together with horizon altitude data deduced from Ordnance Survey maps.[74] The pattern of declinations obtained shows that although some are oriented upon sunset at winter solstice, including the aligned passages of two passage tombs which occur in association with a single ring cairn at Balnuaran of Clava itself (NH 757444) (Fig. 8.6a),[75] the simplest consistent explanation of the orientations of the group as a whole is lunar (Fig. 8.6b). Burl himself felt that the evidence favoured a mixture of solar and lunar symbolism, and it is certainly true that his declinations from the passage tombs, whose passages define an orientation more precisely than the entrances of the ring cairns, fall neatly into three groups closely corresponding to the setting (or, in three cases, rising)[76] positions of the moon at its major and minor standstill limits and, between these, that of the setting solstitial sun.[77] However, a simpler explanation that would fit all the data would be orientation upon (say) the full moon at midsummer, at arbitrary points in the lunar node cycle.[78] In either case, though, this group appears to provide reasonably convincing evidence for intentional lunar, as opposed to solar, symbolism.[79]

Most of the Clava cairns are surrounded by a circle of standing stones, whose graded heights reflect the general orientation preference of the group, the tallest stones generally being to the SSW. This invites a comparison with the recumbent stone circles that are found further south-east in Aberdeenshire, and is one reason why Burl has postulated that the Clava cairns were precursors of the RSCs, suggesting that in this region there were at least some strong elements of continuity in the transition to the sorts of monument, suggestive of public ceremonial and without necessarily any direct association with burial, that seem to characterise the later Neolithic. In fact, Early Bronze Age dates have been obtained recently at four Clava cairns (see Archaeology Box 2), thus undermining the chronological assumptions upon which this particular argument is based, but the structural similarities between the two types of monument are still evident.

These dates also mean that our discussion of Clava cairns (and this applies equally to wedge tombs, which also yield dates generally in the latter part of the third millennium BC—see Archaeology Box 2) has swept us rather a long way forward in time. In order to focus on the earliest of the stone circles, we must step back again.

## In tune with the cosmos: circles, landscape and astronomy in the Late Neolithic

The great monumental constructions of the Late Neolithic appear to mark the culmination of attempts to manipulate the ways in which particular places could be encountered and experienced. Ditch-and-bank enclosures, timber rings, and stone circles and avenues all served to set spaces apart from the

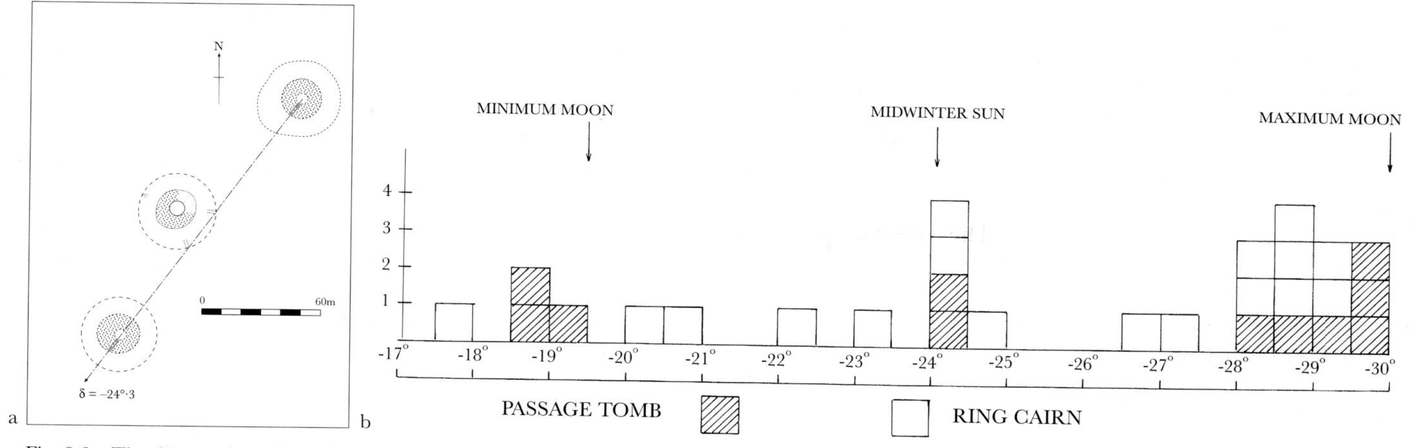


Fig. 8.6 The Clava cairns of Inverness-shire.
a. Plan of Balnuaran of Clava, adapted from Thom 1966, fig. 11.
b. Declinations of the Clava cairns. First published as Burl 1981, fig. 7.5 (the lunar targets have been shifted to take account of parallax—see Astronomy Box 2).

rest of the world. At the great henges built in the third millennium BC, 'this constraint upon bodily movement and physical experience was still further elaborated, with the construction of concentrically-ordered spaces, defined by ditches and timber uprights, linked by avenues and screened by facades of posts'.[80] The epitome of these developments occurred in the Avebury area, where the Sanctuary, Silbury Hill, the great henge itself with its interior stone settings, and the avenues were all constructed within two or three kilometres of one another between about 3000 and 2300 BC.[81] Surprisingly perhaps, little is evident—or at least has been convincingly suggested—in the way of overt astronomical symbolism amongst these monuments.[82]

But low-precision astronomical alignments found repeatedly amongst henges, stone and timber rings give numerous hints that astronomy was very much a part of ceremonial tradition and practice in the Late Neolithic, at least in certain places at certain times.[83] It is regrettable that there have been relatively few systematic studies of groups of monuments from this period, but some small-scale studies do exist, an example being Burl's work on some of the great 'open' stone circles of Cumbria and south-west Scotland. Burl argues that these represent one of the earliest regional groups of stone circles, dating from around 3200 BC.[84] Several of the largest examples were constructed in and around the mountainous region of the Lake District, often well away from fertile valleys (Fig. 8.7), and they appear to be connected with axe production and distribution, possibly serving as meeting and trading places.[85] Nonetheless the rings were evidently laid out with considerable care and some consistency. All have 'entrances', usually formed by placing two extra stones alongside adjacent circle stones so as to make a rectangle; some, like Long Meg and her Daughters, have tall additional stones standing just outside the circle; and some non-circular rings, like Castlerigg, have tall stones marking their longest axis. Seeking to control the many possible alignments that could be used to support almost any speculation, Burl considers only those from the ring centre to the entrance, to the tallest stone in the ring, and to outlying standing stones where present.[86] He concludes that these monuments were designed to encapsulate two alignments, one to a cardinal point and one to sunrise or sunset at an important time of year.

The relevant azimuth and declination data given descriptively by Burl are summarised in Table 8.1, quoted here simply to the nearest degree.[87] It is certainly true, without exception, that the orientations listed by Burl either fall near to (within 8° of) a cardinal direction or else yield declinations between −24° and −16°, and a combination of the two occurs at at least three rings—Brats Hill, Castlerigg, and Long Meg and her Daughters. On the other hand, there is a worrying inconsistency in the use of entrances, tallest stones or outliers, which must raise concerns about data selection.[88] Burl interprets the three declinations close to −16° as alignments upon sunrise or sunset on the mid-quarter day in early November corresponding to the later Celtic festival of Samhain. He is unable, however, to explain the −22° obtained at Swinside, suggesting that 'if, instead, the line had passed through the southern pair of portal stones the . . . declination would have been −24°·6, close to midwinter sunrise'.[89] This special pleading becomes unnecessary if it is postulated that the spread of declinations between −24° and −16° simply represents 'a concern for the sun at the darker quarters of the year'[90] without seeking specific precursors for the Celtic calendar. Bearing in mind that the exact position of the ring centre is only meaningful as a geometrical construct, all the more complex in the case of non-circular rings, and that there is no archaeological evidence to suggest that the exact centre had any special significance, the simpler explanation seems the safer one.

These cardinal and solar alignments were imprecise, as even Thom himself recognised. To him, Castlerigg was a symbolic observatory, a marvel of design 'controlled by the desire to indicate the rising or setting positions of the sun and moon at important times',[91] but one in which the alignments were never intended for refined observations.[92] Thom felt that astronomical symbolism at such monuments was integrated with their geometry, Castlerigg itself being a Type A flattened circle which had 'been made to fit the astronomical requirements'.[93] The complex geometrical constructions proposed by Thom did not withstand statistical reappraisals in the 1970s (see Statistics Box 5), and his astronomical contentions at Castlerigg can never be proved, but simply looking at the orientation of the longer axes of oval and elliptical rings does hint at wider evidence that astronomy and geometry were sometimes woven together at a much more basic level—just one way perhaps in

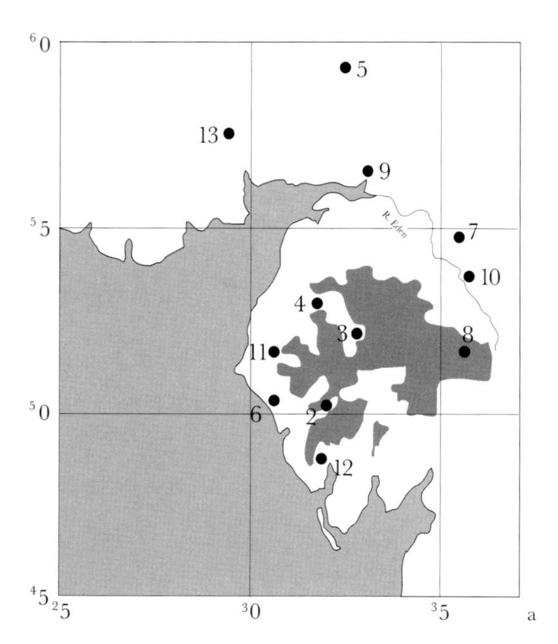


which the design of a timber or stone ring could be related to its place within the cosmos.

One monument often quoted in this context is Woodhenge, whose six concentric rings of timber postholes discovered in the 1920s—adjacent to the great henge at Durrington Walls and only 3 km from Stonehenge—are elliptical in shape with their major axes aligned in the north-east upon midsummer sunrise.[94] Here, however, the exact nature of the timber construction or constructions whose uprights were placed in the holes is unknown—many believe that it was a roofed building rather than a set of rings of free-standing timber uprights[95]—and we should not ignore the fact that the orientation of the entrance to the surrounding ditch-and-bank enclosure is rather different from that of the building, or timber rings, erected inside it (Burl, forthcoming). It is undoubtedly better to view this monument in a broader contextual setting and we shall return to the Stonehenge area shortly.

The longer axes of stone rings may provide more convincing evidence. For example, given the other intimations of low-precision cardinal alignments amongst the great Cumbrian

b

Fig. 8.7 Great stone circles in Cumbria.
a. Location of great stone circles in and around the Lake District, after Burl 1988a, fig. 7.1. The numbers refer to the reference numbers in Table 8.1.
b. The circle at Swinside (Sunkenkirk), Cumbria, viewed from the north-west, . opposite to the entrance.
c. Long Meg and her Daughters.

c

circles, the fact that the long axis of the oval-shaped Twelve Apostles near Dumfries is aligned on azimuth 8°/188° may be interpreted as another case in point.[96] Elsewhere, a number of instances have been quoted where the long axis of oval and elliptical stone rings aligns with a horizon solar event. Some, like the Druid's Circle, Gwynedd (SH 723746) and other rings in north and central Wales, align with sunrise and sunset at some point during the year, and further investigation might reveal whether such axes not only fall within the solar arc but also within a more restricted range of declinations.[97] In other cases the long axis alignment is solstitial. An important example is the elliptical ring at Cultoon, Islay (IS28 in List 2), excavated by Euan MacKie in the 1970s, whose major axis is aligned upon winter solstice sunset; and it may not be coincidental that it is also oriented upon a distant mountain peak in Ireland.[98] Systematic regional studies of the longer axes of stone rings might well reward further study, as indeed might studies of outlier orientations[99] and even the entrance orientations of ditch-and-bank enclosures and henges.

Horizon astronomical events might well have been important when choosing the location of a stone circle, even if no structural indication of the event in question is now evident, or indeed ever existed. This possibility has to be approached not through architectural design and orientation but by examining the astronomical potential of different points in the landscape. An early investigation along these lines was published by John Barnatt and Stephen Pierpoint in 1983. It concerned Machrie Moor on the Isle of Arran (around NR 9132), where six stone circles of different designs are found within 300 m of one another, possibly representing a development that lasted several centuries in an area separate from settlement and cultivation immediately to the west.[100] By taking measurements from points at 100 m intervals over a wide area, the investigators showed that the locations actually chosen seemed to have unusually many astronomical events (solstitial sunrise or sunset, or moonrise or moonset at a lunar standstill limit) in prominent horizon notches.[101] The investigation raises familiar questions of the selection of 'significant' astronomical events and 'prominent' notches as well as the permitted errors in determining whether a given event did coincide with a given horizon feature,[102] but it is important for raising more general questions and techniques concerning what we might call 'highly astronomically charged' points in the landscape.

The example of Cultoon reminds us that astronomical associations are unlikely to have been the only symbolic influence upon the location and orientation of stone circles. A number of scattered examples have been noted of rings being aligned upon prominent natural features in the landscape. For example, one of the larger stones of the Seven Stones of Hordron (Hordron Edge) in Derbyshire (SK 215868) stands in line with the sharply peaked Win Hill.[103] Other rings are located where natural features coincide with astronomical events, such as Nine Stone Close in Derbyshire (SK 225625), from which the moon at the southern major standstill limit sets behind the gritstone crag of Robin Hood's Stride to the SSW, between 'two stubby piles of boulders jutting up at either end of its flat top like the head and pricked-up ears of a wrinkled hippopotamus'.[104] Clearly, in any systematic study it would be unwise to consider astronomical potential in isolation from prominent features in the landscape.

In the case of entrances and outliers, we should be cautious about following the usual assumption that it was necessarily the exact geometrical centre of a ring that was in some sense the 'observing point'—the unique position in relation to which orientations and alignments might have been perceived as significant by the people who built and used these monuments. In most cases the only distinction that is obvious from the material record is the general one between demarcated interior space and exterior space. Admittedly some stone circles did contain centre stones—the practice evolving, probably independently, in places as far apart as Cornwall, Shropshire, Galloway, the Isle of Lewis, and County Cork[105]—but few other signs have been found of the existence of markers, or indeed of any other special activity having taken place, at or close to the geometrical centres of rings. A notable exception is a setting of four stones, carefully placed and levelled, enclosing an area approximately 2 m square in the centre of the Stones of Stenness in Orkney,[106] but this seems to have been constructed around an upright timber post,[107] inviting the suggestion that far from being a viewing place it actually contained a gnomon or marker to be viewed.[108]

The only features within the interiors of stone circles from which it has seriously been suggested that the view outwards was important are the so-called 'coves', curious sentry-box-shaped settings of three tall stones[109] which may have been intended to resemble the entrance forecourts of chambered tombs.[110] Three examples occur within (but not at the centres of) circle-henges well dispersed geographically: at Cairnpapple in West Lothian (NS 987717),[111] at Arbor Low in Derbyshire (SK 160636), and in the north circle at Avebury. A fourth, while associated with the three stone circles at Stanton Drew in Somerset (ST 601631), is located some 150 m to the west of the southernmost of them. The fifth and last, Beckhampton, once stood to the west of Avebury in the Beckhampton Avenue.[112] There is no consistency in their orientation: the five coves face (or faced) E, SSW, NE, SSE, and SE respectively.[113] Yet despite this diversity it has been suggested that the Beckhampton Cove was deliberately aligned towards midwinter sunrise[114] and, more tentatively, that three others faced major standstill limiting positions of the moon: Avebury towards most northerly moonrise,[115] Stanton Drew towards most southerly moonrise, and Arbor Low towards most southerly moonset.[116]

While statistically unverifiable, these hypothetical associations do represent the first tentative evidence of symbolic pre-eminence being assigned to the moon as opposed to the sun, at least architecturally, in association with stone circles. It should certainly come as no surprise that both the sun and moon, and indeed other celestial bodies, might not have been of great sacred importance or that, say, ceremonies taking place in stone circles might not have been timed in relation to the moon as well as the sun.[117] Indeed, the more extensive, systematic archaeoastronomical studies of the north-eastern Scottish recumbent stone circles described in chapter five appear to provide strong evidence of lunar, rather than solar, symbolism. The RSCs were apparently built up to about 2000 BC, and this immediately suggests the possibility that, while retaining the custom of demarcating a circular space, they reflect a very different ideological tradition from most other stone circles in Britain and Ireland, and from contemporary henges;[118] they share this distinctive orientation custom with the Clava

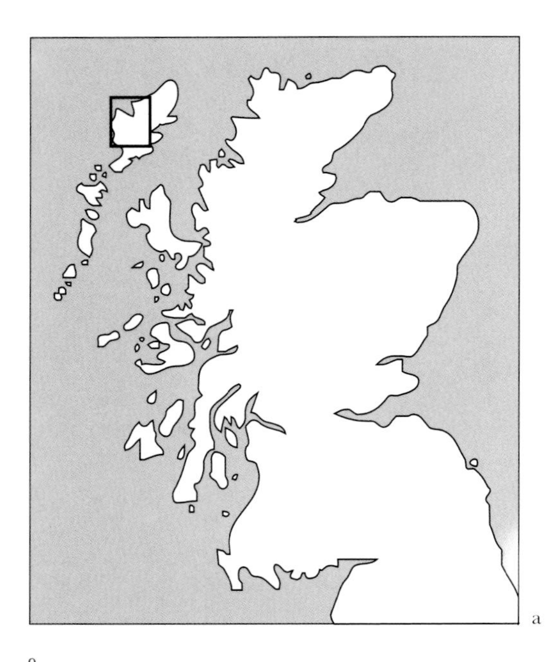


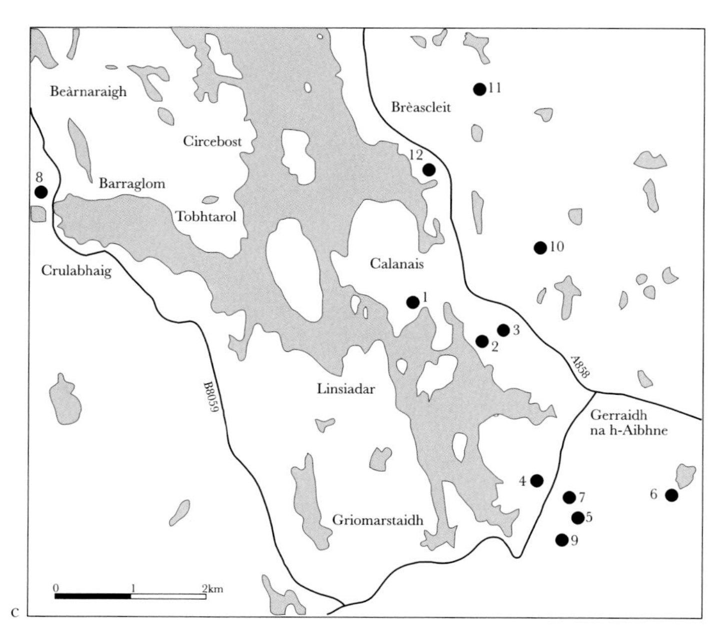


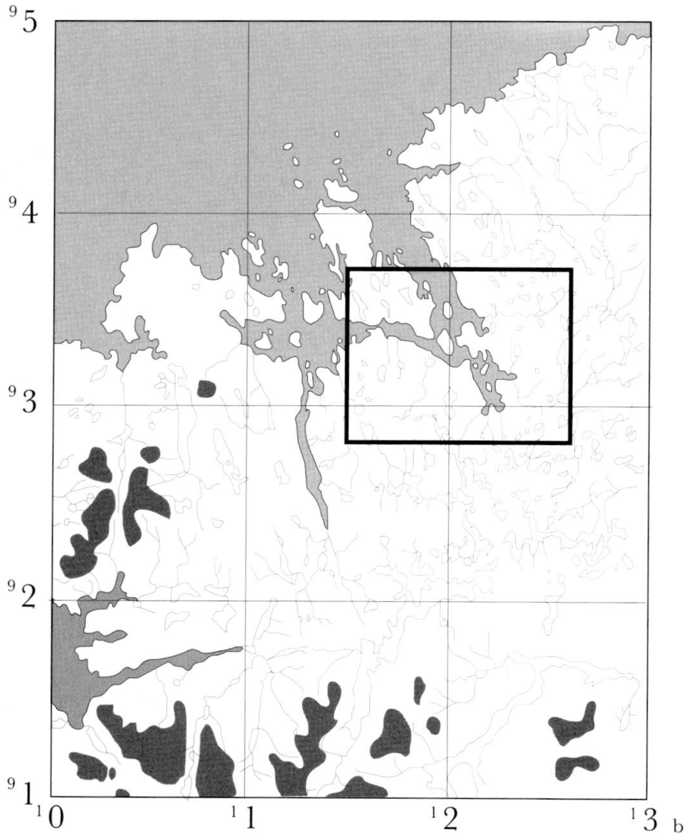


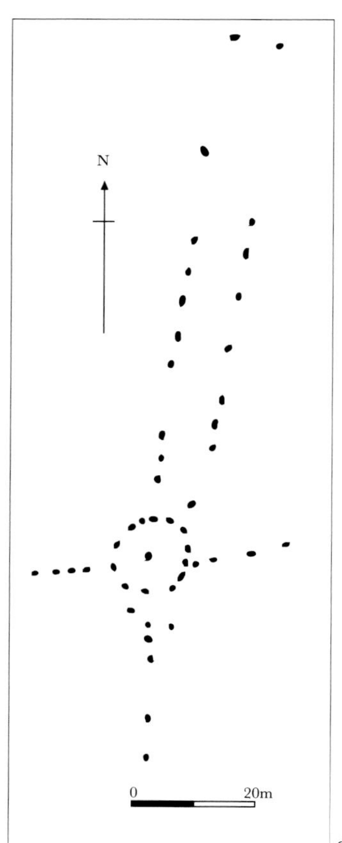


Fig. 8.8 Callanish, Isle of Lewis.
a,b. The location of Callanish.
c. Monuments in the vicinity of Callanish, after Ashmore 1995, 12. The site numbers correspond to the alternative names 'Callanish I', 'Callanish II', etc., most of which are given in List 2 (LH 8–24). For information on the remainder, mostly of more doubtful status, see Ruggles 1984a, 28–9, 46, 50, 53.
d. Plan of the standing stones of Callanish, after *ibid.*, 6 (scale added).
e. The standing stones of Callanish: aerial view from the NNW. The outcrop of Cnoc an Tursa is clearly visible beyond the end of the southern row.

cairns. It is possible, of course, that more detailed archaeoastronomical investigations of regional groups of stone circles elsewhere, of henges, or of other Late Neolithic enclosures might uncover evidence of a widespread lunar symbolism in earlier times, but the evidence currently available offers little to suggest that this might turn out to be the case. Instead, one wonders whether the RSCs represent a relatively late, and quite distinctive, tradition that resulted in most burial and open circular monuments facing south-west or westwards. The even later axial stone circles of south-west Ireland, which were still being constructed well into the Bronze Age, certainly around 1600 BC and possibly as late as 1200 BC (see chapter five), might even represent a late remnant of a related tradition where the subtleties of relationship to the moon and other landscape features were no longer important.

Claimed associations between stone circles and the moon have received a good deal of attention in two other specific cases. The first is Calanais, more usually known in the angli-

e

cised form as Callanish, on the west coast of the remote Isle of Lewis in the Outer Hebrides. The famous standing stones of Callanish (LH16 in List 2) stand grandly on a promontory overlooking East Loch Roag (Fig. 8.8). A small but spectacular ring of tall, slender standing stones of Lewisian gneiss, with an even taller (4·5 m-high) central stone and additional radial lines and avenue forming the shape of a distorted Celtic cross, this monument is widely acclaimed as the 'Stonehenge of the North'.[119] The circle itself was probably erected around 2900 BC, prior to the construction of the tiny chambered tomb inside it.[120] In the vicinity are several smaller stone circles and other settings of standing stones.[121] The most obvious orientation feature of Callanish is a prepossession with the cardinal directions, not just indicated by an entrance, an axis or an outlier but represented physically in stone in what was probably a set of later additions to the existing circle. The fact that not all these cardinal indications, if such they were, are particularly precise—the avenue to the north, for example, is actually oriented roughly 10°/190°[122]—has triggered a variety of more specific stellar explanations, none of which stands up to scrutiny for a number of reasons of the sort described in chapter two.[123]

The possible lunar significance of the southern orientation of the Callanish avenue, which is roughly aligned upon moonset at the southern major standstill limit, was emphasized by Hawkins and Thom,[124] although much debate was generated by the fact that both these authors claimed precise alignments upon the distant mountain of Clisham on Harris, whereas in fact Clisham cannot actually be seen along the line of the stones owing to the rocky outcrop of Cnoc an Tursa immediately to the south.[125] In fact, as viewed from the north end of the avenue, the midsummer full moon in a year close to major standstill is seen to skim along the horizon, disappear behind Cnoc an Tursa, and then reappear briefly at ground level within the silhouette of the stone circle.[126] Perhaps the intention was to establish a dramatic relationship between the monument, the landscape and the heavens in this way,[127] one that would only be enacted on certain rare and auspicious occasions.

It has been suggested that Callanish and the neighbouring monuments formed a 'complex' in which ranges of sacred hills were framed in relation to lunar rising and setting positions on the horizon.[128] These ideas are intriguing and contrast absolutely with attempts to determine the astronomical potential of the Callanish monuments according to a strictly objective code of practice in the 1970s.[129] In one way, they represent the epitome of contextual reasoning, integrating monuments, landscape and sky, but it is a contextual reasoning that pays little attention to negative data and questions of data selection.

The second famous case of claimed associations between a stone circle and the moon is Stonehenge.

## Stonehenge and its landscape

The account up to this point has been broad-based and necessarily cursory. It has concentrated upon burial and ceremonial monuments because these have been the focus of most work to date, and it has emphasized repeated trends visible in the archaeological record because therein lie the most compelling evidence of common practices that merit explanation. In order to give the flavour of a fuller account in a local context quite different from the fringes of western Scotland examined in the North Mull Project, and in a social landscape of considerably greater complexity, we turn once more to Stonehenge. The architectural developments at and around the monument, and the changing context of the wider physical and social environment, are summarised for reference in Archaeology Boxes 3 and 4.

Thinking chronologically, the first question we might ask is whether there was anything special about the spot chosen for the first ditch-and-bank enclosure built on the site of Stonehenge in about 2950 BC. In the absence of intervening woodland it would just have been visible from the causewayed enclosure at Robin Hood's Ball,[130] but too far to the south for its direction to have had any association with the rising sun or moon.[131] It is also unlikely that the siting of this first earthwork monument bore any relation to the Stonehenge cursus to the north, even though the sarsen monument is prominently visible today from the eastern part of the latter.[132] A possibility is that the enclosure was placed so as to be viewed from a distance[133]—it is on a visibility 'high spot' at the centre of a number of 'nested bowls'[134]—but it was merely one of several circular enclosures built in the Wessex downlands at around this time (cf. p. 157) and a comparison with the siting of the others could help to establish whether this reflects a common pattern of preference. On the other hand, it is possible that this special property of its location helped to single it out for further development later on.

The fact that the enclosure had entrances facing north-east and south has given rise to a comparison with the great Cumbrian stone circles, with their two alignments, one in a cardinal direction and one within the solar arc.[135] However, the fact that the azimuth of the axis through the north-east entrance was well to the north of the solar arc and the presence, at least at first before it was blocked off, of a possible further entrance to the SSW seem to argue against this. Fig. 8.9a shows the orientations of the entrances as viewed from the centre of the ring, and gives the horizon declinations estimated to the nearest degree. The north-eastern entrance actually spans approximately the range of horizon where the moon sometimes rises, but never the sun (i.e. the declination range +24° to +28°). That this was intentional is a possibility, although this again raises worries about whether people in prehistory necessarily shared our prepossession with the geometrical centre.[136] Focusing instead on the axis through the north-eastern entrance, this is centred on a declination of about +27°, sufficiently close to the major standstill limit that the moon would only be seen to rise on this axis on occasional favourable days during a period of about five years in every 18·6-year cycle.[137]

Taken in isolation this orientation cannot provide persuasive evidence of any intentional connection between the orientation of the north-east entrance and the moon. Evidence from other middle Neolithic enclosures might cast further light on the question. But the idea of such a connection does not end there. There has been a great deal of detailed speculation on the issue, due largely to the presence of an array of small postholes discovered in the causeway across the ditch by Hawley during his excavations in the 1920s.[138] This began in the 1960s with Peter Newham's assertion that wooden stakes were erected to mark moonrises over several 18·6-year cycles, and hence to set up an exact orientation upon the northern lunar limit.[139] The idea has been accepted by a succession of later commentators,[140] but in fact the technicalities do not withstand close scrutiny, quite apart from any archaeological considerations.[141] The positions of the postholes are shown in Fig. 8.9b. Four larger postholes in a line a few metres outside the entrance—the so-called 'A' holes—have also been suggested as forewarners of northernmost moonrise, but the mere presence of this line of posts does not in itself prove that they were used for lunar observation.[142]

In fact, the balance of the archaeological evidence assigns the entrance postholes to Phase 2, suggesting that they were added some time after the ditch-and-bank enclosure was built,[143] perhaps in an attempt to restrict access down to two narrow 'entrance corridors',[144] or even to block it off completely, the interior being accessed solely via the remaining entrance in the south.[145] Interestingly, the two proposed passageways line up with gaps between the 'A' postholes further

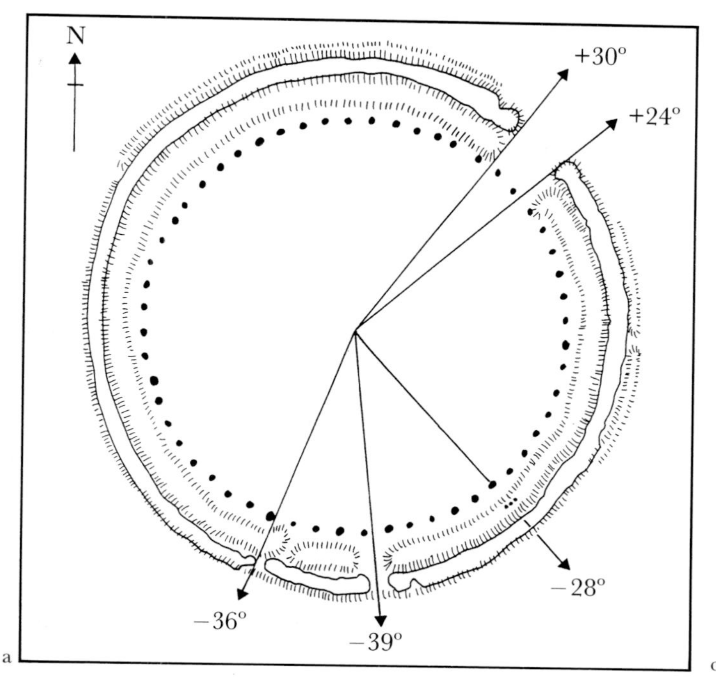


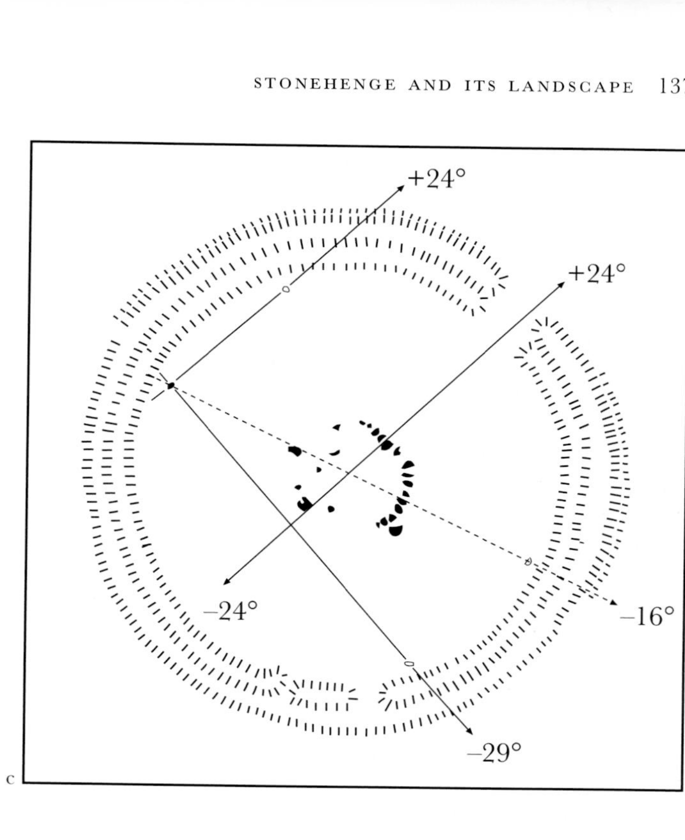


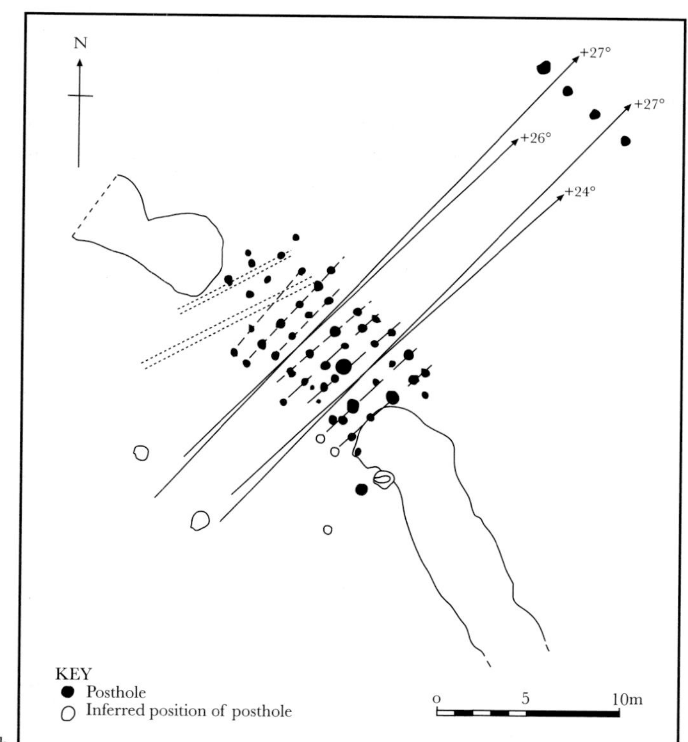


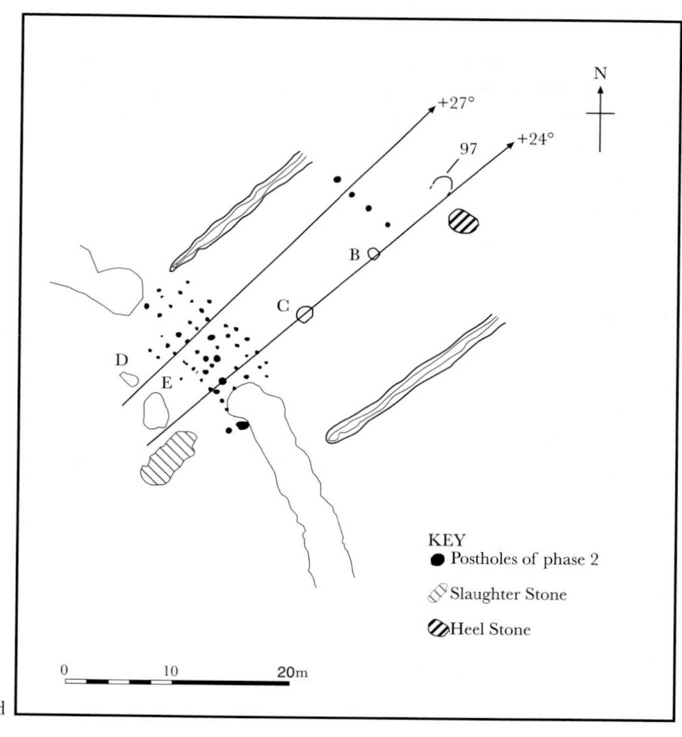


Fig. 8.9 Stonehenge: orientations and astronomical potential. The arrows and 'indicated declinations' are merely intended to help the reader to assess the astronomical potential of a given direction, not to imply that observations were necessarily made of horizon phenomena, nor that they were made from a certain position along the arrow as 'sightline'. For commentary see the text. The azimuths have been judged from plans in Cleal *et al.* 1995, and a horizon altitude of 0°·5 has been assumed, which should result in a declination within 0°·5 of the true value since the Stonehenge horizon altitude is always between 0° and 1°.

a. Stonehenge 1/2. First published as Ruggles 1997, fig. 4, based on Cleal *et al.* 1995, fig. 256. For the raw data, including azimuths, see Ruggles 1997, table 5.

b. The north-eastern entrance during Phase 2. First published as Ruggles 1997, fig. 5, based on Cleal *et al.* 1995, fig. 68. For the raw data, including azimuths, see Ruggles 1997, table 6.

c. Stonehenge 3i/3a. First published as Ruggles 1997, fig. 6, based on Cleal *et al.* 1995, fig. 256. For the raw data, including azimuths, see Ruggles 1997, table 7.

d. The north-eastern entrance during Phase 3. The earlier stakeholes are also shown. First published as Ruggles 1997, fig. 7, based on Cleal *et al.* 1995, fig. 156. For the raw data, including azimuths, see Ruggles 1997, table 8.

out, and are in the same orientation as the Phase 1 axis, yielding a horizon declination of around +27°. These lines do not, however, pass through the centre of the monument, so that if observations were made to the north-east from the interior at this time, it may not have been from the geometric centre.[146] Viewing the passageways from the centre yields orientations further to the east, the right-hand one lining up on the midsummer sunrise. The possibilities are shown in Fig. 8.9b.

A few other things may add support to the idea of lunar significance in Phases 1 and 2. For example, it has been argued that the distribution of deposits of bone, antler and pottery in the Aubrey Holes and in the enclosure ditch and bank reveals spatial patterning relating to the moon (and/or the sun).[147] Also, three small postholes discovered by Atkinson under (or through) the bank on the south-east side of the monument yield a declination of −28° from the centre, which is close to southern major moonrise (Fig. 8.9a).[148] This and subsequent activity in this sector of the site raises the possibility of a sacred significance being attached to the directions of both the northern and southern limits of moonrise. But none of the evidence is compelling.[149]

This situation changes in the Late Neolithic. Stonehenge up to this time was not exceptional, a simple ditch-and-bank enclosure in a mixed landscape within which a succession of wooden structures were erected, presumably for ceremonial purposes. Perhaps other timber rings were put up after the Aubrey circle had been forgotten.[150] As the third millennium progressed there was a marked increase in pastoral and arable farming, accompanied by considerable social change, with an ever-growing labour force that could evidently be mobilised for large construction projects. The henge tradition had spread to Wessex by this time: a major communal focus in the area was undoubtedly the great henge built at Durrington Walls, some 3 km to the ENE; Woodhenge was built adjacent to it, and another henge was constructed at Coneybury, a little over 1 km to the ESE.

It is at this time that strong evidence for astronomical symbolism appears, and this is associated with the sun. The solstitial orientation of Woodhenge has already been mentioned; Coneybury had a similar orientation towards midsummer sunrise.[151] At Stonehenge itself, the transformation of the monument into stone was accompanied by a shift of several degrees in its axis to bring this into line with summer solstice sunrise to the north-east and winter solstice sunset in the south-west.[152]

The details are not straightforward, particularly because the archaeological sequencing remains unclear in many parts of the site (see Archaeology Box 3). Nonetheless, whatever the exact nature of the bluestone setting of Stonehenge 3i, we can be reasonably confident that its axis of symmetry yields a declination of +24° to the north-east and −24° to the south-west (Fig. 8.9c), with the Altar Stone placed on this axis.[153] If this is contemporary with Stonehenge 3a, and stone 97 did indeed stand as a companion to the Heel Stone, then this axis passed between them. 'If the main purpose was to bracket the rising midsummer sun, as seems likely, then the positioning of the two stones here would have achieved that aim'.[154] On the other hand, if the Altar Stone was the focus of attention and the Heel Stone and its companion marked the ceremonial entrance to the monument, it is certainly just as plausible, and arguably more so, that the alignment of particular symbolic value was that of the Altar Stone with the direction of midwinter sunset in the south-west.[155]

The other structure that may have been built at around this time is the Station Stone rectangle. This anomalous setting, which may have been influenced by Breton styles of architecture,[156] is aligned in one direction with the Phase 3i axis, but in the other (NW–SE) upon declination ±29°, corresponding roughly to moonrise at the southern major standstill limit (full in summer) and moonset at the northern major limit (full in winter).[157] This raises the question of whether, given that the solstitial alignment was intentional, the lunar one passed unnoticed, or whether the fact that the moon came very occasionally to rise or set in the direction at right angles to the axis was actively exploited,[158] perhaps perpetuating an earlier interest in the southernmost rising moon in the south-east.[159] Possible astronomical interpretations of the diagonals are even more tentative, but have been suggested in an attempt to explain why the station stones were placed in a rectangle and not a square.[160] A problem here is that the diagonal lines were probably obscured, at least partially, by the bluestones, and certainly would have been blocked by the later sarsens.[161]

If one thing is clear, it is that from around 2500 BC Stonehenge became pre-eminent. Apparently, the effort invested in the construction first of the bluestone, and then of the sarsen monument, and then in their continued remodelling, represented an attempt by those in control of this prosperous region to legitimise their position through a massive and unchallengeable construction occupying a central place in the natural and supernatural scheme of things and serving as a permanent testimony to past achievements. If so, it paid off well. While other Late Neolithic centres of activity fell into disuse, including even the mighty Durrington Walls, Stonehenge remained a place of major importance, a ceremonial focus that endured for something like a thousand years.[162] Its symbolic power was expressed in a variety of ways:[163] in its size, shape and texture; in the materials used (including the various different non-local stones); in the positions where commemorative offerings and other deposits were placed; in what was seen and experienced by people approaching the monument, especially later, along the avenue; in the rituals that took place in the interior, and who and what was excluded; and possibly even in how its architecture reflected its position in the wider landscape.[164]

In this wider expression of symbolic power, the solstitial axis evidently played a major part, although the details are confused. The so-called 'Slaughter Stone' and stone *E*, erected in the entrance (see Fig. 8.9d), are important in this context. It is unlikely that they predated the sarsen circle. Radiocarbon dates from two antler picks in the stonehole of *E* imply otherwise;[165] the Slaughter Stone was carefully worked;[166] and it is difficult to imagine getting the sarsens in past the two entrance stones if they were already in place (there is historical evidence that the Slaughter Stone still stood in 1666).[167] Burl argues that the entrance stones were deliberately placed in line with the Heel Stone and its companion so as to produce a 'corridor' down which the light of the rising midsummer sun would have shone directly into the interior of the monument.[168] In this confined space, now mostly cut off from the view of the outside world—and with most of its view of the outside world cut off—by the sarsen stones,[169] the effect would have been extremely dramatic for the privileged few allowed to witness it.

Unfortunately the archaeological evidence, while not ruling out this interpretation, does permit a range of others. For example, the relative chronology is uncertain, and there are some reasons to suggest that the Heel Stone's companion had already been removed by the time the Slaughter Stone and stone *E* were even erected.[170] In addition, it is clear that a third entrance stone (*D*) stood to the north-west of the Slaughter Stone and *E*,[171] giving rise to the suggestion that the stones formed a double entrance reflecting the two entrance corridors suggested in the earlier timber phase.[172] While this interpretation does not effect the orientation of the 'solar corridor' it does tend to diminish its apparent importance.[173] Finally, the idea that the earthwork entrance was deliberately widened in order to bring it into line with the new axis and to match the width of the avenue—as Atkinson, the excavator, believed[174]—is now disputed;[175] under the new interpretation it seems strange that the avenue, especially emphasized by having the 'solar corridor' running centrally down it, should continue to be half blocked by the ditch where it meets the Stonehenge enclosure.[176]

In 1982 Richard Atkinson summarised a review of the astronomy of Stonehenge by saying that 'only the alignment of the Avenue on the summer solstice sunrise . . . can be accepted with confidence. All other interpretations are open to doubt or to alternative explanations'.[177] Despite all the new ideas put forward in recent years, Atkinson's remark remains true in broad terms. The obvious conclusion is that the axial orientation was changed in around 2500 BC so as to incorporate the solar alignment, perhaps as part of the process of legitimising the place of the monument at the centre of the cosmos and hence reinforcing its symbolic power. If this was the case, then it is hard to believe that other astronomical alignments—perhaps involving the sun at other times of year; perhaps involving the moon, planets or stars—were not also exploited at Stonehenge. It is also quite possible that fortuitous associations between the architecture of the monument and prominent astronomical events—totally unintended and unappreciated by those who planned and constructed some part of the monument—were subsequently noticed and 'used'. However, because of the vast number of different possibilities it seems hopeless to try to identify, simply from the material record, particular associations as possibly meaningful.[178] Only strong contextual arguments might give any justification for attaching possible significance to some astronomical alignments while ignoring a great many others.

Whether or not the entrance stakeholes and stones represent deliberate attempts to restrict access to the interior,[179] it seems clear that rebuilding the monument in stone reflected an inexorable process of restricting 'movement, operation and observation' both at Stonehenge and in the surrounding landscape.[180] The sudden intrusion of sunlight into the inner sanctum of the sarsen monument at midsummer dawn, created by the corridor of pairs of stones to the north-east, could only be appreciated from the interior itself and was a highly exclusive spectacle.[181] Astronomical observations were almost certainly made in earlier times, but they were probably accessible to all: during the time of Stonehenge 1 or 2, sunrise or moonrise (say on a particular ceremonial occasion could have been observed by many hundreds of people, not just those within the enclosure.

Possibly, architectural embellishments at Stonehenge served to manipulate what began as the simple observation of summer solstice sunrise, transforming a communal sacred event into one accessible only to a favoured few.[182] On the other hand perhaps the emphasis on the sun was new, the manifestation of a new dominant ideology replacing an older one focused upon the moon.[183] The archaeoastronomical evidence does not favour either idea conclusively. Despite the enthusiasm of some recent authors for the idea, the direct evidence to support the idea of predominantly lunar symbolism in Phases 1 and 2 is not strong, and neither, as we saw earlier, is the indirect evidence from the Wessex long barrows of a practice of predominantly lunar orientation dating back to the Early Neolithic. Perhaps it is safest simply to conclude that sacred 'knowledge' relating to celestial events was increasingly controlled at and around Stonehenge through the Neolithic, paralleling the wider tendency that increasingly placed control in the hands of a privileged few, but that at present it is difficult to be more specific.

## The Bronze Age and beyond

The onset of the Early Bronze Age brings sharp contrasts between the Stonehenge area, where it seems that existing ideologies were successfully perpetuated and adapted so as to legitimise and maintain existing structures of political power, and many other regions of Britain and Ireland where the ceremonial traditions of the Late Neolithic appear to have broken down completely. Around Stonehenge, we see the appearance of great displays of personal prestige and wealth, in the form of richly accompanied burials within round barrows. Curiously perhaps, there is little to suggest overt astronomical symbolism in Wessex at this time, apart from that preserved in Stonehenge itself.[184] While many round barrows were located so as to be prominently visible from Stonehenge,[185] there is no evidence that astronomical considerations were important.[186] Burl has pointed out that the bodies within Wessex round barrows have a predominantly cardinal point orientation,[187] but there is nothing in the data to indicate any more specific astronomical symbolism. And finally, despite much astronomical attention being focused recently upon a magnificent gold lozenge discovered during explorations of the Bush Barrow in 1808 (Fig. 8.10),[188] there is really no convincing reason for believing it to be anything other than a particularly fine decorative artefact.[189]

Early Bronze Age astronomical traditions are actually far more evident in certain areas of Scotland and Ireland where Neolithic practices of public monument construction and use appear to have been perpetuated, albeit on a much smaller scale. The monuments concerned, such as small stone circles and short stone rows, seem to have been ceremonial foci serving relatively small groups of people. They are open: there is little or no evidence for any physical restriction on access, and what is visible from within a circle or beside a row is generally also visible over a wide area, although the spot itself may have special properties. Because similar monuments were constructed in considerable numbers in certain regions, we have the opportunity to study repeated trends and gain reliable

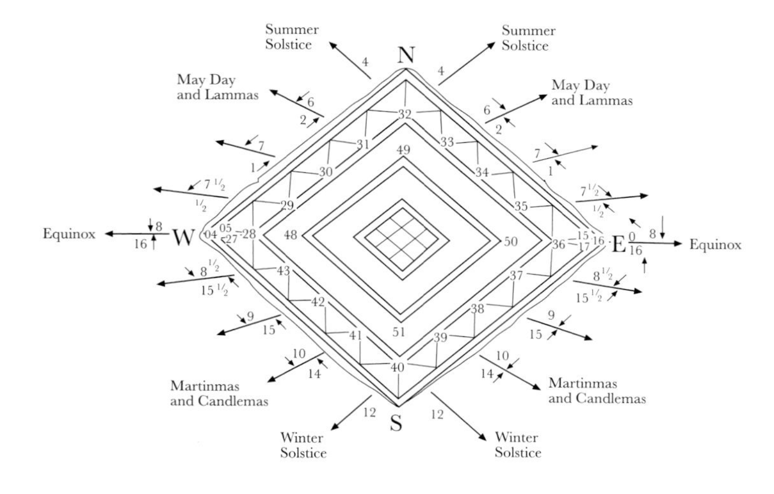


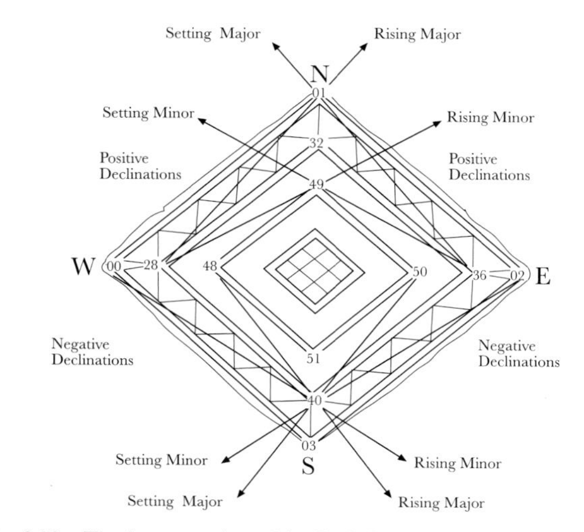


Fig. 8.10 The interpretation of the Bush Barrow gold lozenge as a solar and lunar calendar by A. S. Thom *et al.* (after Antiquity 62 (1988), 492–502, figs 1, 2 and 4).

insights into elements of common practice that were both deliberate and meaningful. As we have seen in chapters five to seven, such studies have begun to reveal a surprising wealth of detail. It is not unreasonable to suggest that the associations, astronomical and otherwise, between these monuments and the natural world around them were framed from a straightforward concern to keep human actions in harmony with nature and the cosmos.

Amongst the Scottish and south-west Irish rows, as we have seen, there are overall patterns of location and orientation that, although different between the two regions, both seem predominantly to be related to the moon as well as to terrestrial features such as prominent hilltops. A similar conclusion has been reached with regard to the recumbent stone circles (RSCs) of eastern Scotland; but the Irish axial stone circles (ASCs), while mimicking the RSCs in many aspects of their architectural design and in their general south-westerly orientation, show very little consistency in their locations in the landscape or in their astronomical potential, lunar or otherwise. The well preserved example at Drombeg stands as an isolated instance of an apparently clear solstitial orientation. If the Irish ASC tradition derives somehow from the Scottish RSCs but with differences of symbolic association (see chapter five), then this raises the question of how the ASCs relate to the stone rows of the same region; a problem indeed, because the latter seem to exhibit so clearly the symbolic associations on the moon and prominent hills that are absent in the former. Then there are the wedge tombs, the main period of whose construction seems to be earlier than that of the circles and rows according to the available radiocarbon evidence.[190] It seems to be generally assumed that these reflect solar symbolism,[191] leading to suggestions on the one hand that monuments where solar and lunar rituals took place coexisted[192] and on the other that a common emphasis on the setting sun provides evidence for ritual continuity.[193] All of this is questionable on the available evidence, and it would clearly be of great interest to undertake an integrated archaeological and archaeoastronomical investigation in a suitable locality, perhaps resembling the North Mull project, in order to provide a richer contextual picture that would help to clarify and progress these ideas.

Longer stone rows, various forms of which are found in scattered concentrations in Caithness, northern Ireland, the Welsh borders, and south-west England, as well as in Brittany,[194] generally seem to have been built somewhat earlier than the short rows. Burl estimates that they were constructed in the last two or three centuries of the third millennium BC and the first two or three centuries of the second,[195] although others suggest that their construction was a little earlier.[196] Few of the British and Irish examples have been studied astronomically, a conspicuous exception being Merrivale (SX 556748), one of several double rows on Dartmoor which received attention as a possible high-precision lunar observatory in the 1970s.[197] Generally, it is hard to conceive of any precise astronomical alignments along rows which are long and sinuous, but if they were built to serve as processional ways[198] less precise astronomical correlations might still seem a possibility. However, in Dartmoor, where over seventy single and double rows are concentrated,[199] their orientations are spread virtually all round the compass.[200]

Rock carvings are a feature of the Neolithic that extended well into the Bronze Age. Ranging from superb decorative art such as is found in the Boyne Valley tombs through to simple cup and cup-and-ring marks, 'petroglyphs' are found on a variety of natural outcrops around Britain and Ireland, as well as on the stones of stone monuments. We have mentioned (p. 129 and note 48) how, for example, the rayed circle in the roof-box corbel at Newgrange and a variety of similar designs have been interpreted as symbolic representations of the sun. When it comes to cup-and-ring marks, for which little or no contextual evidence is available, such attempts to ascribe direct meaning to the designs are even more speculative. A more productive approach may be to examine the spatial context of various designs. In a series of recent publications Bradley has examined correlations between structural components of petroglyphic designs and their location, concluding that they played a fundamental role in the organisation of the prehistoric landscape, helping to mark and characterise important natural places.[201] Burl has pointed out that markings on the sarsen stones at Stonehenge occupy cardinal orientations from the centre, hinting at the sorts of consideration that may have governed the positioning of petroglyphs on the small scale, within a monument.[202]

Astronomical associations are potentially important in the context of such arguments. As we have noted, some Boyne valley carvings may be placed in relation to the play of sunlight and shadow at certain times, although the likelihood of fortuitous coincidences is high and many or all may be unintentional. The location of carvings on solar alignments has been noted from Scotland to Brittany,[203] but more compelling evidence may only come from examining their orientations at groups of monuments,[204] and once again systematic studies are needed.[205] In one such study, a strong correlation has been noted between the orientations of cup markings at recumbent stone circles and the horizon rising and setting positions of the moon.[206]

An indication that certain types of round barrow in southern Britain might be worthy of attention in this context is provided by the bell-barrow at Crick in Gwent (ST 4844 9026), an isolated example of this type of barrow located well away from the main group in Wessex. Just two of the boulders surrounding the mound bear cup marks: one, the largest, has twenty-three cups on its outer face and from the centre is in the direction of midwinter sunrise; and the other, which bears seventeen marks on its upper surface, is in the direction of sunrise in early May and August (Fig. 8.11a).[207] It is especially interesting that the cup marks on the largest stone are concentrated towards the solstitial sunrise position itself, with a sharp cut-off beyond that position, as might be expected if they were positioned by 'lining up' with sunrise on days around the solstice (Fig. 8.11b).[208] The interpretation is not without its difficulties, however,[209] and in any case we would not wish to argue for meaningful correlations between the positions of cup marks and the sun on the basis of the evidence from this monument alone; nonetheless, it does provide an intimation of solar symbolism involving the position of petroglyphs well into the Bronze Age that should encourage broader, more systematic, investigations.

The types of investigation that have characterised archaeoastronomy in Britain and Ireland since the 1970s, as well as before, have seldom looked beyond the Middle Bronze Age.[210] This is undoubtedly due in no small measure to the lack in later times of the sorts of conspicuous monument that have served to stimulate, but also to constrain, modes of thought regarding prehistoric astronomy.[211] There *is* alignment evidence from the Iron Age,[212] but art and artefacts may also reveal a great deal. Miranda Green has argued that during the Bronze Age we begin to see evidence of complex rituals and ceremonies associated with a cult of the sun, and that these were perpetuated all over Europe in the Iron Age. Support for this idea comes in the form of sun symbols found over and over again on armour, jewellery, and small votive objects such as personal talismans.[213] Representations of the crescent moon are also found, for example, on coins[214] and on some anthropomorphic hilted short swords, a distinctive and specialised type of weapon made in 'Celtic' Europe during the La Tène phase of the later Iron Age.[215] It has recently been suggested that combinations of circles and crescents inlaid in some of these short swords, rather than representing the sun and moon, actually represented the moon at different phases, and that these short swords were used by a specialist religious class in practices associated with calendrical ceremonies.[216]

On the other hand, the much-quoted idea of a single 'Celtic calendar', with its eight seasonal festivals dividing the solar year into eight exactly equal parts,[217] stands on much weaker foundations than is generally assumed. First, the extent to which common 'Celtic' practices were widespread through north-western Europe in the Iron Age, and to which we can speak of a (timeless) Celtic tradition that perpetrated through into medieval times, is arguable enough in itself.[218] Second, the historical evidence for the Celtic calendrical festivals is actually very

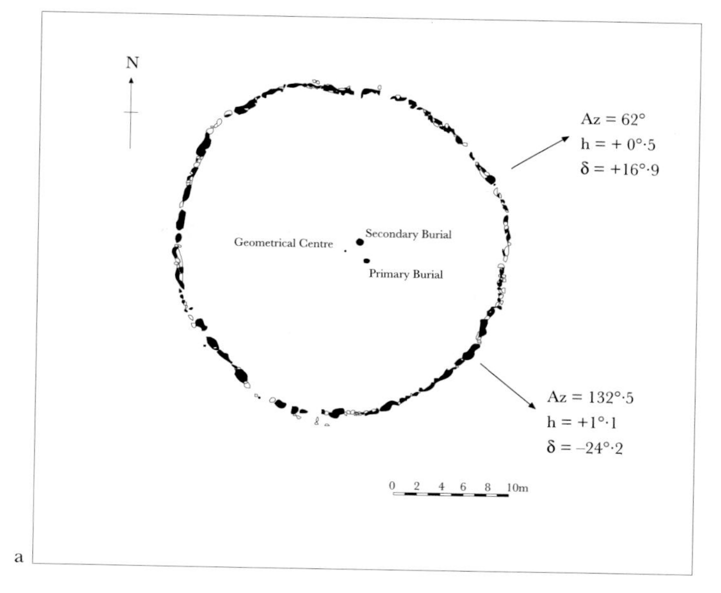


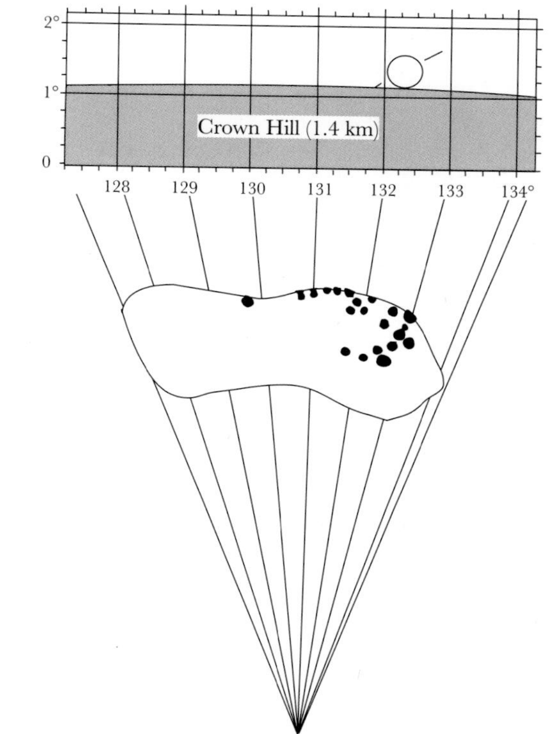


Fig. 8.11 Cup marks on the bell-barrow at Crick, Gwent.
a. Plan of the barrow, showing the positions of the two cup-marked stones (after Powell 1995, fig. 3).
b. The alignment through the south-eastern cup-marked stone (after Powell 1995, fig. 5).

weak.[219] It is true that solar alignments in earlier megalithic monuments such as those already discussed have often been quoted in support of these ideas, but the arguments can become dangerously circular. Third, even some of those who support the idea of a Celtic calendar and have attempted detailed constructions of it concede that in all likelihood the eight festivals were determined in practice by rules of thumb invoking a combination of solar and lunar observations rather than by counting off exact intervals of days.[220] Finally, the idea that a Celtic calendar of this nature left its mark in the famous bronze inscriptions at Coligny and Villards d'Héria in France[221] is only supported by indirect arguments.[222] And while there is certainly evidence that pre-existing seasonal festivals survived the onset of Christianity, going on to influence and help shape medieval and even modern folk traditions,[223] it is possible that these 'pagan' festivals may themselves date no earlier than Roman times.[224] Radical approaches may be needed if strong new insights are to emerge into astronomical elements of religion and ritual in the pre-Roman Iron Age.

## Overview: continuity and change

All human societies have a sky, and the things that people perceive there form an integral part of the world in which they live and of which, since Palaeolithic times, they have striven to make sense. Because of this it is hardly surprising to find widespread intimations of astronomical—or, at least, solar and lunar—symbolism within the material record in prehistoric Britain and Ireland. There is every reason to believe that further investigation will be fruitful in revealing many more. Yet even from the very patchy evidence briefly surveyed in this chapter, no clear picture emerges of overall astronomical development.[225] In one way this should not be surprising: particular sets of circumstances can tend to give rise to particular local practices, beliefs, and ideologies; and these can change rapidly when circumstances change. Thus at Flag Fen in Cambridgeshire the gradual encroachment of the sea in the Later Bronze Age, and the resulting competition for resources and struggle for land and power, appears to have given rise to a water cult where many highly valuable prestige objects were ritually cast into the waters.[226] The Neolithic and Bronze Age are characterised by such huge transformations in subsistence methods and economy and in the nature of the environment, together with continual social change and upheaval, and show such considerable variation from one region of Britain or Ireland to another at different times, that it would be surprising indeed to discover any simple, overall, pattern of developments in cognition, concerning the sky or anything else.

On the other hand, it can be argued that public ritual has a distinctive character, and that it tends to change on a slow timescale, emphasising continuity even in the face of considerable social change or innovation.[227] As we have seen, there is certainly evidence to support this in prehistoric Britain, for example in the persistence of the circular form of Stonehenge, including its transformation from timber into stone and its continued use for a millennium thereafter.[228] In somewhere as distant, and as different, from Wessex as the Scottish islands there is still every reason to believe that 'certain aspects of place changed little over millennia while dramatic changes in language, material culture and power relations swept through'.[229] Certainly this is encouraging for the prehistorian, since if certain ritual practices are less transitory, they are more likely to be discernible in the archaeological record. This also provides further justification for taking an interest in common practice amongst regional groups of monuments.

It has to be said that the available evidence gives little support for two general ideas concerning astronomy in prehistoric Britain and Ireland that have received a great deal of attention in the past and continue to receive tacit support. The first is that there was a shift from predominantly lunar to predominantly solar symbolism in the Late Neolithic. On the contrary, the most convincing evidence of lunar symbolism is found first amongst the Scottish recumbent stone circles, later amongst the Clava cairns, in both cases associated with apparently anomalous patterns of south-westerly orientation, and finally in the Bronze Age stone rows of western Scotland and south-west Ireland. But it is also fair to say that more work is needed before the idea that Early Neolithic long barrow orientations were associated with the moon rather than the sun or anything else can finally be dismissed.

The second idea is that a calendar involving eight-fold divisions of the year was in use during the Neolithic and Bronze Age, resembling—and perhaps even being a direct precursor of—the later Celtic calendar. According to this view, the importance not only of the solstices but also of the equinoxes and mid-quarter days was reflected in various architectural alignments from Early Neolithic times onwards. However, with the refutation of the statistical evidence in support of Thom's 'megalithic calendar', the idea only finds support in highly selective lists of monumental orientations: scattered instances of alignments upon sunrise and sunset on days such as 1 May (which, it is said, corresponds to the later Beltane), that seem to have been emphasized because these calendar dates are thought of as significant targets by modern investigators in the first place.[230] It finds no support in any systematic studies of the evidence from regional groups of monuments.[231] Given that there is not even any strong support for the idea of a single 'Celtic calendar' in the Iron Age, and of the festival dates being fixed by counting equal intervals of approximately forty-six days, the evidence is weak indeed. Attractive as it may be to envisage thin threads connecting Early Neolithic practices to folk traditions surviving into modern times, arguments for continuity on this scale simply have to make too many speculative assumptions at every stage.

On the other hand repeated astronomical trends are evident in the archaeological record, especially amongst the orientations of local groups of broadly contemporary monuments, and they reveal strong local traditions operating from time to time: examples include the strong orientation preferences of various regional groups of chambered tombs; the low-precision cardinal and solar orientations amongst the large stone circles in Cumbria and south-west Scotland; and the solstitial orientations of enclosures and henges in the Stonehenge area, including the axis of the sarsen monument at Stonehenge itself. The material evidence is also beginning to reveal some persistent traits relating to astronomy which, while changing through time, reveal basic continuities of tradition that seem to persist in an otherwise complex picture of continual change. The clearest example is the apparent development of a distinctive tradition of westerly- and south-westerly-facing tombs and ceremonial monuments incorporating predominantly lunar symbolism in parts of Scotland and Ireland in the Early Bronze Age. Finally, where astronomical associations have formed part of the wider studies of monuments in the landscape, as in the case of the Dorset cursus and the stone rows of northern Mull, we are beginning to obtain useful evidence that moves us towards a better understanding of the role of astronomical symbolism in defining the place of a monument in the perceived cosmos (see chapter nine).

The picture is still sketchy, not least because intentional alignments can not be convincingly demonstrated on the basis of a few scattered examples. Ironically, over the last two decades, as archaeoastronomers have become steadily more wary of individual alignments, stressing over and over again the importance of studying common trends amongst groups, so mainstream archaeologists have become more willing to consider individual alignments as part of broader interpretations. This has its merits, but also considerable dangers. Certainly it is clear that a great deal more systematic work could be undertaken on local groups of monuments, and also that there is considerable potential for broader studies of monuments and places in the landscape that are prepared to consider celestial associations as part of the wider picture. However, before launching headlong into new archaeoastronomical research programmes it is necessary to pause and take a considered look at exactly what we might hope to achieve by doing so. Questions of interpretation are critical here, and these relate intimately to wider questions of theory development and method in prehistoric archaeology as a whole. In the last part of the book we examine some of the wider issues and look to the future.

# Part III FUTURE DIRECTIONS

# 9

# Wider Issues

If we would improve the intellect, first of all, we must ascend: we cannot gain real knowledge on a level; we must generalize, we must reduce to method, we must have a grasp of principles, and group and shape our acquisitions by them.

John Henry [Cardinal] Newman, 1852[1]

There can be little doubt that astronomy is the oldest science in the world.

Patrick Moore, 1996[2]

Anthropology is not short of facts but simply of anything intelligent to do with them. The notion of 'butterfly-collecting' is familiar within the discipline and serves to characterize the endeavours of many ethnographers and failed interpreters, who simply amass neat examples of curious customs arranged by area, or alphabetically, or by evolutionary order, whatever the current style may be.

Nigel Barley, 1983[3]

## The big question: so what?

A divide between archaeoastronomy and mainstream archaeology most evidently still exists. Archaeoastronomical investigations still too often proceed from astronomical questions rather than archaeological ones, a case in point being a recent interpretation of Stonehenge and other prehistoric monuments in Wessex by a prominent historian of science in which earthen long barrows, cursuses, earth-cut figures and white horses, avenues, henges, stone circles, and of course Stonehenge itself, all receive a stellar, solar or lunar interpretation as part of a broad theme of astronomical development.[4] On the other hand, archaeologists who sweepingly criticise archaeoastronomy sometimes seem to show a bewildering ignorance of the archaeoastronomical literature since 1980,[5] as in fact do others who tentatively accept some of its earlier conclusions.[6] Yet others make statements that are unjustified,[7] simply wrong,[8] or even nonsensical.[9]

It is certainly true that in so far as archaeoastronomy consists of approaching prehistoric monuments looking for astronomical alignments,[10] then it is at best misleading and at worst completely valueless. The simple reason is that many factors could have influenced a monument's orientation and position in the landscape, and while we should not ignore orientations, as archaeologists have often done in the past, we should certainly study them open-mindedly, not starting from the assumption that astronomy is the (sole or primary) motivation.[11]

There may certainly be value in including orientation and alignment (azimuth and declination) data within archaeological inventories,[12] but for this work astronomers are not needed so much as archaeologists trained to measure azimuths and altitudes and calculate declinations as part of their armoury of field techniques.[13] Even so, the simple collection of data, while 'posing a threat to no-one',[14] is generally of limited interest; what is interesting is their interpretation.[15] And the process of interpretation must proceed with considerable caution. Even demonstrably 'non-random' patterns in declination distributions are not necessarily caused by astronomy,[16] and certainly not by astronomy alone. Such patterns can only really start to be properly understood in a wider interpretative context.[17] This point, once accepted, implies that it is clearly better to start from archaeological goals and look at astronomical questions as and when (or indeed, if) they arise. If there is to be interdisciplinary co-operation between archaeologists and astronomers, as many have argued,[18] it should really aim at the full exchange of ideas and integrated work programmes to address common questions and goals.[19] But this is only likely to happen if the archaeologists are already convinced that archaeoastronomy is likely to answer questions of interest to them.[20]

It is an understatement to say that this has not always been the case in the past. Despite the development of world archaeoastronomy beyond 'alignment studies' since the 1970s,[21] an American anthropologist could still write in 1992 that 'archaeologists see archaeoastronomers as answering questions that, from a social scientific standpoint, no-one is asking'.[22] From an archaeological point of view, the fact that a monument points at (say) midsummer sunrise—even if there are convincing reasons for believing this to be intentional—is in itself no better than a potsherd or other artefact, a collector's item of little interest unless it has the potential to tell us something about the people who made it.[23] On the other hand, Alexander Marshack's discovery of possible early lunar 'calendrical' notations (see chapter eight), although controversial, is of great interest to archaeologists because such evidence could give insights into 'the psychological, cognitive and conceptual processes involved in early forms of human symbolic behaviour and "visual thinking"'.[24]

This line of argument leads directly to the question of whether archaeoastronomy[25] should exist as such at all, despite the apparatus of conferences, journals, and, recently, learned societies,[26] that has come into being. The answer may not be self-evident,[27] but on the other hand it may not be obvious to the non-anthropologist why we should not wish to know more about prehistoric astronomy *per se*, purely out of anthropological interest. Fortunately, this level of argument aside, there are compelling reasons for wishing to focus upon astronomical practice in order to approach a range of more general cultural questions, and it is upon these that we shall concentrate in what follows.[28]

If we move away from the pursuit of evidence of astronomy *per se* towards 'the study of the practice *and use* of astronomy',[29] we quickly find, from examples ancient and modern, from all parts of the world, and from the most technologically primitive hunter-gatherer groups through to complex urbanised states, that a society's view of and beliefs about the sky are inextricably linked to the realm of politics, economics, religion, and ideology.[30] Thus

> From ancient Babylonia to China, Mexico, and Peru, from empire and city state to tribe, astronomical information was gathered, recorded, and used by those whose interests lay as much in the spheres of status enforcement and political ideology as in predicting rainy seasons or planning agricultural schedules. By making natural phenomena appear liable to social manipulation in the form of cosmological myth, cleverly aligned architecture, and appropriately timed religious ritual, the elite used the predictive value of astronomical knowledge as an impressive display and justification of their power and prestige.[31]

One of the most obvious advantages of the study of astronomical practice within the study of human societies as a whole is the quality of the basic physical data. We can directly reconstruct important components of the night sky at any place on earth and at any time during the last several millennia, to within known margins of error. This can be said of few other aspects of a past society's natural environment.[32] The fact that many aspects of the sky are similar, especially at similar latitudes, allows the anthropologist to investigate how the same immutable resource within the natural environment is perceived, integrated into different systems of classification and belief, and manipulated to different or similar ideological ends.[33] Some argue that we should seek cultural correlates for ancient astronomy, hoping that studying astronomy in a variety of cultural contexts will give general insights into processes of cultural development and change.[34]

If we go even further and try to comprehend the *meaning* of astronomy to ancient peoples,[35] then we soon discover that we must strive to understand how they conceived of themselves in relation to what might be called their 'natural world'—their physical and human environment, of which, for almost everyone on earth, celestial objects and events form an integral part.[36] In one sense, every culture makes an appraisal of its own environment by selecting certain aspects of its physical surroundings—such as other people, animals, mountains, meteorological and celestial phenomena—and assigning meaning to them and to the relationships that they regard as existing between them.[37] The resulting framework of understanding—'cosmology', 'world-view' or 'cosmovisión'[38]—is likely to bear very little resemblance to the principles of Linnaean classification that underpin Western ('rational' or 'objective') scientific thought, with its roots in ancient Greece and Babylonia. It is much less likely to classify objects and phenomena into categories familiar to us, such as celestial and terrestrial, sacred and mundane, animate and inanimate, empirically real (the notion of objective reality being part of the Western world-view in the first place) or 'fantastical'.[39] Thus, for example, 'Old Star', chief protector of the inhabitants of the sacred *He* world of the Barasana of the Colombian Amazon, is at once a short trumpet, Orion, the fierce thunder jaguar, and a human warrior.[40]

It is also much more likely to envisage direct relationships between celestial and other phenomena. The Barasana, for example, link the 'Caterpillar Jaguar' constellation to the appearance of living caterpillars, which form an essential part of their diet. Regarded as the 'father of caterpillars', the Caterpillar Jaguar is believed to be responsible for the increase in their numbers as he rises higher and higher in the sky at dusk. To us, the correlation can be seen to be due to the fact that the constellation is in the eastern sky at dusk at the time of year when caterpillars pupate and come down from the trees on which they feed. In Barasana thought there exists a direct correspondence between two entities that to us are quite distinct: a constellation and its position in the sky and an important part of the seasonal food supply.[41] This is not to say that we should abandon the principles of observation and analysis derived from the Western scientific viewpoint in order to try to understand other world-views in terms meaningful to ourselves; merely that we must recognise that the world-views themselves are fundamentally different.[42]

Although celestial aspects may have formed only a small part of complex, integrated frameworks of understanding, much of which are lost to us, they assume a very special importance in our attempts to comprehend something about those frameworks for three basic reasons. First, as we have already said, recurrent phenomena in another people's sky are directly accessible to us, a rare privilege in a prehistoric context. Second, because the sky is immutable, celestial objects and events can *only* be used metaphorically, not directly. Finally, astronomical elements are almost invariably of great importance in human world-views:

> A basic feature of traditional rituals and cosmologies [is] their astrological insistence that good fortune on earth can be ensured only by keeping human action fundamentally in tune with observable astronomical events. 'On earth as it is in heaven.' Again and again, we find this belief that the template for the ancestral 'Way' or 'Law' lies in the skies.[43]

Once we tackle questions of people's perceptions of their natural environment as a whole, it would be unreasonable *not* to consider their perceptions of objects and events in the sky.[44]

## ARCHAEOLOGY BOX 8

### MODES OF EXPLANATION

The ultimate challenge facing the archaeologist is to answer the question 'why?'. In seeking to explain some event or pattern of events in the past, it is necessary to move beyond the mere observation and recording of the archaeological record. A body of theory must be established, in order to clarify the nature (and level) of understanding we are trying to achieve. Methods of archaeological enquiry must then be developed that are focused towards advancing that understanding.[1]

Since the 1960s, a great deal of thought has gone into the establishment of suitable theoretical foundations for archaeology. Traditional methods, extrapolating directly from artefact assemblages to archaeological 'cultures',[2] gave way in the 1970s to the 'processual' methods of the so-called 'New Archaeology',[3] and subsequently to a variety of 'post-processual' approaches reflecting different philosophical standpoints, such as neo-Marxist, post-structuralist, hermeneutic, and feminist. In the resulting interactions, ideas are continuing to develop both from processual and post-processual standpoints. 'Theoretical archaeology' has become a burgeoning field.[4]

#### Levels of explanation: the aim

The New Archaeology laid great stress upon raising the level of generality of archaeological explanations.[5] In doing so, it emphasised social process, and in the early days attempted to establish laws that would explain changes in social organisation in terms of social, economic, and environmental factors.[6]

The quest for non-trivial universal laws of human behaviour was soon found to be unproductive; for one thing, the actions of individuals are very much dependent upon factors such as ideology and historical context that do not adhere to purely functional, ecologically deterministic, explanations.[7] Post-processualists argued that we should not pursue 'absolute' process models that apply to societies as a whole, although some concede that we might still seek generalisations that concern individual behaviour in certain contexts.[8] In other words, we can still strive to recognise and appreciate general tendencies that result simply from our being human—to appreciate some of 'the unique qualities and capabilities of conception, and action based on that conception, that we all share'.[9]

In recent years there has been a shift in emphasis away from trying to understand how societies were organised ('social archaeology') towards trying to comprehend what was in people's minds ('cognitive archaeology').[10] A particular focus of attention is how systems of thought and belief—ideologies—are expressed in ritual practice, art, and so on. New Archaeologists sought direct correlations between patterns in the material record and patterns of human activity,[11] but since the mid-1970s it has been recognised that material objects are 'meaningfully constituted' in the sense that their presence and their relationships to other material objects in any particular cultural context reflect specific factors, such as ideological considerations, that are not directly knowable.[12] From a post-processual point of view, one of the main aims of archaeologists is to unlock some of the meanings that particular objects expressed to particular people in particular contexts.

#### Advancing understanding: the method

Processual archaeologists stress the importance of scientific methodology, and in the past tended to follow the Popperian ideal where the material record is used to 'test' general hypotheses.[13] There are a number of problems with this procedure if adopted too simplistically in an archaeological context, perhaps the most obvious being that new hypotheses and new data do not cleanly precede each other, but must develop in some kind of loop or 'hermeneutic circle'.[14] To many post-processualists, however, the whole approach is an anathema because it views the past as objective reality that can be assessed according to Western scientific principles, reflecting an attitude tantamount to Western cultural imperialism.[15] Taken to the extreme, the conclusion is that all interpretative frameworks are personal and there can be no clear-cut, let alone universal, methodology for guiding empirical research itself, merely restraints on the way in which ideas and empirical data are developed alongside each other.[16] If, however, we operate (say) within the framework of Western thought and wish to communicate our ideas to others within the same framework with whom we wish to build up some level of common understanding, then there is no reason why we should not be able to agree on certain methodological principles.

The post-processual agenda has focused attention on cultural meaning. Material objects expressed symbolic meanings to the people who used them; by 'reading' the signs and patterns we should be able to reconstruct some of those meanings.[17] The problem is to know the 'language' in which they are expressed, a real problem given that the form of material symbols in different contexts, like words in unrelated languages, generally seems arbitrary in relation to their meanings.[18] Only if we believe that it is possible to generalise about human behaviour in certain contexts, or we are prepared to assume some 'universal grammar of human understanding', do we have any hope of unravelling such meanings.[19]

Generalisation, then, is still very much on the agenda. Without general theories that apply to many situations, we lose the ability to extrapolate from the limited evidence available in any particular instance in order (say) to explain why a particular social group developed in a certain way, or to suggest what particular material objects might have meant to particular people. Those who stress historical specificity to the point of arguing that no generalisations are possible must confine themselves to finding ways of making their interpretations more sensitive to the social and historical context of the particular problem they are studying.

How, then, do we proceed? How can we begin to try to understand a world-view operating in another cultural context?[45] One approach starts by recognising a human propensity to express and reinforce the perceived qualities of the natural world, using verbal or visual metaphors, something that may be manifested for example in myth, in activities we might see as sacred (ritual) or mundane, or publicly displayed in architecture.[46] The hope for the prehistorian is that we can recognise cosmological metaphors preserved in the material record by detecting distinctive combinations of symbols used to express them.[47]

The main problem, of course, is exactly how we proceed from objects and patterns in the archaeological record to make deductions about concepts in prehistoric people's minds. Here it is crucial not simply to proceed intuitively. If we are to extrapolate from the material remains in a more disciplined way, it is essential to develop a broader theoretical framework that suggests more general relationships between human thought and material remains. In order not to become embroiled here in the many intensively debated issues of archaeological theory towards which this inexorably leads, we shall merely be explicit about our overall approach for using the material record to increase our understanding on some level. Stated as blandly as possible this is as follows: we start with a set of initial ideas and we then wish to study the material record to see whether it tends to confirm or strengthen, or else to refute or weaken, those ideas.[48] By leaving open (at least for the moment) the question of the level of generality or specificity of the initial ideas in question, and also the question of whether the 'we' refers to a collection of scholars believing that they share some level of common understanding or an individual investigator believing that he or she is acting essentially alone in a cultural milieu, the paradigm is general enough to encompass a range of both processual and post-processual approaches (see Archaeology Box 8). Nonetheless, it is sufficient to give us a framework for organising the ideas that follow.

We begin, then, by examining ways of improving our initial ideas, starting with an attempt to examine and deconstruct concepts that are specifically related to the Western tradition, and proceeding to the thematic generation of new ideas. In the following chapter we proceed to look at what we can study in the material record to test these ideas, and suggest fieldwork priorities. Finally, we come to the question of how to determine the extent to which the material data really do strengthen or weaken our ideas, giving particular consideration to the place of quantitative measures, and statistical analysis and inference.

## Casting aside the baggage

Perceptions of the sky are culture-specific. What one society might perceive as significant might arouse little or no interest in another.[49] Thus, just because people brought up in the Western tradition generally feel that the brightest stars are the most prominent and hence the most notable or consequential, others would not necessarily do so: the Borana of southern Ethiopia, for example, attach key importance in their calendar to the relatively dim stars in our constellation of Triangulum,[50] and many traditional peoples in the southern hemisphere attach great significance and meaning to dark patches in the Milky Way.[51] This did not stop ethnographers around the beginning of the twentieth century struggling to relate Andean perceptions of the sky to constellations in the European tradition,[52] but this is of no help whatsoever in trying to understand the Andean world-view.

Many archaeoastronomers (and even some archaeologists) continue to exhibit an extraordinary unwillingness to challenge concepts which, with a little thought, can readily be seen to be part of our Western 'cultural baggage'. These concepts are very unlikely to be helpful—indeed, they are quite likely to be highly misleading—in trying to understand people in prehistory. Questioning some of the assumptions that are generally made, and in particular identifying concepts that are potentially meaningless outside the context of the Western tradition, is a necessary preliminary to improving our understanding of meanings that operate within the context of other world-views. The list that follows is by no means exhaustive, but illustrates some of the interpretative pitfalls that we must strive to avoid.

On the most fundamental level, indigenous notions of space and time are context-rich, unlike the Western abstractions.[53] One consequence of this is that the nature of objects in the natural world and their location in space and time may not be seen as separate categories in the way that we conceive of them.[54] To give a celestial example, the people of the Andean village of Misminay do not give stars and planets fixed names but names by which they are associated with a position in space and a position in time, and this affects the whole way in which relationships between particular objects and the motions of the heavens are conceived.[55]

A key concept in archaeoastronomy is that of direction. Yet we should not assume that, outside the Western framework of thought, people will inevitably think of direction as point-azimuth.[56] Anthropological work amongst modern Yucatec Maya villagers suggests that the four 'directions' that are a common feature of world-views both in pre-Columbian Mesoamerica and indigenous cultures in the US South-West,[57] were generally conceived as (what we would see as) azimuth ranges centred upon (what we would see as) the cardinal points.[58] Other examples of directions conceived as azimuth ranges can be found in Europe in the Middle Ages.[59] The implication is that if, say, we observe in the archaeological record a set of orientations falling within an azimuth range centred roughly upon due east, we should consider that they might have been conceived as pointing in the same direction; it may be unproductive to look for anything more subtle, specific or precise in our own terms.[60]

This also has a strong potential impact on the interpretation of 'alignments', a notion that in itself may not be meaningful outside the Western cultural context. A sacred mountain or astronomical event might have been conceived to be 'in the same direction as' an architectural feature, and hence symbolically associated with it, even if in our view they are oriented many degrees apart.[61] This is not to say that alignment data—that is, data involving the orientations of structures and how these relate to celestial objects or events visible at or close above the horizon—should be avoided; it is merely that we must be cautious in interpretation. It is convenient to retain the word 'alignment', for example in phrases such as 'the Phase 3 axis of Stonehenge is roughly aligned upon midsummer sunrise', but merely as a guide to what could have been seen in a certain direction of possible interest, either on the horizon or above it, and to what other effects (such as shadows) might have occurred at certain times.

A final cautionary word on directions is the following. Even in contexts where the concept of a direction as point-azimuth exists, we can not assume that those directions that *we* consider of particular significance, such as the cardinal directions, are necessarily of special importance in the other conceptual system. The Chorti Maya of present-day Yucatan, for example, regard the rising and setting directions of the sun on the day of zenith passage as 'east' and 'west', although these are actually several degrees away from our cardinal points.[62]

A question that the author continues to be asked frequently, even occasionally by prehistorians, is 'which alignments at [some monument or set of monuments, most usually Stonehenge] were significant?'. The problem, of course, is that whatever meaning might have been attached to the motions of the heavenly bodies has to be approached through concepts that might make sense within a non-Western world-view, not through any absolute notion of 'significance'. In considering prehistoric astronomy it is all too common to apply a 'recipe book' of solar and lunar horizon targets which consists, say, of the rising or setting sun at the solstices and equinoxes and the most northerly or southerly rising or setting moon at one of the standstill limits (cf. Fig. 1.20).[63] Anthony Aveni has characterised this as the 'Thom paradigm' and particularly criticised its application in pre-Columbian America.[64] Even in prehistoric Europe it has to be avoided at all costs.

One way to proceed, as we have seen, is simply to let the data speak for themselves through declinations;[65] this avoids prejudging the nature of possible horizon targets at all, but proceeds to focus on purely statistical arguments prior to any attempt at interpretation.[66] This 'classic', or 'green',[67] archaeoastronomical approach may be appropriate for certain types of dataset (such as initial investigations of groups of similar monuments) and, as we have seen in this book, can address a limited range of questions. However, if we are really to try to understand what intentional astronomical alignments might have meant to the people who constructed and used them, then we need to focus on particular horizon astronomical events and think more deeply about ways in which their symbolic 'encoding' in architecture might have imbued them with meaning.

Some events certainly do achieve recognition and importance in a wide range of cultural contexts, the most obvious being the rising and setting of the sun at the solstices.[68] But even here, we must move on from viewing the four solstitial rising and setting positions as abstractions—a set of phenomena similar in nature inevitably to be considered together as one package. The very fact of when they occur—at opposite times of the year and day—endows them with different inherent qualities that will tend to invoke different sets of associations and meanings. Winter solstice occurs at a time of year when days are cold, food is short, and the sun may need to be 'turned round' in order to create longer days; summer solstice, on the other hand, is a time of first fruits and celebration.[69] Sunrise often has connotations of rebirth; sunset of death.[70] The place on the horizon where the solstitial sun is seen to rise or set may be imbued with sacred significance.[71] Where we can try to interpret a range of cult objects and imagery,[72] or where we are lucky enough to have relevant historical or ethnographic evidence, we may begin to appreciate much richer sets of meanings.

The equinox, however, is a concept whose meaning and importance are far from obvious outside the framework of Greek geometrical astronomy that underlies the Western scientific tradition.[73] To us, the vernal and autumnal equinoxes are precisely defined in terms of the places where, and the times when, the sun crosses the celestial equator (see Astronomy Box 8). From the perspective of a different world-view, it is not impossible that people attributed significance to such events as the day when the sun rises (or sets) exactly half-way along the horizon between the solstices, or to the day halfway in time between the solstices, or even the day on which sunrise and sunset occur opposite to each other;[74] but each type of explanation suffers from difficulties at three levels.

First, and most fundamentally, they depend upon concepts whose meaning and significance are themselves questionable outside the Western framework. It is not, for example, self-evident that a mid-point dividing something into two precisely equal parts, either in space or time, should have been of particular importance; this only seems natural to us because we view space and time as abstract 'axes'. Second, even if one or more of these dates were significant to certain groups of prehistoric people, they would not generally correspond exactly to our equinox in any case, so that even if alignments upon significant sunrises or sunsets were incorporated in ceremonial architecture, this would not manifest itself in the material record as a set of alignments clustering closely around our equinox (see Astronomy Box 8). Third, these 'half-way' dates bear no relationship whatsoever *on the conceptual level* to the modern astronomer's equinox. Even if any of them was convincingly shown to have been of importance in prehistoric times, this would still not provide evidence that 'our' equinox was conceived and observed.[75]

Despite all this, claims of equinoctial alignments are still encountered frequently, not only in the archaeoastronomical literature but also in mainstream archaeological accounts.[76] This is wholly different from the quite reasonable suggestion, widely attested in other contexts, that significance might have been attached to the position of sunrise or sunset at particular times of year, perhaps corresponding to calendrical festivals;[77] the point is that there is no reason to suppose that our equinox would occupy any special place in the range of possible festivals. Perhaps we would do better to try to understand the meaning of possible solar alignments to the people who built them by simply eliminating the word 'equinox' once and for all from the archaeoastronomer's vocabulary.

The lunar standstill limits (see Astronomy Box 4) have been widely discussed as possible horizon targets for architectural alignments in prehistoric Britain. The lunar standstills tend to be treated, explicitly or implicitly, as events that prehistoric people might have sought to track down through careful observation and measurement.[78] One problem with this is that the concept of a standstill only makes sense when visualised in terms of graphs of the moon's declination plotted against time, as is clear from the many explanations of the concept in books, including this one (see Fig. 1.19). Another is that the amount of ethnohistoric or ethnographic evidence showing that the standstills were conceptualised as such outside Western culture is precisely zero.[79]

This does not mean that the changes in the moon's motions over the lunar node cycle might not have been appreciated in various ways in prehistoric times,[80] nor that this might not have been reflected in the design of ceremonial architecture—resulting, for example, in the apparent existence of deliberate, low-precision alignments upon the lunar standstill limits. The point is that the way these associations were conceived and understood by prehistoric people may be very different from the way in which we might choose to describe them in terms convenient to ourselves.[81]

For example, special significance might have been ascribed to the part of the horizon where the moon, but not the sun, could sometimes reach, and the times when this happened. Then again, the full moon is a distinctive event each month when the moon rises around sunset, is prominent all night long and sets around sunrise. Because of the interplay between the monthly north–south motions of the moon, the time of year and the lunar phase cycle, the rising and setting positions of successive *full* moons progress up and down the eastern and western horizons respectively over an annual cycle. The full moon nearest the summer solstice rises and sets furthest south and that nearest the winter solstice rises and sets furthest north. The full moon nearest one or other solstice in each year is a significant time for a number of indigenous peoples around the world.[82]

The northerly and southerly limits have very different inherent qualities. One obvious difference is that around midwinter the presence or absence of a near-full moon makes the difference between a long dark night and an illuminated night. A set of architectural associations with (say) the rising full moon around midwinter would be perceived by us as a set of alignments towards the north-east falling roughly between the northerly major and minor standstill limits with (if no account were taken of the 18·6-year lunar node cycle) concentrations towards both ends, i.e. towards the standstill limits themselves.[83] It does not seem unreasonable to suggest that special significance might have been attached to those periods of a few years, occurring once or twice in each generation, when the period of moonlight close to full moon in wintertime was unusually long.[84] A set of particular associations with midwinter moonrise or set during these periods would manifest itself as a set of alignments close to the northern major limit. The southern full moon lacks these attributes; whether or not there is a near-full moon on a short night close to midsummer, astronomical twilight ensures that the night is never very dark. We would have to follow a different line of argument to provide the beginnings of a convincing explanation of sets of alignments clustering particularly around the southern major limit. Surprisingly, this crucial difference between the northern and southern lunar limits is rarely noted.

Before leaving the topic of the 'Thom paradigm', it should be mentioned that it has tended to close people's eyes to other possible targets for astronomical alignments, the most obvious being the brighter planets, and particularly Venus.[85] The reason is that because the motions of the planets are both more complicated than those of the sun or moon, and also confined to roughly the same ranges of declination, so putative planetary alignments are difficult to distinguish from solar or lunar ones. If we are interested in interpreting what we find in the archaeological record, we should be led by what seem the most plausible explanations, not by whether they are easy to test.

## Lateral thinking

Having revealed some of the assumptions that we should avoid making, how can we start to theorise about what, say, the orientation of a recumbent stone circle upon the southern moon might have meant to the people who built or used the monument in prehistory, in the hope ultimately of beginning to understand something about their world-view? We need to consider the meanings that might have been conveyed to different people at different times by architectural features of public monuments, including possible astronomical symbolism.[86] This will include thinking about what the location of a monument might have meant before and after it was built, the possible significance of the form it took and the materials of which it was made, why it might have been important to make (or not to make) certain modifications, and so on. Where can we look to generate plausible ideas?

The most obvious place to look is at the material record itself—at what has already been discovered from

## ASTRONOMY BOX 8

### CONCEPTUALISING THE EQUINOXES

The equinoxes are often assumed to offer horizon targets of comparable importance to the solstices in terms of possible ritual or practical significance; targets that might have been indicated with precision. This is a quite unjustifiable assumption.

The equinoxes as defined by us represent the mid-points between the solstices in the sense that the sun's centre passes the celestial equator at declination 0°. Unlike the solstices, however, they are not marked in any special way by the physical phenomena of sunrise or sunset. The rising or setting sun itself does not mark this special point; instead it races past it at a daily rate in declination of some 24′.

So how likely is it that people without our grounding in modern western science or its progenitors might have conceived of anything in any way directly related to the equinox as conceived by us? Anyone relying upon observations of the rising or setting sun has only two obvious ways to conceptualise and determine the mid-point between the solstices: spatially, by bisecting the positions of the directions of the summer and winter solstices; or temporally, by counting the number of days between the solstices and halving the difference.

The problem with the spatial method is that the process would only yield declination 0° if the horizon were completely level and flat; otherwise the spatial mid-point could be several degrees away from the equinox as defined by us (Fig. 9.1). Furthermore, it is difficult to envisage a method of determining this mid-direction to any great precision. Finally, there seems little justification for making the assumption that prehistoric people should have conceptualised the space between the solstices in such an abstract way that its mid-point was of special significance.

The temporal method, which is the basis of Thom's postulated solar calendar (see Astronomy Box 5), involves a similar assumption about the temporal mid-point. It also presupposes a sufficient level of numeracy to count, record, recall, and manipulate (halve) numbers up to at least 183,[1] and that the solstices themselves had already been determined precisely to the day, which as we have already noted can not be achieved by direct observation. And even if a procedure of counting 91 or 92 days forwards from a solstice had been developed and was followed strictly, the declination of the 'equinoctial' setting sun so determined might have been anywhere from +0°8′ to +0°31′, depending on the year.[2]

Another day that might be deemed significant is that on which the length of night is equal to the length of day. But this will not correspond to our equinox because of the twilight periods before sunrise and after sunset. In any case, the point at which night can be deemed to have started is arbitrary[3] and poorly defined: how rapidly it gets dark after sunset is highly dependent upon weather conditions, whether the moon is up, and so on. It could possibly be argued that significance might have been attached to the day when the time from sunset to sunrise equalled that from sunrise to sunset; but this presupposes that prehistoric people had developed devices capable of measuring periods of time, both by night and by day, to a precision of a minute or two, as well as ways to record and compare the results. And even if they

Fig. 9.1 Possible 'mid-way' concepts yielding dates approximating to the modern concept of the equinox, for a hypothetical eastern horizon within Britain or Ireland. The spatial mid-point between sunrise at the two solstices (sun symbol) is at *S*; while sunrise halfway in time between the solstices may be anywhere within the shaded area *T*. Sunrise on the day when the sun rises and sets in opposite directions will generally be different again, depending on the altitude of the western horizon. Sunrise at the equinox, corresponding to a declination of 0°, is *E*. First published as Ruggles 1997, fig. 2.

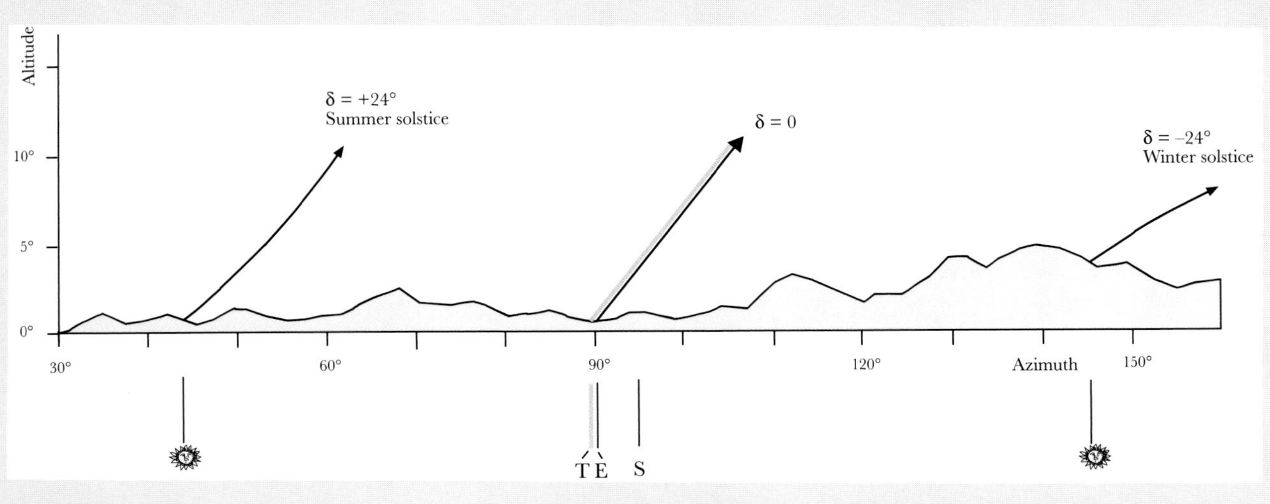

did, the result would not generally correspond to our equinox because of the altitude of the horizon and the effects of refraction.

Yet another day that has been suggested as possibly significant is that on which the rising of the sun occurs exactly opposite to its setting, but once again this is not equivalent to our concept of the equinox and will not generally correspond to it because the exact date when this event would occur is dependent upon the horizon altitudes in the east and west at a particular place.

In short, the equinox is a concept unlikely to have any meaning from an earth-based perspective within a non-Western world view. Even if such people felt it important to determine and mark the (spatial or temporal) mid-point between the solstices this is unlikely to have been done to very great precision. This is not to deny the existence of alignments upon sunrise or sunset at particular times of year, perhaps on the occasion of calendrical festivals, as is widely attested from different cultures around the world; nor that the mid-point between the solstices might have had particular significance in certain cultures, as some claim it did for the Celts (but see pp. 141–2). But there is no *a priori* reason why the half-way point between the solstices should occupy a special place in the range of possible festivals and, even where it did, the mid-point would not correspond conceptually—or, in most cases, actually—to our equinox.

archaeological investigations. Having formulated some fresh ideas on the basis of some existing archaeological data,[87] we will inevitably seek to acquire and/or examine more data in order to 'test' them, and this will result in our modifying our existing ideas and quite possibly formulating new ones. The ever-present danger here is that of circular argument[88] and the aim is to ensure that traversing this loop results in the quality and explanatory power of the ideas, and the mass of useful archaeological data, successfully developing in parallel.[89]

Nonetheless, the material record is limited—as we have already seen, a great many meanings are simply irrecoverable—and will often be misleading: there may, for example, be obvious evidence of the ideology of a powerful élite but little or no trace of co-existing folk ideologies.[90] We need different ways of enriching the set of initial ideas that can be 'tested' by looking afresh at existing, or else by seeking new, archaeological data. The aim may be to raise the level of generality of the ideas being generated, or to find ways that make those initial interpretations more sensitive to the social and historical context of the particular problem we are studying (see Archaeology Box 8). Either way, a valuable tool is that of cultural analogy or 'parallel'.

In questioning long-held assumptions, analogies can be used very valuably to open up a wider range of interpretative possibilities and to suggest more relevant questions to be asking.[91] In focusing on meanings we are doing something different from seeking any form of direct correlation between, say, aspects of astronomical practice and aspects of social organisation.[92] We are trying, rather, to broaden the range of possible interpretations of symbolic associations—say, between celestial and terrestrial features[93]—that we might find in the material record. Historical or modern indigenous peoples may offer particularly valuable insights.[94]

A third source of ideas, one that has generated a good deal of interest amongst archaeologists in recent years, is phenomenology. The basis of this approach is that individual experience lies at the heart of (say) how monuments were used,[95] or how the landscape was perceived.[96] Our personal experience as human beings is relevant to understanding that of others, even in other cultural contexts; even in prehistory.[97] This is especially true in trying to appreciate the visual effect of an astronomical event, such as midsummer sunrise at Stonehenge.[98]

> One should make a point of being there. Celestial events take place in context. When you see what actually happens in the landscape, you learn a great deal more about it. It is, of course, possible to determine the alignment of a building with instruments, but nuances of light and shadow, details of the horizon, difficulties in practical observation, and many more subjective aspects of a site can only be appreciated in person.[99]

Similarly, by walking about in the modern landscape we can try to appreciate patterns of relationship between monuments and the topographic elements of the landscape as experienced by people living in that landscape in prehistoric times.[100]

There are many ways in which ideas can be generated through personal experience. To give a rather different example, most of us share the perception that the moon appears particularly spectacular when near to (say, within one or two degrees of) the horizon, and especially when it is close to small and distinct horizon features, such as distant mountains, buildings or trees. This is doubtless due in part to the well-known effect that the proximity of the moon to terrestrial features makes it seem larger to the eye; in part because the moon's light reflected in the atmosphere shines around objects on the horizon; and partly perhaps because there is a feeling of the celestial and the terrestrial coming together, being in direct association. We cannot say that any prehistoric individual necessarily shared such feelings; but the idea deserves serious consideration in contextual interpretations. If we postulate that the horizon moon was significant at all, we should certainly give serious consideration to the possibility that it had particular symbolic significance whenever it appeared close to (perhaps some way above) a particular part of the horizon or a particular horizon feature, rather than narrow-mindedly assuming that special significance could only have been attached to the actual moment of rising or setting.

As well as generating new ideas, phenomenology can be

used to widen the range of interpretative possibilities, as is the case with cultural analogy. In chapters five to eight the idea was discussed that certain stone circles and rows were oriented with respect to 'prominent' natural features such as distant hill- and mountain-tops on the horizon. Prominence is of course a subjective concept, and there is no reason to assume that anyone else's idea of what constitutes a prominent (significant) feature is the same as ours.[101] An exercise was undertaken in 1991 using freshly-arrived teams of Earthwatch volunteers, largely ignorant of background work on astronomy in prehistoric Britain and Ireland, who were asked to give their personal assessments of the relative prominence of features on the skyline from positions close to the Glengorm standing stones on Mull.[102] Only about a third thought that distant mountain peaks were the most prominent skyline feature. Most laid much greater emphasis upon the false summits of hill slopes close by, or upon skyline trees or peaks that happened to align with other trees, peaks or rivers in the foreground.

Generating ideas that are both interesting and testable, ones that will enable us to extrapolate from objects and patterns found in the archaeological record to aspects of world-view and ideology, must involve some degree of generalisation (Archaeology Box 8). The isolated insight that the builders of Glengorm stone row might have chosen to orient it upon Ben More and the rising moon has no great value. But if this suggests something about human behaviour in general, or human responses to some situation in general, then it becomes much more interesting and challenging. The really stimulating archaeological questions involving astronomy relate it to broader themes.

## Themes for the future

Investigating the ways in which astronomical knowledge was apparently used in different prehistoric contexts can certainly provide useful insights relevant to themes in social archaeology. Many questions immediately suggest themselves. Was astronomy a matter of public or private knowledge? Did it instil or reinforce privilege and power? Did common elements of ritual practice, such as those reflected in the design of stone rows and circles in certain regional groups, merely reflect interaction between politically independent social groups or are they indicative of wider political control? And how and why did things change with time?

As we have already mentioned, celestial phenomena and events, duly manipulated by those with the ability to predict them, can be used to provide an impressive display and justification of élite power and prestige.[103] Conversely, unpredicted or unpredictable events in the sky may undermine the authority of an élite who have persuaded others that they have powers and influence over the celestial sphere, and thus cause social and political instability.[104] Astronomical and calendrical knowledge can be a crucial political resource, as is equally evident from small, modern indigenous groups[105] and from large states such as the Inca empire.[106] In short, archaeoastronomical evidence is undeniably relevant to a range of interesting social questions, as we have hopefully begun to demonstrate in chapter eight. In addition, if we accept that public ritual tends to change on a slow timescale, its shifting contexts can help to highlight social change.[107] Conversely, discontinuities of ritual tradition, as manifested by clear changes in the patterns of astronomical symbolism incorporated in public monuments, may indicate significant social upheaval.

But perhaps what makes prehistoric astronomy most challenging and exciting is its relevance to our broader understanding of developments in human thought. For example, astronomy is intimately related to people's conceptualisation of space and time, and the study of astronomical symbolism in the archaeological record can give valuable insights into the latter.[108] However, caution is needed. It is certainly true that the conceptualisation of time (raising questions such as whether time is conceived as cyclical or linear), is a complex and interesting theme in itself,[109] and it is equally true that astronomical periodicities are an obvious way for people to measure units of time. Much has been written on the early development of calendars in pre-literate societies, from simple tally notations recording the days or lunar months through to institutionalised self-consistent systems of marking time.[110] However, a number of assumptions are often made: one is that there is an almost inevitable progression from lunar, through to luni-solar and eventually solar calendars; another that a seasonal calendar is necessarily a solar calendar.[111] These are questionable outside a Western world-view and indeed can be countered by a number of modern examples.[112] Indeed, the very notion of the development of units of measurement of time[113] seems to assume an abstract (context-free) conception of time that is unlikely be appropriate in a non-Western world-view.[114]

It is better to recognise at the outset that people try to make sense of the passage of time by classifying and categorising events in relation to their personal or shared experience rather than viewing time as an abstract entity.[115] In historical and modern indigenous societies we see ample evidence of the various ways in which people strive to harmonise human activities with cycles of events in the natural world. Ritual performances may have a key role in reaffirming the natural (cosmic) order, as they did, for example, in ancient Mesoamerica.[116] The same is true of traditional Hopi villages in the US south-west,[117] much mentioned in the context of discussions of prehistoric astronomy because of the precision with which solar horizon observations are used to pinpoint ceremonial events to within a day or two in the calendar year.[118] In fact, it would be simplistic to think that the Hopi simply had a sacred calendar related to crop-planting activities that was regulated by horizon sun observations. At the village of Walpi, observations of the sun rising and setting at the solstices are actually superfluous to the ceremonial calendar,[119] but are important because they mark the four directions that are sacred in Hopi world-view. For the Hopi, these are not geometrical abstractions but empirical realities with rich symbolic associations,[120] and the points on the horizon behind which the sun rises and sets at the solstices as viewed from Walpi are themselves sacred places which are visited at certain times of the year for prayer sticks to be offered to the sun.[121] All this forms part of a complex cycle of ceremonial activity that provides a constant reaffirmation of the Hopi view of the world and their place within it.[122]

Space, like time, was not perceived in the abstract, and if we are better to understand non-Western world-views it is more helpful to think in terms of a context-rich concept of 'place'[123] giving rise to 'sacred geographies' in which landscapes were charged with meaning (see Archaeology Box 7). Most indigenous world-views express an intimate, spiritual connection between people and their natural environment, the land and sky. By associating special places (such as prominent landmarks) with particular celestial features and with sacred and secular practices, through myth and oral tradition, the power and vitality of being in a particular place—perhaps at a particular time—is reinforced.[124] The oral tradition of the Lakota people of South Dakota, for example, leads them to undertake an annual progression of movement, involving subsistence and ceremonial activity, through the famous Black Hills that is seen to be related to, and in tune with, the path of the sun through specific constellations. The constellations themselves are directly associated with particular landmarks.[125] The Hopi and Lakota examples show how considerations of place and time may be interrelated in complex ways, and demonstrate again how events in the sky form an integral part of the wider picture.

In recent years there has been a great deal of interest in interpreting prehistoric landscapes in a similar way (Archaeology Box 7). Identifying sacred places themselves may not be easy, except where they came to be marked by monuments, offerings, or rock art,[126] but it is also arguable that 'features of the natural landscape may be held to have provided a symbolic resource of the utmost significance to prehistoric populations'[127] and that associations between monuments and places such as hilltops were also highly important.[128] In the context of such ideas, possible associations between places and horizon or near-horizon astronomical events assume great potential significance, and in view of the importance of the sky in non-Western world-views it would be absurd not even to consider them.

> Astronomical phenomena were not privileged over ancestral monuments or landscape features. We are not witnessing evidence for scientific observations of the heavens, so much as a perceived unity of sky and land, past and present, all being manipulated to bring more and more emphasis onto particular spaces and places. This would tend to heighten the significance of whatever transactions and performances took place there. At the same time, it would also limit access to these spaces in terms of both direction and timing, and would contribute to the way in which the space was experienced by promoting the impression that it stood at an axial point of an integrated cosmos.[129]

There has been much discussion recently of the importance of perceptions gained not just by existing in, but by moving through, the prehistoric landscape. This has stemmed from questioning the conventional view (Archaeology Box 5) that Neolithic people generally lived in fixed settlements, suggesting instead that in much of Britain and Ireland the early herders and cultivators of the Neolithic may have tended to move seasonally about the landscape according to the availability of different resources at different times of the year, following traditional paths,[130] much as the earlier Mesolithic hunter-gatherers may have done.[131] Conspicuous monuments would then have marked reference points in these patterns of movement,[132] and 'part of the sense of place [would have been] the action of approaching it from the "right" (socially prescribed) direction'.[133] In fact, people did not move solely through the landscape but through the whole cosmos as they perceived it. We should therefore take into account the sky, and especially its different configurations that might have regulated and defined such movement, just as it continues to do amongst the Lakota.[134] It is certainly tempting to interpret certain linear monuments, such as cursuses and stone avenues, as actually marking paths of particular ceremonial importance, and Tilley has concluded that 'the experience of walking along [the Dorset cursus] was an essential ingredient in its meaning'.[135] Once again, we should ask whether such movement was most propitious at certain times, and whether astronomical alignments might give us clues as to when those times were.

Studies of non-Western societies in the modern world reveal many examples of the symbolic expression of relationships between celestial and other objects which, if paralleled in prehistoric times, could have left their mark in features of the location and design of both sacred and domestic structures, as well as in the organisation of household and settlement space. Perceived cosmic order is reflected in the layout of traditional buildings such as Navajo hogans,[136] Hopi kivas,[137] and Skidi Pawnee earth lodges[138] in the US south-west, in Amazonian roundhouses,[139] and in the layout of villages in the Andes[140] and in central South America,[141] to name but a few examples from the Americas alone. The case of the Yucatec Maya village of Yalcobá, where the structure of the cosmos is reflected in a whole variety of aspects of social behaviour,[142] gives a glimpse into the sheer complexity that we may encounter when we try to begin to appreciate symbolic schemes as an integrated whole. Nonetheless, if similar practices in prehistoric times were sufficiently widespread they may have left tangible archaeological evidence.

Such evidence is being found amongst domestic structures in the Iron Age. Roundhouses in Britain have their entrances predominantly facing east or south-east, and the same may be true of their counterparts in Ireland.[143] Particular concentrations of orientations are evident within about 10° of due east and around the direction of winter solstice sunrise.[144] A similar trend is evident in the orientations of entrances to hillforts.[145] Added to this, it has been suggested that spatial patterns of activity, such as preferences for certain places when disposing of the dead, of animals, or of other objects, may have reflected symbolic or cosmological principles:[146] similarly, certain activities may only have been appropriate in particular quadrants of roundhouses.[147]

What, though, of the design of earlier monuments such as stone circles and rows? It has been suggested that a stone or timber circle served to demarcate the space within it, separating it from the rest of the world, imbuing it with special meaning. A tradition of 'magic circles', sacred places whose interior offered protection from the devil, persisted in medieval times.[148] One reason why a circle—or at least a ring—was so often the shape chosen may be that it mimicked the sun, the bringer of warmth which alone always appears as a full circle.[149] Another is that it acted as a microcosm, modelling the known world with its surrounding circular horizon: 'a universe in miniature where beneficent forces could be gathered and focused . . .'.[150] It has even been suggested that the space within the sarsen Stonehenge and the landscape around it were conceptually divided into four quarters demarcated by the solstitial axes,[151] resembling the quadripartition common in native American world-views. All this suggests the importance of a fixed place, reaffirmed perhaps by symbolic relationships between the place itself and the cosmos (land and sky) surrounding it. Linear monuments, on the other hand, give the feeling of having been more related to patterns of movement, and not necessarily along them. It has been suggested, for example, that certain short stone rows marked boundaries of (sacred) space that had to be crossed, implying that the movement most charged with symbolic meaning was a movement across the alignment.[152] These are speculative ideas, of course, but they lead to suggestions, in ways that seem plausible in the context of non-Western world-views, as to why the orientations of stone rows or circle entrances in relation to prominent landscape features or astronomical events, or the positioning of a monument such that a prominent mountain was in line with an astronomical event at an important time, might have been important.

We return, then, to orientations and 'astronomical alignments', but considered now in wider frameworks of contextual ideas. Orientation clearly mattered—at least amongst the monuments in those local groups where strong overall trends are evident—just as it did in later communities throughout the world.[153] Why might it have been important to orient the axis or entrance of monument upon an astronomical event at or near the horizon? One possibility is that this was one way in which to harmonise the monument, or the place where it was located, with the cosmos: 'By incorporating into its structure an important astronomical alignment, those who built it made those developments appear to be part of the functioning of nature'.[154] Perhaps, in small communities, astronomical alignments simply helped to affirm a monument as being at 'the centre of the world';[155] but in other cases they may have had more to do with making its power impossible to challenge and thereby affirming ideological structures and political control.[156]

Astronomical alignments, as opposed to symbolic references to permanent features in the landscape, had the special property that the occurrence of an astronomical event, such as the appearance of the sun or moon in line with a monument, could also have served to place that monument in time, empowering it—giving it special significance—on certain (regular) occasions.[157] This is particularly easy to imagine in the case of stone rows aligned upon solstitial sunrise or sunset, and Tim Robinson gives a marvellous description of the visual effect of midwinter sunset at the Gleninagh stone row in Connemara (L 815552).[158] 'Time, in our everyday experience, does not consist of such moments; they are as rare in the general flux as grains of gold in the gravel of our Connemara streams.'[159] While it seems unduly speculative to picture stone circles as refuges for communities during times of solstitial 'cosmic crisis'[160] it is certainly possible that, by performing appropriate ceremonial activities as dictated by spectacular cosmic events associated with a monument, people kept their activities in tune with the workings of the universe as they understood it. The importance of being in a certain place at a certain time is illustrated in modern times by the strength of purpose with which new-age travellers try to converge each year upon Stonehenge at summer solstice.

Yet it may be a mistake to categorise orthostatic stone monuments as ceremonial places for the living in direct contrast to cairns and tombs—obviously places for the dead. Clear associations exist between these different types of monuments, such as that between RSCs and ring cairns (see chapter five), which suggests that no such rigid dichotomy existed in the minds of prehistoric people. A rather different framework of interpretation is suggested by analogies from modern Madagascar and elsewhere, in which the inherent qualities of stone lead to strong and enduring metaphorical associations with ancestors, whereas only perishable materials, such as wood, are suitable as building materials for places used by the living.[161] One possibility currently being considered very seriously by some archaeologists is that some of the best known prehistoric stone circles were places seen to be inhabited by the ancestors, paralleling in their form of construction the timber circles that co-existed with them elsewhere in the landscape, or had once preceded them, and were or had been used for ceremonial purposes by the living. Parts of the landscape surrounding stone monuments such as Stonehenge 3 or Avebury may even have become 'domains of the ancestors', rarely visited by the living.[162] In the context of such ideas, solar and lunar alignments incorporated into stone monuments might well reflect strong conceptual links between sun and moon worship and the ancestors, and give us some valuable insights into perceived relationships between the living and the dead.[163]

The horizon—the place where the earth meets the heavenly vault—is of widespread mythical significance,[164] just as caves are often considered sacred because they are where the world comes into contact with the powers of the underworld, and they become 'places of emergence' or 'points of origin'.[165] But once again we should avoid being too narrow-minded in our ideas, for example in concentrating on the horizon itself as we perceive it. We have mentioned that the moon appears larger when it is close to the horizon. As a result it may be perceived as closer,[166] and so we should explore how distance was perceived in other cultural contexts and ask whether special significance might have been attached to the moon 'when it was near'. Might this explain the apparent interest in Bronze Age western Scotland and south-west Ireland in the southern moon near the major standstill limit, when the moon never rises far above the southern horizon?

Finally, we should at least consider other possibilities, for example that monuments might have done anything *but* point directly at sacred places. Taboos on pointing directly at anything sacred (at least, by people themselves) are certainly encountered amongst modern indigenous peoples,[167] and we should certainly not ignore the possibility that some prehistoric constructions might actually have avoided direct reference to (being within sight of, being oriented upon) sacred places or events.

A point that is often forgotten because of the prepossession of archaeoastronomy with 'alignment studies' is that there is a third dimension: if astronomical symbolism was incorporated in a monument, it need not have been restricted to horizontal orientations.[168] And even so-called alignments upon astronomical events near the horizon might have derived their significance more from their effects in light and shadow than with observations of the sun or moon against the distant skyline. We have seen how sunrise on days around midsummer would have produced a profound effect within the sarsen circle at Stonehenge, and even the moon close to the horizon can cast pronounced shadows.[169] In the archaeoastronomical literature, there has been a good deal of discussion concerning shadow and light phenomena. Examples include the 'equinox hierophany' at the Temple of Kukulcan ('El Castillo') at Chichen Itza, which attracts tens of thousands of visitors each year;[170] the famous 'sun dagger' at Fajada Butte in Chaco Canyon[171] and many lesser known effects of sun-and-shadow upon rock-art designs;[172] and 'zenith tubes' in Mesoamerica.[173] Recently, there have been computer simulations of light and shadow effects amongst the Stonehenge sarsens.[174] As to the meaning and significance of such phenomena, we can perhaps gain some insight from the documented sky traditions of the Chumash Indians of California: 'In the play of winter solstice sunlight upon symbolic rock art, the shaman saw the sign of cosmic order. . . . The shrines themselves were sacred because they were the places where cosmic order was revealed.'[175]

If, however, we are to move seriously beyond merely discovering and describing shadow and light phenomena—if instead we are to gain useful insights into prehistoric metaphysics—then wider perspectives and richer contextual arguments must be brought to bear. We could, for example, explore meanings associated with light, reflectivity, and shiny or brilliant materials. These appear to have formed part of multifaceted concepts widespread throughout pre-Columbian America, where understanding their significance has the potential to give us new insights into native American world-views as well as a completely new indigenous perspective on the European conquest of the Americas.[176] To these Amerindians, the sun, moon, meteorological phenomena, metals, minerals, shells, water, pelts, feathers, and semen (the list is far from exhaustive) all revealed their inner sacredness—their 'spiritual essence'—by displaying light at their surface. (In contrast, the absence of light, as during eclipses, or the presence of 'unnatural' light, from unusual sources including meteors and comets, was associated with sickness, catastrophe and destruction.) We could go on to examine the role played by the symbolism of light in people's perceptions of the cosmos and sacred space, for example in emphasising shiny elements in the landscape such as stones and water.[177] More specifically, the fact that the soft white light of quartz or mica was often associated with the moon, in Mesoamerica and elsewhere,[178] suggests new interpretations of the possible meanings of Bronze Age stone circles and rows, where quartz scatters are often found in association with monumental architecture apparently aligned upon the moon itself (see chapters five and seven).

A rather different approach is to consider the practicalities of monument construction when constrained by desired symbolic relationships with the natural world. Exploring this theme in relation to various details in the design of the three cairns at Balnuaran of Clava, Richard Bradley concludes that structural risks were taken in order to conform to cosmological requirements, resulting in a 'creative tension between cosmology and engineering'. For example, because taller kerbstones needed to be used on the south-west side of each cairn—reflecting a principle of height gradation towards the south-west which was important and enduring enough to be perpetuated from the (probably) earlier recumbent stone circle tradition—the tallest part of the chamber wall of one of the passage tombs had to be built on top of the lowest of the orthostats, bringing a real danger of collapse (indeed, the central ring cairn did collapse, quite possibly for this sort of reason).[179] In examining how structural integrity may have been compromised by the need to adhere to a powerful symbolic system, we may be able to use details of engineering and construction to cast light on prevailing perceptions of the cosmic order.

Returning finally to the general theme of developments in human thought, there is one topic of considerable interest that often crops up in the context of astronomy and deserves special mention: namely where we might find the earliest signs of human enquiry into the causes of natural phenomena, 'scientific explanation' in its widest sense. Ironically in view of the 'science v. symbolism' debates in archaeoastronomy in the early 1980s (see chapter four), answers are not to be found by searching for proto-Western science but in the very symbolism that was once seen as its antithesis. If we see non-Western systems of thought as substituting (superstitious) myth, ritual, and ceremony for (rational) explanation, we fail to recognise that the conceptual structures underlying them themselves constitute mechanisms for explanation that are perfectly coherent and logical in their own terms. Correspondences, such as that between the Barasana Caterpillar Jaguar constellation and earthly caterpillars, have explanatory power within a non-Western world-view, and symbolic expressions of such correspondences—such as that between quartz and moonlight—possess the power to express perceived reality in such a framework. Astronomical symbolism incorporated, for example, in monumental architecture reflected the structure of the world—the cosmos—as prehistoric communities in Britain and Ireland understood it. That understanding was their science, and by striving to understand symbolic associations in the material record aspects of it may begin to be revealed to us.[180]

In sum, astronomical symbolism is relevant to any attempt to interpret prehistoric perceptions of cosmic order. Astronomy occupied an essential place in the ways people ordered the world, defining places within it and pathways through it, and thereby came to 'understand' and control it. This observation suggests a rich variety of contextual ideas for which evidence can be sought in the archaeological record. It is tempting to speak of astronomy as the oldest science, but in doing so we may fail to recognise that attempting to make sense of what was perceived in the sky formed an inextricable part of trying to understand the natural world in an integrated way. The external, 'objective' view of the world underlying modern astronomy is in sharp contrast to the internalised, contextually rich nature of most non-Western world-views.

# 10
# Looking Further

It is easy to argue . . . for a multidisciplinary approach to every scientific and technological problem: it is not so easy to orchestrate the contributions of research specialists from very different scientific backgrounds.

John Ziman, 1994[1]

I object on general principle to the notion of giving full professors well paid vacations to persue [sic] arcane interests. . . .

Grant reviewer, quoted by Anthony F. Aveni, 1981[2]

When astronomers deal more with the cultural context of sites in an area, and when archaeologists deal more with the ritual aspects of the same sites, we will begin to see progress in the maturation of archaeoastronomy as a truly integrated discipline.

W. James Judge, 1987[3]

## Seeking the evidence

Several authors, from a range of disciplines—a group who would disagree fundamentally with one another on many matters of detail—are nonetheless united in feeling, in common with Hawkes more than thirty years earlier,[4] that Stonehenge was a cosmic temple where the sun, moon and stars were indeed observed. According to historian of astronomy John North,[5] Stonehenge 'was in a very real sense a cosmos, a geometrically ordered monument aligned upon the universe of stars, Sun and Moon, and an embodiment of the spiritual forces they represented to most of mankind'; and the statement is enthusiastically endorsed by archaeologist Colin Renfrew.[6] Geographer Rodney Castleden suggests that the sarsen circle was dedicated to sky gods and the ring of Aubrey Holes to earth deities, the symbolism of solar and lunar alignments serving to honour mythical events in the interaction of these deities.[7] To archaeoastronomer Anthony Aveni, Stonehenge was 'a consecrated space for watching the sky . . . more closely allied with theater than with exact science'.[8] Such ideas are plausible enough—perhaps even compelling—given our wider knowledge of the ways in which monuments and buildings carry rich cosmological symbolism in many indigenous societies, but we should not lose sight of the need to assess the extent to which the material record actually provides direct support for them. In view of the discussion in chapter eight, not all would agree with Aveni that 'the sun and moon alignments [at Stonehenge] have withstood the test of time'.[9] While we would not wish to dismiss ideas such as these without good reason, we also need good reason for continuing to support them.

Likewise, statements that have emanated from phenomenological studies, for example 'that a desire that prominent Tors be visible from the [stone] circles [of Bodmin Moor] played a major role in their precise location is evident from a consideration of a number of specific instances'[10] provoke basic questions of data selection similar to those that caused us to question Thom's work in chapter two. At Stonehenge and in its environs, as well as elsewhere in Britain and Ireland, there is a pressing need to examine further evidence on the location and design of monuments in relation to the contemporary landscape, but in a systematic way. This will enable us to question, and ultimately improve, a range of ideas about the ways in which symbolic relationships between monuments and the surrounding terrain and sky reflected contemporary world-views.

Regional groups of similar monuments have particular potential in this regard because features of their design which are found repeatedly, and likewise repeated aspects of their location in the contemporary landscape, may reveal common elements of ritual tradition (and hence world-view) that can be identified with some confidence as something that was intentional and meaningful. Despite the fact that they have long been ignored by most mainstream archaeologists, analyses of orientation and possible astronomical associations may be particularly useful in this regard. It has been more common amongst archaeologists to examine the spatial distributions of different sets of monuments, often in relation to settlement, land use, or ecological resources,[11] or to examine their interrelationships with one another, such as intervisibility;[12] but any inferences made are critically dependent upon the assumption that the present distribution of monuments is a true reflection of the past one. Decay and destruction, through natural erosion and human activity, can be highly dependent upon factors such as modern land use which varies from one area to another, and can badly distort the picture (see Archaeology Box 6). This problem is far less critical in the sort of analysis where we are concerned with the relationship of monuments to immutable astronomical phenomena or physical features in the landscape; here conclusions can be more reliably based upon the sample of monuments that remain.[13]

Recumbent stone circles and short stone rows have received particular archaeoastronomical attention up to now, but there are many other groups of monuments that merit investigation, from the Early Neolithic through to the middle of the Bronze Age. For example, there is a clear need for people to investigate the possible existence of astronomical alignments in the forerunners of the circles and rows, the chambered tombs and earthen long barrows of Britain and Ireland. As we have seen in chapter eight, the orientations of certain regional groups reveal clear trends over relatively wide areas, and there is an obvious need for new surveys to improve upon the low-precision data that are generally all that is currently available. The very nature of the long barrows seems to place limits upon the degree of precision that might be meaningful: several of the Wessex examples are over 100 m long[14] and curved, and some contour round hills with the result that the axial orientations of the two ends are many degrees different from each other. Even restricting attention to the entrance end, it seems meaningless to attempt to measure azimuths to a precision greater than about ±2°.[15] Nonetheless, investigation of regional preferences in orientation or location at this level has the potential to reveal possible topographic or particular astronomical influences upon the orientations,[16] and while these may still be crude in relation to what has been suggested in the Late Neolithic and Early Bronze Age, recognising them could greatly enhance our understanding of later developments. In respect of astronomical conclusions it is necessary, even at this level of precision, to take account of horizon altitude, since every one degree of uncertainty in altitude corresponds to almost as much in declination,[17] and this implies new fieldwork. This is certainly critical in order to address seriously the key question of whether the principal associations were with the sun or moon. In the case of well preserved chambered tombs, it may be possible—and meaningful—to define orientations to 1°, and suitable azimuth data have been published for some regional groups.[18] Most other regional groups are covered by inventories containing published plans,[19] but many—especially the older ones—do not contain reliable north-points.[20] In any case, new fieldwork would be required in order to obtain reliable declination data taking account of horizon altitude, and to explore more subtle relationships with features in the natural landscape.

Cursus monuments are also of considerable interest. The Dorset cursus has received particular attention but the fact that many interpretations are specific to this monument opens a series of questions reminiscent of the controversies surrounding Ballochroy. We should see whether common elements might be more reliably revealed by more systematic studies of these enigmatic monuments. There is good potential for this: previously undetected cursuses have been discovered at a high rate in recent years, particularly using aerial photography, and excavations have been undertaken at a number of them. Over eighty examples are now known in England alone, with perhaps fifty or sixty more elsewhere in Britain and Ireland.[21]

Similarly, the only real hope of tackling the controversial question of whether the orientation of the north-east entrance to the bank and ditch enclosure at Stonehenge 1 was astronomically determined, by reference to the moon or anything else, may be to examine the orientations of similar, contemporary circular enclosures in the vicinity.[22] A suitable catalogue, including aerial photographs, is being compiled by English Heritage (David Batchelor, priv. comm., 1997) and may facilitate such a study. And, moving through to the Late Neolithic and Bronze Age, there are many groups of monuments where archaeoastronomical surveys could produce evidence relevant to a range of thematic ideas. The five-stone circles and stone pairs in Cork and Kerry are important for confirming whether trends apparent in the larger circles and longer rows extended to smaller monuments[23] and may help us to recognise developments in space and time; important too in this context are the wedge tombs (see chapter eight). Certainly, the orientations of all of these monuments are strongly clustered.[24] A concentration of short stone rows in the north of Ireland is particularly important in view of possible continuities of tradition between western Scotland and southern Ireland.[25] Other regional groups particularly worthy of further investigation include the great stone circles of Cumbria and south-west Scotland,[26] concentrations of henge monuments in southern and north-eastern England and elsewhere,[27] the Clava cairns of Inverness-shire,[28] the stone pairs and 'four-posters' of Perthshire,[29] the kerb-cairns of western, central and eastern Scotland,[30] and possibly even groups of single, oriented standing stones in some areas such as Argyll.[31]

Of course, there are dangers in basing interpretations upon simple typological classifications, especially where there is the possibility of miscategorisations owing to degradation of the monuments: this is particularly evident in the case of single standing stones, stone pairs, and short stone rows.[32] We must also be highly cautious about assuming that similarity of form implies similarity of function, meaning or use.[33] The largest examples of their respective classes—monuments such as Newgrange, Maes Howe, the Dorset Cursus, Balnuaran of Clava, and Stonehenge—may be atypical, perhaps incorporating a weight of symbolism not found elsewhere (including more careful alignments); and if we consider them only as part of a larger group we may miss something vital.[34] Neither should we forget that many monuments changed form in antiquity, and certainly we must not lose sight of the need to accumulate more conventional archaeological evidence of construction sequences at individual monuments and chronology and change in regional contexts. Nonetheless, the very existence of recognisable common local trends in monumental architecture—and of the repeated expression of similar relationships between monuments and the surrounding world or cosmos in the form of orientations or alignments upon celestial bodies or prominent topographic features in the landscape—gives us important information relevant to our wider ideas about social or cognitive development. The totality of the evidence available to the prehistorian is so sparse that it would be absurd not to pay attention to these vital clues.

In this context, it would be unduly restrictive to confine our attention to orientations. For example, it has been suggested that the use of geometric forms based on a standard unit of length formed an important element of the ritual symbolism at Newgrange and the other Boyne valley tombs.[35] Furthermore, the sacred order may have been reflected not just in the design of monuments and their location within the landscape, but also in the spatial disposition of various activities, and hence of formal deposits, within the monuments themselves.[36] We have seen how scatters of white quartz found at RSCs are concentrated in an area adjacent to the recumbent stone. Astronomical considerations may have defined places within a monument that had particular sacred significance, as has been suggested to explain the peculiar spatial patterns of deposition found at Stonehenge.[37] We should certainly examine the spatial distribution of intra-site activity with astronomical considerations in mind. This is a largely neglected area of enquiry that deserves a good deal more attention, although it is dependent upon evidence from excavation.

Relationships between monuments may also have been important, and not only between contemporary ones. Conspicuous monuments remained in the landscape long after they had ceased to be used for the purposes for which they were originally built.[38] Once constructed, they became 'indelible marks on the landscape', part of the established order, influencing people's understanding of the world for generations to come.[39] There is evidence that later monuments were located and oriented with reference to earlier ones, one example being the Dorset cursus,[40] another being the stone alignments in the regions of Carnac (see Fig. 7.7) and Saint-Just, Brittany.[41] In consequence, the idea no longer seems far-fetched that a monument might have been located so that an important astronomical event might take place behind an existing, perhaps much older, monument. It follows then, that—despite the critiques of Thom's 'inter-site' alignments of high astronomical precision—deliberate, meaningful astronomical alignments between monuments of very different dates might actually exist. Fresh statistical investigations suggest that there may even be evidence of orientation trends and astronomical alignments amongst the inter-site data from the 1984 reassessment of 300 western Scottish monuments.[42] This merits further investigation.

In order to explore a wider range of thematic ideas such as those developed in chapter nine, it is too restrictive to limit our data to structural orientations and alignments. Other approaches might be more enlightening, provided they can generate systematic data. For example, it has been suggested in another context that certain ritual monuments were deliberately positioned in places highly visible but inaccessible from settlements, in order to place them on the threshold of consciousness.[43] Such 'perceived marginality' and its possible importance in the prehistoric landscape deserves a good deal more attention, especially as there is mounting evidence attesting to the importance of places being hidden or partially hidden. The 'elbow' in the Stonehenge Avenue to the north-east of the sarsen monument draws attention to the possible importance of 'blind spots' in the landscape.[44] Silbury Hill comes in and out of vision as one walks along the Avebury Avenue.[45] The five northernmost stone rows on Mull are located at the very limit of visibility for Ben More (chapter seven), and some Cork–Kerry axial stone circles like Gortanimill (ASC11) were placed just to the north-east of a ridge, while moving a mere ten metres or so beyond the axial stone to the south-west reveals a wide distant horizon. Programmes of work might be devised to generate systematic field data relating specifically to such issues, not ignoring the possibility that relationships between places and astronomical phenomena—just as much a part of the perceived cosmos as prominent landscape features or ancestral monuments—were hidden or restricted from view in various ways.

As we have seen in chapter nine, there is every reason to expect that cosmological symbolism was reflected just as much in domestic constructions as in public monuments. Where evidence exists on the location and layout of settlements, it should receive just as much attention as that from conspicuous monuments. After all, 'the same Neolithic people lived in houses within the settlements, participated in or watched various ceremonies and rites of passage, including funerals'.[46] The recent discovery of a Neolithic village at Barnhouse in Orkney, close to Maes Howe and the Ring of Stenness,[47] together with the already famous villages at Skara Brae and Rinyo, make Orkney a natural focus for studies of the principles of order and classification that might have determined spatial arrangements within a particular sample of Neolithic houses.[48] For instance, the orientations of central hearths in Orcadian houses clearly separate into four wide bands centred roughly upon the four intercardinal directions.[49] This is strongly suggestive of cosmological principles involving a four-directional symbolism, and it is possible that the four directions were conceptually associated with the solstitial directions at their centres.[50]

On a more cautionary note, the Barnhouse excavations have revealed a house-like building 'of "monumental" proportions' structurally similar to Maes Howe but whose entrance is oriented towards sunrise at the summer solstice. This suggests a dichotomy between the symbolism related to light and warmth for the living and cold and darkness for the dead.[51] The idea may seem attractive, but in our enthusiasm to embrace this important new evidence from domestic architecture we should not leap to conclusions that ignore the existing data from the other Orkney tombs. The south-westerly orientation of Maes Howe itself is, after all, anomalous in itself; most Orkney tombs face towards the eastern half of the compass, and those that do not are generally exceptional in other ways.[52]

As we move our point of focus from the monuments themselves to people in the landscape, we encounter the suggestion that in some parts of Britain and Ireland seasonal patterns of movement remained the norm well into, even perhaps throughout, the Neolithic (see chapter nine). In this context, monuments are likely to have formed key points of reference, and petroglyphs may also have played a major role in marking and characterising important places.[53] It is even possible that patterns of movement persisted well into the Bronze Age in some marginal areas such as Mull, where widespread arable cultivation was ruled out by the rugged landscape and poor soils, and fishing and livestock rearing almost certainly remained the main means of subsistence. Various ideas could be explored by examining monument orientations and archaeoastronomical evidence in relation to other archaeological evidence. Are consistent astronomical orientations only to be found in those later monuments constructed by people living in fixed settlements, as part of establishing and reinforcing their permanent 'sense of place'? Or might monuments built by people who only visited them at certain times of year symbolically reflect that fact in their design, perhaps in their alignment upon with the rising or setting of the sun or full moon at the appropriate time?[54] It is even conceivable that such evidence could inform questions of when and how transitions from seasonal movement to fixed settlement occurred in different regions. We should certainly seek it.

To judge by the modern examples described in chapter nine, we should give some consideration to the idea that certain seasonal movements through the landscape in prehistoric times might have been correlated with, and hence regulated by, observations of the different configurations of the sky:

> What did the sky look like as people stood at certain points in the landscape, or moved around in certain ways? Was it important, perhaps, to be in a certain place, or to move in certain ways, at a particular time, when the celestial configuration was right? Was it important, perhaps, sometimes to approach Stonehenge along the avenue at night, when the stars in the sky would have been as prominent as earthly signs of ancestors during the day?[55]

Phenomenology could be useful for exploring astronomical factors in patterns of movement and approach and for generating initial ideas. For example, observing the sunset phenomenon at the Dorset Cursus (chapter eight) may have been just one of a sequence of dramatic effects intended to be experienced by someone moving along the interior of the monument.[56] If they then proceeded from the Bottlebush terminal along to the south-west in the dusk of an early December evening, what stars would they have seen setting in line with the monument? The problem here is that simply by walking about in the modern landscape, one is unable to visualise the different configurations of the sky at various times of day and year in the past. While it may be possible to do so back in the office using one of the now commonly available computer programs for this purpose, the real need is to visualise monuments, topography, and sky together. In the future, this should become increasingly possible using virtual reality modelling,[57] but even then initial ideas will need to be investigated further by the more systematic acquisition of data, which is likely to involve field survey.

In the context of certain questions raised by the prehistoric record in Britain and Ireland there is a role for reassessments of historical and ethnographic evidence further afield and even for fresh fieldwork. One key question that is raised by the conclusions reached in chapters five to seven is why associations with the moon close to the horizon might have been so particularly important to the builders of certain groups of Late Neolithic and Bronze Age monuments. While there is widespread interest in the moon even amongst hunter-gatherer groups around the world,[58] there is little direct evidence of any indigenous group orienting architecture in relation to its motions near to the horizon.[59] One suggestion as to why the southernmost limits of the moon's motions might have been particularly significant to certain groups of people in prehistoric Scotland, as implied by the evidence from the recumbent stone circles and the western Scottish rows, is that the moon when close to the major standstill limit is seen to scrape along just above the southern horizon—a rare and dramatic event which could perhaps have assumed great importance.[60] This is a function of the latitude of Britain and Ireland in general and Scotland in particular.[61] It would clearly be of great interest to discover evidence of a similar interest in the rising and setting moon amongst traditional communities at similar latitudes, such as in north-eastern Europe, central Russia and Siberia.[62]

Finally, there is a need to compensate for the over-attention on earlier prehistory by stimulating studies in the Later Bronze Age and Iron Age. Here we have a potentially rich body of evidence concerning possible cosmological referents in houses, settlements, and even forts, including structure orientations and spatial patterns of activity. Furthermore, the material record is supplemented by historical and literary evidence. We should also consider the cycles of time whereby people arranged their daily lives. The relentless pursuit of 'the' Celtic calendar has tended to obscure the fact that some form or forms of seasonal calendar are certainly to be expected in Iron Age times, especially if we accept that cosmological concerns do seem to have exerted a strong influence upon architecture and the use of space. While the Coligny calendar is unlikely to reflect a single Celtic calendar, it *is* possible that it is based upon traditional, oral knowledge that was committed to writing for the first time in the third century AD.[63] And there is certainly evidence (albeit almost entirely indirect) that the Druids, the principal mediators between the natural and the supernatural worlds in Celtic times,[64] had considerable practical knowledge of astronomy and calendrics.[65] Last of all, hitherto unsuspected patterns of continuity of material tradition from earlier times through to the Iron Age, as have been suggested recently in the case of timber circles,[66] imply that one should not retain an entirely closed mind about the possibility of some sort of continuity from even older sacred or calendrical traditions. An intriguing example in this context is Altar wedge tomb in Co. Cork (V 859302)[67], which is oriented directly upon Mizen Peak, a distinctive pyramidal peak[68] over which the sun would have set at around the time of Samhain;[69] excavations have revealed evidence of ritual activity at the site stretching from around 2000 BC until after the year 0.[70] Despite all the prejudices and misunderstandings of recent decades, all these matters deserve further, careful investigation.

The foregoing suggestions are by no means intended as a comprehensive priority list. They do, however, make apparent the need for ideas concerning astronomy to be considered as an integral part of our efforts to formulate viable theories about cognitive developments in prehistory. There is also a clear need for archaeoastronomical evidence to be considered as an important element in—perhaps even in some cases the main focus of—specific investigations setting out to test those theories.

## Beyond the green and the brown

Up to this point we have discussed ways of improving some of our initial ideas concerning prehistoric world-views, and suggested some priorities for generating field data (with particular emphasis upon archaeoastronomical field data) that will be helpful in assessing and further developing those ideas. The last major issue that remains is precisely *how* we should use the material record to modify and develop our ideas: the question of methodology.

We can clarify this slightly by restating our overall approach[71] as follows: to develop ways of determining the extent to which the material evidence reinforces or dilutes existing ideas. This retains the idea of 'testing' our ideas against the archaeological record[72] but rejects any thought that we should expect a yes/no answer—'proof' or 'disproof' of a given

## STATISTICS BOX 7

### CLASSICAL AND BAYESIAN APPROACHES

There are two distinct areas of interest in statistical studies: the need to describe a practical situation and the need to prescribe action in the context of that situation. The former is termed statistical inference (since information is used to infer, through a probability model, a description of a practical situation) and the latter is termed decision making. The Classical and the Bayesian viewpoints represent two fundamentally different approaches to statistical inference.[1]

In Classical inference and decision-making we test a working (basic, null) hypothesis in order to decide whether to accept or reject it (Statistics Box 4). The classical approach dictates that a concept of probability can only be adequately defined, and a statistical theory developed, for information obtained (potentially at least) in a repetitive situation—the observed outcomes from what are assumed to be independent repetitions of a situation under identical circumstances. It provides no statistical procedures capable of processing prior information accumulated from past (or external) experience.

The Bayesian school, on the other hand, maintains that inferences are made by combining an assessment of a state of knowledge and a practical situation. Probability is viewed as expressing a (personal) degree of belief, and a (personal) state of knowledge is expressed in terms of the probability distribution of the parameters of a model 'prior' to the current situation being taken into account, which is then modified by sample data arising from the current situation. The inferences are expressed (solely) by the posterior distribution of those same parameters. No prescription is given as to what action might be taken as a result.[2]

#### PROBABILITY AND ODDS VIEWED THE BAYESIAN WAY

The description of probability and odds given in Statistics Box 1 was expressed in a classical framework. Bayesian statisticians have a different view of these concepts.[3] Probability in a Bayesian framework expresses a degree of belief, somewhere from 0 (expressing certainty, say, that an event will *not* happen) to 1 (certainty that it will). How, though, can a specific probability figure be placed upon someone's (generally somewhat ill-defined) prior feelings about the likelihood of the event?

One way is to use a betting analogy. I will presumably only bet on a horse at given odds if I think it has a sufficiently good chance of winning—in other words, if my prior belief in the likelihood of its winning exceeds some critical level. A probability of $1/n$ can be defined as the level above which I would consider it worth taking the risk of betting on a horse at odds of $(n - 1)$ to 1 against. This corresponds to the value in the Classical view (cf. Statistics Box 1) but now represents a personal degree of belief that might take into account a host of 'external' factors such as my background knowledge of the horse's form, the state of the course, or even the jockey's astrological chart.

Another difference is that, when probability is viewed in a Classical way, the probability of an event such as throwing a six with a die is independent of what has gone before. If a die has come down 6 six times in a row, the probability of it happening again is still $1/6$. However, if I roll a die and obtain a six over and over again, I might start to question the implicit assumption that the die is fair. The probability of a six on the next throw, defined as my prior belief in obtaining a six, might well steadily increase from $1/6$ up towards 1.

#### THE BAYESIAN PROCEDURE

The Bayesian procedure essentially comprises three steps. The first is to express one's state of knowledge about some domain of interest in the form of the probability distribution of the parameters of a suitable model. This is known as the *prior probability density*[4] or 'prior'. The second step is to develop a model that expresses how likely the observed data are to arise for each set of possible values of the unknown parameters. This is known as the *likelihood function* (or 'likelihood'). The final step is to combine the two mathematically using Bayes' theorem, to obtain a revised probability distribution of the parameters of the model given the observed data. This *posterior probability density* or 'posterior' expresses how the prior picture of the world should be modified in the light of the data.[5]

The Bayesian approach has two features that are particularly attractive in the archaeological context. First, data of many different kinds may be considered in the formulation of a prior, and indeed must be if they have contributed to the archaeologist's belief about a particular issue. (Classical statistics, on the other hand, lacks any means of taking into account 'background' or 'corroborating' evidence relating to the hypothesis being tested.) Second, data may legitimately be reused to examine the nature of the posterior obtained when the same data are investigated in the light of different prior ideas. A further feature is that knowledge may be accumulated, by using the posterior obtained after the consideration of one set of data as a subsequent prior.

Until recently, the Bayesian approach has suffered from two major drawbacks. The first is that the formulation of suitable quantifications for the prior and likelihood is often far from easy. Fortunately, a number of archaeological applications in recent years have produced a variety of case studies that inform future efforts,[6] although each problem still has to be approached on its own merits.[7] The other drawback was that it was simply not feasible to compute most posterior distributions, which were often mathematically complex. The situation has now improved greatly owing to the advent of new computational techniques such as 'Gibbs sampling'[8] and of powerful machines on which to implement them.

The Bayesian method is certainly appropriate to questions relating to astronomy in prehistoric Britain and Ireland, and work on the development of suitable case studies is in progress.

theory.[73] If we work within the Western tradition or similar modern systems of thought, and wish to justify our conclusions to others working within the same framework, then our techniques must certainly be 'scientific' in the broad sense that they must be systematic and self-critical.[74] But to what extent should they be quantified? What is the place of statistical analysis and inference?

For this purpose, classical statistical inference is inappropriate.[75] This is because it is predicated on a paradigm of repeated experimentation which makes a series of assumptions—that the hypothesis precedes the data; that new, independent sets of data can be repeatedly acquired on which to test a hypothesis; and that each dataset must be used once only and then thrown away—each of which is clearly inapplicable in the context of most archaeological investigations.[76] One, not altogether unreasonable, reaction is to reject the use of quantitative procedures for this purpose, arguing that their use should be limited to helping us to recognise patterning in the archaeological record and to specify its nature.[77]

Methodological debates within archaeoastronomy have reflected some of the wider arguments within archaeology as a whole.[78] In the 1980s, the principal divide was between the 'green' approach, developed mainly in Britain and characterised by its preoccupation with statistical techniques for assessing supposed astronomical alignments, and a 'brown' approach, developed mainly in the Americas in the context of a much richer cultural record, led mainly by historical and written evidence, where data from architectural alignments were used to support and corroborate, rather than to generate, new ideas.[79] More recently, the particular emphasis on alignments has been questioned and they are now generally seen in proper perspective as merely one aspect of the evidence relating to astronomical practice that might be retrievable from the material record.[80] Despite this, many archaeoastronomers still seem to view any attempt at quantitative assessment with suspicion, regarding this as inevitably 'positivistic' (cf. Archaeology Box 8).[81]

The way forward here, as in archaeology as a whole, must surely lead from the middle ground where contextual ideas are developed using the widest possible range of pre-existing (but, from a formal statistical point of view, ultimately subjective) knowledge and serious attempts are made to 'test' these ideas (in the broadest sense) by examining patterns in the material record.[82] Since coherence in design may not represent coherence of purpose or meaning, we need a methodology capable of dealing with variation (i.e. allowing us to put forward contextual theories to account for it, and seeing whether the record strengthens or weakens those theories) without resorting to selecting the evidence that fits and ignoring any that does not.[83]

In recent years, a technique has emerged that appears to allow us to assess quantitatively the extent to which certain material evidence strengthens or weakens particular ideas (or beliefs) without compromising our ability to use whatever means we wish to generate those ideas in the first place. The Bayesian approach differs from the classical approach in several fundamental ways (see Statistics Box 7). Starting from a view of probability as degree of belief, which releases it from the paradigm of repeated experimentation, it permits statistical inference about a particular set of empirical data to take account of 'prior information' accumulated from (subjective) experience not directly related to the data in hand, and allows the same data to be used many times over, to examine their effect upon different sets of prior beliefs.[84]

The investigator begins by attempting to express his or her prior beliefs about the likely values of the parameters of some model. The Bayesian method gives a mathematical way of combining this with a set of observed data in order to produce a posterior probability distribution which expresses how the investigator's beliefs should be updated in the light of the data (for more details, and references, see Statistics Box 7). How this is interpreted is an archaeological question; it can not be mathematically prescribed.

For example, we might wish to assess a set of recumbent stone circles in the light of our prior beliefs about the class of associations between ritual monument orientations and the moon (and possibly other factors such as prominent hills). If these beliefs could be expressed in terms of a prior probability distribution on declination (i.e. in terms of a one-parameter model), we would then need to calculate the posterior distribution and display the results.[85] (The principle is essentially the same as the cumulative probability idea underlying the 'curvigrams' used earlier in this book; however, they would no longer necessarily contain neat gaussian humps.) Furthermore, we could examine the data in the light of a number of different priors—that astronomy was irrelevant; that the moon passing close to the horizon was important, but no more; that the southern full moon, rising or setting close to the summer solstice, was particularly important—producing a graph of the posterior distribution in each case. What conclusions were then drawn would be a matter for the people viewing them.

It is important to understand the role being suggested for Bayesian statistics here, especially in view of earlier claims that attempted (and failed) to justify Bayesian inference as a tool for the inductive assessment of competing theories.[86] That role is, essentially, to provide a quantitative method for evaluating data in the light of current knowledge (and beliefs).[87] A recent review by Aveni[88] highlights some common misconceptions about the Bayesian approach, to which it is informative to respond here.[89] First, it is not an attempt to provide a quantitative expression of states of knowledge that is justifiable in any absolute sense. The point is that by developing a quantitative description that satisfies the investigator concerned, that investigator (and others) can judge the extent to which the empirical data strengthen or weaken particular ideas by studying the posterior description that results. But those judgements themselves are ultimately subjective.

Second, there is no need to try to evaluate quantitatively those other sources of information that may have influenced our prior ideas, such as the reliability or relevance of informants' accounts, nor the constraints on their selection. Indeed, one should not. These are part of the complex processes, generally undertaken in the context of the very different frameworks of current thought and practice within diverse disciplines such as history, art history or social science, that lead to a given investigator's prior state of knowledge or belief; it is that state, and not how it was arrived at, that the investigator concerned needs to describe.[90]

Remarkably, perhaps, we seem to have come full circle, from criticising the inherent biases in the work of Thom and others, through seeking rigour and objectivity, to recognising the shortfalls of that approach and discovering the need to readmit subjectivity as part of a controlled approach which involves a 'continuous dialectic between ideas and empirical data'.[91] How the exploration loop is controlled is obviously critical, lest we simply fall once again into all the traps identified towards the beginning of this book.[92] The Bayesian approach promises to reintroduce a quantitative element appropriate, and necessary, to this process.

Perhaps the most important lessons to be learned overall are that (except on the most basic level) different interpretations are always possible, even given the same empirical data, and that personal convictions should never go beyond statements such as 'it seems possible that', 'it seems very likely that', and 'the evidence clearly seems to indicate that' (although Bayesians would be prepared to put probability values on such statements). While it may sound more enticing to say, with absolute conviction, that people in prehistory did this, or thought that, or that the purpose and meaning of particular monuments was this—and this is something of which archaeoastronomers and archaeologists alike have been guilty over the years—it is ultimately misleading, and should immediately raise our suspicions.[93] Better, and more rewarding by far, is to seek to assess existing ideas against new empirical evidence, and to seek to raise interesting and exciting new possibilities. Through these twin goals we can hope to inch towards genuinely better understandings of the past in our own terms, whether personal or shared with others of similar background, rather than arriving with misplaced confidence at completely indefensible conclusions.

## Drawing together the strands

In forty years of research and fieldwork on megalithic monuments, Alexander Thom produced over six hundred drawings and index cards, filled over one hundred notebooks, and published three books and several dozen research papers.[94] As their cataloguer remarks,[95] few other individuals in recent times have generated such a wealth of material and information in pursuing an interest outside their chosen profession.

Thom's contribution to the study of orthostatic stone settings bears witness to his own remarkable range of skills, both theoretical and practical, as well as to his sheer enthusiasm and determination. That much of this effort, while producing a valuable data resource, was ultimately misguided resulted from Thom's total ignorance of (and lack of interest in) the framework of ideas being developed by archaeologists during this time. Thom was not alone in this; archaeoastronomy has been beset for many years by 'otherwise sane and apparently sober physical scientists with respectable academic positions [who, from an archaeological perspective,] appear to lose all critical ability when utilizing archaeological data'.[96] The much broader range of skills needed to appreciate and address the archaeological and anthropological issues, as well as to develop and carry out astronomically related research to answer interesting archaeological questions, is even rarer.

One thing that is evident is the value of truly interdisciplinary discourse and broad-based skills. This is not only important in research specialities such as archaeoastronomy that cut across fields; it is particularly relevant in a wider sense in a political climate where goals and achievements tend to be assessed in a context of increasing compartmentalisation, where interdisciplinary awareness and tolerance are increasingly strained, and where scientific and technical skills and achievements are becoming increasingly undervalued by those without a scientific training. For these reasons archaeoastronomy is increasingly being seen as a valuable topic area in training in secondary schools[97] and in public education[98] as well as at undergraduate level, where it can be a vehicle within interdisciplinary programmes for learning to appreciate other students' very different disciplinary[99] and well as cultural[100] perspectives. It also has considerable potential as a means of teaching science awareness and skills in an indigenous context.

In the past, investigations of prehistoric astronomy, when they weren't being blatantly ethnocentric, tended to be positivistic in approach, grounded in an abstract, detached, context-free view of the past. Times have changed; and in the meantime archaeologists themselves, whether adopting processual or post-processual standpoints, have become increasingly concerned with cognition, ideology and meaning.[101] As we begin to develop broader perspectives on the possible significance of what have hitherto been identified as astronomical 'alignments'; as our explanations of human activity in the landscape attempt to take greater account of people's perceptions of the world around them, and symbolic ways of representing and reinforcing their understanding of it; and as we begin to recognise that the sky was an integral part of the cosmos as perceived by most non-Western peoples, so the study of astronomically-related elements in the material record begins to assume considerable relevance in modern archaeology.[102] Archaeoastronomy—or whatever other term might be used to describe it[103]—as well as coming of age, may at last have found its time: not simply as the accumulation of a body of knowledge on ancient astronomy and calendrics, but as a valuable part of our general inquiry into past human societies and developments in human cognition.

> Archaeoastronomers have begun to ask a new set of broader questions . . . Why did ancient societies develop astronomy? What role did astronomy play in their . . . view of the world? What accounts for variations in the development of astronomy? What universal questions can be perceived from cross-cultural comparisons? Does the ability to speculate on the abstract world really depend upon the achievement of a certain level of social, economic, cultural, religious, or scientific development in all societies? These questions have more to do with cosmological and ideological rather than scientific astronomical systems. . . . They raise the level of discussion and expand it to areas where it can be addressed meaningfully by all scholars of antiquity.[104]

Astronomy is an integral part of cosmology, in the anthropological sense. In this, the Maya world-view is typical:

> Understanding the symbolism of a culture often begins by bearing witness to the complex behavior of the things and phenomena of that segment of the world-view we call 'natural'. For Maya symbolism specifically, this means we are obligated to know the life cycle of the toad, the stingless bee, and the maize plant, to name but a few of the entities that we, in our unfortunate wisdom, separate from the rest of nature and relegate to the zoological and botanical realms. We must also be able to follow the course of the sun, the stars, and the intricate movement of Venus, matters that we choose to label astronomy.[105]

But let us finish by returning to Stonehenge and prehistoric Wessex. Here too 'to ignore [astronomy] in attempts to understand people's perceptions of the Stonehenge landscape is to impose another twentieth-century agenda, one rooted in the backlash to astronomical overload in the 1960s and 1970s; an understandable one maybe, but an unreasonable one nonetheless.'[106] In the meantime, Stonehenge continues to be the very icon of archaeoastronomy, appearing on the covers and dust jackets of a variety of academic books on this subject as well as ancient astronomy in general,[107] while astronomy continues to be the very bane of many archaeologists' existence.[108] This situation will change as archaeologists become better trained in archaeoastronomical field techniques and archaeoastronomers, if they continue to exist as such, become more familiar with the archaeological issues that must motivate their work if it is ultimately to have any value. In this way, one can hope that the quest to uncover ancient achievements as measured in our own terms will be abandoned once and for all, but that there will be a great deal more exploration of a window giving unique insights into the changing ways in which prehistoric people made sense of themselves and their environment.

# Appendix
# Horizon Survey and Data Reduction Techniques

This appendix contains guidelines for determining the azimuth, altitude and declination of points on the horizon as viewed from particular locations, and for generating properly scaled horizon 'profiles' of the sort used in this book. It addresses questions relating to data capture, data reduction, and presentation of the results. This is generally what is meant by 'archaeoastronomical fieldwork', although the term carries the unfortunate, and not always inaccurate, overtone that an overriding astronomical motivation is being assumed even before the evidence is assessed.[1] It may also be taken to imply, incorrectly, that the use of such field techniques is confined to specialist archaeoastronomers as opposed to being available to all archaeologists. Orientation, and even declination, data may certainly be useful in wider interpretative contexts, as is clear from some of the discussions in chapters nine and ten.[2] More generally, data on skyline and visible geographical features are increasingly being seen as an important part of the archaeological survey of orthostatic monuments, providing valuable data on the relationship of those monuments to the surrounding landscape.[3]

As with all archaeological data obtained in the field, orientation, horizon survey and astronomical data should be presented definitively, so that any interpretations arising from them can be independently assessed. Like landscape survey but unlike excavation, archaeoastronomical field techniques are non-destructive, but often the risk of monument degradation or landscape change due to development or afforestation, that would make it impossible or difficult to capture similar data in the foreseeable future, present an additional argument for exactitude.

In view of this it is generally meaningless to make statements like 'from Malcolm's Monument the solstitial sun sets over High Hill' or 'Lionel's Alignment is oriented upon the equinoctial setting sun'. Even when backed up with photographs, such accounts beg questions of observing position, precision, exact date of the year, the fact that the sun's motions have altered slightly over the millennia, and so on; they also make a number of implicit interpretative assumptions, for example that the sun rather than any other astronomical body was the intended 'target', and that the solstices and equinoxes were events meaningful to those who constructed and used the monument. Neither is it sufficient to give diagrams, with or without setting suns or moons, if these diagrams are unscaled.[4] Best practice must involve the measurement and the presentation of azimuths, altitudes and declinations from locations defined to an appropriate precision: at this level it is essential to let the data speak for themselves.

## Preparations

### *What are the goals?*

As with any archaeological fieldwork, it is essential to determine, before a programme of horizon survey commences, its precise aims within the context, say, of a set of research questions or the preparation of an inventory. A clear set of objectives should be identified, from which an appropriate work strategy must be designed. The word 'strategy' is used rather than 'programme' because there will inevitably be uncertainties in what can be achieved owing to the weather conditions encountered at the time of fieldwork and also—if it has not been possible to reconnoitre the area beforehand—because of the possibility of horizons being totally or partially obscured by vegetation or man-made constructions such as modern buildings. The extent to which these are likely to be a problem depends upon a number of factors following from the precise objectives, such as whether measurements are to be made from one or more locations close to one another, perhaps as part of the work relating to an excavation or small-scale field survey project, or at a set of locations spread over a considerable area, as in studies of orientation patterns and location patterns within the wider landscape (e.g. key locations plus controls), or amongst a group of scattered monuments. One point to bear in mind is whether the strategy can be justified afterwards: one's colleagues may be convinced by the argument that a horizon profile was not surveyed because of obstruction by modern buildings, or that a monument was not visited because it was too remote and could not be located in the time available, but not if they suspect that the decision was affected by whether an alignment was considered to point anywhere 'interesting'.

### *Prior generation of ideas in the field*

If doing so is at all feasible, it may well be useful to visit a particular location prior to any actual survey work being carried out. If the research objectives are already clear, the aim may simply be to reconnoitre the area in order to help develop an optimal fieldwork strategy; a rather different aim may be to help generate new ideas worth testing with quantitative measurements. There are evident dangers in developing theories 'on the fly', devoid of social context, but phenomenological approaches exploring possible symbolic relationships between human constructions and activities and natural features in the landscape and objects in the sky, including where they meet on the horizon, may be well worth while.[5] If an approach of this kind is to achieve its full potential then it may be necessary to stay in the landscape for a considerable time, or to return to it several times, walking through it, approaching places from different directions,[6] examining them in different lighting conditions—at different times of day, at dawn and dusk, and during the night—and looking at all of the horizon from different places rather than focusing on a small part of it as viewed, say, along the principal axis of a single conspicuous monument. Some would see the generation of ideas in this way as an end in itself; but otherwise, a clear set of objectives for testing those ideas should follow.

All this implies a clear separation between the development and testing of ideas that many would feel to be unrealistic. In fact, it is pointlessly restrictive, as well as totally unrealistic, to pretend that the field surveyor is a brainless automaton and to insist that new ideas cannot be generated in the course of survey fieldwork. Equally, anyone generating ideas involving orientation and astronomy is likely at least to want to get out a magnetic compass to test them in broad terms as they go along. The overriding criterion must be that by the time the results of the fieldwork are eventually reported, it must be possible to present a clear set of objectives, a clear strategy, and show how it was followed. In particular, it must be possible to demonstrate that one has not simply selected evidence that tends to support a particular theory while ignoring other evidence that does not.

### *Prior decisions: precision and instrumentation*

A key decision is the precision to which azimuths and altitudes need to be determined. This will be set by the research objectives and limited by such factors as the nature of any architectural alignments that might have astronomical significance and the state of the structures concerned.[7] It will affect the instrumentation needed (or even the decision of whether to go into the field at all). Generally speaking, it is worth considering measuring azimuths and altitudes to a precision rather greater than that which is actually needed and will ultimately be quoted; this makes allowance for unforeseen errors and uncertainties. But in no case should this 'uncertainty factor' exceed ten times (one order of magnitude): otherwise, a good deal of time is wasted and one risks obscuring any intentional astronomical correlations with meaningless detail.

It may not be necessary to make survey measurements in the field: data of low precision can be reconstructed from plans, maps and/or digital elevation data used in the context of a GIS rather than being obtained from first-hand surveys, and this may be perfectly adequate for some purposes. Factors such as the accuracy of north-points on site plans, or the distance of relevant horizon profiles, may be critical, so the decision of whether to proceed in this way may not be a straightforward one. It might, however, be useful as an alternative if a period of fieldwork is badly affected by adverse weather. In addition, even map- or GIS-based conclusions may need verification by 'ground-truthing'.

Another method, which might be useful if a site is visited at night and there is no time or equipment available to undertake a survey, is to photograph a part of the horizon together with a set of stars higher up in the sky. A short exposure is necessary so that stars appear as points rather than drawn-out trails: with reasonably fast film a short exposure (e.g. 1/30 sec) might be sufficient. The date and time must be noted. If the brightest stars can be identified, then their declinations and right ascensions,[8] and hence their azimuths and altitudes, can be determined using one of the many star mapping computer programs currently available. If there are enough stars to calibrate the photograph, azimuth and altitudes can then be deduced for points on the horizon. While quick on site, the data reduction process can be long and messy, and photographic distortion may be a problem given the potentially wide field of view.

*Compass and Clinometer.* For measuring azimuths to a precision of about 1°, it may be sufficient to use a prismatic compass that allows the user to view the horizon, a vertical index line, and (by means of a reflector) the measurement scale all at the same time, superimposed on one another. These are small and light to carry and quick and easy to use, but like all magnetic compasses have the disadvantage that measurements are notoriously unreliable unless a careful procedure is followed. The reason for this is that the difference between magnetic north and true north changes over the years and may be subject to local anomalies. As a result, all compass readings must be calibrated by observations of visible landmarks (such as church spires) whose azimuth can be obtained sufficiently accurately from maps. In addition, it should be standard practice to observe all magnetic bearings in both directions along an observation line where this is possible. This can lead to the detection of anomalous bearings from one end of the line.

For archaeoastronomical purposes it is also necessary to measure altitudes to a similar precision.[9] It is essential not to fall into the trap, as do many published accounts, of quoting azimuths to a precision of one degree but failing to quote altitudes, which may be uncertain by several degrees. A suitable portable instrument is the Abney Level, which allows the user simultaneously to view the horizon, a horizontal index line, and (by means of a reflector assembly) a spirit bubble which is adjusted until level. When this has been done, the altitude is read from a semi-circular scale with a vernier attachment. Modern pocket-sized combined compass–clinometers are also now available.

Even if a theodolite is the preferred survey instrument, having a prismatic compass and clinometer to hand may be useful as an alternative if weather conditions or other problems prevent its effective use.

*Theodolite.* Using a theodolite is often the only way to obtain demonstrably reliable readings, and minute-of-arc precision or better is easily possible; but, once again, a careful procedure must be followed on site. Against the advantage of reliability is the disadvantage that theodolites, together with their tripods, are relatively heavy so that carrying them to a site, setting them up and using them generally requires two people, although a fit and competent person can manage on their own. In addition, setting up and using the instrument at each location takes a considerable time—generally a minimum of one hour, even in perfect circumstances and conditions with minimal data to collect.

An obvious factor to take into account in choosing a theodolite is its weight and transportability. There are several small models with lightweight aluminium tripods which are best suited to reaching remote locations, ranging down to the tiny Kern DKM1 (now out of production) which can be fitted into a backpack along with many other instruments and accessories. Amongst the most useful accessories are sun filters, which allow direct observation of the sun rather than by projection; corner eyepieces, which are necessary with some designs of theodolite for high-altitude observations such as sun sightings close to noon; and measurement scale illuminators for observations at night.

The choice between using a theodolite or a compass and clinometer, either at a particular location or on a fieldwork expedition as a whole, may well depend upon practical considerations such as the number of investigators and their state of fitness, and the distance apart and accessibility of survey locations given the time available. In what follows, we shall assume that a theodolite is being used.

*Timing devices.* When using a theodolite to measure azimuths it is necessary to determine the direction of true north to the requisite precision in order to orient the horizontal circle on the theodolite. This can be done by taking timed observations of the azimuth of the sun, which requires a means of accurately determining the time.[10] The sun moves through the sky at a rate of about 1° every four minutes.[11] The resultant rate of change in its azimuth depends on the time of day—generally speaking, the higher the sun is in the sky, the greater the rate of change in its azimuth—but at the latitude of Britain and Ireland it will never exceed about one minute [of arc] every two seconds [of time].[12] An ordinary digital watch should be sufficient for the purpose; these are preferable to analogue watches on which it is easy to misread the minute hand. An alternative strategy is to use a radio receiver tuned to a continuous time-signal or a hand-held (pocket) GPS receiver.[13]

## Data capture in the field

The golden rule of archaeoastronomical fieldwork in Britain and Ireland is to be aware of the weather. If at all possible, be prepared to work any or all hours in order to take advantage of the most suitable weather conditions. Successful theodolite survey of horizons depends critically upon two things: that the horizon profiles of interest are clearly visible (at least when being photographed and surveyed), and that there is a spell of clear sunshine, preferably lasting at least ten minutes, to permit the necessary sun-azimuth observations to be made. (Another method is to use stars, such as Polaris, at night, but clear weather is still needed.) Spells of changeable weather are often better than long periods of high pressure since the latter often leads to lasting atmospheric haze, whereas clear spells following showers may bring excellent visibility. Periods of hyperactivity may be necessary, possibly interspersed with long spells when patience, and a good book or a detailed interest in the local flora and fauna, are a distinct advantage. If it is possible to take a portable computer into the field, then data reduction work can be undertaken at the field base on days when survey work is thwarted by bad weather.

### *Equipment inventory*

A typical field inventory might include the following: theodolite (with or without EDM[14] attachment), tripod, supplementary lenses and filters including sun filter, illuminator; digital watch or chronometer, backup watch and radio receiver for time signals, or GPS; field notebooks and pens/pencils; large-scale maps for locating monuments, smaller-scale maps for identifying features in the landscape and their approximate distances, protractor for determining angles on maps; torch and batteries; camera(s) and lenses, camera tripod; chalk for scratching RO on nearby object where no distant object is available (e.g. in forest); survey pegs and ranging poles for determining survey locations relative to each other and fixed, recognisable points; steel or cloth tapes or EDM reflector for ground-plan measurements where necessary; sun cream, rain protection, insect repellent, etc.

### *What needs to be done first?*

Before commencing a site visit, it is advisable to calibrate any watch(es) using a telephone time signal. In order to ensure that the time shown by a watch has not been accidentally altered during fieldwork, this also needs to be done as soon as possible afterwards, as a check. It is perfectly acceptable to apply a correction to all watch readings when reducing the data rather than trying to make the watch read the exact time; indeed, this may be preferable, since it serves as a reminder that watch readings may always be subject to a correction.

In ideal conditions, it is possible to follow a standard procedure such as that described below. However, in Britain and Ireland, especially in the north and west, conditions are seldom completely ideal. A strategy is needed if weather conditions leave something to be desired, both to optimise the chances of a successful survey and to achieve the greatest efficiency overall. Nevertheless, disappointments can still happen; hours of careful surveying may end up being of little use for the want of a sun-azimuth calibration. Several of my own survey assistants will recall hours spent on site at the end of the day waiting for the sinking sun to break through clouds just for a minute or two so as to allow us to obtain some sun readings before returning to base.

Hence, before setting out, especially for a remote location, assess the chances of successfully obtaining photographs of, and survey data from, the relevant horizon profiles and of making direct sightings of the sun. On arrival, check horizon visibility and cloud cover in relation to the position of the sun. Bear in mind that a suitable theodolite position must be chosen and the theodolite fully set up before any meaningful readings can be taken, whereas a photograph can be taken almost immediately. So if the visibility of any profile is rapidly deteriorating, take suitable photographs immediately; if bad conditions persist, it may still be possible to use a few surveyed points to calibrate the photograph.[15] Once the theodolite is set up, it may be urgent to obtain sun-azimuth readings, or readings of points on a deteriorating horizon profile, or both.

Throughout the subsequent site work, keep an eye on unsurveyed profiles and (until sufficient sun-azimuth readings have been obtained) the sky. Be prepared to switch to taking sun-azimuth readings immediately the sun comes out. If a mountain top emerges from cloud, take a photograph and then attempt to survey the summit while it is still visible. If it proves impossible to complete the survey, the only alternative to abandoning the survey completely is to leave a survey peg to mark the spot and to return on another day.[16]

## *A 'standard procedure'*

Aside from all the variations and constraints imposed by weather conditions, the following standard procedure has worked successfully for me over many years.

*Positioning the theodolite.* Given the fieldwork objectives, identify the key position(s) from which it is desirable to know what astronomical events could be seen at the horizon.[17] The precision with which these points need to be defined in relation to the questions being asked can be determined using the following rule of thumb: a lateral movement of 1 m will shift the azimuth of a point 1 km away by about 0°·05, or 3 arc minutes.[18] Doubling the lateral shift doubles the azimuth shift, whereas doubling the distance of the observed object halves the azimuth shift. The same rule applies to the effect of vertical movement upon the altitude.

If possible, determine a single position at which the theodolite may be placed in order to obtain all the necessary data. The following criteria must be satisfied:

- The horizon profiles must be unobscured from this point.[19]
- The sun must be unobscured when sun-azimuth readings are taken. It is embarrassing if, when the time comes to take these readings, the sun as viewed through the theodolite telescope has just disappeared behind a close-by building or tree.
- It must be possible to set up the tripod stably at this point. Firm but penetrable ground is ideal.

Try to choose a position that minimises the parallax corrections that need to be applied to azimuth and altitude readings taken from the theodolite in order to estimate those from the appropriate key positions. Bear in mind that the nearer the horizon, the smaller the lateral shift in the observing position to have a significant effect. The rule of thumb already stated applies.[20]

It is not always possible to acquire all the necessary data from a single theodolite position. If the instrument has to be moved from one place to another it is useful, and often essential, to set up a number of 'control points' whose positions relative to one another are determined during the course of the survey. All the points need to be identified at the outset and marked by a survey peg for the duration of the entire survey, so they can be measured from other points as required. Special measurements are made in order to determine the positions of the control points relative to one another, and this obviates the need to undertake repeated sun-azimuth observations at each theodolite position, which is useful if the sun becomes completely obscured by cloud. On the other hand, setting up at more than one position adds considerably to the time spent on a survey and this needs to be weighed up carefully against the potential benefits. In addition, if care is not taken, it can introduce serious systematic errors.

*The reference object.* After setting up the tripod and levelling the theodolite, it is necessary to identify and survey a suitable reference object (RO). This should be fixed, easily locatable but sufficiently distant to be precisely positionable in the theodolite crosshairs. Fixed points on distant buildings (corners of windows, tops of chimneys, etc.) are often ideal; they are generally preferable to prominent features on the distant horizon, because of the propensity of these to disappear from view if the weather changes. Occasionally, where for example a monument is surrounded by trees but a theodolite reading is necessary to determine an orientation, it may be necessary to resort to using tiny pencil or chalk crosses on nearby tree trunks.

Locate the RO in the telescope and note the horizontal (H) and vertical (V) circle readings to the required precision, and whether these were obtained on left or right face.[21] It is good practice to repeat this on both faces, both in order to check instrument errors and to prevent the occurrence of a gross error through misreading or misrecording,[22] unless there are more urgent things to be done.

Repeat RO readings between major sets of other readings, immediately before and after sets of sun-azimuth readings, immediately after it is suspected that something affecting the PB–Az correction—the correction that needs to be applied to horizontal circle readings ('plate bearings') to convert them into azimuths (see below)—may have occurred,[23] and immediately before the theodolite is dismantled.

When setting up the theodolite at a location where it has been set up before (e.g. after abandoning an earlier day's work because of bad weather), use the same RO as before.

*Sun-azimuth readings.* A period of strong sunshine is needed at this stage. *Under no circumstances look directly through the theodolite at the sun, as this could cause permanent damage to your eyes, even if the sun is close to the horizon.* If you are using the projection method, point the telescope roughly at the sun, hold a piece of plain paper or card a few centimetres from the eyepiece and adjust the eyepiece focus until the crosshairs are clearly visible superimposed on the solar image. If you are using a sun filter, point the telescope roughly at the sun and adjust until the light of the sun shines directly down the telescope. Then fit the filter. The sun should be visible within the field of view.

It is at this point that the presence of two people is most desirable. Person *A* views the sun and shouts out when one of its limbs (edges) crosses the vertical crosshair. Person *B* notes the time. In the meantime, person *A* immediately measures the alidade bubble tilt and then, without adjusting the telescope, takes the horizontal circle reading. The result is a set of four measurements: watch time (WT); limb ($p$ = preceding or $f$ = following); tilt (which is measured in divisions, or seconds of arc, towards ($t$) or away ($a$) from the vertical circle; and the horizontal circle (LH/RH) reading. These measurements constitute one sun-azimuth reading. Ideally, a group of twelve readings should be taken, to minimise the possible errors: three of each solar limb on each face. When well practised, the process should take no more than ten to fifteen minutes.

*Horizon profiles.* In general, it is a sector of horizon that is likely to be of interest in the context of an investigation, rather than a single point. For each such sector, begin by sketching the horizon profile in the field notebook sufficiently carefully so as to allow surveyed points to be identified unambiguously from the accompanying photographs. Take the photographs at this time, noting photograph numbers, lens focal lengths etc. Short focal lengths, and especially wide-angle lenses, produce distortions that may be significant for higher-precision work. Against this, long focal-length lenses may require several photographs to cover a single stretch of surveyed horizon, which would make them difficult to calibrate using surveyed points. Take photographs from the appropriate key position, rather than the theodolite, wherever possible. Survey readings are then corrected before being used to calibrate the photograph.

Survey the azimuth and altitude of horizon points as required (cf. Fig. 1.7). If this information is needed to calibrate photographs, at least three unambiguous points (and preferably rather more) are needed within each photograph. If the declinations of particular points (such as hill summits or junctions between hills) are of particular interest then survey these points directly, but also survey points around them. If the horizon is relatively featureless and unambiguously defined points are few and far between, include some points that could not be reliably relocated, and mark them accordingly.

If possible, identify horizon profiles on site. Naming the hills and other features that make up the horizon can be surprisingly difficult to do after the event, even if estimated distances are marked in the field notebook. As well as appropriate Ordnance Survey maps, a compass of the type that allows angles to be plotted on maps, and even a protractor, can be helpful to have on site for this purpose. Another benefit of this exercise is that one can make sure that there are no more distant profiles lurking unseen in the clouds behind measured ones.

*Local ground-plan (station description).* It is important to record, to an appropriate precision, the point(s) from which theodolite measurements are made, and from which photographs are taken, in relation to fixed and relocatable points such as identifiable parts of a prehistoric monument. In many cases pacing and rough estimates of the orientation will be sufficient, but in other cases theodolite and tape/EDM measurements will be helpful in constructing enough of a ground-plan for this purpose.

*Tie-in to other control points.* Where the theodolite is placed at one of several control points, you may need to determine the azimuth and possibly the distance of some or all of the other control points. The latter will only be possible if the points concerned are sufficiently close for steel tape to be used, or if your theodolite has an EDM attachment. There are various strategies for achieving a complete 'tie-in'.[24] If the terrain is hilly then altitudes should also be measured, in order to determine the difference in elevation between the current theodolite station and each remote control point.[25]

### *Data recording on site*

It is not desirable to use standard data input forms for this sort of survey data, although this is standard practice for many other types of archaeological data obtained in the field. The reason is that it is vital to keep a chronological record of theodolite readings; if for any reason the PB–Az correction changes during the course of a survey, a chronological record makes it possible to locate the event and to make the appropriate corrections before and after, thereby avoiding potentially serious systematic errors. Since the exact order in which the different types of measurement are made will vary from place to place according to conditions, best practice is to record everything chronologically in a permanent field notebook, although it is definitely helpful to follow standard presentation formats for sun-azimuth measurements, profile measurements and so on.

Fig. X.1 is based on a typical page from a field notebook at a site where the standard procedure was followed. Note the watch calibration prior to reaching the survey location, and the format of the reference object, sun-azimuth and horizon profile readings, which are shown in a recommended layout. In the profile sketch, note the use of dots to specify non-relocatable points: in this case only points 3 and 5 were considered relocatable.

## Data reduction

In most archaeoastronomical fieldwork to date, data reduction has tended to take place as a separate process following a period of fieldwork. Specially programmed survey equipment could, however, enable some data reduction to take place on site. Certainly, portable computers can enable much of the data reduction work to be completed while in the field. This has several advantages, one of which (if it suits the overall objectives) is that the strategy for subsequent site work can be influenced by the results obtained so far.

### *Calculating declinations*

The basic formula for converting readings of azimuth and altitude into declinations was given in Astronomy Box 2. In order to convert 'raw' theodolite readings of horizon points into useful data, the data reduction process is a little more complicated, and may involve several stages, as follows.

*Determine plate-bearing errors.* Calculate the theodolite plate bearing (PB) errors from pairs of horizontal circle readings on both faces taken before, after and throughout a period of fieldwork. These are only likely to be of concern in the highest-precision work; otherwise, the pairs of RO readings taken as a matter of course will generally serve as a check that such errors are indeed negligible, or else supply enough information for them to be taken into account in the subsequent data reduction.[26]

*Calculate the (PB–Az) correction.* Every time the theodolite is set up, it is necessary to determine the correction that needs to be applied to horizontal circle readings (plate bearings) in order to transform them into true azimuths. If this (PB–Az) correction is to be determined from sun-azimuth observations, ephemeris data must be obtained from which the true azimuth of the sun at the time of the observation can be calculated so that this can be compared with the scale reading (plate bearing) obtained.[27] Sequences of up to twelve sets of readings (each consisting of the time, limb, tilt and plate bearing) allow various errors to be taken into account and should prevent any possibility of gross error by making obvious any anomalies (e.g. due to misrecording of readings).

If the orientation of the horizontal circle changes for any reason in the course of a theodolite session, it will be necessary to apply different PB–Az corrections to readings taken before

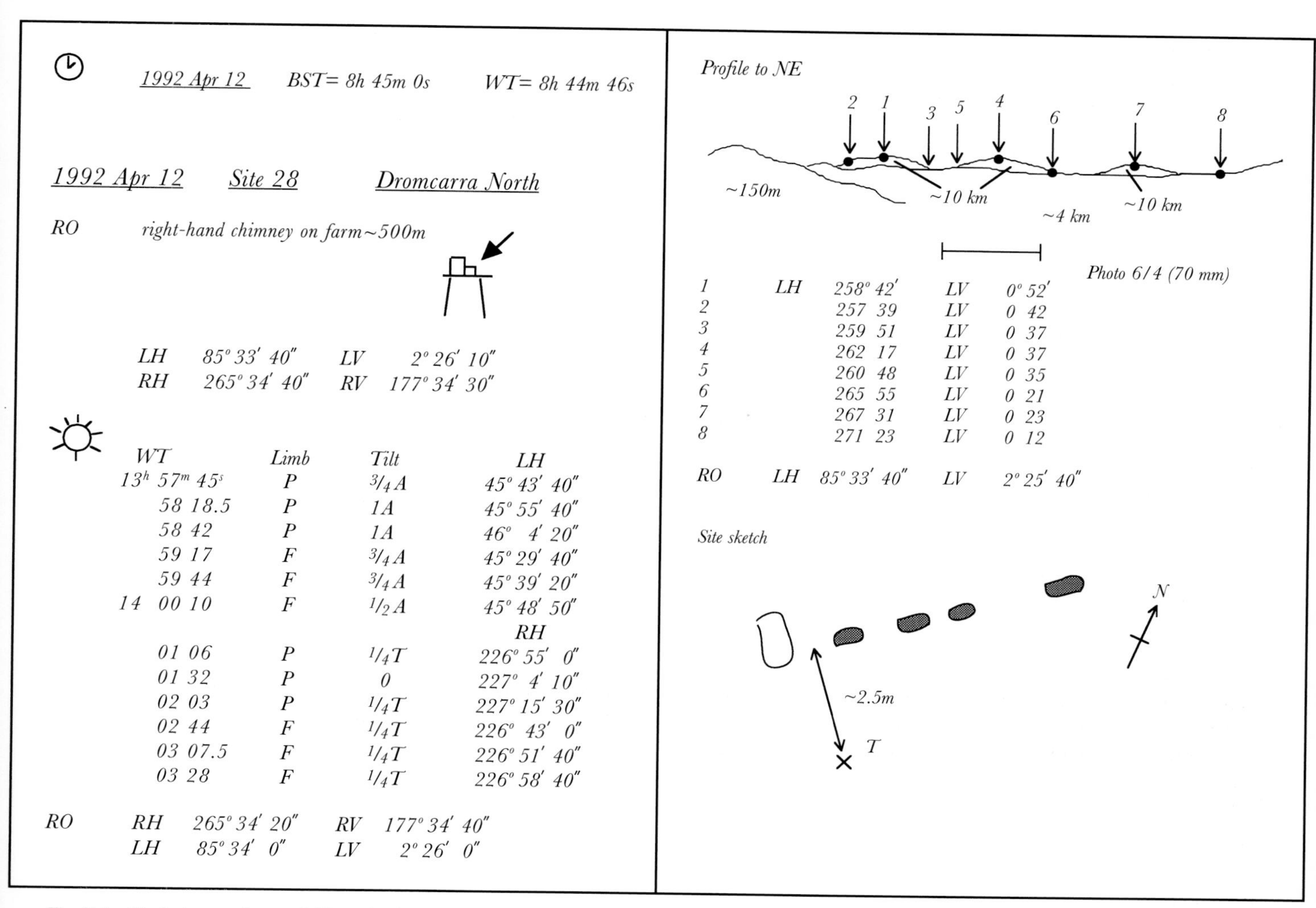


Fig. X.1 Typical pages from a field notebook.

and after the event causing the change. In general, only one set of sun-azimuth readings will be available, giving the PB–Az correction either before or after the event; but the change in the PB–Az correction caused by the occurrence, and hence the actual correction both before and after, can be determined from a comparison of RO readings in the two time periods.

*Correct from theodolite position to key position.* If the corrections involved are significant, it will be necessary to transform the azimuth and altitude readings of horizon points to obtain those that would have been measured from the relevant key position(s). The correction to be applied (called 'eccentric position correction') will depend upon the distance of the horizon point.

*Calculate declinations.* The declinations of horizon points as viewed from one or more key positions can now be deduced.

Various computer programs are available to help in the reduction process, depending on the accuracy sought. For example, DECPAK is a suite of three DOS programs written by the author to assist in calculating declinations from field data.[28] The first, PBERRS, calculates the relevant plate-bearing errors from sets of pairs of horizontal scale readings. The second, STIMES, calculates the PB–Az correction from a sun-azimuth observation, or set of sun-azimuth observations, together with the relevant ephemeris data. The third program, GETDEC, calculates the declination of a horizon point or points from recorded theodolite readings. In order to determine the corrections that must be applied, the user must also supply the plate-bearing errors (if significant), the PB–Az correction, the location of any key positions relative to the theodolite position, and the distance of any horizon points where a transformation of the surveyed data to a different key position is required.

Where awkward corrections and high accuracy are not required, declinations can be obtained more simply from azimuth, altitude and latitude data using spreadsheets. Declinations can also be estimated from map data rather than survey data: the GETDEC program has facilities for this, and it can also be done using a GIS.[29]

Since developments involving computer software and communications technology can be rapid and unexpected, further details are not given here but up-to-date information will be posted on the author's Web site which, at the time of writing, is located at http://www.le.ac.uk/archaeology/rug/. It is anticipated that the DECPAK programs and associated documentation, or updates of them, will be available by

*Reduction of Site 28: Dromcarra North* *NGR* $^{1}278^{0}680$

*Mean PB-Az from sun azimuth observations= −142° 14′*

*PBRO throughout = 85° 34′*

*Profile to NE:*

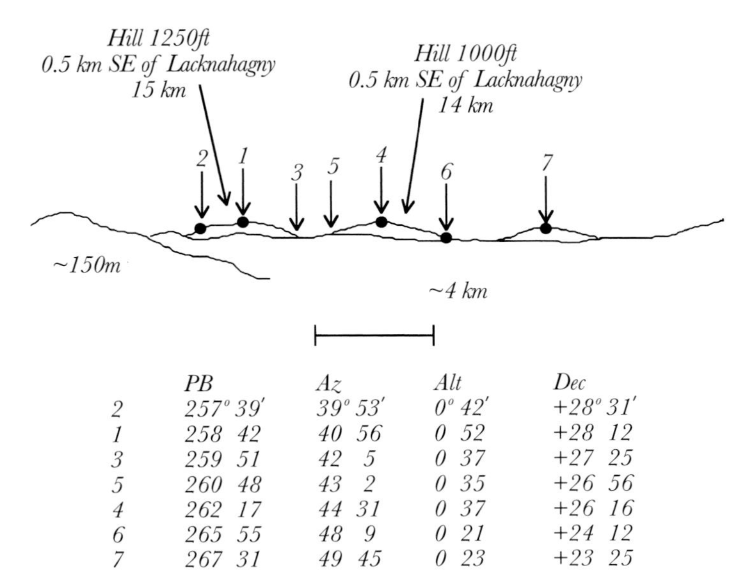


| | PB | Az | Alt | Dec |
|---|---|---|---|---|
| 2 | 257° 39′ | 39° 53′ | 0° 42′ | +28° 31′ |
| 1 | 258 42 | 40 56 | 0 52 | +28 12 |
| 3 | 259 51 | 42 5 | 0 37 | +27 25 |
| 5 | 260 48 | 43 2 | 0 35 | +26 56 |
| 4 | 262 17 | 44 31 | 0 37 | +26 16 |
| 6 | 265 55 | 48 9 | 0 21 | +24 12 |
| 7 | 267 31 | 49 45 | 0 23 | +23 25 |

Fig. X.2 The reduction of the field data from Fig. X.1.

downloading from this Web site, as will suitable spreadsheets and other relevant information.

Fig. X.2 shows the reduction of the field data from Fig. X.1, again in a recommended format.

## *Integrating survey data and horizon profile photographs*

One of the most reliable ways in which to produce a horizon profile plot properly scaled in both azimuth and altitude is to use a photograph (or series of overlapping photographs) calibrated using surveyed points that are easily and unambiguously identifiable. A prerequisite is that the photographs are taken with a sufficiently long lens that problems of distortion can be assumed to be negligible.

One method is to digitise the photograph(s) so as to produce a preliminary plot in which the scale can be assumed to be linear (i.e. a certain distance represents 1 degree of arc throughout the photograph and in any direction) but the actual scale is unknown, as is the extent to which the plot might need be rotated in order to make the azimuth scale horizontal and the altitude scale vertical. Effectively, the plot exists in an $(x, y)$–co-ordinate system[30] and needs to be converted (by scaling, linear transformation and rotation) into a new, $(X, Y)$–co-ordinate system where $X$ represents azimuth and $Y$ represents altitude. For each of the surveyed points, both $(x, y)$ and an estimate of $(X, Y)$ (the surveyed azimuth and altitude) is known. By performing a least-squares fit, the 'best' consistent transformation can be determined overall and the entire plot can be converted accordingly into azimuth-altitude co-ordinates.

The plots in the western Scottish project, published in 1984 (cf. Fig. 3.2), were produced using computer programs to perform the least-squares fit, to convert the digitised plots to azimuth and altitude, and to plot the resulting profiles with lines of constant declination shown.[31]

## *Presentation of the results*

Two principles that will enhance good practice in presenting horizon astronomical surveys are not to quote results to a precision greater than either the archaeological questions or the measurements taken in the field merit, and to present the data with as little interpretative baggage as possible. These principles can be illustrated both in relation to astronomical 'indications' shown on site plans, and to horizon plots themselves.

*Site plans.* Where horizon 'indications' are shown on site plans, it is important simply to quote the azimuth and altitude, and/or the declination, thus allowing the reader to assess directly the astronomical potential of a given direction. This practice is evident throughout the work of Thom and has generally been followed in this book. While selection clearly operates in choosing certain directions to be highlighted while others are not, at least this is explicit. A common practice which fails to give a definitive account of the data, and is potentially misleading, is that of quoting assumed targets without supplying the azimuth and altitude, or else the declination, data (as we have done ourselves in Fig. 8.4). It is not possible from such a figure to determine the 'goodness of fit' to the claimed targets. Worse still is where azimuth or declination figures are given for the assumed target rather than the actual alignment, as was done in *Stonehenge Decoded*[32] (cf. Fig. 1.22a; compare this, and particularly the alignment from Stone 93 to Stone 91, with Fig. 8.9c).[33] Such an approach is highly misleading and is to be avoided.

Of course, the 'neutral' presentation of the data in this sense does not prevent specific astronomical interpretations being put forward in accompanying commentaries. For a summary of the declinations of prominent astronomical events at different epochs see Ruggles 1997, tables 1–4.

*Horizon profile diagrams.* In addition to clearly marked azimuth and altitude scales, it is certainly helpful to indicate lines of constant declination at regular intervals (cf. Fig. 3.2), which saves readers having to work these out for themselves. On the other hand, with the exception perhaps of the solstitial sun,[34] the common practice of marking the rising or setting sun or moon can be misleading, since this often singles out a mere one of many possible (and inherently equally likely) rising or setting positions. This is less serious in the case of the sun if the relevant declination, or century plus calendar date, is clearly marked, but can never be justified in the case of the moon, whose complex motions (see Astronomy Boxes 4 and 8) mean that emphasizing any single rising or setting position—even a theoretical standstill limit—can never be justified on *a priori* grounds.[35]

In addition to the profile itself, the position from which the profile will appear as presented should be identified in the accompanying text. How closely this position must be defined will depend upon the distance of the horizon, according once again to the rule of thumb that a 1 m lateral shift will alter the azimuth of a point 1 km away by about 3′.

It can also be helpful to show the mean direction or range of directions 'indicated' by a structural alignment if there is one. In the western Scottish project of 1984, 'indicated azimuth ranges' and wider, 'adjacent azimuth ranges' were identified in each case according to strict criteria (see chapter three) and identified on each profile diagram (see Fig. 3.2). An alternative method is to indicate a mean direction, or the edges of a range of directions, on an accompanying site plan.

Although others before him had made inspired suggestions,[36] it was Alexander Thom's work from the 1950s onwards that set the agenda for the representation of data from horizon astronomical surveys. The principles described here largely follow that work.

# Reference lists of monuments

## Sources used in cross-referencing monuments

The six reference lists that follow give background information on monuments that have been grouped together for the purposes of the astronomical and statistical investigations described in the text. The lists are not intended as archaeological inventories in their own right but do include basic descriptions and cross-references to further archaeological data. The information given is as accurate as possible but should not be taken as definitive: the definitive sources are the inventories cited. Other sources are indicated as follows:

- A — RCAHMS 1988 [Mid-Argyll and Cowal]: entry number.
- B — Barnatt 1989 [stone circles], vol. 2: entry number.
- BC — Burl 1976 [stone circles], appendix 1 (gazetteer): reference number. Figures in square brackets refer to the revised reference number in the forthcoming 2nd edn, provisionally entitled *The Stone Circles of Britain, Ireland and Brittany*, anticipated publication date 2000. Information kindly supplied by Aubrey Burl, priv. comms, 1998.
- BG — Burl 1995 [stone circles guide]: entry number.
- BR — Burl 1993 [stone rows], gazetteer: page number.
- C — RCAHMS 1911 [Caithness]: entry number.
- D — Cuppage 1986 [Dingle peninsula]: entry number.
- DM — RCAHMS 1920 [Dumfries]: entry number.
- E — Ordnance Survey Archaeological Record number for an English monument (this information is now held by RCHME)
- EC — Power 1994 [East and South Cork]: entry number.
- EL — RCAHMS 1924 [East Lothian]: entry number.
- F — RCAHMS 1933 [Fife]: entry number.
- FP — Burl 1988b [four-posters], gazetteer: reference number.
- G1 — RCAHMS 1912 [Galloway 1: Wigtown]: entry number.
- G2 — RCAHMS 1914 [Galloway 2: Kirkcudbright]: entry number.
- H — Henshall 1963; 1972 [Scottish chambered tombs]: reference number.
- I — RCAHMS 1984 [Jura, Islay, Colonsay and Oronsay]: entry number.
- J — Jack Roberts, *Exploring West Cork*, Key Books, Skibbereen, 1988: chapter/entry number.
- K — RCAHMS 1971 [Kintyre]: entry number.
- L — RCAHMS 1975 [Lorn]: entry number.
- M — RCAHMS 1980 [Mull, Tiree, Coll and North Argyll]: entry number.
- MC — Power 1997 [Mid Cork]: entry number.
- N — National Monuments Record of Scotland (NMRS): record number.
- O — RCAHMS 1928 [Outer Hebrides, Skye and the Small Isles]: entry number.
- OC — Ó Nualláin 1984a [stone circles in Cork and Kerry]: catalogue number
- OR — Ó Nualláin 1988 [stone rows in southern Ireland]: catalogue number.
- P — RCAHMS 1994 [south-east Perth]: page number.
- Q — Burl 1980 [recumbent stone circles], appendix 1: reference number.
- R — Roberts 1996 [stone circles of Cork and Kerry: an astronomical guide], entry number.
- R3 — Ruggles 1996 [3-stone rows in Cork and Kerry]: reference number.
- R4 — Ruggles 1994a [4- to 6-stone rows in Cork and Kerry]: reference number.
- S — RCAHMS 1963 [Stirling]: entry number.
- SK — O'Sullivan and Sheehan 1996 [South Kerry]: entry number.
- T — Thom 1967 and unpublished site list: reference number.
- U — Ruggles 1984a, site reference number. Included in List 1 and List 6 where the monument is listed there but not in List 2.
- W — Ordnance Survey Archaeological Record number for a Welsh monument (this information is now held by RCAHMW)
- WC — Power 1992 [West Cork]: entry number.
- X — *DES*: year and page number.[1]
- Z — RCAHMS 1946 [Orkney and Shetland]: entry number.

# List 1

*Monuments of High-Precision Lunar Significance According to Thom*

This is a reference list of British megalithic monuments claimed to be of high-precision lunar significance by Thom, either because they incorporate high-precision lunar alignments included in the analyses at Levels 2, 3 and 4, or because they mark the extrapolation distance (see chapter two). Cross-references are given to monument reference numbers and names used in this book as well as to information in other lists and gazetteers. Information is given for each 'site' as defined by Thom, but is divided into parts where this corresponds to more than one monument as defined by other sources.

The following information, as available, is given for each entry, except that items 4–8 are omitted where the monument is also included in List 2:

1. Reference number (L1, etc.: the reference number used in Ruggles 1981 is given in brackets) and name.
2. Alternative names if any.
3. National Grid reference (NGR), generally quoted to 10 m.
4. Cross-references to the monument in other sources.
5. Date of most recent visit by the author ("NV" indicates not visited).
6. Brief description and other information.
7. Reference of published excavation report(s).
8. Reference of published site plan(s).

## *Scotland*

L1, Wormadale Hill (Shetland). HU 4052 4651. Z(1503); N(HU44NW06); T(Z3/4). Aug 81. Standing stone 2·4 m high, leaning by about 20° to the sw.

L2 (1), Brodgar (Orkney). Brogar. T(O1/1). Sep 90.

(*a*) Ring of Brodgar. HY 2945 1335. Z(875); N(HY21SE01); B(1:1); BG(183); BC(Orkney1[1]). Class II henge with rock-cut ditch and interior circle of stones up to 4·2 m high, 104 m in diameter. Exc: Renfrew 1979, ch. 5. Plans: Thom and Thom 1978a, 122; see also Ritchie 1988 and refs therein.

(*b*) Comet Stone. HY 2963 1331. Z(877); N(HY21SE13); FP(Orkney1). Standing stone 1·8 m high, together with the stumps of two companions, set in a low mound.

(*c*) Fresh Knowe; Mound *B* (Thom). HY 2960 1339. Z(882); N(HY21SE12). Elongated mound some 35 m long (nw to se) by about 25 m wide and 4 m high, probably a despoiled chambered tomb.

(*d*) Salt Knowe; Mound *N* (Thom). HY 2927 1328. Z(884); N(HY21SE14). Roughly circular mound some 18 m in diameter and 6 m high.

(*e*) Mound *A* (Thom). HY 2951 1348. Z(883); N(HY21SE11). Roughly circular mound some 19 m in diameter, formerly containing two cists.

(*f*) Mounds *J*, $L_2$, *L* (Thom). HY 2954 1323. Z(885); N(HY21SE16). Three mounds in a nw–se alignment. *J* is about 1 m high and 8 m across, $L_2$ is almost ploughed flat, and *L* is about 0·5 m high and 8 m across.

(*g*) Mound *M* (Thom). HY 2941 1328. N(HY21SE15). Mound some 2 m high and 10 m across north to south.

(*h*) Mound *K* (Thom). HY 2946 1313. N(HY21SE21). Mound some 0·5 m high and 6 m across.

L3, Stenness (Orkney). Stones of Stenness. HY 3067 1252. Z(876); N(HY31SW02); B(1:2); BG(184); BC(Orkney2[2]); T(O1/2). Sep 90. Class I henge with rock-cut ditch and interior circle of stones up to 5·7 m high, 31 m in diameter. Exc: Ritchie 1976. Plan: Curtis 1988, 373.

L4 (2), Dirlot (Caithness). ND 1226 4860. C(165); N(ND14NW06); BR(222, 243); T(N1/17). Sep 79. Grid of small stones, apparently forming a fan-shaped arrangement with approximately fifteen rows up to 45 m in length. Plans: Thom 1971, fig. 9.5; see also Myatt 1988, 294.

L5 (3), Battle Moss (Caithness). Loch of Yarrows. ND 3130 4405. C(570); N(ND34SW22); BR(222, 243); T(N1/7). Aug 79. Six or seven parallel or slightly fanned rows of small stones some 40 m in length. Plans: Thom 1971, fig. 9.7; Ruggles 1981, fig. 4.5; see also Myatt 1988, 290 and 291.

L6 (4), Camster (Caithness). T(N1/14). Aug 79. ND 2602 4377. C(573); N(ND24SE03); BR(222, 243). A number of small stones, possibly the remains of a fan-shaped arrangement of six or more rows up to 30 m in length.

L7 (5), Mid Clyth (Caithness). Hill o'Many Stanes. ND 296 384. C(292); N(ND23NE06); BR(222, 243); T(N1/1). Aug 79. Fan-shaped grid of stones with approximately twenty-three rows up to 45 m in length. Plans: Thom 1967, fig. 12.12; 1971, fig. 9.1; see also Myatt 1988, 283.

L8 (6), Guidibest (Caithness). Guidebest; Latheronwheel. ND 1802 3511. C(279); N(ND13NE03); B(2:11); BG(139); BC(Caithness5[11]); T(N1/13). Aug 79. Remains of a stone circle, diameter approx. 57 m. Plan: Thom, Thom and Burl 1980, 324.

L9 (9), Callanish (Lewis). Callanish I. NB 2130 3300. *See LH16 in List 2.*

L10 (10), Cnoc Ceann a' Gharaidh (Lewis). Callanish II; Loch Roag. NB 2220 3260. *See LH19 in List 2.*

L11 (11), Airigh nam Bidearan (Lewis). Callanish V. NB 2342 2989. *See LH24 in List 2. Also SSR4 in List 6.*

L12 (7), Dursainean (Lewis). Dursainean NE; Allt na Muilne. NB 5281 3340. *See LH29 in List 2.*

L13 (8), Lower Bayble (Lewis). Clach Stein. NB 5165 3174. *See LH31 in List 2.*

L14 (12), Barpa nam Feannag (N Uist). NF 8567 7207. O(238); N(NF87SE13); H(UST7); T(H3/6); U(UI17). Jul 81. Chambered long tomb. Plan: Henshall 1972, 501.

L15 (13), South Clettraval (N Uist). NF 7501 7118. *See UI22(b) in List 2.*

L16 (14), Unival (N Uist). Leacach an Tigh Chloiche. NF 8003 6685. *See UI28 in List 2.*

L17 (15), Dulnainbridge (Inverness). NJ 0117 2468. N(NJ02SW04); T(B7/3). Aug 79. Three standing stones, up to about 2·5 m high, spaced about 100 m apart and about 15° off line.

L18 (16), Esslie the Greater (Kincardine). Esslie sw; Esslie South. NO 7171 9159. Recumbent stone circle; internal ring cairn. *See RSC90 in List 3.*

L19 (17), Kempston Hill (Kincardine). NO 8767 8947 and 8760 8942. N(NO88NE23); BR(223, 265); T(B3/5). Aug 79. Two standing stones, some 2·5 m and 3 m high, 80 m apart.

L20 (18), Corogle Burn (Angus). Glen Prosen. NO 3488 6017. N(NO36SW02); B(7:20); BC(Angus5[8]); BR(222, 263); FP(Angus8); T(P3/1). Aug 79. Four-poster and outlying pair of standing stones, now both fallen. The westernmost two stones of the four-poster and the outlying pair are roughly in line. Plans: Thom 1966, fig. 20; Burl 1988b, 108.

L21 (19), Fowlis Wester (Perth). NN 9239 2492. N(NN92SW01); B(7:32, 7:33); BG(203); BC(Perth25a,b [28a,b]); BR(224, 267); T(P1/10). Aug 79. Ring of small stones, 5·5 m to 7 m in diameter (Thom's *W*); excavated stonehole of large standing stone (Thom's *L*) some 6 m E of *W*; kerb-cairn (Thom's *E*) some 15 m to E of *L*; standing stone (Thom's *M*) some 10 m NE of *E*; three stones of uncertain status (Thom's *A*, *B* and *C*) some 25 m S of *E*. Exc: Young 1943. Plans: *ibid.*, figs 1, 2, 4; Thom and Thom 1978a, fig. 13.5; Thom, Thom and Burl 1990, 324.

L22 (20), Comrie (Perth). Tullybannocher. NN 7548 2247. N(NN72SE07); B(7:18); BC([Perth17]); BR(224, 249); FP(Perthshire28); T(P1/8). Sep 79. Two standing stones approx. 1·9 m high, 6 m apart. Four stones were originally recorded here, but the configuration is uncertain. Possibly they formed a row; it has also been suggested that they represent the remains of a four-poster. Plan: Thom, Thom and Burl 1990, 323. *Also SSR11 in List 6.*

L23 (21), Muthill (Perth). Dalchirla. NN 8244 1588. N(NN81NW03); B(7:e); BR(224, 266); T(P1/1). Aug 79. Two standing stones, 2·3 m and 1·3 m high, 2·4 m apart.

L24 (22), Lundin Links (Fife). NO 4048 0271. F(379); N(NO40SW01); B(7:39); BG(148); BC(Fife4[5]); FP(Fife1); T(P4/1). Aug 79. Three large standing stones between 4 m and 5 m high; a fourth has been removed since the late eighteenth century. Possible, but unlikely, four-poster. Plans: Thom 1971, fig. 5.5; Burl 1988b, 126.

L25 (23), Kell Burn (East Lothian). Kingside. N(NT66SW02); T(G9/13). Jun 79.

(*a*) NT 6434 6419. Kingside School. EL(246); B(8:18); BC([EastLothian2]); FP(EastLothian2). Setting of small stones of indeterminate date, possibly the remains of a four-poster, now destroyed.

(*b*) NT 643 642. EL(220). Remains of oval enclosure wall, now destroyed.

Plan: Thom 1966, fig. 10(c).

L26 (24), Balinoe (Tiree). Balemartin. NL 9731 4258. *See CT9 in List 2.*

L27 (25), Quinish (Mull). Mingary. NM 4134 5524. *See ML2 in List 2. Also SSR19 in List 6.*

L28 (26), Lochbuie (Mull). NM 617 251. *See ML28 in List 2.*

L29 (27), Taoslin (Mull). Bunessan. NM 3973 2239. *See ML30 in List 2.*

L30 (28), Kintraw (Mid-Argyll). NM 8305 0498. A(63); N(NM80SW01); T(A2/5); U(AR5). May 85. Cairn 15 m in diameter; kerb-cairn 7 m in diameter; standing stone 4·0 m high; enclosure, possibly prehistoric. Exc: Simpson 1967 [cairns]; Cowie 1980 [stone re-erection]. Plans: Simpson 1967, fig. 1; see also Fig. 1.13 in this volume.

L31 (29), Kilmartin (Mid-Argyll). Temple Wood + Nether Largie. NR 827 978. *See AR13 in List 2. Also AP3 and AP4 in List 6 supp. A.*

L32 (30), Ballymeanoch (Mid-Argyll). Duncracaig. NR 8337 9641. *See AR15 in List 2. Also SSR30 in List 6 and AP5 in List 6 supp. A.*

L33 (31), Dunadd (Mid-Argyll). NR 838 935. *See AR27 in List 2.*

L34 (32), Achnabreck (Mid-Argyll). NR 856 900. *See AR31 in List 2.*

L35 (33), Knockrome (Jura). NR 549 715. *See JU4 in List 2.*

L36 (34), Craighouse (Jura). Carragh a'Ghlinne. NR 5128 6648. *See JU6 in List 2. Also SSR33 in List 6.*

L37 (35), Ballinaby (Islay). NR 221 673. *See IS15(a) in List 2.*

L38 (36), Stillaig (Cowal). May 81.

(*a*) Cnoc Pollphail, Low Stillaig. NR 9316 6835. A(207); N(NR96NW05); BR(222, 263); T(A10/5). Two standing stones, respectively 2·9 m high and a stump 0·6 m high, 7 m apart. Plan: RCAHMS 1988, 131.

(*b*) Creag Loisgte, Low Stillaig. NR 9352 6778. A(209); N(NR96NW08); T(A10/6). Standing stone 1·9 m high, 670 m SE of (*a*).

L39 (37), Escart (Kintyre). NR 8464 6678. *See KT5 in List 2. Also SSR36 in List 6.*

L40, Skipness (Kintyre). NR 9063 5876. N(NR95NW01); U(KT6). May 81. A rounded boulder 1·0 m across; not an antiquity.

L41 (38), DUNSKEIG (Kintyre). Clach Leth Rathad. NR 7624 5704. *See KT8 in List 2.*

L42 (39), TARBERT (Gigha). Carragh an Tarbert. NR 6555 5227. *See KT12 in List 2.*

L43 (40), BEACHARR (Kintyre). Beacharra. NR 6926 4330. *See KT15 in List 2.*

L44 (41), BEINN AN TUIRC (Kintyre). Arnicle; Crois Mhic Aoida. NR 7349 3506. *See KT23 in List 2.*

L45 (42), HIGH PARK (Kintyre). NR 6950 2572. *See KT29 in List 2.*

L46 (43), CAMPBELTOWN (Kintyre). Balegreggan. NR 7238 2123. *See KT36 in List 2.*

L47 (44), KNOCKSTAPPLE (Kintyre). NR 7026 1240. *See KT41 in List 2.*

L48 (45), HAGGSTONE MOOR (Wigtown). T(G3/2).
- (*a*) *C* [Thom]. NX 067 727. N(NX07SE29). NV. Possible cairn.
- (*b*) *M* [Thom]. NX 065 726. N(NX07SE27). Jun 79. Prostrate standing stone, more than 1·7 m long.
- (*c*) NX 0577 7251. N(NX07SE12); B(8:k); BC(Ayr5[–]). NV. Though reported as a possible stone circle, this is actually the kerb-stones of a denuded cairn.
- (*d*) Long Tom; *L* [Thom]. NX 0814 7186. G1(49); N(NX07SE03). Jun 79. Standing stone 1·7 m high.
- (*e*) Old Park of the Gleick; *S* [Thom]. NX 0603 7160. N(NX07SE28). Jun 79. Stone 0·8 m high, probable grounder for field wall.
- (*f*) Taxing Stone; *T* [Thom]. NX 0623 7097. G1(47); N(NX07SE01). Jun 79. Standing stone 1·8 m high.

Plan: Thom 1971, fig. 6.17 [does not include (*c*)].

L49 (46), LAGGANGARN (Wigtown). NX 2225 7166. G1(282); N(NX27SW04); B(8:p); BC(Wigtown3[4]); BR(224, 267–8); T(G3/3). May 79. Two cross-incised standing stones probably dating to the Early Christian period, but possibly the survivors of a prehistoric stone circle. A number of nearby stones are not antiquities.

L50 (47), BLAIR HILL (Kirkcudbright). The Thieves. NX 4041 7160. G2(367); N(NX47SW02); B(8:30); T(G4/2). May 79. Two standing stones and a prostrate monolith, within a roughly elliptical earth bank. Possibly the remains of a stone ring. Plan: Thom 1967, fig. 6.12; Thom, Thom and Burl 1980, 278.

## *England and Wales*

L51 (48), CASTLERIGG (Cumbria). NY 2913 2364. B(9:8); BG(18); BC(Cumberland8[8a]); BR(214, 227); T(L1/1). May 79. Stone circle, a flattened ring between 30 m and 33 m in diameter. Plan: Thom, Thom and Burl 1980, 28.

L52, BLAKELEY MOSS (Cumbria). Blakeley Raise. NY 0600 1402. E(NY01SE01); B(9:3); BG(15); BC(Cumberland3[3]); BR(214, 227); T(L1/16). Sep 81. Stone circle 17 m in diameter. Reconstructed, possibly inaccurately; possibly completely bogus. Plan: Thom, Thom and Burl 1980, 52.

L53 (49), BURNMOOR (Cumbria). BG(16); T(L1/6). May 79. Overall plan: Burl 1995, 40.
- (*a*) Low Longrigg; *A* and *B* (Thom). NY 1725 0276. E(NY10SE02); B(9:24, 9:25); BG(16c); BC(Cumberland24[20a,b]). Two stone circles, diameters between 15 m and 22 m (*A*, to the NE) and 15 m (*B*, to the SW), about 20 m apart. Plan: Thom, Thom and Burl 1980, 36.
- (*b*) White Moss; *C* and *D* (Thom). NY 1726 0239. E(NY10SE01); B(9:33, 9:32); BG(16B); BC(Cumberland29[27a,b]). Two stone circles, diameters 17 m (*C*, to the WSW) and 16 m (*D*, to the ENE), about 30 m apart. They are some 370 m S of (*a*). Plan: Thom, Thom and Burl 1980, 38.
- (*c*) Brats Hill; *E* (Thom). NY 1735 0233. E(NY10SE01); B(9:5); BG(16A); BC(Cumberland4[4]). Stone circle, a flattened ring between 30 m and 32 m in diameter, located about 100 m SE of (*b*). Plan: Thom, Thom and Burl 1980, 40.

L54 (50), KIRKSANTON (Cumbria). Giants' Graves. SD 1361 8110. BR(214, 250); T(L1/11). May 79. Two standing stones, some 3·0 m and 2·2 m high, placed 5 m apart.

L55 (51), BOROUGHBRIDGE (N Yorkshire). Devil's Arrows. SE 391 665. BR(217, 245); T(L6/1). Sep 79. 175 m-long row of three stones, up to 6·9 m high. Former companion to centre stone is now removed.

L56 (52), PARC-Y-MEIRW (Dyfed). SM 9988 3591. W(SM93NE12); BR(225, 240); T(W9/7). Nov 79. 40 m-long row of eight standing stones up to 3·4 m high, four now prostrate. Plan: Thom, Thom and Burl 1990, 362.

L57 (53), THE SANCTUARY (Wiltshire). SU 118 680. B(15:10); BG(83); BC(Wiltshire6[9]); T(S5/2). Nov 79. Concentric rings of timber posts, probably supports for a sequence of huts. Later, two concentric stone circles, the outer (at least) free-standing. Destroyed in the eighteenth century. Markers show timber postholes and stoneholes excavated in 1930. Exc: Cunnington 1931. Plan: Thom, Thom and Burl 1980, 124.

L58 (55), ALTARNUN, Cornwall. T(S1/2). Nov 79.
- (*a*) Nine Stones. SX 2361 7815. E(SX27NW29); B(14:1); BG(9); BC(Cornwall1[1]). Stone circle, 15 m in diameter.
- (*b*) SX 2384 7824. E(SX27NW27). Line of stones running away from (*a*) to the ENE.
- (*c*) SX 2340 7791. E(SX27NW28). Line of stones running away from (*a*) to the SW.

The two radial lines of stones mark a parish boundary, and hence are probably modern.

# List 2

*Monuments in Western Scotland used in an Independent Study of 'Megalithic Astronomy'*

This is a reference list of those monuments included in an independent study of 'megalithic astronomy' carried out between 1975 and 1981 and published in 1984.[1] It was constructed specifically in the course of a project that aimed to assess independently the astronomical potential of settings of standing stones using strict selection criteria (see chapter three). Monuments were considered, and included or excluded, according to those criteria solely for the purposes of a statistical reappraisal. The list is not intended to be complete in any other sense and should not be treated as an archaeological inventory.

Cross-references are given to monument reference numbers and names used in this book as well as to information in other lists and gazetteers. These include ones not available at the time of the original project, such as the RCAHMS 1988 inventory of mid-Argyll and Burl's gazetteers of stone rows and 'four-posters'. However, no new entries have been added, and none of the original entries has been excluded, in the light of this new material.

Information is given for each 'site' as defined in the study, but is divided into parts where this corresponds to more than one monument as defined by other sources.

The following information is given for each entry:

1. Reference number and name used in Ruggles 1984a.
2. Alternative names if any.
3. National Grid reference (NGR), generally quoted to 10 m.
4. Cross-references to the monument in other sources.
5. Date of most recent visit by the author ('NV' indicates not visited).
6. Brief description and other information. Note that this is not meant to be complete or precise but merely to give an indication of the nature of the monument. For more complete information the reader is referred to the inventories cited.
7. Reference of published excavation report(s).
8. Reference of published site plan(s).

## *Lewis*

LH1, Port of Ness. Clach Stein. NB 5349 6417. O(19); N(NB56SW12); FP(Lewis1). Aug 79. Two standing stones, 1·5 m and 1·0 m high, some 3 m apart. Possibly the remains of a four-poster.

LH5, Ballantrushal. Clach an Trushal. NB 3755 5377. O(16); N(NB35SE01); T(H1/12). Jul 81. Standing stone more than 5·5 m high.

LH6, Carloway. Clach an Tursa. NB 2041 4295. O(87); N(NB24SW01); BR(223, 255); T(H1/16, earlier H1/8). Jul 81. A standing stone 2·5 m high together with two large prostrate stones, both split in two and originally about 4 m long. Possibly the remains of a 5 m-long row. *Also SSR3 in List 6.*

LH7, Kirkibost (Great Bernera). Callanish XV. NB 1775 3459. X(76:59). Jul 81. Prostrate stone, 3·5 m long, which appears to have stood at its sw end.

LH8, Bernera Bridge (Great Bernera). Cleiter; Great Bernera; Callanish VIII. NB 1642 3424. O(86); N(NB13SE02); B(3:g); BR(223; 265); T(H1/8, earlier H1/7); X(76:57). Jul 81. Two standing stones, about 2 m and 3 m high, a small erect slab under 1 m high, and a prostrate stone 2·5 m long which was re-erected in 1985. Subsequently, its true socket hole was established. Situated on a cliff edge, it has been suggested that this arc of stones represents the remains of a stone circle about 20 m in diameter whose other half has fallen away, but this can now be ruled out. Exc: see Curtis 1988, 366–8. Plan: Curtis 1988, 366.

LH10, Beinn Bheag. Callanish XI; Airigh na Beinne Bige. NB 2223 3568. B(3:c); X(76:58). May 88. Possible standing stone 1·5 m high, together with other stones of uncertain status submerged in the peat, described by Ponting and Ponting 1981, 82–6. Plans: Tait 1978 (but see Ponting and Ponting 1981, 107, note 2); *ibid.*, 85.

LH16, Callanish. Tursachan Callanish; Callanish I. NB 2130 3300. O(89); N(NB23SW01); B(3:2); BG(185); BC(Lewis3[Hebrides8]); BR(223, 231, 248); H(LWS3); T(H1/1). May 88. 13 m-diameter ring of large standing stones, heights about 3 m to 4 m, surrounding a cairn and single stone standing to 4·5 m; five radial lines of standing stones, two forming an avenue approx. northwards and one each approx. east, west and south. Exc: Ashmore 1995, 27–36. Plans: Tait 1978; Ponting and Ponting 1981, 80; Curtis 1988, 355; see also Fig. 8.8 in this volume. *Also L9 in List 1.*

LH18, Cnoc Fillibhir Bheag. Callanish III. NB 2250 3269. O(91); N(NB23SW02); B(3:6); BG(188); BC(Lewis4

[Hebrides11]); T(H1/3); X(76:57). May 88. A 17 m-diameter ring of standing stones up to 2·5 m in height, surrounding four more standing stones which appear to be the remains of an internal ring. Plans: Tait 1978; Ponting and Ponting 1981, 81; Curtis 1988, 359.

LH19, Cnoc Ceann a' Gharaidh. Callanish II; Loch Roag. NB 2220 3260. O(90); N(NB23SW03); B(3:5); BG(187); BC(Lewis8[Hebrides10]); T(H1/2). May 88. Ring, between 18 m and 22 m in diameter, of which five stones up to 2·5 m high remain standing. Plans: Tait 1978 (but see Ponting and Ponting 1981, 107, note 2); Ponting and Ponting 1981, 81; Curtis 1988, 357. *Also L10 in List 1.*

LH21, Ceann Hulavig. Callanish IV. NB 2297 3042. O(93); N(NB23SW04); B(3:4); BG(186); BC(Lewis7[Hebrides9]); T(H1/4); X(76:57). May 88. Five large standing stones from 2 m to 3·5 m high in a 13 m-diameter ring surrounding a cairn. Plans: Tait 1978; Ponting and Ponting 1981, 81; Curtis 1988, 360.

LH22, Cul a' Chleit. Callanish VI. NB 2465 3034. O(95); N(NB23SW07); B(3:i); BC(Lewis5[Hebrides12]); BR(223, 265); T(H1/6); X(76:57). Jul 75. Two standing stones, 1·5 m and 0·8 m high, some 10 m apart upon a small rocky knoll.

LH24, Airigh nam Bidearan. Callanish V. NB 2342 2989. O(94); N(NB22NW01); B(3:b); BC(Lewis2[–]); BR(223, 255); T(H1/5). Jul 81. Five small upright slabs about 0·5 m high, three of which are close (within 10 m) and in line. Possibly the remaining part of a field or enclosure wall. Plans: Tait 1978; Ponting and Ponting 1981, 84; Thom, Thom and Burl 1990, 230. *Also L11 in List 1 and SSR4 in List 6.*

LH27, Newmarket. NB 4132 3565. N(NB43NW04); B(3:13). Jul 81. Three small erect stones about 1 m high and spaced about 6 m apart, possibly the remains of a stone circle about 45 m in diameter.

LH28, Priests Glen. Laxdale. NB 4111 3519. O(56); N(NB43NW01); B(3:10); BC(Lewis10[Hebrides16]). Jul 81. Three prostrate stones, between 1·5 m and 2 m long, possibly the remains of a ring about 50 m in diameter.

LH29, Dursainean NE. Allt na Muilne. NB 5281 3340. N(NB53SW07); T(H1/15). Jul 81. Possible standing stone 1·7 m high. *Also L12 in List 1.*

LH31, Lower Bayble. Clach Stein. NB 5165 3174. O(57); N(NB53SW05); T(H1/14). Aug 79. Large prostrate monolith broken into two parts each approximately 1·5 m long, apparently a fallen standing stone. *Also L13 in List 1.*

LH33, Sideval. Loch Seaforth. NB 2781 1662. N(NB21NE01); B(3:9); BC(Lewis9[hebrides5]). Aug 79. Two standing stones about 1·5 m high and other prostrate stones which appear to form the remains of a stone circle about 17 m in diameter.

## *Harris*

LH36, Horgabost. Nisabost; Clach Mhic Leoid. NG 0408 9727. O(135); N(NG09NW04); BR(223 [listed under Lewis], 255); T(H2/2). Aug 79. Standing stone 3·5 m high. Two small slabs some 2·5 m to the west are probably the kerbstones of a cairn.

LH37, Scarista. Borvemore. NG 0202 9392. O(136); N(NG09SW02); B(3:f); BC(Harris1[Hebrides5]); T(H2/3). Aug 79. Standing stone 2·0 m high and two prostrate stones in a triangular setting.

## *North Uist*

UI6, Borve (Berneray). Cladh Maolrithe. NF 9122 8068. O(133); N(NF98SW07); T(H3/1). Jul 81. Standing stone 2·5 m high.

UI9, Newtonferry. Crois Mhic Jamain. NF 8937 7818. O(243); N(NF87NE08); T(H3/22). Jul 81. Two standing stones, 1·5 m and 0·5 m high, set into the summits of two mounds some 6 m apart.

UI15, Maari. NF 8645 7292. N(NF87SE23). Jul 81. Standing stone 2·2 m high, leaning by 45°.

UI19, Blashaval. Na Fir Bhreige. NF 8875 7176. O(246); N(NF87NE14); BR(223, 255); T(H3/8). Jul 81. Row of three stones standing only to about 0·5 m above present peat level, spaced at intervals of about 25 m and 35 m. According to Burl (1993, 255) the westernmost stone is prostrate, but this was not the case in 1981.

UI22, South Clettraval. Jul 77.
- (*a*) NF 7496 7139. O(233); H(UST12). Clyde-group chambered long tomb. Exc: W. Lindsay Scott, 'The chambered cairn of Clettraval, North Uist', *PSAS* 69 (1935), 480–536.
- (*b*) NF 7503 7120. O(256); N(NF77SE13); T(H3/3). Standing stone 1·5 m high. *Also L15 in List 1.*
- (*c*) Tigh Chloiche (W). NF 7516 7101. O(234); N(NF77SE14); H(UST28); T(H3/4). Hebridean-group chambered tomb.

UI23, Toroghas. Fir Bhreige. NF 7700 7029. O(255); N(NF77SE12); BR(223, 266); T(H3/5). Jul 81. Two stones standing to about 1·0 m above present peat level and about 35 m apart.

UI24, Marrogh. Tigh Chloiche (E). NF 8324 6952. O(242); N(NF86NW02); T(H3/13). Jul 79. Possible standing stone 1·0 m high.

UI26, Beinn a'Charra. NF 7863 6909. O(253); N(NF76NE01); T(H3/9). Jul 81. Standing stone 2·8 m high, leaning to the south.

UI28, Unival. Leacach an Tigh Chloiche. NF 8003 6685. O(228); N(NF86NW04); H(UST34); T(H3/11). Jul 79. Hebridean-group chambered tomb; standing stone 3·0 m high, 7 m to the sw. Exc: W. Lindsay Scott, 'The chamber tomb of Unival, North Uist', *PSAS* 82 (1947), 1–49. Plan: Moir *et al.* 1980, 41. *Also L16 in List 1.*

UI29, Loch na Buaile Iochdrach. NF 8031 6651. O(254); N(NF86NW17). Jul 79. Possible standing stone 1·0 m high.

UI31, Claddach Kyles. Jul 79.
- (*a*) Clach Mhor à Chè. NF 7700 6620. O(252); N(NF76NE02); T(H3/12). Standing stone 2·5 m high.
- (*b*) Dùn na Càrnaich. NF 7700 6617. O(231); H(UST16). Ruined chambered tomb, possibly of Clyde group, some 20 m south of (*a*).

UI33, Ben Langass. Sornach Coir Fhinn; Pobull Fhinn. NF 8427 6502. O(250); N(NF86NW07); B(3:11); BG(190);

BC(NUist5[Hebrides21]); BR(223, 227–8); T(H3/17). Jul 81. A 35 m-diameter ring of stones up to 2·1 m in height. Plans: RCAHMS 1928, fig. 141; Thom, Thom and Burl 1980, 310; Curtis 1988, 367.

UI35, Cringraval W. NF 8116 6447. O(251); B(3:7) ['Cringravel']; BC(NUist3[Hebrides19]). Jul 81. Five stones, one standing 0·7 m high and the others prostrate and up to 1·5 m long, in the eastern arc of a ring about 35 m in diameter.

UI37, Loch a'Phobuill. Sornach a'Phobuill; 'Sornach Coir Fhinn' [Thom]. NF 8289 6302. O(249); N(NF86SW28); B(3:8); BG(189); BC(NUist4[Hebrides20]); T(H3/18). Jul 81. A 40 m-diameter ring of stones up to 1·0 m in height. Plans: Thom 1966, fig. 14; Thom, Thom and Burl 1980, 312; Curtis 1988, 369.

UI40, Carinish. NF 8321 6021. O(248); N(NF86SW01); B(3:3); BC(NUist2[Hebrides18]). Jul 81. The remains of a stone circle about 40 m in diameter, through the middle of which runs the A865 road. Only four standing stones now remain to its north and one to its south. Plan: Curtis 1988, 371.

### *Benbecula*

UI44, Hacklett. NF 8445 5381. O(355); N(NF85SW03); T(H4/6). NV. A prostrate cup-marked stone 3·0 m long, apparently a fallen standing stone.

UI46, Stiaraval. Rueval. NF 8142 5315. T(H4/4). Jul 81. Possible standing stone 1·2 m high.

### *South Uist*

UI48, Stoneybridge. Crois Chnoca Breaca. NF 7340 3366. O(408); N(NF73SW03); H(UST34b); T(H5/2). Jul 81. Standing stone 2·5 m high atop a 1 m-high mound.

UI49, Beinn a'Charra. An Carra. NF 7703 3211. O(407); N(NF73SE01); T(H5/1). Jul 81. Standing stone some 5·0 m high.

UI50, Sligeanach Kildonan. Ru Ardvule. NF 7273 2860. O(406); N(NF72NW03); BR(224, 256); T(H5/3); X(77:18). Jul 81. Standing stone located amongst shifting sand dunes, which had become completely buried by 1979, its tip being marked by a wooden post. Two prostrate stones some 90 m to the nnw, 2·0 m and 1·8 m long and lying one on the other, were documented in 1914 and are shown on Thom's plan made in 1949, but were not located by monuments inspectors in 1965 and have not been found since. Plan: Thom, Thom and Burl 1990, 260.

### *Barra*

UI57, Borve. NF 6527 0144. O(461); N(NF60SE10); BR(222, 264); T(H6/1). Jul 81. Two stones 8 m apart, one standing to a height of 1·5 m above present ground level, the other fallen with only its tip showing above the machair. Plan: Thom, Thom and Burl 1990, 262.

UI58, Beul a'Bhealaich. NF 684 008. T(H6/2). Jul 81. A large prostrate stone 4·5 m long, possibly a fallen standing stone.

UI59, Brevig. NL 6890 9903. O(460); N(NL69NE01); BR(222, 264); T(H6/3). Jul 79. Two stones some 5 m apart, one standing to a height of 2·5 m, the other prostrate and broken into two. The other components of alignments noted by Thom appear to be natural rocks and outcrops. Plan: Thom, Thom and Burl 1990, 263.

UI60, Ben Rulibreck (Vatersay). Cuithe Heillanish. NL 6277 9389. O(462); N(NL69SW02); T(H6/4). Jul 81. Standing stone 1·7 m high, now forming the west jamb of the entrance of an old enclosure.

### *Ardnamurchan*

NA1, Branault. Cladh Chatain. NM 5268 6950. M(99); N(NM56NW02); BR(223, 265) [listed under Mull]. Jun 81. Standing stone 2·2 m high together with a stone 0·4 m high which may represent the stump of another. The two are only 1 m apart.

NA3, Camas nan Geall. Cladh Chiarain. NM 5605 6184. M(263); N(NM56SE02). Jun 81. Standing stone 2·3 m high, probably erected in prehistoric times but decorated with motifs of later date and now standing adjacent to a burial ground.

### *Morvern*

NA7, Beinn Bhan. NM 6591 4926. M(93); N(NM64NE08). Jun 81. Standing stone 1·8 m high.

### *Coll*

CT1, Acha. Loch nan Cinneachan. NM 1860 5674. M(50); N(NM15NE17). Jul 79. Standing stone 1·0 m high some 20 m from two possible cairns.

CT2, Totronald. NM 1665 5594. M(120); N(NM15NE15); BR(222, 264); T(M3/1). Jul 79. Two standing stones, 1·5 m and 1·4 m high, some 14 m apart and oriented across the line joining them.

CT3, Breachacha. NM 1519 5329. M(94); N(NM15SE15). Jul 79. A large prostrate slab some 10 m to the west is probably not a fallen standing stone.

CT4, Caolas. NM 122 532. M(95); N(NM15SW16). Jul 79. Possible standing stone about 0·8 m high; another stone 1 m long and 0·5 m high, situated some 6 m to its se, is probably an outcrop.

### *Tiree*

CT5, Caoles. NM 0776 4833. M(96); N(NM04NE14); T(M4/1). Jul 79. Standing stone 3·0 m high, now leaning at about 45° to the nw.

CT7, Hough. M(107); BG(231). Jul 79.

(*a*) Hough nne; Moss A. NL 9588 4518. N(NL94NE20); B(4:9); BC(Tiree3a[1a]). Stone ring about 40 m in diameter.

(*b*) Hough ssw; Moss B. NL 9580 4505. N(NL94NE23); B(4:10); BC(Tiree3b[1b]). Stone ring about 40 m in diameter, about 150 m sw of (*a*), centre to centre.

CT8, BARRAPOLL. NL 9468 4300. M(91); N(NL94SW11); T(M4/3). Jul 79. Standing stone 1·5 m high.

CT9, BALINOE. Balemartin. NL 9731 4258. M(89); N(NL94SE04); T(M4/2). Jul 79. Standing stone 3·5 m high. *Also L26 in List 1.*

## *Mull*

ML1, GLENGORM. NM 4347 5715. M(105); N(NM45NW02); BR(223, 255); T(M1/7). Jun 91. The remains of a row of three stones, all around 2 m in height, one in its original position. One of the other two, both prostrate in 1882, one has been re-erected close to its original position but the other has been dragged out of line and re-erected to form a triangular setting. Exc: Ruggles and Martlew 1989, S143–8; Martlew and Ruggles 1996, 121–2. Plans: Ruggles and Martlew 1989, fig. 3; Martlew and Ruggles 1996, fig. 3. *Also SSR18 in List 6.*

ML2, QUINISH. Mingary. NM 4134 5524. M(111); N(NM45NW05); BR(223, 248, 255); T(M1/3). Jun 91. A standing stone 2·8 m high and three prostrate stones, which appear to have formed a 10 m-long row. A fifth stone is documented. Plan: Ruggles 1981, fig. 4.4. *Also L27 in List 1 and SSR19 in List 6.*

ML4, BALLISCATE. Sgriob-Ruadh; Tobermory. NM 4996 5413. M(90); N(NM45SE01); B(4:a); BC(Mull1[–]); BR(223, 255); T(M1/8). Jul 91. Three-stone row about 5 m long, the end stones 2·5 m and 1·8 m in height, the central one fallen, 2·8 m long. Plan: RCAHMS 1980, fig. 39. *Also SSR20 in List 6.*

ML6, LAG. Mornish. NM 3626 5331. M(109); N(NM35SE23); BR(223, 265). Jul 91. Standing stone 1·6 m high; prostrate stone 1·4 m long which appears to have stood only about 1 m away from it. Two further small standing stones some 15 m to the SE, noted by James Orr ('Standing stones and other relics in Mull', *Transactions of the Glasgow Archaeological Society* 9 (1937), 128–34, pp. 132–3), are the remains of a more recent boundary wall.

ML7, CILLCHRIOSD. NM 3773 5348. M(98); N(NM35SE05); T(M1/2). Jul 91. Standing stone 2·6 m high.

ML8, CALGARY. Frachadil. NM 3849 5231. M(104); N(NM35SE22); BR(223, 265). Jul 91. Two prostrate stones 2·8 m and 2·6 m long, lying close together.

ML9, MAOL MOR. Dervaig A; Dervaig NNW. NM 4355 5311. M(101(1)); N(NM45SW05); B(4:m); BC(Mull4[–]); BR(223, 248); T(M1/4). Jun 90. A 10 m-long row of four stones, three standing all about 2 m high; the fourth fallen, 2·4 m long. Plans: Thom 1966, fig. 8(e); RCAHMS 1980, fig. 41; Thom, Thom and Burl 1990, 267. *Also SSR21 in List 6.*

ML10, DERVAIG N. Dervaig B; Dervaig Centre. NM 4390 5202. M(101(2)); N(NM45SW04; 10); B(4:f); BC(Mull2[–]); BR(223, 248); T(M1/5). Jun 91. Two standing stones 2·5 m and 2·4 m high, and three prostrate stones, which appear to have formed an 18 m-long five-stone row. Some 250 m to the SE, and also in the alignment, are an erect stone 1·0 m high and other stones shown in Thom's plans, but which are probably natural boulders built into an old field wall. Plans: Thom 1966, fig. 7; RCAHMS 1980, fig. 42; Thom, Thom and Burl 1990, 268. *Also SSR22 in List 6.*

ML11, DERVAIG S. Dervaig C; Dervaig SSE; 'Glac Mhór'. NM 4385 5163. M(101(3)); N(NM45SW07); BR(223, 248); T(M1/6). Jun 91. A 5 m-long row of three stones between 1·0 m and 1·3 m in height, two appearing to have been broken off. Plan: RCAHMS 1980, fig. 43. *Also SSR23 in List 6.*

ML12, ARDNACROSS. NM 5422 4915. M(10); N(NM54NW03); BR(223, 255); T(M1/9). Jul 91. One standing and five prostrate stones, the remains of two parallel three-stone rows respectively 12·5 m and 10·0 m long, together with three kerb-cairns. Exc: Martlew and Ruggles 1993; 1996, 125–8. Plans: Martlew and Ruggles 1993, fig. 16; Ruggles and Martlew 1993, fig. 2; Martlew and Ruggles 1996, fig. 8. *Also SSR24 and SSR25 in List 6.*

ML13, TENGA. Loch Frisa. NM 5040 4632. M(117); N(NM54NW04); B(4:30); BC(Mull5[2]); T(M1/10). Jun 90. Four stones between 0·8 m and 1·8 m high in a trapezoidal formation, in an area of deep peat, which may represent the remains of a stone ring of diameter about 35 m, though this is far from certain. Plan: RCAHMS 1980, fig. 48.

ML14, TOSTARIE. NM 3917 4561. M(119); N(NM34NE03); T(M1/11). Jul 91. Standing stone 1·7 m high.

ML15, KILLICHRONAN. Torr nam Fiann. NM 5401 4193. M(108); N(NM54SW01); T(M2/15). Jul 91. Standing stone 2·4 m high, now leaning by about 70° from the vertical.

ML16, GRULINE. M(106); BR(223, 265). Jul 91.
- (*a*) NM 5437 3977. N(NM53NW03); T(M2/16). Standing stone 2·4 m high.
- (*b*) NM 5456 3960. N(NM53NW01); T(M2/1). Standing stone 2·3 m high, some 250 m SE of (*a*).

ML17, ORMAIG (Ulva). NM 4256 3928. M(122); N(NM43NW03). Jul 91. Standing stone 2·7 m high, prostrate in 1979 but now re-erected.

ML18, CRAGAIG (Ulva). 'Cragaid'. NM 4028 3901. M(100); N(NM43NW09); BR(223, 265). Jul 91. Two standing stones, 1·6 m and 1·3 m high but both apparently broken off, placed about 4 m apart.

ML19, DISHIG. NM 4969 3574. M(102); N(NM43NE03). Aug 76. Possible standing stone 1·0 m high.

ML21, SCALLASTLE. NM 6999 3827. M(113); N(NM63NE02); BR(223, 255); H(MUL1c). May 85. A standing stone 1·2 m high and two prostrate stones 1·7 m and 2·0 m long, which apparently formed a 5 m-long three-stone row. *Also SSR26 in List 6.*

ML23, DUART. Barr Leathan. May 85.
- (*a*) NM 7260 3425. M(92, part); N(NM73SW02); T(M2/2, part). Standing stone 2·5 m high, leaning to the west.
- (*b*) NM 7275 3425. M(92, part); N(NM73SW04); T(M2/2, part). Setting of six boulders, possibly the disturbed remains of some sort of prehistoric structure, but possibly natural.

ML24, PORT DONAIN. NV.
- (*a*) NM 7368 2932. M(52); N(NM72NW04). Prostrate slab 4·1 m long, possibly a fallen standing stone, lying close to the perimeter of a 6 m-diameter kerb cairn. Plan: RCAHMS 1980, fig. 29.
- (*b*) NM 737 292. M(4); H(MUL1). Clyde-group chambered long tomb, about 100 m to the SE of (*a*).

ML25, ULUVALT. Uluvalt I. Jun 91.
(*a*) NM 5469 3004. M(121(3)); N(NM53SW02, part); H(unnumbered: 1972, p. 467). Reported remains of a chambered tomb, but probably more modern in origin.
(*b*) NM 5463 3002. M(121(2)); N(NM53SW02, part); BR(223, 248–9). Immediately west of (*a*), three prostrate stones between 2·0 m and 2·3 m long, possibly the remains of a 10 m-long three-stone row. A boulder placed within this alignment is considered an upright stone by RCAHMS (1980) who list this monument as a four-stone row. *Also SSR27 in List 6.*
(*c*) NM 546 300. M(121(4)). Prostrate stone 2·1 m long some 50 m west of (*b*), possibly a fallen standing stone.
(*d*) NM 5468 2996. M(121(1)); N(NM52NW03). Standing stone 1·9 m high, some 80 m south of (*b*), traditionally identified as one of a series of marker stones erected along the pilgrim route to Iona [see T. M'Laughlin, 'Notice on monoliths in the Island of Mull', *PSAS* 5 (1865), 46–52, p. 49] and not certainly prehistoric.
Plan: RCAHMS 1980, fig. 49.

ML27, ROSSAL. Breac Achadh. NM 5434 2820. M(54); N(NM52NW06). Aug 76. Standing stone 2·0 m high. Traditionally identified as a pilgrim route marker [see ML25(*d*)] and not certainly prehistoric.

ML28, LOCHBUIE. BG(182); T(M2/14). Jun 87.
(*a*) *M* [RCAHMS]; *D* [Thom]. NM 6163 2543. M(110, part); N(NM62NW03). Standing stone 2 m high, located 350 m NNW of (*c*).
(*b*) NM 6141 2524. M(49); N(NM62NW04). Remains of a kerb-cairn, located about 390 m WNW of (*c*).
(*c*) *A–J* [RCAHMS]; circle and *A* [Thom]. NM 6178 2512. M(110, part); N(NM62NW01); B(4:14); BC(Mull3[1]). Stone circle, 12 m in diameter; standing stone 0·9 m high, 5 m to the SE of the circle.
(*d*) *K* [RCAHMS]; *B* [Thom]. NM 6175 2507. M(110, part); N(NM62NW02). Standing stone 3 m high, about 40 m SW of (*c*).
(*e*) *L* [RCAHMS]; *C* [Thom]. NM 6169 2506. M(110, part); N(NM62NW05). Standing stone 2·2 m high, originally higher, about 100 m SW of (*c*).
Plans: RCAHMS 1980, figs 46 and 47; Thom, Thom and Burl 1980, 320. *Also L28 in List 1.*

ML30, TAOSLIN. Bunessan. NM 3973 2239. M(116); N(NM32SE01); T(M2/8). May 85. Standing stone 2·1 m high. Traditionally identified as a pilgrim route marker [see ML25(*d*)] and not certainly prehistoric. *Also L29 in List 1.*

ML31, UISKEN. Druim Fan; Am Fan. NM 3916 1961. M(103); N(NM31NE02); T(M2/10). Aug 76. Standing stone 2·2 m high.

ML33, ARDALANISH. NM 3784 1888. M(88); N(NM31NE01); BR(223, 265); T(M2/9). May 85. A standing stone 1·9 m high and a prostrate stone at least 2·3 m long, located 11 m apart. The standing slab is roughly oriented along the line joining the two stones.

ML34, SUIE. Dail na Carraigh. NM 3706 2185. M(61); N(NM32SE07); BR(223, 265); H(unnumbered: 1972, p. 467); T(M2/7). Aug 76. Two standing stones 2·0 m and 1·1 m high, and 1·2 m apart, situated on the NE side of a large ruined cairn some 25 m in diameter. A row of stones referred to by Thom in fact consists of the remaining grounders of an old field dyke.

ML35, TIRGHOIL. Ross of Mull. NM 3531 2240. M(118); N(NM32SE06); T(M2/6). May 85. Standing stone 2·4 m high. Traditionally identified as a pilgrim route marker [see ML25(*d*)] and not certainly prehistoric.

ML37, POIT NA H-I. Torr Mor. NM 3251 2216. M(112); N(NM32SW02); T(M2/5). Aug 76. Standing stone 2·4 m high.

ML39, ACHABAN HOUSE. NM 3133 2331. M(87); N(NM32SW01); T(M2/4). Aug 76. Standing stone 2·5 m high. Traditionally identified as a pilgrim route marker [see ML25(*d*)] and not certainly prehistoric.

### *Appin*

LN1, ACHARRA. NM 9866 5455. L(110); N(NM95SE03); T(M7/1). Jun 81. Standing stone some 3·5 m high.

LN2, INVERFOLLA. NM 9583 4503. L(118); N(NM94NE01); T(M7/2). Jun 81. Fallen standing stone about 3·5 m in length.

### *Benderloch*

LN3, BARCALDINE. NM 9636 4214. L(111); N(NM94SE03); BR(222, 263); T(M8/2). Jun 81. Standing stone 1·7 m high, with a second stone 1·6 m high moved to its side and propped up against it.

LN5, ACHACHA. L(10); N(NM94SW09); T(M8/1). May 81.
(*a*) NM 9436 4076. Cairn between 4 m and 5 m in diameter, with a kerb of large boulders.
(*b*) NM 9445 4075. Standing stone 2·5 m high, situated 90 m to the east of (*a*).

LN7, BENDERLOCH N. NM 9062 3865. L(113); N(NM93NW03, 10); B(4:31). May 81. Standing stone 1·5 m high, possibly all that remains of a stone circle.

LN8, BENDERLOCH S. May 81. NM 9033 3802. L(112); N(NM93NW09, 28); T(M8/4). Standing stone 2·1 m high. Documentary evidence suggests that there were originally more stones here.

### *Lorn*

LN14, TAYNUILT. Taynuilt 2. NN 0120 3115. L(122); N(NN03SW14). May 81. Standing stone 1·2 m high which appears originally to have been higher.

LN15, BLACK LOCHS. NM 9155 3124. N(NM93SW04). May 81. Possible standing stone 1·4 m high.

LN17, STRONTOILLER. Loch Nell. T(A1/2). May 81. Thom, Thom and Burl 1980, 140.
(*a*) NM 9067 2914. L(120); N(NM92NW08); B(4:25); BG(124); BC(Argyll1[6]). Ring of rounded boulders some 20 m in diameter. Plans: RCAHMS 1975, fig. 27; Thom, Thom and Burl 1980, 140.
(*b*) Clach na Carraig. NM 9076 2897. L(78, part); N(NM92NW02). Standing stone about 4·0 m high, some 190 m to the SE of (*a*).
(*c*) NM 9078 2896. L(78, part); N(NM92NW07). Ring

cairn 5 m across, 12 m ESE of (*b*). Exc: Graham Ritchie, 'Excavation of a cairn at Strontoiller, Lorn, Argyll', *Glasgow Arch. J.* 2 (1971), 1–7.

LN18, CLENAMACRIE. NM 9250 2854. L(114); N(NM92NW01); BR(222, 254). May 81. Standing stone 1·5 m high together with two large boulders whose status is uncertain.

LN22, DUACHY. Loch Seil. NM 8014 2052. L(116); N(NM82SW01); BR(222, 254); T(A1/4). Jun 81. 5 m-long row of three stones up to 2·8 m high, the central one now leaning at about 70° from the vertical. Some 40 m to the east is the 0·3 m-high stump of a further stone, which stood 2·5 m high until 1963. Plans: Thom 1966, fig. 10(a); Thom, Thom and Burl 1990, 98. *Also SSR28 in List 6.*

## *Mid-Argyll*

AR2, SLUGGAN. May 85.
- (*a*) NM 8395 0769. A(77(1)). Cairn 23 m in diameter and 3 m high.
- (*b*) NM 8405 0762. N(NM80NW06); T(A2/2, part). A small earthfast slab, possibly the remains of a standing stone, surrounded by field clearance material, 30 m ENE of (*c*).
- (*c*) NM 8401 0761. A(77(2)). Small cairn 7 m in diameter and 0·5 m high, some 110 m SE of (*a*).
- (*d*) NM 8403 0757. A(77(3)); N(NM80NW04); T(A2/2, part). Standing stone 2·5 m high, 30 m SSE of (*c*).

AR3, BARBRECK. NM 8315 0641. A(200); N(NM80NW19); B(4:4); BR(222, 263); T(A2/3). May 81. Aligned pair of standing stones, 2·5 m and 1·5 m in height, some 3 m apart. A further standing stone 2·6 m high, oriented roughly parallel to the first two, is situated some 23 m to the east, perpendicular to the alignment. It is surrounded by three small erect slabs, each about 0·6 m high, forming the central parts of three sides of a rectangle roughly 4 × 3 m. Plans: Patrick 1979, S80; RCAHMS 1988, 130. *Also AP1 in List 6 supp. A.*

AR6, SALACHARY. NM 8405 0403. A(225); N(NM80SW16); BR(222, 254) [NGR given erroneously as NM 847 040]; T(A2/26). May 85. Three standing stones that appear to have formed a row about 4 m long. Two stood to a height of about 2·5 m, but one now leans at about 70° to the vertical. The third is prostrate, about 3 m long. Plan: RCAHMS 1988, 137. *Also SSR29 in List 6.*

AR7, TORRAN. NM 8788 0488. A(230); N(NM80SE37); T(A1/8). Jun 79. Standing stone 3·3 m high, with crosses pecked on both faces.

AR8, FORD. NM 8668 0333. A(215); N(NM80SE42); T(A2/22). Jun 79. Standing stone some 3 m high; a second stone had disappeared by 1900.

AR9, GLENNAN N. Creagantairbh Beag. NM 8595 0157. A(208); N(NM80SE29); T(A2/23). May 85. 2 m-high stump of a standing stone. The top part, which broke off in 1879 and lies adjacent, is some 4 m long, and the original height appears to have been almost 5 m.

AR10, GLENNAN S. Glennan. NM 8573 0113. A(216); N(NM80SE28); T(A2/24). May 85. Standing stone 2·2 m high.

AR12, CARNASSERIE. NM 8345 0080. A(202); N(NM80SW22); BR(222, 263); T(A2/6). May 85. Aligned pair of standing stones, 2·6 m and 2·7 m high, some 2·5 m apart. *Also AP2 in List 6 supp. A.*

AR13, KILMARTIN. T(A2/8). Jun 88.
- (*a*) Temple Wood; Slockavullin. NR 8263 9783. A(228); N(NR89NW06); B(4:26, 4:27); BG(125); BC(Argyll2[7a,b]). Stone circle between 12 m and 13 m in diameter, including burial cairns and other deposits; site of second stone circle, 25 m to NE. Exc: Scott 1989; see also RCAHMS 1988, no. 228. Plans: Thom, Thom and Burl 1980, 144 and 146; RCAHMS 1988, 141 and 142; Scott 1989, fig. 3.
- (*b*) Nether Largie. A(222); BR(222, 264).
  - (*b1*) *M* [RCAHMS]; $S_6$ [Thom]. NR 8279 9774. N(NR89NW44). Standing stone 1·8 m high, located about 170 m SE of (*a*) and about 100 m to the NNW of the pair at (*b2*) and roughly in line with them.
  - (*b2*) *L* and *K* [RCAHMS]; $S_2$ and $S_3$ [Thom]. NR 8284 9764. N(NR89NW03, part). Pair of standing stones around 2·8 m high, placed about 4 m apart in a NNW–SSE alignment. *Also AP3 in List 6 supp. A.*
  - (*b3*) *F* and *G–J* [RCAHMS]; $S_1$ [Thom]. NR 8283 9761. N(NR89NW03, part); B(4:28) ['Temple Wood—site 3']. Standing stone 2·8 m high, extensively cup-marked, flanked by four small erect slabs up to 0·8 m high forming the central parts of the sides of a rectangle some 3 × 2 m. Located about 30 m SSW of (*b2*).
  - (*b4*) *C–E* [RCAHMS]; *Q* [Thom]. NR 8283 9760. N(NR89NW45); B(4:29) ['Temple Wood—site 4']. Three small erect slabs up to 0·9 m high forming the central parts of three sides of a rectangle some 5 × 3 m, located about 5 m SSW of (*b3*).
  - (*b5*) *A* and *B* [RCAHMS]; $S_4$ and $S_5$ [Thom]. NR 8281 9758. N(NR89NW03, part). Pair of standing stones around 2·8 m high, placed about 4 m apart in a NNW–SSE alignment. Located about 30 m SSW of (*b3*). *Also AP4 in List 6 supp. A.*
  - (*b6*) NR 8252 9761. N(NR89NW73). Buried stump of a standing stone uncovered by excavation in 1973. Located about 300 m west of (*b3*). Exc: see Gerald S. Hawkins, *Mindsteps to the Cosmos*, Harper and Row, New York, pp. 98–102.

  Plans: Thom 1971, 46; Hawkins (*op. cit.*), 100; Patrick 1979, S80; RCAHMS 1988, 136; Thom, Thom and Burl 1990, 106; see also Fig. 2.9 in this volume.

*Also L31 in List 1.*

AR15, BALLYMEANOCH. Duncracaig. NR 8337 9641. A(199); N(NR89NW14); BR(222, 248, 263); T(A2/12). Jun 79. 15 m-long row of four slabs up to 4 m high; aligned pair of slabs up to 3 m high and 3 m apart, parallel to the row and some 40 m to the SW; site of a holed stone, about 2·8 m high, some 20 m to the WNW of the pair. There is a kerb cairn 30 m to the NE of the row and a henge about 130 m to the SSW. Exc [holed stone]: John W. Barber, 'The excavation of the holed-stone at Ballymeanoch, Kilmartin, Argyll', *PSAS* 109 (1978), 104–11. Plans: Thom 1971, 52; Barber, *op. cit.*, 105; RCAHMS 1988, 128; Thom, Thom and Burl 1990, 110. *Also L32 in List 1, SSR30 in List 6 and AP5 in List 6 supp. A.*

AR16, ROWANFIELD. Poltalloch. NR 8205 9585. A(224); N(NR89NW47); T(A2/11). May 81. Standing stone 2·6 m high.

AR17, DUNTROON. NR 8034 9561. N(NR89NW57). May 85. Standing stone 1·3 m high.

AR18, CRINAN MOSS. NR 8083 9410. A(211); N(NR89SW05); BR(222, 240, 263); T(A2/7). Jun 79. Five upright stones protruding through peat, up to about 1 m high, irregularly placed. Plan: RCAHMS 1988, 132.

AR19, INVERARY. NN 0947 0905. A(217); N(NN00NE06); T(A2/1). May 85. Standing stone 2·8 m high.

AR23, LECHUARY. An Car. NR 8757 9550. A(194); N(NR89NE08); T(A2/17). May 85. Standing stone 2·9 m high, now leaning to the NE.

AR24, LOCH NA TORRNALAICH. NR 8548 9509. N(NR89NE10); T(A2/16). May 81. A prostrate slab 2·5 m long, probably fallen naturally from the adjacent crags.

AR25, TORBHLARAN N. NR 8639 9449. A(229); N(NR89SE03); T(A2/15). May 85. Standing stone 2·1 m high.

AR27, DUNADD. A(212); T(A2/13). May 81.
- (*a*) NR 8397 9343. N(NR89SW25). Large prostrate stone 4·2 m long, which was erect in 1867 and faced ENE.
- (*b*) NR 8386 9362. N(NR89SW35). Standing stone 1·4 m high, some 220 m NNW of (*a*).

*Also L33 in List 1.*

AR28, DUNAMUCK I. Dunamuck N. NR 8471 9290. A(213(1)); N(NR89SW28); BR(222, 254); T(A2/21). May 85. 7 m-long row of three standing stones, the end ones both being some 2·5 m high, and the central one, which has fallen, some 3·5 m long. Plan: RCAHMS 1988, 133. *Also SSR31 in List 6.*

AR29, DUNAMUCK II. Dunamuck Mid. NR 8484 9248. A(213(2)); N(NR89SW27); BR(222, 263–4); T(A2/14). May 85. Aligned pair of standing stones some 3·5 m and 2·5 m high, situated some 8 m apart. Plan: RCAHMS 1988, 133. *Also AP6 in List 6 supp. A.*

AR30, DUNAMUCK III. Dunamuck S. NR 8484 9233. A(213(3)); N(NR89SW24); BR(222, 264); T(A2/20). May 81. Two prostrate slabs, 4·1 m and 3·1 m long, lying 3·2 m apart.

AR31, ACHNABRECK. BR(222, 263); T(A2/19). May 85.
- (*a*) NR 8554 9018. A(193); N(NR89SE13). Prostrate stone some 4·5 m long.
- (*b*) Stane Alane. NR 8563 8993. A(226); N(NR88NE16). Standing stone some 2·5 m high, about 250 m to the SSE of (*a*) and roughly oriented upon it.

*Also L34 in List 1.*

AR32, OAKFIELD. Auchendarroch. NR 8572 8852. A(223); N(NR88NE15). Jul 79. Standing stone 1·7 m high.

AR33, KILMORY CASTLE. NR 8674 8652. A(220); N(NR88NE14). Jul 79. Standing stone 2·1 m high.

## *Jura*

JU1, TARBERT. East Loch Tarbert. T(A6/5). Jun 79. Plan: Thom, Thom and Burl 1990, 157.
- (*a*) NR 6062 8231. I(122); N(NR68SW01). Standing stone some 2·5 m high.
- (*b*) NR 6089 8221. I(328); N(NR68SW02). Standing stone 1·8 m high, some 290 m to the ESE of (*a*), cross-incised and standing within a burial ground, but possibly erected in prehistoric times.

JU2, KNOCKROME N. NR 5505 7192. I(109(3)); N(NR57SE02). Jun 79. Possible standing stone 0·9 m high.

JU3, ARDFERNAL. NR 5601 7171. I(75); N(NR57SE01); BR(223, 254–5, part); T(A6/4, part). Jun 79. Standing stone 1·2 m high. Plans: see JU4.

JU4, KNOCKROME. BR(223, 254–5, part); T(A6/4, part). Jun 79. Plans [including JU3]: Thom 1966, fig. 10(b); Thom, Thom and Burl 1990, 154.
- (*a*) NR 5503 7148. I(109(1)); N(NR57SE03). Standing stone 1·4 m high.
- (*b*) NR 5484 7144. I(109(2)); N(NR57SW03). Standing stone 1·5 m high, some 200 m to the west of (*a*).

*Also L35 in List 1.*

JU5, LEARGYBRECK. NR 5387 7128. I(111); N(NR57SW02). Jun 79. Possible standing stone 1·3 m high.

JU6, CRAIGHOUSE. Carragh a'Ghlinne. NR 5128 6648. I(83); N(NR56NW02); BR(223, 248); T(A6/6). Jun 79. 5 m-long four-stone row, one standing 2·4 m high and three prostrate. Plan: RCAHMS 1984, fig. 64D. *Also L36 in List 1 and SSR33 in List 6.*

JU7, SANNAIG. NR 5184 6480. I(116); N(NR56SW04); BR(223, 255); T(A6/3). Jun 79. The remains of a 5 m-long three-stone row, comprising a prostrate stone 2·5 m long, a standing stone 2·2 m high and a 0·3 m-high stump together with what appears to be its upper portion. Plan: RCAHMS 1984, fig. 70E. *Also SSR34 in List 6.*

JU8, STRONE. NR 5078 6375. I(120); N(NR56SW03); BR(223, 265); T(A6/2). Jun 79. A standing stone 3·0 m high and a prostrate stone 2·6 m long, some 2 m apart. Plans: Thom 1966, fig. 8(g); Thom, Thom and Burl 1990, 153.

JU9, CAMAS AN STACA. NR 4641 6477. I(81); N(NR46SE01); T(A6/1). Jun 79. Standing stone 3·5 m high.

## *Islay*

IS3, BEINN A' CHUIRN. Clach an Tiampain. NR 3475 6978. I(86); N(NR36NW01); T(A7/2). Jun 79. Standing stone 1·2 m high.

IS4, FINLAGGAN. NR 3927 6856. I(97); N(NR36NE03); BR(223, 265); T(A7/24). Jul 78. Standing stone 2·0 m high; a second stone stood here in the seventeenth century.

IS5, SCANISTLE. NR 4108 6724. I(119); N(NR46NW06). Jul 78. Standing stone 1·2 m high.

IS6, BEINN CHAM. NR 3492 6793. I(80); N(NR36NW02); BR(223, 264); T(A7/3). Jun 79. Standing stone 1·2 m high; adjacent to it is a loose stone 1·0 m long which may originally have stood in the vicinity also.

IS7, BALLACHLAVIN. Baile Tharbhach. NR 3636 6762. I(78); N(NR36NE16); T(A7/4). Jun 79. Standing stone 1·6 m high.

IS11, KNOCKLEAROCH. NR 3989 6483. I(108); N(NR36SE01); BR(223, 265). Jul 78. Two standing stones, 1·7 m and 1·5 m high and both leaning to the south, some 2·5 m apart.

IS12, MULLACH DUBH. NR 4037 6410. I(113); N(NR46SW01); T(A7/10). Jun 79. Standing stone 1·2 m high.

IS15, BALLINABY. BR(223, 254); T(A7/5). Jun 79.
- (*a*) NR 2200 6719. I(79, part); N(NR26NW13). Standing stone 4·9 m high. *Also L37 in List 1.*
- (*b*) NR 2210 6738. I(79, part); N(NR26NW14). Stump of a standing stone, 2 m high, 220 m NE of (*a*).
- (*c*) NR 222 675. N(NR26NW15). Although it has been identified as the possible stump of a third standing stone, this stone 140 m NE of (*b*) appears to be a natural outcrop.

IS18, FORELAND HOUSE. NR 2692 6429. I(98); N(NR26SE03). Jul 78. Standing stone some 2·0 m high.

IS19, UISGEANTSUIDHE. NR 2938 6335. I(124); N(NR26SE03); T(A7/20). Jul 78. Standing stone some 2·5 m high.

IS21, KNOCKDON. NR 3360 6423. I(107); N(NR36SW12); T(A7/8). Jul 78. Standing stone 1·5 m high.

IS23, GARTACHARRA. NR 2527 6137. I(99); N(NR26SE08); T(A7/13). Jul 78. Standing stone 2·7 m high.

IS24, CARN MOR. Cnoc nan Guaillean. NR 2241 6041. I(91); N(NR26SW32); T(A7/12). Jul 78. Standing stone 2·7 m high.

IS25, CNOC THORNASAIG. Cnoc a'Charraigh. NR 2317 6000. I(90); N(NR26SW23). Jul 78. Prostrate slab about 4 m long, which is probably a fallen standing stone.

IS26, DROIGHNEACH. NR 2106 5938. I(102); N(NR25NW11); T(A7/14). Jul 78. Standing stone 2·7 m high.

IS28, CULTOON. NR 1956 5697. I(94); N(NR15NE01); B(4:6); BG(164); BC(Islay2[2]); T(A7/15); X(75:08). Jul 78. Elliptical stone ring with diameter varying from 35 m to 40 m. Exc: MacKie 1981, 116–28. Plans: MacKie 1977, 93; 1981, 118.

IS31, KELSAY. NR 1901 5561. I(104); N(NR15NE08). Jul 78. Possible standing stone 1·2 m high.

IS34, ARDTALLA. NR 4658 5457. I(77); N(NR45SE16); X(76:09). Jul 78. Probable standing stone 1·3 m high.

IS35, CLAGGAIN BAY. NR 4618 5372. I(88); N(NR45SE02); T(A7/21). Jun 79. A massive boulder some 1·7 m high, possibly erected in antiquity.

IS36, TRUDERNISH. NR 4630 5290. I(123); N(NR45SE09). Jun 79. Standing stone 1·8 m high.

IS37, ARDILISTRY. NR 4426 4919. I(76); N(NR44NW27); B(4:1); BC(Islay1[1]); FP(Islay1). NV. Four stones no more than 0·5 m high, possibly the remains of a four-poster. Plans: RCAHMS 1984, fig. 63D; Burl 1988b, 128.

IS38, CNOC RHAONASTIL. Clachan Ceann Ile. NR 4369 4832. I(85); N(NR44NW06); BR(223, 264–5). Jun 79. Two stones, standing 1·5 m and 0·7 m high, some 10 m apart, of doubtful antiquity.

IS39, LAGAVULIN N. 'Lagavullin'. NR 3954 4621. I(110); N(NR34NE09); BR(223, 265). Jun 79. A standing stone 3·5 m high and a prostrate slab 3·7 m long, some 2 m apart, which seem to have formed an aligned pair. *Also AP7 in List 6 supp. A.*

IS41, ACHNANCARRANAN. Laphroaig. NR 3895 4607. I(74); N(NR34NE07); BR(223, 254); T(A7/19). Jul 78. A 7·5 m-long three-stone row. The end stones are 2·9 m and 2·7 m high; the central one is fallen and about 3·0 m long. Plan: RCAHMS 1984, fig. 63A. *Also SSR35 in List 6.*

IS42, KILBRIDE. NR 3838 4657. I(106); N(NR34NE05); BR(223, 265); T(A7/18). Aug 78. Standing stone 2·7 m high; a sunken stone at least 2·5 m long some 3 m to its SW may be a fallen companion, but this is far from certain.

IS46, PORT ELLEN I. NR 3715 4559. I(115); N(NR34NE12). Jul 78. Standing stone 4·3 m high.

IS47, KINTRA. Druim an Stuin; Carragh Bhan. NR 3283 4781. I(84); N(NR34NW07); T(A7/17). Jul 78. Standing stone 2·2 m high. Within 3 m are two stones, no more than 0·3 m high, which might be the stumps of two companions. They do not form a row.

IS49, CORNABUS. Cnoc Ard. NR 3264 4600. I(93); N(NR34NW11). Jul 78. Standing stone 1·5 m high.

IS50, KINNABUS. Glac a'Charraigh. NR 2975 4315. I(101); N(NR24SE02). Jul 78. Standing stone originally 1·9 m high, whose top was broken off between June 1977 and July 1978.[2]

## *Knapdale*

AR37, BARNASHAIG. 'Barnashalg'. NR 7298 8640. A(201); N(NR78NW01); T(A3/4, part). May 85. Standing stone 3·5 m high.

AR38, UPPER FERNOCH. Tayvallich. T(A3/4). May 85.
- (*a*) NR 7280 8614. A(231); N(NR78NW06). Standing stone some 2·5 m high, now leaning by about 20° to the east.
- (*b*) NR 7269 8594. N(NR78NW07); BR(222, 264). A group of stones up to 2·5 m long, some 200 m to the south of (*a*), marked as 'Ring A' and 'Ring B' on Thom's plan, but most likely natural.

Plans: Thom 1966, fig. 7; Thom, Thom and Burl 1990, 127.

AR39, LOCHEAD. NR 7772 7801. A(221); N(NR77NE04); T(A3/9). May 85. Standing stone some 2·0 m high.

KT1, CRETSHENGAN. NR 7072 6689. A(210); N(NR76NW03); T(A3/5). May 85. A standing stone 1·6 m high, leaning by about 15° from the vertical. It has been wedged into a small rock cleft, which suggests that it may not be prehistoric.

KT2, CARSE. Loch Stornoway. A(204); N(NR76SW01); BR(222, 263); T(A3/6). Jun 79.
- (*a*) NR 7414 6166. Standing stone 2·3 m high, about 110 m west of (*b*).
- (*b*) NR 7425 6163. Aligned pair of standing stones 3·2 m and 2·4 m high, 2·5 m apart. Plan: RCAHMS 1988, 131. *Also AP8 in List 6 supp. A.*

KT3, ARDPATRICK. Achadh-Chaorann. NR 7573 6014. A(192); N(NR76SE03); T(A3/7). Jun 79. Standing stone 2·1 m high.

KT4, AVINAGILLAN. NR 8391 6746. A(197); N(NR86NW01); T(A3/8). May 85. Standing stone 1·9 m high.

*Kintyre*

KT5, ESCART. NR 8464 6678. K(143); N(NR86NW02); BR(222, 248); T(A4/1). Jun 79. A 15·5 m-long five-stone row. The stones are up to 3·3 m high, and the line is rather sinuous. There is some evidence to suggest that there may originally have been more stones here. Plans: Thom 1971, 60; RCAHMS 1971, fig. 34. *Also L39 in List 1 and SSR36 in List 6.*

KT8, DUNSKEIG. Clach Leth Rathad. NR 7624 5704. T(A4/23). Jun 79. Two stones, 1·0 m and 0·4 m in height, about 6 m apart. Possibly the grounders of a modern field wall. *Also L41 in List 1.*

KT9, LOCH CIARAN. NR 7802 5479. K(150); N(NR75SE01). May 85. Standing stone 1·8 m high.

KT10, BALLOCHROY. NR 7309 5241. K(57); N(NR75SW03); BR(222, 254); T(A4/4). May 81. A 5 m-long row of three stones, some 3·5 m, 3·0 m and 2·0 m in height, the last of which appears to have been broken off at the top. Some 35 m away, in the alignment, is a cist that was once covered by a cairn. The row once contained a further cairn and standing stone. Plans: Thom 1966, fig. 9; 1971, 37; Thom, Thom and Burl 1990, 132; see also Fig. 1.5 in this volume. *Also SSR37 in List 6.*

KT12, TARBERT (Gigha). Carragh an Tarbert. NR 6555 5227. K(136); N(NR65SE22); T(A4/17). May 85. Standing stone 2·3 m high, leaning by about 20°. *Also L42 in List 1.*

KT14, RHUNAHAORINE. NR 7141 4893. N(NR74NW01). May 81. Erect stone 1·1 m high of uncertain status.

KT15, BEACHARR. Beacharra. NR 6926 4330. K(3, 134); N(NR64SE02); H(ARG27); T(A4/5). May 85. Clyde-group chambered long tomb; standing stone 5·0 m high, 30 m to the south. *Also L43 in List 1.*

KT19, SOUTH MUASDALE. Carragh Muasdale. NR 6792 3914. K(153); N(NR63NE20); BR(222, 264); T(A4/6). May 81. Standing stone some 3.0 m high. Some 12 m to its WSW is a 1·1 m-high stone that is possibly the stump of a companion.

KT21, BARLEA. Glenbarr. NR 6616 3707. K(132); N(NR63NE18); T(A4/7). May 85. Standing stone 1·8 m high.

KT23, BEINN AN TUIRC. Arnicle; Crois Mhic Aoida. NR 7349 3506. K(130); N(NR73SW07); T(A4/9). May 81. Standing stone 1·8 m high, now leaning by about 20°. *Also L44 in List 1.*

KT24, TIGHNAMOILE. NR 6641 2898. N(NR62NE03). May 81. Erect stone 0·8 m high of uncertain status.

KT25, DRUMALEA. NR 6609 2756. K(142); N(NR62NE19). Jun 79. Fallen stone 1·8 m long, which was standing in 1900.

KT27, CLOCHKEIL. NR 6577 2445. K(137); N(NR62SE15); BR(222, 254). Jun 79. 5·5 m-long row of three stones, one 1·9 m high, one 1·4 m high, and one fallen, some 2·0 m long. *Also SSR38 in List 6.*

KT28, SKEROBLINGARRY. Skeroblin Cruach; Pobull Burn. NR 7094 2701. K(152); N(NR72NW13); T(A4/11). May 85. Standing stone 1·5 m high.

KT29, HIGH PARK. NR 6950 2572. K(148); N(NR62NE18); T(A4/2). Jun 79. Standing stone 3·0 m high. *Also L45 in List 1.*

KT31, CRAIGS. NR 6902 2362. K(139); N(NR62SE05). May 85. Standing stone 2·5 m high.

KT32, GLENCRAIGS S. NR 6932 2354. K(144); N(NR62SE06); T(A4/13). May 85. Standing stone some 2·0 m high.

KT35, GLENLUSSA LODGE. Peninver. NR 7614 2541. K(146); N(NR72NE13); T(A4/10). May 81. Standing stone 2·3 m high, now set into a stone wall.

KT36, CAMPBELTOWN. Balegreggan. NR 7238 2123. K(131); N(NR72SW03); T(A4/14). Jun 79. Standing stone 4·0 m high. *Also L46 in List 1.*

KT37, STEWARTON. NR 6995 1982. K(147); N(NR61NE12). Jun 79. Standing stone which stood to a height of 1·8 m in 1930 but is now fallen and broken into two pieces.

KT39, MINGARY. NR 6533 1940. K(34); N(NR61NE07). Jun 79. Standing stone 1·4 m high, standing by the outer of two banks surrounding a cairn. Plan: RCAHMS 1971, fig. 23.

KT40, LOCHORODALE. NR 6657 1546. K(151); N(NR61NE10). May 85. Standing stone some 1·5 m high, now leaning by about 50° from the vertical, which may have been moved to its present position in modern times.

KT41, KNOCKSTAPPLE. NR 7026 1240. K(149); N(NR71SW10); T(A4/19). Jun 79. Standing stone 3·2 m high. *Also L47 in List 1.*

KT42, CULINLONGART. NR 6517 1192. K(140); N(NR61SE04); T(A4/16). Jun 79. Standing stone 2·1 m high.

KT44, SOUTHEND. Brunerican. NR 6976 0787. K(135); N(NR60NE06). Jun 79. Standing stone 2·7 m high.

# List 3

*Recumbent Stone Circles in North-Eastern Scotland*

This is a reference list of recumbent stone circles in north-east Scotland, a group of monuments whose possible astronomical significance was studied in the early 1980s by Burl and Ruggles (Burl 1980; Ruggles 1984c; Ruggles and Burl 1985). The reference numbers follow the last two of these papers, as does the status assignation of the monuments as certain, probable, possible and unlikely RSCs. All four of these categories are included in the list. Where monuments are still considered to be unlikely RSCs the site reference number is shown in square brackets. Further information, where available, is given in the notes.

Supplementary List A gives details of a related monument identified since the work that was published in 1985. Supplementary List B details four more monuments included in the NMRS at the time of writing as RSCs or possible RSCs.

The following information is given for each monument in the main list:

1. Reference number and name used in Ruggles 1984c.
2. Alternative names if any.
3. National Grid reference (NGR), generally quoted to 100 m.
4. Cross-references to the monument in other sources.
5. Status as RSC if not certain: prob[able], poss[ible] or unl[ikely] according to Ruggles and Burl 1985.
6. Destr[oyed or unrecognisable] according to Burl 1976.
7. Date of most recent visit by the author ('NV' indicates not visited).
8. Brief description:
   - Diameter or estimate of original diameter to nearest metre (maximum and minimum values are given for non-circular rings).
   - Condition of recumbent stone and flankers (included only where verified at first hand by the author):
     - East flanker: 'E' = standing (or leaning or earthfast stump), 'e' = fallen, '–' = moved or removed.
     - Recumbent stone: 'R' = standing (or leaning), 'r' = fallen, '–' = moved or removed.
     - West flanker: 'W' = standing (or leaning or earthfast stump), 'w' = fallen, '–' = moved or removed.
   - Number of other circle stones still standing, or present as earthfast stumps.

   For more complete information the reader is referred to the notes where present or to the cross-references cited.
9. Reference of published excavation report(s).
10. Reference of published site plan(s).

*Moray*

RSC1, Innesmill. Urquhart. NJ 289 640. B(5:26); N(NJ26SE07); BG(180); BC(Moray6[9]); T(B5/1). Prob RSC. Jul 81. Desc: diam 34 m; – – –; cir 5. Plan: Thom, Thom and Burl 1980, 240.

*Banff*

RSC2, St Brandan's Stanes. NJ 608 611. B(6:83); N(NJ66SW01); BC(Banff11[14]). Prob RSC. Destr. NV.

RSC18, Harestane. NJ 664 438. B(6:49); N(NJ64SE01); BC(Banff4[4]). Prob RSC. Destr. NV. Desc: diam 18 m?

RSC23, Rothiemay. Milltown. NJ 550 487. B(6:82); N(NJ54NE06); BG(134); BC(Banff10[13]); Q(16(A)); T(B4/4). Jun 81. Desc: diam 28 m; – R –; cir 4. Plan: Thom, Thom and Burl 1980, 238.

*Aberdeen*

RSC3, Clochforbie. Gray Stone. NJ 797 586. B(6:25); N(NJ75NE01); BC(Aberdeen29[28]). Aug 81. Desc: diam ?; – r –; cir 0.

RSC4, Cortes. NJ 99 59. B(6:k); N(NJ95NE05); BC(Aberdeen34[33]). Poss RSC. NV.

RSC5, Berrybrae. NK 028 572. B(6:13); N(NK05NW02); BG(95); BC(Aberdeen11[10]); Q(24(A)). Jun 81. Desc: diam 13 × 11 m; E R W; cir 2. Exc: Burl 1979a, 26–31.

RSC6, Netherton. NK 043 572. B(6:68); N(NK05NW03); BC(Aberdeen83[86]); Q(31(B)). Aug 81. Desc: diam 17 m; E R W; cir 7.

RSC7, Strichen. NJ 937 545. B(6:91); N(NJ95SW02); BG(116); BC(Aberdeen102[105]); T(B1/1). Aug 81. Desc:

diam 13 m; E R W; cir 4. Exc: Hampsher-Monk and Abramson 1982. Plan: Thom, Thom and Burl 1980, 156.

RSC8, Auchcorthie. NJ 925 523. B(6:107); Coles 1904, no. 17. Prob RSC. Destr. NV.

RSC9, Gaval. NJ 981 515. B(6:45); N(NJ95SE03); BC(Aberdeen53[53]). Aug 81. Desc: diam ?; – – –; cir 1.

RSC10, Auchmachar. NJ 948 502. B(6:5); N(NJ95SW11); BC(Aberdeen4[4]). Aug 81. Desc: diam 15 m?; E r w; cir 1.

RSC11, Loudon Wood. NJ 962 497. B(6:61); N(NJ94NE01); BG(108); BC(Aberdeen73[74]); Q(10(A)). Jun 81. Desc: diam 19 m; e R W; cir 2.

RSC12, Aikey Brae. NJ 959 471. B(6:1); N(NJ94NE04); BG(90); BC(Aberdeen1[1]); Q(8(A)). Jun 81. Desc: diam 17 × 13 m; E R w; cir 3.

RSC13, Hatton. NK 050 364. B(6:134); BC(Aberdeen57[58]). Poss RSC.[1] Destr. NV.

RSC14, Culsh. Hill of Culsh. NJ 881 483. B(6:123); N(NJ84NE02); BC(Aberdeen39[38]). Poss RSC.[2] Destr. NV. Desc: diam 9 m?

RSC15, Auchmaliddie. Muckle Stone. NJ 881 448. B(6:6); N(NJ84SE01); BC(Aberdeen5[5]). Aug 81. Desc: diam ?; – r w; cir 0.

RSC16, Cairn Riv. Carlin Stone. NJ 674 466. B(6:19); N(NJ64NE04); BG(93B); BC(Aberdeen21[20]). Prob RSC.[3] NV. Desc: diam 29 m?

RSC17, Corrydown. NJ 707 445. B(6:28); N(NJ74SW11); BC(Aberdeen33[32]); Q(39(C)). Aug 81. Desc: diam 23 m?; – R w; cir 0.

RSC19, Pitglassie. NJ 686 434. B(6:75); N(NJ64SE08); BC(Aberdeen88[91]); Q(37(C)). Aug 81. Desc: diam 18 m?; – R –; cir 0.

RSC20, Mains of Hatton. Charlesfield. NJ 699 425. B(6:62); N(NJ64SE06); BC(Aberdeen74[75]); Q(38(C)); T(B1/25). NV. Desc: diam 20 m. Plan: Thom, Thom and Burl 1980, 188.

RSC21, Rappla Burn. NJ 727 405. B(6:hh); N(NJ74SW07); BC(Aberdeen91[–]) ['Rapplabum']. Prob RSC. Destr. NV.

RSC22, Rappla Wood. Burreldales. NJ 736 402. B(6:81); N(NJ74SW06); BC(Aberdeen92[94]). Prob RSC. NV. Desc: diam 15 m?

RSC24, Arnhill. The Ringing Stone; Haddock Circle. NJ 532 455. B(6:3); N(NJ54NW12); BG(92); BC(Aberdeen3[3]); Q(49(C)). Jul 81. Desc: diam 18 m?; – R –; cir 0.

RSC25, Yonder Bognie. NJ 601 458. B(6:106); N(NJ64NW15); BG(122); BC(Aberdeen118[122]); Q(28(B)); T(B1/23). Aug 81. Desc: diam 24 m; e R W; cir 4. Plan: Thom, Thom and Burl 1980, 184.

RSC26, Cairnton. NJ 585 446. B(6:20); N(NJ54SE01); BC(Aberdeen22[22]); Q(43(C)). Jul 81. Desc: diam ?; – R W; cir 0.

RSC27, Frendraught. Auchaber. NJ 632 403. B(6:41); N(NJ64SW05); BC(Aberdeen50[50]). Destr. NV. Desc: diam 26 m?

RSC28, Upper Third. NJ 677 394. B(6:155); N(NJ63NE03); BC(Aberdeen109[114]). Poss RSC.[4] Destr. NV.

RSC29, Gingomyres. Hill of Milleath. NJ 467 429. B(6:46); N(NJ44SE09); BC(Aberdeen54[54]). Destr. NV. Desc: diam 18 m?

RSC30, Huntly. NJ 529 399. B(6:55); N(NJ53NW01); BC(Aberdeen63[64]). Poss RSC.[5] Jun 81. Desc: diam 14 m?; – – –; cir 2.

RSC31, Nether Dumeath. NJ 425 378. B(6:67); N(NJ43NW07); BC(Aberdeen82[85]). Prob RSC.[6] Destr. NV. Desc: diam 12 m?

RSC32, Upper Ord. NJ 483 270. B(6:99); N(NJ42NE06); BC(Aberdeen108[112]). Prob RSC. Aug 81. Desc: diam 23 m?; – – –; cir 2.

RSC33, Corrstone Wood. Mains of Druminnor. NJ 510 271. B(6:27); N(NJ52NW02); BC(Aberdeen32[31]); Q(47(C)); T(B1/21). Aug 81. Desc: diam ?; e r W; cir 0.

RSC34, Hillhead. Clatt, Hillhead. NJ 528 265. B(6:118); N(NJ52NW10); BC(Aberdeen28[27b]). Poss RSC.[7] Destr. NV. Desc: diam 23 m?

RSC35, Bankhead. Clatt, Bankhead. NJ 529 269. B(6:117); N(NJ52NW25); BC(Aberdeen27[27a]). Poss RSC.[7] Destr. NV.

RSC36, Holywell. Sunken Kirk. NJ 549 271. B(6:53); N(NJ52NW06); BC(Aberdeen61[62]). Prob RSC.[7] Destr. NV. Desc: diam 24 m?

RSC37, Ardlair. Holywell. NJ 552 279. B(6:2); N(NJ52NE04); BG(91); BC(Aberdeen2[2]); Q(2(A)); T(B1/18). Aug 81. Desc: diam 12 m; E R W; cir 1. Plan: Thom, Thom and Burl 1980, 182.

RSC38, Culsalmond. Colpy. NJ 641 326. B(6:122); N(NJ63SW01); BC(Aberdeen38[37]). Prob RSC. Destr. NV.

RSC39, Candle Hill. NJ 599 299. B(6:23); N(NJ52NE10); BC(Aberdeen23[23]); Q(45(C)). Jun 81. Desc: diam 13 m; e r w; cir 1.

RSC40, Inschfield. NJ 623 293. B(6:57); N(NJ62NW06); BC(Aberdeen66[67]); Q(48(C)); T(B1/14). Jun 81. Desc: diam 27 m; E r –; cir 0. Plan: Thom, Thom and Burl 1980, 178.

RSC41, Stonehead. Whitebrow. NJ 601 287. B(6:89); N(NJ62NW05); BC(Aberdeen100[103]); Q(50(C)). Jun 81. Desc: diam ?; E R W; cir 0.

RSC42, Dunnideer. NJ 608 284. B(6:35); N(NJ62NW04); BC(Aberdeen45[44]). Aug 81. Desc: diam ?; – R –; cir 0.

RSC43, Wantonwells. NJ 619 273. B(6:100); N(NJ62NW02); BC(Aberdeen110[115]); Q(20(A)); T(B1/12). Aug 81. Desc: diam ?; e R –; cir 0. Plan: Thom, Thom and Burl 1980, 174.

RSC44, Braehead. NJ 592 255. B(6:15); N(NJ52NE06); BC(Aberdeen15[14]); Q(34(B)). Prob RSC. Aug 81. Desc: diam ?; – R –; cir 0.

RSC45, Broomend. Husband Hillock; Auchleven. NJ 621 245. B(6:d); N(NJ62SW02); BC(Aberdeen17[16]). Poss RSC.[8] Destr. NV.

RSC46, Loanend. NJ 604 242. B(6:60); N(NJ62SW01); BC(Aberdeen69[70]); Q(35(B)). Aug 81. Desc: diam ?; – R –; cir 1.

RSC47, DRUIDSTONE. NJ 615 222. B(6:34); N(NJ62SW04); BC(Aberdeen43[42]); Q(46(C)). Aug 81. Desc: diam 17 m?; e – w; cir 3.

RSC48, COTHIEMUIR WOOD. NJ 617 198. B(6:29); N(NJ61NW01); BG(98); BC(Aberdeen35[34]); Q(18(A)). Jul 81. Desc: diam 21 × 19 m?; E R W; cir 4.

RSC49, OLD KEIG. NJ 596 193. B(6:73); N(NJ51NE02); BG(111); BC(Aberdeen86[89]); Q(21(A)). Jun 81. Desc: diam 26 m?; E R W; cir 1. Exc: Childe 1933.

RSC50, CROOKMORE. NJ 586 187. B(6:121); N(NJ51NE16); BC(Aberdeen36[35]). Poss RSC.[9] Destr. NV.

RSC51, CORRIE CAIRN. NJ 552 205. B(6:i); N(NJ52SE13); BC(Aberdeen31[30]). Poss RSC. NV. Desc: diam 19 m?

RSC52, DRUIDSFIELD. Druidstone. NJ 578 177. B(6:33); N(NJ51NE01); BC(Aberdeen42[41]); Q(36(C)). Jun 81. Desc: diam 15 m?; E – W; cir 0.

RSC53, NETHER BALFOUR. NJ 539 172. B(6:146); N (NJ51NW05); BC(Aberdeen79[82]). Prob RSC.[9] Destr. NV.

RSC54, NORTH STRONE. NJ 584 138. B(6:71); N(NJ51SE02); BG(110); BC(Aberdeen85[88]); Q(29(B)). Jun 81. Desc: diam 20 × 19 m?; E r w; cir 3.

RSC55, OLD RAYNE. NJ 679 280. B(6:74); N(NJ62NE01); BG(112); BC(Aberdeen87[90]); Q(15(A)); T(B1/13). NV. Desc: diam 26 m? Plan: Thom, Thom and Burl 1980, 176.

RSC56, HATTON OF ARDOYNE. NJ 659 268. B(6:50); N(NJ62NE07); BG(103); BC(Aberdeen58[59]); Q(25(A)). Jun 81. Desc: diam 24 m; E R –; cir 3.

RSC57, MILL OF CARDEN. NJ 693 260. B(6:143); N(NJ62NE04); BC(Aberdeen77[79]). Poss RSC.[10] Aug 81. Desc: diam ?; – – –; cir 0.

RSC58, NEW CRAIG. NJ 746 297. B(6:69); N(NJ72NW03); BC(Aberdeen84[87]); Q(12(A)). Jun 81. Desc: diam ?; E R W; cir 0.

RSC59, LOANHEAD OF DAVIOT. NJ 747 288. B(6:59); N(NJ72NW01); BG(106); BC(Aberdeen70[71a]); Q(14(A)); T(B1/26). Jul 81. Desc: diam 21 m; E R W; cir 8. Exc: Kilbride-Jones 1935. Plan: Thom, Thom and Burl 1980, 190.

RSC60, SHELDON OF BOURTIE. NJ 823 249. B(6:84); N(NJ 82SW01); BC(Aberdeen94[96]); T(B1/8). Prob RSC. NV. Desc: diam 33 m. Plan: Thom, Thom and Burl 1980, 166.

RSC61, KIRKTON OF BOURTIE. NJ 801 249. B(6:58); N(NJ82SW02); BG(105); BC(Aberdeen67[68]); Q(11(A)); T(B1/7). Jul 81. Desc: diam 22 m?; E R –; cir 2. Plan: Thom, Thom and Burl 1980, 164.

RSC62, BALQUHAIN. NJ 735 241. B(6:11); N(NJ72SW02); BG(94); BC(Aberdeen10[9]); Q(9(A)); T(B1/11). NV. Desc: diam 20 m? Plan: Thom, Thom and Burl 1980, 172.

RSC63, EASTER AQUORTHIES. NJ 732 208. B(6:37); N(NJ72SW12); BG(102); BC(Aberdeen47[46]); Q(13(A)); T(B1/6). Jul 81. Desc: diam 20 m; E R W; cir 9. Plan: Thom, Thom and Burl 1980, 162.

RSC64, CHAPEL O'SINK. Westerton. NJ 706 189. B(6:h); N(NJ71NW04); BC(Aberdeen26[26]); T(B1/16). Prob RSC. Jun 81. Desc: diam ?;[11] – – –; cir 0. Plan: Thom, Thom and Burl 1980, 180.

RSC65, NETHER COULLIE. NJ 710 156. B(6:66); N(NJ71NW11); BC(Aberdeen81[84]). Prob RSC.[12] Jun 81. Desc: diam ?; – – –; cir 1.

RSC66, CASTLE FRASER. NJ 715 125. B(6:24); N(NJ71SW03); BG(97); BC(Aberdeen25[25]); Q(19(A)); T(B2/3). Aug 81. Desc: diam 20 m; E R W; cir 4. Plan: Thom, Thom and Burl 1980, 198.

RSC67, SOUTH LEY LODGE. Leylodge. NJ 767 132. B(6:87); N(NJ71SE03); BC(Aberdeen98[101]); Q(44(C)); T(B2/14). Jun 81. Desc: diam 30 m?; E R W; cir 0. Plan: Thom, Thom and Burl 1980, 218.

RSC68, SOUTH FORNET. NJ 782 109. B(6:86); N(NJ71SE01); BC(Aberdeen97[100]); Q(42(C)). Prob RSC.[13] Jun 81. Desc: diam 27 m; E – W; cir 0.

RSC69, NETHER CORSKIE. NJ 748 096. B(6:147); N(NJ70NW03); BC(Aberdeen80[83]). Poss RSC.[14] Jun 81. Desc: diam ?; E – W; cir 0.

RSC70, WESTER ECHT. NJ 739 083. B(6:101); N(NJ70NW02); BC(Aberdeen112[118]). Poss RSC.[15] Jun 81. Desc: diam 29 m?; – – –; cir 3.

RSC71, MIDMAR KIRK. NJ 699 064. B(6:64); N(NJ60NE03); BG(109); BC(Aberdeen76[78a]); Q(22(A)); T(B2/17). Jun 81. Desc: diam 17 m; E R W; cir 5. Plan: Thom, Thom and Burl 1980, 222.

RSC72, SUNHONEY. NJ 716 057. B(6:92); N(NJ70NW55); BG(117); BC(Aberdeen103[106]); Q(23(A)); T(B2/2). Jun 81. Desc: diam 25 m; E r W; cir 9. Plan: Thom, Thom and Burl 1980, 196.

RSC73, WHITEHILL WOOD. Tillyfourie Hill. NJ 643 135. B(6:103); N(NJ61SW03);[16] BG(120); BC(Aberdeen115[121]); Q(17(A)); T(B2/18). Jul 81. Desc: diam 22 m?; e R W; cir 2. Plan: Thom, Thom and Burl 1980, 224.

RSC74, AULD KIRK O'TOUGH. NJ 625 092. B(6:4); N(NJ60NW01); BC(Aberdeen6[6]); Q(27(B)). Aug 81. Desc: diam 25 m?;[17] – – –; cir 0.

RSC75, TOMNAGORN. NJ 651 077. B(6:95); N(NJ60NE01); BG(118); BC(Aberdeen104[108]); Q(32(B)); T(B2/16). Jun 81. Desc: diam 22 m; E R W; cir 4. Plan: Thom, Thom and Burl 1980, 220.

RSC76, BALNACRAIG. NJ 603 035. B(6:10); N(NJ60SW05); BC(Aberdeen9[8]). Aug 81. Desc: diam 14 m; – R –; cir 2.

RSC77, TOMNAVERIE. NJ 486 034. B(6:96); N(NJ40SE01); BG(119); BC(Aberdeen105[109]); Q(26(A)); T(B2/9). Jun 81. Desc: diam 17 m; e r w; cir 4. Plan: Thom, Thom and Burl 1980, 210.

RSC78, BLUE CAIRN. NJ 411 063. B(6:c); N(NJ40NW04); BC(Aberdeen13[12]); Q(33(B)). Unl RSC.[18] Aug 81. Desc: diam 23 m; – R –; cir 0.

[RSC79, FORVIE]. Sands of Forvie. NK 011 263. B(6:r); N(NK02NW03); BC(Aberdeen49[49]); T(B1/27). Unl RSC.[19] NV. Desc: diam 20 m. Exc: W. Kirk, 'Prehistoric sites at the sands of Forvie, Aberdeenshire: a preliminary examination', *Aberdeen University Review*, 35 (1953), 150–71.

RSC80, HILL OF FIDDES. NJ 935 243. B(6:52); N(NJ92SW01); BC(Aberdeen60[61]). Aug 81. Desc: diam 14 m?; – R W; cir 0.

RSC81, Potterton. NJ 952 163. B(6:76); N(NJ91NE07); BC(Aberdeen89[92]). Aug 81. Desc: diam ?; e R w; cir 0.

RSC82, Mundurno. NJ 940 131. B(6:145); N(NJ91SW05); BC(Aberdeen78[81]). Poss RSC.[20] Aug 81. Desc: diam ?; – – –; cir 1.

RSC83, Dyce. Tyrebagger. NJ 860 132. B(6:36); N(NJ81SE11); BG(101); BC(Aberdeen46[45]); Q(5(A)); T(B2/1). Jun 81. Desc: diam 18 m; E R W; cir 8. Plan: Thom, Thom and Burl 1980, 194.

RSC84, Binghill. NJ 855 023. B(6:14); N(NJ80SE16); BC(Aberdeen12[11]); Q(41(C)). Aug 81. Desc: diam 10 m; – R w; cir 2.

*Kincardine*

RSC85, Old Bourtreebush. Aquhorthies N. NO 903 961. B(6:72); N(NO99NW02); BG(173); BC(Kincardine13[16]); T(B3/2). Jun 81. Desc: diam 31 × 23 m?; – – –; cir 4. Plan: Thom, Thom and Burl 1980, 228.

RSC86, Aquhorthies. Aquhorthies S. NO 901 963. B(6:7); N(NO99NW01); BG(165); BC(Kincardine1[2]); Q(3(A)); T(B3/1). Jun 81. Desc: diam 23 m; – R W; cir 5. Plan: Thom, Thom and Burl 1980, 226.

RSC87, Cairnfauld. NO 754 941. B(6:18); N(NO79SE01); BG(166); BC(Kincardine2[3]); T(B2/11). Prob RSC. Aug 81. Desc: diam 23 m; – – –; cir 4.

[RSC88, Tilquhillie]. NO 722 941. B(6:94); N(NO79SW10); BC(Kincardine16[19]). Unl RSC.[21] NV.

RSC89, Esslie the Lesser. Esslie NE. NO 722 921. B(6:40); N(NO79SW01); BG(170); BC(Kincardine9[12]); Q(6(A)); T(B2/5). Prob RSC. Aug 81. Desc: diam 13 m; – r W; cir 4. Plan: Thom, Thom and Burl 1980, 202.

RSC90, Esslie the Greater. Esslie SW. NO 717 916. B(6:39); N(NO79SW02); BG(169); BC(Kincardine8[11]); Q(4(A)); T(B2/4). Aug 81. Desc: diam 25 × 23 m; E R W; cir 5. Plan: Thom, Thom and Burl 1980, 200. *Also L18 in List 1.*

RSC91, Garrol Wood. NO 723 912. B(6:43); N(NO79SW08); BG(171); BC(Kincardine10[13]); Q(1(A)); T(B2/6). Aug 81. Desc: diam 18 × 15 m; e r W; cir 7. Plan: Thom, Thom and Burl 1980, 204.

RSC92, Raes of Clune. Clune Wood. NO 795 950. B(6:79); N(NO79SE02); BC(Kincardine15[18]); Q(7(A)); T(B3/7). Aug 81. Desc: diam 17 m; E R W; cir 3. Plan: Thom, Thom and Burl 1980, 232.

[RSC93, Cotbank of Barras]. Mitton Hill. NO 827 791. B(6:l); N(NO87NW01); BC(Kincardine5[7]). Unl RSC.[22] Destr. NV. Desc: diam 18 m?

RSC94, The Camp, Montgoldrum. NO 817 772. B(6:22); N(NO87NW05); BC(Kincardine4[5]); Q(40(C)). Prob RSC.[23] Aug 98. Desc: diam 24 m; – r –; cir 0. Plan: Barclay and Ruggles, forthcoming.

RSC95, Millplough. NO 819 754. B(6:65); N(NO87NW06); BC(Kincardine12[15]). Prob RSC. Destr.[24] Aug 98.

*Angus*

RSC96, Colmeallie. NO 565 781. B(6:26); N(NO57NE03); BC(Angus4[7]); Q(30(B)). Aug 98. Desc: diam 15 m?; E r w; cir 5. Plan: Barclay and Ruggles, forthcoming.[25]

RSC97, Newbigging. NO 541 693. B(6:148); N(NO56NW03); BC(Angus9[12]). Prob RSC. Destr. NV. Desc: diam 17 m?

*Perth*

[RSC98, Coilleachur]. Urlar. NN 849 464. B(7:15); N(NN84NW24); BC(Perth11[14]). Unl RSC.[26] NV. Desc: diam 49 m?

[RSC99, Fortingall South]. Fortingall Church. NN 746 469. B(7:30); N(NN74NW03); BG(202c); BC(Perth24[27c]); T(P1/6). See also Burl 1988b, 175. Unl RSC.[27] NV. Desc: diam 23 m?. Exc: Derek Simpson, 1970, unpublished, but see Burl 1988b, 168–75. Plan: Thom, Thom and Burl 1980, 336; Burl 1988b, 174.

## SUPPLEMENTARY LIST A: related monuments

*Kincardine*

The Cloch. NO 7817 6788. N(NO76NE01); BC ([Kincardine6]). Aug 98. A large recumbent stone adjacent to a ring cairn. In the vicinity of the recumbent are a number of smaller stones, possibly kerbstones. See Barclay and Ruggles forthcoming.

## SUPPLEMENTARY LIST B: additional monuments identified as RSCs and possible RSCs in the NMRS

*Banff*

Ley. Gaul Cross N. NJ 535 639. B(6:130); N(NJ56SW11); BC(Banff3a[3a]). A single fallen stone, itself destroyed between 1967 and 1991, was reported in 1905 as representing the remains of a stone circle. This was possibly an RSC.

Bellman's Wood. NJ 605 504. B(6:12); N(NJ65SW04); BC(Banff1[1]). Two fallen stones. They have been interpreted as flankers but Barnatt (1989, 272) suggests the monument is a ruined four-poster.

*Aberdeen*

Leslie Parish. Leslie. NJ 59 24. B(6:141); N(NJ52SE24); BC(Aberdeen68[69]). Stone circle, possibly an RSC. Completely destroyed.

*Perth*

Moncreiffe House. NO 133 193, moved to 136 193. B(7:41); N(NO11NW11); BC(Perth32[40]); T(P1/20). Exc and plans: Margaret E. C. Stewart, 'The excavation of a henge, stone circles and metal working area at Montcrieffe, Perthshire', *PSAS*, 115 (1985), 125–50. Pre-excavation plan: Thom, Thom and Burl 1980, 350. Moved and re-erected prior to the construction of the A90 road. Despite being suggested as a possible RSC by its excavator (*ibid.*, 135–7), the excavated evidence provides little justification for this.

# List 4

*Axial Stone Circles of More than Five Stones in Counties Cork and Kerry*

This is a reference list of axial stone circles of more than five stones, often referred to as 'multiple-stone circles', in Counties Cork and Kerry. The possible astronomical significance of this group of monuments has been studied recently (for a preliminary report see Ruggles and Prendergast 1996; a full report is in preparation).

The monument reference numbers are ordered from grid north to grid south. In naming ASCs, the convention is used that directions, when spelt in full, e.g. Derreenataggart West, are part of townland names: abbreviated directions, e.g. Knocks S, are used to distinguish different ASCs within the same townland.

The list includes probable and possible ASCs as well as certain ones. Some recently documented examples of ASCs that were not accommodated in the original numbering scheme are given reference numbers such as ASC21a. Monuments considered unlikely to be ASCs although listed as such elsewhere have been relegated to a supplementary list and are not considered further.

The following information is given for each monument in the main list:

1. Reference number and name.
2. Alternative names if any.
3. National Grid reference (NGR), generally obtained from Ó Nualláin 1984a and quoted to 100 m. Where more precise grid references are available, as in Power 1992, they are quoted in preference.
4. Cross-references to the monument in other sources.
5. Status as ASC if not certain: prob[able], or poss[ible].
6. Destr[oyed] or unrecognisable, or reconst[ructed].
7. Date of most recent visit by the author ('NV' indicates not visited).
8. Brief description, verified at first hand by the author where visited, otherwise taken or estimated from published information:
   - Diameter or estimate of original diameter (through the stone centres rather than the internal or external edges) to the nearest 0·5 m. Maximum and minimum values are given for non-circular rings.
   - Estimate of original number of stones in circle.
   - Condition of axial stone and portals (included only where verified at first hand by the author):
     - Left portal (as viewed from outside the circle in the portal-axial direction): 'L' = standing (or leaning or earthfast stump), 'l' = fallen, '–' = moved or removed.
     - Axial stone: 'A' = standing (or leaning), 'a' = fallen, '–' = moved or removed.
     - Right portal: 'R' = standing (or leaning or earthfast stump), 'r' = fallen, '–' = moved or removed.
   - Number of other circle stones still standing, or present as earthfast stumps ['cir'].
   - Alternatively (where the identification of the axial and/or portal stones is considered uncertain) total number of stones remaining standing, or present as earthfast stumps ['rem'].

   For more complete information the reader is referred to the cross-references cited.
9. Reference of published excavation report(s).
10. Reference of published site plan(s).

*Kerry*

ASC1, Lissyviggeen. V 998 906. OC(38); BG(350); BC(Kerry12[13]); R(100). Apr 94. Desc: diam 3·5 m; st 7; L A R; cir 4. Plan: Ó Nualláin 1984a, fig. 10.

ASC13, Killowen. V 924 716. OC(40); BC(Kerry10[11]); R(98). Prob ASC. Destr. NV. Plan: Ó Nualláin 1984a, fig. 12.

ASC14, Kenmare. V 906 707. OC(41); BG(349); BC(Kerry9[10]); R(94). Poss ASC.[1] Apr 94. Desc: diam 17·0 × 15·5 m; st 15; L A R; cir 10. Plan: Ó Nualláin 1984a, fig. 6.

ASC15, Gurteen. W 006 698. OC(43); BG(348); BC(Kerry7[9]); R(96). Apr 94. Desc: diam 10·5 m; st 11; L A R; cir 7. Plan: Ó Nualláin 1984a, fig. 7.

ASC16, Doughill. V 962 696. OC(42); BC([Kerry2]); R(97). Destr. NV. Desc: diam *c.* 9·0 m; st 17? Plan: Ó Nualláin 1984a, fig. 13.

ASC18, Dromod. V 544 692. SK(381); OC(39); BC(Kerry4[6]); R(99). Poss ASC. NV. Desc: diam *c.* 4·0 m; st 7?; rem 3. Plans: Ó Nualláin 1984a, fig. 8; O'Sullivan and Sheehan 1996, fig. 78.

ASC19, Lackaroe. V 958 687. Paul Walsh, priv. comm., 1997; BC([Kerry12]). NV. Desc: diam 8·0 m; st 13;[2] – – R; cir 8.

ASC20, Lohart. Tuosist. V 823 663. OC(44); BG(351); BC(Kerry16[14]); R(93). Poss ASC. Reconst. NV. Desc: diam 11·0 m; st ≥12.

ASC21, Dromroe. V 881 657. OC(45); BG(347); BC(Kerry5[7]); R(92). Apr 94. Desc: diam 10·0 m; st 13; l A R; cir 8. Plan: Ó Nualláin 1984a, fig. 7.

ASC21a, Dromagorteen. V 958 653. Paul Walsh, priv. comm., 1997; BC([Kerry3]). NV. Desc: diam *c.* 10·0 m; st 13; rem 6.

ASC24a, Uragh. V 825 630. NV. E. Twohig, 'Two stone circles at Uragh, Co. Kerry', *JKAHS*, 20(1987), 111–18, pp. 114–16; BC([Kerry19a]). Desc: diam *c.* 10 × 8 m; st 11; L A R; cir 5. Plan: Twohig, *ibid.*, fig. 3.

ASC28, Drombohilly Upper. V 790 607. OC(48); BG(346); BC(Kerry3[5]); R(90). Apr 94. Desc: diam 9·5 × 8·5 m; st 11?; L – R; cir 7. Plan: Ó Nualláin 1984a, fig. 2.

ASC30, Cashelkeelty. V 747 575. OC(47); BG(344); BC([Kerry1b]); R(88). NV. Desc: diam 17·0 m; st 11/13; rem 2. Exc: Lynch 1981, 64–9, 76–97.

ASC33, Shronebirrane. Drumminboy. V 754 554. OC(46) [NGR given erroneously as V 735 554]; BG(353); BC(Kerry6[18]); R(89). Apr 94. Desc: diam 7·5 m; st 13?; L A –; cir 6. Plan: Ó Nualláin 1984a, fig. 2.

### *Cork*

ASC2, Lissard. Reim na Gaoithe. W 582 903. MC(6448); OC(1); BC(Cork66[74]); R(3). Poss ASC. Destr. NV. Destroyed between 1963 and 1970.

ASC3, Gowlane North. W 484 856. MC(6444); OC(5); BC(Cork39[53]); R(14). Apr 94. Desc: diam *c.* 7·0 m; st 9; L A R; cir 3. Plan: Ó Nualláin 1984a, fig. 3.

ASC4, Glantane East N. W 282 840. MC(6441); OC(2); BG(319, part); BC(Cork37[43a]); R(4). Prob ASC. NV. Desc: diam 5·0 m; st 13?; – – –; cir 6. Plan: Ó Nualláin 1984a, fig. 9.

ASC5, Carrigagulla. Carrigagulla SW. W 371 834. MC(6435); OC(4); BG(310); BC(Cork15[16b]) ['Carrigagulla B'] [NGR given erroneously as W 389 822]; R(11). Apr 94. Desc: diam 8·5 m; st 17; L A R; cir 12. Plan: Ó Nualláin 1984a, fig. 4.

ASC6, Glantane East S. W 289 834. MC(6440); OC(3); BG(319, part); BC([Cork43b]); R(5). Apr 94. Desc: diam 5·5 m; st 11; L A R; cir 7. Plan: Ó Nualláin 1984a, fig. 1.

ASC7, Kilmartin Lower. W 451 824. MC(6446); OC(6); BC([Cork60]); R(15). Apr 94. Desc: diam 3·5 m; st 7; L A R; cir 2. Plan: Ó Nualláin 1984a, fig. 9.

ASC8, Oughtihery. Keel Cross. W 415 801. MC(6449); OC(8); BG(334) ['Oughtihery SE']; BC(Cork42[78a]); R(20). Prob ASC. Apr 94. Desc: diam 3·0 m; st 7; – A R; cir 3. Plan: Ó Nualláin 1984a, fig. 8.

ASC9, Kilboultragh. W 320 757. MC(6445); OC(7), BC(Cork43[59]); R(23). Destr. NV. Desc: diam *c.* 9·0 m; st 9? Plans: Ó Nualláin 1984a, figs 8, 13.

ASC10, Gortanacra. W 204 756. MC(6442); OC(9); BC([Cork47]); R(27). Apr 94. Desc: diam 8·5 m; st 13; L A R; cir 5. Plan: Ó Nualláin 1984a, fig. 1 [stone immediately to west of axial stone actually stands].

ASC11, Gortanimill. W 208 741. MC(6443); OC(10); BG(320); BC(Cork38[48]); R(29). Apr 94. Desc: diam 7·5 m; st 9; L A R; cir 6. Plan: Ó Nualláin 1984a, fig. 4.

ASC12, Carrigaphooca. W 295 734. MC(6436); OC(11); BC(Cork16[19]); R(31). NV. Desc: diam ≥5·5 m; st ?; rem 3. Plans: Ó Nualláin 1984a, figs 4, 14.

ASC17, Teergay. W 291 694. MC(6450); OC(12); BG(338); BC(Cork71[84]); R(33). Apr 94. Desc: st 9; L A R; cir 4. Plan: Ó Nualláin 1984a, fig. 3.

ASC22, Currabeha N. Shandagan. W 410 642. MC(6438); OC(14); BC(Cork70[31a]); R(38). NV. Desc: diam *c.* 8·5 m; st 13?; rem 5. Plan: Ó Nualláin 1984a, fig. 2.

ASC23, Currabeha S. W 411 640. MC(6439); OC(15); BC(Cork25[31b]); R(39). Apr 94. Desc: diam 9·0 m; st 13; L A R; cir 7. Plan: Ó Nualláin 1984a, fig. 4.

ASC24, Coolaclevane. W 287 639. MC(6437); OC(13); BC(Cork20[24]) ['Coolclevane']; R(37). Apr 94. Desc: diam 8·0 m; st 9; L A R; cir 2. Plans: Ó Nualláin 1984a, figs 3, 14.

ASC25, Knocknaneirk. W 370 626. MC(6447); OC(19); BC(Cork46[66b]) ['Knocknaneirk A']; R(41). Apr 94. Desc: diam 10·5 m; st 9; L A R; cir 4. Plan: Ó Nualláin 1984a, fig. 1.

ASC26, Derrynafinchin. W 0476 6219. WC(51); OC(16); BC(Cork27[33]); R(77). Apr 94. Desc: diam *c.* 8·5 m; st 11?; – A –; cir 3. Plan: Ó Nualláin 1984a, fig. 5.

ASC27, Coolmountain. W 1919 6083. WC(48); OC(17); BC(Cork21[25]). NV. Desc: diam 9·0 m?; st 11?; rem 2. Plans: Ó Nualláin 1984a, figs 8, 13.

ASC29, Gortroe. W 2582 6055. WC(56); OC(18); BG(322); BC(Cork34[52]) ['Gertroe']; R(74). Poss ASC. NV. Desc: diam 8·5 m; st ?; rem 8. Plan: Ó Nualláin 1984a, fig. 9.

ASC31, Ardgroom Outward NE. Ardgroom NE. W 7188 5646. WC(42); OC(20); BG(302B); BC([Cork3a]); R(86). Apr 94. Desc: diam ?; st ?; – A –; cir 2. Plan: Ó Nualláin 1984a, fig. 8.

ASC32, Maughanaclea. W 1047 5645. WC(59); OC(23); BG(329) ['Maughanaclea ENE']; BC(Cork59[75b]) ['Maughanaclea A']; R(71); J(5/36). Apr 94. Desc: diam 11·5 m; st 13; l A R; cir 5. Plan: Ó Nualláin 1984a, fig. 5.

ASC34, Ardgroom Outward SW. Ardgroom SW. W 7087 5534. WC(41); OC(21); BG(302A); BC(Cork3[3b]) ['Ardgroom']; R(85). Apr 94. Desc: diam 7·0 m; st 11; rem 9.[3] Plan: Ó Nualláin 1984a, fig. 9.

ASC35, Breeny More. W 0508 5526. WC(45); OC(24); BG(305); BC([Cork9]); R(68); J(5/27). Apr 94. Desc: diam *c.* 14·0 m; st ?; L A R; cir 0. Plan: Ó Nualláin 1984a, fig. 6.

ASC36, Dromkeal. W 0031 5445. WC(53); BC([Cork39]). NV. Desc: 9·0 m; st 13?; rem 5.

ASC37, Cappanaboul. W 0340 5324. WC(46); OC(22); BC(Cork9[14]) ['Brinny More']; R(67); J(5/25). Apr 94. Desc: diam 10·5 × 9·5 m; st 13; L A –; cir 6. Plan: Ó Nualláin 1984a, fig. 5.

ASC37a, KEALAGOWLANE. V 876 523. Paul Walsh, priv. comm., 1998. NV. Desc: diam *c.* 8·0 m; st ?; rem 4.

ASC38, COULAGH. V 6322 4926. WC(49); OC(25); BC([Cork26]); R(84). NV. Desc: diam 8·5 m?; st ?; rem 2. Plan: Ó Nualláin 1984a, fig. 2.

ASC39, CURRADUFF. V 7414 4837. WC(211). Destr. NV.[4]

ASC40, DERREENATAGGART WEST. V 6653 4628. WC(50); OC(26); BG(316); BC(Cork26[32]); R(83). Apr 94. Desc: diam 8·0 m; st 15?; L A R; cir 6. Plan: Ó Nualláin 1984a, fig. 2.

ASC41, KNOCKS N. W 2994 4564. WC(57); OC(28); BC(Cork53[68a]) ['Knocks B']; R(56; 57); J(6/16, part). NV. Desc: diam *c.* 10·0 × 9·0 m; st 11; L – R; cir 6.[5] Plan: Ó Nualláin 1984a, fig. 3.

ASC42, GARRYGLASS. Durraghalicky. W 2228 4522. WC(55); OC(27); BC(Cork32[41]); R(59); J(6/5). Destr. NV. Desc: diam ?; st 11; rem 1.

ASC43, KNOCKS S. W 3022 4429. WC(58); OC(29); BC(Cork52[68b]) ['Knocks A']; R(58); J(6/16, part). Apr 94. Desc: diam *c.* 8·5 m; st 9?; L A –; cir 3. Plan: Ó Nualláin 1984a, fig. 8.

ASC44, MAULATANVALLY. W 2633 4421. WC(60); OC(30); BG(331); BC(Cork61[76]); R(53); J(6/11). Apr 94. Desc: diam *c.* 9·5 m; st 11; L A –; cir 5. Plan: Ó Nualláin 1984a, fig. 4.

ASC45, TEMPLEBRYAN NORTH. W 3890 4371. WC(62); OC(31); BG(339); BC(Cork72[85]); R(44); J(1/16). Apr 94. Desc: diam *c.* 9·5 m; st 9; rem 4. Plan: Ó Nualláin 1984a, fig. 4.

ASC45a, AHAGILLA. W 333 435. WC(3525); BC([Cork1]); R(46) ['Bealad']. NV. Desc: diam ?; st ?; L A –; cir 1.

ASC46, CARRIGAGRENANE. W 2541 4322. WC(47); OC(33); BG(308); BC(Cork12[16b]); R(51); J(6/12). Apr 94. Desc: diam 8·5 m; st 19?; L A R; cir 9. Plan: Ó Nualláin 1984a, fig. 3.

ASC47, BALLYVACKEY. W 3438 4267. WC(43); OC(35); BC(Cork4[4]) ['Ballvackey']; R(45); J(1/19). Apr 94. Desc: diam 8·5 m; st 9?; L A –; cir 5. Plan: Ó Nualláin 1984a, fig. 3.

ASC48, REANASCREENA SOUTH. Reenascreena South. W 2639 4106. WC(61); OC(34); BG(335); BC(Cork65[81]); R(49); J(6/13). Apr 94. Desc: diam 10·0 × 9·5 m; st 13; L A R; cir 10. Exc: E. M. Fahy, 'A recumbent stone circle at Reenascreena South, Co. Cork', *JCHAS* 67 (1962), 59–69. Plan: Ó Nualláin 1984a, fig. 11.

ASC49, GORTEANISH. V 860 396. WC(215); BC([Cork49]). NV. Desc: diam *c.* 8·0 m; st 11?; rem 4.

ASC50, DUNBEACON. V 9271 3921. WC(54); OC(32); BG(318); BC(Cork31[40]); R(80); J(4/50). NV. Desc: diam 8·5 m; st 11; rem 6. Plan: Ó Nualláin 1984a, fig. 1.

ASC51, BOHONAGH. W 3073 3686. WC(44); OC(36); BG(304); BC(Cork8[8]); R(47); J(1/33). Sep 98. Desc: diam 8·5 m; st 13; L A R; cir 3. Exc: E. M. Fahy, 'A stone circle, hut and dolmen at Bohonagh, Co. Cork', *JCHAS* 66 (1961), 93–104. Plan: Ó Nualláin 1984a, fig. 1.

ASC52, DROMBEG. W 2465 3516. WC(52); OC(37); BG(317); BC(Cork30[36]); R(48); J(2/5). Sep 98. Desc: diam 9·5 m; st 17; L A R; cir 11. Exc: E. M. Fahy, 'A recumbent stone circle at Drombeg, Co. Cork', *JCHAS* 64 (1959), 1–27. Plan: Ó Nualláin 1984a, fig. 1.

## SUPPLEMENTARY LIST:

monuments considered unlikely ASCs, although listed as 'recumbent stone circles' (or possible recumbent stone circles) of more than 5 stones in Burl 1976

### *Cork*

DERRYLAHAN. Knockeenadara. R 816 087. BC(Cork51[–]). This is a form of cairn or kerb-circle.

KNOCKAUNAVADDREEN. R 512 048. OC(appendix, Cork 3); BC(Cork45[63]) ['Knockane']. Probably the remains of a kerb-circle (i.e. a kerb of contiguous uprights).

CUPPAGE. R 777 041. OC(appendix, Cork 4); BC(Cork23[30]) ['Cuppoge']. Probably the remains of a kerb-circle.

CLONLEIGH. W 67 49. BC(Cork19[–]). At the NGR given by Burl there is a boulder-burial and adjacent standing stone.

# List 5

*Short Stone Rows in Counties Cork and Kerry*

This is a reference list of short stone rows of three to six stones in Counties Cork and Kerry, a group of monuments whose possible astronomical significance has been studied recently by this author (Ruggles 1994a; 1996).

The monument reference numbers are ordered from grid north to grid south. In naming short stone rows, the convention is used that directions, when spelt in full, e.g. Glantane East, are part of townland names: abbreviated directions, e.g. Carrigagulla W, are used to distinguish different short stone rows within the same townland.

The list includes probable and possible short stone rows as well as certain ones. Some recently documented examples of short stone rows that were not accommodated in the original numbering scheme are given reference numbers such as CKR7a. Monuments considered unlikely to be short stone rows although listed as such elsewhere have been relegated to a supplementary list and are not considered further.

The following information is given for each monument in the main list:

1. Reference number and name.
2. Alternative names if any.
3. National Grid reference (NGR), generally obtained from Ó Nualláin 1988 and quoted to 100 m. Where more precise grid references are available, as in Power 1992, they are quoted in preference.
4. Cross-references to the monument in other sources.
5. Status as stone row (defined as three or more stones) if not certain: prob[able], or poss[ible].
6. Destr[oyed or unrecognisable], or reconst[ructed].
7. Date of most recent visit by the author ('NV' indicates not visited).
8. Brief description, verified at first hand by the author where visited: length or estimate of original length to nearest 0·5 m; no. of stones still standing; no. of stones remaining (standing or prostrate); estimate of original no. of stones in row.

   For more complete information the reader is referred to the cross-references cited.
9. Reference of published excavation report(s).
10. Reference of published site plan(s).

*Kerry*

CKR1, Beal Middle. Q 885 475. OR(57); BR(221, 247) ['Beale Middle']; R4(57). Apr 91. Desc: len ≥9·5 m; st 5, 6, 6. Plan: Ó Nualláin 1988, fig. 30.

CKR2, Feavautia. R 065 235. BR(247); R4(U1). Apr 91. Desc: len *c.* 7·5 m ; st 3, 4, 4.

CKR4, Cloonsharragh. Q 511 128. D(52); OR(58); BR(221, 247); R4(58). Apr 91. Desc: len ≥7·0 m; st 3, 4, 4.[1] Plans: Cuppage 1986, fig. 25; Ó Nualláin 1988, fig. 18.

CKR5, Ballygarret. Q 686 099. D(51); OR(59); BR(221, 253); R3(59). May 93. Desc: len *c.* 8·0 m; st 2, 3, 3. Plans: Cuppage 1986, fig. 23; Ó Nualláin 1988, fig. 18.

CKR7a, Knockrower West. R 073 006. Patricia O'Hare, 'A stone alignment at Knockrower West', *The Kerry Magazine*, no. 7 (1996), 10–12. NV. Desc: len *c.* 8·0 m; st 2, 3, 3. Plan: *ibid.*, p. 10.

CKR8, Ardamore. Q 522 000. D(50); OR(60); BR(221, 253); R3(60). May 93. Desc: len 7·5 m; st 3, 3, 3. Plans: Cuppage 1986, fig. 24; Ó Nualláin 1988, fig. 17.

CKR8a, Scrahanagullaun. W 141 947. Walsh 1997, 12. NV. Desc: len 7·0 m; st 2, 3, 3.

CKR19, Gortnagulla. V 568 836. SK(113); OR(61); BR(221, 253); R3(61). May 93. Desc: len *c.* 4·5 m; st 2, 3, 3. Plans: Ó Nualláin 1988, fig. 17; O'Sullivan and Sheehan 1996, fig. 27e.

CKR25, Curragh More. V 801 820. SK(100); OR(62); BR(221, 253); R3(62). May 93. Desc: len 2·5 m; st 3, 3, 3. Plans: Ó Nualláin 1988, fig. 19; O'Sullivan and Sheehan 1996, fig. 24.

CKR28, Dromteewakeen. V 763 808. SK(108); OR(63); BR(221, 253); R3(63). May 93. Desc: len 5·5 m; st 2, 2, 3. Exc: John Sheehan, 'Dromteewakeen: stone row and possible boulder burial', in I. Bennett, *Excavations 1989: Summary Accounts of Archaeological Excavations in Ireland,* Wordwell, Dublin, 1990, 30. Plans: Ó Nualláin 1988, fig. 14; O'Sullivan and Sheehan 1996, fig. 26c.

CKR36, Derrylicka. V 784 754. SK(104). NV. Desc: 4·5m; st 2, 3, 3.

CKR38, Gortacloghane. V 761 741. SK(112); OR(65); OC(appendix, Kerry 4); BR(221, 247); R4(65). Apr 91.

Desc: len *c.* 6·5 m?; 3, 4(5?), 4(5?). Plans: Ó Nualláin 1988, fig. 32; O'Sullivan and Sheehan 1996, fig. 30.

CKR41, Fermoyle. V 453 724. SK(110); OR(176); BR(221, 253); R3(176). Prob SR. NV. Desc: len ?; st 2, 2, 3. Plans: Ó Nualláin 1988, fig. 35; O'Sullivan and Sheehan 1996, fig. 26d.

CKR43, Derrineden NE. V 573 717. SK(103); BR(247); R4(U2). Apr 91. Desc: len 6·0 m; st 4, 4, 4. Plan: O'Sullivan and Sheehan 1996, fig. 27c.

CKR44, Derrineden SW. V 569 714. SK(102); BR(253). NV. Desc: len *c.* 3·5 m, st 2, 3, 3. Plan: O'Sullivan and Sheehan 1996, fig. 26b.

CKR44a. Dromatouk NW. V 950 713. Paul Walsh, priv. comm., 1998. Poss SR.[2] NV. Desc: len 10·0 m; st 3, 3, 3. Plan: Ó Nualláin 1984b, fig. 24.

CKR45, Dromatouk SE. V 952 711. OR(66); BR(221 ['Dromatouk NE'], 253); R3(66). May 93. Desc: len 4·0 m; st 3, 3, 3. Exc: Lynch 1981, 97–103. Plan: Ó Nualláin 1988, fig. 17.

CKR46, Doory. V 545 709. SK(105); OR(64); BR(221, 253); R4(64). May 93. Desc: len 22·5 m; st 4, 4, 4.[3] Plans: Ó Nualláin 1988, fig. 31; O'Sullivan and Sheehan 1996, fig. 27a.

CKR47, Dromod. Cloughane. V 552 704. SK(107); OR(156) [listed as stone pair]; BR(221, 262) [listed as stone pair]. NV. Desc: len *c.* 7·0 m; st 2, 3, 3. Plans: Ó Nualláin 1988, fig. 54; O'Sullivan and Sheehan 1996, fig. 27d.

CKR49, Dromkeare. V 541 684. SK(106); BR(221, 247); R4(U3). Apr 91. Desc: len 7·5 m; st 4, 4, 4. Plan: O'Sullivan and Sheehan 1996, fig. 27b.

CKR54, Eightercua. V 512 647. SK(109); BR(221, 247); R4(U4). Apr 91. Desc: len 8·5 m; st 4, 4, 4. Plans: Ó Nualláin 1988, fig. 62; O'Sullivan and Sheehan 1996, fig. 28.

CKR54a. Derrysallagh. V 874 645. Paul Walsh, priv. comm., 1998. Poss SR. NV. Desc: len 4·5 m; st 2, 3, 3.

CKR55, Kildreelig. V 409 636. SK(114); OR(67); BR(221, 247); R4(67). NV. Desc: len 5·0 m; st 4, 4, 4. Plans: Ó Nualláin 1988, fig. 33; O'Sullivan and Sheehan 1996, fig. 26a.

CKR57, Derreenauliff. V 634 626. SK(101); BR(253). NV. Desc: len 3·0 m; st 2, 3, 3. Plan: O'Sullivan and Sheehan 1996, fig. 25.

CKR59, Garrough. V 559 609. SK(111); OR(68); BR(221, 247); R4(68). Apr 91. Desc: len *c.* 11·0 m; st 2, 4, 4. Plans: Ó Nualláin 1988, fig. 33; O'Sullivan and Sheehan 1996, fig. 29.

CKR66, Cashelkeelty. V 748 575. OR(69); OC(93); BR(221, 247); R3(69). May 93. Desc: len >6·5 m; st 3, 3, 4. Exc: Lynch 1981, 64–9, 76–97. Plans: Ó Nualláin 1984a, fig. 17; 1988, fig. 19.

## *Cork*

CKR3, Cloghvoula. R 157 143. OR(1); BR(219, 246); R4(1). Apr 91. Desc: len *c.* 7·0 m; st 3, 4, 4. Plan: Ó Nualláin 1988, fig. 25.

CKR3a. Knockaclarig. R 174 130. Paul Walsh, priv. comm., 1998. Prob SR. NV. Desc: len 7·5 m; st 4, 4, 4.

CKR6, Knocknanagh East. R 228 057. OR(2); BR(220, 247); R4(2). Apr 91. Desc: len 7·5 m; st 1, 4, 4.[4] Plan: Ó Nualláin 1988, fig. 12.

CKR7, Tooreenglanahee. R 173 022. OR(3) [NGR given erroneously as V 147 207]; BR(220, 253); R3(3). NV. Desc: len ?; st 1, 1, 3.

CKR9, Gneeves. W 470 928. OR(4); BR(220, 252); R3(4). Apr 92. Desc: len 8·0 m; st 3, 3, 3. Plan: Ó Nualláin 1988, fig. 8.

CKR10, Garrane. Garrane NW. W 479 910. OR(5); BR(219, 247); R4(5). Apr 91. Desc: len *c.* 14·5 m?; st 3, 4, 4. Plan: Ó Nualláin 1988, fig. 24.

CKR11, Kippagh. W 227 882. MC(6521); OR(6); BR(220, 252); R3(6). Apr 92. Desc: len 5·0 m; st 1, 3, 3.[5] Plan: Ó Nualláin 1988, fig. 7.

CKR12, Beenalaght. An Seisar. W 485 873. OR(12); BR(219, 246); R4(12). Apr 91. Desc: len 11·0 m; st 5, 6, 6. Plan: Ó Nualláin 1988, fig. 29.

CKR13, Cloghboola More. W 277 872. MC(6510); OR(7); BR(219, 252); R3(7). Apr 92. Desc: len 4·5 m; st 3, 3, 3. Plan: Ó Nualláin 1988, fig. 15.

CKR14, Tullig. Kerryman's Table. W 320 872. MC(6529); OR(8); OC(appendix, Cork 6); BR(220, 247); R4(8). Apr 91. Desc: len 5·0 m; 4, 4, 4. Plan: Ó Nualláin 1988, fig. 20.

CKR14a. Lyradane. W 554 872. MC(6532). Poss SR. Destr. NV. Desc: len *c.* 3·0 m?; st 0, 0, 3.

CKR15, Curraghdermot. W 8664 8688. EC(3679). NV. Desc: len 5·5 m; st 1, 3, 3.

CKR16, Garryduff. W 9142 8536. EC(3680); OR(174); BR(220, 252); R3(174). NV. Desc: len 5·0 m; st 3, 3, 3. Plan: Ó Nualláin 1988, fig. 34.

CKR17, Cloghboola Beg. W 305 852. OC(53); BR(219, 252). Poss SR. NV. Desc: len ?; st 0, 3, 3. Plan: Ó Nualláin 1984a, fig. 19.

CKR18, Pluckanes North. W 535 846. Paul Walsh, 'A "most interesting set of stones" at Pluckanes North, Co. Cork', *JCHAS* 103 (1998), 141–51; MC(6525); BR(253). Destr. NV. Desc: len *c.* 7·5 m; st 0, 1, 4. Plan: Somerville 1923, 200.

CKR20, Glantane East. W 278 830. MC(6519); OR(9); BR(220 ['Glantane W'], 252); R3(9). Apr 92. Desc: len 5·5 m; st 3, 3, 3. Plan: Ó Nualláin 1988, fig. 7.

CKR21, Knocknagappul. W 345 830. MC(6522); OR(173); BR(220, 252); R3(173). NV. Desc: len ?; st 1, 3, 3. Plan: Ó Nualláin 1988, fig. 35.

CKR22, Barrahaurin. W 455 830. MC(6503); OR(13); BR(219, 246); R4(13). Destr. NV. Desc: len 6·5 m; st 0, 0, 5.

CKR23, Carrigagulla W. W 371 829. MC(6507); OR(10); BR(219, 252) ['Carrigagulla NE']; R3(10). Apr 92. Desc: len 5·5 m; st 3, 3, 3. Plan: Ó Nualláin 1988, fig. 9.

CKR24, Carrigagulla E. W 384 829. MC(6508); OR(11); BR(219, 252) ['Carrigagulla SW']; R3(11). NV. Desc: len ?; st 2, 3, 3. Plan: Ó Nualláin 1988, fig. 16.

CKR26, Ballideenisk. W 7225 8179. EC(3678), OR(14), BR(219, 252); R3(14). NV. Desc: len *c.* 8·5 m?; st 1, 3, 3. Plan: Ó Nualláin 1988, fig. 27.

CKR27, Dooneens. W 384 814. MC(6517); OR(19); BR(219, 252); R3(19). NV. Desc: len ?; st 1, 3, 3. Plan: Ó Nualláin 1988, fig. 28.

CKR29, Newcastle. W 575 804. OR(98); BR(220, 252–3). See also Ruggles 1996, note 5. Poss SR. Reconst? NV. Desc: len *c.* 10·0 m; st 3, 3, 2/3?. Plans: Somerville 1923, 200; Ó Nualláin 1988, fig. 47.

CKR30, Cabragh N. Cabragh A. W 278 798. MC(6505); OR(16); BR(219, 246); R4(16). Apr 91. Desc: len 8·0 m; st 4, 4, 4. Plan: Ó Nualláin 1988, fig. 26.

CKR31, Carrigonirtane. W 277 798. MC(6509); BR(252). NV. Desc: len 4·0 m; st 2, 3, 3.

CKR32, Cabragh S. Cabragh B. W 278 792. MC(6506); OR(17); BR(246); R4(17). May 93. Desc: len *c.* 10·0 m; st 4, 6, 6. Plan: Ó Nualláin 1988, fig. 29.

CKR33, Coolacoosane. W 323 790. MC(6512); OR(18); BR(219, 252); R3(18). Apr 92. Desc: len 5·5 m; st 3, 3, 3. Plan: Ó Nualláin 1988, fig. 9.

CKR34, Derrynasaggart. W 181 781. MC(6516); OR(15); BR(219, 252); R3(15). Apr 92. Desc: len 6·5 m; st 3, 3, 3. Plan: Ó Nualláin 1988, fig. 11.

CKR34a. Leadawillin. W 395 773. MC(6523). NV. Desc: len 7·0 m; st 3, 4, 4.

CKR34b. Curraghnalaght. W 621 772. MC(6531). Poss SR. Destr. NV. Desc: len ?; st 0, 0, 3.

CKR35, Coolgarriff. W 412 769. MC(6514); OR(20); BR(246); R4(20). Apr 91. Desc: len ≥5·0 m; st 2, 4, 4.[6] Plan: Ó Nualláin 1988, fig. 20.

CKR37, Coolavoher. W 192 753. MC(6513); OR(21); BR(219, 252); R3(21). Apr 92. Desc: len 3·0 m; st 3, 3, 3. Plan: Ó Nualláin 1988, fig. 9.

CKR39, Bealick. Laught Malon. W 351 729. MC(6504); OR(24); BR(219, 252); R3(24). NV. Desc: len ≥6·5 m; st 2, 2, 3. Plan: Ó Nualláin 1988, fig. 16.

CKR40, Reananerree. W 204 728. MC(6526); OR(22); BR(220, 247) ['Reanerre']; BG(336); R4(22). Apr 91. Desc: len 7·0 m; st 6, 6, 6. Plan: Ó Nualláin 1988, fig. 30.

CKR42, Gortyleahy. W 320 717. MC(6520); OR(23); BR(220, 252); R3(23). Apr 92. Desc: len 4·5 m; st 3, 3, 3. Plan: Ó Nualláin 1988, fig. 11.

CKR48, Rooves Beg. W 451 702. MC(6527); OR(25); BR(220['Roovesmore NW'], 247 ['Roovesmore']); R4(25). Apr 91. Desc: len *c.* 12·5 m?; st 1, 6, 6. Plan: Ó Nualláin 1988, fig. 31.

CKR50, Cloonshear Beg. W 266 682. MC(6511); OR(27); BR(219, 252); R3(27). Apr 91. Desc: len 5·0 m; st 3, 3, 3. Plan: Ó Nualláin 1988, fig. 8.

CKR51, Dromcarra North. W 278 679. MC(6518); OR(28); BR(219, 246); R4(28). Apr 92. Desc: len *c.* 9·5 m; st 4, 5, 5. Plan: Ó Nualláin 1988, fig. 23.

CKR52, Turnaspidogy. W 188 666. MC(6530); OR(26); BR(220, 253); R3(26). NV. Desc: len *c.* 5·0 m?; st 2, 3, 3. Plan: Ó Nualláin 1988, fig. 13.

CKR53, Rossnakilla. W 324 657. MC(6528); OR(29); BR(220, 253); R3(29). Apr 92. Desc: len 7·5 m; st 3, 3, 3. Plan: Ó Nualláin 1988, fig. 7.

CKR56, Derrynagree. W 140 627. MC(6515); OR(32); BR(219, 252); R3(32). NV. Desc: len *c.* 4·5 m; st 3, 3, 3. Plan: Ó Nualláin 1988, fig. 14.

CKR58, Monavaddra. W 196 624. MC(6524); OR(34); BR(220, 252); R3(34). NV. Desc: len 3·5 m; st 3, 3, 3. Plan: Ó Nualláin 1988, fig. 9.

CKR60, Castlenalacht. W 4864 6087. WC(141); OR(36); BR(219, 246); R4(36). Apr 91. Desc: len 13·5 m; st 4, 4, 4. Plan: Ó Nualláin 1988, fig. 22.

CKR61, Farrannahineeny. W 2153 6081. WC(151); OR(35); BR(219, 246); R4(35). Apr 91. Desc: len ≥8·0 m; st 4, 5, 5. Plan: Ó Nualláin 1988, fig. 25.

CKR62, Piercetown. Carrigaline. W 6908 5926. EC(3681); OR(37); BR(220, 247);[7] R4(37). NV. Desc: len 8·5 m; st 5, 5, 5.[8] Plan: Ó Nualláin 1988, fig. 25.

CKR63, Maughanasilly. Ahil More. W 0440 5851. WC(160); OR(33); BR(220, 247); R4(33); J(5/29). Apr 91. Exc: Lynch 1981, 69–71, 107–11; 1999. Desc: len 6·0 m; st 5, 5, 5.[9] Plan: Ó Nualláin 1988, fig. 23.

CKR64, Canrooska. V 9353 5821. WC(140); OR(30); OC(76); BR(219, 252); R3(30). May 93. Desc: len *c.* 6·5 m; st 2, 3, 3. Plans: Ó Nualláin 1984a, fig. 20; 1988, fig. 21.

CKR65, Canrooska/Currakeal.[10] Currakeal. V 9359 5812. WC(146); OR(31); BR(219, 252); R3(31). May 93. Desc: len 2·5 m; st 3, 3, 3. Plan: Ó Nualláin 1988, fig. 7.

CKR67, Behagullane. W 2765 5665. WC(139); OR(117); BR(219, 246); R4(117). Apr 91. Desc: len 6·5 m; st 2, 4, 4. Plan: Ó Nualláin 1988, fig. 42.

CKR68, Dromdrasdil. W 1723 5581. WC(147); OR(41); BR(219, 246); R4(41). Apr 91. Desc: len 5·5 m; st 3, 4, 4.[11] Plan: Ó Nualláin 1988, fig. 20.

CKR69, Derrynacaheragh. W 1815 5526. WC(166). Poss SR. NV. Desc: len 6·5 m; 3?, 4?, 5?

CKR70, Ardrah. W 0710 5448. WC(138); OR(40); BR(219, 246); R4(40). Apr 91. Desc: len 4·5 m; st 4, 4, 4. Plan: Ó Nualláin 1988, fig. 23.

CKR71, Coomleagh East. W 1195 5364. WC(144); OR(39); BR(219, 252); R3(39); J(5/39). Poss SR.[12] NV. Desc: len ?; st 1, 2, 3. Plan: Ó Nualláin 1988, fig. 13.

CKR72, Kilcaskan. V 817 523. WC(155) [NGR given erroneously as V 7495 5060]; OR(38): BR(220, 252); R3(38). May 93. Desc: len 5·0 m; st 3, 3, 3. Plan: Ó Nualláin 1988, fig. 10.

CKR73, Cullenagh. W 1536 5220. WC(145); OR(42); BR(219, 252); R3(42); J(5/41). Apr 92. Desc: len 9·0 m; st 3, 3, 3. Plan: Ó Nualláin 1988, fig. 8.

CKR74, Eyeries. V 6457 5043. WC(148); OR(43); BR(219, 252); R3(43). NV. Desc: len 5·0 m; st 2, 3, 3. Plan: Ó Nualláin 1988, fig. 11.

CKR75, Leitry Lower. W 1380 4971. WC(158); OR(49) [NGR given erroneously as V 138 488]; BR(220, 252); R3(49). May 93. Desc: len 3·5 m; st 3, 3, 3. Plan: Ó Nualláin 1988, fig. 10.

CKR76, Foildarrig. V 6812 4674. WC(152); OR(46); BR(219, 252); R3(46). NV. Desc: len 8·0 m?; st 1, 3, 3. Plan: Ó Nualláin 1988, fig. 10.

CKR77, Derrymihin West. V 6963 4611. WC(165); BR(246); R4(U5). Destr. NV. Desc: len 6·0 m?; st 0, 0, 5.[13] Plan: Somerville papers, University College, Cork.

CKR78, Cullomane West. W 0177 4598. WC(142); BR(252). NV. Dec: len 6·0 m; st 2, 3, 3.

CKR79, Knockawaddra. W 2703 4598. WC(157); OR(50); BR(220 ['Knockawaddra W'], 252); R3(50). May 93. Desc: len 7·0 m; st 3, 3, 3. Plan: Ó Nualláin 1988, fig. 9.

CKR80, Scartbaun. W 0018 4594. WC(163); OR(47); BR(220, 253); R3(47). Apr 92. Desc: len 4·0 m; st 3, 3, 3. Plan: Ó Nualláin 1988, fig. 7.

CKR81, Keilnascarta. V 9933 4515. WC(154); OR(48); BR(220 ['Keilnascarta NW'], 252); R3(48); J(5/10). Apr 92. Desc: len 6·5 m; st 3, 3, 3. Plan: Ó Nualláin 1988, fig. 14.

CKR82, Fanahy. V 6512 4487. WC(149); OR(44); BR(219, 252); R3(44). NV. Desc: len ?; st 1, 3, 3. Plan: Ó Nualláin 1988, fig. 15.

CKR83, Knockoura. V 631 447. WC(3533); BR(252). NV. Desc: len ?; st 3, 3, 3.[14]

CKR84, Clonglaskan. V 6454 4453. WC(164); OR(45); BR(219, 252); R3(45). Prob SR. NV. Desc: len 10·5 m?; st 1, 3, 3. Plan: Ó Nualláin 1988, fig. 28.

CKR85, Knockatlowig. W 3206 4435. WC(156); OR(51); BR(220, 252); R3(51). NV. Desc: len *c.* 7·0 m?; st 2, 3, 3. Plan: Ó Nualláin 1988, fig. 13.

CKR86, Maulinward. V 9911 4381. WC(161); OR(52) [NGR given erroneously as V 975 438]; BR(220, 252); R3(52). May 93. Desc: len 4·5 m; st 3, 3, 3. Plan: Ó Nualláin 1988, fig. 27.

CKR87, Farranamanagh. V 8255 3846. WC(150); OR(175); BR(219, 252); R3(175); J(4/58). NV. Desc: len ?; st 1, 3, 3. Plan: Ó Nualláin 1988, fig. 34.

CKR88, Dromnea. V 8409 3824. WC(167); BR(252). Destr. NV. Desc: len 3·0 m; st 0, 0, 3.

CKR89, Lissaclarig West. W 040 367. WC(159); OR(53); BR(220, 252); R3(53). NV. Desc: len ?; st 2, 3, 3. Plan: Ó Nualláin 1988, fig. 15.

CKR90, Murrahin North. W 029 365. WC(162); OR(54); BR(220, 252); R3(54); J(4/1). NV. Desc: len ?; st 2, 3, 3. Plan: Ó Nualláin 1988, fig. 16.

CKR91, Gurranes. Three Ladies; Five Fingers. W 1746 3148. WC(153); OR(55); BR(220; 247); R4(55); J(2/28). Sep 98. Desc: len ≥11·5 m; st 3, 4(5?), 5?. Plan: Ó Nualláin 1988, fig. 12.

CKR92, Comillane. V 9748 2306. WC(143); OR(56); BR(219, 252); R3(56); J(3/5). NV. Desc: len ?; st 2, 3, 3. Plan: Ó Nualláin 1988, fig. 13.

## SUPPLEMENTARY LIST:
monuments considered unlikely short stone rows, although listed as such in Burl 1993.

### *Kerry*

Curraduff. Q 703 088. D(54); BR(221, 253). Destr. Insufficient evidence that this was a stone row (Paul Walsh, priv. comm., 1997).

Clogher. Q 314 033. D(53); BR(221, 253). Destr. Insufficient evidence that this was a stone row (Paul Walsh, priv. comm., 1997).

# List 6

*Scottish Short Stone Rows*

This is a reference list of Scottish short stone rows compiled from various published lists backed up by data from the National Monuments Record of Scotland (NMRS). Supplementary List A gives details of aligned pairs of slabs, as defined in the independent study of 'megalithic astronomy' described earlier in this book,[1] listed within the study area for that project. The monument reference numbers are ordered from grid north to grid south within each geographical area. Parallel rows in close proximity are listed as separate entries, e.g. SSR24 and 25.

The main list includes probable and possible short stone rows as well as certain ones. Supplementary List B details monuments considered unlikely to be short stone rows or which have been rejected outright (e.g. because of the length criterion, see p. 108)—despite being included by other authors (and in particular Burl 1993)—together with the reasons for this.

The following information, as available, is given for each entry in the main list and Supplementary List A, except that where the monument is also included in List 1 or List 2 only the Burl 1993 reference is given under item 4, and items 7, 9 and 10 are omitted completely:

1. Reference number and name.
2. Alternative names if any.
3. National Grid reference (NGR), generally quoted to 10 m.
4. Cross-references to the monument in other sources.
5. Status as stone row (defined here in the wider sense, to include aligned pairs) if not certain: prob[able], or poss[ible].
6. Destr[oyed or unrecognisable].
7. Date of most recent visit by the author ('NV' indicates not visited).
8. Brief description, verified at first hand by the author where visited: length or estimate of original length to nearest 0·5 m; no. of stones still standing; no. of stones remaining (standing or prostrate); estimate of original no. of stones in row. For more complete information the reader is referred to the cross-references cited.
9. Reference of published excavation report(s).
10. Reference of published site plan(s).

*Shetland*

SSR1, Hamna Voe. The Giant's Stones. HU 243 806. Z(1358); N(HU28SW03); BR(224, 256); T(Z3/2). Poss SR. NV. Desc: len ?; st 2, 2, 3. Two standing stones, 2·4 m and 1·8 m high, 20 m apart on a WSW–ENE line. A third stood in 1774,[2] but whether it was in line is unclear. There are several smaller stones in the vicinity. Plans: Thom and Thom 1978a, 164; Thom, Thom and Burl 1980, 366; 1990, 348.

*Orkney*

SSR2, St Tredwell's Chapel (Papa Westray). HY 497 509. Z(543); N(HY45SE24); BR(223, 255). Poss SR. Destr. NV. Desc: len ?; st 0, 0, 3. Three standing stones (one fallen) were reported here in 1701. Whether they were in line is uncertain.

*Lewis*

SSR3, Carloway. Clach an Tursa. NB 2041 4295. BR(223, 255). Prob SR. Desc: len 5·0 m; st 1, 3, 3. *See LH6 in List 2.*

SSR4, Airigh nam Bidearan. Callanish V. NB 2342 2989. BR(223, 255). Poss SR. Desc: len 9·0 m; st 3, 3, 3. *See LH24 in List 2. Also L11 in List 1.*

*Skye*

SSR5, Sornaichean Coir Fhinn. Loch Eyre. NG 414 526. O(638); N(NG45SW04); BR(224, 256) [NGR given erroneously as NG 414 563]; T(H7/3). Prob SR. Jul 75. Desc: len c. 8·0 m?; st 2, 2, 3. Two standing stones, both 1·7 m high, 4·0 m apart in a NW–SE line. According to Burl (Thom, Thom and Burl 1990, 265) there were once three stones equally spaced and in line.

SSR6, Clachan Erisco. Borve. NG 451 480. O(636); N(NG44NE01); BR(224); T(H7/5). Jul 75. Desc: len 8·5 m; st 3, 3, 3. A row of three stones up to 1·8 m high, oriented NW–SE.

*Perth*

SSR7, East Cult. NO 0725 4216. P(34, 152). N(NO04SE02); BR(224, 266–7) [listed as stone pair]; T(P2/7). NV. Desc:

len *c.* 23 m; st 2, 3, 3. Plan: Thom, Thom and Burl 1990, 334.

SSR8, GALLOWHILL WOOD. NO 1681 3604. P(34, 152); N(NO13NE20). Prob SR. NV. Desc: len *c.* 7·0 m, st 0, 3, 3. Three large prostrate stones that probably formed a row oriented roughly NNE–SSW.

SSR9, ST. MARTINS. Cupar Stone. NO 1592 3122. P(152); N(NO13SE09); B(7:46); BC(Perth39[–]); FP(Perth25). Poss SR. Desc: len c. 4·0 m, st 1, 3, 3. Three stones that possibly formed a row running NNE–SSW.

SSR10, COMMONBANK. Kilspindie. NO 1749 2484. P(34, 152); N(NO12SE23); BR(224, 255–6); FP(Perth9); X(64:44, 73:44)[3]. NV. Desc: len ?; st 0, 3, 3. Three recumbent boulders in an ENE–WSW alignment, situated 8·5 m east of a probable four-poster.

SSR11, COMRIE. Tullybannocher. NN 7548 2247. BR(224, 249). Poss SR. Desc: len ?; st 2, 2, 4?. *See L22 in List 1.*

SSR12, ST MADOES STONES. NO 197 210. N(NO12SE20); BR(224, 256). Prob. SR. NV. Desc: len 2·5 m; st 2, 3, 3. A pair of stones 1·6 m and 1·5 m high together with a small recumbent boulder forming a row oriented NNW–SSE.

SSR13, CRAGGISH. NN 763 207. N(NN72SE08); BR(224, 255). Poss SR. Destr. NV. Desc: len ?; st 0, 0, 3?. Site of 'several' standing stones, now destroyed.

SSR14, COWDEN. Roman Stone. NN 7744 2064. N(NN72SE05); BR(224, 255). Poss SR. NV. Desc: len 2·5 m; st 3, 3, 3. One standing stone 1·9 m high with 'two adjacent squat stones'.

SSR15, DOUNE. Glenhead Farm. NN 755 004. N(NN70SE03); BR(224, 249); T(P1/2). NV. Desc: len 8·5 m; st 3, 4, 4. Plan: Thom, Thom and Burl 1990, 322.

### *Stirling*

SSR16, BLANEFIELD. Dumgoyach; Duntreath. NS 5328 8072. S(58); N(NS58SW03); BR(224, 249); T(A11/2); X(72:38–9). Poss SR. NV. Desc: len 17·0 m?; st 1, 5, 5. Exc and plan: MacKie 1973.

SSR17, MIDDLETON. NS 561 766. S(63); N(NS57NE18); BR(224, 249). Destr. NV. Desc: len ?; st 0, 0, 4.

### *Mull*

SSR18, GLENGORM. NM 4347 5715. BR(223, 255). Desc: len [originally] 9·0 m; st 3 [1 moved], 3, 3. *See ML1 in List 2.*

SSR19, QUINISH. Mingary. NM 4134 5524. BR(223, 248, 255). Desc: len 10·0 m; st 1, 4, 5.[4] *See ML2 in List 2. Also L27 in List 1.*

SSR20, BALLISCATE. Sgriob-Ruadh; Tobermory. NM 4996 5413. BR(223, 255). Desc: len 5·0 m; st 2, 3, 3. *See ML4 in List 2.*

SSR21, MAOL MOR. Dervaig A; Dervaig NNW. NM 4355 5311. BR(223, 248). Desc: len 10·0 m; st 3, 4, 4. *See ML9 in List 2.*

SSR22, DERVAIG N. Dervaig B; Dervaig Centre. NM 4390 5202. BR(223, 248). Desc: len 18·5 m; st 2, 5, 5. *See ML10 in List 2.*

SSR23, DERVAIG S. Dervaig C; Dervaig SSE; Glac Mhór. NM 4385 5163. BR(223, 248). Desc: len 5·0 m; st 3, 3, 4. *See ML11 in List 2.*

SSR24, ARDNACROSS (northern row). NM 5422 4915. BR(223, 255). Desc: len 12·5 m; st 0, 3, 3. *See ML12 in List 2.*

SSR25, ARDNACROSS (southern row). NM 5422 4915. BR(223, 255). Desc: len 10·0 m; st 1, 3, 3. *See ML12 in List 2.*

SSR26, SCALLASTLE. NM 6999 3827. BR(223, 255). Prob SR. Desc: len *c.* 5·0 m?; st 1, 3, 3. *See ML21 in List 2.*

SSR27, ULUVALT. Uluvalt I, Barr Leathan. NM 5463 3002. BR(223, 248–9). Poss SR. Desc: len *c.* 10·0 m; st 1, 4, 4. *See ML25(b) in List 2.*

### *Lorn*

SSR28, DUACHY. Loch Seil. NM 8014 2052. BR(222 [appears twice, as Duachy and Loch Seil], 254). Desc: len 5·0 m; st 3, 3, 3. *See LN22 in List 2.*

### *Mid-Argyll*

SSR29, SALACHARY. Bealach Mor. NM 8405 0403. BR(222, 254). Desc: len *c.* 4·0 m; st 2, 3, 3. *See AR6 in List 2.*

SSR30, BALLYMEANOCH (NE row). Duncracaig. NR 8337 9641. BR(222, 248). Desc: len 15·0 m; st 4, 4, 4. *See AR15 in List 2. Also L32 in List 1.*

SSR31, DUNAMUCK I. Dunamuck N. NR 8471 9290. BR(222, 254). Desc: len 7·0 m; st 2, 3, 3. *See AR28 in List 2.*

### *Cowal*

SSR32, INVERYNE. NR 915 749. A(218); N(NR79SW14); BR(222, 254). NV. Desc: len 4·0 m; st 3, 3, 3. Plan: RCAHMS 1988, 134.

### *Jura*

SSR33, CRAIGHOUSE. Carragh a'Ghlinne. NR 5128 6648. BR(223, 248). Desc: len 5·0 m; st 1, 4, 4. *See JU6 in List 2. Also L36 in List 1.*

SSR34, SANNAIG. NR 5184 6480. BR(223, 255). Desc: len 2·5 m; st 2, 3, 3. *See JU7 in List 2.*

### *Islay*

SSR35, ACHNANCARRANAN. Laphroaig. NR 3895 4607. BR(223, 254). Desc: len 7·5 m; st 3, 3, 3. *See IS41 in List 2.*

### *Kintyre*

SSR36, ESCART. NR 8464 6678. BR(222, 248). Desc: len 15·5 m; st 5, 5, >5?. *See KT5 in List 2. Also L39 in List 1.*

SSR37, BALLOCHROY. NR 7309 5241. BR(222, 254). Desc: len 5·0 m; st 3, 3, 4. *See KT10 in List 2.*

SSR38, CLOCHKEIL. NR 6577 2445. BR(222, 254). Desc: len 5·5 m; st 2, 3, 3. *See KT27 in List 2.*

SSR39, MACHRIHANISH. Cnocan Sithein. NR 6446 2065. BR(222, 248). Poss SR. Destr. NV. Desc: len ?, st 0, 0, 4?.

A row of stones was reported in 1831 to the north of the cairn [N(NR62SW02); BR(222, 248); T(A4/15); U(KT38)] and aligned with its central cist.

*Bute*

SSR40, STRAVANAN BAY. Largizean Farm. NS 0846 5535. N(NS05NE07); BR(222, 254); T(A9/7). NV. Desc: len 6·0 m; st 3, 3, 3. Plan: Thom, Thom and Burl 1990, 163.

*Wigtown*

SSR41, TORHOUSEKIE E. Torhouse. NX 3837 5650. G1(534); N(NX35NE12); BR(224, 256); T(G3/7, part). NV. Desc: len 5·0 m; st 3, 3, 3. Plans: Thom, Thom and Burl 1980, 274; 1990, 202.

SSR42, TORHOUSEKIE W. NX 381 565. G1(532); N(NX35NE25); BR(224, 256). Poss SR. Destr. NV. Desc: len *c.* 4·5 m; st 0, 0, 3.

SSR43, DRUMTRODDAN. NX 3644 4430. G1(231); N(NX34SE02); BR(224, 256); T(G3/12). NV. Desc: len 13·0 m; st 2, 3, 3. Plan: Thom, Thom and Burl 1990, 205.

*Dumfries*

SSR44, DYKE. Moffat; 'Three Stannin' Stanes'. NT 084 036. DM(426); N(NT00SE17); BR(222, 254); T(G6/4). NV. Desc: len 8·5 m; st 3, 3, 3. See also Thom, Thom and Burl 1990, 218.

## SUPPLEMENTARY LIST A: aligned pairs in regions covered by Ruggles 1984a

*Mid-Argyll*

AP1, BARBRECK. NM 8315 0641. BR(222, 263). Desc: len 5·0 m; st 2, 2, 2. *See AR3 in List 2.*

AP2, CARNASSERIE. NM 8345 0080. BR(222, 263). Desc: len 2·5 m; st 2, 2, 2. *See AR12 in List 2.*

AP3, NETHER LARGIE (*LK*). NR 828 977. BR(222, 264 [Kilmartin (a)]). Desc: len 4·0 m; st 2, 2, 2. The smaller stone *M* to the NW is roughly in line with *L* and *K*, but it is about 100 m away, so the three stones would not be considered a short stone row in any case. *See AR13(b2) in List 2. Also part of L31 in List 1.*

AP4, NETHER LARGIE (*AB*). NR 8281 9758. BR(222, 264 [Kilmartin (c)]). Desc: len 4·0 m; st 2, 2, 2. *See AR13(b5) in List 2. Also part of L31 in List 1.*

AP5, BALLYMEANOCH (SW row). Duncracaig. NR 8337 9641. BR(222, 263). Desc: len 3·0 m, st 2, 2, 2. *See AR15 in List 2. Also L32 in List 1.*

AP6, DUNAMUCK II. Dunamuck Mid. NR 8484 9248. BR(222, 263–4). Desc: len 8·0 m; st 2, 2, 2. *See AR29 in List 2.*

*Islay*

AP7, LAGAVULIN N. 'Lagavullin'. NR 3954 4621. BR(223, 265). Desc: len *c.* 2·0 m; st 1, 2, 2. *See IS39 in List 2.*

*Knapdale*

AP8, CARSE. Loch Stornoway. NR 7425 6163. BR(222, 263). Desc: len 2·5 m; st 2, 2, 2. *Also KT2(b) in List 2.*

## SUPPLEMENTARY LIST B: monuments considered unlikely short stone rows, or rejected as short stone rows, although listed as such in Burl 1993

*Orkney*

BRAESIDE (Eday). Carrick. HY 564 371. Z(211); N(HY53NE05); BR(223, 255). Three prostrate stones up to 1·4 m long were reported here in 1928. Whether they were actually in line is not clear, and the monument is now destroyed, but in any case the furthest were some 90 m apart.

*Caithness*

BROUGHWIN. Broughwin Centre W [Burl]; Broughwin II (part) [Myatt]. ND 311 408. N(ND34SW24); BR(222, 248). Myatt (1988, 287) reported 'a row of five small upright stones running along the top of a ridge' but according to the NMRS all the small stones in this area are natural.

*Harris*

HORGABOST. Nisabost; Clach Mhic Leoid. NG 0408 9727. BR(223 [listed under Lewis], 255). The two small slabs to the west of the standing stone are probably the kerbstones of a cairn. *See LH36 in List 2.*

*Skye*

CLACH ARD. Uig Bay. NG 3943 6284. O(637); N(NG36SE02); BR(224, 256); T(H7/1). Re-erected standing stone atop a flat-topped cairn. Earlier this century two rough boulders were reported some 2·0 m to the NE, 0·5 m apart, but only one now remains. There is no evidence that these represent the remains of a stone row.

*North Uist*

BAYS LOCH (Berneray). Cnoc na Greana. NF 9230 8187. O(134); N(NF98SW02); BR(223, 248) [listed under Lewis]; H(HRS2); U(UI4). If prehistoric at all, this is probably the remains of the chamber of a cairn.

BLASHAVAL. Na Fir Bhreige. NF 8875 7176. BR(223, 255). This alignment of three stones stretches over 59 m. *See UI19 in List 2.*

SKEALTRAVAL. NF 853 707. N(NF87SE29); BR(223, 249); X(86:52). This alignment of six prostrate stones and boulders is some 150 m long.

*South Uist*

SLIGEANACH KILDONAN. Ru Ardvule. NF 7273 2860. BR(224, 256).[5] This possible alignment of three stones is some 90 m long. *See UI50 in List 2.*

### *Perth*

DUNRUCHAN STONES. NN 790 168 – 792 174. N(NN71NE01, 'B'–'E'); BR(224, 249). NV. This alignment of four stones stretches over some 640 m.

### *Mull*

'DERVAIG D'. NM 442 519. N(NM45SW10); BR(223 ['Dervaig E'], 255); X(57:9). The alleged stone row here in fact consists of natural boulders built into an old field wall.

### *Lorn*

CLENAMACRIE. NM 9250 2854. BR(222, 254). The status of the two boulders here is uncertain. *See LN18 in List 2.*

### *Jura*

KNOCKROME–ARDFERNAL. NR 548 714 – 560 717. BR(223, 254–5). These three stones form an alignment some 1,200 m long.[6] *See JU3 and JU4 in List 2.*

### *Islay*

BALLINABY. NR 220 672. BR(223, 254). Even if the three stones were in line, which is not clear, and the third stone was placed between the extant pair, the row would still have been *c.* 220 m long. Thomas Pennant's comment in 1772 that the three stones were 'nearly equidistant' might imply an alignment of some twice this length, but if Pennant's third stone is rightly identified with the rock outcrop at NR 222 675 (IS15(*c*) in List 2), as has been suggested, then the three-stone alignment is almost certainly spurious in any case.[7] *See IS15 in List 2. Also L37 in List 1.*

### *Knapdale*

UPPER FERNOCH. Tayvallich. NR 7269 8594. N(NR78NW07); BR(222, 254); T(A3/4, part); U(AR38). Three alleged prostrate standing stones which are almost certainly natural rock outcrops (see Ruggles 1984a, 161–2 (AR38)).

### *Kintyre*

MEALDARROCH. NR 877 680. N(NR86NE36); BR(222, 254). NV. According to the NMRS, the 'three stones standing in a straight line' described here in 1919[8] are not antiquities.

MACHARIOCH. NR 7365 0927. K(11); N(NR70NW05); BR(222, 248); H(ARG34); T(A4/20); U(KT43). Probable remains of a chambered cairn.

### *Ayr*

BALLANTRAE. Garleffin standing stones. NX 0873 8172. N(NX08SE01); BR(222, 254); T(G1/4). Plan: Thom, Thom and Burl 1990, 199. The alleged stone row here is merely an apparently fortuitous alignment of stones over 150 m long amongst a scatter of eight stones.

# TABLES

**Table 2.1 Thom's lunar alignments: summary data.**

This table presents background information about monuments claimed by Thom to incorporate deliberate lunar alignments, together with a summary of the number of such alignments at each 'level' (of precision).

Column headings:

1 Reference no. in List 1
2 Thom's site reference no.
3 Number of lines included in the Level 2 reassessment
4 Number of lines included in the Level 3 reassessment
5 'y' if an extrapolation length is allegedly marked at the site
6 No. of lines included in the Level 4 reassessment
7 Brief description, generally following Ruggles 1981, 204–5
8 Authenticity of monument (A = no doubt; B = some doubt; C = serious doubt; Z = not an antiquity)
9 Notes

Key to Column 7 (Brief description)

C Stone circle or ring
H Henge
R(..) Row consisting of .. [N.B. An aligned pair of standing stones is listed as a two-stone row. Two standing stones are only considered an aligned pair if they are *both* slabs (i.e. stones that are sufficiently wide and flat that a direction of orientation is determinable) oriented along the line of centres, or if there is evidence that a third standing stone, which has subsequently disappeared, originally stood in line with them.]
X 'Four-poster'
G Setting of small erect slabs forming the centres of the four sides of a rectangle
M Standing stone
S Stone (unspecified)
T Chambered tomb
U Cairn or mound
F.. Fallen..
P.. Possible..
../.. Surrounding
Y.. Remains of..
Z.. Site of..
n.. Several..

Key to Column 9 (Notes)

a Described by Thom (1971, 70) as a four-stone alignment, but all the stones recorded by Thom are part of an enclosure wall. The remains of a possible four-poster (Burl 1988b, 125 (East Lothian 2)) are merely noted as 'four stones in a rectangle nearby' in Thom's plan (Thom 1966, fig. 10(c)).
b Not certainly prehistoric. See Ruggles 1981, 164 (site 11).
c Listed by Thom (1967, table 12.1), as 'A, 4M', but see Atkinson 1981, 208 (site 46). See also Jane Murray, 'The stone circles of Wigtownshire', *TDGNHAS* 56 (1981), 18–30.
d The proposed indication is provided by one of two lines of stones leading radially away from the stone circle. It marks a parish boundary and is almost certainly modern. See Ruggles 1981, 164 (site 55).
e Not certainly prehistoric. See Atkinson 1981, 207 (site 27).
f Not certainly prehistoric. See Ruggles 1981, 164 (site 7).
g Natural boulder, not an antiquity. See Ruggles 1982b, S32 (line 32).
h Not certainly prehistoric. See Atkinson 1981, 207–8 (site 38).
j There is uncertainty about the accuracy of the reconstruction of this monument, and a possibility that it is completely bogus. See Thom, Thom and Burl 1980, 53; Ruggles 1982b, S33–4 (line 43).

| 1 | 2 | 3 | 4 | 5 | 6 | 7 | 8 | 9 |
|---|---|---|---|---|---|---|---|---|
| L1 | Z3/4 | — | — | | 1 | M | A | |
| L2 | O1/1 | — | — | | 9 | H/C, M, 2YM, PT, nU | A | |
| L3 | O1/2 | — | — | | 1 | H/C | A | |
| L4 | N1/17 | — | 2 | y | — | nR(nS) | A | |
| L5 | N1/7 | — | — | y | — | nR(nS) | A | |
| L6 | N1/14 | — | — | y | — | nR(nS) | A | |
| L7 | N1/1 | — | 2 | y | 2 | nR(nS) | A | |
| L8 | N1/13 | 1 | — | | — | YC | A | |
| L9 | H1/1 | 1 | 2 | | 1 | C/U, M, 5R(nM) | A | |
| L10 | H1/2 | 1 | — | | — | C | A | |
| L11 | H1/5 | 1 | 4 | | 1 | R(3S), 2S | B | b |
| L12 | H1/15 | 1 | — | | — | PM | B | f |
| L13 | H1/14 | 1 | — | | — | FM | A | |
| L14 | H3/6 | 1 | — | | — | T | A | |
| L15 | H3/3 | 1 | — | | — | M | A | |
| L16 | H3/11 | — | 2 | | — | T, M | A | |
| L17 | B7/3 | 1 | — | | — | 3M | A | |
| L18 | B2/4 | 1 | — | | — | C | A | |
| L19 | B3/5 | 1 | — | | — | 2M | A | |
| L20 | P3/1 | 1 | 1 | y | 1 | X, 2FM | A | |
| L21 | P1/10 | — | 1 | y | 1 | C, ZM, U, M, 3S | A | |
| L22 | P1/8 | 1 | — | | — | 2M[YPX] | A | |
| L23 | P1/1 | 1 | — | | — | 2M | A | |
| L24 | P4/1 | — | 1 | | 1 | R(2M), M | A | |
| L25 | G9/13 | 1 | — | y | — | PX, ZR(4S) | C | a |
| L26 | M4/2 | 1 | — | | — | M | A | |
| L27 | M1/3 | — | 1 | | — | R(M, 3FM), ZM | A | |
| L28 | M2/14 | 1 | — | | — | C, 4M, U | A | |
| L29 | M2/8 | 1 | — | | — | PM | B | e |
| L30 | A2/5 | — | 2 | | 1 | 2U, M | A | |
| L31 | A2/8 | 2 | 4 | y | 4 | 2C, M, 2R(2M), G/M, G, ZM | A | |
| L32 | A2/12 | 1 | 2 | | 2 | R(4M), R(2M), ZM | A | |
| L33 | A2/13 | — | 1 | | 1 | FM, M | A | |
| L34 | A2/19 | 1 | — | | — | FM, M | A | |
| L35 | A6/4 | 1 | 1 | | 1 | 2M | A | |
| L36 | A6/6 | 1 | — | | — | R(M, 3FM) | A | |
| L37 | A7/5 | — | — | | 2 | R(2M, ZM) | A | |
| L38 | A10/6 | 1 | 1 | y | 1 | 2M, M | A | |
| L39 | A4/1 | 1 | 1 | | 1 | R(5M) | A | |
| L40 | — | — | — | | 1 | S | Z | g |
| L41 | A4/23 | — | — | | 1 | 2PM | C | h |
| L42 | A4/17 | — | 2 | | 1 | M | A | |
| L43 | A4/5 | — | 1 | | 1 | T, M | A | |
| L44 | A4/9 | — | 2 | | 1 | M | A | |
| L45 | A4/2 | — | 2 | | 2 | M | A | |
| L46 | A4/14 | — | 2 | | 2 | M | A | |
| L47 | A4/19 | — | — | | 1 | M | A | |
| L48 | G3/2 | — | 2 | | 1 | 2M, FM, PU, S | A | |
| L49 | G3/3 | 2 | — | | — | R(2PM) | C | c |
| L50 | G4/2 | 1 | — | | — | 2M, FM[YPC] | A | |
| L51 | L1/1 | 1 | — | | — | C | A | |
| L52 | L1/16 | — | — | | 1 | C | B | j |
| L53 | L1/6 | 3 | — | | — | 5C | A | |
| L54 | L1/11 | 1 | — | | — | 2M | A | |
| L55 | L6/1 | 1 | — | | — | R(3M) | A | |
| L56 | W9/7 | 1 | 1 | | 1 | R(4M, 4FM) | A | |
| L57 | S5/2 | 1 | — | | — | C/C etc. | A | |
| L58 | S1/2 | 1 | — | | — | C, 2A(nS) | C | d |

## Table 2.2 Thom's lunar alignments: a reassessment

Independent field data and notes from a reassessment of Thom's 'Level 2' data undertaken in 1979.

Column headings:
1 Reference number in List 1
2 Thom's site reference number
3 Nature of indication
4 Horizon foresight?
5 Authenticity of monument (see Table 2.1; A = no doubt; B = some doubt; C = serious doubt)
6 'Intrinsic' status of alignment (A = reasonable; B = somewhat dubious; C = very dubious; Z = dismissed)
7 Comments on intrinsic status
8 Declination obtained by Thom, in degrees
9 Declination obtained independently by author, where available
10 Source of value quoted in column 9
11 Difference from relevant standstill limit, in degrees
12 Standard deviation deemed appropriate, in degrees (from Ruggles 1981, table 4.1)
13 Reasons for larger standard deviation than assumed by Thom
14 Notes

Key to Column 3 (Nature of indication)
Feature codes follow Table 2.1, column 7, and are listed in order along the indication, e.g.
M-T Standing stone to chambered tomb;
T-S Chambered tomb to 'stone' (unspecified).
Subscript "O", e.g. $M_O$, is used where the indication is in the direction of orientation of a slab.
$C_C$ denotes the *centre* of a circle or ring
$C_D$ denotes the stone at one end of the major axis (maximum diameter) of a flattened or elliptical stone ring, according to Thom's interpretation of its geometry
L(..) denotes that the indication is a line consisting of more than five features of the given type

Key to Column 7 (Comments on intrinsic status)
d The claimed indication is along the avenue looking south (Thom 1971, 68–9). However, the southern horizon is in fact obscured by a local outcrop (Cooke *et al.* 1977, 120). See also Ruggles 1981, 166 (site 9).
e The type of indication given in Thom 1967, table 8.1 is 'alignment to stone'. The indication is actually between two non-intervisible standing stones about 100 m apart. A third stone between them is about 15° off line. See Ruggles 1981, 167 (site 15).
f The foresight 'stone' is not an antiquity and could not be located. See Ruggles 1981, 167.
g Two standing stones remain but there were once at least four, probably in the form of a four-poster. Hence it is unlikely that the orientation of the line joining them was of any particular significance in the original structure. See Atkinson 1981, 207 (site 20).
j The indication is from the centre to two stoneholes *N*1 and *N*2 which lie just outside the outer stone circle (*A*) to the north-west (see Cunnington 1931, 329 and pl. 1). However, it is likely that they were associated with the two radial lines of stoneholes further round to the west which are thought to represent the beginning of the avenue to Avebury. See Ruggles 1981, 165 (site 53).
k Thom's 'outlier' is apparently a stone of doubtful authenticity a mere 3 m outside the stone circle. See Ruggles 1981, 167–8 (site 16); Atkinson 1981, 206 (site 16).
m The central stone of the row was originally one of a pair. See Atkinson 1981, 208 (site 51).
n Thom's 'outlier' is apparently one of a group of stones of uncertain status about 50 m from the circle centre. See Ruggles 1981, 168 (site 6).
r The three stones were probably once part of a stone circle. Hence it is unlikely that the orientation of the line joining the two stones still standing was of any particular significance in the original structure. See Atkinson 1981, 208 (site 47).

Key to Column 10 (Source of value quoted in column 9)
b Ruggles 1981, table 4.1.
G Ruggles 1984a, fig. 5.8 and table 11.2 (line 30).
J *Ibid.*, fig. 5.13 and table 11.2 (line 49).
K *Ibid.*, fig. 7.4 and table 11.2 (line 120).
N *Ibid.*, fig. 7.11 and table 11.2 (line 143).
S *Ibid.*, fig. 8.6 and table 11.2 (line 166).
T *Ibid.*, fig. 8.7 and table 11.2 (line 168).
W Ruggles 1985, tables 3 and 4.
Y Ruggles 1984a, fig. 10.3 and table 11.2 (line 244).

Key to Column 13 (Reasons for larger standard deviation than assumed by Thom)
c The exact position of the circle/ring centre is uncertain.
l The outlying stoneholes are not quite in line with monument centre.
o The outlier is close to the ring centre.
q The foresight subtends a wide angle.
s The stones are close together.
t The foresight subtends about 0°·6 in azimuth, despite this being one of six lines picked out by Thom (1967, marked 'P' in table 10.1; see p. 121) as having particularly high precision. It appears that he was sighting onto one of the remaining kerbstones. See Ruggles 1981, 171 (site 8).
u One or more standing stones are fallen; their original position is uncertain.
z The circles are close together.
L The longer faces of the standing stone are irregular and point in somewhat different directions.
O See also Ruggles 1981, 171 (site 24).
Q The foresight is well below the horizon.
U The row is not straight.
V See also Ruggles 1981, 171 (site 30).
X The southernmost stone forms the horizon, giving an altitude uncertainty of 0°·6 depending on whether the base or tip of the stone is taken as the foresight. See Ruggles 1985, S115, note *o*.

Key to Column 14 (Notes)
a The foresight is a ruined Hebridean-type passage tomb, Garrabost (Henshall 1972, no. LWS6). This is Thom's site H1/13 ('Dursainean'), not H1/15, as stated in Thom 1967, table 8.1.
h The declination quoted by Thom is that of a horizon foresight some 4° to the left of the alignment of the stones. This was deduced by Ruggles (1981, table 4.1, comment 'i') and later confirmed in a remark by Thom in Thom, Thom and Burl 1990, 323.
i The declination quoted by Thom appears to be that of a horizon foresight, although not noted as such in Thom 1967 table 8.1, as is known to have been the case at site L22.
p The foresight is a ruined Hebridean-type chambered tomb, Unival (site L16 in List 1). This is Thom's site H3/11 ('Leacach an Tigh Chloiche').
v Thom (1967, table 8.1) lists this as a row of three stones, but the central stone appears not to be original. See Atkinson 1981, 207 (site 21).
w The flat faces of the backsight are in fact oriented some 10° to the left of the foresight (Ruggles 1981, 171 (site 36)). The foresight stone is one of a pair (site L38(a) in List 1).
x The backsight is actually an aligned pair of oriented stones, but the suggested indication is not in the direction of the alignment.
D Thom (1967, tables 8.1 and 12.1) lists this as a row of three stones, but there is no trace of a third, and no evidence for further stones apart from a possible adjacent stone circle, now destroyed. See Atkinson 1981, 208 (site 50).
E The first two stones in the line form part of a four-poster (see List 1).
F The foresight stone is one of a pair (site LH22 in List 2).
H The backsight is not one stone but a three-stone row. The distant foresight, site L10, is not in the direction of orientation of the row.
I The backsight is the easternmost stone of the pair.
M The indication is from outlier *L* to outlier *J* (see List 1).
P The indication is from the south-west circle to standing stone *F* [$S_1$] (see List 1).
R The indication is along the line *KFA* [$S_3S_1S_4$]. All these stones are themselves oriented in a different direction.

| 1 | 2 | 3 | 4 | 5 | 6 | 7 | 8 | 9 | 10 | 11 | 12 | 13 | 14 |
|---|---|---|---|---|---|---|---|---|---|---|---|---|---|
| L8 | N1/13 | $C_C$-S | | A | B | n | −29·7 | −29·7 | b | +0·25 | 0·2 | | |
| L9 | H1/1 | L($M_O$) | | A | Z | d | −30·2 | | | | | | |
| L10 | H1/2 | $C_C$-M | | A | A | | −18·8 | −19·5 | G | +0·08 | 0·1 | | F |
| L11 | H1/5 | M-$C_C$ | | B | A | | +27·8 | +27·4 | J | −0·77 | 0·1 | | H |
| L12 | H1/15 | M-T | | B | A | | −19·3 | | | +0·28 | 0·3 | q | a |
| L13 | H1/14 | M-T | | A | A | | +28·5 | | | +0·33 | 0·2 | t | a |
| L14 | H3/6 | T-S | | A | Z | f | −29·8 | | | | | | |
| L15 | H3/3 | M-T | | A | A | | −19·2 | −19·1 | b | +0·48 | 0·1 | | p |
| L17 | B7/3 | M-M | | A | Z | e | −19·5 | | | | | | |
| L18 | B2/4 | $C_C$-S | | A | C | k | +18·4 | +18·6 | b | +0·66 | 0·7 | oc | i |
| L19 | B3/5 | M-M | | A | A | | −19·8 | | | −0·22 | 0·2 | | |
| L20 | P3/1 | M-M-M-M | | A | A | | −29·9 | −29·9 | b | +0·05 | 0·2 | u | E |
| L22 | P1/8 | M-M | | A | C | g | +18·2 | +20·9 | b | +2·96 | 0·5 | s | h |
| L23 | P1/1 | M-M | | A | A | | +18·7 | | | +0·76 | 1·5 | s | v |
| L25 | G9/13 | S-S-S-S | | C | Z | f | −19·7 | | | | | | |
| L26 | M4/2 | $M_O$ | y | A | A | | −30·4 | −31·1 | K | −1·15 | 1·0 | LO | |
| L28 | M2/14 | M-M | | A | A | | +18·2 | | | +0·26 | 0·1 | | M |
| L29 | M2/8 | P$M_O$ | y | B | A | | +28·6 | +28·0 | N | −0·17 | 0·4 | L | |
| L31 | A2/8 | $C_C$-M | | A | A | | −20·1 | | | −0·52 | 0·2 | cQ | P |
| L31 | A2/8 | M-M-M | | A | A | | −30·3 | −29·4 | S | +0·55 | 0·2 | i | R |
| L32 | A2/12 | M-M-M-M | | A | A | | +28·2 | +28·8 | T | +0·63 | 0·6 | UV | |
| L34 | A2/19 | $M_{?O}$-$M_O$ | | A | A | | −29·9 | −29·8 | W | +0·15 | 0·2 | X | |
| L35 | A6/4 | $M_O$ | y | A | A | | −30·4 | −30·5 | b | −0·55 | 0·7 | L | I |
| L36 | A6/6 | M-M-M-M | y | A | A | | −20·0 | | | −0·42 | 1·5 | u | |
| L38 | A10/6 | $M_O$-M | y | A | A | | +27·9 | +27·5 | b | −0·67 | 0·1 | | w |
| L39 | A4/1 | M-M-M-M-M | y | A | A | | −29·7 | −29·4 | Y | +0·55 | 0·3 | U | |
| L49 | G3/3 | M-S | | C | Z | f | −30·4 | | | | | | x |
| L49 | G3/3 | M-S | | C | Z | f | −19·6 | | | | | | x |
| L50 | G4/2 | M-M | | A | C | r | +28·5 | +27·7 | b | −0·47 | 0·3 | s | |
| L51 | L1/1 | $C_D$-$C_D$ | | A | A | | −29·8 | −30·4 | b | −0·45 | 0·2 | | i |
| L53 | L1/6 | $C_C$-$C_C$ [C to D] | | A | A | | +18·3 | +17·8 | b | −0·14 | 0·6 | z | |
| L53 | L1/6 | $C_C$-$C_C$ [E to C] | | A | A | | +17·8 | +17·4 | b | −0·54 | 0·2 | c | |
| L53 | L1/6 | $C_C$-$C_C$ [E to D] | | A | A | | +27·5 | +26·7 | b | −1·47 | 0·2 | c | |
| L54 | L1/11 | M-M | | A | A | | −30·2 | −30·0 | b | −0·05 | 0·3 | s | D |
| L55 | L6/1 | M-M-M | | A | B | m | −30·7 | | | −0·75 | 0·2 | U | |
| L56 | W9/7 | L(M) | | A | A | | +17·8 | | | −0·14 | 0·1 | | |
| L57 | S5/2 | $C_C$-$M_O$-$M_O$ | y | A | B | j | +28·4 | +28·5 | b | +0·33 | 0·5 | l | |
| L58 | S1/2 | $C_C$-L(S) | | C | A | | +17·5 | | | −0·44 | 0·1 | | |

**Table 2.3 The evidence for high-precision lunar foresights**

This table lists the declinations of forty putative foresights at thirty-four sites, identified in Thom 1971, table 7.1, and presents some notes relevant to their reassessment.

Column headings:
1 Reference number in List 1
2 Thom's site reference number
3 Nature of backsight
4 Nature of indication (for key, see Table 2.2)
5 Nature of foresight
6 Page reference of horizon profile diagram in Thom 1971
7 Authenticity of monument (see Table 2.1; A = no doubt; B = some doubt; C = serious doubt)
8 Status of line (A = reasonable; B = somewhat dubious; C = very dubious; Y = no genuine indication exists, so dismissed; Z = dismissed out of hand)
9 Comments on the status of the line
10 Declination obtained by Thom, in degrees and minutes
11 Difference from relevant standstill limit, in minutes
12 Notes
13 Difference from relevant standstill limit of horizon notches and bottoms of dips only
14 Notes on the data given in column 13, if not as per Ruggles 1981 (table 4.4).

Key to Column 3 (Nature of backsight)
See brief description codes in Table 2.1, plus
V Hill summit at the uphill end of the stone rows

Key to Column 5 (Nature of foresight)
D Bottom of dip (Thom and Thom 1980b, type Ia or IIIa).
N Notch or hill junction (Thom and Thom 1980b, type I or III).
S Section of hill-slope parallel to the moon's rising or setting path (Thom and Thom 1980b, type II or IIa).
T Hilltop, or point on convex hill-slope tangential to setting path (Thom and Thom 1980b, type IV).

Key to Column 9 (Comments on the status of the line)
a There is no foresight. The horizon is close, flat and featureless. See Ruggles 1981, 175 (site 18).
b The foresight can not be seen from the backsight.
e There is no indication.
m The indications are formed by small stones in the vicinity of the standing stone planned by Thom (1971, 56–7) but which are not antiquities. For details see Ruggles 1982b, S32–3 (line 36).
n The indication is some degrees from the foresight.

Key to Column 12 (Notes)
c See Cooke *et al.* 1977, 120; Ruggles 1981, 166; Ruggles 1982b, S28 (line 14).
d See Cooke *et al.* 1977, 121; Ruggles 1981, 178 (site 11); Ruggles 1982b, S28 (line 15). Thom's profile diagram was calculated. For a surveyed profile diagram see Ruggles 1984a, fig. 5.12 (line 45).
f See Thom 1971, 95–7; Ruggles 1981, 178 (site 2).
g The possible former existence of further rows, which would have indicated the foresights, is postulated by Thom (1971, 92). See Ruggles 1982b, S28 (lines 12–13).
h The profile here is where the moon would rise when it set on the horizon indicated by the aligned pair to the south-west (Thom 1971, 55–6), but no direct indication of the foresight exists at all. See Ruggles 1982b, S29 (line 18). According to Burl (1988b, 127) the indication is from the north-west to the south-east stone, but this direction is some 30° away from the foresight.
i The backsight is the fallen standing stone *M* (site L48(*a*) in List 1). The foresight is the tip of the Mull of Kintyre ($A_2$).
j The backsight is the standing stone 'Long Tom' (site L48(*c*) in List 1). Viewed from here a possible cairn (site L48(*b*)) is visible beneath the tip of the Mull of Kintyre, but a feature more than 1° to the right ($A_1$) is considered to be the foresight.
k See also Ruggles 1981, 178 (site 45).
l See Ruggles 1982b, S33 (lines 39–40).
o The indication formed by the original alignment of stones seems to have been some 20° or more to the left of the foresight. See Ruggles 1981, 179 (site 25). For a site plan see *ibid.*, fig. 4.4. For a surveyed profile diagram see Ruggles 1984a, fig. 7.4 (line 122).
p The alignment from the fallen stone to the small standing stone (from L33(*a*) to L33(*b*), see List 1) is some 3° to the right of the foresight (for a surveyed profile diagram see Ruggles 1984a, fig. 8.10 (line 180)). See also Ruggles 1982b, S30–1 (line 26). The fallen stone was not oriented towards the foresight but originally faced WSW–ENE, according to Campbell and Sandeman (1961, no. 175).
q The proposed indication is the flat face of the standing stone adjacent to a large chambered tomb. See Ruggles 1981, 182 (site 14). See also Moir *et al.* 1980. For an independent surveyed profile diagram see Ruggles 1981, fig. 4.3a.
r The foresight is a kerb-cairn. See Atkinson 1981, 206–7 (site 19); Ruggles 1982b, S29 (line 17).
s The foresight is Thom's small notch $A_1$ (Thom 1971, fig. 5.1). For an independent surveyed profile diagram see Ruggles 1981, fig. 4.3c.
t The backsight is standing stone *F* (Thom's $S_1$). Its south-west (flatter) face is actually oriented about 7° to the right of the foresight, and the notch is above a point between the two circles at Temple Wood (Ruggles 1982b, S30 (line 20)).
u The backsight is the setting *C–E* (Thom's 'group *Q*'). The indication is the south-west circle at Temple Wood, which sits centrally beneath the notch (Ruggles 1982b, S30 (line 21)).
v The backsight is the aligned pair *BA* (Thom's $S_5S_4$). The alignment is actually oriented about 4°·5 to the right of the foresight, and the south-west circle at Temple Wood is about 3°–4° to its right (Ruggles 1982b, S30 (line 22)).
w The indication is along standing stones *KFA* (Thom's $S_3S_1S_4$). See also Ruggles 1982b, S30 (line 23). For an independent surveyed profile diagram see Ruggles 1984a, fig. 8.6 (lines 164 and 166).
y For a description see Ruggles 1982b, S31 (line 27). For an independent surveyed profile diagram see Ruggles 1981, fig. 4.3b.
z The indication is from the standing stone through the centre of the kerb-cairn to its south-west (Thom 1969, 9). For more details see Ruggles 1982b, S29 (line 19).
E See Ruggles 1982b, S32 (line 34).
G See *ibid.*, S32 (line 35).
H See *ibid.*, S33 (lines 37–8).

Key to Column 14 (Notes on the data in column 13)
x For the north-west profile, only the values from setting *C–E* (group *Q*) have been included, the others being unindicated.
F These data were not included in Ruggles 1981, table 4.4 but are included here in the light of *ibid.*, 179, note 1. They have been generated afresh from unpublished survey data obtained in 1981. They are consistent with Ruggles 1983, table 5 (omitting lines of type V) but with a systematic difference of −2′ due to additional corrections applied in Ruggles 1983 to match those applied by the Thoms at Level 4.

| 1 | 2 | 3 | 4 | 5 | 6 | 7 | 8 | 9 | 10 | 11 | 12 | 13 | 14 |
|---|---|---|---|---|---|---|---|---|---|---|---|---|---|
| L4 | N1/17 | V | — | D | 96 | A | Y | e | −29 19·1 | −16·8 | f | — | |
| L4 | N1/17 | V | — | S | 96 | A | Y | e | −28 48·5 | +13·8 | f | | |
| L7 | N1/1 | V | — | S | 93 | A | Y | e | −29 26·6 | −24·3 | g | — | |
| L7 | N1/1 | V | — | S | 93 | A | Y | e | −29 08·7 | −6·4 | g | | |
| L9 | H1/1 | M | L($M_O$) | D | 69 | A | Z | b | −29 27·1 | −24·8 | c | — | |
| L9 | H1/1 | M | L($M_O$) | T | 69 | A | Z | b | −29 11·6 | −9·3 | c | | |
| L11 | H1/5 | M | M-M-M | T | 69 | B | Z | b | −29 26·3 | −24·0 | d | — | |
| L11 | H1/5 | M | M-M-M | T | 69 | B | Z | b | −29 20·5 | −18·2 | d | | |
| L11 | H1/5 | M | M-M-M | D | 69 | B | Z | b | −28 58·1 | +4·2 | d | | |
| L11 | H1/5 | M | M-M-M | D | 69 | B | Y | n | −28 35·4 | +26·9 | d | | |
| L16 | H3/11 | M | $M_O$ | D | 70 | A | B | | −29 29·5 | −27·2 | q | −34, −5, | |
| L16 | H3/11 | M | $M_O$ | D | 70 | A | B | | −29 04·6 | −2·3 | q | +7, +15 | |
| L20 | P3/1 | M | M-M-M-M | — | — | A | Z | a | −28 55 | +7·3 | | — | |
| L21 | P1/10 | U | U-$M_O$ | N | 54 | A | B | | +29 18·8 | +16·5 | r | +19 | |
| L24 | P4/1 | M | — | T | 55 | A | Y | e | −18 59·7 | −15·1 | h | — | |
| L27 | M1/3 | M | M-M-M-M | T | 67 | A | Y | n | −29 26·9 | −24·6 | o | — | |
| L30 | A2/5 | M | M-U | N | 39 | A | B | z | −18 40·6 | +4·0 | | −33, −29, −20, −3, | F |
| L30 | A2/5 | M | M-U | S | 39 | A | B | z | −18 20·3 | +24·3 | | −2, +5, +10 | |
| L31 | A2/8 | M | $M_O$ | N | 46 | A | Y | n | +28 56·5 | −5·8 | st | +1, +2; | x |
| L31 | A2/8 | G | C | N | 46 | A | A | | +29 02·5 | +0·2 | su | −17, −16, | |
| L31 | A2/8 | M | M-M | N | 46 | A | Y | n | +29 19·1 | +16·8 | sv | +14, +19 | |
| L31 | A2/8 | M | M-M-M | N | 46 | A | A | | −28 48 | +14·3 | w | | |
| L32 | A2/12 | M | M-M-M-M | S | 52 | A | A | | +29 13·6 | +11·3 | | +19, +32 | |
| L32 | A2/12 | M | M-M-M-M | S | 52 | A | A | | +29 28·5 | +26·2 | | | |
| L33 | A2/13 | M | M-M | S | 64 | A | Y | n | +29 17·0 | +14·7 | p | — | |
| L35 | A6/4 | M | $M_O$ | T | 65 | A | A | | −29 25·9 | −23·6 | y | −26, −24 | |
| L38 | A10/6 | M | M-M | S | 66 | A | A | | +28 52·4 | −9·9 | | — | |
| L39 | A4/1 | M | M-M-M-M-M | S | 60 | A | A | | −28 39·3 | +23·0 | | +20 | |
| L42 | A4/17 | M | — | N | 62 | A | Y | e | +28 34·6 | −27·7 | E | — | |
| L42 | A4/17 | M | — | S | 62 | A | Y | e | +29 26·3 | +24·0 | E | | |
| L43 | A4/5 | M | — | N | 61 | A | Y | e | +29 12·0 | +9·7 | G | — | |
| L44 | A4/9 | S | S-M | D | 57 | A | Y | m | −28 37·3 | +25·0 | | — | |
| L44 | A4/9 | S | S-S | T | 57 | A | Y | m | +18 18·0 | −26·6 | | | |
| L45 | A4/2 | M | — | S | 60 | A | Y | e | −29 26·4 | −24·1 | H | — | |
| L45 | A4/2 | M | — | S | 60 | A | Y | e | −28 45·7 | +16·6 | H | | |
| L46 | A4/14 | M | — | D | 62 | A | Y | e | −28 53·8 | +8·5 | l | — | |
| L46 | A4/14 | M | — | T | 62 | A | Y | e | −28 44·0 | +18·3 | l | | |
| L48 | G3/2 | M | — | N | 72 | A | Y | e | +18 17·8 | −26·8 | ik | — | |
| L48 | G3/2 | M | — | S | 72 | A | Y | e | +19 09·9 | +25·3 | jk | | |
| L56 | W9/7 | M | L(M) | S | 73 | A | A | | +18 19·7 | −24·9 | | −29, −24, −8, +7, +23 | |

**Table 3.1 Selected indications: summary data from the study of 300 western Scottish sites, 1975–81**

Column headings:
1 Reference number in List 2
2 Site classification (see text)
3 Number of on-site lines in category A (reliably surveyed); category B (less reliably surveyed); category C (calculated); category L (dismissed because horizon is local); and category W (dismissed because IAR is too wide).

| 1 | 2 | 3 | | | | |
|---|---|---|---|---|---|---|
| | | A | B | C | L | W |
| LH5 | 5 | 0 | 0 | 0 | 0 | 2 |
| LH6 | 2 | 0 | 0 | 0 | 0 | 2 |
| LH8 | 3 | 1 | 0 | 0 | 1 | 0 |
| LH10 | 4 | 0 | 1 | 0 | 1 | 0 |
| LH16 | 1 | 5 | 0 | 2 | 3 | 0 |
| LH22 | 3 | 1 | 0 | 1 | 0 | 0 |
| LH24 | 3 | 2 | 0 | 0 | 0 | 0 |
| LH29 | 6 | 1 | 0 | 0 | 1 | 0 |
| LH36 | 5 | 1 | 0 | 0 | 1 | 0 |
| LH37 | 3 | 0 | 0 | 1 | 1 | 0 |
| UI6 | 5 | 2 | 0 | 0 | 0 | 0 |
| UI9 | 3 | 1 | 0 | 0 | 1 | 0 |
| UI19 | 1 | 1 | 0 | 0 | 1 | 0 |
| UI23 | 3 | 1 | 0 | 0 | 1 | 0 |
| UI28 | 5 | 1 | 0 | 0 | 1 | 0 |
| UI29 | 6 | 0 | 0 | 0 | 2 | 0 |
| UI31 | 5 | 1 | 0 | 0 | 1 | 0 |
| UI46 | 6 | 2 | 0 | 0 | 0 | 0 |
| UI49 | 5 | 1 | 0 | 1 | 0 | 0 |
| UI57 | 3 | 2 | 0 | 0 | 0 | 0 |
| UI59 | 4 | 2 | 0 | 0 | 0 | 0 |
| NA1 | 4 | 1 | 1 | 0 | 0 | 0 |
| NA3 | 5 | 1 | 0 | 0 | 1 | 0 |
| CT1 | 5 | 0 | 0 | 1 | 1 | 0 |
| CT2 | 3 | 2 | 0 | 0 | 0 | 0 |
| CT9 | 5 | 1 | 0 | 0 | 1 | 0 |
| ML2 | 2 | 1 | 0 | 0 | 1 | 0 |
| ML4 | 2 | 1 | 0 | 1 | 0 | 0 |
| ML6 | 4 | 0 | 0 | 0 | 2 | 0 |
| ML7 | 5 | 1 | 0 | 0 | 1 | 0 |
| ML8 | 5 | 0 | 0 | 0 | 0 | 2 |
| ML9 | 1 | 0 | 0 | 1 | 1 | 0 |
| ML10 | 2 | 0 | 1 | 0 | 1 | 0 |
| ML11 | 1 | 1 | 0 | 0 | 1 | 0 |
| ML12 | 2 | 2 | 0 | 0 | 0 | 0 |
| ML16 | 3 | 0 | 2 | 0 | 0 | 0 |

| 1 | 2 | 3 | | | | |
|---|---|---|---|---|---|---|
| | | A | B | C | L | W |
| ML18 | 3 | 1 | 0 | 0 | 1 | 0 |
| ML21 | 2 | 0 | 0 | 0 | 2 | 0 |
| ML25 | 3 | 2 | 0 | 0 | 0 | 0 |
| ML30 | 6 | 1 | 0 | 0 | 1 | 0 |
| ML31 | 5 | 2 | 0 | 0 | 0 | 0 |
| ML33 | 3 | 1 | 0 | 0 | 1 | 0 |
| ML34 | 3 | 0 | 0 | 0 | 0 | 2 |
| LN7 | 5 | 0 | 0 | 1 | 1 | 0 |
| LN18 | 4 | 0 | 1 | 0 | 1 | 0 |
| LN22 | 1 | 1 | 1 | 0 | 0 | 0 |
| AR2 | 5 | 0 | 0 | 1 | 1 | 0 |
| AR3 | 1 | 2 | 0 | 0 | 0 | 0 |
| AR6 | 2 | 1 | 0 | 0 | 1 | 0 |
| AR9 | 5 | 0 | 1 | 0 | 1 | 0 |
| AR10 | 5 | 0 | 0 | 0 | 2 | 0 |
| AR12 | 1 | 0 | 0 | 0 | 2 | 0 |
| AR13 | 1 | 2 | 1 | 3 | 2 | 0 |
| AR15 | 1 | 0 | 3 | 1 | 0 | 0 |
| AR16 | 5 | 0 | 2 | 0 | 0 | 0 |
| AR19 | 5 | 0 | 0 | 0 | 0 | 2 |
| AR23 | 5 | 0 | 0 | 0 | 2 | 0 |
| AR25 | 5 | 0 | 0 | 0 | 2 | 0 |
| AR27 | 4 | 0 | 1 | 0 | 1 | 0 |
| AR28 | 2 | 0 | 2 | 0 | 0 | 0 |
| AR29 | 1 | 0 | 2 | 0 | 0 | 0 |
| AR30 | 5 | 0 | 0 | 0 | 0 | 2 |
| AR31 | 4 | 0 | 0 | 0 | 2 | 0 |
| AR32 | 5 | 0 | 0 | 2 | 0 | 0 |
| AR33 | 5 | 0 | 0 | 2 | 0 | 0 |
| AR39 | 5 | 0 | 0 | 0 | 0 | 2 |
| JU1 | 4 | 1 | 0 | 0 | 1 | 0 |
| JU4 | 3 | 1 | 1 | 0 | 0 | 0 |
| JU6 | 2 | 0 | 0 | 0 | 0 | 2 |
| JU7 | 2 | 0 | 0 | 2 | 0 | 0 |
| JU8 | 4 | 0 | 0 | 0 | 2 | 0 |
| JU9 | 5 | 2 | 0 | 0 | 0 | 0 |

| 1 | 2 | 3 | | | | |
|---|---|---|---|---|---|---|
| | | A | B | C | L | W |
| IS3 | 5 | 1 | 0 | 0 | 1 | 0 |
| IS4 | 5 | 1 | 0 | 0 | 1 | 0 |
| IS5 | 5 | 0 | 0 | 0 | 0 | 2 |
| IS7 | 5 | 0 | 0 | 0 | 2 | 0 |
| IS11 | 3 | 1 | 0 | 0 | 1 | 0 |
| IS15 | 3 | 0 | 0 | 0 | 2 | 0 |
| IS19 | 5 | 1 | 0 | 0 | 1 | 0 |
| IS23 | 5 | 1 | 0 | 0 | 1 | 0 |
| IS24 | 5 | 0 | 0 | 0 | 0 | 2 |
| IS35 | 6 | 0 | 0 | 0 | 0 | 2 |
| IS38 | 4 | 0 | 0 | 2 | 0 | 0 |
| IS39 | 3 | 1 | 0 | 0 | 1 | 0 |
| IS41 | 2 | 0 | 0 | 1 | 1 | 0 |
| IS42 | 3 | 0 | 0 | 0 | 0 | 2 |
| IS47 | 5 | 0 | 0 | 0 | 0 | 2 |
| KT2 | 3 | 2 | 1 | 2 | 1 | 0 |
| KT3 | 5 | 2 | 0 | 0 | 0 | 0 |
| KT4 | 5 | 0 | 0 | 1 | 1 | 0 |
| KT5 | 1 | 0 | 0 | 2 | 0 | 0 |
| KT8 | 5 | 1 | 0 | 0 | 1 | 0 |
| KT10 | 1 | 0 | 1 | 0 | 1 | 0 |
| KT12 | 5 | 0 | 0 | 1 | 1 | 0 |
| KT19 | 4 | 0 | 0 | 1 | 1 | 0 |
| KT21 | 5 | 0 | 0 | 0 | 2 | 0 |
| KT23 | 5 | 1 | 0 | 0 | 1 | 0 |
| KT27 | 2 | 1 | 1 | 0 | 0 | 0 |
| KT28 | 5 | 1 | 0 | 0 | 1 | 0 |
| KT29 | 5 | 1 | 0 | 0 | 1 | 0 |
| KT31 | 5 | 1 | 0 | 0 | 1 | 0 |
| KT32 | 5 | 0 | 0 | 0 | 1 | 1 |
| KT35 | 5 | 0 | 0 | 2 | 0 | 0 |
| KT36 | 5 | 0 | 0 | 0 | 2 | 0 |
| KT39 | 5 | 2 | 0 | 0 | 0 | 0 |
| KT40 | 6 | 0 | 0 | 0 | 2 | 0 |
| KT41 | 5 | 2 | 0 | 0 | 0 | 0 |
| KT44 | 5 | 2 | 0 | 0 | 0 | 0 |

**Table 3.2 Indicated horizons: summary of data from the study of 300 western Scottish sites, 1975–81**

The data on azimuth and declination limits have been taken from Ruggles 1984a, table 11.2. One adjustment has been made, to line 129, where the calculated profile has been revised and the declination limits have been altered by up to 0°·6. The mean altitude and maximum and minimum declination data have been obtained from the profile diagrams in Ruggles 1984a or from raw survey data.

Column headings:

1 Line number in Ruggles 1984a
2 Site reference number in Ruggles 1984a. For the site name cross-refer to List 2
3 Details of the indication
4 On-site line classification[1]
5 Inter-site line classification[2]
6 Reliability of the data (A = surveyed; B = extrapolated from surveyed points or unreliable for other reasons, e.g. large parallax correction; C = calculated)
7 Lesser azimuth at limit of IAR, quoted to the nearest 0·2 degrees
8 Greater azimuth at limit of IAR, quoted to the nearest 0·2 degrees
9 Mean altitude within IAR, quoted to the nearest 0·2 degrees
10 Lesser declination at limit of IAR, quoted to the nearest 0·1 degrees
11 Greater declination at limit of IAR, quoted to the nearest 0·1 degrees
12 Minimum or maximum declination within IAR where different, quoted to the nearest 0·1 degrees
13 Lesser azimuth at limit of AAR, quoted to the nearest 0·2 degrees
14 Greater azimuth at limit of AAR, quoted to the nearest 0·2 degrees
15 Mean altitude within AAR, quoted to the nearest 0·2 degrees
16 Lesser declination at limit of AAR, quoted to the nearest 0·1 degrees
17 Greater declination at limit of AAR, quoted to the nearest 0·1 degrees
18 Minimum or maximum declination within AAR where different, quoted to the nearest 0·1 degrees

| 1 | 2 | 3 | 4 | 5 | 6 | 7 | 8 | 9 | 10 | 11 | 12 | 13 | 14 | 15 | 16 | 17 | 18 |
|---|---|---|---|---|---|---|---|---|---|---|---|---|---|---|---|---|---|
| 1 | LH7 | I/s to LH8 | | B1 | A | 251·6 | 251·8 | 0·6 | −9·5 | −9·3 | | 250·6 | 252·8 | 0·6 | −10·1 | −8·9 | |
| 2 | LH8 | *ba* (to SW) | 3a | | A | 219·4 | 223·6 | 1·3 | −23·1 | −21·6 | | 217·4 | 225·6 | 1·4 | −23·8 | −20·6 | |
| 3 | LH8 | I/s to LH7 | | B1 | A | 71·6 | 71·8 | 1·0 | 10·0 | 10·1 | | 70·6 | 72·8 | 1·0 | 9·5 | 10·7 | |
| 4 | LH10 | *ab* (to SE) | 4a | | B | 146·0 | 147·0 | 0·0 | −26·7 | −26·5 | | 145·0 | 148·0 | 0·0 | −27·1 | −26·1 | |
| 5 | LH10 | I/s to LH16 | | A2 | A | 194·8 | 195·0 | 0·5 | −30·7 | −30·5 | | 193·8 | 196·0 | 0·6 | −30·9 | −29·7 | |
| 6 | LH10 | I/s to LH18 | | A1 | A | 170·4 | 170·6 | 0·7 | −31·1 | −31·0 | | 169·4 | 171·6 | 0·8 | −31·2 | −30·6 | |
| 7 | LH10 | I/s to LH19 | | A1 | A | 176·0 | 176·2 | 0·3 | −31·9 | −31·9 | | 175·0 | 177·2 | 0·3 | −32·1 | −31·9 | |
| 8 | LH10 | I/s to LH21 | | A3 | A | 167·6 | 167·8 | 0·6 | −30·8 | −30·7 | | 166·6 | 168·8 | 0·8 | −30·8 | −30·6 | |
| 9 | LH10 | I/s to LH22 | | A3 | B | 151·4 | 151·6 | 0·2 | −28·0 | −27·9 | | 150·4 | 152·6 | 0·2 | −28·2 | −27·6 | |
| 10 | LH10 | I/s to LH24 | | B3 | A | 164·2 | 164·4 | 0·6 | −30·3 | −30·2 | | 163·2 | 165·4 | 0·7 | −30·4 | −30·0 | |
| 11 | LH16 | Av W to N | 1a | | A | 7·6 | 10·8 | 1·4 | 32·1 | 32·5 | | 5·6 | 12·8 | 1·4 | 32·1 | 32·5 | |
| 12 | LH16 | Av E to N | 1a | | A | 9·8 | 12·0 | 1·6 | 32·3 | 32·5 | | 7·8 | 14·0 | 1·5 | 31·9 | 32·5 | |
| 13 | LH16 | S row to N | 1a | | A | 359·4 | 0·0 | 1·0 | 32·4 | 32·4 | | 358·4 | 1·0 | 1·0 | 32·3 | 32·5 | |
| 14 | LH16 | W row to W | 1a | | A | 266·0 | 268·6 | 0·7 | −1·8 | −0·6 | | 264·0 | 270·6 | 0·6 | −2·8 | 0·3 | |
| 15 | LH16 | W row to E | 1a | | C | 86·0 | 88·6 | 0·6 | 0·9 | 2·3 | | 84·0 | 90·6 | 0·6 | −0·1 | 3·3 | |
| 16 | LH16 | E row to W | 1a | | C | 258·8 | 259·6 | 1·0 | −5·2 | −5·0 | | 257·8 | 260·6 | 1·1 | −5·5 | −4·6 | |
| 17 | LH16 | E row to E | 1a | | A | 78·8 | 79·6 | 0·9 | 5·9 | 6·3 | | 77·8 | 80·6 | 0·8 | 5·3 | 6·7 | |
| 18 | LH16 | I/s to LH10 | | A2 | A | 14·8 | 15·0 | 1·4 | 31·6 | 31·7 | | 13·8 | 16·0 | 1·4 | 31·3 | 31·9 | |
| 19 | LH16 | I/s to LH18 | | A1 | A | 99·8 | 100·8 | 1·0 | −5·1 | −4·6 | | 98·8 | 101·8 | 1·1 | −5·6 | −4·1 | |
| 20 | LH16 | I/s to LH19 | | A1 | A | 109·6 | 111·4 | 0·4 | −11·1 | −10·2 | | 107·8 | 113·2 | 0·4 | −12·1 | −9·3 | |
| 21 | LH16 | I/s to LH21 | | A3 | A | 143·0 | 143·2 | 0·5 | −24·9 | −24·8 | | 142·0 | 144·2 | 0·4 | −25·3 | −24·6 | |
| 22 | LH16 | I/s to LH22 | | A1 | A | 124·0 | 124·6 | 0·2 | −17·7 | −17·4 | | 123·0 | 125·6 | 0·2 | −18·1 | −16·9 | |
| 23 | LH16 | I/s to LH24 | | B3 | A | 141·6 | 141·8 | 0·4 | −24·5 | −24·5 | | 140·6 | 142·8 | 0·4 | −24·8 | −24·0 | |
| 24 | LH18 | I/s to LH10 | | A2 | A | 350·2 | 350·4 | 1·5 | 32·5 | 32·5 | | 349·2 | 351·4 | 1·5 | 32·3 | 32·5 | |
| 25 | LH18 | I/s to LH16 | | A1 | A | 278·8 | 283·6 | 0·2 | 4·5 | 7·0 | | 276·8 | 285·6 | 0·3 | 3·6 | 8·0 | |
| 26 | LH18 | I/s to LH19 | | A1 | A | 247·2 | 251·0 | 1·4 | −10·9 | −9·1 | | 245·2 | 253·0 | 1·3 | −11·9 | −8·0 | |
| 27 | LH18 | I/s to LH21 | | A1 | A | 164·2 | 164·4 | 1·0 | −29·9 | −29·9 | | 163·2 | 165·4 | 1·0 | −30·1 | −29·8 | |
| 28 | LH18 | I/s to LH24 | | B2 | A | 157·8 | 158·2 | 0·4 | −29·3 | −29·2 | | 156·8 | 159·2 | 0·5 | −29·3 | −28·9 | |
| 29 | LH19 | I/s to LH10 | | A3 | A | 356·0 | 356·2 | 1·6 | 32·9 | 33·0 | | 355·0 | 357·2 | 1·6 | 32·8 | 32·9 | Max 33·0 |
| 30 | LH19 | I/s to LH22 | | A1 | A | 128·8 | 129·2 | 0·4 | −19·6 | −19·4 | | 127·8 | 130·2 | 0·4 | −20·0 | −18·9 | |
| 31 | LH19 | I/s to LH24 | | B2 | A | 152·0 | 152·2 | 0·8 | −27·4 | −27·4 | | 151·0 | 153·2 | 0·7 | −27·8 | −27·2 | |
| 32 | LH21 | I/s to LH10 | | A3 | C | 347·6 | 347·8 | 0·7 | 31·3 | 31·4 | | 346·6 | 348·8 | 0·7 | 31·1 | 31·5 | |
| 33 | LH21 | I/s to LH16 | | A1 | A | 322·6 | 324·0 | 0·1 | 24·4 | 24·8 | | 321·2 | 325·4 | 0·1 | 23·8 | 25·4 | |
| 34 | LH21 | I/s to LH18 | | A2 | A | 344·0 | 344·4 | 0·6 | 30·5 | 30·6 | | 343·0 | 345·4 | 0·6 | 30·2 | 30·8 | |
| 35 | LH21 | I/s to LH22 | | A2 | A | 88·6 | 89·0 | 1·0 | 1·0 | 1·3 | | 87·6 | 90·0 | 1·0 | 0·3 | 2·0 | |

*Continued*

| 1 | 2 | 3 | 4 | 5 | 6 | 7 | 8 | 9 | 10 | 11 | 12 | 13 | 14 | 15 | 16 | 17 | 18 |
|---|---|---|---|---|---|---|---|---|---|---|---|---|---|---|---|---|---|
| 36 | LH21 | I/s to LH24 | | B2 | A | 134·8 | 135·4 | 1·2 | −21·2 | −21·0 | | 133·8 | 136·4 | 1·2 | −21·7 | −20·5 | |
| 37 | LH22 | *ba* (to NNE) | 3a | | C | 33·0 | 35·8 | 0·8 | 25·8 | 26·5 | | 31·0 | 37·8 | 0·8 | 25·3 | 27·3 | |
| 38 | LH22 | *ab* (to SSW) | 3a | | A | 213·0 | 215·8 | 1·2 | −25·8 | −24·5 | | 211·0 | 217·8 | 1·1 | −26·6 | −24·0 | |
| 39 | LH22 | I/s to LH10 | | A3 | C | 331·4 | 331·6 | 0·6 | 27·7 | 27·8 | | 330·4 | 332·6 | 0·5 | 27·2 | 28·2 | |
| 40 | LH22 | I/s to LH16 | | A1 | A | 304·2 | 304·8 | 0·1 | 16·8 | 17·1 | | 303·2 | 305·8 | 0·1 | 16·4 | 17·5 | |
| 41 | LH22 | I/s to LH19 | | A3 | A | 308·8 | 309·2 | 0·0 | 18·8 | 19·0 | | 307·8 | 310·2 | 0·1 | 18·4 | 19·5 | |
| 42 | LH22 | I/s to LH21 | | A2 | A | 268·0 | 268·4 | 1·2 | −0·3 | −0·2 | | 267·0 | 269·4 | 1·1 | −1·0 | 0·1 | |
| 43 | LH22 | I/s to LH24 | | B2 | A | 243·8 | 247·8 | 0·8 | −13·1 | −11·0 | | 241·8 | 249·8 | 0·9 | −13·9 | −10·0 | |
| 44 | LH24 | *bcd* (to NNW) | 3c | | C | 342·0 | 344·0 | 0·4 | 29·9 | 30·5 | | 340·0 | 346·0 | 0·3 | 29·5 | 30·9 | |
| 45 | LH24 | *dcb* (to SSE) | 3c | | A | 162·0 | 164·0 | 1·0 | −29·6 | −29·2 | | 160·0 | 166·0 | 1·0 | −29·6 | −29·2 | |
| 46 | LH24 | I/s to LH10 | | B3 | C | 344·2 | 344·4 | 0·6 | 30·5 | 30·6 | | 343·2 | 345·4 | 0·5 | 30·3 | 30·8 | |
| 47 | LH24 | I/s to LH16 | | B3 | A | 321·4 | 322·0 | −0·1 | 23·7 | 24·0 | | 320·4 | 323·0 | −0·1 | 23·3 | 24·4 | |
| 48 | LH24 | I/s to LH18 | | B2 | A | 337·8 | 338·2 | 0·3 | 29·0 | 29·2 | | 336·8 | 339·2 | 0·3 | 28·8 | 29·3 | |
| 49 | LH24 | I/s to LH19 | | B2 | A | 332·0 | 332·2 | 0·1 | 27·3 | 27·4 | | 331·0 | 333·2 | 0·1 | 26·9 | 27·7 | |
| 50 | LH24 | I/s to LH21 | | B2 | A | 313·4 | 314·8 | 0·0 | 20·8 | 21·4 | | 312·0 | 316·2 | 0·0 | 20·3 | 21·8 | |
| 51 | LH24 | I/s to LH22 | | B2 | A | 65·2 | 65·6 | 0·6 | 12·8 | 13·0 | | 64·2 | 66·6 | 0·6 | 12·3 | 13·4 | |
| 52 | LH29 | Flat face to NW | 6 | | A | 305·4 | 307·2 | 0·1 | 17·4 | 18·3 | | 303·6 | 309·0 | 0·1 | 16·5 | 19·0 | |
| 53 | LH36 | Flat face to WNW | 5a | | A | 284·4 | 289·0 | 0·0 | 6·9 | 9·6 | | 282·4 | 291·0 | 0·0 | 5·8 | 10·9 | |
| 54 | LH36 | I/s to LH37 | | A3 | A | 207·6 | 207·8 | 1·2 | −27·3 | −27·2 | | 206·6 | 208·8 | 1·4 | −27·4 | −26·8 | |
| 55 | LH37 | *ba* (to WNW) | 3b | | C | 302·0 | 305·0 | −0·1 | 15·8 | 17·0 | | 300·0 | 307·0 | −0·1 | 14·8 | 18·1 | |
| 56 | LH37 | I/s to LH36 | | A3 | A | 27·6 | 27·8 | 1·6 | 29·3 | 29·4 | | 26·6 | 28·8 | 1·7 | 29·3 | 29·5 | |
| 57 | UI6 | Flat face to WNW | 5a | | A | 293·0 | 294·0 | −0·2 | 11·3 | 11·8 | | 292·0 | 295·0 | −0·2 | 10·8 | 12·3 | |
| 58 | UI6 | Flat face to ESE | 5a | | A | 113·0 | 114·0 | 0·1 | −12·8 | −12·3 | | 112·0 | 115·0 | 0·1 | −13·3 | −11·8 | |
| 59 | UI6 | I/s to UI9 | | A3 | A | 212·0 | 212·4 | 0·3 | −27·0 | −26·8 | | 211·0 | 213·4 | 0·3 | −27·2 | −26·4 | |
| 60 | UI9 | *ba* (to SW) | 3a | | A | 229·0 | 232·4 | 0·5 | −20·6 | −18·9 | | 227·0 | 234·4 | 0·5 | −21·1 | −18·0 | |
| 61 | UI9 | I/s to UI6 | | A3 | A | 32·0 | 32·4 | 1·0 | 27·4 | 27·6 | | 31·0 | 33·4 | 0·9 | 26·9 | 27·9 | |
| 62 | UI15 | I/s to UI19 | | A2 | C | 111·8 | 112·8 | −0·2 | −12·6 | −12·2 | | 110·8 | 113·8 | −0·2 | −13·1 | −11·9 | |
| 63 | UI19 | *abc* (to WNW) | 1a | | A | 288·8 | 290·6 | 2·3 | 11·7 | 12·5 | | 287·0 | 292·4 | 2·1 | 10·7 | 12·9 | |
| 64 | UI19 | I/s to UI15 | | A2 | C | 292·0 | 292·6 | 1·7 | 12·9 | 13·0 | | 291·0 | 293·6 | 1·7 | 12·6 | 13·3 | |
| 65 | UI22 | I/s to UI26 | | A3 | A | 115·2 | 115·6 | 0·1 | −13·8 | −13·5 | | 114·2 | 116·6 | 0·0 | −14·3 | −13·2 | |
| 66 | UI22 | I/s to UI31 | | A3 | A | 153·4 | 153·8 | −0·2 | −29·6 | −29·3 | | 152·4 | 154·8 | −0·2 | −30·0 | −28·8 | |
| 67 | UI23 | *ba* (to E) | 3a | | A | 94·0 | 96·0 | 1·1 | −2·8 | −1·2 | | 92·0 | 98·0 | 1·0 | −4·1 | 0·3 | |
| 68 | UI23 | I/s to UI26 | | A1 | A | 120·4 | 120·8 | 0·2 | −16·1 | −15·9 | | 119·4 | 121·8 | 0·3 | −16·3 | −15·5 | |
| 69 | UI23 | I/s to UI28 | | A1 | A | 133·6 | 133·8 | 0·0 | −22·3 | −22·2 | | 132·6 | 134·8 | 0·0 | −22·7 | −21·5 | |
| 70 | UI23 | I/s to UI31 | | A3 | A | 175·2 | 175·4 | −0·1 | −33·0 | −32·9 | | 174·2 | 176·4 | 0·0 | −33·0 | −32·7 | |
| 71 | UI26 | I/s to UI22 | | A1 | A | 295·2 | 295·6 | 0·2 | 12·9 | 13·1 | | 294·2 | 296·6 | 0·2 | 12·2 | 13·7 | |
| 72 | UI26 | I/s to UI23 | | A2 | A | 300·4 | 300·8 | 0·7 | 16·0 | 16·2 | | 299·4 | 301·8 | 0·7 | 15·5 | 16·7 | |
| 73 | UI26 | I/s to UI28 | | A1 | A | 143·8 | 144·2 | 0·2 | −26·1 | −25·9 | | 142·8 | 145·2 | 0·2 | −26·6 | −25·3 | |
| 74 | UI26 | I/s to UI31 | | A3 | A | 204·8 | 205·2 | −0·6 | −30·4 | −30·3 | | 203·8 | 206·2 | −0·6 | −30·6 | −30·0 | |
| 75 | UI28 | Flat face to SSE | 5a | | A | 154·8 | 156·4 | −0·1 | −30·0 | −29·8 | | 153·2 | 158·0 | −0·2 | −30·6 | −29·4 | |
| 76 | UI28 | I/s to UI23 | | A3 | A | 313·6 | 313·8 | 0·3 | 21·5 | 21·6 | | 312·6 | 314·8 | 0·3 | 21·0 | 22·0 | |
| 77 | UI28 | I/s to UI26 | | A2 | A | 323·8 | 324·2 | 0·3 | 25·4 | 25·6 | | 322·8 | 325·2 | 0·3 | 25·1 | 25·9 | |
| 78 | UI28 | I/s to UI33 | | A3 | A | 108·8 | 109·2 | 0·2 | −10·5 | −10·2 | | 107·8 | 110·2 | 0·1 | −11·1 | −9·9 | |
| 79 | UI28 | I/s to UI35 | | B2 | C | 150·0 | 150·2 | −0·3 | −28·6 | −28·5 | | 149·0 | 151·2 | −0·3 | −28·9 | −28·3 | |
| 80 | UI28 | I/s to UI37 | | A3 | C | 138·4 | 138·8 | −0·2 | −24·5 | −24·3 | | 137·4 | 139·8 | −0·3 | −25·1 | −24·0 | |
| 81 | UI31 | Flat face to ESE | 5a | | A | 107·4 | 108·8 | 0·8 | −9·3 | −9·1 | | 106·0 | 110·2 | 0·8 | −9·9 | −8·6 | |
| 82 | UI31 | I/s to UI22 | | A3 | C | 333·4 | 333·8 | 1·2 | 29·4 | 29·6 | | 332·4 | 334·8 | 1·2 | 29·1 | 29·9 | |
| 83 | UI31 | I/s to UI23 | | A3 | A | 355·2 | 355·4 | 1·2 | 33·1 | 33·1 | | 354·2 | 356·4 | 1·2 | 33·0 | 33·1 | |
| 84 | UI31 | I/s to UI26 | | A3 | A | 24·8 | 25·2 | 1·2 | 29·9 | 30·0 | | 23·8 | 26·2 | 1·2 | 29·8 | 30·2 | |
| 85 | UI31 | I/s to UI35 | | B3 | C | 107·6 | 108·2 | 0·8 | −9·2 | −9·1 | | 106·6 | 109·2 | 0·9 | −9·3 | −8·9 | |
| 86 | UI33 | I/s to UI28 | | A1 | A | 288·8 | 289·2 | 0·6 | 10·1 | 10·4 | | 287·8 | 290·2 | 0·6 | 9·4 | 11·2 | |
| 87 | UI33 | I/s to UI37 | | A2 | A | 210·0 | 211·0 | −0·2 | −28·4 | −28·0 | | 209·0 | 212·0 | −0·2 | −28·7 | −27·7 | |
| 88 | UI35 | I/s to UI28 | | B1 | A | 330·0 | 330·2 | 1·0 | 28·2 | 28·3 | | 329·0 | 331·2 | 1·0 | 27·8 | 28·8 | |
| 89 | UI35 | I/s to UI31 | | B3 | A | 287·6 | 288·2 | −0·2 | 8·7 | 9·0 | | 286·6 | 289·2 | −0·2 | 8·1 | 9·5 | |
| 90 | UI35 | I/s to UI37 | | B2 | A | 125·0 | 126·0 | 0·2 | −18·7 | −18·2 | | 124·0 | 127·0 | 0·2 | −19·3 | −17·7 | |

*Continued*

| 1 | 2 | 3 | 4 | 5 | 6 | 7 | 8 | 9 | 10 | 11 | 12 | 13 | 14 | 15 | 16 | 17 | 18 |
|---|---|---|---|---|---|---|---|---|---|---|---|---|---|---|---|---|---|
| 91 | UI35 | I/s to UI40 | | B3 | C | 149·2 | 149·6 | −0·2 | −28·3 | −28·1 | | 148·2 | 150·6 | −0·2 | −28·6 | −27·8 | |
| 92 | UI37 | I/s to UI28 | | A1 | A | 318·4 | 318·8 | 0·8 | 23·9 | 24·1 | | 317·4 | 319·8 | 0·8 | 23·3 | 24·7 | |
| 93 | UI37 | I/s to UI33 | | A2 | A | 30·0 | 31·0 | 1·4 | 28·3 | 28·7 | | 29·0 | 32·0 | 1·4 | 27·9 | 29·0 | |
| 94 | UI37 | I/s to UI35 | | B2 | C | 305·0 | 306·0 | 0·4 | 17·8 | 18·3 | | 304·0 | 307·0 | 0·4 | 17·4 | 18·7 | |
| 95 | UI40 | I/s to UI35 | | B3 | A | 329·2 | 329·6 | 0·4 | 27·4 | 27·5 | | 328·2 | 330·6 | 0·4 | 27·0 | 28·0 | |
| 96 | UI46 | Flat face to NW | 6 | | A | 303·6 | 304·4 | −0·2 | 16·6 | 17·0 | | 302·6 | 305·4 | −0·2 | 16·1 | 17·4 | |
| 97 | UI46 | Flat face to SE | 6 | | A | 123·6 | 124·4 | −0·2 | −18·2 | −18·0 | | 122·6 | 125·4 | −0·2 | −18·8 | −17·5 | |
| 98 | UI48 | I/s to UI49 | | A3 | A | 108·4 | 108·6 | 3·0 | −7·5 | −7·4 | | 107·4 | 109·6 | 3·0 | −8·3 | −6·4 | |
| 99 | UI48 | I/s to UI50 | | A3 | C | 182·8 | 183·2 | −0·2 | −33·3 | −33·3 | | 181·8 | 184·2 | −0·2 | −33·4 | −33·3 | |
| 100 | UI49 | Flat face to NE | 5a | | A | 51·2 | 52·0 | 3·4 | 22·4 | 22·4 | | 50·2 | 53·0 | 3·4 | 22·1 | 22·6 | |
| 101 | UI49 | Flat face to SW | 5a | | C | 231·2 | 232·0 | −0·2 | −20·6 | −20·2 | | 230·2 | 233·0 | −0·2 | −21·0 | −19·7 | |
| 102 | UI49 | I/s to UI48 | | A1 | C | 288·4 | 288·6 | −0·2 | 9·1 | 9·2 | | 287·4 | 289·6 | −0·2 | 8·6 | 9·7 | |
| 103 | UI49 | I/s to UI50 | | A3 | C | 226·0 | 226·4 | −0·2 | −22·8 | −22·6 | | 225·0 | 227·4 | −0·2 | −23·2 | −22·2 | |
| 104 | UI50 | I/s to UI48 | | A1 | C | 2·8 | 3·2 | 0·0 | 32·3 | 32·3 | | 1·8 | 4·2 | 0·0 | 32·2 | 32·3 | |
| 105 | UI50 | I/s to UI49 | | A3 | C | 46·0 | 46·4 | 0·6 | 22·1 | 22·2 | | 45·0 | 47·4 | 0·6 | 21·8 | 22·6 | |
| 106 | UI57 | *ab* (to NE) | 3b | | A | 33·4 | 37·6 | 4·4 | 30·0 | 30·9 | | 31·4 | 39·6 | 4·4 | 29·8 | 31·4 | |
| 107 | UI57 | *ba* (to SW) | 3b | | A | 213·4 | 217·6 | 5·6 | −21·4 | −21·0 | | 211·4 | 219·6 | 5·4 | −21·7 | −20·8 | |
| 108 | UI59 | *ab* (to WNW) | 4a | | A | 298·6 | 300·6 | 10·0 | 23·8 | 24·7 | | 296·6 | 302·6 | 10·4 | 23·3 | 25·9 | |
| 109 | UI59 | *ba* (to ESE) | 4a | | A | 118·6 | 120·6 | −0·2 | −16·8 | −15·8 | | 116·6 | 122·6 | −0·2 | −17·6 | −14·8 | |
| 110 | NA1 | *ab* (to NNW) | 4a | | B | 328·0 | 329·6 | 0·8 | 28·1 | 28·4 | Max 28·9 | 326·4 | 331·2 | 0·6 | 27·2 | 29·1 | |
| 111 | NA1 | *ba* (to SSE) | 4a | | A | 148·0 | 149·6 | 2·6 | −25·8 | −25·5 | | 146·4 | 151·2 | 2·6 | −26·4 | −25·4 | |
| 112 | NA3 | Flat face to NNW | 5a | | A | 328·0 | 331·0 | 6·2 | 33·3 | 34·9 | | 326·0 | 333·0 | 6·4 | 33·1 | 35·5 | |
| 113 | CT1 | Flat face to S | 5a | | C | 178·0 | 181·4 | −0·2 | −34·1 | −34·1 | | 176·0 | 183·4 | −0·2 | −34·1 | −34·0 | |
| 114 | CT2 | *ab* (to NNE) | 3a | | A | 17·8 | 19·0 | 0·4 | 31·2 | 31·5 | | 16·6 | 20·2 | 0·4 | 30·7 | 31·8 | |
| 115 | CT2 | *ba* (to SSW) | 3a | | A | 196·6 | 200·2 | −0·2 | −32·5 | −32·1 | | 197·8 | 199·0 | −0·2 | −32·6 | −31·9 | |
| 116 | CT2 | I/s to CT3 | | A2 | A | 204·6 | 205·0 | −0·2 | −30·7 | −30·6 | | 203·6 | 206·0 | −0·2 | −31·0 | −30·4 | |
| 117 | CT3 | I/s to CT2 | | A1 | A | 24·6 | 25·0 | 0·4 | 29·8 | 29·9 | | 23·6 | 26·0 | 0·4 | 29·9 | 30·2 | Min 29·7 |
| 118 | CT7 | I/s to CT8 | | A2 | B | 204·0 | 205·2 | 0·8 | −30·0 | −29·7 | | 202·8 | 206·4 | 0·6 | −30·3 | −30·2 | Max −29·6 |
| 119 | CT8 | I/s to CT7 | | A2 | C | 24·0 | 25·2 | 0·0 | 29·7 | 29·9 | | 22·8 | 26·4 | 0·0 | 29·3 | 30·3 | |
| 120 | CT9 | Flat face to SSW | 5a | | A | 195·0 | 196·6 | 1·4 | −31·0 | −31·0 | Min −31·1 | 193·4 | 198·2 | 1·4 | −30·8 | −30·6 | Min −31·1 |
| 121 | ML1 | I/s to ML9 | | A3 | C | 175·0 | 175·6 | 1·4 | −32·2 | −32·1 | | 174·0 | 176·6 | 1·4 | −32·3 | −32·0 | |
| 122 | ML2 | *dcba* (to SSE) | 2c | | A | 166·0 | 170·0 | 1·8 | −31·5 | −30·8 | | 164·0 | 172·0 | 1·8 | −31·4 | −30·5 | Min −31·5 |
| 123 | ML2 | I/s to ML7 | | A3 | C | 240·0 | 241·0 | 1·0 | −15·4 | −14·9 | | 239·0 | 242·0 | 1·0 | −15·9 | −14·5 | |
| 124 | ML4 | *abc* (to N) | 2a | | C | 4·4 | 6·2 | 0·8 | 33·2 | 34·0 | | 2·6 | 8·0 | 0·8 | 32·8 | 33·8 | Max 34·2 |
| 125 | ML4 | *cba* (to S) | 2a | | A | 184·4 | 186·2 | 5·0 | −28·6 | −28·4 | | 182·6 | 188·0 | 5·0 | −28·7 | −28·2 | |
| 126 | ML7 | Flat face to SE | 5a | | A | 132·6 | 133·4 | 2·2 | −20·5 | −20·2 | | 131·6 | 134·4 | 2·2 | −20·9 | −19·6 | |
| 127 | ML7 | I/s to ML2 | | A3 | A | 60·0 | 61·0 | 0·8 | 15·8 | 16·3 | | 59·0 | 62·0 | 0·8 | 15·5 | 16·8 | |
| 128 | ML9 | *abcd* (to NNW) | 1a | | C | 341·0 | 343·0 | 0·2 | 30·9 | 31·6 | | 339·0 | 345·0 | 0·2 | 30·2 | 31·9 | |
| 129 | ML9 | I/s to ML1 | | A3 | C | 355·0 | 355·6 | −0·2 | 32·4 | 32·5 | | 354·0 | 356·6 | −0·2 | 32·3 | 32·5 | |
| 130 | ML10 | *abcde* (to NNW) | 2a | | B | 328·6 | 331·0 | 0·2 | 27·5 | 28·7 | | 326·6 | 333·0 | 0·2 | 26·7 | 29·8 | |
| 131 | ML11 | *abc* (to SSE) | 1a | | A | 156·4 | 157·8 | 1·8 | −29·0 | −28·9 | | 155·0 | 159·2 | 2·0 | −29·3 | −28·5 | |
| 132 | ML12 | *abc* (to NNE) | 2c | | A | 26·0 | 29·2 | 2·0 | 30·4 | 31·3 | | 24·0 | 31·2 | 2·0 | 29·8 | 31·8 | |
| 133 | ML12 | *cba* (to SSW) | 2c | | A | 206·0 | 209·2 | 7·0 | −23·2 | −22·2 | | 204·0 | 211·2 | 7·0 | −23·5 | −21·6 | |
| 134 | ML15 | I/s to ML16 | | A2 | C | 163·4 | 167·2 | 5·4 | −27·0 | −27·0 | | 161·4 | 169·2 | 5·4 | −27·0 | −26·8 | |
| 135 | ML16 | *ba* (to NW) | 3a | | B | 307·6 | 308·6 | 3·4 | 22·3 | 23·0 | | 306·6 | 309·6 | 3·4 | 21·8 | 23·7 | |
| 136 | ML16 | *ab* (to SE) | 3a | | B | 127·6 | 128·6 | 2·6 | −18·3 | −17·7 | | 126·6 | 129·6 | 2·6 | −18·6 | −17·1 | |
| 137 | ML16 | I/s to ML15 | | A2 | C | 343·4 | 347·2 | 3·6 | 35·3 | 35·9 | | 341·4 | 349·2 | 3·6 | 34·9 | 36·2 | |
| 138 | ML18 | *ba* (to ENE) | 3a | | A | 66·6 | 67·6 | 6·4 | 17·5 | 18·0 | | 65·6 | 68·6 | 6·4 | 16·9 | 18·8 | |
| 139 | ML25 | *abc* (to NW) | 3c | | A | 316·4 | 317·6 | 14·8 | 36·8 | 37·2 | Max 37·3 | 315·2 | 318·8 | 14·6 | 35·8 | 37·4 | |
| 140 | ML25 | *cba* (to SE) | 3c | | A | 136·4 | 137·6 | 6·4 | −18·5 | −17·9 | | 135·2 | 138·8 | 6·4 | −18·9 | −17·4 | |
| 141 | ML25 | I/s to ML27 | | B2 | A | 191·6 | 192·0 | 3·0 | −30·0 | −29·9 | | 190·6 | 193·0 | 3·0 | −30·1 | −29·8 | |
| 142 | ML27 | I/s to ML25 | | B2 | C | 11·6 | 12·0 | 5·4 | 38·0 | 38·1 | | 10·6 | 13·0 | 5·4 | 37·7 | 38·3 | |
| 143 | ML30 | Flat face to NNW | 6 | | A | 328·2 | 330·4 | 0·0 | 27·5 | 28·5 | | 326·2 | 332·4 | 0·0 | 27·0 | 28·9 | |
| 144 | ML31 | Flat face to NE | 5a | | A | 49·4 | 50·2 | 2·4 | 22·6 | 23·0 | | 48·4 | 51·2 | 2·4 | 22·2 | 23·4 | |
| 145 | ML31 | Flat face to SW | 5a | | A | 229·4 | 230·2 | 0·2 | −21·3 | −21·0 | | 228·4 | 231·2 | 0·2 | −22·0 | −20·7 | |

*Continued*

| 1 | 2 | 3 | 4 | 5 | 6 | 7 | 8 | 9 | 10 | 11 | 12 | 13 | 14 | 15 | 16 | 17 | 18 |
|---|---|---|---|---|---|---|---|---|---|---|---|---|---|---|---|---|---|
| 146 | ML33 | *ba* (to WNW) | 3b | | A | 281·6 | 283·2 | 3·0 | 8·8 | 9·5 | | 280·0 | 284·8 | 3·0 | 7·9 | 10·8 | |
| 147 | LN7 | Flat face to N | 5a | | C | 0·0 | 4·0 | 1·2 | 34·0 | 34·5 | | 358·0 | 6·0 | 1·2 | 33·7 | 34·8 | |
| 148 | LN18 | *ab* (to W) | 4a | | B | 270·2 | 273·8 | 2·0 | 1·7 | 3·1 | | 268·2 | 275·8 | 2·0 | 1·2 | 4·0 | |
| 149 | LN22 | *abc* (to NNW) | 1a | | A | 326·6 | 329·0 | 6·2 | 32·7 | 34·8 | | 324·6 | 331·0 | 6·2 | 31·7 | 35·9 | |
| 150 | LN22 | *cba* (to SSE) | 1a | | B | 146·6 | 149·0 | 7·2 | −21·5 | −21·2 | | 144·6 | 151·0 | 7·2 | −21·8 | −20·6 | |
| 151 | AR2 | Flat face to N | 5a | | C | 356·0 | 358·0 | 7·2 | 40·9 | 41·0 | | 354·0 | 0·0 | 7·4 | 40·7 | 41·6 | |
| 152 | AR2 | I/s to AR3 | | A2 | C | 213·6 | 214·4 | 0·0 | −28·2 | −27·7 | | 212·6 | 215·4 | 0·0 | −28·7 | −27·3 | |
| 153 | AR3 | *ab* (to N) | 1b | | A | 6·4 | 7·6 | 5·2 | 38·5 | 38·6 | | 5·2 | 8·8 | 5·2 | 38·4 | 38·7 | |
| 154 | AR3 | *ba* (to S) | 1b | | A | 186·4 | 187·6 | 3·2 | −30·7 | −30·5 | Min −30·8 | 185·2 | 188·8 | 3·2 | −30·7 | −30·5 | Min −30·8 |
| 155 | AR3 | I/s to AR2 | | A2 | C | 33·6 | 34·4 | 2·8 | 29·7 | 30·1 | | 32·6 | 35·4 | 2·8 | 29·1 | 30·6 | |
| 156 | AR6 | *abc* (to N) | 2a | | A | 356·4 | 357·8 | 1·6 | 35·0 | 35·3 | | 355·0 | 359·2 | 1·6 | 34·9 | 35·2 | Max 35·3 |
| 157 | AR7 | I/s to AR8 | | A2 | C | 214·6 | 215·0 | 2·2 | −25·4 | −25·2 | | 213·6 | 216·0 | 2·2 | −25·8 | −24·8 | |
| 158 | AR8 | I/s to AR7 | | A2 | C | 34·6 | 35·0 | 2·2 | 29·0 | 29·2 | | 33·6 | 36·0 | 2·2 | 28·5 | 29·5 | |
| 159 | AR9 | Flat face to ENE | 5a | | B | 69·4 | 71·0 | 2·6 | 12·4 | 13·9 | | 67·8 | 72·6 | 2·6 | 11·6 | 14·6 | |
| 160 | AR9 | I/s to AR10 | | A2 | C | 203·4 | 203·8 | 6·8 | −24·0 | −24·0 | | 202·4 | 204·8 | 6·8 | −24·1 | −23·9 | |
| 161 | AR10 | I/s to AR9 | | A2 | C | 23·4 | 23·8 | 1·8 | 32·3 | 32·4 | | 22·4 | 24·8 | 1·8 | 32·2 | 32·5 | |
| 162 | AR13 | $S_2S_3S_6$ | 1a | | A | 329·0 | 329·8 | 5·8 | 33·8 | 34·2 | | 328·0 | 330·8 | 5·8 | 33·5 | 34·3 | |
| 163 | AR13 | $S_5S_1S_2$ | 1a | | C | 20·2 | 21·2 | 2·0 | 33·0 | 33·2 | | 19·2 | 22·2 | 2·0 | 32·8 | 33·2 | |
| 164 | AR13 | $S_2S_1S_5$ | 1a | | C | 200·2 | 201·2 | 1·0 | −30·9 | −30·7 | | 199·2 | 202·2 | 0·8 | −31·1 | −30·6 | |
| 165 | AR13 | $S_4S_1S_3$ | 1a | | C | 26·0 | 26·8 | 2·8 | 32·4 | 32·5 | | 25·0 | 27·8 | 2·8 | 32·3 | 32·5 | |
| 166 | AR13 | $S_3S_1S_4$ | 1a | | B | 206·0 | 206·8 | 1·0 | −29·5 | −29·5 | Max −29·3 | 205·0 | 207·8 | 0·8 | −29·9 | −29·6 | Max −29·3 |
| 167 | AR13 | $S_5S_4$ | 1b | | A | 322·2 | 322·8 | 4·6 | 30·3 | 30·5 | | 321·2 | 323·8 | 4·6 | 29·8 | 31·0 | |
| 168 | AR15 | *abcd* (to NW) | 1a | | B | 320·8 | 324·0 | 3·0 | 28·3 | 29·3 | | 318·8 | 326·0 | 3·0 | 27·3 | 29·9 | |
| 169 | AR15 | *dcba* (to SE) | 1a | | B | 140·8 | 144·0 | 1·8 | −25·8 | −23·7 | | 138·8 | 146·0 | 1·8 | −26·5 | −22·7 | |
| 170 | AR15 | *ef* (to NNW) | 1b | | B | 333·4 | 335·4 | 3·0 | 32·7 | 33·2 | | 331·4 | 337·4 | 3·0 | 32·2 | 33·6 | |
| 171 | AR15 | *fe* (to SSE) | 1b | | C | 153·4 | 155·4 | 0·8 | −30·2 | −29·6 | | 151·4 | 157·4 | 0·8 | −30·8 | −29·0 | |
| 172 | AR15 | I/s to AR18 | | B3 | C | 224·6 | 225·0 | 1·0 | −22·8 | −22·6 | | 223·6 | 226·0 | 1·0 | −23·3 | −22·0 | |
| 173 | AR16 | Flat face to NW | 5a | | B | 316·8 | 319·6 | 4·0 | 27·2 | 28·8 | | 314·8 | 321·6 | 4·0 | 26·3 | 29·5 | |
| 174 | AR16 | Flat face to SE | 5a | | B | 136·8 | 139·6 | 0·8 | −24·7 | −23·5 | | 134·8 | 141·6 | 0·8 | −25·5 | −22·4 | |
| 175 | AR16 | I/s to AR17 | | A2 | C | 258·8 | 259·4 | 0·8 | −5·9 | −5·4 | | 257·8 | 260·4 | 0·8 | −6·5 | −4·8 | |
| 176 | AR17 | I/s to AR16 | | A2 | C | 78·8 | 79·4 | 1·0 | 6·5 | 6·9 | | 77·8 | 80·4 | 1·0 | 5·9 | 7·6 | |
| 177 | AR17 | I/s to AR18 | | B2 | C | 158·6 | 159·6 | 2·2 | −29·7 | −29·5 | | 157·6 | 160·6 | 2·2 | −29·8 | −29·5 | |
| 178 | AR18 | I/s to AR15 | | B3 | A | 44·6 | 45·0 | 2·4 | 25·1 | 25·3 | | 43·6 | 46·0 | 2·2 | 24·8 | 25·9 | |
| 179 | AR18 | I/s to AR17 | | B2 | C | 338·6 | 339·6 | 2·4 | 33·1 | 33·5 | | 337·6 | 340·6 | 2·4 | 32·8 | 33·7 | |
| 180 | AR27 | *ab* (to NNW) | 4a | | B | 327·4 | 329·2 | 1·6 | 29·3 | 29·9 | | 325·6 | 331·0 | 1·6 | 28·7 | 30·6 | |
| 181 | AR28 | *abc* (to NNW) | 2a | | B | 344·2 | 345·6 | 3·2 | 35·3 | 35·6 | | 342·8 | 347·0 | 3·2 | 34·9 | 35·9 | |
| 182 | AR28 | *cba* (to SSE) | 2a | | B | 164·2 | 165·6 | 2·2 | −31·0 | −30·6 | | 162·8 | 167·0 | 2·2 | −31·5 | −30·1 | |
| 183 | AR28 | I/s to AR29 | | A2 | B | 159·4 | 160·0 | 2·6 | −29·4 | −29·2 | | 158·4 | 161·0 | 2·6 | −29·8 | −28·8 | |
| 184 | AR28 | I/s to AR30 | | A2 | B | 164·0 | 164·6 | 2·2 | −30·8 | −30·5 | | 163·0 | 165·6 | 2·2 | −31·0 | −30·2 | |
| 185 | AR29 | *ab* (to NW) | 1b | | B | 314·0 | 318·4 | 2·0 | 24·2 | 26·3 | | 312·0 | 320·4 | 2·0 | 23·3 | 26·6 | |
| 186 | AR29 | *ba* (to SE) | 1b | | B | 134·0 | 138·4 | 3·6 | −21·5 | −19·5 | | 132·0 | 140·4 | 3·6 | −21·9 | −18·4 | |
| 187 | AR29 | I/s to AR28 | | A2 | C | 339·4 | 340·4 | 1·8 | 33·1 | 33·3 | | 338·4 | 341·4 | 1·8 | 32·5 | 33·3 | |
| 188 | AR30 | I/s to AR28 | | A2 | C | 344·0 | 344·6 | 1·8 | 34·0 | 34·1 | | 343·0 | 345·6 | 1·8 | 33·7 | 34·5 | |
| 189 | AR32 | Flat face to WNW | 5a | | C | 287·0 | 289·0 | 4·6 | 13·1 | 14·3 | | 285·0 | 291·0 | 4·6 | 11·9 | 15·4 | |
| 190 | AR32 | Flat face to ESE | 5a | | C | 107·0 | 109·0 | 3·2 | −8·1 | −6·8 | | 105·0 | 111·0 | 3·2 | −9·2 | −5·6 | |
| 191 | AR33 | Flat face to WNW | 5a | | C | 299·0 | 302·0 | 2·8 | 18·2 | 19·3 | | 297·0 | 304·0 | 2·8 | 17·2 | 20·3 | |
| 192 | AR33 | Flat face to ESE | 5a | | C | 119·0 | 122·0 | 2·0 | −15·6 | −14·2 | | 117·0 | 124·0 | 2·0 | −16·5 | −13·2 | |
| 193 | JU1 | *ba* (to WNW) | 4b | | A | 285·2 | 285·6 | 2·4 | 10·1 | 10·5 | | 284·2 | 286·6 | 2·4 | 9·3 | 11·4 | |
| 194 | JU2 | I/s to JU3 | | B2 | A | 98·6 | 99·0 | 2·4 | −3·3 | −2·9 | | 97·6 | 100·0 | 2·4 | −3·8 | −2·4 | |
| 195 | JU3 | I/s to JU2 | | B2 | C | 278·6 | 279·0 | 4·8 | 8·6 | 8·9 | | 277·6 | 280·0 | 4·8 | 7·6 | 9·7 | |
| 196 | JU3 | I/s to JU4 | | A2 | B | 253·4 | 254·0 | 4·0 | −6·0 | −5·8 | | 252·4 | 255·0 | 4·0 | −6·3 | −5·3 | |
| 197 | JU3 | I/s to JU5 | | B2 | B | 255·8 | 256·2 | 3·6 | −5·1 | −4·8 | | 254·8 | 257·2 | 3·6 | −5·5 | −4·5 | |
| 198 | JU4 | *ba* (to ENE) | 3a | | B | 73·0 | 73·2 | 1·8 | 10·6 | 10·8 | | 72·0 | 74·2 | 1·8 | 10·2 | 11·3 | |
| 199 | JU4 | *ab* (to WSW) | 3a | | A | 253·0 | 253·2 | 5·4 | −5·1 | −5·0 | | 252·0 | 254·2 | 5·4 | −5·4 | −4·7 | |
| 200 | JU4 | I/s to JU3 | | A1 | B | 73·4 | 74·0 | 1·8 | 10·2 | 10·5 | | 72·4 | 75·0 | 1·8 | 10·2 | 11·1 | |

*Continued*

| 1 | 2 | 3 | 4 | 5 | 6 | 7 | 8 | 9 | 10 | 11 | 12 | 13 | 14 | 15 | 16 | 17 | 18 |
|---|---|---|---|---|---|---|---|---|---|---|---|---|---|---|---|---|---|
| 201 | JU4 | I/s to JU5 | | B2 | A | 257·0 | 257·4 | 4·8 | −3·3 | −3·1 | | 256·0 | 258·4 | 4·8 | −3·9 | −2·9 | |
| 202 | JU5 | I/s to JU3 | | B2 | B | 75·8 | 76·2 | 0·8 | 8·2 | 8·3 | | 74·8 | 77·2 | 1·0 | 8·0 | 8·7 | |
| 203 | JU5 | I/s to JU4 | | B2 | B | 77·0 | 77·4 | 1·4 | 7·9 | 8·1 | | 76·0 | 78·4 | 1·4 | 7·4 | 8·3 | |
| 204 | JU7 | *abc* (to NNE) | 2a | | C | 18·0 | 22·0 | 0·8 | 31·6 | 33·3 | | 16·0 | 24·0 | 1·0 | 30·7 | 34·2 | |
| 205 | JU7 | *cba* (to SSW) | 2a | | C | 198·0 | 202·0 | −0·2 | −33·1 | −32·2 | | 196·0 | 204·0 | −0·2 | −33·5 | −31·5 | |
| 206 | JU9 | Flat face to NNW | 5a | | A | 334·0 | 334·8 | 4·2 | 34·0 | 34·5 | | 333·0 | 335·8 | 4·2 | 33·7 | 35·0 | |
| 207 | JU9 | Flat face to SSE | 5a | | A | 154·0 | 154·8 | 0·0 | −31·1 | −30·8 | | 153·0 | 155·8 | 0·0 | −31·3 | −30·7 | |
| 208 | IS3 | Flat face to WSW | 5a | | A | 256·8 | 257·4 | 0·8 | −7·0 | −6·6 | | 255·8 | 258·4 | 1·0 | −7·5 | −6·0 | |
| 209 | IS4 | Flat face to WNW | 5a | | A | 296·4 | 297·2 | 4·4 | 18·2 | 18·6 | | 295·4 | 298·2 | 4·4 | 17·5 | 19·0 | |
| 210 | IS5 | I/s to IS12 | | A3 | C | 189·2 | 189·4 | 1·2 | −32·9 | −32·9 | | 188·2 | 190·4 | 1·2 | −33·0 | −32·7 | |
| 211 | IS6 | I/s to IS7 | | A2 | A | 98·8 | 99·0 | −0·2 | −5·7 | −5·6 | | 97·8 | 100·0 | −0·2 | −6·3 | −4·9 | |
| 212 | IS6 | I/s to IS11 | | A3 | A | 118·4 | 118·6 | 0·2 | −15·8 | −15·7 | | 117·4 | 119·6 | 0·2 | −16·5 | −15·0 | |
| 213 | IS7 | I/s to IS6 | | A1 | A | 278·8 | 279·0 | 2·4 | 6·8 | 6·9 | | 277·8 | 280·0 | 2·4 | 6·3 | 7·3 | |
| 214 | IS7 | I/s to IS11 | | A3 | A | 124·6 | 124·8 | 0·6 | −18·7 | −18·6 | | 123·6 | 125·8 | 0·6 | −19·1 | −17·8 | |
| 215 | IS7 | I/s to IS12 | | A3 | A | 127·6 | 128·0 | 0·8 | −19·8 | −19·6 | | 126·6 | 129·0 | 0·8 | −20·2 | −19·3 | |
| 216 | IS11 | *ab* (to ENE) | 3a | | A | 58·4 | 61·8 | 2·0 | 16·6 | 18·6 | | 56·4 | 63·8 | 2·0 | 16·0 | 19·9 | |
| 217 | IS11 | I/s to IS6 | | A3 | C | 298·4 | 298·6 | 0·4 | 15·5 | 15·6 | | 297·4 | 299·6 | 0·4 | 15·0 | 16·1 | |
| 218 | IS11 | I/s to IS7 | | A3 | A | 304·6 | 304·8 | 0·2 | 18·4 | 18·5 | | 303·6 | 305·8 | 0·2 | 18·2 | 19·0 | |
| 219 | IS12 | I/s to IS5 | | A3 | C | 9·2 | 9·4 | −0·2 | 32·7 | 32·8 | | 8·2 | 10·4 | −0·2 | 32·9 | 33·0 | Min 32·7 |
| 220 | IS12 | I/s to IS7 | | A3 | C | 307·6 | 308·0 | 0·2 | 19·7 | 19·9 | | 306·6 | 309·0 | 0·2 | 19·2 | 20·4 | |
| 221 | IS19 | Flat face to SSE | 5a | | A | 168·4 | 169·6 | 0·4 | −33·7 | −33·6 | | 167·2 | 170·8 | 0·4 | −33·8 | −33·5 | |
| 222 | IS19 | I/s to IS23 | | A3 | A | 240·6 | 241·0 | 1·0 | −15·5 | −15·3 | | 239·6 | 242·0 | 1·0 | −16·0 | −14·7 | |
| 223 | IS23 | Flat face to E | 5a | | A | 92·4 | 94·0 | 1·2 | −1·6 | −0·8 | | 90·8 | 95·6 | 1·0 | −2·5 | −0·3 | |
| 224 | IS23 | I/s to IS19 | | A3 | A | 60·6 | 61·0 | 0·8 | 16·2 | 16·3 | | 59·6 | 62·0 | 0·8 | 15·5 | 17·3 | |
| 225 | IS28 | I/s to IS31 | | B2 | C | 198·0 | 198·4 | −0·2 | −33·2 | −33·1 | | 197·0 | 199·4 | −0·2 | −33·4 | −32·9 | |
| 226 | IS31 | I/s to IS28 | | B1 | A | 17·6 | 18·8 | 1·0 | 32·8 | 33·0 | | 16·4 | 20·0 | 1·0 | 32·5 | 33·2 | |
| 227 | IS35 | I/s to IS36 | | B2 | A | 168·2 | 168·6 | 0·6 | −33·4 | −33·3 | | 167·2 | 169·6 | 0·6 | −33·5 | −33·2 | |
| 228 | IS36 | I/s to IS35 | | B2 | A | 348·2 | 348·6 | 2·4 | 35·5 | 35·6 | | 347·2 | 349·6 | 2·4 | 35·5 | 35·7 | |
| 229 | IS38 | *ab* (to ENE) | 4b | | C | 58·0 | 59·2 | 0·2 | 16·6 | 17·3 | | 56·8 | 60·4 | 0·2 | 16·0 | 17·8 | |
| 230 | IS38 | *ba* (to WSW) | 4b | | C | 238·0 | 239·2 | 1·0 | −16·7 | −16·0 | | 236·8 | 240·4 | 1·0 | −17·5 | −15·4 | |
| 231 | IS39 | *ab* (to E) | 3b | | A | 82·4 | 86·6 | 0·4 | 1·8 | 4·3 | | 80·4 | 88·6 | 0·4 | 0·7 | 5·2 | |
| 232 | IS39 | I/s to IS41 | | A2 | A | 252·4 | 253·0 | 0·8 | −9·4 | −9·1 | | 251·4 | 254·0 | 0·8 | −9·8 | −8·7 | |
| 233 | IS41 | *cba* (to S) | 2a | | C | 168·4 | 169·8 | −0·2 | −34·5 | −34·3 | | 167·0 | 171·2 | −0·2 | −34·6 | −34·0 | |
| 234 | KT2 | *ba* (to S) | 3a | | C | 177·4 | 177·8 | 1·4 | −32·9 | −32·9 | | 176·4 | 178·8 | 1·4 | −32·9 | −32·9 | |
| 235 | KT2 | *ac* (to WNW) | 3a | | A | 284·0 | 284·6 | 3·4 | 10·6 | 10·9 | | 283·0 | 285·6 | 3·4 | 9·9 | 11·5 | |
| 236 | KT2 | *ca* (to ESE) | 3a | | B | 104·0 | 104·6 | 1·2 | −7·5 | −7·1 | | 103·0 | 105·6 | 1·2 | −8·3 | −6·5 | |
| 237 | KT2 | *bc* (to WNW) | 3a | | A | 281·8 | 283·2 | 3·2 | 9·0 | 10·0 | | 280·4 | 284·6 | 3·2 | 8·1 | 10·9 | |
| 238 | KT2 | *cb* (to ESE) | 3a | | C | 101·8 | 103·2 | 1·2 | −6·6 | −5·7 | | 100·4 | 104·6 | 1·2 | −7·5 | −4·7 | |
| 239 | KT3 | Flat face to NW | 5a | | A | 316·6 | 317·0 | 2·8 | 26·5 | 26·6 | | 315·6 | 318·0 | 2·8 | 26·0 | 27·0 | |
| 240 | KT3 | Flat face to SE | 5a | | A | 136·6 | 137·0 | 2·2 | −22·5 | −22·3 | | 135·6 | 138·0 | 2·2 | −23·0 | −21·8 | |
| 241 | KT3 | I/s to KT8 | | B1 | A | 167·4 | 167·6 | 1·8 | −31·8 | −31·8 | | 166·4 | 168·6 | 1·8 | −31·9 | −31·8 | |
| 242 | KT4 | Flat face to S | 5a | | C | 170·4 | 171·6 | 2·8 | −31·3 | −30·9 | | 169·2 | 172·8 | 2·8 | −31·5 | −30·5 | |
| 243 | KT5 | *abcde* (to NNW) | 1a | | C | 27·4 | 28·6 | 2·2 | 31·4 | 32·0 | | 26·2 | 29·8 | 2·2 | 30·6 | 32·4 | |
| 244 | KT5 | *edcba* (to SSW) | 1a | | C | 207·4 | 208·6 | 0·8 | −29·5 | −29·3 | | 206·2 | 209·8 | 0·8 | −29·9 | −29·1 | |
| 245 | KT8 | *ba* (to SE) | 5b | | A | 128·4 | 129·2 | 1·6 | −19·9 | −19·3 | | 127·4 | 130·2 | 1·4 | −20·5 | −18·9 | |
| 246 | KT8 | I/s to KT3 | | B2 | A | 347·4 | 347·6 | 0·8 | 33·7 | 33·7 | | 346·4 | 348·6 | 0·8 | 33·7 | 33·9 | |
| 247 | KT10 | *cba* (to SW) | 1a | | B | 222·0 | 224·0 | 0·0 | −25·5 | −24·5 | | 220·0 | 226·0 | 0·0 | −26·3 | −23·7 | |
| 248 | KT12 | Flat face to N | 5a | | C | 4·8 | 6·4 | 0·8 | 34·4 | 34·4 | | 3·2 | 8·0 | 0·8 | 34·3 | 34·5 | |
| 249 | KT19 | *ab* (to WSW) | 4b | | C | 250·0 | 250·6 | −0·2 | −11·8 | −11·3 | | 249·0 | 251·6 | −0·2 | −12·4 | −10·7 | |
| 250 | KT23 | Flat face to SW | 5a | | A | 221·6 | 225·4 | 0·2 | −25·3 | −23·5 | | 219·6 | 227·4 | 0·2 | −26·1 | −22·7 | |
| 251 | KT27 | *abc* (to NE) | 2a | | A | 47·4 | 49·4 | 3·0 | 24·1 | 25·0 | | 45·4 | 51·4 | 3·0 | 23·0 | 25·9 | |
| 252 | KT27 | *cba* (to SW) | 2a | | B | 227·4 | 229·4 | 0·0 | −22·9 | −22·2 | | 225·4 | 231·4 | 0·0 | −24·1 | −21·3 | |
| 253 | KT27 | I/s to KT39 | | A3 | A | 181·8 | 182·0 | 1·6 | −33·3 | −33·3 | | 180·8 | 183·0 | 1·6 | −33·4 | −33·2 | |
| 254 | KT28 | Flat face to ENE | 5a | | A | 67·0 | 72·0 | 2·6 | 11·7 | 15·0 | | 65·0 | 74·0 | 2·6 | 10·4 | 16·4 | |
| 255 | KT29 | Flat face to ENE | 5a | | A | 56·8 | 59·6 | 2·0 | 18·2 | 19·7 | | 54·8 | 61·6 | 2·0 | 17·3 | 20·9 | |

*Continued*

| 1 | 2 | 3 | 4 | 5 | 6 | 7 | 8 | 9 | 10 | 11 | 12 | 13 | 14 | 15 | 16 | 17 | 18 |
|---|---|---|---|---|---|---|---|---|---|---|---|---|---|---|---|---|---|
| 256 | KT31 | Flat face to WNW | 5a | | A | 296·4 | 298·0 | 0·8 | 14·9 | 15·7 | | 294·8 | 299·6 | 0·8 | 14·0 | 16·2 | |
| 257 | KT31 | I/s to KT37 | | A3 | A | 163·2 | 163·4 | 1·4 | −31·9 | −31·8 | | 162·2 | 164·4 | 1·4 | −32·2 | −31·5 | |
| 258 | KT31 | I/s to KT39 | | A3 | C | 218·0 | 218·2 | 1·8 | −25·1 | −25·1 | | 217·0 | 219·2 | 2·0 | −25·2 | −24·5 | |
| 259 | KT32 | I/s to KT31 | | A2 | A | 283·8 | 284·2 | 0·2 | 7·4 | 7·6 | | 282·8 | 285·2 | 0·2 | 7·0 | 8·1 | |
| 260 | KT32 | I/s to KT39 | | A3 | A | 220·8 | 221·0 | 1·8 | −24·0 | −23·8 | | 219·8 | 222·0 | 1·8 | −24·3 | −23·4 | |
| 261 | KT35 | Flat face to WNW | 5a | | C | 288·4 | 289·8 | 2·0 | 11·7 | 12·5 | | 287·0 | 291·2 | 2·0 | 10·9 | 13·4 | |
| 262 | KT35 | Flat face to ESE | 5a | | C | 108·4 | 109·8 | 0·0 | −11·4 | −10·7 | | 107·0 | 111·2 | 0·0 | −12·2 | −10·0 | |
| 263 | KT36 | I/s to KT37 | | A2 | B | 236·6 | 236·8 | 1·6 | −17·1 | −17·0 | | 235·6 | 237·8 | 1·6 | −17·8 | −16·3 | |
| 264 | KT37 | I/s to KT31 | | A3 | A | 343·2 | 343·4 | 1·4 | 34·0 | 34·0 | | 342·2 | 344·4 | 1·4 | 33·8 | 34·3 | |
| 265 | KT37 | I/s to KT36 | | A2 | A | 56·6 | 56·8 | 1·6 | 19·2 | 19·4 | | 55·6 | 57·8 | 1·6 | 18·3 | 20·1 | |
| 266 | KT37 | I/s to KT39 | | A3 | A | 261·6 | 261·8 | 1·0 | −4·3 | −4·1 | | 260·6 | 262·8 | 1·0 | −4·8 | −3·5 | |
| 267 | KT39 | Flat face to NW | 5a | | A | 313·4 | 315·6 | 0·0 | 22·4 | 23·5 | | 311·4 | 317·6 | 0·0 | 21·3 | 24·4 | |
| 268 | KT39 | Flat face to SE | 5a | | A | 133·4 | 135·6 | 1·2 | −22·9 | −22·2 | | 131·4 | 137·6 | 1·2 | −23·4 | −21·6 | |
| 269 | KT39 | I/s to KT27 | | A3 | A | 1·8 | 2·0 | 0·0 | 34·0 | 34·0 | | 0·8 | 3·0 | 0·0 | 33·8 | 34·1 | |
| 270 | KT39 | I/s to KT31 | | A3 | C | 38·0 | 38·2 | 1·0 | 27·2 | 27·3 | | 37·0 | 39·2 | 1·0 | 26·8 | 27·7 | |
| 271 | KT39 | I/s to KT32 | | A3 | A | 40·8 | 41·0 | 1·0 | 25·8 | 25·9 | | 39·8 | 42·0 | 1·0 | 25·4 | 26·5 | |
| 272 | KT39 | I/s to KT37 | | A3 | A | 81·6 | 81·8 | 0·2 | 4·3 | 4·4 | | 80·6 | 82·8 | 0·2 | 3·7 | 5·0 | |
| 273 | KT41 | Flat face to NW | 5a | | A | 320·0 | 324·0 | 0·8 | 26·4 | 27·5 | | 318·0 | 326·0 | 0·8 | 25·8 | 28·0 | |
| 274 | KT41 | Flat face to SE | 5a | | A | 140·0 | 144·0 | 1·2 | −26·8 | −24·8 | | 138·0 | 146·0 | 1·2 | −27·7 | −23·9 | |
| 275 | KT44 | Flat face to N | 5a | | A | 8·4 | 9·8 | 1·6 | 35·2 | 35·5 | | 7·0 | 11·2 | 1·6 | 35·3 | 35·8 | Min 35·2 |
| 276 | KT44 | Flat face to S | 5a | | A | 188·4 | 189·8 | 0·0 | −34·9 | −34·7 | | 187·0 | 191·2 | 0·0 | −35·1 | −34·5 | |

**Table 5.1 Axial orientations of recumbent stone circles**

Centre Line and Perpendicular Line principal axis orientations of recumbent stone circles, from fieldwork in 1981. Values quoted by previous authors, or deduced from site plans published by them, are included for comparison.

Column headings:

1 Reference number in List 3

2 Centre Line azimuth to nearest half-degree, from fieldwork by this author in 1981

- * Significant uncertainty exists owing to the state of the monument (e.g. in determining the centre of the ring, or the long axis of the recumbent stone)
- (. . .) Value deduced from groundplan rather than directly measured, so some error is possible
- [. . .] Azimuth determined by prismatic compass only
- {. . .} No azimuth data are available from the 1981 fieldwork; the figure represents the value considered most reliable from the existing publications of other authors (Ruggles 1984c, S72)

3 Perpendicular Line azimuth to nearest half-degree, from fieldwork by this author in 1981. Annotations as in column 2

4 Azimuth quoted by Burl (1980, appendix I)

5 Azimuth quoted by Burl (fieldwork subsequent to the publication of Burl 1980, first published in Ruggles 1984c, table 2, col. 9)

6 Azimuth from Thom in Thom, Thom and Burl 1980

7 Azimuth from Lockyer 1909

8 Azimuth from Keiller 1934, 12

9 Azimuth from Coles 1900; Fred R. Coles, 'Stone circles in the N.E. of Scotland. Inverurie district', *PSAS* 35 (1901), 187–248; 'Stone circles in the N.E. of Scotland. Aberdeenshire', *PSAS* 36 (1902), 488–581; Coles 1903; 1904; Fred R. Coles, 'Stone circles in Aberdeenshire', *PSAS* 39 (1905), 206–18

10 Azimuth from another author (JC = J. Craig, 'The stone circles of the Ladieswell of Balronald and of Knocksoul', *Aberdeen University Review* 33 (1950), 428–30; KJ = Kilbride-Jones 1935; Og = A. Ogston, *The Prehistoric Antiquities of the Howe of Cromar*, Third Spalding Club, Aberdeen, 1931; GC = Childe 1933)

| 1 | 2 | 3 | 4 | 5 | 6 | 7 | 8 | 9 | 10 |
|---|---|---|---|---|---|---|---|---|---|
| RSC3 | — | [219]* | — | — | — | — | — | 201 | |
| RSC5 | 231·0 | 231·0 | 232 | 232 | — | 241 | — | 206 | |
| RSC6 | 185·0 | 185·0 | 198 | 198 | — | — | — | 174 | |
| RSC7 | 161·0* | {158} | — | 158 | 158 | — | — | — | |
| RSC10 | — | [201·]* | — | — | — | — | — | 206 | |
| RSC11 | (194·0) | (190·0) | 192 | 190 | — | 192 | 189 | 203 | |
| RSC12 | 185·5* | 185·5 | 184 | 184 | — | 184 | 189 | 198 | |
| RSC15 | — | [181]* | — | — | — | — | — | 188 | |
| RSC17 | {186}* | [196]* | 180 | — | — | — | — | 186 | |
| RSC19 | — | {170} | 170 | — | — | — | 170 | — | |
| RSC20 | {165}* | — | 173 | — | 165 | — | — | 166 | |
| RSC23 | 216·5* | 202·5* | 197 | 213 | 197 | — | — | 230 | |
| RSC24 | 188·5* | 195·5* | 203 | — | — | — | — | 208 | |
| RSC25 | 182·5* | 173·5 | 168 | 170 | 155 | — | 175 | 175 | |
| RSC26 | — | 182·0* | 189 | — | — | — | 189 | 183 | |
| RSC33 | — | (210·5)* | 197 | — | — | — | — | 204 | |
| RSC37 | 159·0* | 147·0 | 160 | — | 160 | 151 | — | 163 | |
| RSC39 | 163·5 | — | 195 | 161 | — | — | — | 184 | |
| RSC40 | — | 202·0* | 201 | 198 | 215 | 201 | — | 221 | |
| RSC41 | — | 214·0 | 217 | 208 | — | 218 | 202[1] | 206 | |
| RSC42 | — | [191]* | — | — | — | — | — | 195 | |
| RSC43 | — | {202} | 202 | 203 | 202 | 202 | — | 201 | |
| RSC44 | — | {202}* | 204 | — | — | 203 | — | 201 | |
| RSC46 | — | {225}* | 229 | 229 | — | — | 221 | 202 | |
| RSC47 | {199}* | — | 196 | — | — | — | — | 199 | |
| RSC48 | (200·0) | (196·0) | 198 | — | — | 199 | — | 170 | |
| RSC49 | 212·5 | (209·5) | 229 | 205 | — | 202 | — | 212 | 211 (GC) |
| RSC52 | — | 170·0* | 168 | 168 | — | — | 168 | — | |
| RSC54 | 180·0 | 191·5* | 185 | 187 | — | — | — | 156 | |
| RSC55 | {195} | — | 196 | — | 196 | — | — | 199 | |
| RSC56 | 219·5 | 236·5 | 234 | 216 | — | 237 | — | 203 | |
| RSC58 | — | 202·5 | 195 | 201 | — | 201 | 202 | 204 | |
| RSC59 | (200·0) | (187·0) | 196 | — | 196 | 198 | 202 | 180 | 196 (KJ) |
| RSC61 | — | (199·0) | 194 | 192 | 194 | 195 | — | 210 | |
| RSC62 | {190} | {190} | 189 | 184 | 189 | — | — | 175 | |
| RSC63 | (195·5) | (195·5) | 196 | 195 | 196 | — | 180 | 195 | |
| RSC66 | 203·0 | 203·0 | 202 | 200 | 202 | 201 | — | 185 | |
| RSC67 | — | 197·0 | 194 | — | 208 | 194 | — | 180 | |
| RSC68 | — | 183·5* | 188 | — | — | 188 | — | — | |
| RSC69 | — | {180} | — | — | — | — | — | 180 | |
| RSC71 | 231·0 | 226·0 | 230 | 228 | 230 | 227 | — | 251 | |
| RSC72 | 231·0 | 224·5 | 231 | 230 | 231 | 233 | — | 227 | |
| RSC73 | (203·0) | (187·0) | 197 | 196 | 197 | — | — | 193 | |
| RSC75 | 202·5 | (204·0) | 199 | 200 | — | 195 | 202 | 219 | |
| RSC76 | [224]* | — | — | — | — | — | — | — | |
| RSC77 | 235·5 | — | 235 | 232 | 235 | — | — | 235 | 235 (Og) |
| RSC78 | [207]* | — | 201 | — | — | — | 216 | — | 201 (JC); 193 (Og) |
| RSC80 | — | [206] | — | — | — | — | — | 164 | |
| RSC81 | — | ? | — | — | — | — | — | — | |
| RSC83 | 178·5 | 178·5 | 180 | — | 180 | — | — | 193 | |
| RSC84 | (187·0) | (194·0) | 187 | — | — | — | — | 187 | |
| RSC86 | 174·0 | 174·0 | 172 | — | 172 | — | — | 175 | |
| RSC89 | 183·5* | — | 180 | — | 180 | — | 180 | — | |
| RSC90 | 176·0 | (183·0) | 176 | — | 176 | 185 | 180 | 172 | |
| RSC91 | (157·5) | 150·5 | 155 | — | 155 | — | — | 160 | |
| RSC92 | (183·0) | (168·0) | 181 | — | 181 | — | — | 171 | |
| RSC94 | 179·0* | — | 186 | — | — | — | — | 186 | |
| RSC96 | 202·5* | — | 190 | 190 | — | — | — | — | |

**Table 5.2 Indicated horizons at recumbent stone circles: Centre Line data**

Azimuths, altitudes and declinations of the horizon ranges demarcated by the recumbent and flankers, as determined during fieldwork in 1981. The assumed observing position is the geometrical centre of the ring. The data have been taken or deduced from Ruggles 1984c, table 3.[1]

The quoted 'East flanker' azimuth is that of the inside (right-hand) edge of the flanker as viewed from the direction of the ring interior. The 'West flanker' azimuth is that of the inside (left-hand) edge of that flanker.

Column headings:

1 Reference number in List 3
2 East flanker azimuth, quoted to the nearest 0·5 degrees
3 Recumbent Left azimuth, quoted to the nearest 0·5 degrees
4 Recumbent Centre azimuth, quoted to the nearest 0·5 degrees
5 Recumbent Right azimuth, quoted to the nearest 0·5 degrees
6 West flanker azimuth, quoted to the nearest 0·5 degrees
7 Mean altitude within indication, quoted to the nearest 0·2 degrees
8 East flanker declination, quoted to the nearest 0·2 degrees
9 Recumbent Left declination, quoted to the nearest 0·2 degrees
10 Recumbent Centre declination, quoted to the nearest 0·2 degrees
11 Recumbent Right declination, quoted to the nearest 0·2 degrees
12 West flanker declination, quoted to the nearest 0·2 degrees
13 Notes

Key to Column 13 (Notes)

a Some or all of the azimuth values were deduced from a surveyed groundplan rather than measured directly, increasing the possible error in the quoted azimuth and declination values.
h Some or all of the altitude values were deduced from 1-inch Ordnance Survey maps rather than measured directly, increasing the possible error in the quoted altitude and declination values.
o The azimuth data have been taken from the surveys of other authors.
u There is significant uncertainty owing to the poor state of preservation of the monument.

| 1 | 2 | 3 | 4 | 5 | 6 | 7 | 8 | 9 | 10 | 11 | 12 | 13 |
|---|---|---|---|---|---|---|---|---|---|---|---|---|
| RSC5 | | 215·5 | 231·0 | 243·0 | 247·0 | 0·8 | | −25·6 | −19·4 | −13·8 | −12·0 | |
| RSC6 | 176·0 | 177·0 | 185·0 | 193·5 | 194·0 | 1·0 | −32·0 | −32·0 | −31·6 | −30·8 | −30·8 | |
| RSC7 | 143·5 | | 161·0 | | 179·0 | 1·6 | −24·0 | | −29·4 | | −31·6 | u |
| RSC11 | | 183·0 | 194·0 | 204·0 | 204·0 | 0·8 | | −32·6 | −30·8 | −29·0 | −29·0 | ah |
| RSC12 | 167·5 | 170·5 | 185·5 | 202·5 | | 0·8 | −32·0 | −32·4 | −32·2 | −28·4 | | au |
| RSC17 | | 179·5 | 186·0 | 192·5 | | 0·6 | | −32·4 | −32·2 | −31·8 | | oh |
| RSC20 | | 158·5 | 165·0 | 171·5 | | 0·2 | | −30·4 | −31·6 | −32·4 | | oh |
| RSC23 | | 207·5 | 216·5 | 225·0 | | 1·2 | | −27·6 | −24·8 | −21·6 | | u |
| RSC24 | | 181·5 | 188·5 | 196·5 | | 1·2 | | −31·8 | −31·2 | −29·8 | | u |
| RSC25 | | 169·5 | 182·5 | 190·5 | 193·5 | 1·2 | | −31·0 | −31·8 | −31·0 | −30·6 | u |
| RSC37 | 142·0 | 147·0 | 159·0 | 172·0 | 174·0 | 1·6 | −24·8 | −26·0 | −29·0 | −30·6 | −31·0 | u |
| RSC39 | | 148·0 | 163·5 | 182·0 | | 0·6 | | −26·6 | −31·0 | −32·8 | | |
| RSC47 | | 189·0 | 199·0 | 209·0 | | 0·4 | | −32·6 | −30·8 | −28·0 | | oh |
| RSC48 | 189·0 | 190·5 | 200·0 | 208·0 | 212·5 | 0·8 | −31·6 | −31·6 | −30·2 | −28·4 | −26·8 | ah |
| RSC49 | 196·5 | 200·0 | 212·5 | 225·0 | 225·0 | 1·0 | −30·8 | −30·2 | −26·8 | −21·4 | −21·4 | a |
| RSC54 | 174·5 | 176·5 | 180·0 | 184·5 | 185·0 | 0·6 | −32·2 | −32·6 | −32·6 | −32·6 | −32·6 | |
| RSC55 | | 186·0 | 195·0 | 204·0 | | 3·6 | | −28·6 | −27·8 | −26·6 | | oh |
| RSC56 | 215·5 | 215·5 | 219·5 | 224·0 | | 1·0 | −25·4 | −25·4 | −24·2 | −22·6 | | |
| RSC59 | 190·0 | 190·0 | 200·0 | 211·0 | 212·0 | 0·4 | −32·0 | −32·0 | −30·8 | −27·6 | −27·2 | ah |
| RSC62 | | 179·0 | 190·0 | 201·0 | | 2·8 | | −30·0 | −29·4 | −28·4 | | oh |
| RSC63 | 181·0 | 185·0 | 195·5 | 206·5 | 208·0 | 0·6 | −33·6 | −32·6 | −31·0 | −28·8 | −28·4 | a |
| RSC66 | 196·5 | 196·5 | 203·0 | 209·0 | 210·0 | 1·6 | −29·8 | −29·8 | −28·4 | −27·2 | −27·2 | |
| RSC71 | 213·5 | 218·5 | 231·0 | 245·0 | 245·5 | 3·8 | −21·8 | −20·0 | −16·8 | −11·6 | −11·6 | |
| RSC72 | 219·5 | 220·5 | 231·0 | 241·5 | 242·0 | 4·4 | −21·0 | −20·6 | −16·2 | −11·0 | −10·8 | |
| RSC73 | | 196·5 | 203·0 | 210·0 | 211·5 | 2·6 | | −28·6 | −27·8 | −25·8 | −25·2 | ah |
| RSC75 | 194·0 | 194·0 | 202·5 | 209·5 | | 1·6 | −30·6 | −30·6 | −28·8 | −27·2 | | |
| RSC77 | | 223·0 | 235·5 | 246·5 | | 1·8 | | −22·2 | −17·0 | −10·8 | | |
| RSC78 | | 199·0 | 207·0 | 215·0 | | 5·0 | | −27·4 | −25·0 | −19·8 | | oh |
| RSC83 | 166·5 | | 178·5 | | 195·5 | 0·4 | −32·2 | | −32·6 | | −31·4 | ah |
| RSC84 | | 175·0 | 187·0 | 199·5 | | 0·6 | | −32·6 | −32·4 | −30·4 | | ah |
| RSC86 | | 167·5 | 174·0 | 183·0 | 183·0 | −0·2 | | −33·4 | −33·6 | −33·4 | −33·4 | h |
| RSC89 | | 174·0 | 183·5 | 192·5 | | 2·4 | | −31·0 | −30·4 | −30·2 | | au |
| RSC90 | 167·5 | 169·5 | 176·0 | 183·5 | 184·0 | 3·2 | −28·8 | −28·8 | −29·6 | −31·0 | −31·0 | |
| RSC91 | 143·0 | 143·0 | 157·5 | 172·5 | 172·5 | 2·2 | −24·2 | −24·2 | −28·6 | −30·6 | −30·6 | a |
| RSC92 | 171·0 | 174·0 | 183·0 | | 193·5 | 2·8 | −29·6 | −29·8 | −31·0 | | −29·0 | ah |
| RSC94 | | 172·5 | 179·0 | 185·0 | | 0·0 | | −33·8 | −33·6 | −33·4 | | au |
| RSC96 | 190·5 | | 202·5 | | 219·0 | 6·6 | −26·2 | | −24·8 | | −18·0 | |

**Table 5.3 Indicated horizons at recumbent stone circles: Perpendicular Line data**

Azimuths, altitudes and declinations of the horizon ranges demarcated by the recumbent and flankers, as determined during fieldwork in 1981. The assumed observing position is 10 m behind the recumbent stone. The data have been taken or deduced from Ruggles 1984c, table 4.[1]

The quoted 'East flanker' azimuth is that of the inside (right-hand) edge of the flanker as viewed from the direction of the ring interior. The 'West flanker' azimuth is that of the inside (left-hand) edge of that flanker.

For column headings and key to the notes, see Table 5.2.

| 1 | 2 | 3 | 4 | 5 | 6 | 7 | 8 | 9 | 10 | 11 | 12 | 13 |
|---|---|---|---|---|---|---|---|---|---|---|---|---|
| RSC5 | | 223·0 | 231·0 | 237·0 | 239·0 | 0·8 | | −22·8 | −19·4 | −16·6 | −15·6 | a |
| RSC6 | 176·0 | 177·0 | 185·0 | 193·5 | 194·0 | 1·0 | −32·0 | −32·0 | −31·6 | −30·8 | −30·8 | |
| RSC7 | | 150·5 | 158·0 | 165·5 | | 1·6 | | −26·4 | −28·6 | −30·4 | | o |
| RSC11 | | 181·0 | 190·0 | 198·0 | 198·0 | 0·6 | | −32·8 | −31·4 | −30·4 | −30·4 | ah |
| RSC12 | | 174·5 | 185·5 | 197·5 | | 0·8 | | −32·6 | −32·2 | −29·6 | | a |
| RSC17 | | 188·5 | 196·0 | 203·5 | | 0·6 | | −31·8 | −31·2 | −29·4 | | oh |
| RSC23 | | 191·5 | 202·5 | 213·0 | | 1·2 | | −31·2 | −28·8 | −26·2 | | u |
| RSC24 | | 187·5 | 195·5 | 202·5 | | 1·8 | | −31·4 | −30·0 | −27·4 | | u |
| RSC25 | | 165·5 | 173·5 | 179·5 | 181·5 | 1·4 | | −30·0 | −31·2 | −31·2 | −31·6 | a |
| RSC26 | | 172·0 | 182·0 | 190·0 | 190·5 | 1·0 | | −31·6 | −32·2 | −31·4 | −31·4 | au |
| RSC33 | | 198·0 | 210·5 | 222·0 | | 1·4 | | −29·0 | −27·4 | −22·8 | | ahu |
| RSC37 | 138·5 | 141·0 | 147·0 | 155·0 | 155·5 | 1·2 | −23·6 | −24·4 | −26·0 | −28·4 | −28·6 | |
| RSC40 | 190·5 | 191·0 | 202·0 | 213·5 | | 1·4 | −31·4 | −31·4 | −29·0 | −25·4 | | u |
| RSC41 | 198·5 | 206·0 | 214·0 | 225·5 | 226·5 | 2·6 | −28·6 | −26·2 | −23·6 | −21·2 | −20·8 | |
| RSC43 | | 193·0 | 202·0 | 211·0 | | 1·6 | | −30·2 | −28·6 | −26·6 | | oh |
| RSC44 | | 193·0 | 202·0 | 211·0 | | 3·2 | | −29·4 | −26·6 | −25·0 | | oh |
| RSC46 | | 213·5 | 225·0 | 236·5 | | 3·2 | | −23·2 | −19·4 | −16·0 | | oh |
| RSC48 | 185·0 | 187·0 | 196·0 | 204·0 | 208·0 | 1·0 | −31·6 | −31·6 | −31·0 | −29·2 | −28·4 | ah |
| RSC49 | 192·0 | 196·0 | 209·5 | 223·0 | 223·0 | 1·0 | −31·6 | −31·0 | −27·8 | −22·4 | −22·4 | a |
| RSC52 | 157·5 | | 170·0 | | 181·0 | 1·8 | −28·3 | | −31·0 | | −31·2 | u |
| RSC54 | 186·0 | 188·5 | 191·5 | 195·5 | 197·0 | 0·8 | −32·2 | −31·8 | −31·6 | −31·4 | −30·8 | u |
| RSC56 | 231·0 | 231·0 | 236·5 | 242·5 | | 1·0 | −19·6 | −19·6 | −16·6 | −13·8 | | |
| RSC58 | 193·5 | 194·0 | 202·5 | 211·5 | 211·5 | 0·6 | −31·6 | −31·4 | −30·0 | −27·0 | −27·0 | a |
| RSC59 | 178·5 | 178·5 | 187·0 | 196·5 | 197·5 | 0·2 | −33·2 | −33·2 | −32·8 | −31·2 | −31·0 | ah |
| RSC61 | 183·0 | 183·0 | 199·0 | 216·5 | | 1·4 | −30·2 | −30·2 | −29·6 | −26·2 | | a |
| RSC62 | | 179·0 | 190·0 | 201·0 | | 2·8 | | −30·0 | −29·4 | −28·4 | | oh |
| RSC63 | 182·5 | 185·5 | 195·5 | 205·5 | 207·0 | 0·6 | −33·6 | −32·6 | −31·0 | −29·0 | −28·6 | a |
| RSC66 | 197·0 | 197·0 | 203·0 | 208·5 | 209·5 | 1·6 | −29·8 | −29·8 | −28·4 | −27·2 | −27·4 | |
| RSC67 | 189·5 | 191·0 | 197·0 | 200·0 | 204·5 | 1·8 | −31·0 | −30·8 | −29·6 | −29·0 | −27·8 | |
| RSC68 | 176·5 | | 183·5 | | 190·5 | 0·2 | −33·0 | | −33·0 | | −32·6 | u |
| RSC71 | 213·0 | 215·5 | 226·0 | 237·0 | 237·0 | 4·6 | −22·0 | −20·8 | −18·4 | −14·4 | −14·4 | |
| RSC72 | 212·0 | 212·0 | 224·5 | 237·5 | 237·5 | 4·8 | −22·4 | −22·4 | −18·8 | −13·2 | −13·2 | a |
| RSC73 | | 180·5 | 187·0 | 194·0 | 195·5 | 2·6 | | −31·4 | −30·0 | −29·0 | −28·8 | ah |
| RSC75 | 199·0 | 199·0 | 204·0 | 210·0 | | 1·6 | −29·6 | −29·6 | −28·4 | −27·0 | | a |
| RSC83 | 168·5 | | 178·5 | | 193·0 | 0·6 | −32·4 | | −32·6 | | −31·4 | ah |
| RSC84 | | 188·0 | 194·0 | 200·0 | | 0·8 | | −32·4 | −31·4 | −30·4 | | ah |
| RSC86 | | 166·5 | 174·0 | 181·5 | | −0·4 | | −33·2 | −33·6 | −33·8 | | ah |
| RSC90 | 176·5 | 176·5 | 183·0 | 190·0 | 190·0 | 2·8 | −29·6 | −29·6 | −30·8 | −30·0 | −30·0 | a |
| RSC91 | 144·0 | 144·0 | 150·5 | 157·5 | 157·5 | 2·0 | −24·6 | −24·6 | −26·4 | −28·6 | −28·6 | a |
| RSC92 | 158·0 | 160·5 | 168·0 | | 177·0 | 2·8 | −27·4 | −28·2 | −29·2 | | −31·0 | ah |

**Table 5.4 Hilltops within indicated horizons at recumbent stone circles**

Azimuths, altitudes and declinations of single conspicuous hilltops contained within RSC indicated horizon ranges, as determined during fieldwork in 1981 and subsequent data reduction. The data have been taken from Ruggles and Burl 1985, table 6. Collecting these data was not an original objective of the fieldwork project, so hills were not always identified on site. Thus while the azimuth, altitude and declination data should be reliable the identifications and distances quoted 'are unchecked and should be treated as provisional' (cf. *ibid.*, S46).

Column headings:
1 Reference number in List 3
2 Name of hill
3 Distance (km)
4 Azimuth of hill summit, to the nearest 0·1 degrees
5 Altitude of hill summit, to the nearest 0·1 degrees
6 Declination of hill summit, to the nearest 0·1 degrees
7 Difference between azimuth of hill summit and Centre Line azimuth, in degrees
8 Difference between azimuth of hill summit and Perpendicular Line azimuth, in degrees
9 Status of measurements (S = Surveyed profile; P = Partial survey, with large parallax errors, together with calculated profile; C = Calculated profile; V = Profile surveyed, but more distant hill suspected and calculated)

| 1 | 2 | 3 | 4 | 5 | 6 | 7 | 8 | 9 |
|---|---|---|---|---|---|---|---|---|
| RSC11 | Hill of Dens | 4 | 191·0 | 1·0 | −31·2 | 4 | 1 | C |
| RSC17 | Hill of Rothmaise | 11 | 189·0 | 0·8 | −31·7 | 3 | 7 | C |
| RSC19 | Hill of Rothmaise | 10 | 177·5 | 0·4 | −32·5 | ? | 7 | C |
| RSC20 | Gordonstown Hill | 3 | 156·0 | 0·4 | −29·6 | 9 | ? | C |
| RSC23 | Hillhead of Avochie | 2 | 202·5 | 1·6 | −28·6 | 14 | 0 | S |
| RSC24 | Tap o'Noth | 17 | 195·5 | 1·4 | −30·1 | 7 | 0 | V |
| RSC25 | Hill of Foudland | 13 | 178·5 | 1·6 | −31·2 | 4 | 5 | V |
| RSC33 | Morven | 27 | 209·5 | 1·2 | −27·2 | ? | 1 | C |
| RSC37 | Knock Saul | 6 | 150·5 | 1·6 | −26·7 | 9 | 3 | S |
| RSC40 | Satter Hill | 8 | 201·0 | 1·4 | −29·2 | ? | 1 | C |
| RSC41 | Hill of Flinder | 1 | 212·0 | 3·8 | −23·6 | ? | 2 | S |
| RSC43 | Satter Hill | 5 | 203·5 | 1·8 | −28·1 | ? | 2 | C |
| RSC44 | Knock Saul | 3 | 205·0 | 4·2 | −25·4 | ? | 3 | C |
| RSC46 | Spur of Satter Hill | 1 | 217·5 | 4·2 | −21·7 | ? | 7 | C |
| RSC49 | Mount Keen | 37 | 209·5 | 1·0 | −27·5 | 3 | 0 | C |
| RSC52 | Benaquhallie | 9 | 162·5 | 2·0 | −29·3 | ? | 8 | S |
| RSC54 | Mill Maud | 7 | 189·5 | 1·2 | −31·6 | 9 | 2 | S |
| RSC55 | Oxen Craig | 5 | 197·0 | 4·2 | −27·2 | 2 | ? | C |
| RSC58 | Mount Battock | 49 | 203·0 | 0·6 | −29·7 | ? | 1 | P |
| RSC59 | Hill at 733219 | 7 | 191·0 | 0·6 | −31·8 | 9 | 4 | S |
| RSC63 | Hill of Fare | 20 | 196·5 | 0·8 | −30·8 | 1 | 1 | P |
| RSC66 | Hill of Fare | 10 | 201·0 | 2·0 | −28·8 | 2 | 2 | S |
| RSC71 | Hill to south | 1 | 217·0 | 6·0 | −20·2 | 14 | 9 | S |
| RSC72 | Blackyduds | 3 | 226·0 | 4·8 | −17·9 | 5 | 2 | S |
| RSC73 | Green Hill | 4 | 191·0 | 3·4 | −29·0 | 12 | 4 | C |
| RSC77 | Lochnagar | 30 | 234·0 | 1·8 | −17·3 | 2 | ? | S |
| RSC83 | Brimmond Hill | 4 | 184·5 | 1·6 | −31·4 | 6 | 6 | C |
| RSC89 | Shillofad | 3 | 180·5 | 3·0 | −30·2 | 3 | ? | P |
| RSC90 | Shillofad | 3 | 171·0 | 4·0 | −28·8 | 5 | 12 | P |
| RSC91 | Spur at 738886 | 3 | 150·0 | 2·4 | −26·1 | 8 | 1 | S |
| RSC92 | Strathgyle | 2 | 171·5 | 3·2 | −29·6 | 12 | 3 | C |
| RSC96 | Craig of Shanno | 2 | 193·0 | 6·8 | −25·4 | 9 | ? | S |

**Table 5.5 Cup marks at recumbent stone circles**

Azimuths of groups of cup marks from the ring centres, together with declinations of the horizon above them. The data have been taken from Ruggles and Burl 1985, table 9.

Column headings:
1 Reference number in List 3
2 RSC name
3 Number of cup marks on the circle stone east of the east flanker
4 Number of cup marks on the east flanker
5 Number of cup marks on the recumbent stone
6 Number of cup marks on the west flanker
7 Number of cup marks on the circle stone west of the west flanker
8 Azimuth of the cup marks from the ring centre, to the nearest 0·5 degrees
9 Horizon altitude above the cup marks, to the nearest 0·1 degrees
10 Declination of the horizon above the cup marks, to the nearest 0·1 degrees
11 Comments

Key to Column 11 (Comments)
a The ring centre has been determined from Thom's plan (Thom, Thom and Burl 1980, 238) in which no absolute azimuth is given. The azimuth was determined during fieldwork in 1981.
b It is possible that these cup marks are natural (Keiller 1934, 20).
c The cup marks are on a chockstone under the east end of the recumbent (J. Ritchie, 'Cupmarks on the stone circles and standing stones of Aberdeenshire and part of Banffshire', *PSAS*, 52 (1918), 86–121, p. 99).
d The cup marks are at the extreme western end on the outer face.
e The azimuth is taken from Thom's plan (Thom, Thom and Burl 1980, 190). It is the azimuth of his line *BC*; although *B* and *C* are spurious the line fortuitously goes through the relevant circle stone.

| 1 | 2 | 3 | 4 | 5 | 6 | 7 | 8 | 9 | 10 | 11 |
|---|---|---|---|---|---|---|---|---|---|---|
| RSC23 | Rothiemay | | | 119 | | | 202·5 | 1·4 | −28·4 | |
| RSC23 | Rothiemay | | | | | 2 | 167·0 | 0·5 | −31·0 | a |
| RSC24 | Arnhill | | | 2 | | | 195·5 | 1·6 | −29·6 | b |
| RSC44 | Braehead | | | 4 | | | 193·0 | 2·6 | −29·1 | c |
| RSC46 | Loanend | | | 2 | | | 225·0 | 3·6 | −19·1 | |
| RSC48 | Cothiemuir Wood | | | 3–4 | | | 204·0 | 0·8 | −29·2 | d |
| RSC59 | Loanhead of Daviot | 12 | | | | | 140·0 | 0·2 | −24·2 | e |
| RSC62 | Balquhain | | | | | 25 | 232·0 | 0·5 | −18·9 | |
| RSC62 | Balquhain | | | 1 | | | 190·0 | 3·0 | −29·4 | |
| RSC62 | Balquhain | | 4 | | | | 172·0 | 3·0 | −29·3 | |
| RSC72 | Sunhoney | | | 31 | | | 224·5 | 5·0 | −18·8 | |

**Table 5.6 Indicated horizons at axial stone circles**

Azimuths, altitudes and declinations of the horizon ranges demarcated by the axial stone and portals, as determined during fieldwork in 1994. The assumed observing position is 5 m outwards from the portals. The data have been taken from Ruggles and Prendergast 1996, table 2.

The quoted 'Portal Left' azimuth is that of the inside (right-hand) edge of the left portal as viewed in the direction towards the axial stone. The 'Portal Right' azimuth is that of the inside (left-hand) edge of that portal.

Column headings:
1 Reference number in List 4
2 Portal Left azimuth, quoted to the nearest 0·5 degrees
3 Axial Left azimuth, quoted to the nearest 0·5 degrees
4 Axial Centre azimuth, quoted to the nearest 0·5 degrees
5 Axial Right azimuth, quoted to the nearest 0·5 degrees
6 Portal Right azimuth, quoted to the nearest 0·5 degrees
7 Mean altitude within indication, quoted to the nearest 0·2 degrees
8 Portal Left declination, quoted to the nearest 0·2 degrees
9 Axial Left declination, quoted to the nearest 0·2 degrees
10 Axial Centre declination, quoted to the nearest 0·2 degrees
11 Axial Right declination, quoted to the nearest 0·2 degrees
12 Portal Right declination, quoted to the nearest 0·2 degrees
13 Notes

Key to Column 13 (Notes)
a All of the indicated azimuths are subject to uncertainty because of the lack of one portal stone.
b Some or all of the indicated azimuths were obtained by offset, because nearby vegetation partially obscures the horizon as viewed from the circle, and they are therefore subject to uncertainty.
c Some or all of the indicated declinations were obtained by map calculation, because nearby vegetation partially obscures the horizon as viewed from the circle, and they are therefore subject to uncertainty.
d The relevant indicated azimuth has been obtained by estimating the upright position of the leaning or fallen portal.
e The mean axial direction has been estimated by taking the line joining the mid-point of the flankers to the mid-point of the portals.
f The mean axial direction has been estimated as that at right angles to the flat face of the axial stone. Since no stones remain on the north-eastern side of the circle, the observing position has been taken as 12 m from the axial stone.
g All of the indicated azimuths are subject to uncertainty because the portal and entrance stones were overgrown with thick vegetation that could not be removed at the time of our visit.
h The indicated azimuths are subject to considerable uncertainty because of the lack of both portal stones.
i The indicated azimuths have been obtained by estimating the upright position of the leaning or fallen axial stone.
j The indicated horizon is formed by outcroppings immediately adjacent to the circle (within 20 m), and it is meaningless to quote a precise altitude. An approximate figure of 5° has been assumed, and the deduced declinations are subject to considerable uncertainty.
k Trees cover part of the indicated horizon, so the altitude of the horizon points in question has been estimated.

| 1 | 2 | 3 | 4 | 5 | 6 | 7 | 8 | 9 | 10 | 11 | 12 | 13 |
|---|---|---|---|---|---|---|---|---|---|---|---|---|
| ASC1 | 211·0 | 211·5 | 214·0 | 216·5 | 217·0 | 2·4 | −29·0 | −29·4 | −28·4 | −28·4 | −28·4 | |
| ASC3 | 200·0 | 200·0 | 202·5 | 205·5 | 205·5 | 0·8 | −35·2 | −35·2 | −34·4 | −33·6 | −33·6 | |
| ASC5 | 240·5 | 242·0 | 246·0 | 250·0 | 252·0 | 4·5 | −14·0 | −13·2 | −11·0 | −8·5 | −7·5 | c |
| ASC6 | 189·0 | 189·0 | 191·0 | 193·0 | 193·0 | 1·4 | −36·4 | −36·4 | −36·0 | −36·0 | −36·0 | b |
| ASC7 | 238·0 | 240·0 | 245·5 | 250·5 | 252·0 | 2·8 | −17·2 | −15·8 | −12·6 | −9·6 | −8·6 | |
| ASC8 | — | 224·0 | 229·0 | 234·0 | 234·0 | 5·2 | — | −22·0 | −19·4 | −17·0 | −17·0 | ak |
| ASC10 | 199·0 | 199·5 | 202·5 | 205·5 | 207·0 | 0·6 | −35·2 | −35·0 | −34·8 | −33·4 | −33·0 | |
| ASC11 | 196·0 | 196·5 | 199·5 | 202·5 | 204·0 | 5 | −32 | −32 | −31 | −30 | −30 | dj |
| ASC14 | 257·5 | 263·5 | 265·5 | 267·5 | 272·5 | 1·4 | −7·2 | −3·2 | −2·0 | −0·6 | +2·8 | bc |
| ASC15 | 188·0 | 194·0 | 198·5 | 203·0 | 210·0 | 8·6 | −25·4 | −26·8 | −27·6 | −28·0 | −27·6 | |
| ASC17 | 218·0 | 221·0 | 222·5 | 224·0 | 228·0 | 3·8 | −26·0 | −24·4 | −23·8 | −23·0 | −21·0 | |
| ASC21 | 259·5 | 263·5 | 266·5 | 269·5 | 274·0 | 4·2 | −1·8 | −0·8 | +1·2 | +2·6 | +5·6 | d |
| ASC23 | 207·0 | 207·0 | 210·0 | 213·5 | 214·0 | 0·4 | −33·4 | −33·4 | −32·6 | −31·2 | −31·0 | d |
| ASC24 | 241·5 | 247·0 | 249·0 | 250·5 | 257·0 | 0·6 | −17·0 | −14·2 | −12·6 | −11·8 | −7·8 | |
| ASC25 | 228·5 | 234·5 | 237·0 | 239·5 | 245·5 | 1·2 | −23·6 | −20·6 | −18·8 | −17·4 | −14·0 | b |
| ASC26 | — | 197·5 | 200·5 | 203·5 | — | 0·0 | — | −36·8 | −36·0 | −35·0 | — | h |
| ASC28 | 202·0 | — | 206·5 | — | 211·0 | 5·4 | −29·0 | — | −28·6 | — | −28·0 | de |
| ASC31 | — | 192·0 | 197·5 | 203·0 | — | 7·4 | — | −28·0 | −29·0 | −29·6 | — | f |
| ASC32 | 212·5 | 220·0 | 223·0 | 226·0 | 230·0 | 8·0 | −21·8 | −20·8 | −19·8 | −18·4 | −16·8 | d |
| ASC33 | 253·0 | 253·0 | 257·0 | 261·0 | — | 22·8 | +6·4 | +6·4 | +10·2 | +14·2 | — | a |
| ASC34 | 203·5 | — | 210·5 | — | 218·0 | 5·2 | −28·8 | — | −27·6 | — | −25·4 | e |
| ASC35 | 194·0 | 196·5 | 199·5 | 202·5 | — | 1·0 | −36·2 | −35·8 | −35·2 | −34·4 | — | abi |
| ASC37 | 266·5 | 268·0 | 270·0 | 272·5 | — | 1·2 | −1·6 | −0·4 | +0·6 | +2·0 | — | a |
| ASC40 | 264·0 | 264·0 | 271·5 | 279·0 | 279·0 | 4·4 | +0·4 | +0·4 | +4·2 | +8·6 | +8·6 | |
| ASC43 | 208·5 | 213·0 | 215·5 | 218·0 | — | 3·2 | −30·4 | −28·6 | −27·8 | −26·4 | — | a |
| ASC44 | 248·5 | 254·0 | 256·5 | 259·5 | — | 4·0 | −10·4 | −6·8 | −5·4 | −3·4 | — | a |
| ASC46 | 203·5 | 207·0 | 210·0 | 213·5 | 214·5 | 2·0 | −33·4 | −32·0 | −30·6 | −29·4 | −29·0 | g |
| ASC47 | 255·5 | 258·5 | 261·5 | 264·0 | — | 2·0 | −7·8 | −5·8 | −4·0 | −2·6 | — | ab |
| ASC48 | 242·5 | 246·0 | 248·5 | 251·0 | 252·5 | −0·2 | −17·0 | −15·2 | −14·4 | −12·2 | −11·4 | b |
| ASC51 | 262·5 | 265·0 | 268·0 | 271·0 | 273·0 | 2·6 | −3·0 | −1·4 | +0·6 | +3·4 | +3·8 | |
| ASC52 | 223·0 | 222·5 | 227·0 | 231·0 | 230·5 | 3·0 | −24·6 | −25·0 | −23·0 | −20·2 | −20·4 | |

**Table 5.7 Hilltops within indicated horizons at axial stone circles**

Azimuths, altitudes and declinations of single conspicuous hilltops contained within ASC indicated horizon ranges, as determined during fieldwork in 1994 and subsequent data reduction. The data have been taken from Ruggles and Prendergast 1996, table 3.

The first part of the table concerns horizons in the direction from the portals to the axial stone. The second part, for comparison, contains similar data for the opposite direction.

Column headings:
1 Reference number in List 4
2 Name of hill
3 Distance (km)
4 Azimuth of hill summit, to the nearest 0·1 degrees
5 Altitude of hill summit, to the nearest 0·1 degrees
6 Declination of hill summit, to the nearest 0·1 degrees
7 Difference between azimuths of hill summit and ASC axis, in degrees
8 Status of measurements (S = Surveyed profile; C = Calculated profile)

| 1 | 2 | 3 | 4 | 5 | 6 | 7 | 8 |
|---|---|---|---|---|---|---|---|
| ASC25 | Hill 1797ft at W150602 | 14 | 253·1 | 1·3 | −9·7 | 4 | S |
| ASC38 | Sugarloaf Mountain | 16 | 267·8 | 1·5 | −0·4 | 2 | S |
| ASC40 | Knockgour, N end of ridge | 4·5 | 268·5 | 5·5 | +3·3 | 3 | S |
| ASC43 | Unnamed hill | 1·5 | 221·3 | 3·5 | −24·9 | 6 | S |
| ASC44 | Carrigradda, N end of ridge | 1·5 | 256·4 | 4·1 | −5·3 | 0 | S |
| ASC3 | Torc Mountain | 8 | 210·8 | 3·3 | −29·1 | 3 | S |
| ASC16 | Knocknagullin | 15 | 264·5 | 1·5 | −2·5 | 1 | C |
| ASC38 | Nowen Hill | 11 | 89·4 | 2·3 | +2·0 | 1 | S |
| ASC17 | Crohane | 14 | 17·3 | 2·0 | +37·8 | 1 | C |
| ASC34 | Unnamed hill | 2·0 | 78·9 | 6·9 | +12·3 | 2 | S |

**Table 6.1 Indicated horizons at Cork–Kerry short stone rows**

Azimuths, altitudes and declinations of the horizon ranges indicated in each direction along the rows, as determined during fieldwork between 1991 and 1993. The data have been taken from Ruggles 1994a, table 2 and 1996, table 2. Where an apparent direction of indication has been ascertained from the stone height gradation, the data in that direction are shown in bold type; otherwise, no apparent direction of indication could be identified.

The first part of the table contains data from monuments identified as four- to six-stone rows during 1991–3, and hence investigated during the first stage of the project; the second part contains data from monuments identified as three-stone rows.

Column headings:
1 Reference number in List 5
2 Axial orientation (to 'NE'), quoted to the nearest degree
3 Axial orientation (to 'SW'), quoted to the nearest degree
4 Minimum Azimuth (to 'NE'), quoted to the nearest degree
5 Maximum Azimuth (to 'NE'), quoted to the nearest degree
6 Mean altitude ('NE' indication), quoted to the nearest 0·2 degrees
7 Maximum Declination ('NE' indication), quoted to the nearest 0·2 degrees
8 Minimum Declination ('NE' indication), quoted to the nearest 0·2 degrees
9 General notes (horizon to 'NE')
10 Notes on Lynch's result where significantly different (horizon to 'NE')
11 Minimum Azimuth (to 'SW'), quoted to the nearest degree
12 Maximum Azimuth (to 'SW'), quoted to the nearest degree
13 Mean altitude ('SW' indication), quoted to the nearest 0·2 degrees
14 Minimum Declination ('SW' indication), quoted to the nearest 0·2 degrees
15 Maximum Declination ('SW' indication), quoted to the nearest 0·2 degrees
16 General notes (horizon to 'SW')
17 Notes on Lynch's result where significantly different (horizon to 'SW')

Key to Column 2 (Axial orientation)
* Labelling the stones *a*, *b*, etc. from the south-west, the line *acd* (i.e. from the outlier to the SW-most pair of Ó Nualláin's three-stone row, *b* being the possible fallen stone), yields an azimuth range of roughly 32°–37°, whereas the NE-most pair *de* yield 45°–48°, so the mean figure of 40° may be misleading.

Key to Columns 9 and 16 (General notes on horizons)
a Horizon less than 500 m distant, so not surveyed.
b Horizon local and obscured by a forestry plantation adjacent to the stones, so not surveyed.
c Horizon probably local and obscured by a forestry plantation adjacent to the stones, so not surveyed.
d Distant horizon obscured by nearby field walls and vegetation, so not surveyed.
e The indicated azimuth range here is very wide, owing to the sinuous nature of the alignment.
f The right-hand end of the indication was not surveyed, owing to tall vegetation on the local horizon (see Ruggles 1994a, table 1). The declination figure quoted is an estimate.
g The north-east horizon was not surveyed owing to obscuration by field walls and gorse some 20 m from the stones. In this direction is land near the tip of the peninsula leading to Rough Point, some 12 km distant. This has an altitude close to that of the sea horizon. Consequently, an altitude of −0°·3 has been assumed in order to provide the declination estimates given.
h Some error is possible owing to large extrapolations between surveyed points.
B The 'fine outlook to the east and north' noted by Ó Nualláin (1988, no. 50) is now obscured by a forestry plantation adjacent to the stones, so was not surveyed. The tabulated values have been obtained by calculation.
C The indicated azimuth range has been estimated using the positions of the stoneholes of the two stones removed in about 1987. They are consistent with Ó Nualláin's plan (1988, fig. 7).
F The left-hand end of the indication is close (about 300 m) and was not surveyed. The declination figure quoted is an estimate.
G Some error is possible owing to poor visibility conditions and the lack of an accurate sun-azimuth determination during the site survey.
H The lower precision is due to large extrapolations between surveyed points.
S Although in earlier publications the stone heights in this row were considered to be graded towards the north-east, this is only true of the middle three stones; the end stones break the pattern. I am grateful to Ann Lynch (*priv. comm.*, 1997) for pointing this out.

Key to Columns 10 and 17 (Notes on Lynch's result where significantly different)
j Lynch appears to have transposed the horizon altitudes in the two directions along this row.
k Lynch quotes a much higher altitude (8°·4).
l Lynch quotes a rather higher azimuth, of 244°·2.
m Lynch quotes a rather higher azimuth (48°·8) and a much higher altitude (9°·4). At azimuths between 45° to 51° we obtain altitudes around 2°·6.
n Lynch quotes a rather lower azimuth (52°·2/232°·2) and an altitude of 0°·0 in both directions.
o Lynch quotes an azimuth of 36°·2, corresponding to *acd* (see note * above) and a rather higher altitude of 3°·0.
p Lynch quotes an azimuth of 216°·2, corresponding to *dca* (see note * above) and a lower altitude of 0°·0. At azimuths between 215° to 222° we obtain altitudes between 3°·0 and 3°·2.
q Lynch quotes a higher altitude of 2°·0.
r Lynch quotes an azimuth of 222°·4 and an altitude of 0°·2. This azimuth is close to the summit of Nowen Hill (azimuth 222°·5), for which we obtain an altitude of 1°·6.
s Lynch quotes a rather lower azimuth (39°·6/219°·6).
t Lynch quotes a rather lower altitude of −0°·6.
K Lynch quotes a much lower altitude of 3°·0.
L Lynch quotes a higher altitude of 6°·9.
M Lynch quotes an azimuth of 29°·0/209°·0.
N Lynch does not provide data for this monument.
P Lynch quotes an azimuth of 33°·5/213°·5.
Q Lynch quotes an impossibly low altitude of −2°·8.
R Lynch quotes a higher altitude of +1°·4.

| 1 | 2 | 3 | 4 | 5 | 6 | 7 | 8 | 9 | 10 | 11 | 12 | 13 | 14 | 15 | 16 | 17 |
|---|---|---|---|---|---|---|---|---|---|---|---|---|---|---|---|---|
| CKR1 | **38** | 218 | **35** | **40** | **0·2** | **+29·4** | **+27·6** | | | — | — | — | — | — | a | |
| CKR2 | 38 | 218 | 36 | 40 | 0·2 | +29·4 | +27·6 | | N | 216 | 220 | 1·2 | −28·6 | −27·4 | | N |
| CKR3 | 0 | **180** | 355 | 5 | 0·2 | +37·4 | +37·2 | | N | **175** | **185** | **3·0** | **−35·0** | **−34·8** | | N |
| CKR4 | **60** | 240 | **58** | **63** | **−0·2** | **+18·2** | **+15·4** | g | j | 238 | 243 | 7·4 | −13·2 | −9·8 | | j |
| CKR10 | **36** | 216 | **33** | **38** | **0·2** | **+30·8** | **+28·6** | | k | — | — | — | — | — | b | |
| CKR14 | 62 | **242** | — | — | — | — | — | a | | **241** | **243** | **1·8** | **−16·2** | **−15·0** | | l |
| CKR30 | **43** | 223 | **42** | **44** | **2·8** | **+29·6** | **+28·6** | | m | — | — | — | — | — | a | |
| CKR32 | 55 | 235 | 54 | 57 | 1·4 | +22·4 | +20·2 | | n | 234 | 237 | 0·4 | −21·4 | −19·4 | | n |
| CKR35 | 39 | **219** | — | — | — | — | — | c | | **216** | **222** | **0·0** | **−30·4** | **−27·8** | | N |
| CKR38 | 38 | 218 | 35 | 41 | 4·2 | +35·2 | +30·2 | | N | 215 | 221 | 2·6 | −28·6 | −25·4 | | N |
| CKR40 | 33 | **213** | 32 | 35 | 0·6 | +31·6 | +30·4 | | | **212** | **215** | **0·6** | **−31·8** | **−30·0** | | |
| CKR43 | 34 | **214** | — | — | — | — | — | a | | **213** | **216** | **0·4** | **−31·2** | **−30·2** | | N |
| CKR46 | **40*** | 220 | **32** | **48** | **2·0** | **+33·4** | **+25·6** | | o | 212 | 228 | 2·4 | −28·8 | −23·4 | | p |
| CKR49 | 23 | **203** | 20 | 26 | 4·2 | +40·0 | +37·0 | | | **200** | **206** | **2·0** | **−33·8** | **−32·4** | | |
| CKR51 | 44 | 224 | 41 | 46 | 0·6 | +28·2 | +25·4 | | q | — | — | — | — | — | a | |
| CKR54 | 47 | **227** | 45 | 49 | 1·0 | +26·2 | +24·6 | | | **225** | **229** | **0·8** | **−25·8** | **−24·0** | | |
| CKR59 | 42 | **222** | 40 | 43 | 11·4 | +38·2 | +36·8 | h | N | **220** | **223** | **−0·3** | **−29·2** | **−27·8** | | N |
| CKR60 | **57** | 237 | **—** | **—** | **—** | **—** | **—** | d | | 233 | 241 | 0·2 | −22·0 | −17·8 | | N |
| CKR61 | 41 | **221** | 39 | 44 | 2·0 | +30·6 | +26·8 | | | **219** | **224** | **1·2** | **−28·6** | **−25·8** | | r |
| CKR63 | 45 | 225 | 42 | 49 | 1·6 | +28·4 | +25·4 | S | s | 222 | 229 | 1·2 | −26·6 | −23·4 | | s |
| CKR91 | **63** | 243 | **53** | **72** | **0·2** | **+21·8** | **+10·6** | ef | t | 233 | 252 | 0·4 | −21·6 | −11·6 | e | |
| CKR5 | 58 | **238** | 54 | 61 | 0·2 | +20·8 | +16·8 | | N | **234** | **241** | **4·8** | **−17·6** | **−13·8** | | N |
| CKR6 | **46** | 226 | **42** | **49** | **2·4** | **+29·8** | **+25·0** | G | | 222 | 229 | 0·0 | −27·6 | −24·0 | G | |
| CKR9 | **94** | 274 | **89** | **98** | **2·4** | **+2·4** | **−3·2** | | | 269 | 278 | 2·0 | +0·8 | +6·2 | | |
| CKR11 | 356 | **176** | 354 | 358 | −0·2 | +36·8 | +37·2 | C | N | **174** | **178** | **14·6** | **−23·6** | **−23·0** | C | N |
| CKR13 | 21 | **201** | 18 | 24 | 1·0 | +36·2 | +34·8 | | N | **—** | **—** | **—** | **—** | **—** | a | |
| CKR19 | 57 | **237** | 53 | 61 | 11·0 | +30·6 | +25·8 | | N | **233** | **241** | **0·8** | **−21·2** | **−17·4** | | N |
| CKR20 | 62 | **242** | 56 | 68 | 4·4 | +23·6 | +17·2 | F | j | **236** | **248** | **2·8** | **−18·4** | **−11·0** | | j |
| CKR23 | 28 | 208 | 26 | 30 | 1·8 | +35·2 | +33·8 | | Q | 206 | 210 | −0·2 | −34·4 | −33·0 | | R |
| CKR25 | **87** | 267 | **79** | **96** | **3·2** | **+10·0** | **−2·8** | | N | — | — | — | — | — | a | |
| CKR28 | **52** | 232 | **45** | **58** | **10·0** | **+36·0** | **+25·2** | | K | 225 | 238 | 3·6 | −21·8 | −16·8 | | L |
| CKR33 | 24 | 204 | 11 | 38 | 2·8 | +40·4 | +30·4 | e | | — | — | — | — | — | a | |
| CKR34 | **106** | 286 | **102** | **111** | **−0·2** | **−7·8** | **−13·8** | | N | — | — | — | — | — | a | |
| CKR37 | 44 | 224 | — | — | — | — | — | a | | 220 | 228 | 1·8 | −27·0 | −22·8 | | N |
| CKR42 | 59 | **239** | — | — | — | — | — | a | | **234** | **243** | **0·4** | **−20·8** | **−16·4** | | N |
| CKR45 | 47 | 227 | 43 | 50 | 1·2 | +27·4 | +24·6 | | | 223 | 230 | 2·4 | −25·4 | −21·0 | | |
| CKR50 | 54 | **234** | 53 | 55 | 0·4 | +21·8 | +20·6 | | j | **233** | **235** | **1·4** | **−21·0** | **−19·8** | | j |
| CKR53 | 42 | 222 | 40 | 44 | 1·8 | +29·6 | +28·0 | | | 220 | 224 | 1·8 | −26·8 | −25·0 | | |
| CKR64 | 78 | **258** | 77 | 80 | 2·8 | +10·0 | +8·2 | | N | **257** | **260** | **2·2** | **−6·6** | **−4·6** | | N |
| CKR65 | 62 | **242** | 57 | 68 | 3·0 | +22·6 | +15·6 | | N | **237** | **248** | **2·0** | **−18·2** | **−12·0** | | N |
| CKR66 | **83** | 263 | **81** | **86** | **2·6** | **+7·2** | **+4·2** | | | 261 | 266 | 2·4 | −2·8 | −1·6 | | |
| CKR72 | 23 | 203 | 20 | 25 | 15 | +50 | +48 | H | j | 200 | 205 | 0·4 | −35·8 | −34·4 | | j |
| CKR73 | 30 | **210** | 28 | 31 | 0·4 | +33·0 | +32·0 | | | **—** | **—** | **—** | **—** | **—** | a | |
| CKR75 | 45 | **225** | 43 | 47 | 3·0 | +29·2 | +27·4 | | P | **223** | **227** | **0·2** | **−27·4** | **−25·4** | | P |
| CKR79 | **33** | 213 | **31** | **35** | **−0·2** | **+31·2** | **+30·0** | B | M | 211 | 215 | 0·4 | −32·4 | −30·4 | | M |
| CKR80 | 42 | **222** | — | — | — | — | — | a | | **219** | **224** | **0·4** | **−28·4** | **−26·8** | | N |
| CKR81 | 59 | **239** | — | — | — | — | — | a | | **236** | **241** | **0·8** | **−20·0** | **−17·2** | | N |
| CKR86 | 37 | 217 | 27 | 48 | 2·2 | +35·6 | +26·2 | | j | 207 | 228 | 4·6 | −27·6 | −22·2 | | j |

### Table 6.2 Hilltops within indicated horizons at Cork–Kerry short stone rows

Azimuths, altitudes and declinations of hilltops within the indicated horizon ranges in each direction along the rows, as determined during fieldwork between 1991 and 1993 and subsequent data reduction. The data have been taken from Ruggles 1994a, table 3 and 1996, table 3. Where the hilltop is in the apparent direction of indication as ascertained from the stone height gradation, the azimuth, altitude and declination data are shown in bold type. Otherwise, the direction concerned is opposite to this, or else no apparent direction of indication could be identified.[1]

On selection criteria, see the main text. Note particularly that 'hilltop' here means the point of highest altitude as viewed from the monument, not necessarily the point of highest elevation (i.e. the summit).

The first part of the table contains data from monuments identified as four- to six-stone rows during 1991–3, and hence investigated during the first stage of the project; the second part contains data from monuments identified as three-stone rows.

Column headings:

1 Reference number in List 5
2 Direction ('NE' or 'SW')
3 Name of hill
4 Distance, to the nearest 0·1 km if under 5 km, to the nearest km otherwise
5 Azimuth of hilltop, to the nearest 0·1 degrees
6 Altitude of hilltop, to the nearest 0·1 degrees
7 Declination of hilltop, to the nearest 0·1 degrees
8 Notes

Key to Column 8 (Notes)

a Calculated owing to poor visibility at time of survey.
b Calculated owing to obscuration of the distant horizon (see chapter six, note 25).
c Whether Nowen Hill is within the indicated azimuth range is uncertain because no precise determination of the orientation of the row could be obtained (see chapter six, note 24).
d Calculated (see chapter six, note 25). Whether Milane Hill is in fact visible is uncertain, and this entry must be regarded as tentative.
e Small error possible. Extrapolated between surveyed points.
f The value quoted corresponds to that point on the sea horizon directly above the highest point of Two-Headed Island.
g In Ruggles 1994a, table 3, the information for this row (number 55) has been confused: the data in columns 4–8 ascribed to the NE indication actually refer to the SW one, and vice versa. The mistake has also been carried through to Ruggles and Burl 1995, table 2.

| 1 | 2 | 3 | 4 | 5 | 6 | 7 | 8 |
|---|---|---|---|---|---|---|---|
| CKR1 | NE | Slievecallan (NW peak) | 40 | **38·5** | **0·3** | **+28·2** | |
| CKR1 | NE | Slievecallan (SE peak) | 40 | **39·7** | **0·3** | **+27·7** | |
| CKR2 | NE | Hill, 883ft, at R188397 | 20 | 36·2 | 0·2 | +29·2 | |
| CKR2 | NE | Hill, 916ft, at R197387 | 20 | 40·0 | 0·2 | +27·6 | |
| CKR4 | SW | Point on ridge between Brandon Peak and Brandon Mountain | 4·8 | 242·7 | 8·0 | −9·8 | |
| CKR10 | NE | Coolfree Mountain | 32 | **37·4** | **0·3** | **+29·0** | |
| CKR14 | SW | Mullaganish | 11 | **241·8** | **1·8** | **−15·7** | |
| CKR30 | NE | Musherabeg | 4·5 | **43·3** | **2·8** | **+28·9** | |
| CKR32 | SW | Knockboy | 32 | 237·1 | 0·7 | −19·4 | |
| CKR35 | SW | Milane Hill | 36 | **220·1** | **0·1** | **−28·6** | a |
| CKR40 | SW | Douce Mountain | 15 | **214·6** | **1·0** | **−30·0** | |
| CKR43 | SW | Knockstooka | 14 | **214·9** | **0·7** | **−30·2** | |
| CKR46 | NE | Colly | 14 | **45·5** | **2·4** | **+27·4** | |
| CKR49 | SW | Beenarourke | 9 | **204·5** | **2·2** | **−32·4** | |
| CKR51 | NE | Hill, 1250 ft, 0·5 km SE of Lacknahagny | 15 | 40·9 | 0·9 | +28·2 | |
| CKR54 | NE | Knocknacusha | 22 | 48·6 | 1·3 | +24·9 | |
| CKR54 | SW | Reenearagh | 6 | **227·5** | **1·1** | **−24·1** | |
| CKR59 | SW | Two-Headed Island | 7 | **221·4** | **−0·3** | **−28·5** | f |
| CKR60 | SW | Hill, 550 ft, at W447582 | 4·7 | 234·2 | 0·3 | −21·4 | |
| CKR61 | SW | Nowen Hill | 11 | **222·5** | **1·6** | **−26·1** | |
| CKR63 | NE | Hill 1017 ft, at W099632 | 7 | 49·0 | 2·1 | +25·5 | e |
| CKR67 | SW | Milane Hill | 13 | 236·9 | 1·1 | −19·2 | d |
| CKR68 | SW | Nowen Hill | 4·1 | **226·6** | **5·5** | **−20·5** | c |
| CKR70 | SW | Seefin | 29 | **238·2** | **0·2** | **−19·4** | b |
| CKR91 | NE | Hill, 565 ft, 2 km ENE of Glandore | 8 | **54·0** | **0·4** | **+21·3** | g |
| CKR91 | SW | Hill, 531 ft, 2 km WSW of Castletownshend | 2·6 | 237·5 | 1·0 | −19·1 | g |
| CKR5 | SW | Hill, 656 ft, at Q656081 | 3·5 | **237·7** | **5·2** | **−14·8** | |
| CKR6 | NE | Knockmulanane | 7 | **42·7** | **3·7** | **+29·9** | a |
| CKR19 | NE | Been Hill, ridge to SE of summit | 2·0 | 57·5 | 11·3 | +28·6 | |
| CKR19 | SW | Benlee | 11 | **236·9** | **1·3** | **−18·9** | |
| CKR25 | NE | Brassel Mountain | 3·0 | **85·0** | **5·5** | **+7·3** | |
| CKR37 | SW | Hill 1265 ft at W162727 | 3·9 | 227·5 | 2·2 | −23·0 | |
| CKR42 | SW | Hill 1797 ft at W151601 | 20 | **234·6** | **1·0** | **−20·4** | |
| CKR50 | SW | Hill 1797 ft at W151601 | | **233·7** | **1·7** | **−20·3** | |
| CKR66 | NE | Droppa | 8 | **84·3** | **2·9** | **+5·6** | |
| CKR72 | NE | Slopes of unnamed hill, 1714 ft, 1 km SW of Glenkeel Lake/Lough | 1·0 | 22·3 | 15·4 | +49·5 | |
| CKR72 | SW | Hill 795 ft, 2·5 km from Sheep's Head | 20 | 201·3 | 0·5 | −35·3 | |
| CKR75 | SW | Mount Kiel | 13 | **225·8** | **0·4** | **−25·7** | |
| CKR79 | NE | Green Hill | 18 | **34·7** | **0·1** | **+30·2** | b |
| CKR80 | SW | Mount Corin | 9 | **219·6** | **0·9** | **−28·1** | |
| CKR86 | NE | Point on ridge, V983448 | 1·3 | 34·2 | 2·7 | +33·1 | |

**Table 6.3 Indicated horizons at western Scottish short stone rows**

Azimuths, altitudes and declinations of the horizon ranges indicated in each direction along the rows. The data have been taken from various surveys undertaken by the author between the mid-1970s and the mid-1990s. The widths of the indicated ranges are not always compatible from one dataset to the next and minor adjustments have been made to fit where necessary, but this has had little or no effect on the resulting declinations.

The first part of the table contains data from short rows of at least three stones, the second part contains data from aligned pairs.

Column headings:
1 Reference number in List 6
2 Axial orientation (northwards), quoted to the nearest degree
3 Axial orientation (southwards), quoted to the nearest degree
4 Distance (northerly indication), quoted to the nearest 0·5 km
5 Minimum Azimuth (northerly indication), quoted to the nearest degree
6 Maximum Azimuth (northerly indication), quoted to the nearest degree
7 Mean altitude (northerly indication), quoted to the nearest 0·5 degrees
8 Maximum Declination (northerly indication), quoted to the nearest 0·5 degrees
9 Minimum Declination (northerly indication), quoted to the nearest 0·5 degrees
10 Source (northerly indication)
11 Reliability (northerly indication)
12 Distance (southerly indication), quoted to the nearest 0·5 km
13 Minimum Azimuth (southerly indication), quoted to the nearest degree
14 Maximum Azimuth (southerly indication), quoted to the nearest degree
15 Mean altitude (southerly indication), quoted to the nearest 0·5 degrees
16 Minimum Declination (southerly indication), quoted to the nearest 0·5 degrees
17 Maximum Declination (southerly indication), quoted to the nearest 0·5 degrees
18 Source (southerly indication)
19 Reliability (southerly indication)

Key to Columns 10 and 18 (Sources)
Line number in Ruggles 1984a (azimuths and declinations are taken from the IAR data in table 11.2; for altitude data see the figures; for distances see table 11.2 or tables 5.1–10.1), or
a Ruggles 1985, table 3.
b Ruggles 1988b, table 9.1.
c Ruggles and Martlew 1989, S147 and fig. 4.
d Martlew and Ruggles 1993, fig. 18.
f Newly calculated: the azimuths given in Ruggles 1985 correspond to *MLK* (Thom's $S_6$-$S_2$-$S_3$) rather than *LK*.
i The azimuth values adopted are in line with those suggested in the opposite direction following the excavation (Martlew and Ruggles 1993, fig. 18). Declination values have been estimated accordingly from unpublished field data.

Key to Columns 11 and 19 (Reliability)
A Reliably surveyed
B Less reliably surveyed
C Calculated
For notes relating to reliability of the data see the original sources given in columns 10 and 18.
e The IAR azimuths have been adjusted to correspond to the indication *KL* rather than *KLM* (i.e. Thom's $S_3$-$S_2$-$S_6$). Declination values have been estimated from Ruggles 1984a, fig. 8.5 (no. 162).
g The IAR azimuths have been adjusted to correspond with the more recent measurements in the other direction.
h The azimuth values adopted are in line with those suggested in the opposite direction following the excavation (Martlew and Ruggles 1993, fig. 18). Declination values have been estimated from Ruggles 1984a, fig. 7.8 (no. 132) and unpublished field data.

| 1 | 2 | 3 | 4 | 5 | 6 | 7 | 8 | 9 | 10 | 11 | 12 | 13 | 14 | 15 | 16 | 17 | 18 | 19 |
|---|---|---|---|---|---|---|---|---|---|---|---|---|---|---|---|---|---|---|
| SSR4 | 343 | 163 | 8·0 | 342 | 344 | 0·5 | +30·0 | +30·5 | 44 | C | 16·5 | 162 | 164 | 1·0 | −29·5 | −29·0 | 45 | A |
| SSR18 | 336 | 156 | 0·0 | — | — | — | — | — | — | — | 2·0 | 154 | 158 | 2·0 | −29·5 | −27·5 | c | A |
| SSR19 | 348 | 168 | 0·0 | — | — | — | — | — | — | — | 8·0 | 166 | 170 | 2·0 | −31·5 | −31·0 | 122 | A |
| SSR20 | 5 | 185 | 13·5 | 4 | 6 | 0·5 | +33·0 | +34·0 | 124 | C | 1·5 | 184 | 186 | 5·0 | −28·5 | −28·5 | 125 | A |
| SSR21 | 342 | 162 | 47·0 | 341 | 343 | 0·0 | +31·0 | +31·5 | 128 | C | 0·5 | 161 | 163 | 2·0 | −30·0 | −29·5 | b | B |
| SSR22 | 330 | 150 | 3·0 | 329 | 331 | 0·0 | +27·5 | +28·5 | 130 | B | 0·5 | 149 | 151 | 3·5 | −26·0 | −25·0 | a | B |
| SSR23 | 337 | 157 | 0·5 | — | — | — | — | — | — | — | 8·0 | 156 | 158 | 2·0 | −29·0 | −29·0 | 131 | A |
| SSR24 | 12 | 192 | 8·0 | 11 | 14 | 1·5 | +33·5 | +33·5 | i | B | 2·0 | 191 | 194 | 6·5 | −26·5 | −26·0 | d | A |
| SSR25 | 28 | 208 | 8·0 | 22 | 34 | 2·0 | +29·0 | +32·5 | 132 | Bh | 2·0 | 202 | 214 | 7·0 | −24·0 | −20·0 | d | A |
| SSR26 | 302 | 122 | 0·5 | — | — | — | — | — | — | — | 1·0 | 119 | 125 | 2·0 | −16·5 | −14·0 | a | A |
| SSR27 | 317 | 137 | 3·5 | 316 | 318 | 15·0 | +37·0 | +37·5 | 139 | A | 4·0 | 136 | 138 | 6·5 | −18·5 | −18·0 | 140 | A |
| SSR28 | 328 | 148 | 1·0 | 327 | 329 | 6·5 | +32·5 | +35·0 | 149 | A | 1·0 | 147 | 149 | 7·0 | −21·5 | −21·0 | 150 | B |
| SSR29 | 357 | 177 | 2·0 | 356 | 358 | 1·5 | +35·0 | +35·5 | 156 | A | 1·0 | 176 | 178 | 2·0 | −32·0 | −31·5 | a | A |
| SSR30 | 322 | 142 | 4·0 | 321 | 324 | 3·0 | +28·5 | +29·5 | 168 | B | 3·0 | 141 | 144 | 2·0 | −26·0 | −23·5 | 169 | B |
| SSR31 | 346 | 166 | 1·5 | 345 | 348 | 3·0 | +35·5 | +36·0 | 181 | Bg | 1·5 | 165 | 168 | 2·0 | −31·5 | −30·5 | a | B |
| SSR34 | 20 | 200 | 16·5 | 18 | 22 | 0·5 | +31·5 | +33·5 | 204 | C | Sea | 198 | 202 | 0·0 | −33·0 | −32·0 | 205 | C |
| SSR35 | 349 | 169 | 1·0 | — | — | — | — | — | — | — | >70 | 168 | 170 | 0·0 | −34·5 | −34·5 | 233 | C |
| SSR36 | 28 | 208 | 3·0 | 27 | 29 | 2·5 | +31·5 | +32·0 | 243 | C | 9·5 | 207 | 209 | 1·0 | −29·5 | −29·5 | 244 | C |
| SSR37 | 43 | 223 | 0·5 | — | — | — | — | — | — | — | 12·0 | 222 | 224 | 0·0 | −25·5 | −24·5 | 247 | B |
| SSR38 | 48 | 228 | 3·0 | 47 | 49 | 3·0 | +24·0 | +25·0 | 251 | A | >45 | 227 | 229 | 0·0 | −23·0 | −22·0 | 252 | B |
| AP1 | 7 | 187 | 3·5 | 6 | 8 | 5·0 | +38·5 | +38·5 | 153 | A | 4·0 | 186 | 188 | 3·0 | −30·5 | −30·5 | 154 | A |
| AP2 | 347 | 167 | 1·0 | — | — | — | — | — | — | — | 0·0 | 163 | 171 | 2·0 | −32·0 | −30·0 | a | B |
| AP3 | 329 | 149 | 2·5 | 332 | 333 | 5·5 | +34·5 | +34·5 | 162 | Ae | 0·5 | 152 | 153 | 3·0 | −27·0 | −27·0 | f | C |
| AP4 | 322 | 142 | 2·0 | 321 | 323 | 4·5 | +30·0 | +30·5 | 167 | Ag | 0·5 | 141 | 143 | 3·0 | −24·0 | −23·0 | a | C |
| AP5 | 334 | 154 | 4·0 | 333 | 335 | 3·0 | +32·5 | +33·0 | 170 | B | 5·5 | 153 | 155 | 0·5 | −30·0 | −29·5 | 171 | C |
| AP6 | 319 | 139 | 1·0 | 317 | 321 | 2·0 | +26·0 | +26·5 | 185 | Bg | 1·0 | 137 | 141 | 3·5 | −22·5 | −21·0 | a | B |
| AP7 | 85 | 265 | 36·0 | 82 | 87 | 0·5 | +2·0 | +4·5 | 231 | A | 0·5 | — | — | — | — | — | — | — |
| AP8 | 358 | 178 | 0·5 | — | — | — | — | — | — | — | 1·5 | 177 | 178 | 1·5 | −33·0 | −33·0 | 234 | C |

**Table 6.4 Hilltops within indicated horizons at western Scottish short stone rows**

This is a provisional list of azimuths, altitudes and declinations of hilltops within the indicated horizon ranges in each direction along the rows, where hilltops have been defined in accordance with the criteria used for the Cork–Kerry rows. Thus 'hilltop' means the point of highest altitude as viewed from the monument, not necessarily the point of highest elevation (i.e. the summit).

The list has been compiled simply by examining the profile diagrams in Ruggles 1984a, which are not available for all the relevant indications, and therefore should not be regarded as definitive. Peaks anywhere inside the AAR (as defined on p. 70) have been considered.

Aligned pairs are not included in this table.

Column headings:
1 Reference number in List 6
2 Direction ('N' or 'S')
3 Name of hill
4 Distance, to the nearest 0·1 km if under 5 km, to the nearest km otherwise
5 Azimuth of hilltop, to the nearest 0·1 degrees
6 Altitude of hilltop, to the nearest 0·1 degrees
7 Declination of hilltop, to the nearest 0·1 degrees
8 Line number in Ruggles 1984a
9 Reliability of the data
10 Notes

Key to Column 9 (Reliability)
A Reliably surveyed
B Less reliably surveyed
C Calculated
For notes relating to reliability of the data see Ruggles 1984a.

Key to Column 10 (Notes)
c This profile is only partially visible from the row, because of intervening ground (see Ruggles 1981, 178; 1984a, fig. 5.12). Thom (1971, 69) gives a different azimuth (163°·0) for the summit of Mór Monach, obtained by calculation.
f These data are obtained from surveys undertaken during 1987–91.
j This profile has been recalculated using computer programs that can easily convert between the British and Irish National Grids.
x This peak is included although it is just outside the AAR.

| 1 | 2 | 3 | 4 | 5 | 6 | 7 | 8 | 9 | 10 |
|---|---|---|---|---|---|---|---|---|---|
| SSR4 | N | Hill at 221361 | 6 | 344·0 | 0·6 | 30·5 | 44 | C | |
| SSR4 | S | Mór Mhonadh | 16 | 162·0 | 1·2 | −29·4 | 45 | A | c |
| SSR19 | S | Cruachan Ceann a'Ghairbh | 7 | 165·7 | 1·9 | −30·6 | 122 | B | |
| SSR20 | N | Beinn na Leathaid | 13 | 4·1 | 1·3 | 34·2 | 124 | A | f |
| SSR20 | S | | 1·3 | 186·9 | 5·1 | −28·2 | 125 | A | |
| SSR21 | N | Orval | 47 | 343·8 | 0·3 | 31·7 | 128 | C | |
| SSR23 | S | Beinn Bhuidhe | 8 | 159·5 | 2·1 | −29·3 | 131 | A | x |
| SSR25 | S | Meall na Caorach | 1·9 | 210·0 | 7·0 | −21·9 | 133 | A | |
| SSR27 | N | Ben More | 3·7 | 317·0 | 15·0 | 37·3 | 139 | A | |
| SSR29 | N | Meall Reamhar | 2·2 | 358·1 | 1·9 | 35·3 | 156 | A | |
| SSR30 | N | Garbh Sron | 4·0 | 320·4 | 3·1 | 28·3 | 168 | B | |
| SSR36 | N | Hill at 863695 | 3·2 | 27·8 | 2·5 | 31·9 | 243 | C | |
| SSR36 | S | Sheirdrim Hill | 10 | 206·9 | 0·9 | −29·6 | 244 | C | |
| SSR37 | S | Cara Island | 13 | 222·5 | 0·1 | −25·0 | 247 | A | |
| SSR38 | S | Knocklayd | 48 | 226·5 | 0·4 | −23·0 | 252 | C | j |

**Table 7.1 Ben More and the northernmost stone rows on Mull**

Azimuth, altitude and declination of the summit of Ben More as viewed from the vicinity of each of the rows. The data are taken from Ruggles and Martlew 1992, table 9.

Column headings:
1 Reference number in List 6
2 Name of the stone row
3 Orientation of the row, to the nearest degree
4 Azimuth of the summit, to the nearest 0·1 degrees
5 Altitude of the summit, to the nearest 0·1 degrees
6 Declination of the summit, to the nearest 0·1 degrees
7 Difference between the row orientation and the azimuth of the summit, in degrees
8 Status of measurements

Key to Column 8 (Status of measurements)
S Surveyed profile
C Calculated profile

| 1 | 2 | 3 | 4 | 5 | 6 | 7 | 8 |
|---|---|---|---|---|---|---|---|
| SSR18 | Glengorm | 156° | 156°·0 | 2°·0 | −28°·6 | 0 | S |
| SSR19 | Quinish | 168° | 149°·6 | 2°·0 | −26°·7 | 18 | C |
| SSR21 | Maol Mor | 162° | 152°·3 | 2°·1 | −27°·5 | 10 | C |
| SSR22 | Dervaig N | 150° | 151°·9 | 2°·2 | −27°·2 | 2 | C |
| SSR23 | Dervaig S | 157° | 151°·3 | 2°·3 | −26°·9 | 6 | C |

**Table 8.1 Orientations of 'Cumbrian' stone circles**

Entrance, outlier and tallest stone orientations as viewed from the centre of twelve great circles in and around Cumbria. Azimuths and declinations are quoted to the nearest degree. Reference numbers and page numbers refer to Burl's account (Burl 1988a), where the figures given are quoted. Some were obtained in turn from Thom 1967, table 8.1.

| Ref. no. | Name | Location | NGR | Thom ref. | Entrance az (°) | Entrance dec (°) | *Page no.* | Tallest stone az (°) | Tallest stone dec (°) | *Page no.* | Outlier az (°) | Outlier dec (°) | *Page no.* |
|---|---|---|---|---|---|---|---|---|---|---|---|---|---|
| 1 | Ballynoe | Co. Down | J 481404 | — | 264 | | *197* | 357 | | *199* | | | |
| 2 | Brats Hill | Cumbria | NY 173023 | L1/6E | | | | 178 | | *199* | | −16 | *198* |
| 3 | Castlerigg | Cumbria | NY 292236 | L1/1 | 0 | | *199* | 127 | −16 | *198* | | | |
| 4 | Elva Plain | Cumbria | NY 176317 | L1/2 | | | | 262 | | *199* | | | |
| 5 | Girdle Stanes | Dumfries | NY 254961 | G7/5 | | | | (E?) | | *199* | | | |
| 6 | Grey Croft | Cumbria | NY 034024 | L1/10 | | | | | | | 354 | | *199* |
| 8 | Gunnerkeld | Cumbria | NY 568178 | L2/10 | 358 | | *199* | | | | | | |
| 10 | Long Meg and her Daughters | Cumbria | NY 571373 | L1/7 | | | | 86 | | *199* | 223 | −24 | *196–7* |
| 11 | Studfold | Cumbria | NY 040224 | L1/14 | | | | (N) | | *199* | | | |
| 12 | Swinside | Cumbria | SD 172883 | L1/3 | 129 | −22 | *198* | 6 | | *199* | | | |
| 13 | Twelve Apostles | Dumfries | NY 947794 | G6/1 | | | | 241 | −16 | *198* | | | |
| | Mayburgh | Cumbria | NY 523284 | — | 92 | | *199* | | | | | | |

# Notes

If a particular book or article is frequently mentioned in the notes, then only the author's name, date of publication and page number(s) are cited. Full details will be found in the bibliography. As an example, in the introduction, note 2, 'Heggie 1981a' refers the reader to Heggie, Douglas C. (1981), *Megalithic Science: Ancient Mathematics and Astronomy in Northwest Europe*, London: Thames and Hudson, in the bibliography. Where a work appears just once or very rarely, the full reference is given in the notes, as in the introduction, note 1.

If a journal or publisher's name appears frequently in the notes or bibliography, then it is abbreviated and the full name is given in the list preceding the bibliography. If it appears only once or very rarely, then its name is given in full where it occurs.

Notes to the Boxes are listed separately after the notes to the main text, on pp. 259–66.

PREFACE AND ACKNOWLEDGEMENTS

1 The remark (Anthony F. Aveni and Giuliano Romano, 'Orientation and Etruscan ritual', *Antiquity*, 68 (1994), 545–63, p. 545) was actually made in the more general context of studies of orientation.

2 Chapter 2 is the most technical, and the astronomical boxes there are only needed for a full understanding of that chapter.

SOME CONVENTIONS

1 Aitken 1990; see also Renfrew and Bahn 1996, ch. 4.

2 For calibration curves see Minze Stuiver and Bernd Becker, 'High-precision decadal calibration of the radiocarbon time scale, AD 1950–6000 BC', *Radiocarbon*, 35(1) (1993), 35–65. On the CALIB computer program for generating this information, see Stuiver and Reimer 1993.

3 Julian D. Richards and Nick S. Ryan, *Data Processing in Archaeology*, CUP, Cambridge, 1985, 20.

4 Whyte and Paul 1997, 10.

5 *Ibid.*

6 This follows Ruggles 1981, 154, 168.

7 The county names used may not always reflect current administrative entities: for example, the historical county of Kincardineshire—between 1974 and 1996 the Kincardine and Deeside District of the Grampian Region—is now subsumed within Aberdeenshire Council.

8 The six- and eight-figure grid references given in the main text and site reference lists are intended to be accurate to the nearest 100 m and 10 m respectively. Thus, for example, a grid reference of 327191 is meant to imply that the actual easting is between 32,650 m and 32,749 m east of the western side of the relevant 100 km grid square, and the actual northing between 19,050 m and 19,149 m north of the southern side. However, other sources, including at least some SMRs (see Archaeology Box 1), would use the same grid reference to imply that the point lies within the 100 m grid square whose bottom-left hand point is 327191, so that actual easting is between 32,700 m and 32,799 m and the actual northing between 19,100 m and 19,199 m. It has not always been possible to disentangle these conventions, but as the grid references in question are mainly intended for use in locating monuments with the aid of Ordnance Survey maps, this is not considered a serious problem.

INTRODUCTION

1 C. P. Snow, 'The two cultures', *New Statesman and Nation*, 52 n.s. (6 October 1956), 413–14.

2 Heggie 1981a, 8.

3 Julian Thomas, Review of Ruggles 1988a, *Antiquity* 63 (1989), 183–4.

4 Examples are 'Haughey's faith in ancestors confirmed', *Irish Times*, 22 December 1987, 7; 'Newgrange solstice sun shines for its top fan', *Irish Independent*, 22 December 1987, 1.

5 I am grateful to Frank Prendergast for bringing this episode to my attention.

6 The connection of these Druidic ceremonies with Stonehenge has absolutely no justification in historical fact. Not only is there no connection between the modern 'Ancient Order of Druids' and the ancient Celtic Druids; the Celtic Druids themselves lived over a millennium after Stonehenge fell into disuse in the Bronze Age. On the Celtic Druids see Miranda Green, *Exploring the World of the Druids*, Thames and Hudson, London, 1997; Andrew P. Fitzpatrick, *Who Were the Druids?*, Weidenfeld and Nicolson, London, 1997. On the development, from the seventeenth century onwards, of ideas linking the Druids and Stonehenge see Smiles 1994, ch. 5; Green, *Exploring the World of the Druids*, ch. 9; on modern Druids see *ibid.*, ch. 10. See also Chippindale 1994, 172–3.

7 'Mr Lyon Thomson was stationed at Lark Hill, two miles from Stonehenge, during the war, and noticed that one hut in a row was missing, though the foundations were laid. Inquiry showed that a hut had originally been built there for forty men, but had been demolished as it would have prevented the midsummer sunrise being seen from Stonehenge'. Contribution to the discussion on Somerville 1923, *Archaeologia* 73 (1923), 224. See also Chippindale 1994, 175.

8 For an amusing account of the various midsummer activities at Stonehenge during the 1960s and 1970s see Chippindale 1994, ch. 16. The festival became an increasingly worrying problem for those responsible for the preservation of ancient monuments in the area and was eventually outlawed in 1985. More than ten years later a four-mile (6 km) pedestrian-free 'exclusion zone' was still being imposed around the site each year. See also Barbara Bender, 'Stonehenge—contested landscapes (medieval to present-day)', in Bender 1993, 245–79.

9 Newgrange's excavator Michael O'Kelly records that '. . . a belief existed in the neighbourhood that the rising sun, at some unspecified time, used to light up the three-spiral stone . . . in the end recess' (O'Kelly 1982, 123), a tradition that his wife had dismissed as an 'old wives' tale' until witnessing the phenomenon for herself (Claire O'Kelly, *Illustrated Guide to Newgrange*, 2nd edn, John English, Wexford, 1971, 94–5; this remark is omitted from the 3rd edn). See also Burl 1987b, 7–8.

10 Burl 1982, 146, quoting Janet and Colin Bord, *The Secret Country: An Interpretation of the Folklore of Ancient Sites in the British Isles*, Paladin, St Albans, 1978. On the long barrow itself and other folklore associated with it, see Leslie V. Grinsell, 'Somerset barrows, part 2: north and east', *Proceedings of the Somerset Archaeological and Natural History Society*, 115 (1971), special supp., 44–137, pp. 68, 88 [Wrington II].

11 For example, children with rickets were passed, always against the sun, through the holed stone at Men-an-Tol in Cornwall (Burl 1981c, 197). Diseased people would walk three times deasil (sunwise) around the standing stones of Stenness in the Orkney islands and a similar procedure took place on the Hebridean island of Eigg around a stone called Martin Dessil (Leslie V. Grinsell, *Folklore of Prehistoric Sites in Britain*, David and Charles, Newton Abbot, 1976, 25).

12 'Unearthed: a sun temple that puts Stonehenge in the shade', *The Independent on Sunday*, 17 February 1991, 8. On this monument, near Godmanchester, see chapter eight.
13 'Stones may be meteor mapper', *The Times Higher Education Supplement*, 15 December 1995, 44.
14 'Waiting for the solstice sun to set in Connemara' (p. 2) and 'Seeing light at the end of the tunnel at Newgrange' (p. 7).
15 Barry W. Cunliffe (ed.), *The Oxford Illustrated Prehistory of Europe*, OUP, Oxford, 1994.
16 In fact, these are both quite speculative interpretations that have attracted a good deal of discussion and controversy. On Palaeolithic tally-marks see chapter eight; on Dacian sanctuaries see Schlosser and Cierny 1996, 101–2.
17 First published in the USA in 1965, this was published in London a year later and as a Fontana paperback in 1970 (Hawkins and White 1970). In the literature there are numerous commentaries on Hawkins's work and the archaeologists' reaction to it; for one of the most recent see Aveni 1997, 64–71.
18 The two scientific articles upon which the popular book is based were published in the previous two years (Hawkins 1963, 1964) and are reprinted as appendices in the book.
19 The name 'HAL' was derived from 'IBM' by taking the previous letters (see Jerome Agel, *The Making of Kubrick's 2001*, Signet, New York, 1970). See also David G. Stork (ed.), *HAL's Legacy: 2001's Computer as Dream and Reality*, MIT Press, Cambridge Mass., 1997.
20 See also Chippindale 1994, 230–1.
21 A typical account is the following: 'On the 20th of June, we excavated a small mound of earth near the large barrow at Sterndale, opened in September, 1846, but failed to discover an interment. When, to occupy the afternoon, we worked a little in the large barrow, where, in 1846, a bronze dagger was found, and made two cuttings to no purpose, as we observed only remains of animal bones and pottery, some of which was of the Romano-British period, and doubtless belonged to a late interment that was found near the surface some years before by men getting stone.' Thomas Bateman, *Ten Years' Diggings in Celtic and Saxon Grave Hills*, J. R. Smith, London, 1861, 67. For an example from a hundred years earlier still see Thomas Pennant, *A Tour in Scotland, 1769*, 3rd edn, Warrington, 1774, 138–40, who describes incursions into several cairns in the Grampian Region of Scotland. See also Hayman 1997, ch. 11.
22 On the rise of professionalism in archaeology between 1920 and 1960, see Bahn 1996, ch. 6, esp. 198–210.
23 Paul Screeton, *Quicksilver Heritage—The Mystic Leys: Their Legacy of Ancient Wisdom*, Thorsons, Wellingborough, 1974, 13. Ley lines had in fact arisen as far back as 1921: at this time Herefordshire businessman Alfred Watkins first noticed that prehistoric, historic and natural features could time after time be connected up by perfectly straight lines, which he called 'leys'. See Alfred Watkins, *Early British Trackways*, Watkins Meter Co., Hereford, 1922; *The Old Straight Track*, Methuen, London, 1925. Watkins's idea of ancient straight trackways brought an immediate amateur following, but created little interest amongst professional archaeologists, and ley lines faded quietly from view. They might have been forgotten completely but for their resurrection in the 1960s as lines of earth energy.
24 'When on leys in the countryside, you will become aware of senses outside the range of scientifically accepted ones, where intuition may reign and subtle power may be perceived.' Screeton, *Quicksilver Heritage*, 46.
25 For a detailed discussion both of ley lines and of the academic arguments against them, see Tom Williamson and Liz Bellamy, *Ley Lines in Question*, World's Work, Tadworth, 1983. See also Hutton 1991, 118–32; Hayman 1997, chs 21–4. For an archaeological perspective see also Burl 1979a, 78–83. For an argument between an archaeologist and a ley-enthusiast see Aubrey Burl and John Michell, 'The great debate. Living leys or laying the lies?', *Popular Archaeology* 4(8) (1983), 13–18. For statistical examinations of chance alignments in a random distribution of points see M. J. O'Carroll, 'On the probability of general and concurrent alignments of randomly distributed points', *Science and Archaeology* 21 (1979), 37–40; D. G. Kendall and W. S. Kendall, 'Alignments in two-dimensional random sets of points', *Advances in Applied Probability* 12 (1980), 380–424; S. R. Broadbent, 'Simulating the ley hunter', *JRSS* A143 (1980), 109–40.
26 Hawkins was not himself motivated by alternative archaeology.
27 For example, a drawing in the *Ley Hunter* magazine, issue 74 (1977), p. 12, shows a standing stone being used to 'tap' the 'earth currents', the energy release being related to the lunar phase cycle.
28 John Michell, *A Little History of Astro-archaeology: Stages in the Transformation of a Heresy*, Thames and Hudson, London, 1977. The quote is from the cover blurb of the edition published in the same year by Penguin Books under the title *Secrets of the Stones: The Story of Astro-archaeology*. This work also draws extensively on the work of Alexander Thom and others, see below.
29 Atkinson 1966. The article had the matchless title 'Moonshine on Stonehenge'.
30 Atkinson 1979. The first edition was published in 1956.
31 Richard J. C. Atkinson, 'Decoder misled?' (Review of *Stonehenge Decoded*), *Nature* 210 (1966), 1302.
32 Atkinson 1966, 216. Atkinson was also irritated (*ibid.*, 213) by the unacknowledged paraphrasing of passages from his own guidebook on Stonehenge and Avebury (Richard J. C. Atkinson, *Stonehenge and Avebury and Neighbouring Monuments: an Illustrated Guide*, HMSO, London, 1959).
33 Inigo Jones, *The Most Notable Antiquity of Great Britain, Vulgarly Called Stone-Heng, on Salisbury Plain, Restored*, London, 1655. Cf. Aveni 1997, 58–9.
34 Hawkes 1967, 174.
35 These include Evan Hadingham, *Circles and Standing Stones*, Heinemann, London, 1975; Lancaster Brown 1976; Wood 1978; and Cornell 1981.
36 They also described in detail the ideas of C. A. ['Peter'] Newham, Fred Hoyle and others concerning Stonehenge (see chapter one for further discussion) and those of Alexander Thom and others that followed. They tended to be sympathetic to Newham and Hoyle but attitudes to Thom differed.
37 For example Cornell 1981, 71–2.
38 A noteworthy exception is Renfrew 1976, 242–4. While describing Hawkins's theory elsewhere as 'very ill-substantiated' (*ibid.*, 249) Renfrew nonetheless concludes that 'his arguments for at least some of the alignments are now widely accepted, and it is clear that Stonehenge was used to observe the motions of the moon as well as the sun: it was indeed an observatory'.
39 'Many years ago the editors of *Scientific American* told me that there were two magically appealing subjects at least one of which they invariably tried to include in every issue: archaeology and astronomy. The romance of space and the mystery of the past!' Owen Gingerich, foreword to Aveni 1980, xi.
40 Snow first identified and discussed the divide between the 'two cultures'—science and 'traditional' culture (arts and literature)—in an article in the *New Statesman* in 1956 (see note 1).
41 Smiles 1994, ch. 8.
42 Martin described the stone circles of Brodgar and Stenness in Orkney as places of pagan sacrifice, with the sun being worshipped in the larger circle (Brodgar) and the moon in the smaller. Martin Martin, *A Description of the Western Islands of Scotland*, 2nd edn, London, 1716, 365. The theme is also evident in William Stukeley's interpretation of Avebury (see Burl 1979b, 48–9) and John Toland's of Callanish in Lewis (John Toland, *A Critical History of the Celtic Religion and Learning, Containing an Account of the Druids*, Lockington, Hughes, Harding and Co., London, 1820 [originally published 1726], 122–4).
43 Stukeley is generally accredited with the discovery of the alignment of the axis of Stonehenge upon midsummer sunrise. He first mentioned this in a manuscript dated 1723 (Cardiff Public Library MS 4.253), writing more famously seventeen years later that 'the intent of the founders of Stonehenge, was to set the entrance full north east, being the point where the sun rises, or nearly, at the summer solstice' (William Stukeley, *Stonehenge. A Temple Restor'd to the British Druids*, London, 1740, 56). See also Stuart Piggott, *William Stukeley: An Eighteenth-Century Antiquary* (revised edn), Thames and Hudson, London, 1985, 87–92.
44 Wood first proposed a lunar explanation for the Heel Stone: '. . . the altar belonging to the temple of the moon at Stantondrue is situated north eastward from that fabrick, the same as the great pillar before the front of Stonehenge is situated north eastward from that edifice; each work is placed upon ascending ground; and in each work a phase of the new moon is pointed out.' (John Wood, *Choir Gaure*, Oxford, 1747, 80–1). He also attempted to relate the number of sarsen uprights to alternating lunar months of 30 and 29 days (*ibid.*, 84–5).
45 Smith was the first to propose a systematic astronomical explanation for the layout of Stonehenge, a complex one involving the sun, moon, planets, and even the signs of the zodiac. John Smith, *Choir Gaur: the Grand Orrery of the Ancient Druids, Commonly Called Stonehenge, on Salisbury Plain, Astronomically Explained*, Salisbury, 1771. But see, e.g., Castleden 1994, 60.
46 H. du Cleuziou, instructed in 1874 to make an official plan of the Carnac stone rows in Brittany, suggested that they were oriented towards solar risings and settings at the solstices and equinoxes. See Giot 1988, 323.

47 Lewis, in a number of papers published between 1877 and 1914, expressed the notion that astronomical orientations might be incorporated in many megalithic monuments. See particularly Lewis 1892. He also pointed out (*ibid.*, 142) that horizon features might play a significant role, a possibility that would receive a great deal of attention more than half a century later in the work of Alexander Thom.

48 In fact, this idea was expressed as early as 1808 by James Headrick, in an editor's footnote on pp. 213–15 of George Barry, *History of the Orkney Isles*, 2nd edn, London, 1808. The idea may in fact be attributable to Joseph Banks: see Averil M. Lysaght, 'Joseph Banks at Skara Brae and Stennis, Orkney, 1772', *Notes and Records of the Royal Society of London*, 28 (1974), 221–34, p. 232, where Headrick's footnote is reproduced in full. Note that the version of Barry's *History* published by Mercat Press, Edinburgh in 1975 is a facsimile of the first edition of 1805, and does not contain the footnote.

49 There are many discussions of the historical development of these ideas. See, for example, Heggie 1981a, 83–4; Burl 1976, 14–17, on stone circles; Burl 1993, 12–14, with particular reference to Callanish; and Chippindale 1994, ch. 5, for the wider context of the early development of these ideas, and particularly the development of ideas associating megalithic monuments with Druidical practices.

50 J. Norman Lockyer, *The Dawn of Astronomy*, Cassell, London, 1894.

51 Lockyer 1909, v.

52 See especially Boyle Somerville, 'Prehistoric monuments in the Outer Hebrides and their astronomical significance', *JRAI*, 42 (1912), 23–52; Somerville 1923; 1927; A. P. Trotter, 'Stonehenge as an astronomical instrument', *Antiquity* 1 (1927), 42–53.

53 Burl 1993, 16.

54 See, for example, Burl 1993, 14–16, 84–6. For a statistical critique see Heggie 1981a, 179–81.

55 Indeed, one can find precursors of ley lines in the work of Lockyer (e.g. Lockyer 1909, 412–13).

56 Burl 1988a, 176, makes the point that most discoveries about stone circles have been made by non-archaeologists. In 1940, despite decades of research by surveyors, engineers, astronomers and so on, archaeologist Grahame Clark could still say of stone circles: 'what they signify no man can say' and 'in all essentials the great stone circles retain their mystery'. J. G. D. Clark, *Prehistoric England*, Batsford, London, 1940, 103 and 116.

57 Childe 1940, 109.

58 Stuart Piggott, *British Prehistory*, OUP, Oxford, 1949, 119.

59 Hawkes 1962, 168.

60 Thom 1955.

61 Cf. Archibald S. Thom, *Walking in all the Squares*, Argyll Publishing, Glendaruel, 1995, 175, 201, 217, 314. Somerville—a naval captain in 1912, later to become a Rear-Admiral—was a keen amateur astronomer and surveyor.

62 Childe 1955.

63 Childe 1940, 145.

64 *Ibid.*, 138.

65 Childe 1955, 293.

66 Hoyle 1966. Hoyle's ideas were subsequently developed and refined, and eventually published in two books: Hoyle 1972, 19–54; and Hoyle 1977.

67 These included Alexander Thom, 'The geometry of Megalithic Man', *Mathematical Gazette* 45 (1961), 83–93; 'The egg-shaped standing stone rings of Britain', *Archives Internationales d'Histoire des Sciences* 14 (1961), 291–300; 'The megalithic unit of length', *JRSS* A125 (1962), 243–51; 'The larger units of length of Megalithic Man', *JRSS* A127 (1964), 527–33; Thom 1966 and Thom 1969 [*Vistas in Astronomy*]; Alexander Thom, 'The lunar observatories of Megalithic Man', *Nature* 212 (1966), 1527–8; 'Megaliths and mathematics', *Antiquity* 40 (1966), 121–8; 'The metrology and geometry of cup and ring marks', *Systematics* 6 (1968), 173–89; 'The geometry of cup-and-ring marks', *Transactions of the Ancient Monuments Society* 16 (new series) (1969), 77–87. A number of popular articles also appeared in *New Scientist* magazine: Alexander Thom, 'Megalithic geometry in standing stones', *New Scientist* 21 (1964), 690–1; 'Observatories in ancient Britain', *New Scientist* 23 (1964), 17–19; 'Timekeeping with standing stones', *New Scientist* 32 (1966), 719–21; 'Prehistoric observatories', *New Scientist* 38 (1968), 32–5; and a further paper appeared in the *Journal of the British Astronomical Association*: Alexander Thom, 'Observing the moon in megalithic times', *JBAA* 80 (1970), 93–9.

68 Thom 1967.

69 Although he had travelled extensively throughout Scotland, England, and Wales, and later made many visits to Brittany, he never made any systematic investigation of monuments in Ireland, only ever making one trip there (Thom 1988, 9).

70 Richard J. C. Atkinson, Review of Thom 1967, *Antiquity* 42 (1968), 77–8.

71 See Glyn Daniel, editorial, *Antiquity* 55 (1981), 88; also Thom 1988, 9.

72 Thom 1971.

73 David G. Kendall, Review of Thom 1971, *Antiquity* 45 (1971), 310–13, p. 310.

74 Heggie 1972, 48. Heggie modified his view significantly in the subsequent decade (cf. Heggie 1981a).

75 In particular, these included several publications on the Carnac monuments, where Thom had returned several times since Glyn Daniel's original invitation: Thom and Thom 1971; Alexander Thom and Archibald S. Thom, 'The Carnac alignments', *JHA* 3 (1972), 11–26; Alexander Thom, 'The uses of the alignments at Le Ménec, Carnac', *JHA* 3 (1972), 151–64; Thom and Thom 1973; Alexander Thom and Archibald S. Thom, 'The Kerlescan cromlechs', *JHA* 4 (1973), 168–73; 'The Kermario alignments', *JHA* 5 (1974), 30–47; Thom 1974; Thom, Thom and Thom 1974; 1975; Thom and Thom 1975; Thom, Thom and Gorrie 1976; Alexander Thom, Archibald S. Thom, and T. R. Foord, 'Avebury (1): a new assessment of the geometry and metrology of the ring', *JHA* 7 (1976), 183–92; Alexander Thom and Alexander S. Thom, 'Avebury (2): the West Kennet Avenue', *JHA* 7 (1976), 193–7; Thom and Thom 1977.

76 Thom 1971, 9–11.

77 Hawkes 1967, 174.

78 Atkinson 1975, 51.

79 Fleming 1975.

80 For accounts of this, see O'Kelly 1978, 111–12; O'Kelly 1982, 123–5.

81 The results were published in a paper in *Nature* (Patrick 1974).

82 Barber 1973.

83 MacKie 1974.

84 MacKie 1977, 21.

85 Renfrew 1976, 261–3. The first edition was published in 1973.

86 Atkinson 1975, 51.

87 This is the title of ch. 8 of Thom 1967.

88 Heggie 1972.

89 Michael A. Hoskin, 'The fourth session, on "Megalithic Astronomy, Fact or Speculation?"', part of a longer editorial article on IAU Commission 41 at Grenoble, *JHA* 7 (1976), 219–23, pp. 219–20.

90 A collection of papers under the heading 'The place of astronomy in the ancient world' appeared in *PTRS*, A276 (1974), 1–276. A number of them are cited in this book.

91 Elizabeth Chesley Baity, 'Archaeoastronomy and ethnoastronomy so far', *Curr. Anth.* 14 (1973), 389–449.

92 See, for example, Anthony F. Aveni (ed.), *Archaeoastronomy in Pre-Columbian America*, UTP, Austin, 1975; (ed.), *Native American Astronomy*, UTP, Austin, 1977.

93 The term was actually coined by MacKie in 1969 at a talk given at the University of Glasgow (see Euan W. MacKie, *An Archaeological View of Neolithic Astronomy*, Hunterian Museum, Glasgow, 1970) though it appears to be used to mean the dating of archaeological structures by astronomical methods. It achieved more widespread use following the publication of Baity's 1973 article (see note 91), which included it in the title. The alternative term 'astro-archaeology', coined by Hawkins in 1965 (Hawkins and White 1970, 155 [p. 121 in the original 1965 edn]) found favour amongst some American commentators who used it to denote the more restricted study of astronomical alignments typical of megalithic astronomy (see for example Aveni 1981, 4; also Aveni 1994, 26), but was unpopular in Europe, mostly perhaps because it had proceeded to acquire connotations with the archaeological fringe. More recently, it has been pointed out that the term 'archaeoastronomy', defined as the study of ancient astronomy based on the fullest possible range of both written and unwritten evidence, is itself misleading precisely because the range of evidence is much wider than just the archaeological; thus the term 'cultural astronomy' has been suggested to embrace both archaeoastronomy and its twin discipline of ethnoastronomy (see chapter ten, note 103).

94 The *Archaeoastronomy* supplement to the *Journal for the History of Astronomy* was first published in 1979. *Archaeoastronomy*, the bulletin, later the journal, of the Center for Archaeoastronomy in College Park, Maryland, was first published in 1977.

95 Christopher Chippindale, 'Stonehenge astronomy: anatomy of a modern myth', *Archaeology* 39(1) (1986), 48–52.

96 Edwin C. Krupp, 'Introduction', in Edwin C. Krupp (ed.), *In Search of Ancient Astronomies*, Doubleday, New York, 1977, xiii–xvii, p. xiii.

97 Glyn E. Daniel, 'Megalithic monuments', *Scientific American* 243 (1980), 64–76, p. 71. In an *Antiquity* editorial the following year he added '. . . our sad view of those who make our megaliths into observatories [is that] they are deluded men' (*Antiquity*, 55 (1981), 87).
98 Burl 1970, 1976, and many subsequent publications considered elsewhere in this book.
99 MacKie 1977.
100 Stuart Piggott, Review of MacKie 1977, *Antiquity*, 52 (1978), 62–3.
101 For example, Fred Hoyle's interpretations of Stonehenge were criticised strongly by Gordon Moir, an applied mathematician, in 1979: see Moir 1979. The surveyor Jon Patrick published a critique of Thom's interpretation of a Scottish site in 1979: see Patrick 1979. A further critique of some of Thom's surveys appeared in Moir *et al.* 1980.
102 The conference led to the publication of a mixed volume of papers under the title *Astronomy and Society in Britain During the Period 4000–1500 BC* (Ruggles and Whittle 1981).
103 Heggie 1981a.
104 A major reason for this was that in the Americas there is a demonstrable degree of cultural continuity between the pre-Colonial past and the indigenous present. The 'Old World' and 'New World' proceedings of the conference were published as two separate volumes: Heggie 1982a and Aveni 1982, respectively.
105 *The Times*, 8 July 1994, p. 8. Ironically, in a 1997 advertisement for English Heritage membership (*Heritage Today* magazine, no. 37, March 1997, 13), Stonehenge is presented prominently as 'the world's oldest calendar', even though the small print qualifies this by adding 'considered by many to be . . .' A balanced view is in fact presented in a 1997 English Heritage publication, Souden 1997, 118–27.
106 Hugh Thurston, *Early Astronomy*, Springer-Verlag, Berlin, 1994.
107 Chippindale 1994, 276.
108 *Ibid.*, 230.
109 *Ibid.*, 263.
110 North 1996.
111 Hayman 1997.
112 This is implied rather than stated explicitly. The fuller quote is: 'in the late seventies, my first contact with [alternative archaeology] was through ley hunting which, alongside astronomy, was at that time at the height of its popularity. Although I quickly rejected anything I considered to be a non-intellectual notion of the past . . .' (*ibid.*, xiii).
113 Bradley 1984, 77.

CHAPTER 1

1 Galileo Galilei, *Dialogue on the Two Great World Systems*, transl. Thomas Salusbury, rev. Giorgio de Santillana, University of Chicago Press, Chicago, 1953, p. 398.
2 Thom 1954, 403.
3 Atkinson 1979, 96 [p. 89 in the original 1956 edn].
4 O'Kelly 1989, 50–2, 97–100.
5 Seán P. Ó Ríordáin and Glyn E. Daniel, *New Grange and the Bend of the Boyne*, Ancient Peoples and Places 40, Thames and Hudson, London, 1964, 21.
6 For the definitive description of the site and a detailed report of the excavation, see O'Kelly 1982.
7 Burnt material from within the tomb, apparently undisturbed since the tomb was being built, yielded uncorrected dates of 2475 ± 45 uncal BC (GrN-5462-C) and 2465 ± 40 uncal BC (GrN-5463) (O'Kelly 1982, 230–1). On radiocarbon dates and their calibration see ahead to Statistics Box 3.
8 Claire O'Kelly, 'Corpus of Newgrange art', Part V of O'Kelly 1982, 146–85.
9 O'Kelly estimated that if the circle had been complete and the stones fairly regularly spaced, then it would have consisted of some 35 to 38 stones. However, his excavations could find no trace of the stoneholes of the 'missing' stones. See O'Kelly 1982, 66, 79.
10 P. David Sweetman, 'A late Neolithic/Early Bronze Age pit circle at Newgrange, Co. Meath', *PRIA* 85C (1985), 195–221, pp. 208–9.
11 'New Grange . . . so far as I can make out, is oriented to the Winter Solstice' (Lockyer 1909, 430). Lockyer did not, *pace* Brennan 1983, 32, 'draw attention to this fact on a scientific level'.
12 O'Kelly 1982, 123–4. For a more detailed description with diagrams see Brennan 1983, 72–81.
13 O'Kelly 1982, 93–6 and fig. 52.
14 A quantitative response to the question may be couched in terms of a probability (see Statistics Box 1) that the phenomenon could have occurred fortuitously. However, it is notoriously difficult to derive a reliable estimate of this probability. Any attempt to do so inevitably rests upon a number of detailed, and often questionable, assumptions. Several general issues relating to this question will be tackled later in this chapter and in the book. For differing points of view on Newgrange see Patrick 1974; Heggie 1981a, 213, who estimates the probability that the solstitial phenomenon was not deliberate as a mere 'one in ten or fifteen' and hence 'not really significant enough to excite much interest'; and Ray 1989. Elsewhere, Ray is reported as estimating the probability that the solstitial phenomenon was not deliberate as about 1 in 300, which is strongly supportive of deliberate orientation ('Newgrange: the oldest observatory in the world, say scientists', *Irish Times*, 21 June 1988).
15 O'Kelly 1978, 113.
16 *Ibid.*, 94.
17 Patrick 1974.
18 Ray 1989.
19 *Ibid.*, 343.
20 O'Kelly's excavations uncovered unburnt bones, apparently from two individuals, together with fragments of burnt bone from at least three, and possibly a good many, more (T. P. Fraher, 'The human skeletal remains', part of Appendix B to O'Kelly 1982, 197–205, pp. 200 and 205).
21 O'Kelly 1982, 98 refers to this as the 'closing stone'. It was wedged in front of the entrance and behind the decorated entrance stone, which remains in situ.
22 O'Kelly 1982, 96.
23 Lynch 1973, 152.
24 O'Kelly 1982, 123. See also Burgess 1980, 52.
25 See O'Kelly 1982, 96–8. O'Kelly himself was confident he had restored the passage walls to their likely original height. The roof-box corbels had slipped behind the orthostats.
26 'It is difficult not to see this as a deliberate and successful attempt to incorporate the midwinter sunrise as a significant element in the planning and use of the monument.' Colin Renfrew, foreword to O'Kelly 1982, 7–8, p. 8.
27 Burl 1983, 26. Aubrey Burl has elaborated on this general theme in a number of publications, e.g. Burl 1981a; 1987a.
28 RCAHMS 1971, no. 57.
29 *Ibid.*
30 Thom 1954.
31 Thom 1971, 36.
32 Thom 1971, 13–14.
33 The exact figure depends upon the time of day at which the sun reaches its limiting declination, the event that actually defines the solstice.
34 In the remainder of this chapter, and in the following chapter, we shall continue to quote some declinations to a precision of one arc minute in order to derive a better understanding of past ideas and debates, and of the lessons that can be learned from them. Thereafter, throughout the remainder of the book, declinations will be quoted to at most the nearest 0°·1 (six arc minutes), since recent work on variable refraction (see Astronomy Box 3) suggests that higher precision is unjustified.
35 Thom 1954 (see also 1967, fig. 12.2) quoted a declination of $-23°53'$ for the centre of the sun setting so that its upper limb just clips the end of Cara Island and $+23°54'$ for the setting path with the sun's upper limb twinkling down the slope of Corra Bheinn. An independent investigation, assuming existing models for refraction uncertainties and working to high precision, obtained a postulated date of use of 1580 ± 100 BC for the Cara Island foresight and 1640 ± 70 BC for Corra Bheinn (Bailey *et al.* 1975).
36 This sort of 'astronomical dating' is usually unacceptable in an isolated case, because the chances are too great of being able to fit a spurious high-precision astronomical explanation to a horizon feature by choosing a suitable date for its supposed use. See chapter two for similar problems with stellar and high-precision lunar interpretations. For lower-precision astronomical explanations, where changes in the motions of the sun over the centuries have negligible effect, the possibility of astronomical dating does not arise.
37 See MacKie 1974, 177; Heggie 1981a, 192; Ruggles 1984a, fig. 10.5.
38 MacKie 1974, 177.
39 Heggie 1981a, 190. This in turn reveals the danger that the close coincidence in the declinations of the two foresights might have come about because of the *a posteriori* selection of the Corra Bheinn declination from amongst a rather wider range of possibilities. In deriving their postulated date of 1640 ± 70 BC for the Corra Bheinn foresight, Bailey *et al.* 1975 assume that the upper limb of the midsummer sun should clip the summit of the hill. An alternative assumption, that the limb of the sun should just appear further down the slope (to an observer by stone *b*) would yield a date closer to 1250 BC.

40 MacKie 1974, 177.
41 *Ibid.*; see also J. G. Scott, 'The *Clyde* Cairns of Scotland', in Powell *et al.* 1969, 175–222, p. 198.
42 Wood 1978, 90.
43 J. L. Campbell and Derick Thomson, *Edward Lhuyd in the Scottish Highlands, 1699–1700*, OUP, Oxford, 1963, pl. v(a). See also Burl 1979a, 66; 1980, 192.
44 Heggie 1981a, 192; see also Clive Ruggles, 'Prehistoric astronomy: how far did it go?', *New Scientist* 90 (1981), 750–3, p. 752.
45 Burl 1983, 7–11.
46 Patrick 1981, 215–17. Cf. chapter nine, note 8.
47 RCAHMS 1988, no. 63.
48 Cowie 1980.
49 Simpson 1967.
50 RCAHMS 1988, no. 63.
51 Thom 1967, 156.
52 *Ibid.*, 155. See also Thom 1971, 13–14 and 37.
53 Thom 1971, 39.
54 Thom 1988, 6.
55 Thom 1971, 39–40.
56 Thom 1969, 7.
57 MacKie 1974, 180–1; see also MacKie 1988, 210.
58 MacKie 1974, 178–85; see also MacKie 1977, 84–92.
59 It would not, however, have been conclusive. It is possible, for example, that a platform might have been built merely to aid the passage of cattle along the side of the gorge (R. B. K. Stevenson, 'Kintraw again', *Antiquity*, 56 (1982), 50–1).
60 MacKie 1974, 181–3; see also MacKie 1977, 88.
61 W. C. Krumbein, 'Preferred orientation of pebbles in sedimentary deposits', *Journal of Geology* 47 (1939), 673–706.
62 J. S. Bibby, 'Petrofabric analysis', *PTRS* A276 (1974), 191–4. This forms an appendix to MacKie 1974.
63 MacKie 1974, 184. See also MacKie 1977, 89–91.
64 MacKie 1977, 92.
65 See especially Thomas McCreery, 'The Kintraw stone platform', *Kronos* 5(3) (1980), 71–9.
66 e.g. MacKie 1981, 116.
67 MacKie *et al.* 1985, 160.
68 Patrick 1981, 213.
69 MacKie 1981, 140–1.
70 Ruggles 1984b, 240.
71 McCreery *et al.* 1982.
72 Heggie 1981a, 190.
73 Patrick 1981, 213.
74 McCreery *et al.* 1982, 189.
75 Thom 1971, 37.
76 *Ibid.* In relation to the method itself the reader's attention is drawn once more to the problems of variable refraction (see Astronomy Box 3). Even Thom (*ibid.*) recognises that the procedure would be 'bedevilled by refraction changes from evening to evening'. Because of atmospheric effects on the sun's apparent position it is misleading to suggest, even if the visibility of the col itself behind the intervening ridge were not in dispute, that the platform could have been used as a 'permanent observing point from which the day of the midwinter solstice was regularly checked' (MacKie 1974, 184).
77 Patrick 1981, 211; McCreery *et al.* 1982, 187–8.
78 For example, a different interpretation is suggested by Burl (1976, 199; 1983, 42–4).
79 RCAHMS 1988, no. 364; NMRS, no. NR99NE10.
80 P. Fane Gladwin, 'Discoveries at Brainport Bay, Minard, Argyll: an interim report', *The Kist* 16 (1978), 1–15.
81 *Ibid.*; see also RCAHMS 1988, no. 364.
82 See MacKie 1981, 131.
83 *Ibid.*, 134.
84 MacKie 1988, 213.
85 MacKie 1981, 131–4; see also MacKie *et al.* 1985.
86 A date of 1060 ± 80 uncal BC (GU 1705) was obtained for charcoal associated with flints found on top of a buried sandy soil behind the back platform (MacKie *et al.* 1985, 159).
87 MacKie 1981, 134.
88 MacKie *et al.* 1985, 159.
89 See MacKie 1988, 214 and table 8.1.
90 Hoyle (1966, 270–1) made a similar point with regard to solar and lunar alignments at Stonehenge (see below).
91 Reversing the argument, MacKie (1988, 229–30) has even used the Brainport Bay evidence in support of the general principle of precise solstitial foresights, arguing that at this particular site, assuming that they discovered rather than chose the alignment, they would have had to make do with the notch that was already present and not quite at the solstitial position.
92 Fane Gladwin 1985.
93 *Ibid.*, 17.
94 Its status as a fallen standing stone is disputed by the RCAHMS (1988, no. 364; NMRS, no. NR99NE10). A 'Car Park Stone' referred to by RCAHMS, and also considered by them to be a natural boulder, is not mentioned either by Gladwin (*ibid.*) or MacKie (1981; 1988; MacKie *et al.* 1985) in the context of calendrical alignments.
95 MacKie *et al.* 1985. See also MacKie 1988, 217.
96 MacKie *et al.* 1985, fig. 4.
97 MacKie 1988, 220.
98 Euan W. MacKie, P. Fane Gladwin and Archie E. Roy, 'Brainport Bay: a prehistoric calendrical site in Argyll', *AAJ* 8 (1985), 53–69, fig. 6.
99 MacKie 1988, 218–20.
100 The declination of the notch itself is $+0°\,9'$ (MacKie *et al.* 1985; see also MacKie 1988, 218), a value greater by the semidiameter of the sun, which is about $16'$.
101 MacKie 1988, 223.
102 *Ibid.*, 230.
103 MacKie 1977, 10–12; see also MacKie 1981, 113.
104 We shall turn to more general issues of theory and procedure in chapters nine and ten.
105 A classic work on the megalithic tombs of Brittany is Jean l'Helgouac'h, *Les Sépultures Mégalithiques en Armorique*, Université de Rennes, Rennes, 1965. Another on the megalithic monuments of north-western France is Glyn Daniel, *The Prehistoric Chamber Tombs of France: a Geographical, Morphological and Chronological Survey*, Thames and Hudson, London, 1960. For a more recent interpretation see Patton 1993. General descriptions and guide books include Pierre-Roland Giot, *Les Alignements de Carnac*, Ouest France, Rennes, 1983 and Burl 1985.
106 Burl 1993, 134–46. The book as a whole presents an integrated account and catalogue of the stone rows of Britain, Ireland and Brittany.
107 *Ibid.*, 135. See also Thom and Thom 1971, 149; Alexander Thom, 'Moving and erecting the menhirs', *PPS* 50 (1984), 382–4.
108 Thom and Thom 1971; Thom, Thom and Gorrie 1976. See also Thom and Thom 1978a, 98–108.
109 Hadingham 1981, 36. The process of laying out the traverses was not easy, since much of the area is now covered in dense woodland and gorse (Motz 1988, 19).
110 Atkinson 1975, 43.
111 Thom and Thom (1971) made no attempt to estimate the probability that eight suitable backsight candidates could have arisen by chance, given the total number of monuments in the area, and this was criticised by Jon D. Patrick and C. J. Butler, 'On the interpretation of the Carnac menhirs and alignments by A. and A. S. Thom', *IARF*, 1(2) (1974), 29–39, p. 30. Atkinson (1975, 44–5) did make the attempt and concluded that Thom's results could not be attributed to chance, but Peter R. Freeman, 'Carnac probabilities corrected', *JHA* 6 (1975), 219, identified errors in Atkinson's argument and concluded to the contrary. However, such arguments are of limited value since large numbers of monuments to the south of the great menhir have been destroyed since Neolithic times by the rising sea (Hadingham 1981, 36).
112 *Ibid.*, 37–9.
113 Robert L. Merritt and Archibald S. Thom, 'Le Grand Menhir Brisé', *Arch. J.*, 137 (1980), 27–39, argue that the great menhir did indeed stand, and consider various reasons for its subsequent collapse. However, R. Hornsey, 'The Grand Menhir Brisé: megalithic success or failure?', *Oxford J. Arch.* 6(2) (1987), 185–217 concludes that the stone never stood, but dropped and shattered during its erection.
114 See also notes 14 and 111.
115 For a detailed discussion of the structural history of Stonehenge, see Cleal *et al.* 1995. For other accounts of the history of Stonehenge, with interpretation, see Burl 1987a; Bradley 1991a; Castleden 1993; and various papers in Cunliffe and Renfrew 1997.
116 Atkinson 1979, 68–101. The first edition was published by Hamish Hamilton in 1956.
117 Cleal *et al.* 1995, 465–70. In brief, Atkinson's Stonehenge I is replaced by the new Stonehenge 1 and 2; the arrival of the bluestones, which brings this period to a close, is now dated between 2600 and 2500 BC rather than around 2100 BC. His Stonehenge II and III are replaced by Stonehenge 3, where a more complex sub-phasing, with considerable uncertainties remaining, has now been identified both within the interior of the monument and from the entrance outwards to the north-east. The

idea that the avenue was only completed down to the River Avon at a late phase, known as Stonehenge IV (RCHM(E) 1979, 11), has now been refuted (Cleal *et al.* 1995, 533–4).

118 For a well illustrated general introduction see Souden 1997.

119 Amongst the many accounts and discussions of Stonehenge astronomy, see especially Heggie 1981a, which gives a critical discussion from the point of view of an astronomer; and Burl 1987a, which covers the topic in a wider archaeological context. The first edition of Chippindale's *Stonehenge Complete* (Thames and Hudson, London, 1983) discusses Stonehenge astronomy in some detail (pp. 216–35), but this is considerably cut down in ch. 14 of the 1994 edition (Chippindale 1994). See also Castleden 1993, 18–27.

120 Atkinson 1979, 30; see also Heggie 1981a, 196; Castleden 1993, 60. In any case, there is no firm evidence of any structure marking an observing position at the centre of the site. As Atkinson (1979, 96) remarked: 'The assumed line of sight along the axis is not marked positively in any way, . . . [but] by the mid-points of a number of empty spaces between pairs of upright stones'.

121 In order to distinguish between the sun's position on the solstice and (say) three days before or after, the observer would need to be able to distinguish a movement no greater than 2′ in declination (see Astronomy Box 3), i.e. about 3′ in azimuth. But for a foresight at a distance of 75 m, a transverse movement of only 5 cm would shift the apparent azimuth of the foresight by about 3′. See also Atkinson 1979, 96.

122 Castleden (1987, 130) suggests that the Heel Stone could have been used to pinpoint thirteen days before and after the solstice. But at these times the daily change in declination, about 6′ according to formula (A3.2) of Astronomy Box 3, is still equivalent to a transverse movement of only about 15 cm for a foresight 75 m distant. See also Hoyle 1972, 36–40; 1977, 59; Heggie 1981a, 196.

123 In order for a spot marked on the ground to suffice as an indication of where to stand, it must be sufficient to define the observing position to within, say, 1 m. This being the case, in order to achieve a resolution of 15′, half the sun's diameter, a foresight at least 200 m distant is needed; for a resolution of 3′ the distance must be at least 1 km. The distance required, $d$, increases in proportion to the allowable error in the observing position, $p$, and in roughly inverse proportion to the resolution required, $\theta$, according to the formula $d = p / \tan\theta$ where $d$ and $p$ are in the same units.

124 The lines are tabulated in Hawkins and White 1970, 143, 169 and shown in fig. 14. Some had already been identified by authors such as Lockyer and by Newham: see Heggie 1981a, 196 for details. Newham's work had been prevented from publication by a series of accidents: see Lancaster Brown 1976, 103–8; 'Newham and Stonehenge', *AA* no. 4 (*JHA* 13) (1982), S73–4.

Fig. 1.22a shows a simpler scheme, including only those alignments upon the solstices and lunar limits, which follows Hawkins and White 1970, fig. 11 (see also Hawkins 1963).

125 Hawkins and White (1970, 172) actually give 0·00006, but they made an arithmetical error and the answer by their own formula should be 0·000006. See Heggie 1981a, 148.

126 For example, given that the line from Station Stone 93 to Station Stone 94 points to midsummer sunrise, then 92–91 does so too, not because the two pairs were independently aligned upon an astronomical target, but because they form the opposite sides of a rectangle. Furthermore, because the precision of alignment being considered is only about one degree and the horizon is fairly level, a line between a pair of points indicating midsummer sunrise in one direction will inevitably indicate midwinter sunset in the other. Thus lines 93–94, 94–93, 92–91 and 91–92 should really be counted as one 'hit' upon an astronomical target, not four.

127 Perhaps the most notorious manifestation of this is the fact that only one of a line of four entrance postholes was included (Atkinson 1966, 214).

128 Atkinson considered it likely that holes *F*, *G* and *H* were tree-root holes rather than stoneholes (Atkinson 1966, 215; see also Cleal *et al.* 1995, 288–9). To this can be added the fact that the Station Stones and Heel Stone probably belong to Stonehenge 3 (see Archaeology Box 3).

129 Hawkins's targets included the 'equinox moon' at declination ±5°, but see Heggie 1981a, note 43 to p. 95, on p. 239; see also Hoyle 1966, 275. Eight equinoctial sun and 'equinox moon' alignments were included in the total of 24, but similar probability arguments apply even if these are excluded.

130 Different methods were proposed by Hawkins 1964 (see also Hawkins and White 1970, 177–83) and Fred Hoyle, 'Stonehenge—an eclipse predictor', *Nature*, 211 (1966), 454–6 (see also Hoyle 1977, ch. 5). See Heggie 1981a, 101–4 for a discussion.

131 Heggie 1981a, 103.

132 The methods predict 'eclipse seasons' or 'danger periods', which recur with a period of roughly 173 days and last for about three weeks, when it is possible that one or more eclipses (solar or lunar) may occur. All that is actually certain is that an eclipse cannot occur outside these periods. It is by no means certain that a lunar eclipse, if it occurs, will be a 'proper' umbral one; penumbral eclipses often pass unnoticed. Even if it is umbral, it will only be visible at best from that half of the earth where the moon is up in the sky at the time. A solar eclipse, if that occurs, will most likely be partial, and once again most partial solar eclipses pass unnoticed. It is highly unlikely that a total solar eclipse (which may well be what many people have in mind when they hear of these ideas) will be visible from a given spot such as Stonehenge. 'A priest could hardly gain credit by successes that might be ten years apart' (Hawkes 1967, 176). For the technical background see Thom 1971, 17–19. See also Heggie 1981a, 103.

133 Sadler 1966; see also Heggie 1981a, 104.

134 Two relatively recent examples are the idea that they were used to help determine the exact length of the lunar synodic (phase-cycle) month (C. T. Daub, 'The Aubrey Holes revisited', *QJRAS*, 34 (1993), 563–5) and the idea that they were used as a device for calculating areas of land (T. H. Kirk, 'The significance of the 56 holes of the Aubrey Circle at Stonehenge', *QJRAS*, 34 (1993), 567–8).

135 Cf. Chippindale 1994, 223–5.

136 Burl 1969, 7–8; 1981b, 22–3 and table 1, which is reproduced in Burl 1987a, table 8.

137 The fact that it is possible to invent other schemes to predict eclipse seasons using different numbers of pits in the circle (Sadler 1966, 1120; R. Colton and R. L. Martin, 'Eclipse cycles and eclipses at Stonehenge', *Nature*, 213 (1967), 476–8, p. 478; and even see Hawkins and White 1970, 185; for an overview see Heggie 1981a, 205) shows how easy it is to fit this sort of theory to a given number of pits and reinforces the general conclusion that eclipse prediction was not the actual purpose of the Aubrey Holes.

138 E.g. Richard F. Brinckerhoff, 'Astronomically-oriented markings on Stonehenge', *Nature*, 263 (1976), 465–9; A. D. Beach, 'Stonehenge I and lunar dynamics', *Nature*, 265 (1977), 17–21. For a rejoinder see Richard Atkinson, 'Interpreting Stonehenge', *Nature*, 265 (1977), 11.

139 Thom, Thom and Thom 1974, 86–8; 1975. See also Thom and Thom 1978a, 151–62.

140 Of the four proposed foresights one, a mound at Hanging Langford, is part of a complex dating to the Late Iron Age or Romano-British period; one, an earthwork at Gibbet Knoll, is probably the remains of a Civil War gun-battery; one, a mound at Figsbury Rings, overlies modern ridge and furrow cultivation; and the fourth, Peter's Mound, has been shown by excavation to be a First World War military rubbish dump (Atkinson 1981, 209; see also Atkinson 1982, 114). The Thoms also identified the top of Chain Hill as the site of a further lunar foresight, but found no surviving candidate for the foresight itself. The Thoms identified these putative foresights by examining the eight possible lunar sightlines radiating out from Stonehenge (Thom and Thom 1978a, 151), as was the case with the putative backsights for Le Grand Menhir Brisé. See also Castleden 1993, 25.

141 Burl 1987a; Castleden 1993.

142 Pitts 1981. Interestingly, R. S. Newall had suggested in 1929 ('Stonehenge', *Antiquity*, 3 (1929), 75–88, p. 84) that the Heel Stone might have had a companion, on the grounds of symmetry about the axis. See also Archaeology Box 3.

143 Atkinson 1966, 215.

144 Cleal *et al.* 1995, 102–7, 152–3. This does not, of course, prove Hawkins's or Hoyle's arguments.

145 Newham 1993, 35–8 [1972, 23–5]. See Burl 1979a, 65; 1987a, 144.

146 In 1973 they were assumed to date to the Late Neolithic (Lance Vatcher and Faith Vatcher, 'Excavation of three post-holes in Stonehenge car park', *WANHM*, 68 (1973), 57–63), but the presence of pine, and subsequent radiocarbon dating (A. J. Walker, R. S. Keyzor and R. L. Otlet, 'Harwell radiocarbon measurements V', *Radiocarbon*, 29 (1987), 78–99, pp. 78–9) indicated dates back in the eighth millennium BC. See Cleal *et al.* 1995, 526–7; also Castleden 1993, 28–9.

147 Castleden 1993, 30, 52.

148 Box entitled 'Dead snails and Stonehenge astronomy', in Christopher Chippindale, 'Life around Stonehenge', *New Scientist*, 101 (1984), 12–17, p. 13.

149 Castleden 1993, 92.

150 See, for example, the passage written by Thom and quoted in Chris Jennings, 'Megalithic landscapes', in Ruggles 1988a, 155, 172.

151 For examples see chapter seven, note 3.
152 Peter D. Hinge, 'The distribution of visibility from stone alignments: a case study', unpublished MSc thesis, University of Leicester, 1990, 2.
153 Newham 1993, 26 [1972, 18].
154 See note 136.
155 Henshall 1972, 381–3, no. ARN 10.
156 Burl 1981a, 253, 256.
157 *Ibid.*, 254–6. Table 7.1 lists 22 tombs, fig. 7.3 shows 21 (tomb ARN 13 is omitted and ARN 8 contains three tombs) and the text on p. 256 refers to twenty (Carn Ban plus 'nineteen others').
158 Strictly, the orientation pattern of the Arran tombs as a whole leads us to have much less confidence in the Carn Ban alignment being anything other than fortuitous, but we can never say for certain that it was not intentional: particular considerations may have operated at this monument alone. Further consideration of this possibility leads us towards issues that will be addressed in chapter ten.

CHAPTER 2

1 Thom 1981, 54–7.
2 Gingerich 1981, 117.
3 Ferguson 1988, 32.
4 The four levels were first identified in Ruggles 1981, 153–4 and 189. While they present Thom's ideas in a logical progression, they do not entirely reflect the order in which those ideas developed. For instance, a 'Level 2' histogram showing the sun's limbs appears in Thom 1954 (fig. 2).
5 Thom 1955, tables 5 and 6.
6 Thom 1967, table 8.1. The table also contains a line at Burnmoor, Cumbria (Thom's L1/6) simply marked 'meridian' with no declination given.
7 Thom 1967, chs 9 and 10 respectively.
8 Thom 1969. The definitive version was published in book form in 1971 (Thom 1971).
9 Thom 1971, ch. 7.
10 Thom and Thom 1978b; 1980a; Thom 1981.
11 Thom and Thom 1980a, S88; Thom 1981, 38.
12 Ruggles 1981; 1982b; 1983.
13 Heggie 1981b, S18–19.
14 The term 'curvigram' was coined by this author as a convenience (Ruggles 1981, 156).
15 Measured declinations are quoted by Thom to the nearest 0°·1.
16 For Thom's own graphs, see Thom 1955, fig. 8 and Thom 1967, fig. 8.1. These are similar in form but distinguish between rising and setting lines and show all the constituent humps rather than just the cumulative result. They also show the proposed astronomical targets.
17 The area under each constituent hump is 1·0. The *y*-values reflect this and have no other significance. Note especially that comparison from one graph to another on the basis of the y-values will only have meaning if the *x*-axes are calibrated in the same units and the areas under the two sets of humps truly reflect the relative weightings to be assigned to the two sets of indications.
18 The *x*-axis resolution used in displaying these graphs is 0°·2, which is all that is needed in order to illustrate the points made here. In the case of Fig. 2.3c, where the standard deviation of most constituent humps is only 0°·1, this results in a very jagged appearance. For a version of Fig. 2.3c with a much finer *x*-axis resolution, see Ruggles 1984b, fig. 5, which can be directly compared with Thom's own graph (Thom 1967, fig. 8.1).
19 For the form of curve that would be expected given random orientations, see ahead to Fig. 3.4. Whether they are significantly different has, of course, to be determined by suitable statistical analysis. See, for example, Thom 1955, 286–8; Freeman and Elmore 1979; Patrick and Freeman 1988. However, our primary concern here is with data selection questions that affect the validity of such statistical tests in the first place.
20 Cooke *et al.* 1977, 130. The dangers are evident in using supposed stellar alignments to try to provide a construction date, a good example being the various attempts to fit stellar explanations to the avenue and radial rows of stones at Callanish in Lewis (see chapter eight, note 123), all of which yielded dates around 1750 BC. In fact, it is clear that they represent 'no more than the selection by modern researchers of stars which happened to be rising in the right place at the seemingly right time' (Burl 1982, 145), as is confirmed by the archaeological evidence placing the construction of the monument considerably earlier (chapter eight).
21 Thom 1955, 284–8. The analysis found that significantly more stellar 'hits' were obtained for a date of around 2100 BC than for other dates, suggesting that many of the supposed stellar alignments were indeed deliberate and mostly erected around this same date. For a commentary see Heggie 1981a, 162–8.
22 Thom himself addressed this issue in Thom 1967, ch. 13.
23 Bradley E. Schaefer, 'Atmospheric extinction effects on stellar alignments', *AA* no. 10 (*JHA* 17) (1986), S32–42. The implication is that if a structure was aligned upon the first appearance, or last disappearance, of a certain star, the shallow angle at which celestial bodies rise and set at the latitude of Britain or Ireland would generally result in an orientation significantly different from that to be expected if the point of reference was the theoretical point of appearance or disappearance on the horizon (see *ibid.*, S37–41 for quantitative arguments). Where the horizon is higher the problem is lessened.
24 This means that even if a number of alignments were set up at around the same time and oriented upon the appearance or disappearance of the same star, changes in day-to-day atmospheric conditions, and conditions from site to site, would be sufficient to scatter the orientations considerably (*ibid.*, S38).
25 See also Ruggles 1984b, 246–8.
26 Thom 1955, 284.
27 Thom 1967, table 12.1.
28 *Ibid.*, 135.
29 Thom was kind enough to supply the author in the early 1980s with a copy of his full unpublished site list, which was used in the full reassessment described in chapter three.
30 Thom 1967, table 8.1.
31 Examples are the five-stone row (in which four stones now remain) at Quinish, Mull (Thom's M1/3; ML2 in List 2) and the three-stone rows at Dervaig S (Thom's M1/6 'Dervaig C'; ML11 in List 2) and Balliscate (Thom's M1/8 'Tobermory'; ML4 in List 2).
We prefer wherever possible to use the term 'stone row' rather than to follow Thom and speak of 'stone alignments'. This is because the use of the latter term tends to lead to a prejudgement about the function of the monument. In this we follow, for example, Ó Nualláin 1988 and Burl 1993.
32 Examples are the three-stone row at Carloway (Clach an Tursa), Lewis (Thom's H1/16; LH6 in List 2) and the alignment of two slabs at Barbreck, mid-Argyll (Thom's A2/3; AR3 in List 2), a monument that will be mentioned later (p. 76).
33 Ruggles 1981, 156–8.
34 Examination of the unpublished site list does show, however, that a great many sites included there and not carried forward to the general reference list are antiquities marked on nineteenth- or early twentieth-century 1:10560 (6-inch) Ordnance Survey maps which are of doubtful archaeological status or in a hopeless state of repair, and thus were quite legitimately omitted from further consideration (Ruggles 1982a, 93).
35 Many megalithic rings are accompanied by outlying standing stones placed a small distance outside the ring. This type of indication raises the issue of how accurately the centre of the ring can be defined, especially when it is significantly non-circular. With the exception of a few rings with a central standing stone (Burl 1976, 205–8), we usually know of no marker at the centre.
36 'The most difficult part of the whole investigation is to decide when to include a line and when to exclude it. The decision must always be a matter of personal opinion and is influenced by the viewpoint and the other lines with which, at the time, it is being compared' (Thom 1967, 96).
37 Examples are Blashaval (Na Fir Bhreige), North Uist (Thom's H3/8; UI19 in List 2), Maol Mor, Mull (Thom's M1/4 'Dervaig A'; ML9 in List 2); Ardnacross, Mull (Thom's M1/9; ML12 in List 2); *BFL* ($S_5S_1S_2$) at Nether Largie ('Temple Wood'), mid-Argyll (Thom's A2/8; AR13b in List 2); Dunamuck I, Mid-Argyll (Thom's A2/21 'Dunamuck North'; AR28 in List 2); and Escart, Kintyre (Thom's A4/1; KT5 in List 2). See also Heggie 1981a, 160.
38 In his 1967 data table (Thom 1967, table 8.1), Thom did include eighteen indications (Class C, see *ibid.*, 96) for no reason other than that they produce an interesting astronomical declination. However, these were omitted from his own graph (*ibid.*, fig. 8.1) and are omitted from our Fig. 2.3. The most serious questions relate to almost half (112) of the remaining 243 lines which are classified as Class B: 'borderline cases' in which there is a subjective element. These are shown unshaded in Thom's figure, but mixed in with Class A alignments which Thom considered to be fully objective. They are also shown unshaded in Fig. 2.3, but separated from Class A lines so that their effect upon the overall result is clear. For additional commentary see Heggie 1981a, 153.
39 Ruggles 1982a, 94. Similar questions will arise at the higher Levels. On unwritten selection criteria see also Heggie 1981a, 159–62.

40 Ruggles 1981, 168–9.
41 Ruggles 1982a, 94–5.
42 Thom, Thom and Burl 1990, 323. The fact that the quoted declination is that of an indicated foresight is not recorded in Thom 1967, table 8.1.
43 Thom 1971, 11–12; Thom and Thom 1978a, 178. On Castlerigg itself, see Thom 1966, 22, figs 21, 38 and 39.
44 The distant island of Boreray is listed as an indicated foresight with a declination of +8°·8 (Thom 1967, table 8.1).
45 RCAHMS 1928, no. 170; Ruggles 1984a, 46 (UI12). See also Moir 1981, 233.
46 Moir 1981, 226–7; Ruggles 1984a, 48 (AR36).
47 'Another of [Captain Riddel's] imitations was a Bronze Age stone circle which he erected on a knoll a short way up the river. It is now a most realistic affair, correct in lay-out and so weathered as to take in anyone. It should be a warning to all antiquaries.' 'The Glencairn Area', in 'Field meetings', *TDGNHAS* 25 (1948), 182–6, p. 185. See also Moir 1981, 232.
48 Thom 1967, 151.
49 Burl 1988a, 179. See also Moir 1981, 235.
50 Thus for example Thom's Leacach an Tigh Chloiche (H3/11), 'a mixture of open kists and upright stones' which from the astronomical point of view he considered the most important site on North Uist (Thom 1967, 131), is in fact the remains of a large chambered tomb known to archaeologists as Unival (Moir *et al.* 1980).
51 Even the relatively few (sixteen) solstitial alignments listed in Thom's earliest paper on megalithic astronomy (Thom 1954, table 5) include many different types of indication.
52 It also seems to contradict the related idea of Thom that a precisely defined unit of measurement (the megalithic yard) was uniform from the Orkney islands to Brittany to a precision of about 0·1 mm, a point made by Fleming 1975.
53 In addition, the construction of the various monuments in Thom's sample spans over two thousand years of prehistory, and excavations have highlighted instances where monuments have been modified and reused over considerable periods of time (see p. 77), which only serves to reinforce this doubt.
54 Ruggles 1982a, 94.
55 Thom's data are given in Thom 1967, table 10.1 and the result resembles Thom's own graph, *ibid.*, fig. 10.1. We have assumed a standard deviation of 0°·1, with 0°·2 being used for those less accurate lines marked '±' by Thom. To judge by the specimen humps shown at the foot of Thom's graph, these values appear to match those employed by Thom. Three lines marked '±' in Thom 1967, table 8.1 are not marked thus in *ibid.*, table 10.1, but have nonetheless been assumed to be less accurate. For other details concerning the raw data, which explain minor differences between our graph and Thom's, see Ruggles 1981, 159.

The major and minor limiting declinations are taken, following Thom (1967, table 10.1), to be −29°·95, −19°·58, +17°·94 and +28°·17. This corresponds to a date somewhere between 2000 and 1500 BC (see Astronomy Box 6), but given the level of precision concerned, the data are not greatly affected by date.

56 In fact, Thom 1967, fig. 9.2 shows a slight preference for the upper limb at the winter solstice and almost total preference for the upper limb at the summer solstice. There is no definitive bimodal structure amongst the clusters of declinations at other calendrical epoch dates included in the same graph, but then even if there was a preference for one or other solar limb in such observations, it would be blurred out by the inherent uncertainty in each epoch declination and by the fact that two distinct epoch dates fall close together in each case. See also Heggie 1981a, 156–7.
57 In fact, a Kolmogorov–Smirnov test suggests that a normal distribution does provide an adequate fit, at least to the data in Thom 1966, table 4, which represents a subset of the data in Thom 1967, table 10.1. See Heggie 1981a, note 124 to ch. 7.
58 Thom 1967, 165; Alexander Thom, contribution to 'Hoyle on Stonehenge: some comments', *Antiquity* 41 (1967), 95–6.
59 Ruggles 1981, 159–74.
60 See Thom 1967, 121. Thom also cuts off the data further than 0°·8 from the mean and excludes these other lines—i.e. humps whose maxima fall outside the range −0°·8 to +0°·8—from his graph (Thom 1967, fig. 10.1). For the same graph but with the other data included, see Ruggles 1981, fig. 4.1a.
61 The authenticity of a further monument, Corogle Burn (L20), was considered doubtful by Ruggles (1981, 165 (site 18)), but subsequent work (Burl 1988b, 108–9) suggests that it is in fact genuine, comprising a four-poster and outlying pair.
62 Five indications involve chambered tombs or cairns, either as foresights or backsights. Ruggles (1981, 164) also relegated these to lower archaeological status, arguing that such monuments were dubious contenders as components of sightlines since they had an obvious primary function as burial places. This reasoning, however, attempts to impose a single function and meaning on such monuments, and is thus flawed (see p. 89). The symbolic placement of burial monuments within the natural landscape, and in relation to earlier monuments pre-existing within the landscape, is an important topic that will be considered later.
63 The use of 'stones' that are not necessarily antiquities as foresights also raises the question of whether they were identified by the dangerous procedure of searching outwards from a putative backsight in promising directions: see pp. 34–5.
64 In one case (L18), the 'outlier' is also very close to the stone circle, an inherently imprecise method of indication which fails to conform to Thom's own selection criteria for solar or lunar lines (Thom 1967, 94). The other outlier (L8) is more distant but still violates these criteria (Ruggles 1981, 168 (site 6)).
65 This is greater than the figure quoted in Ruggles 1981 because lines involving chambered tombs or cairns have not been assigned lower status for this reason alone (see note 62).
66 As an example, one of the indications (at site L51, Castlerigg in Cumbria) is the major axis (maximum diameter) of a flattened stone circle, according to Thom's interpretation of its geometry (see Thom, Thom and Burl 1980, 28–9). Yet the major axis of Brats Hill (Thom's 'Burnmoor E', site L53(e)), a stone ring in the same region that has a very similar geometrical form (a fact noted by Thom himself: see Thom, Thom and Burl 1980, 40–1), has no astronomical interpretation, lunar or otherwise. The two proposed lunar indications at Burnmoor E are from its centre towards the centres of other stone circles.
67 For details see Ruggles 1981, 168–72 and table 4.1.
68 See also Ruggles 1981, 169.
69 The detailed comments given in Ruggles 1981 have been thoroughly rechecked and updated as necessary.
70 Thom 1971, table 7.1.
71 *Ibid.*, ch. 6. For cross-references see Table 2.3.
72 *Ibid.*, ch. 5.
73 For a horizon feature 30 km distant, a sideways shift in the observer's position of 10 m will cause the azimuth of the foresight to change by approximately 1 arc minute. But for a horizon only 3 km distant, a sideways shift of only 1 m will have the same effect. In fact, in Thom 1971, table 7.1 azimuths and altitudes are only quoted to the nearest arc minute.
74 This terminology follows Cooke *et al.* 1977.
75 A standard deviation of 0′·75 has been used, which appears to match Thom's humps, in all but two cases where a declination is only quoted to 1′ by Thom and a standard deviation of 1′·5 is used. The mean extreme (geocentric lunar) declinations used to calculate the relevant differences are ±29° 2′·3 and ±18° 44′·6.
76 It actually varies between about 14′·7 and 16′·7.
77 See Thom 1971, 45–51.
78 Ruggles 1981, 175.
79 At Nether Largie, for example, the only direct indication of the proposed foresight to the north-west is the alignment from a rectangular setting of small stones 5 m SSW of the large standing stone *F*, which Thom labelled 'Group *Q*', over the south-west circle at Temple Wood. A number of equally (and at least some arguably more) plausible alignments, can easily be identified (see Fig. 2.9d), and the subsequent discovery by excavation of a second stone circle at Temple Wood (Scott 1989) merely increases the possibilities, given that one is prepared to speculate that the Nether Largie stones and either of the nearby Temple Wood circles were contemporary and directly related in the first place.

Regarding the overall figures, two of the four lines at Nether Largie/Temple Wood listed by Ruggles 1981, 178–9 as indicated are in fact unindicated; on the other hand two lines at Kintraw listed as unindicated are in fact indicated (*ibid.*, 198, note 1). Thus the total is unaltered.

80 This fact is obscured by the scarcity of site plans in Thom 1971. In 1971 Thom evidently felt, as he had a few years earlier, that 'if there is a suitable natural foresight which gives a commonly found solar or lunar declination exactly then we are entitled to suspect that there had been a secondary indicator which would have identified the foresight but that it has vanished' (Thom 1967, 94) but no longer felt that 'such a line could only be given a low classification and would not be put on a general histogram' (*ibid.*).
81 Later, Thom and Thom (1980b, S90–1) did precisely this, stating: 'if we

stand at a marked backsight and make careful measurements of the profile of part of the horizon which turns out to contain a significant position, we can assume that we are at a real observing point.' This is quite evidently a circular argument.

82 This is evident for two reasons. First, at such locations there are generally very many horizon features similar to those that have been considered elsewhere by Thom as putative foresights (cf. Burl 1976, 143 on the western horizon at Nether Largie; Gingerich 1981, 117). Second, at this level of precision astronomical targets are numerous: there are nine in each lunar band (see Astronomy Box 7), making a total of seventy-two potential targets on any 360° horizon. Assessing the probabilities formally would involve going out to randomly selected locations and undertaking a careful survey of all the horizon features with the same enthusiasm as would be applied at a real monument, something that few have been prepared to do. However, control data of a similar nature have been collected as part of the North Mull project (see chapter seven).

83 This is difficult using purely statistical methods. We could analyse all equally plausible horizon features around the entire horizon regardless of their astronomical potential; but since no more than a handful could possibly have been used simultaneously, even from the most propitiously placed monument, and since without prejudging the result we can not know which, any significant trends would inevitably be submerged in the loud 'background noise' of random data (Ruggles 1981, 180). The problem is similar to that of detecting foresights close to the solar solstices suitable for 'halving the difference', which was discussed in chapter one. An approach that attempts to overcome this problem by introducing contextual argument is discussed in chapter seven.

84 Ruggles (1981, 186; 1982a, 87–8) describes several more examples. Further evidence is provided by a comparison of the indicated foresights in both Level 2 and Level 3: see *ibid.* (169 and table 4.2).

85 The standard deviation is taken as 1′·5 throughout. For reasons why nothing greater is justified see Ruggles 1981, 183. See also Patrick 1979, S81–2.

86 See particularly McCreery 1979 and Gordon Moir, 'A review of megalithic lunar lines', *NA*, 1(1) (1980), 14–22.

87 For example, at Knockrome (L35) there are two standing stones in line with a third 1 km away at Ardfernal (JU3 in List 2). The long alignment provides a much more precise and convincing indication than the flat faces of the central stone, but appears to have no astronomical significance (Ruggles 1981, 183).

88 An example is the north-west foresight at Nether Largie (Patrick 1979, S78–83).

89 The most notable example here is Parc-y-Meirw (L56), where the proposed foresight is 145 km away in Ireland. McCreery (1979, 50–2) argues that the proposed foresight could probably never have been seen at all. The opposite direction along the row, uphill towards a megalithic tomb, was more probably of interest (Burl 1993, 99).

90 Thom 1971, ch. 8. For other explanations of Thom's extrapolation procedure see Wood 1978, 114–29; McCreery 1980, 7–11; Heggie 1981a, 98–100.

91 Thom 1971, 103–5.

92 Ruggles 1981, 193–4. At Dirlot (L2), for example, the alleged markers of the extrapolation length are stones in a modern stone fence (Atkinson 1981, 206 (site 2)).

93 *Ibid.*, 194.

94 Heggie 1972, 47; McCreery 1980, 12–14. See also Ruggles 1981, 194.

95 Heggie 1972, 47.

96 Ruggles 1982a, 87.

97 In fact, the dataset used by Thom and Thom 1980a and Thom 1981 consisted of all indications they considered to have been reliably measured—a total of forty-two. Detailed reasons are given by Archie Thom (Thom 1981, appendix 1.1) why indications included in earlier analyses were not now suitable. Of these forty-two lines, twenty-three were amongst a sample of twenty-five examined separately in an earlier paper (Thom and Thom 1978b) as being most convincing according to strict terms of reference. Following Ruggles 1982b, S21 we take as the Level 4 dataset all lines in either or both of the Thoms' analyses, forty-four in all.

The full dataset at Level 4 is not reproduced here, since it overlaps considerably with that at Level 3. For a complete list see Ruggles 1982b, table I.

98 This overcomes some of the problems identified in the Level 3 reassessment but fails to address others, and actually creates several more. For full details see Ruggles 1983; for brief notes on the astronomical complications at Level 4 see Astronomy Box 7.

99 In both the analyses of twenty-five lines and of forty-two lines (see note 97), the calculated residuals (differences between measured and expected declinations), and hence the inferred precision of the sightlines, are of the order of 1′ or 2′. In the first case see Thom and Thom 1978b, table 3. In the second see Thom and Thom 1980a, table 1 or Thom 1981, table 1.1.

100 Ruggles 1982b; 1983.

101 For more detailed descriptions and archaeological interpretations of Brodgar and Stenness see Renfrew 1979, ch. 5; Ritchie 1985. On historical accounts of Brodgar, see Ritchie 1988.

102 Thom and Thom 1973; 1975; 1977. See also Thom and Thom 1978a, ch. 10. The quote is from *ibid.*, 123.

103 Ruggles 1982b, S23–7.

104 Thom and Thom 1978a, 137.

105 Excavations of the ring ditch at Brodgar (Renfrew 1979, ch. 5) yielded no useful radiocarbon dates, but the conclusion can be made by comparison with other Class II henges and, indeed, henges in general (see Archaeology Box 2). The Ring of Brodgar itself is central to the whole astronomical interpretation. Two of the nine proposed sightlines involve Fresh Knowe, the Thoms' mound *B*, which is possibly even earlier (see Ruggles 1982b, S24), but the date of all the mounds is uncertain and they could in fact have been constructed as late as 1700 BC.

106 The date of the rock-cut ditch at Stenness has been established by excavation and radiocarbon dating. See Ruggles 1982b, S27 (line 11).

107 The data are taken from Ruggles 1982b, table I, col. 13. This shows 15 class 'A' lines, one of which is at Brodgar.

108 Atkinson 1981, 207–8, site 38.

109 For details see Ruggles 1982b. For further commentary see Ruggles 1984b, 253–5.

110 For details, and the results of an attempt at fairer selection using predefined criteria, see Ruggles 1983, S9–13.

111 See note 11.

112 See in particular Thom and Thom 1978a, ch. 2; 1978b, 174–8; and 1980a.

113 Ruggles 1983, S13–31.

114 Thom and Thom 1978b, 174.

115 See Ruggles 1983, S25 and note 157.

116 Thom and Thom 1978b, 178.

CHAPTER 3

1 Heggie 1981a, 140.

2 Bradley 1984, 77.

3 Ruggles 1984a.

4 *Ibid.*, 20–1.

5 For example, the stone circle and nearer 'outliers' at Lochbuie, Mull (ML28 in List 2) may seem like one site, but what about the circle and standing stone at Strontoiller, Lorn (LN17 in List 2), 200 m apart and hidden from each other by intervening high ground? Similarly, it may seem obvious that each of the stone rows of North Mull—up to 20 m long—should be treated as a single site, but what about the 900 m-long alignment of three standing stones at Ardfernal/Knockrome in Jura (JU3 and JU4 in List 2)? See *ibid.*, 26.

Indeed, the whole concept of a 'site' may be unhelpful and even questionable in the first place (see ahead to Archaeology Box 6).

6 This judgement was strongly influenced by Thom's results at Level 1, so that, at the very least, it promised to reveal what remained at the core of Thom's 'megalithic astronomy' once any preferences owing to unwitting selective bias in Thom's own data were removed.

7 The code of practice used in the project evolved from an initial set of selection criteria developed in 1975 and tested at the group of sites around Callanish, Lewis (Cooke *et al.* 1977). These earlier criteria were modified in order to take account both of unexpected site configurations that were encountered, and of suggestions and criticisms by others (on the difficulties of devising suitable criteria see Heggie 1981a, 141; 1981b, S19). A further guarantee that selection procedures at particular sites were not influenced by astronomical predilections was provided by the fact that the reduction of the site survey data did not commence until the entire programme of fieldwork had been completed in 1981; the results did not start to become available until 1983.

8 For a different attempt to define selection criteria see Hawkins 1968, 48–50; for a critique see MacKie 1977, 98–100.

9 See, e.g., Ruggles 1984b, 255–6.

10 Ruggles 1984a, 21.

11 RCAHMS inventories were being prepared for mainland Argyll and the Inner Hebrides during the 1970s and 1980s, so that up-to-date information for these areas was generally available. During the lifetime of the project two of the Argyll Inventories were already published (RCAHMS

1971; 1975), and pre-publication information was available for two more (RCAHMS 1980; 1984) thanks to the co-operation and generosity of the Commission. Only information for mid-Argyll was lacking, and an extensive site list published by Campbell and Sandeman (1961) was used instead. The site list given in this book (List 2) gives cross-references to the more recent mid-Argyll inventory (RCAHMS 1988); for cross-references to Campbell and Sandeman's list see Ruggles 1984a, table 2.1.

In the Outer Hebrides, however, the only information available from the RCAHMS was collected back in the 1920s (RCAHMS 1928).

12 This archive, held at the time of the project on index cards by the Ordnance Survey office in Edinburgh, provides descriptions of Scottish monuments of all periods organised geographically. The NMRS is now held in electronic form by RCAHMS (see Archaeology Box 1).
13 There are 207 such sites, of which the fullest published list (Thom 1967, table 12.1) contains only eighty-five.
14 Burl 1976, appendix 1.
15 For a discussion of these criteria see Ruggles 1984a, 23 and 26.
16 For the full list see Ruggles 1984a, table 2.1.
17 This was due to the fact that the RCAHMS inventory of the island was being prepared while the project was under way.
18 For full details see *ibid.*, 44–58.
19 Ruggles 1984a, 20–1. As an example, consider the selection criterion: 'the orientations of all lines joining two standing stones at a site are to be included in the analysis (in both directions)'. At a site consisting only of two standing stones this is a good criterion, as these are the most obvious ways in which an orientation at the site might be significant. However now consider a circle of twenty stones: the criterion would have us include from the site some 380 orientations, virtually all without a doubt of no particular significance. See also Heggie 1981a, 151–2.
20 An exception was made where there were more than six structures of the highest existing classification at a site. In this case no indications were considered at the site.
21 For all the details see Ruggles 1984a, 59–64.
22 The most significant departure from Thom's own selection criteria at Level 1 was the decision not to consider any indications involving stone rings. The problem is that most stone rings are not perfectly circular or in a good state of repair, which means that there can be considerable uncertainty in determining the centre unless geometrical assumptions are made. Geometrical hypotheses are formally independent of astronomical ones, and until a consensus has been reached in both areas, it can be highly misleading to make geometrical assumptions in the process of testing astronomical hypotheses (Ruggles 1984a, 61; see also Clive Ruggles, comment on Ellegård 1981, *Curr. Anth.*, 22 (1981), 121–2, p. 122). Quite apart from this, there are very few instances of a stone ring together with a single outlier amongst the western Scottish data. The orientations of stone rings and outliers should, it was felt, form the basis of a separate study based in regions where such monuments are more common. See Ruggles 1984a, 61.
23 Ruggles 1984a, 63.
24 Although examples are found throughout the region, there are three notable concentrations: one, around Callanish in Lewis (sites LH10–24), consisting mainly of stone rings; another, in the flat neck of Kintyre between Machrihanish and Campbeltown (sites KT27–39), consisting almost entirely of single stone slabs; and the last, in North Uist (sites UI22–37), consisting of a greater diversity of types of monument.
25 Ruggles 1984a, 65–7. In brief, foresights more than 6 km distant were dismissed. Those on the horizon as seen from the backsight, or on a ridge with more distant horizon beyond, and subtending at least 1 arc minute, were assigned to Class 1. Class 2 contains all other foresights within 3 km, and Class 3 all others within 6 km.
26 For full details, including on-site lines of lower categories and inter-site lines, see Ruggles 1984a, table 11.1.
27 This was to ensure that the assumed observing positions for different indications had been chosen in a consistent manner, rather than to imply that observations had necessarily been made from that position in prehistoric times (*ibid.*, 59–60).
28 See note 7.
29 Ruggles 1984a, 60.
30 *Ibid.*, 60, 65.
31 *Ibid.*, 65.
32 For details see *ibid.*, sections 5.3, 6.3, 7.3, 8.3, 9.3 and 10.3.
33 *Ibid.*, figs 5.2–5.14, 6.3–6.15, 7.2–7.11, 8.2–8.12, 9.2–9.10, and 10.2–10.12.
34 For curvigrams of the overall azimuth and declination distributions see Ruggles 1984a, figs 12.10, 12.11, 12.14 (which presents the same data as Fig. 3.3) and 12.15.
35 This was felt to be important in order to avoid a subjective decision as to the original probable indication in cases where the direction of the original structure orientation was uncertain within wide bounds (Ruggles 1984a, 307; 1984b, 247).
36 See Ruggles 1984a, 19.
37 Although the classical hypothesis-testing method adopted dictated that a null hypothesis of this form should be considered (see Statistics Box 4), the orientations observed could not in fact have arisen randomly, although they could have arisen through the interaction of factors unrelated to astronomy. There remains a need to test that such processes would in fact lead to a pseudo-random distribution of declinations, for this is not self-evident (see Ruggles 1984a, 17–19). Account must also be taken of the fact that the indications in opposite directions by a single structure are not independent of each other (*ibid.*, 228).
38 Data generated in this way, for example by computer random-number generators, are 'pseudo-random' rather than truly random. See, e.g., Peter R. Freeman, 'How to simulate if you must', in Clive L. N. Ruggles and Sebastian P. Q. Rahtz (eds), *Computer and Quantitative Methods in Archaeology 1987*, BAR (International Series 393), Oxford, 1988, 139–46, p. 142. The whole paper gives useful procedural hints for this sort of simulation.
39 This is elaborated in Statistics Box 4.
40 Ruggles 1984a, 255.
41 The results are given in *ibid.*, table 12.2.
42 *Ibid.*, 303–4. For the derivation of these conclusions see *ibid.*, 254–75.
43 Ruggles 1984b, 264.
44 This was the interpretation favoured by Ruggles (1984a), but it depends critically upon the existence of a significant cut-off below −31°. An independent statistical appraisal of the data, using a cluster analysis method (Patrick and Freeman 1988) suggests that this cut-off may not in fact be significant. If the northern trend represents the cause and the southern one the effect, then an alternative explanation may be that structures were oriented preferentially to point farther along the horizon to the north than the moon (or sun) ever rose or set (Ruggles 1984a, 303). On the other hand, it is also possible that this is simply a reflection of the general preference for orientation around north–south.
45 Of the 261 indications listed in Thom 1967, table 8.1, 114 come from the areas covered by Ruggles 1984a.
46 Patrick and Freeman 1988, 257.
47 See also Ruggles 1984a, 306–8.
48 Ruggles (1984a, 307) concluded that there were no discernible trends at all amongst the inter-site alignments, although Patrick and Freeman (1988, 257–9) found evidence of a greater preference for directions between south and west amongst these data than amongst the on-site indications.
49 An exception is Nether Largie (AR13(b)), which really consists of two aligned pairs and a single slab, all parallel to one another, placed in three lines each some 30 m apart. Two class 1 indications in the sample (lines 163/164 and 165/166 in Table 3.2) represent longer alignments of three stones from Nether Largie, comprising one from each of the shorter lines.
50 For diagrams showing regional and other subsets of the data see Ruggles 1984a, figs 12.2–12.3 and 12.5–12.7.
51 *Ibid.*, 242–3.
52 Ironically, this evidence lay unremarked upon amongst the datasets in some of Thom's early publications. This is evident from Fig. 2.3a, where we display the same data as Thom 1955, fig. 8 but use shading to show the data from stone rows and aligned pairs, so that the unshaded data, from ring centres to outliers, can be more easily ignored. The accumulations at the solar solstices and lunar standstill limits observed by Thom were formed largely by the data from the stone rows and aligned pairs. In other words, they were at the basis of the early evidence that encouraged Thom to proceed to postulate ideas of higher precision observations of the sun and moon. Underlying these later and less sustainable claims was a clear trend that was obscured because the stone rows were never examined by themselves as a coherent group.
53 For a detailed discussion of the evidence on types of monument and indication falling within different preferred declination intervals see Ruggles 1984a, 266–75.
54 Twenty-seven rows, pairs and single flat slabs in Mull and mainland Argyll are oriented in the south upon a declination that the moon can reach at the southern limit of its monthly motions (Ruggles 1984b, 265; for the raw data see Ruggles 1984a, 282). In a further fourteen cases the southern indication was excluded from consideration because the horizon was closer than 1 km. Ten examples fall outside this range, but none of these is a stone row or aligned pair, and in six cases there is

special reason to believe the 'indication' to be spurious (Ruggles 1984a, 282–3). For a diagrammatic representation of these results see Ruggles 1984b, fig. 9.

55 The one author who had proposed prior to 1984 that Ballochroy might have a lunar association was Hawkins, who pointed out that to the north-east the row pointed at the major northern moonrise (Gerald S. Hawkins, *Beyond Stonehenge*, Hutchinson, London, 1973, 250).

56 Ruggles 1984a, 304.

57 Ruggles 1985.

58 Aubrey Burl, Euan W. MacKie and Andrew Selkirk, 'Stone circles again', *Curr. Arch.* 2 [no. 12] (1969), 27–8, p. 27.

59 Gerry G. Bracken and Patrick A. Wayman, 'A Neolithic or Bronze-age alignment for Croagh Patrick', *Cathair na Mart (Journal of the Westport Historical Society)*, 12 (1992), 1–11.

60 Freeman 1982, 46–8.

61 *Ibid.*, 48.

62 See Patrick 1979. The indications at Barbreck that, on grounds of architectural similarity, parallel the putative lunar indications at Nether Largie are without lunar significance. However, Patrick was wrong to imply that there is no lunar alignment at all at Barbreck (*ibid.*, S84), for the aligned pair (Patrick's *A–B*) is oriented close to declination −30° in the south (Ruggles 1984a, 306; see also *ibid.*, fig. 8.3).

63 Burl 1985, 135–6.

64 Thom and Thom (1978a, 110–19) propose that an isolated 6·5 m-high standing stone at the top of the hill of Le Manio, some 8 km to the north-west of the great menhir, is the centre of a second Carnac observatory. But some of the hypothetical backsights are not prehistoric. See Burl 1985, 154–5.

65 Heggie 1981b, S18–19; 1982b, 4.

66 Ruggles 1984b, 256.

67 Many factors may have affected this. In the context of the standing stones of Brittany, Giot (1988, 319–20) emphasizes natural agencies and particularly lightning. Deliberate destruction at later times is another possibility, as seems to have happened at Ardnacross on Mull, where the end stones of each of two three-stone rows were cast down and partially buried (see pp. 113–14). There is extensive evidence, through to modern times, of the removal of stone from these monuments and its re-use as material for field walls and buildings.

68 An example is provided by the data from excavations by Jack G. Scott at Temple Wood (AR13(*a*) in List 2) between 1974 and 1979, which revealed the site to have consisted of not one but two stone circles, and suggested a chronological sequence quite possibly spanning several hundred years during which time a timber setting was replaced by a small stone circle, then a second larger one was built, then cairns were added, then the spaces between the circle stones were filled in with small stone slabs, and finally a bank of stones was built around the circle, covering the cairns (RCAHMS 1988, no. 228; Scott 1989).

69 This led the present author to propose ways in which statistical rigour should precede interpretative reasoning (Ruggles 1984a, 309; 1985, S128–9; 1988b, 246–50), but it is impossible to separate the processes of theory development and data acquisition in this way and greater subtlety is needed. The methodological issues that arise are discussed in a broader context in chapters nine and ten.

70 Heggie 1981a, 141.

71 See also Ruggles 1994b, 500. On objectivity, see chapter nine.

72 This is not true of Bayesian inference, which will be examined in chapter ten.

73 This problem was recognised by Heggie (1981a, 144; 1981b, S20).

74 As an example, consider the well known, if at first surprising, fact that if twenty-three people are chosen at random from a population whose birthdays fall independently with equal probability upon any day in the year, the probability is close to 0·5 that two of them will have the same birthday. Suppose that a particular group of people is our 'archaeological record' and we notice that two of them have a birthday on (say) 24 June. Suppose also that, having spotted this pattern, we form the hypothesis that people are preferentially born on 24 June. If the only set of data available for testing this hypothesis is the group of people that suggested it in the first place, and we proceed to calculate the probability that two people out of twenty-three will fortuitously have their birthday on 24 June, this will be very low indeed (less than 0·002), and the result will seem highly significant. The problem is that we have chosen the '24 June' hypothesis from 365 similar ones.

Freeman (1982, 48–50) concludes that the tendency to choose hypotheses on the basis of the existing data is an inadequacy of methodology in archaeoastronomy, but it is arguably unavoidable, as it is within most of archaeology as a whole. Other ways are needed to develop data and theory while avoiding circular argument. For more on this theme, with references, see chapter ten.

75 See Freeman and Elmore 1979. This was also the approach taken by Ruggles 1984a, see p. 261.

76 Even the Thoms (Alexander Thom and Archibald S. Thom, 'Statistical and philosophical arguments for the astronomical significance of standing stones, with a section on the solar calendar', in Heggie 1982a, 53–82) attempt to distinguish between 'statistical' and 'philosophical' proof, the latter consisting of subjective observations that lend support to the astronomical hypothesis but can not be used in statistical analysis.

77 'The reason for this is straightforward. A single alignment of apparent astronomical significance, unlike an artefact, is ambiguous; it might have arisen through the interaction of factors quite unrelated to astronomy. It is an insufficient, indeed futile, yet nonetheless overworked practice in archaeoastronomy simply to seek out examples of astronomical alignments and to lay great emphasis upon them. Instead, we must first examine enough archaeological structures to permit a statistical examination. By this means we can compare the actual number of astronomical alignments obtained with that which would have been expected by chance (i.e. by the interaction of factors unrelated to astronomy). This immediately raises numerous questions of the fair selection of data, which, if we do not pay full attention to them at the outset, will introduce the possibility of bias, and thus invalidate any statistical conclusions. Only if the data can be seen to have been selected fairly (that is without regard for the astronomical possibilities), and the presence of significantly more astronomical alignments than would have been expected by chance can be demonstrated statistically, can the archaeoastronomer present the archaeologist with reliable evidence that astronomical considerations did affect site design' (Ruggles 1984a, 15).

78 Heggie 1981a, 224 ff.

79 Ruggles 1984a, 16–17.

80 I. J. Thorpe, Comment on Ellegård 1981, *Curr. Anth.* 23 (1982), 220–1, p. 220.

81 See chapter nine.

CHAPTER 4

1 John Simpson, *In the Forests of the Night*, Hutchinson, 1993 (pp. 26–7 in the Arrow Books edn, 1994).

2 Thom 1967, 166.

3 Burl 1979b, v.

4 Thom 1971, 9–10. See also Thom and Thom 1978a, ch. 14.

5 Wickham-Jones 1994, 116–17. By 6000 BC people were living on Rhum, Islay, Jura, Ulva (off Mull) and Arran (*ibid.*, 46–7).

6 Thom 1971, 10.

7 This was sparked off by the following remark by Atkinson, referring to Hawkins's ideas, in a 1965 television documentary 'The Mystery of Stonehenge', screened by CBS Television: 'These people were what I would call "howling barbarians", they were practically savages, and everything we know about them suggests they were quite incapable of this degree of scientific and technological sophistication' (Richard Atkinson, 1965, *MacGraw Hill Film Texts*, quoted by Gerald Hawkins, 'Stonehenge: the clues and the challenge', in Romano and Traversari 1991, 169–75, p. 169). For a critique of the latter article see Ruggles 1995, 84–5.

8 See also Castleden 1993, 196.

9 Fred Hoyle, on the other hand, talked of an intellectual prowess astonishing among primitive farmers (Hoyle 1966, 262), and even went so far as to suggest that the constructors of Stonehenge I (see chapter one) formed an 'isolated group or gene pool with extraordinary mental gifts' (see Hoyle 1966, 274 and Hawkes 1967, 176).

10 Burl 1979a, 82.

11 As limited by variable atmospheric conditions: see Astronomy Box 3.

12 Turton and Ruggles 1978, 592.

13 Thom's involvement was intensely personal, as is evident from several general descriptions (see, for example, Thom 1988; Motz 1988; Archibald S. Thom, *Walking in all of the Squares*, Argyll Publishing, Glendaruel, 1995) and he felt himself increasingly close to 'megalithic man', or 'the boys' as he came to call them (*ibid.*, 222).

14 See Ritchie 1982.

15 However, the book by Heggie with this title (Heggie 1981a) was in fact a detailed and careful scientific appraisal of Thom's ideas.

16 e.g. Renfrew 1976, 260–3.

17 Thom and Thom 1978a, 177. See also Statistics Box 5.

18 For example, Colin Renfrew (*Towards an Archaeology of Mind*, CUP, Cambridge, 1985, 16–17; see also Renfrew and Bahn 1996, 382) shows how, from the discovery from a city in the Indus valley, *c.* 2000 BC, of a scale

pan and a set of stone cubes whose weights are found to be integer multiples of a unit of 0·836 g, one can argue that a succession of concepts were held by the people who made them: a notion of weight or mass; a concept of modular measure; a system of numeration based on multiples of 16; a notion of equivalence; and so on.

19 For simplicity, the term 'Western' will be used henceforth to refer to the 'classical' tradition of scientific thought prevalent in the modern Western world.

20 There is a fundamental difference between the geometry that we perceive and the concepts that might have generated a particular shape. For example, the existence of an elliptical ring does not mean that prehistoric people perceived of the geometrical concept of an ellipse; merely that their procedure for laying out this ring resulted in a shape approximating to one. Thus *saroeak*, described by many Western authors as 'circles' of eight stones linked to pastoral practices in the Basque country, were perceived traditionally not as circles but as octagons (Frank and Patrick 1993; Luix Mari Zaldua Etxabe, *Saroeak Urnietan [Stone Octagons in Urnieta]*, Kulturnieta, Urnieta, 1996; 'Configuration and orientation of *saroe* and *artamugarri* in the Basque country', in Jaschek and Atrio 1997, 95–102).

21 Thaddeus M. Cowan ('Megalithic compound ring geometry', in Ruggles 1988a, 378–91) suggests that many megalithic rings were constructed by two people moving at opposite ends of a piece of rope. This would result in equal-width figures, though perhaps with rounded corners like British and Irish 50 pence coins, a prime example being Grange [The Lios] in Co. Limerick (R 640 410) (see also Burl 1995, no. 356). This idea seems to fit the evidence at a number of monuments. In some cases, complex geometrical interpretations of the figures produced have been proposed by Thom and others, but the intention may simply have been to enclose an area, perhaps reflecting the (circular) shape of the larger world, using a convenient method. Even if a shape closely resembling our circle (e.g. a smooth ring with constant diameter) was intended, we should not take it as self-evident that a radius-and-circumference method would be seen by the builders as the best way of trying to achieve this.

22 Thom 1967, 112.

23 Thom (*ibid.*, 111–12) used a least-squares method to calculate the best fit to an ideal solution where the sixteen periods in his 'megalithic calendar' have an equal number of days and fractions of days. However, his solution could not have seemed any better to prehistoric calendar-makers than other schemes of apportioning the days in each period, since the 'ideal' epoch declinations only have meaning to us who have access to formula (A5.1) of Astronomy Box 5. In any case, the notion of prehistoric people striving to obtain the 'best' solution (*ibid.*, 112) as defined in terms of a least-squares fit is flawed because this type of fit is itself a modern scientific notion. Thus, Thom's best-fit solution actually has one period of twenty-four days, i.e. it achieves the mathematical best fit by having one rather larger error (greater than a whole day) in order to minimise several others. In practice, this might not seem as attractive as a solution in which all errors were within a day of the ideal.

24 Writing in 1977, Heggie concluded that 'megalithic astronomy was a religious, ritual or symbolic activity rather than a scientific or practical one' (Douglas C. Heggie, 'Megalithic astronomy—fact or fiction?', *QJRAS*, 18 (1977), 450–8, p. 456). Following a seminal paper on the subject by Burl (1980), a succession of publications argued for the 'ceremonial' interpretation (Burl 1981a; 1983; Moir 1981; Ellegård 1981; Barnatt and Pierpoint 1983; David Fraser, 'In support of festive astronomy', *SAR* 3 (1984), 16–18). The 'scientific' point of view was most vigorously defended by MacKie (1981; MacKie *et al.* 1985). See also Powell 1994. For a North American perspective see McCluskey 1987.

25 The heliacal rising of a constellation, or more exactly of a star, is its reappearance at dawn after a few weeks of absence each year when it is too close to the sun to be seen. The heliacal setting is its disappearance from the sky at dusk before the annual period of invisibility. These phenomena can be used to determine the time within the solar year to within a few days. For a diagram that can be used to give the approximate dates for 2000 BC, see Thom 1967, fig. 8.2.

26 Hesiod, *Works and Days*, 385–7, 564–617. See Richmond Lattimore (transl.), *Hesiod*, University of Michigan Press, Ann Arbor, 1991 (Ann Arbor paperbacks edn.), 65, 85–91.

27 In fact, Galileo's own observations (see Hoskin 1997a, 122–34) tended to be qualitative, and although he made many new discoveries such as the phases of Venus, the moons of Jupiter, and new faint stars in the constellation of Orion, he paid little attention to the measurement of position as had Ptolemy.

28 Clive L. N. Ruggles and Michael Hoskin, 'Astronomy before History', in Hoskin 1997a, 2–21, pp. 8–9.

As Gingerich (1989, 41) points out, in developing an understanding of the heavenly cycles it is tradition or record-keeping that is important rather than high-precision observing instruments. Similarly, Bradley (1993, 61) argues that the purpose of stone monuments themselves could not have been scientific observation because the movements of the heavenly bodies would need to have been understood before the monuments were built. But the point here is one of precision rather than one of purpose or meaning. Temporary markers would certainly have been needed to determine the appropriate location for more permanent monuments of reasonable precision, but this is true whether their purpose was as solar or lunar observatories (e.g. Thom 1971, 13–14) or ceremonial monuments whose 'carefully chosen position' (Bradley 1993, 62) encapsulated sacred knowledge.

According to Fleming (1975), 'the functions discharged by ceremonial, and the need to have recognised ritual sites and other symbolic points in the landscape, would certainly have taken precedence over the painstaking manœuvres necessary for the setting up of observatories'. This assumes once again a correlation between function and precision.

29 The idea that observations were made by 'advanced thinkers' in a spirit of free enquiry (e.g. Thom 1971, 110–11; see also Hawkins and White 1970, 150–1), begs the question and says nothing of the conceptual framework within which the prehistoric inhabitants of Britain or Ireland might have been making discoveries and developing ideas; instead, the implicit assumption is that their framework is similar to our own.

Historians of science and astronomy have made a number of suggestions. One is that 'scientific' activity is characterised by reducing what is observed to a series of rules (North 1994, xxv). A key attribute of scientific activity is often taken to be the ability to make accurate predictions about future occurrences, the judgement of accuracy being made in the context of own scientific understanding of the world. Modern astronomy undoubtedly has its roots in the merging of two great but distinct traditions that took place following the conquests of Alexander the Great in the fourth century BC: the Ancient Greek penchant for philosophical enquiry and the Babylonian prepossession with recording events systematically in order to make accurate predictions (Hoskin 1997a, ch. 2).

30 The tendency to do this is widespread. Thom and Thom (1978a, 181), for example, imagine 'megalithic man' deducing that lunar eclipses represent the shadow of the earth on the moon and the fact that the earth is spherical.

31 See, for example, Aveni 1992a, ch. 1. Judging the achievements of another culture (including past ones) against our own only makes sense if we believe in 'the notion that only the knowledge we have acquired by walking the particular paths of discovery taken by our [scientific] predecessors has intrinsic value' (*ibid.*, 6).

32 The question of literacy sometimes enters this argument, literacy being seen as a prerequisite for logical thought. For a critique of this point of view see John Halverson, 'Goody and the implosion of the literacy thesis', *Man* 27 (1992), 301–17. 'To insist . . . on a rigid and ultimately subjective dichotomy between (literate) science and (nonliterate) non-science serves only to obscure the nature of what surely ranks as an interesting development in the evolution of scientific thought' (Colin Renfrew, Comment on Ellegård 1981, *Curr. Anth.* 22 (1981), 120–1).

33 To Chippindale (1994, 230), its use demonstrates 'the mistake in the astronomers' vision'. See also I. J. Thorpe, 'Prehistoric British astronomy—towards a social context', *SAR* 2 (1983), 2–10, p. 5. This does not seem to have been a problem elsewhere in the world: see, e.g., Williamson 1984 for a case study in North American archaeoastronomy and McCluskey 1987 for a discussion of the point.

34 On the distinction between the two, see Edmund Leach, *Social Anthropology*, OUP, Oxford, 1982, ch. 7, esp. p. 214. See also chapter nine, note 38. Note, however, that Hawkins's use of the term 'megalithic cosmology' ('Megalithic cosmology', in *Encyclopedia of Cosmology*, ed. Norriss S. Hetherington, Garland Publishing, New York, 1993, 380–7) has nothing to do with the broader anthropological concept.

35 See Ruggles and Saunders 1993b, 2–4.

36 Because perception and use may be so different from one group of people to another, it is arguably better to speak of 'astronomies' rather than 'astronomy', as in McCluskey 1987; 1998; and Ruggles and Saunders 1993a.

37 Thom himself conceived of local experts, even going so far as to construct an imaginary dialogue between an instructor and an apprentice at a lunar observatory (Thom 1971, ch. 10). These experts had 'determination, time, and freedom from distraction' (*ibid.*, 86) and could evidently mobilise a significant workforce in support of their activities: 'At each [lunar] maximum and minimum, parties would be out at all possible

places trying to see the moon rise or set behind high trial poles . . . Then there would ensue the nine years of waiting' (Thom and Thom 1971, 158–9). Lunar extrapolation devices some 200 km away in Caithness would have been known about in detail in western Scotland (Thom 1971, 107). Yet the wider implications for communications and social structure were never discussed by Thom and were evidently of no interest to him.

38 For one of the most succinct comments see Fleming 1975.

39 MacKie 1977. The quotation is from p. 168.

40 For example MacKie (1977, ch. 3), extending an original argument by Thom himself (Thom 1967, 34), argued that there was a close relationship between the megalithic yard and several known European units of measurement from more recent times, such as the Iberian *vara*. See also Euan W. MacKie, *The Megalith Builders*, Phaidon, Oxford, 1977.

41 See, for example, Thorpe 1983, 2–3; Bradley 1984, 77. A specific criticism was that societies, such as the Maya, who possessed astronomer-priests also had writing, while Bronze Age Britons did not (Moir 1981, 223; see also Hutton 1991, 112). On the dangers of using a single analogy, see Orme 1981, 27–8.

42 There are two strands to the argument here. The first is that these were prestige goods, and that they might have been for the use of a privileged few. However, while it is arguable that there was indeed a prestige goods economy in parts of Late Neolithic Britain (e.g. Bradley 1984, 46–57), Grooved Ware pottery should probably be seen as only one of a number of styles of ceramics, quite possibly used initially by high-status individuals for special purposes, but whose significance changed as they were emulated by different manufacturers and eventually incorporated into widespread, everyday domestic use (*ibid.*, 70–3). The other strand of the argument is that the people who used the high-status ceramics for a special purpose were privileged élites that necessarily contained full-time specialists such as astronomer-priests. Here a much broader range of interpretation is possible (e.g. Castleden 1987, 204–17, 260–3).

43 Again, a much broader range of interpretation is possible. For some of the main issues see Bradley 1984, 75–84. For a discussion in the context of the Orkney monuments see Ritchie 1985, 126–30. On Skara Brae in particular, see Anna Ritchie, *Prehistoric Orkney*, Batsford/Historic Scotland, London, 1995, 28–40. On the other hand, a more recently discovered and excavated settlement at Barnhouse, close to the stone rings of Stenness and Brodgar, does seem to contain the habitations of high-status individuals (*ibid.*, p. 40; Richards 1990a; see also chapter ten).

44 There were further criticisms on the grounds of the social implications. For example, 'Thom's proposed calendar is not a farming calendar which subdivides the year according to the needs of agriculture, but a systematic calendar that subdivides the year into (nearly) equal periods. Though such a division may suit a civil bureaucracy, astronomers and accountants (rents became due on quarter days), it fits less well with the husbandry of cattle and corn' (Moir 1981, 230). MacKie, on the other hand, argued that the calendar progressed in three stages from a simple one serving the needs of agriculture to one in which there was a 'professional' concern with esoteric knowledge (MacKie 1988, 211–12, 230).

45 Renfrew 1973, 555, reprinted in Renfrew 1984, 242.

46 Colin Renfrew, 'Islands out of time', in Ray Sutcliffe (ed.), *Chronicle: Essays from Ten Years of Television Archaeology*, BBC, London, 1978, 113–26, p. 114, reprinted in Renfrew 1984, 202. (The remark was actually made in relation to Atkinson's 'barbarians', quoting Atkinson 1979, 165–6; see note 7 above.)

47 For a detailed account of the nature of ritual and society in the Neolithic and Bronze Age accessible to the non-specialist, and with frequent reference to astronomical evidence, see Burl 1981c. On Late Neolithic society see also Castleden 1987, ch. 13. For further details and references to some of the mainstream archaeological sources see Archaeology Boxes 4 and 5.

48 See, e.g., Thorpe 1981. This point was also clearly made in the early 1980s by several papers presented at the 1983 Washington conference on ethnoastronomy, a number of which are due to be published, after an extraordinary delay, in *AAJ*, vols 12 (1998) and 13 (1999).

49 Renfrew 1976, 263 and fig. 53; McCluskey 1977.

50 The Yanomamö of Amazonia, for instance, continually conduct fights, raids, and warfare, believing that the blood of the moon spilled upon their layer of the cosmos, causing them to become fierce (Napoleon A. Chagnon, *Yanomamö: The Last Days of Eden*, Harcourt Brace Jovanovitch, San Diego, 1992, 122–3). The significance of the moon in this context would be lost to any future archaeologist working from the material record alone (Ruggles and Saunders 1993b, 14).

51 This has been suggested in the case of Palaeolithic bone fragments from southern and eastern France. See chapter eight.

52 Amongst the Pueblo Indians as documented in historic times, the sun-priests simply used an identifiable spot close to their village that pro-vided a good view of the relevant horizon (Michael Zeilik, 'The ethnoastronomy of the historic pueblos, 1: Calendrical sun watching', *AA* no. 8 (*JHA* 16) (1985), S1–24, pp. S10–12 and table 4; see also 'Keeping the sacred and planting calendar: archaeoastronomy in the Pueblo southwest', in Aveni 1989a, 143–66, pp. 149–51). Sun-watchers amongst the Mursi use a natural object such as a boulder or tree to mark the spot (Turton and Ruggles 1978, 589). There is no need to mark the relevant horizon foresights used, since these are well known.

53 The Mursi provide an excellent example of a calendar that to us appears completely haphazard because of an institutionalised disagreement about the current month, but which works perfectly adequately within the framework of the society in question. See Turton and Ruggles 1978.

54 See p. 32.

55 One way would have been to use shadow casting at noon. See David A. Allen, 'Solstice determination at noon', *AA* no. 17 (*JHA* 23) (1992), S21–31.

56 Clive Ruggles, 'Archaeoastronomy in Europe', in Walker 1996, 15–27, p. 23.

57 This is a theme that has been emphasized by Burl in a number of publications, e.g. Burl 1981a; 1983; 1987a.

58 'We should not expect past societies to make the same distinctions between the domestic, social and religious as we do now' (Bradley 1984, 77).

59 See chapter nine for examples.

60 See Bewley 1994.

61 Burl 1980, 196. Elsewhere, Burl refers to cup marks as 'symbols of the moon' (Burl 1979a, 62). He also suggests (Burl 1980, 196) that 'it may not be too fanciful to think that they saw, in the litter of quartz that glittered so brilliantly in the moonlight, fragments of the moon itself' (*ibid.*). Support for this speculation comes from some Australian Aboriginal groups for whom rock crystals are not only of celestial origin, but are considered 'solidified light' (Mircea Eliade, *Shamanism: Archaic Techniques of Ecstasy*, Princeton UP, Princeton, 1970, 137–9, 508).

62 Brennan 1983, 127–205. But see p. 129.

63 See Morris 1981, 5.

64 Maurice Bloch, 'Questions not to ask of Malagasy carvings', in Ian Hodder, Michael Shanks, Alexandra Alexandri, Victor Buchli, John Carman, Jonathan Last and Gavin Lucas (eds), *Interpreting Archaeology: Finding Meaning in the Past*, Routledge, London, 1995, 212–15, pp. 213–14. Once we admit such an explanation then interpretations of the carvings are possible, not as literal representations or even abstract art *per se* but on quite another level. Bloch suggests that making the carvings forms part of the process by which Zafimaniry married couples continually 'harden' their houses, themselves seen as entities that need to mature and grow; this is also done through the addition of carefully chosen pieces of wood, and of other furnishings. The process is inseparable from that of building and reinforcing a couple's own relationship, throughout their lives and beyond, when their house will be used by their descendants (*ibid.*).

65 John C. Brandt and Ray A. Williamson, 'The 1054 supernova and native American rock art', *AA* no. 1 (*JHA* 10) (1979), S1–38. For a catalogue see *ibid.*, table 1. The Chaco example is shown in *ibid.*, pl. 5b.

66 R. A. Armitage, M. Hyman, J. Southon, C. Barat and M. W. Rowe, 'Rock-art image in Fern Cave, Lava Beds National Monument, California: not the AD 1054 (Crab nebula) supernova', *Antiquity*, 71 (1997), 715–19.

67 See, for example, Stuart Piggott, 'Concluding remarks', *PTRS* A276 (1974), 275–6.

68 e.g. Heggie 1972, 47–8; Ellegård 1981, 107.

69 According to Hubert H. Lamb (*The Changing Climate*, Methuen, London, 1966, 173), the mean temperature in Britain and Ireland in Neolithic times was some 2°C greater than that today. From this, Ellegård (1981, 107) estimated that on average the skies would still have been cloudy between 50 per cent and 60 per cent of the time, and even a cloudless sky may be hazy near the horizon (Heggie 1981b, S26). Recently, though, the idea of a significant climatic deterioration since the Early Bronze Age has been challenged (Graeme Whittington and Kevin J. Edwards, 'Climate change', in Edwards and Ralston 1997, 11–22, pp. 13–14 and references therein).

70 See, e.g., Evans 1975, ch. 6; Kevin J. Edwards and Kenneth R. Hirons, 'Cereal pollen grains in pre-elm decline deposits: implications for the earliest agriculture in Britain and Ireland', *J. Arch. Sci.*, 11 (1984), 71–80; Roberts 1989, 144–7; Bell and Walker 1992, 164–8. On Scotland in

particular see Edwards and Whittington 1997. See also Archaeology Box 5.

71 In the Stonehenge area in Early Neolithic times, for example, it seems that areas were continually clear-felled and subsequently allowed to regenerate, resulting in a complex and dynamic 'mosaic' of vegetation types (see Cleal *et al.* 1995, 473).

72 An example is the Isle of Mull (see p. 113).

73 There are limitations. Pollen analysis is specific to a particular location, and can only give a general indication of the changing types of vegetation in the vicinity. In addition, pollen evidence only indicates the chronological sequence, and other methods are needed to provide an absolute calibration in time. See Renfrew and Bahn 1996, 224–7.

74 MacKie 1977, 226–8.

75 See particularly Burl 1983, 34–5; Darvill 1987, 189–90; MacKie 1988, 225–9. The remnants of this Celtic calendar, the argument continues, survive today in the form of English and Scottish Quarter Days, the latter actually falling on (or close to) the mid-quarter dates. In fact, Thom himself was aware of the possibility that modern quarter and mid-quarter days corresponded to the eight epochs of his megalithic calendar, and mentioned this in his earliest paper on megalithic astronomy (Thom 1954, 399). The idea of alignments on sunrise and sunset at the quarter and mid-quarter days goes right back to Lockyer (1909, see chs 18 and 24) and Somerville (1923, 197–8).

76 Not least of these is the very idea of 'the' (i.e. a single) Celtic calendar. See pp. 141–2.

77 Diodorus of Sicily, *The Library of History*, II, 47, as translated by C. H. Oldfather, *Diodorus of Sicily* (10 vols), Harvard UP, Cambridge Mass. and Heinemann, London, vol. 2 (1935), 41.

78 *Ibid.*, 39.

79 e.g. Gerald S. Hawkins, 'Stonehenge: a Neolithic computer', *Nature* 202 (1964), 1258–61; Hawkins and White 1970, 215; Hoyle 1977, 103–4; and MacKie 1977, 227. But see also Hawkes 1967, 178.

80 Somerville 1912, 91–2; Burl 1993, 64–5; 1995, 150. On Callanish, see chapter eight.

81 Hawkins and White (1970, 215) take 'the god' to be the moon, but MacKie (1977, 227) takes it to be Apollo, the sun.

82 Hawkins and White 1970, 215.

83 In Oldfather's translation the remainder of the last sentence reads '. . . and for this reason the nineteen-year period is called by the Greeks the "year of Meton"'. The Metonic cycle is described further by Diodorus in *The Library of History*, XII, 36, translated by C. H. Oldfather, *Diodorus of Sicily* (10 vols), Harvard UP, Cambridge Mass. and Heinemann, London, vol. 4 (1946), 447–9. See also Aveni 1997, 89. On the confusion between the two lunar cycles, see Aveni 1987, 14–17.

A recognition of the Metonic cycle would have constituted important practical knowledge enabling the solar and lunar calendars to be reconciled by inserting a fixed pattern of intercalary months over each 19-year (235-lunar month) period (cf. Piggott 1968, 123).

84 One of these is Hoyle (1977, 104), despite offering his own interpretation.

85 Strabo, III, 4, 16, transl. Horace Leonard Jones. See *The Geography of Strabo*, vol. II, Heinemann, London, repr. 1960 [first edition 1923].

86 For example, from the stone circle at Beltany, Co. Donegal (C 254003), a neighbouring hill indicates sunrise at the mid-quarter date in Thom's calendar corresponding to the Celtic festival of Beltane (Somerville 1923, 212). See also Burl 1976, 252–3; 1995, 229–31, no. 340. The problem is that, even if such coincidences are not fortuitous, they may well reflect relatively recent traditions rather than indicating any sort of continuity from the distant past.

87 Thus Burl (1976, 294–7) cautiously suggests that seasonal rites at the Rollright Stones, Oxfordshire (SP 296308) in historical times might be related to earlier practices. 'One wonders what time-thin truths survive' (*ibid.*, 296).

88 Frank and Patrick 1993, 88–9. For a commentary on the evidence from the Basque country see Ruggles 1990, 139–41. For a critique of theories of the history and development of the Indo-European languages see Whittle 1996, 137–40.

89 Heggie (1981a, 219), citing the examples just mentioned, is highly dismissive of these forms of evidence. The historian of science Olaf Pedersen, whose remarks at Commission 41 of the International Astronomical Union at Grenoble in 1976 are reported in *JHA* 7 (1976), 223, warns against using such evidence 'unless carefully considered and controlled'. For an idea of the complexities surrounding the question of 'pagan survivals' from pre-Christian times in Britain and Ireland see Hutton 1991, ch. 8.

90 This use of ethnographic evidence is quite different from putting specific cases forward as 'parallels' that can directly inform archaeological interpretations, as in MacKie's use of the Maya analogy (see note 41 above). 'Analogues are better at informing archaeologists about what they do not know about the past than about what they do know or can expect to know' (Gould 1980, 36).

91 E. Crystal, 'Man and the menhir: contemporary megalithic practice of the Sa'dan Toraja of Sulawesi, Indonesia', in Christopher B. Donnan and C. William Clewlow Jr (eds), *Ethnoarchaeology*, Institute of Archaeology UCLA (Archaeological Survey, Monograph IV), Los Angeles, 1974, 117–28. In contrast, principles of a broader cosmology (see chapter nine), such as male-female duality and sacredness of place, were of central importance (*ibid.*, 122). See also Moir 1981, 237.

92 According to Barber (1973, 37) they 'know of a period called the *Duibhré* when the moon is so low in the sky that it does not rise above the surrounding mountains for a period of several weeks. This period occurs once every nine years, when the moon's declination is at a minimum'. This description does not make astronomical sense as the moon's declination moves rapidly from north to south and back each month, and the time of the most extreme motions (around major standstill) occurs every eighteen to nineteen years, not every nine years (see Astronomy Box 4). However, it seems clear that amongst these farmers there was some consciousness of the consequences of the lunar node cycle.

93 e.g. Burl 1993, 62; North 1994, xxiv–xxv.

94 Turton and Ruggles 1978, 589–90.

95 For a general introduction to the uses of ethnographic analogy in archaeology, see Orme 1981, 1–28. See also Gould 1980.

96 e.g. Ruggles 1984c; 1985.

97 On different modes of explanation of European megalithic monuments see Renfrew and Bahn 1996, 466–7, and references therein.

98 For further possibilities see Parker Pearson and Ramilisonina 1998, 311–13 on standing stones in modern Madagascar.

99 This is evident from some of the monuments that have been excavated, such as the Temple Wood stone circles in Argyll (see chapter three, note 69).

100 Renfrew 1973, 555, reprinted in Renfrew 1984, 242.

101 Colin Renfrew, 'Megaliths, territories and populations', in Sigfried J. de Laet (ed.), *Acculturation and Continuity in Atlantic Europe*, De Tempel (Dissertationes Archaeologicae Gandenses XVI), Brugge, 1976, 198–220, p. 208, reprinted in Renfrew 1984, 180. See also Burgess 1980, 295–6; Parker Pearson 1993, 46; Whittle 1996, 243–8.

102 For a classic article see H. J. Rose, 'Celestial and terrestrial orientation of the dead', *JRAI* 52 (1922), 127–40. For an overview in the context of Saxon graves see Philip Rahtz, 'Grave orientation', *Arch. J.* 135 (1978), 1–14. See also Ruggles 1984b, 276–7.

103 The earthen long barrows of Cranborne Chase are an example (Ashbee 1984, 23–4).

104 Thus chambered tombs in the Scilly Isles are generally set on slopes and frequently, though not always, face downhill (Paul Ashbee, *Ancient Scilly*, David and Charles, Newton Abbot, 1974, 71). Ashbee suggests that this may have been (at least in part) in order to make the monuments conspicuous.

105 This is often suggested, but see Ruggles and Burl 1985, S32.

106 Two early papers on orientation in the mainstream journals *Archaeologia* and *Antiquity* (Somerville 1923; 1927) were in fact concerned only with possible astronomical interpretations of orientation and were precursors of Thom's work.

107 e.g. Ashbee 1984, 21–4 and figs 19–22 on earthen long barrows, Henshall 1963, 104 on the chambered tombs of northern Scotland, and Henshall 1972, 99 on Clyde tombs.

108 Powell *et al.* 1969, 236.

109 An exception was the work of David Fraser on the Orkney chambered tombs (Fraser 1983). See chapter eight.

110 In the context of Stonehenge, see Whittle 1997.

111 For further details and references see chapter eight.

112 On Clava cairns see chapter eight. On recumbent stone circles see chapter five and references therein.

113 For a summary, and references to primary sources, see Ruggles 1994a, S1–2. On axial stone circles see also chapter five, and on stone rows see chapter six.

114 Clive Ruggles, 'Orientation analysis and prehistoric astronomy', in George L. Cowgill, Robert Whallon and Barbara S. Ottoway (eds), *Manejo de datos y Métodos Matemáticos de Arqueología*, International Union of Pre- and Proto-Historic Sciences (UISSP Congress X, Commission IV), Mexico City, 1981, 228–34. See also Ruggles 1982a, 96; 1984a, 17.

115 This is not restricted to horizon astronomy, i.e. orientation upon celestial bodies at their time or rising or setting. For example, a tradition of

orientation upon a particular astronomical body while up in the sky would constrain the declinations measured at the horizon to be lower than that of the body concerned, and so could still produce discernible trends in declination.

116 Note, however, that to determine this for declinations is a non-trivial statistical problem (see pp. 70–5).

117 Ruggles 1982a, 97; 1984b, 281–2.

118 A poignant example of this is provided by the Ashanti of Ghana, some of whom deliberately bury their dead in the opposite direction to that considered the norm, in the belief that the dead turn themselves round after a certain passage of time (Peter J. Ucko, 'Ethnography and archaeological interpretation of funerary remains', *WA*, 1 (1969), 262–80, p. 273). For a discussion see Burl 1981a, 250–1.

119 For example, the predominantly south-easterly orientation of the long barrows of Cranborne Chase may be due more than anything else to the local lie of the land, because the ridges along which they tend to be sited (note 103) are themselves mainly oriented NNW–SSE (Ashbee 1984, 23–4).

120 The problem is that if, for example, we notice surprisingly many declinations between, say, −30° and −28° and choose on this basis to test the hypothesis that the declination −29° was preferred to a precision of 1°, then we are selecting one hypothesis (a particular target and a particular precision) from innumerable possibilities, and selecting it on the basis of the data that will be used to test the hypothesis in the first place. See p. 77. See also Ruggles 1982a, 98–9.

121 See ahead to Archaeology Box 8.

122 A related problem is to decide how many directions will be deemed significant. If too few, then trends of possible importance may be missed. If too many, then so many fortuitous orientations will be considered significant that they will inevitably overwhelm any genuine trends in the data (Ruggles 1984a, 18).

123 At the time, results were emerging to suggest that the short stone rows of western Scotland were preferentially aligned upon the moon, to a low level of precision (see chapter three). In addition, Burl had already suggested that lunar symbolism was important in the recumbent stone circles (Burl 1980). At the very least, many overall trends of easterly or westerly orientation amongst groups of Neolithic tombs suggested general traditions relating to the rising or setting sun or moon.

CHAPTER 5

1 Tim Robinson, *Connemara. Part 1: Introduction and Gazetteer*, Folding Landscapes, Roundstone, Co. Galway, Ireland, 1990, p. 9.

2 Lewis 1892, 142.

3 Ritchie and Ritchie 1981, 54.

4 Burl 1976, 167–8.

5 The southerly orientation of the RSCs was first noticed by Hector Boece in about 1520 (Burl 1980, 193). In the 1960s a single RSC, Strichen (RSC7 in List 3), did exist whose recumbent and flankers were on the north-east side of the circle (Thom 1967, 142), but it was already known (Keiller 1934, 12), and excavation subsequently confirmed, that all the circle stones had been demolished in the nineteenth century and re-erected on the wrong side (Hampsher-Monk and Abramson 1982; see also Ruggles and Burl 1985, S29–30). For an early orientation diagram see Burl 1970, fig. 4.

6 Burl 1979a, 16; Ruggles and Burl 1985, S25–6.

7 Burl (1979a, 22; see also Ruggles and Burl 1985, S28) has suggested that they were 'family' monuments serving groups of no more than twenty or thirty people, although the term 'family' has varied connotations anthropologically. Cf. Whittle 1988, 181, who prefers the term 'special interest group'.

8 Ruggles and Burl 1985, S26–8.

9 Barclay 1997, 136–8.

10 The Berrybrae excavation is described in Burl 1979a, 25–31. For a site plan and photo see Burl 1976, fig. 30 and pl. 21.

11 Burl 1995, 93.

12 Burl 1979a, 29.

13 Ritchie and Ritchie 1981, 60.

14 Shepherd 1989, 123–5.

15 Burl 1970. Amongst other things, the RSCs and Clava cairns have a common pattern of orientation towards the SSW (Burl 1976, 168–70; see also Burl 1983, 28, 36–7). See also Clive Ruggles, 'Ritual astronomy in the Neolithic and Bronze Age British Isles: patterns of continuity and change', in Gibson and Simpson 1998, 203–8, p. 207.

16 Burl 1979a, 34–8.

17 Burl 1980.

18 Ruggles 1984b, 256. See also Ruggles 1982a, 99; 1984a, 19.

19 Burl 1976, 168 and fig. 27.

20 Burl 1980, fig. 1.

21 These criteria were developed following an initial period of two weeks in the field in April of the same year when preliminary visits were made to a number of RSCs in the company of Burl.

22 Ruggles 1984c, S61. The classification information is included in List 3. For a commentary see Ruggles 1984c, table 1, col. 10.

23 Nether Corskie (RSC69) is also listed as destroyed or unrecognisable in Ruggles 1984c, table 1, but two erect stones do survive here (*ibid.*, note *r*). Likewise Tilquhillie (RSC88) where the recumbent and west flanker remain according to Barnatt (1989, 304, site 6:94).

24 Sites RSC4, 8, 51, 79, and 98.

25 Rappla Wood (RSC22) and Tilquhillie (RSC88). The other two of the four sites listed as 'not found' in Ruggles 1984c, table 1 are included here amongst those listed as destroyed or unrecognisable.

26 Sites RSC16, 19, 20, 55, 60, 62, 69, and 81. At three of these—Pitglassie (RSC19), Nether Corskie (RSC69), and Potterton (RSC81)—the form of the monument could still be ascertained, and this information has been included in List 3.

27 Ruggles 1984c, section 4 describes the fieldwork procedure for obtaining the new data. Ruggles 1984b, section 4.3.3 also contains a description of some of the problems involved.

28 For many more examples see the plans in Thom, Thom and Burl 1980, 163–239.

29 Ruggles 1984c, S64.

30 This reproduces the information in Ruggles 1984c, table 2. Although Centre Line and Perpendicular Line azimuths are quoted to 0°·5 in the table, in many cases this figure is subject to considerable uncertainty for one reason or another. Some of the reasons are documented in the table. For a commentary see *ibid.*, section 4.

31 See also Ruggles and Burl 1985, S32.

32 Burl had suggested that Mither Tap (NJ 683224)—a prominent nipple-shaped eminence at the eastern end of the Bennachie Hills, which are situated amidst the central concentration of RSCs—was a focus of attention (cf. Burl 1979a, 9, 20, 24, 134), as was Mormond Hill (NJ 981570) to the north (*ibid.*, 25). According to Hampsher-Monk and Abramson (1982, 16), the sites are 'clustered around the separate peaks of Mither Tap and Mormond Hill'. However, there is no evidence that these particular peaks, if they were important, influenced the orientations in a systematic way: while Mither Tap is conspicuous from thirteen RSCs, not a single one of these monuments is aligned upon it (Ruggles and Burl 1985, S50 and table 7).

33 Such a hypothesis has been put forward to explain similar patterns of orientation in Swedish tombs (Lindström 1997, 119–21).

34 For further details see Ruggles 1984c, S63.

35 For the equivalent diagram using Centre Line data see Ruggles 1984c, fig. 4(a).

36 The zone of avoidance is narrower (little more than 10°) in the Centre Line case, providing some support for the conclusion that it was the direction perpendicular to the recumbent stone, rather than the axis through the centre of the circle, that was of particular symbolic significance. But see Ruggles and Burl 1985, S33–4.

37 There is, on the other hand, no evidence of a preference for particularly distant horizons.

38 Ruggles 1984c, S72–6. Only at Dunnideer (RSC42) does the recumbent appear to have obscured the horizon to the SSW. In a few cases, such as Midmar Kirk (RSC71), all three stones are well below the horizon.

39 In addition, Burl's original interpretation depended critically upon a few sites whose principal axes yielded azimuths around 155° and 205°. See Ruggles and Burl 1985, S34–5.

40 The raw azimuth and declination data may be found in cols 4 and 10 of tables 5.2 and 5.3. For the raw altitude data see Ruggles 1984c, tables 3 and 4.

41 The RSCs concerned are Berrybrae (RSC5), Loanend (RSC46), Hatton of Ardoyne (RSC56), Midmar Kirk (RSC71), and Sunhoney (RSC72).

42 Ruggles 1984c, S76. The azimuths, altitudes and declinations of the horizon at points directly above the centre and ends of the recumbent, and cut by or directly above the inner edges of the flankers, as viewed from these two positions, are given in full in Ruggles 1984c, tables 3 and 4. Graphical representations of the declination data appear in *ibid.*, figs 2 and 3. For further discussion see Ruggles and Burl 1985, S35–9.

43 A total of 49 of the 64 RSCs visited are involved, these being the sites where the ends of the recumbent stone and/or of the inner edges of the flankers were in sufficiently good condition to enable the ends of the indicated horizon ranges to be identified.

44 At Auchquhorthies (RSC86), for example, the maximum declination in

the range is −33°·4 (Centre Line) or −33°·2 (Perpendicular Line); at The Camp (RSC94) it is −33°·4 (Centre Line), and at South Fornet (RSC68) it is −32°·6 (Perpendicular Line).

45 The most marginal case is Hatton of Ardoyne (RSC56), where the minimum declination in the range for Perpendicular Line is −19°·6; however, the Centre Line data for this circle yield a range between −22°·6 and −25°·4.

46 This is clear, for example, at Auchmachar (RSC10), Balquhain (RSC62), Cothiemuir Wood (RSC48), Easter Aquorthies (RSC63), and Whitehill (RSC73). At Old Keig (RSC49) the 50-tonne recumbent seems to have been pulled uphill from the Don Valley 6 km away. In one case, Auchmallidie (RSC15), the 3 m-long recumbent and a 2 m-long flanker were made entirely of quartz. At North Strone (RSC54) the circle stones are of alternating pink and grey granite, whereas the recumbent is of quartz (Ruggles and Burl 1985, S28–9).

47 *Ibid.*, S29. For data on flat-topped recumbents see *ibid.*, table 6, col. 3 and S46–7.

48 Burl 1970, 63.

49 Ruggles and Burl 1985, S28, S55. See Table 5.5 for the azimuth and declination data from the eight RSCs with cup marks considered in the 1981 project.

50 Ruggles and Burl 1985, S28. On the Berrybrae excavation see Burl 1979a, 27–31.

51 Burl 1979a, 21.

52 Burl 1970, 69.

53 This is true with only two exceptions, Easter Aquorthies (RSC63) and Esslie the Greater (RSC90), which are both located just to the north side of a ridge (Ruggles 1984c, S63). The trend is reflected in horizon scans for absolute azimuths (*ibid.*, fig. 1; Ruggles 1984b, fig. 10) and may be interpreted as a consequence of the preference for non-local horizons over the recumbent stone which always had to be placed facing a southerly direction.

54 Ruggles and Burl 1985, S45–7 and table 6. The criteria for identifying a conspicuous hilltop were subjective, though.

55 This theme is explored in more detail in chapter nine.

56 Childe 1940, 100–1. However, since he speaks of heliacal risings, which involve looking at the eastern sky before dawn, it seems that Childe envisaged the observer viewing from outside the circle towards the north-east.

57 It also implies that those such as Lockyer (1909, 378–411) who investigated astronomical possibilities in the direction viewing from the recumbent stone towards the interior of the ring, were missing the true intention of the builders.

58 In those cases where the recumbent stone was more towards the south-west than the south, the light of the rising midsummer sun would have shone directly across the monument to light up the recumbent stone from the north-east at the following dawn. Perhaps this was also of significance.

59 Cf. chapter four, note 61. For other evidence in support of this interpretation see p. 155.

60 'Lunar orientation and the presence of a conspicuous hilltop seem to [have been] separate goals, which could often be achieved together' (Ruggles and Burl 1985, S47).

61 Of course, such patterns must be approached with caution since further cup marks, perhaps well worn and almost imperceptible, may be lurking still to be discovered, and yet others may have disappeared without trace.

62 Ruggles and Burl 1985, S55. In the other four cases—Pitglassie (RSC19), Balnacraig (RSC76), Nether Corskie (RSC69), and Potterton (RSC81)—azimuth figures are unobtainable because the centre of the ring is unknown.

63 Convincing as they may seem, there are too few data for a strong argument to be made on statistical grounds.

64 This raises the question: at how many RSCs could this not be done because no stone coincided with the limiting directions?

65 Cf. Ruggles 1984c, table 1, col., 15. For a discussion see Burl 1970, 65–7; 1976, 168, 172, 179.

66 Burl 1970, 61–5; 1976, 171–2 and fig. 28. He also cites differences in the artefacts (pottery and axe-heads) found at the sites to support this conclusion (cf. Burl 1970, table 1).

67 For instance, the internal cairns may well represent secondary features, something that would be consistent with the Early Bronze Age dates from Berrybrae and the idea that the circles themselves were constructed in the Neolithic (cf. Shepherd 1989, 124). Indeed, if this were not the case the cairns would have made it difficult, if not impossible, to observe the moon in the way we have described.

68 Ruggles and Burl 1985, S50–4.

69 Cf. Ruggles and Burl 1985, S30.

70 *Ibid.*, S29. It has also been suggested (*ibid.*, S43–4), on the basis of cumulative histograms of the azimuth data in Tables 5.2 and 5.3 (*ibid.*, figs 12–15), that there was a desire to place the recumbent stone due south of the ring centre, and this might have conflicted with the desire to place the recumbent and flankers symmetrically in the circle.

71 See note 5.

72 At the following nine RSCs there is no documentary evidence that circle stones ever stood (Ruggles and Burl 1985, S30): Clochforbie (RSC3), Auchmallidie (RSC15), Cairnton (RSC26), Dunnideer (RSC42), Wantonwells (RSC43), Braehead (RSC44), Druidsfield (RSC52), South Ley Lodge (RSC67) and Nether Corskie (RSC69). (At Nether Corskie this assumes that the remaining two standing stones are flankers, which is not certain.) At Potterton (RSC81) the evidence is equivocal (*ibid.*, note 73).

73 There are also some sites where nothing but a single recumbent stone, placed on the edge of a cairn or ring cairn, may have been erected. Examples are Blue Cairn (RSC78), The Camp, Montgoldrum (RSC94), and The Cloch (List 3, supp. list A).

74 The term 'axial stone' was coined by Ó Nualláin (1984a, 3). A total of 111 examples are now known (Walsh 1997, 11).

75 Burl (1976, 213) opens a discussion of the Irish monuments with the simple statement 'The stone circles of Cork and Kerry . . . belong to the RSC tradition of north-east Scotland'. We shall use the term 'axial stone circle' here so as to avoid prejudging the nature of the linkage between the two groups of monuments.

76 Ó Nualláin 1988, 3.

77 As with the RSCs, there are many morphological variations amongst the ASCs: several have central stones, several have (so-called) boulder burials, and some have outlying uprights. For an overview, and a discussion of the similarities and differences between the Irish ASCs and the Scottish RSCs, see Burl 1976, 213–24. See also Seán Ó Nualláin, 'The stone circle complex of Cork and Kerry', *JRSAI*, 105 (1975), 83–131, p. 115; Ó Nualláin 1984a, 3–7.

78 Lynch (1981, 66) obtained a date of 970 ± 60 uncal BC (GrN–9173) from charcoal underlying stony soil used to level the site prior to the construction of the circle, together with another of 715 ± 50 uncal BC (GrN–9172) from charcoal in the stony soil itself, although the latter could be a later intrusion.

79 This view is corroborated by a recent date of 790 ± 80 uncal BC (OxA–2683), corresponding to between about 1100 and 800 BC, obtained from charcoal scraped off pot-sherds found in a central pit cremation at Drombeg (ASC52) (William O'Brien, 'Boulder burials: a Later Bronze Age megalith tradition in south-west Ireland', *JCHAS*, 97 (1992), 11–35, p. 33). Two samples of charcoal from Reanascreena South (ASC48) have yielded similar dates (*ibid.*, 33–4).

80 Relevant azimuth information can be found in col. 4 of Table 5.6. For an orientation plot see Ó Nualláin 1984a, fig. 24. For a comparison with the Scottish RSC orientation data see Burl 1982, 148 and fig. 1.

81 Cf. Burl 1976, 213.

82 Ó Nualláin 1984a, 11–30. He terms these monuments 'multiple-stone circles' to distinguish them from the diminutive five-stone circles, which were not considered in this first reconnaissance. For a list of five-stone circles see *ibid.*, 30–45.

In fact, a number of monuments have come to light in recent years (I am grateful to Paul Walsh of Ordnance Survey Ireland for up-to-date information), and these are included in List 4 for completeness.

83 The circle at Ardgroom Outward SW (ASC34) was included because our interpretation of the evidence—that the supposed axial stone had been misidentified in earlier publications and that the axial stone is in fact missing (see List 4, note 3)—only developed as a result of our site visit and survey in 1994. However, other RSCs with a missing axial stone were not visited, so this one should arguably be removed from the statistical sample for consistency.

84 Surveys were aided by the use of a gyroscopic attachment that obviated the need for breaks in the cloud cover which are a prerequisite for timed observations of the sun (see Appendix, note 10).

85 The raw data are yet to be published.

86 For a plot of indicated declinations in the form of a 'curvigram', see Ruggles and Prendergast 1996, fig. 1. Note that some two-thirds of the thirty-one ASCs for which data are given in Table 5.6 are subject to some degree of uncertainty.

87 For a plot in the form of a 'curvigram', see *ibid.*, fig. 2.

88 For comparison, the data on conspicuous summits in the opposite direction, from the axial stone towards the portals, are also given in Table 5.7.

Such summits only occur at three of the circles, and their declinations, +38°, +12° and +2°, are entirely without significance.

89 e.g. Barber 1973, table 2; Jack Roberts, *The Stone Circles of Cork and Kerry: an Astronomical Guide* [map], Key Books, Skibbereen, 1993; Roberts 1996. Barber (1973, 32–3) presented statistical arguments to show that twelve hits on solar or lunar targets out of thirty possibilities were unlikely to have occurred fortuitously, but these were fallacious (Freeman and Elmore 1979, S90–3 and Heggie 1981a, 183).

90 Barber 1973, 33–4, echoed by Burl 1976, 176–8.

91 The declination limits of Venus are around ±26° (for details see North 1996, appendix 5) so declinations of around −31° and −32° cannot be explained in terms of alignments upon it (*pace* Barber).

92 e.g. Somerville 1923, 210–12; 1930, 72–4; O'Kelly 1989, 235; Ronald Hicks, 'The year at Drombeg', in Aveni 1989a, 470–82, pp. 471–2; Roberts 1996, 8–9.

93 As determined during fieldwork in 1994, the azimuth and altitude of the notch viewed from a position 5 m to the north-east of the portal stones on the axis of symmetry are 227°·2 and 2°·6 respectively, yielding a declination of −22°·9, about 0°·7 greater than that of the upper limb of the setting solstitial sun in the early second millennium BC. In other words, the sun would actually have set somewhat to the left of the notch at the solstice itself, only reaching it about a fortnight before and after (see Astronomy Box 3). The azimuth of the axis of symmetry, as estimated to the nearest 0°·1, is 226°·8 (cf. Table 5.6). Somerville (1930, 73) gives the azimuth of the axis of symmetry as 225° 50′, but see Jon D. Patrick and Peter R. Freeman, 'Revised surveys of Cork-Kerry stone circles', *AA* no. 5 (*JHA* 14) (1983), S50–6, table 2.

94 The data do not even support simple hypotheses relating to the setting sun at random days throughout the year, or even just during the winter half (Barber 1973, 36–7; Heggie 1981a, 183).

95 Burl 1982, 159–60.

96 See chapter six.

97 Ó Nualláin 1988 lists 103 pairs of standing stones in Counties Cork and Kerry: 85 in the main list, pp. 241–50; 12 'anomalous pairs' on pp. 250–2; and 6 in an additional list, pp. 252–3. In addition, several more examples have come to light since Ó Nualláin's publication (Seán Ó Nualláin and Paul Walsh, priv. comms), some of which are listed in new regional inventories such as Power 1992, 1994, and O'Sullivan and Sheehan 1996.

98 There is no single list covering the whole area, but (for example) Power 1992 lists 344 standing stones and possible standing stones in West Cork alone, and Cuppage 1986 lists 69 extant stones, together with 23 cases reported but now destroyed, in the Dingle peninsula.

99 Ó Nualláin 1988, table 5, incorporates data from ninety-nine wedge tombs in Counties Cork and Kerry.

100 Of sixty-nine rows whose orientations are plotted by Ó Nualláin (1988, fig. 2a) (note that a few examples from outside Cork and Kerry are included), all but eleven fall between 20°/200° and 75°/255°, and all without exception fall between 350°/170° and 100°/280°. Of sixty stone pairs (*ibid.*, fig. 2b), all but two fall between 6°/186° and 88°/268°.

101 See Seán Ó Nualláin, 'Stone circles, stone rows, boulder-burials and standing stones', in Michael Ryan (ed.), *The Illustrated Archaeology of Ireland*, Country House, Dublin, 1991, 89–92, p. 91.

102 According to Ó Nualláin (1988, 190) this is true of all 460 or so wedge tombs in the whole of Ireland. For an orientation plot of 81 examples in Counties Cork, Kerry, Limerick and Tipperary see de Valera and Ó Nualláin 1982, fig. 36.

103 Lynch 1981, 65–6.

104 Cf. *ibid.*, 73.

CHAPTER 6

1 Heggie 1981a, 138.

2 Burl 1988a, 200.

3 These parameters can conveniently be taken as defining a short stone row (cf. Burl 1993, 147), thus distinguishing them from long rows and avenues, which do not share the same geographical distribution (see Archaeology Box 2).

4 Burl 1993.

5 For overall distribution maps see Burl 1993, figs 37, 38, and 41.

6 Burl 1993, 147–51. See also Ruggles and Burl 1995, 517–18.

7 At Cashelkeelty (CKR66 in List 5), for example, excavation showed that what now appears as a three-stone row originally comprised four stones (Lynch 1981, 66).

8 On the basis of radiocarbon evidence from the vicinity of three Cork–Kerry rows, Cashelkeelty (CKR66), Maughanasilly (CKR63) and Dromatouk (CKR45), Lynch (1981, 73) suggests that they were constructed within the period 1400–1000 uncal BC, i.e. between roughly 1700 and 1100 BC. Martlew and Ruggles (1996, 126) obtained an even later date of 930 ± 60 uncal BC (OxA-3880), equivalent to about 1250–900 BC, from charcoal within a stonehole at Ardnacross (northern row) in Mull (SSR24 in List 6) (see also chapter seven).

In the 1970s MacKie obtained a much earlier date of 2860 ± 270 uncal BC (GX-2781) from Blanefield in Stirlingshire (SSR16) which suggests construction around the mid-fourth millennium BC (MacKie 1973; see also Euan W. MacKie, *Scotland: an Archaeological Guide*, Faber and Faber, London, 1975, 115; RCAHMS 1994, 34). An editorial note following MacKie 1973, however, points out the possibility that what appears to be a five-stone row is actually the ruined façade of a chambered tomb, an argument that the Early Neolithic dates would seem to support (Burl 1993, 8–9, 249).

9 Ruggles and Burl 1996, 518.

10 Cf. chapter ten, note 38.

11 This is not entirely true, because of the work of Ann Lynch in the late 1970s (Lynch 1982) which will be discussed later, but it is certainly true relatively speaking, in comparison with the Scottish rows.

12 This relative inattention was actually a major motivation for the new fieldwork (Ruggles 1994a, S2).

13 *Ibid.*

14 The main reason for this (apart from the earlier critiques of high-precision astronomical alignments elsewhere) was simply the ways the rows were constructed: 'The stones are not set with their long axes in precise alignment and indeed in many cases serious deviations occur' (Ó Nualláin 1988, 180).

15 For the results of the first season's fieldwork see Ruggles 1994a. Ruggles 1996 presents the data from the three-stone rows and examines them in conjunction with the earlier data.

The division into two categories includes a small number of misidentifications, such as Cashelkeelty, which was classified as a three-stone row. Up-to-date information is given in List 5 and is reflected in Fig. 6.2. However, the original division is retained in the discussion that follows in order to show the results as originally obtained from the smaller sample as well as from the surveyed monuments as a whole. In this context it is emphasized that the rows with four or more stones were not, and should not be, viewed as a different class of monument from their three-stone counterparts, but merely as a sample more likely perhaps to reveal trends closer to the original intention.

16 These comprise sixty-nine in the main list (Ó Nualláin 1988, 231–40) plus four in an additional list (*ibid.*, 252). The investigation was restricted to the counties of Cork and Kerry for logistical reasons.

17 Aubrey Burl, 'A county concordance. The stone rows of Britain, Ireland and western Europe', in Thom, Thom and Burl 1990, 421–540.

18 Burl 1993. For the gazetteer of four- to six-stone rows see pp. 245–9; for three-stone rows see pp. 250–7.

19 Cuppage 1986; Power 1992; 1994; O'Sullivan and Sheehan 1996.

20 See also Paul Walsh, 'In circle and row: Bronze Age ceremonial monuments', in Shee Twohig and Ronayne 1993, 101–13, fig. 11.1.

21 See Ruggles 1994a, table 1 and Ruggles 1996, table 1 together for the complete list.

22 Two—Barrahaurin (CKR22) and Derrymihin West (CKR77)—were known to have been destroyed, and the other eighteen were three-stone rows where fewer than three stones were known to remain standing (see Ruggles 1996, table 1 for a list of these). In addition, Garryduff (CKR16) was omitted because it only appeared in a supplementary list. On the other hand, three out of eleven rows where one stone has fallen but the other two remain standing—Ballygarret (CKR5), Gortnagulla (CKR19), and Canrooska (CKR64)—were also visited because time and logistics happened to permit. The data collection and reduction procedure should have ensured that these data were selected fairly (Ruggles 1996, S55–9). Finally, at Kippagh (CKR11), it was found that only one stone now remained upright, but the stoneholes from the other two stones remained visible, and it was possible to determine the original orientation (*ibid.*, table 1, note *h*).

23 These were Kildreelig (CKR55), Derrynagree (CKR56), Monavaddra (CKR58), and Piercetown (CKR62).

24 At Dromdrasdil (CKR68), the orientation could not be accurately determined because the stones were surrounded by a thicket of holly and brambles. However, a survey was carried out in order to measure a hill summit in the general direction of indication.

25 The rows concerned are Beenalaght (CKR12), Rooves Beg (CKR48), Behagullane (CKR67), and Ardrah (CKR70). At Rooves Beg only one stone remains standing and the exact orientation is uncertain. In the other cases the relevant horizons were either obscured by trees and

buildings, or else too close to warrant a survey (see Ruggles 1994a, table 1).

26 Calculated data from Behagullane and Ardrah are included in Table 6.2.

27 These are Kippagh (CKR11) at 356°/176° and Cloghvoula (CKR3) at 0°/180°.

28 These are Canrooska (CKR64) at 78°/258°, Cashelkeelty (CKR66) at 83°/263°, Curragh More (CKR25) at 87°/267°, Gneeves (CKR9) at 94°/274°, and Derrynasaggart (CKR34) at 106°/286°.

29 The tallest stone is usually at one end of the row (Ó Nualláin 1988, 180), although the stone height gradation is not always monotonic, and there are some anomalous cases, such as the six-stone row at Beenalaght (CKR12) where the stones are graded upwards towards each end (Ó Nualláin 1988, fig. 29). In many cases most or all of the stones appear broken off, so that attempting to identify the original height gradation is little more than pure guesswork (Ruggles 1984a, S9).

30 Azimuth, altitude and declination data in an apparent direction of indication, as determined in this way, are highlighted in the tables using bold type. They are also marked in the figures.

31 Of the twenty rows where an apparent direction of indication was determined in the first stage of the project, it is to the SW in eleven cases and to the NE in nine. The second stage increased the numbers to 24 to the SW and 16 to the NE. Full details, with notes, can be found in columns 9 and 10 of Ruggles 1994a, table 1 and 1996, table 1. The number of SW indications should be reduced by one in both cases to take into account a more recent reassessment of Maughanasilly (see Table 6.1, note *S*).

32 Horizon scan data were collected as an integral part of the project, using the same four distance categories as for the circles (see p. 94).

33 For commentary, see Ruggles 1996, S61, S68. It should be emphasized that these plots represent the 'raw' data, unmodified in any way.

34 Anyway, there are two cases—Cloghboola More (CKR13) and Behagullane (CKR67)—where the horizon is local both ways along the row; it was clearly not crucial here to have a distant horizon along the alignment, even in one direction. The Behagullane row is in a valley where the horizon is close (within 1 km) all the way around, a situation that could easily have been avoided by erecting the stones a few tens of metres away.

35 Thus Somerville (1923, 200) noted summer solstitial sunrise alignments at Pluckanes North (CKR18) and Newcastle (CKR29).

36 Lynch 1982.

37 Cuppage 1986, 5; see also *ibid.*, 37.

38 O'Sullivan and Sheehan 1996, 50 (no. 108), 51 (no. 109), 48 (no. 105) and 52 (no. 114) respectively.

39 While Lynch quotes azimuths, and hence declinations, to a precision of 0°·1, the north point was determined from magnetic measurements (Ruggles 1994a, S2), so the results are only likely to be accurate to within 0°·5 at best (see Appendix). There are also some major discrepancies between some of Lynch's quoted altitudes (which effect the calculated declinations) and the present author's (see Ruggles 1994a, S13 and Table 6.1).

40 Lynch 1982, 212.

41 In Lynch's approach, mean row orientations were taken as the line of best fit to the centroids through still-standing stones.

42 See note 14.

43 On survey procedure see Ruggles 1994a, S7.

44 Where a horizon was less than 500 m distant, or wholly obscured by trees or buildings, it was not surveyed. For details see Table 6.1.

45 See also Ruggles 1994a, S13.

46 This differs somewhat from Ruggles 1994a, fig. 3 and Ruggles 1996, fig. 3, where a lower weighting per degree was used for wider ranges according to a cruder formula.

47 None of these groupings is particularly evident when the data are plotted in the form of a curvigram (see Ruggles and Burl 1995, figs. 1 and 2).

48 For details see Ruggles 1994a, S9.

49 The north-eastern profiles at Beal Middle (CKR1) and Feavautia (CKR2) each contain two hilltops judged equally conspicuous. Details of both are included in the table.

50 This is most clearly seen by examining Ruggles 1994a, table 1 and Ruggles 1996, table 1. The exceptions are Cloonsharragh (CKR4) and Castlenalacht (CKR60).

51 For the same data presented in the form of a curvigram see Ruggles and Burl 1995, figs 3 and 4. In these figures, as in Ruggles 1994a, fig. 4 and Ruggles 1996, fig. 4, the mistaken juxtaposition of the data from the NE and SW indications at Gurranes (see Table 6.2, note *g*) resulted in an indication at declination −19°·1 (−19°) being marked as in the apparent direction of indication rather than opposite to it, and one at +21°·3 (+21°) as in the opposite direction rather than in the apparent direction of indication. These errors have been corrected in Fig. 6.6.

52 Actually, Cashelkeelty was originally a four-stone row—see note 15.

53 Examples of such ceremonies amongst modern indigenous peoples are well enough known: cf. chapter nine, note 82.

54 There are no convincing examples of stone rows in Shetland, Orkney, Caithness, or most of the Outer Hebrides. While a few examples are found in central Scotland (Perth and Stirling) and the south-west (Galloway), the main concentrations are in Argyll.

55 Burl 1993, 245–69. The excluded monuments vary in length from 59 m (Blashaval, N Uist) up to 640 m (the Dunruchan Stones, Perth), and even include Knockrome-Ardfernal, Jura, where three stones in approximate alignment stretch over 1·2 km. For details of these and other sites excluded even though they appear in Burl's gazetteer, see Supp. B to List 6.

56 Of course, the various stones at Callanish were not necessarily erected at a similar time: it is not unlikely that the rows were built considerably later than the internal ring. It is also quite possible that some or all of the rows were shorter at first, more stones being added later. See Ashmore 1995, 34–5. On dating evidence from Callanish see chapter eight, note 120. However, the conclusion is unaffected.

57 A 'slab' was defined as a standing stone that 'has opposite faces both of which are of maximum width at least twice that of the other faces', and an aligned pair corresponds to an indication of class (1b), namely 'two slabs which are, or can reasonably be assumed to have been, oriented along the line joining them' (Ruggles 1984a, 61–2). Considering such aligned pairs as stone rows is consistent with the definition of the latter given in Patrick F. O'Donovan, *Archaeological Inventory of County Cavan*, Stationery Office, Dublin, 1995, 14, as 'two or more standing stones with a shared axis'. The aligned pairs do include *LK* and *AB* at Nether Largie.

58 All rows in List 6 within Mull, Lorn, mid-Argyll, Jura, Islay, and Kintyre are included, with the exception of Craighouse (SSR33) whose orientation was considered too ill-defined to be included in the 1984 project (cf. Ruggles 1984a, table 11.1 [JU6]), and Machrihanish (SSR39) which is destroyed. The 1984 project also covered the Outer Hebrides, but the orientation was too ill-defined to be considered at Carloway (SSR3) and rows in the Uists have been excluded here because of the length criterion, leaving only Airigh nam Bidearan (SSR4).

59 Many single standing stones in Argyll and Mull are wide and flat. For a list see Ruggles 1985, table 2; for their orientations see *ibid.*, table 3.

60 The further horizon is southwards in eleven of the twenty stone rows, northwards in seven cases, and the distances are equal to within 0·5 km in the remaining two. In the case of the seven aligned pairs (the anomalous Lagavulin N aside), the figures are three, three and one respectively (Table 6.3, columns 4 and 12).

61 Amongst the rows, there are eight such hilltops within southwards profiles and seven within northwards ones (Table 6.4).

62 For a combined plot of the Irish and Scottish data presented in the form of a curvigram see Ruggles and Burl 1995, fig. 5. Note, however, that it includes Blashaval and the two NNE–SSW alignments at Nether Largie ('Kilmartin'), all dismissed here as being too long.

63 For a combined plot of the Irish and Scottish data presented in the form of a curvigram see Ruggles and Burl 1995, fig. 6. Note, however, that it includes Blashaval and the two NNE–SSW alignments at Nether Largie ('Kilmartin'), all dismissed here as being too long.

64 A northerly orientation is unusual amongst tombs (see chapter eight), but this group of stone rows may have been different.

65 Examples are Scallastle (SSR26), where the southern declination range is −16°·5 to −14°·0; and Uluvalt (SSR27), where it is −18°·5 to −18°·0. Furthermore, some of the missing data, such as the northerly indication at Scallastle (SSR26), may well yield northern declinations below +24°.

66 The following, now omitted, are all responsible for outlying values: Scallastle, Mull (SSR26), −16°·5 to −14°·0; Uluvalt, Mull (SSR27), −18°·5 to −18°·0; Sannaig, Jura (SSR34), −33°·0 to −32°·0; Achnancarranan, Islay (SSR35), −34°·5; Lagavulin N, Islay (AP7) (no values given in Table 6.3); and Carse, Knapdale (AP8), −33°·0.

67 For the original plots see Ruggles 1985, fig. 3 (lower) and Ruggles 1988b, fig. 9.2. Here the data from Table 6.3 have been used, so that (for example) Achnabreck (see AR31 in List 2) has been omitted. However, the NNE–SSW alignments at Nether Largie have been retained, as they form an important part of the interpretation described.

68 These arguments are developed and the conclusions elaborated in Ruggles 1985 and 1988b.

69 During years close to the 18·6-yearly major standstill, the moon would

approach its extreme southerly declination once a month. The rising and setting moon on these occasions would be most visible when it was close to full, i.e. closest to the summer solstice. See also Astronomy Box 4.

70 At Dunamuck (SSR31/AP6), however, the three-stone row and the aligned pair are the other way round.
71 Ruggles 1985, S125; 1988b, 243.
72 Martlew and Ruggles 1993, S62; 1996, 128.
73 RCAHMS 1994, 34.
74 Aubrey Burl, 'The sun, the moon and the megaliths', *UJ Arch.*, 50 (1989), 7–21. See also Ó Nualláin 1988, 195–7 and references therein.
75 Burl 1993, 147–51.
76 Stewart 1966; Ó Nualláin 1988, 198.
77 Burl 1976, 191–5; Ó Nualláin 1984b, 63–5 and fig. 21; 1988, 198–9.
78 Ó Nualláin 1988, 191–4.

CHAPTER 7

1 Raymond P. Norris, 'Megalithic observatories in Britain: real or imagined?', in Ruggles 1988a, 262–76, pp. 275–6.
2 Thomas 1996, 88.
3 For example, excavations at Stackpole Warren in Dyfed (SR 981950) (D. G. Benson, J. G. Evans, G. H. Williams, T. Darvill, and A. David, 'Excavations at Stackpole Warren, Dyfed', *PPS* 56 (1990), 179–245, esp. 184–5, 189–91) revealed that a single standing stone known as the Devil's Quoit actually stood at the corner of a sizeable artificial platform of small boulders, roughly rectangular in shape. They also showed ample evidence of domestic activity in the vicinity both around the time of construction and later. Other excavations at Cojoux in Ille-et-Vilaine, France (Charles-Tanguy Le Roux, Yannick Lecerf and Maurice Gautier, 'Les megalithes de Saint-Just (Ille-et-Vilaine) et la fouille des alignements du Moulin de Cojou', *Revue Archéologique de l'Ouest*, 6 (1989), 5–29) and at Plas Gogerddan, Dyfed (SN 626835) (K. Murphy, 'Plas Gogerddan', *Archaeology in Wales*, 26 (1986), 29–31) have also shown that unexpected complexities may be revealed by excavating these superficially simple monuments.
4 In certain circumstances, geophysical techniques might be useful in determining the original positions of stones which have fallen or been removed, but they seemed to be ruled out in the case of the northern Mull rows (see below) owing to the extremely shallow soils of this area.
5 Renfrew and Bahn 1996, 67.
6 See p. 59.
7 The terms 'northern Mull' and 'North Mull' refer to the area to the north and west of the narrow neck of land between Salen on the sound of Mull (NM 5743) and Knock at the head of Loch na Keal on the west coast (NM 5439).
8 See Ruggles and Martlew 1989, S139–41; Martlew and Ruggles 1996, 118–20.
9 A series of detailed reports on the project has been published in the *Archaeoastronomy* supplement to *JHA*: see Ruggles and Martlew 1989; Ruggles, Martlew and Hinge 1991; Ruggles and Martlew 1992; Martlew and Ruggles 1993. For general reports see Ruggles and Martlew 1993; Martlew and Ruggles 1996.
10 Ruggles and Martlew 1989, S141–3.
11 The fieldwork was sponsored by Earthwatch and the Center for Field Research, Boston, Mass.
12 Graham Ritchie, 'Early settlement in Argyll', in Ritchie 1997, 38–66, p. 44.
13 Cf. RCAHMS 1980, 12.
14 Cf. Keith D. Bennett, 'A provisional map of forest types for the British Isles 5000 years ago', *Journal of Quaternary Science*, 4 (1989), 141–4, fig. 1, reproduced in Bell and Walker 1992, fig. 6.5, or Ashmore 1996, fig. 6.
15 Ruggles and Martlew 1989, S147; Martlew and Ruggles 1996, 122.
16 None of the rows is intact. At Dervaig S, the only site where all three stones remain standing, at least two appear to have been broken *in situ*, and the top of one has been dragged off to be used as a gatepost.
17 In 1882 (John Duns, 'Notes on North Mull', *PSAS*, 17 (1883), 79–89, p. 83) and again in 1926 (Thomas Hannan, *The Beautiful Isle of Mull*, Robert Grant and Son, Edinburgh, 1926, 191) only a single standing stone was recorded at the site, with two fallen stones lying nearby. A note in Gordon Childe's field notebook of 1942 (held in the NMRS, Edinburgh), however, records two stones standing. The remaining fallen stone was re-erected at some time between then and the 1960s when the present owners arrived at Glengorm Castle. A ring of boulders has also been placed as an embellishment around the stones.
18 The other monuments comprise a four-stone setting at Tenga towards the south end of Loch Frisa (ML13 in List 2); a pair of prostrate stones (Calgary, ML8) and two single standing stones (Lag, ML6 and Cillchriosd, ML7) on the remote Mornish peninsula in the far north-west; a single standing stone at Tostarie (ML14) on the west coast; and a single standing stone at Killichronan (ML15) in the south, close to the narrow neck joining northern Mull to the rest of the island.
19 Ruggles, Martlew, and Hinge 1991, S60–1.
20 The landscape of northern Mull is characterised by lines of broad, flat-topped ridges running NW–SE falling in terraces down to the intervening valleys of the rivers Bellart and Aros (Jermy 1978, 3.4). Quinish, Dervaig N and Dervaig S all lie on the western side of a ridge running inland from Quinish point, Maol Mor being located on its crest; and all four are oriented NNW–SSE. On the other hand, Balliscate and Ardnacross are situated on hillside terraces near the eastern coast overlooking east–west valleys, and yet Balliscate is oriented north–south and the Ardnacross rows NNE–SSW (cf. Ruggles 1984a, 243).
21 This was originally conceived as a test of the 'primary–secondary' orientation hypothesis (see chapter six) at settings of standing stones that were in a particularly poor state of repair (Ruggles and Martlew 1989, S141).
22 *Ibid.*, S144–5; Martlew and Ruggles 1996, 121–2.
23 Ruggles and Martlew 1989, S145–6; Martlew and Ruggles 1996, 122.
24 Ruggles and Martlew 1993, 186; see also Martlew and Ruggles 1996, 125–6.
25 It is possible on the one hand that this event was contemporary with the construction of the kerb-cairns (Martlew and Ruggles 1993, S61); on the other it could have occurred in Medieval times (cf. Martlew and Ruggles 1996, 125–6) or even later.
26 Martlew and Ruggles 1993, S59–60.
27 At Glengorm only a thin layer of topsoil separated the bases of the stones from bedrock and, moreover, a possible cairn adjacent to the standing stones turned out to be of modern origin (Ruggles and Martlew 1989, S146), eliminating the possibility of establishing a chronological relationship between the stones and any prehistoric cairn.
28 See also Martlew and Ruggles 1993, fig. 17.
29 It has been suggested that ards were used only to cut up established turf to allow cultivation with spades, in the manner of the 'ristle' and 'caschrom' used until the twentieth century in Scotland (Fenton 1976, 45). In other words, the ard would leave marks in the subsoil that would not subsequently be disturbed. On the other hand, it is always possible to argue that they could represent 'ritual ploughing', part of a specific ceremonial that preceded the erection of the stones.
30 A date of 930 ± 60 uncal BC (OxA–3880) was obtained from charcoal from the stonehole of the central stone of the northern row (Martlew and Ruggles 1996, 126).
31 This was apparent from the presence of a distinctive dark layer containing many small fragments of charcoal, found right across the site (Martlew and Ruggles 1993, S60–1; 1996, 125). Evidence of clearance by burning had also been found at Glengorm—although here it could not be linked stratigraphically to the stones (*ibid.*)—and at two of the south-west Irish rows, Maughanasilly and Dromatouk, excavated twenty years earlier by Lynch (1981, 71, 73; 1999).
32 Martlew and Ruggles 1993, S61.
33 This setting, damaged by subsequent ploughing, contained two flat stones reminiscent of the side stones of a small burial cist and a further two stones pulled out of position (Martlew and Ruggles 1996, 126 and fig. 8). However, there was no evidence of a burial. Nonetheless it did contain a rare and distinctive decorated copper alloy bracelet (*ibid.*, 126–7 and fig. 9), together with enough charcoal to enable a date to be obtained of 850 ± 65 uncal BC (OxA–3879) (*ibid.*). The presence of the setting was unsuspected prior to the excavation.
34 Evidence of narrow-rig cultivation was found around the southern row, and several of the buried stones across the site had been scored by a metal plough-share (Martlew and Ruggles 1993, S61). There are also 'lazy beds' around the stones (Ruggles and Martlew 1993, 186). These are banks of soil and manure heaped up between roughly parallel ditches or furrows, to increase soil depth and warmth, and to improve drainage. This system of cultivation is known from the Neolithic to the nineteenth century in the Outer Hebrides (Fenton 1976, 7–8). The area around the site is now under pasture.
35 Ruggles and Martlew 1989, S147 and fig. 4; Martlew and Ruggles 1996, 122 and fig. 4.
36 Ben More is situated in the south of the island. At 966 m, it is the highest of a chain of mountains that mark the remnants of a great volcanic caldera (A. R. Woolley and A. C. Jermy, 'Geology', in Jermy and Crabbe 1978, 4.1–4.25, pp. 4.14–4.20), and it forms a prominent landmark from many directions.

37 Ruggles and Martlew 1992, S3. If visible from the row, the summit of Ben More would yield $A = 156°{\cdot}0$, $h = 2°{\cdot}0$, and $\delta = -28°{\cdot}6$. These values are not critically dependent upon the observing position, the declination altering by less than 0°·05 as the observer moves 150 m back along the alignment to the north until the ground falls away again (*ibid.*, S2).
38 This had been noted at Quinish in the first season of fieldwork but was unsuspected at first at the other rows owing to poor weather and local tree cover. It only fully emerged in the third season (Ruggles and Martlew 1993, 189). For details see Ruggles and Martlew 1992, S3. This conclusion is unaltered by any assumptions about vegetation on the intervening hills, which are at some distance in all cases (Ruggles and Medyckyj-Scott 1996, 137).
39 The largest difference occurs at Quinish, where only a single stone stands and the original orientation of the row is in greatest doubt. All the other row orientations are within 10° of the direction of Ben More. The fact that the alignment upon the peak is precise at Glengorm, the only row where the original orientation has been reliably determined by excavation, raises the possibility that the correlation was even closer at the time of construction, but that the current state of the monuments obscures this (Ruggles and Martlew 1992, S3–4).
40 *Ibid.*, S4.
41 Cf. *ibid.*, S8.
42 *Ibid.*, S9.
43 Any purely statistical argument is weakened, however, because we have implicitly selected five out of the seven stone rows by omitting Balliscate and Ardnacross on the east coast. More subjective interpretations of the locations of all seven stone rows follow.
44 See note 18.
45 None is actually on the limit of visibility.
46 Ruggles and Martlew 1992, S10–11.
47 All three stones at Uluvalt are prostrate but the alignment is fairly well defined. See Ruggles 1984a, fig. 7.10, no. 139.
48 It is even tempting to speculate that Ben More was of importance to the builders of Balliscate and Ardnacross, who attempted to align their rows upon it despite the fact that it was invisible behind local horizons to the south and south-west. The azimuth of Ben More from Balliscate is some 15° to the left of the row orientation, and at Ardnacross some 12° to the left of that of the northern row. Even the slab at Cillchriosd is oriented to within 10° of Ben More (Ruggles and Martlew 1993, 195).
49 The areas visible from Ben More, and hence from which the summit of Ben More (or any point in the landscape) is visible, can be determined using the viewshed function itself, although allowance may need to be made for the curvature of the earth (see Ruggles and Medyckyj-Scott 1996, 133) and the height of the observer (cf. Fisher *et al.* 1997, 582–3, and references therein, on viewshed accuracy in general). In order to show the Ben More declinations for the visible areas, this layer can then be overlaid with one where each cell contains the declination of Ben More (Ruggles and Medyckyj-Scott 1996, 135). For an example see *ibid.*, fig. 6–5.
50 As listed by RCAHMS 1980.
51 *Ibid.*, no. 79.
52 For a quantitative analysis of the visible areas from the Bronze Age cairns throughout Mull see Fisher *et al.* 1997.
53 More strictly, sets of control points are chosen by a procedure that generates spatially pseudo-random points within some well-defined region, such as the whole of northern Mull. Certain categories of wholly inappropriate locations could be excluded by restricting the region, e.g. to elevations below a certain limit (Ruggles, Martlew and Hinge 1991, S52–3).
54 For the formal procedure adopted see *ibid.*, S64–7.
55 On the other hand, as we have already noted, their topographic positions are very different (*ibid.*, S55).
56 When they undertook this exercise, the volunteers were unaware of the detailed agenda for the ensuing project. As a result we can be confident that their choices, while subjective, were uninfluenced by any of the symbolic factors subsequently considered, such as the presence of prominent hills or the astronomical possibilities.
57 None was identified at Maol Mor owing to extensive afforestation.
58 For detailed descriptions see *ibid.*, S55–8.
59 The categories, A, B, C, and D, are as used as in the previous two chapters. See p. 94.
60 The distribution of horizons in each distance category between the two samples was compared using a Kolmogorov–Smirnov one-sample test. The null hypothesis that the two samples are taken from the same population—in other words there is no significant difference between the distribution of visibility from an actual monument and that from a control point—was rejected at the 1% level for categories B and D and at the 5% level for category A (Ruggles, Martlew and Hinge 1991, S67–8).
61 *Ibid.*, S68.
62 *Ibid.*, S72–4 (appendix). The rich alluvial soils in the base of this wide glaciated valley attracted agriculture and settlement from Neolithic times onwards.
63 *Ibid.*, S68–9.
64 Cf. *ibid.*, S61.
65 For more details see *ibid.*, S69–70.
66 The local analyses support this conclusion. See *ibid.*, S70–2 and fig. 11.
67 See also p. 152.
68 For a methodological discussion see Ruggles, Martlew, and Hinge 1991, S54.
69 Although the determination was subjective, it was undertaken prior to any survey and hence cannot have been influenced by the astronomical possibilities. For the purposes of this exercise, the orientations of the rows were also disregarded (Ruggles and Martlew 1992, S4). For the details see *ibid.*, S4–6.
70 *Ibid.*, S12. At Ardnacross, however, the orientations of the northern and southern rows are respectively some 50° and 60° away from Beinn Talaidh. One suggestion is that, exceptionally, these rows were oriented upon the setting moon after it had arisen over a prominent peak (Beinn Talaidh and Sgurr Dearg respectively) rather than upon the peak itself (Martlew and Ruggles 1993, S62–3; 1996, 128–9).
71 Ruggles and Martlew 1992, S7.
72 Around the time of major standstill, when the moon would be rising in the vicinity of Ben More, it would rise a short time after sunset. Around minor standstill it would rise before sunset some way to the left of Ben More, moving across and over the mountain in the darkening sky.
73 Ruggles and Martlew 1992, S7–8.
74 See *ibid.*, table 10.
75 This recalls the RSCs further to the east, with their pattern of orientation between SSE and WSW emphasizing exactly the same range of directions, and a variety of evidence indicating a possible association with the moon (chapter five). Cf. Ruggles, Martlew and Hinge 1991, S72.
76 Jermy 1978, 3.4.
77 Although the agricultural potential of the land in northern Mull has decreased since Neolithic times, the island's relief, climate and soil all suggest that it was never very great (Ruggles, Martlew and Hinge 1991, S60).
78 The wide Bellart valley, stretching inland to the south-east from Dervaig, shares these attributes.
79 But see note 29.
80 The exact rising position would have depended upon the time in the 18·6-year node cycle and the number of days from the summer solstice. Cf. Astronomy Box 7.
81 This is because the declination of Ben More in this area is never more than about three degrees above the theoretical major standstill limit. Only in years around the time of major standstill would the midsummer full moon generally have risen further to the right.
82 Cf. Ruggles and Burl 1995, 525.
83 See chapter nine.
84 A related interpretation would be in terms of movement through the landscape, monuments being erected at points where a sacred mountain first comes into view when moving along traditional paths. Similar ideas have been put forward, for example, concerning the locations of rock engravings in relation to Mount Teide on Tenerife (José J. Jiménez, César Esteban, J. Víctor Febles, and Juan A. Belmonte, 'Archaeoastronomy and sacred places in Tenerife (Canary Islands)', in Jaschek and Atrio 1997, 185–90, p. 186).
85 Martlew and Ruggles 1996, 125. There is no evidence that any were deliberately worked.
86 Quartz appears in a range of sizes, from scattered small fragments to large boulders, and has been found in association with, or as a structural element of, a variety of Neolithic and Bronze Age monuments. These include not just stone rings (see many entries in Burl 1995; also cf. Burl 1976, 164) and rows (Burl 1993, 173, 183, 190) but also single standing stones (e.g. Williams 1988, 71), cists (*ibid.*, 87), and a variety of types of tomb and cairn, from Neolithic chambered tombs (e.g. Powell *et al.* 1969, 206) to Bronze Age kerb-cairns (e.g. Graham Ritchie, 'Monuments associated with burial and ritual in Argyll', in Ritchie 1997, 67–94, p. 89). For further examples see O'Brien 1994, 223. A connection has been suggested between quartz associations and early copper mining, but see *ibid.*, 223–5.

87 Here, where a five-stone circle is found in association with a pair of standing stones and a cairn (Ó Nualláin 1984a, no. 54 and fig. 19; 1988, no. 82 and fig. 59), L. S. Gogan ('A small stone circle at Muisire Beag', *JCHAS* 36 (1931), 9–19, p. 19) reported finding 'a great number of quartzite stones . . . chiefly about the portal'.
88 Burl 1993, 152.
89 Interestingly, at Maughanasilly (CKR63) a distinct concentration of quartz pebbles was found at the base of the tallest stone, on its south-west side (Lynch 1999), and the row is aligned with moonrise close to the major limit in the north-east. As at the RSCs (see chapter five), quartz pebbles were scattered in front of a stone over which the moon was apparently viewed.
90 Cf. pp. 88 and 98; also Ruggles and Burl 1995, 526. The Connemara rows, however, are predominantly oriented north–south (Burl 1993, 170, 192).
91 RCAHMS 1994, 34.
92 This is because the row orientations are only correlated with the direction of Ben More at a much lower level of precision. In addition, the fact that the mountain is on the margins of visibility further obfuscates the issue when following this approach: at Glengorm, for example, the peak does not appear in horizon profiles drawn from just behind the stones (Ruggles and Martlew 1989, fig. 4; Martlew and Ruggles 1996, fig. 4). Amongst the earlier data from the 1984 project, Ben More only appears in the horizon profile at Uluvalt (Ruggles 1984a, fig. 7.10, no. 139).
93 Many of them appear in Ruggles 1984a. See also Ruggles and Martlew 1992, S2.
94 Ruggles, Medyckyj-Scott, and Gruffydd 1993, 127; Ruggles and Medyckyj-Scott 1996, 132–4.
95 For example, the ecological and economic factors can be combined into a 'site suitability' layer expressing the suitability of a given location before factors relating to the visibility of prominent hills or astronomical events are taken into account. See Ruggles, Medyckyj-Scott, and Gruffydd 1993, 127–8; Ruggles and Medyckyj-Scott 1996, 134.
96 Finer resolution brings the potential for greater accuracy but may substantially increase processing times, as raster grids contain more cells (Ruggles, Medyckyj-Scott, and Gruffydd 1993, 130).
97 Cf. chapter three, note 60.
98 Martlew and Ruggles 1993, 64; 1996, 130.

CHAPTER 8

1 Richard Gough (ed.), in his editorial additions to *Camden's Britannia*, John Stockdale, London, repr. 1806, 432 [1st edn 1789].
2 Thom 1967, data from tables 12.1, 5.1 and 8.1. The idea of contrasting the former quote and a statement similar to this follows Burl 1988a, 175.
3 Burl 1988a, 202.
4 See, for example, Darvill 1987, ch. 2; Christopher Smith, *Late Stone Age Hunters of the British Isles*, Routledge, London, 1992; for a European perspective see Whittle 1996, ch. 2. On Scotland see Wickham-Jones 1994; on the Western Isles in particular see Armit 1996, ch. 3. On Ireland see Cooney and Grogan 1994, ch. 2.
5 See chapter nine for specific examples.
6 See, e.g., Marshack 1972, chs 3–5, 7, 9–11; 'Cognitive aspects of Upper Paleolithic engraving', *Curr. Anth.*, 13 (1972), 445–77; Marshack 1991.
7 For a critique of Marshack's interpretation and a response see the exchange of views between Francesco d'Errico and Alexander Marshack in *Curr. Anth.*, 30 (1989), 117–18 and 491–500. For commentaries see Anthony F. Aveni, *Empires of Time: Calendars, Clocks and Cultures*, Basic Books, New York, 1989, pp. 66–71; Ruggles 1999. Bone fragments containing incised marks dating from Upper Palaeolithic times are not unknown in Britain: an example from Cheddar, Somerset, dating to *c.* 10,000 BC, contains three linear series of cuts, each broken into widely spaced groups containing four, five or six lines, but the numerology does not readily suggest a lunar explanation (E. K. Tratman, 'A Late Palaeolithic calculator (?), Gough's Cave, Cheddar, Somerset', *Proceedings of the University of Bristol Spelaeological Society*, 14 (1976), 123–9).
8 Parker Pearson 1993, 40. In fact, the activities of Mesolithic people could well have had significant ecological impacts: see, e.g., Paul A. Mellars, 'Fire ecology, animal population and man: a study of some ecological relationships in prehistory', *PPS*, 42 (1976), 15–45; Roberts 1989, 93; Bell and Walker 1992, 154–8.
9 Certainly, the Atlantic fringes of northern and western Europe seem to be characterised throughout the Neolithic by an investment in separate, non-domestic ritual, which is very different from the central European tradition (Hodder 1984, 65 [1992, 74]).
10 Parker Pearson 1993, 41.
11 Hodder 1984, 55–7 [1992, 54–5] and references therein; Magdalena S. Midgley, *The Origin and Function of the Earthen Long Barrows of Northern Europe*, BAR (International Series, 259), Oxford, 1985, 63–74, 186–9; see also Stanisław Iwaniszewski, 'Neolithic uses of time and astronomy: funnel beakers and long barrows in Kuiavia, Poland', in Jaschek and Atrio 1997, 173–84. On the earthen long barrows of southern England see particularly Ashbee 1984, 21–4 and figs 19–22.
12 Their rectangular or trapezoidal shape is the most obvious manifestation of this. See Hodder 1984, 54–6 [1992, 50–4]; Parker Pearson 1993, 42–3.
13 On possible reasons for the remarkable unity of form of the earthen long barrows see, e.g., Julian Thomas, 'Relations of production and social change in the Neolithic of north-west Europe', *Man* 22 (1987), 405–30.
14 Burl 1987a, 26–9.
15 For a distribution map showing all the principal groups of long barrows in Dorset and Wiltshire see Renfrew 1973, fig. 2 [reprinted as Renfrew 1984, 230, fig. 1].
16 Burl 1987a, 28.
17 *Ibid*, 26–9. The quote is from p. 28.
18 Recently, North (1996, ch. 2) has suggested that the long barrows were oriented not upon solar or lunar but upon stellar targets. The main problem with this is the room for manoeuvre in fitting bright stars to orientations when the date is uncertain, especially when a third dimension is introduced by suggesting that the back of the barrow, whose original height can now only be estimated, provided an artificial horizon (North 1994, 2). See also Aveni 1996, 404; Ruggles 1997, 212.
19 According to Burl (1987a, 28) only one barrow is oriented to the north of the lunar arc, whose azimuth limit is around 41°. But in *ibid.*, table 4 four barrows are listed as being oriented NNE, i.e. within the azimuth range 11° to 34°.
20 See chapter ten, note 17. Indeed, Castleden (1993, 42) interprets the same data as solar.
21 e.g. Michael A. Hoskin, Elizabeth Allan, and Renate Gralewski, 'Studies in Iberian archaeoastronomy: (3) Customs and motives in Andalucia', *AA* no. 20 (*JHA* 26) (1995), S41–8.
22 This view is summarised in Thomas 1991b. The quote is from p. 11. For a fuller account see Thomas 1991a, 41–7.
23 Several radiocarbon dates from the cursus itself suggest a date for 'the end of its active maintenance' of around 2600 uncal BC (Bradley in Barrett *et al.* 1991, 51–2), i.e. between around 3400 and 3100 BC. On the dating evidence for the long barrows and other elements of the Dorset Cursus complex, see *ibid.*, 51–3.
24 Penny and Wood 1973. See also Wood 1978, 82–4, 101–3; Castleden 1987, 135–6.
25 Bradley in Barrett *et al.* 1991, 47.
26 Penny and Wood (1973, fig. 14) have given a diagram of the Gussage Cow Down barrow as seen from the Bottlebush Down (which they call the Wyke Down) terminal, annotated with azimuth and altitude lines. It is calculated from OS maps. A theodolite survey by this author in 1997 gave the azimuth and altitude for the left-hand side of the barrow as 230°·5 and 0°·2 respectively, yielding a declination of −23°·9; and 231°·3 and 0°·2 for the right side, yielding −23°·5.
27 Bradley in Barrett *et al.* 1991, 56.
28 Bradley and Chambers 1988, 286.
29 RCHM(E) 1979, 13–5.
30 Castleden 1993, 46. The barrow, now destroyed, is Amesbury 42 on Ashbee's (1984) list. But see Burl 1987a, 45.
31 Despite the impression given, for example, by Bradley and Chambers (1988, 286), Parker Pearson (1993, 62), and Darvill (1996, 254), the alignment is not exactly equinoctial; and indeed Burl (1987a, 44) refers to the orientation as (only) 'rather casually laid out to mark the equinoctial sunrises'. On the matter of supposed equinoctial alignments see pp. 148–9.
32 Bradley and Chambers 1988, 287.
33 In addition, Bradley (in Barrett *et al.* 1991, 56) concludes that 'most of [Penny and Wood's] field evidence is unconvincing'.
34 Bradley in Barrett *et al.* 1991, 56. See also Tilley 1994, 196–7.
35 Bradley and Chambers 1988, 286–7.
36 Bradley in Barrett *et al.* 1991, 56–7. Burl (1987a, 65–71) reaches a similar view in the Stonehenge area.
37 This work is as yet unpublished and this information has been kindly supplied by Fachtna McAvoy, English Heritage (priv. comm., 1996), but Parker Pearson 1993, 65 has a description.
38 The claim is that the posts 'were arranged to form alignments with all 12 most extreme settings and risings of the lunar and solar cycles' together with the equinoxes (David Keys, 'Unearthed: a sun temple that puts

Stonehenge in the shade', *The Independent on Sunday*, 17 February 1991, p. 8).

As at Stonehenge, it is possible to estimate the probability that the observed alignments might have arisen fortuitously. Even making the questionable assumption that the data are mutually independent, and generously assuming (since all the suggested alignments are formed by pairs of posts on opposite sides of the rectangle) that only this type of alignment was considered when formulating the astronomical theory, there are still something like 250 possible alignments at the monument. Given this many random shots at the horizon, the odds of hitting all the targets at least once are at worst about 10 to 1 against and at best better than evens (the exact figure depends upon various assumptions made, such as the precision of the claimed alignments). Thus even without examining the archaeological evidence more closely, as Atkinson did at Stonehenge, it is clear that these alignments do not constitute convincing evidence of the monument having being 'used in a ... sophisticated way ... to calculate astronomical events' (Keys, *ibid.*). See Letters to the Editor, *Independent on Sunday*, 24 February 1991, p. 21.

39 E.g. Krupp 1994, xi. The orientations of some chambered tombs have also been used to support this idea (see note 60 below), as has that of one diagonal of the Station Stone rectangle at Stonehenge (note 160). For a critique of Thom's megalithic calendar see chapter two.

40 See Clive Ruggles, 'Summary of the RAS specialist discussion meeting on Current issues in archaeoastronomy', *The Observatory*, 116, 278–85, p. 285. On the equinox as a concept, see pp. 148–9.

41 Altitude data have been taken into account by the National Almanac Office, who have been working with English Heritage.

42 Particularly Brennan 1983, 82–6.

43 For an overall plan of Dowth see Michael J. O'Kelly and Claire O'Kelly, 'The tumulus of Dowth, County Meath', *PRIA*, 83C (1983), 135–90, fig. 3. For detailed descriptions of the two tombs in the western side of the mound, see *ibid.*, pp. 150–8. See also Eogan 1986, 14–15.

44 O'Kelly and O'Kelly, 'The tumulus of Dowth', 178; Eogan 1986, 178–9. For a plan see *ibid.*, fig. 15.

45 Brennan 1983, 102–7.

46 Brennan 1980, 97–100; 1983, 144. See also Susan M. P. McKenna-Lawlor, 'Astronomy in Ireland from earliest times to the eighteenth century', *Vistas in Astronomy*, 26 (1982), 1–13, pp. 2–7 and fig. 2.

47 An examination of Newgrange using three-dimensional survey, recording and visualisation techniques has uncovered some intriguing interplays between sunlight and shadow and decorative art (Frank T. Prendergast, 'Shadow casting phenomena at Newgrange', *Survey Ireland*, no. 9, 1991, 9–18). In particular, one of the great circle stones (GC1) casts a shadow whose tip at winter solstice would have passed down through the three largest spiral carvings on the decorated entrance stone (K1) at about the time of its (GC1's) construction (*ibid.*, fig. 5a, where it is assumed that the construction of the great circle stones postdated that of the tomb, cf. chapter one, note 10).

48 Brennan 1983, 76; Castleden 1993, 132–3. On the general possibility that many circles and rayed circles in European megalithic art may be sun symbols or representations see Green 1991, 20–8. For a drawing showing all of the decorations on the back corbel of the roof-box see O'Kelly 1982, fig. 52.

49 Brennan (1983, 87–126) also describes a number of other alignments and shadow and light phenomena at cairns in the Loughcrew cemetery and elsewhere: these include the sun at different times of the year shining down tomb passages and/or illuminating carved symbols, and may well reward further investigation. However, there is no attempt to describe the negative data—how many alignments do not fit astronomical explanations, how many decorative symbols are not bathed at solstice or equinox in solar light and shadow. Before such evidence can begin to be taken seriously, an attempt must be made to demonstrate that what are described are highly unlikely to have occurred fortuitously.

50 From an anthropological point of view this is clearly naïve and 'bound to be flawed' (cf. Powell 1994, 86).

51 Renfrew 1985.

52 The actual date of Maes Howe is uncertain—a single radiocarbon date corresponding to around 2500 BC comes from a secondary position in the ditch—but its similarity in structure to the principal building at the nearby settlement of Barnhouse (Richards 1990a; see also chapter ten) suggests that it was built several centuries earlier. For background information on the Orkney chambered tombs see Henshall 1985.

53 e.g. Moir 1981, 223–4; Ritchie 1982, 28; 1985, 127.

54 Burl 1981a, 251. The outer part, however, was oriented much more closely towards the midwinter sunset. For a detailed discussion including a ground-plan and horizon diagram see MacKie 1997.

55 The first mention of this in print appears to have been in a popular book accompanying a television series: Simon Welfare and John Fairley, *Arthur C. Clarke's Mysterious World*, Collins, London, 1980, 93. Shortly afterwards, it was mentioned in the academic literature: e.g. Burl 1981c, 124–6; Moir 1981, 224; Ritchie 1982, 28. As at Newgrange, different interpretations are possible (cf. Lynch 1973, 152): see Davidson and Henshall 1989, 85; Henshall 1985, 113–14.

56 Lynch 1973, 158.

57 These are tombs *L* and *K*. See Michael J. O'Kelly, Frances Lynch and Claire O'Kelly, 'Three passage-graves at Newgrange, Co. Meath', *PRIA*, 78C, 249–352, pp. 263 (site *L*) and 276–7 (site *K*). See also O'Kelly 1982, 125.

58 According to Lockyer (1909, 430–1) the orientation is solstitial, but according to Green (1991, 11) it is towards 'May Day sunrise'. As to whether the latter date is particularly significant see note 60 below. Thomas (1991a, 44) has noted that five postholes found some 4·5 m outside the entrance kerbstones here (Claire O'Kelly, 'Bryn Celli Ddu, Anglesey: a reinterpretation', *Arch. Camb.*, 118 (1969), 17–48, p. 46) are reminiscent of the 'A' holes at Stonehenge (see p. 136), concluding that 'it seems highly likely that [they] record a series of observations upon the rising of some heavenly body in order to ascertain its standstill [i.e. limiting—see Astronomy Box 4] position'. The only possibility is the northern minor limit of the moon, and while the adjacent posts are ranged on the correct side to record the position, say, of the midwinter full moonrise in years before and after the minor standstill, many other interpretations of these posts are doubtless possible.

59 Burl 1983, 26.

60 For example Müller (1970, 104–15), drawing on earlier work by Somerville, describes a number of examples from north-western Europe (for a comment see Heggie 1981a, 184). Following this tradition, Edwin C. Krupp ('Archaeoastronomy', in John Lankford (ed.), *History of Astronomy: an Encyclopedia*, Garland, New York, 1997, 21–30, p. 23) has claimed that Maes Howe, Knowth, Bryn Celli Ddu, and La Roche-aux-Fées and Gavr'inis in Brittany are all oriented on 'season stations of the sun', but it is of course meaningless to list isolated examples selected from many hundreds in order to try to support a particular hypothesis, in this case that the equinoxes and mid-quarter days, as well as the solstices, were significant throughout European prehistory in a continuous tradition lasting from the Early Neolithic right through to modern folk calendars (see also note 39 above). Krupp also includes the Clava cairns in this list, but for the actual evidence see below.

61 For overviews see Burl 1981a; 1983, ch. 5.

62 Burl 1981a, 246.

63 See also Henshall 1972, 99. For a different view see Powell *et al.* 1969, 232–6.

64 See also Davidson and Henshall 1991, 79–80.

65 Orientations of chambered tombs in other regional groups also show strong trends. For example, only twelve of the 114 cairns in Henshall's Orkney–Cromarty group face the western half of the compass (Henshall 1963, 104). An easterly direction also seems to have been preferred, although not universally, by the builders of Maes Howe-type cairns (*ibid.*, 130).

66 Working to a lower level of precision, Fraser (1983) investigated the seventy-six chambered tombs in Orkney, analysing their architectural form, their relationship to other material evidence, their spatial interrelationships and their locations within the landscape, including an examination of their 'orientation of visibility' similar to the analyses of the variation of horizon distance with azimuth described in chapters five and six. He concluded (amongst other things) that the chambered tombs were situated so as to facilitate observations of the sun at various times in the agricultural year. For a commentary see Ruggles 1984b, 272–6. For further work on the island of Eday see David Fraser, 'The orientation of visibility from the chambered cairns of Eday, Orkney', in Ruggles 1988a, 325–36. On the orientations of the Orkney chambered tombs see also Davidson and Henshall 1989, 85–6.

67 Burl 1983, 22. Where we observe a set of orientations spanning the azimuth range between midsummer and midwinter sunrise, we should be careful to distinguish between the idea that it was merely important to orient the tomb within the range of horizon where the sun rises at some time in the year, and the more specific idea that the tombs were oriented upon sunrise on the day of their construction. The latter is not convincing where the distribution of orientations is unimodal and centred upon due east, unless it can also be convincingly argued that tomb construction was concentrated in the spring and/or autumn and tended to avoid times around both midwinter and midsummer (Ruggles 1984b, 277–9).

68 Burl 1981a, 284–6. See also pp. 47–8.
69 Patrick 1975, 13; Burl 1983, 22–4. But if this was the intention the alignments were very imprecise, off target by up to 32°. See also Burl 1981a, 250.
70 Burl 1981a, 256. Actually, two recent, unpublished student projects at the Universities of Reading (Richard Bradley, priv. comm., 1997) and Leicester show little evidence for this: the orientations seem to avoid the most prominent targets.
71 Cf. MacKie 1997, 349–53.
72 However, this is not the case, for example, amongst the Arran tombs (see Burl 1981a, 254–5).
73 This has long been noted; see, e.g., Barber 1973, 26; Burgess 1980, 49.
74 Burl 1981a, 257–65.
75 Somerville 1923, 217–22; Thom 1966, 18; Müller 1970, 109–10; Heggie 1981a, 118. The cairns at Balnuaran of Clava have recently been excavated by Richard Bradley (see Richard J. Bradley, 'Excavations at Clava', *Curr. Arch.*, 13 [no. 148] (1996), 136–42). See also p. 155.
76 Four of the eleven azimuths quoted by Burl (1981a, table 7.3a), including all three less than 180°, are in fact within 8° of due south.
77 In fact, the two passage tombs at Balnuaran of Clava form the middle (solstitial) group, the other nine examples falling into one or other of the lunar groups.
78 Ruggles 1990, 130. This would tend to produce concentrations at either end of the lunar range −30° to −20°, with a sharp cut-off below −30° and a slightly less sharp cut-off above −20°, and lower values in between (Ruggles and Burl 1995, 524–5). The observed data are broadly consistent with this. Note that it does not necessarily follow, in this case, that 'the major-minor cycle of the moon [as such] was known in Neolithic times' (Burl 1981a, 262).
A broader case can in fact be made that the Balnuaran of Clava group were exceptional. They are larger than all the other Clava cairns, are built in a slightly different way, incorporate coloured stonework, and (unlike nearly all the others) are part of a cemetery of at least eight monuments (Richard Bradley, priv. comm., 1997). If this is accepted, the argument becomes much stronger that the builders of the remaining Clava cairns not only oriented them with respect to the moon, but were also aware of, and took into account, the changing position of (say) the midsummer full moon over the nineteen-year cycle, waiting for this to reach a limit before commencing construction.
79 Interestingly, in several cases the roof was too low to allow (sunlight or) moonlight to penetrate along the entire passage to the tomb interior, even when the moon was on the horizon. Cf. Burl 1981a, 262.
80 Thomas 1991b, 11.
81 For an interpretation of the Avebury monuments and landscape in these terms see Julian Thomas, 'The politics of vision and the archaeologies of landscape', in Bender 1993, 19–48, pp. 29–43. On Avebury generally see Burl 1979b; Malone 1989; Peter J. Ucko, Michael Hunter, Alan J. Clark and Andrew David, *Avebury Reconsidered: from the 1660s to the 1990s*, Unwin Hyman, London, 1991.
82 For an account of astronomical speculations at Avebury see Burl 1979b, 200, 215–16. See also *ibid.*, 72, and 131–2 on Silbury Hill. One possible exception is the orientation of the cove (see below).
83 Burl 1982, 154–5 lists examples from all over Britain and Ireland. See also, for example, Hutton 1991, 117–18.
84 Burl 1988a, 176, 184. The group includes the circle at Ballynoe, Co. Down.
85 *Ibid.*, 181–6.
86 *Ibid.*, 197.
87 Of the twelve rings shown in Fig. 8.7a, two (no. 7: Grey Yauds, Cumbria [NY 544486] and no. 9: Lochmaben, Dumfries [NY 312659]) have been completely destroyed. A further three (Elva Plain, Gunnerkeld, and Studfold) are in poor condition and the relevant figures may not be reliable (Burl 1988a, 197). A further example, Mayburgh, is mentioned in the text but not shown in Burl's figure.
88 The first step in addressing these concerns would be to obtain, where available, and include the 'missing' data in Table 8.3. Interestingly, the orientations of a similar variety of special features at timber circles may indicate that the cardinal directions were important at a number of these monuments also (Gibson 1994, 200).
89 Burl 1988a, 198.
90 *Ibid.*
91 Thom and Thom 1978a: 22. For details of the proposed indications see Thom 1966, 22, figs 21 and 39.
92 Thom 1971, 12.
93 Thom 1967, 145–50; the quote is from p. 145.
94 Maud E. Cunnington, *Woodhenge*, Simpson and Co., Devizes, 1929, pl. 3; Thom 1966, fig. 32; 1967, fig. 6.16; Burl 1987a, 123–7.
95 For a summary see Richards 1991, 96–8.
96 Thom, Thom and Burl 1980, 288–9; Burl 1988a, 199–200.
97 Aubrey Burl, 'Stone circles: the Welsh problem', *CBA Report no. 35*, 1985, 72–82. For plans of Druid's Circle see Thom 1967, fig. 6.18; Burl 1976, 270; Thom, Thom and Burl 1980, 372–3 (this is Thom's Penmaenmawr, W2/1). This ring is given special emphasis by Burl (1995, 177) because it aligns with sunset close to May Day, corresponding to the later Celtic festival of Beltane, but as with the Cumbrian circles one must examine a broader sample before jumping to such specific conclusions. In fact, according to Burl, several rings in the region do have their long axis aligned ESE-WNW, which may indicate a clustering around sunset on this date (Burl, 'Stone circles: the Welsh problem', 81–2; 1995, 177).
98 MacKie 1977, 92–4; 1981, 116–28.
99 The possible astronomical significance of outliers as viewed from the ring centres is one that has received attention ever since Thom included such alignments in his earliest data sets (see chapter two), but apart from the Cumbrian circles no discernible regional trends have been sought or discovered (cf. Burl 1979b, 154, 214).
100 Burl 1995, 114–18; Barnatt 1989, 247–50 (nos 4:15 to 4:22). Excavations have been undertaken since Barnatt and Pierpoint's work: see Alison Haggarty, 'Machrie Moor, Arran: recent excavations at two stone circles', *PSAS* 121 (1991), 51–94. 'Circle I' (NR 912 324), for example, consisted of two timber rings surrounding a horseshoe-shaped timber setting facing north-west, probably erected early in the fourth millennium BC, later replaced by an eleven-stone circle.
101 Barnatt and Pierpoint 1983, 108–13. For a more detailed commentary see Ruggles 1984b, 279–81.
102 Cf. Clive Ruggles, comment on Barnatt and Pierpoint 1983, *SAR* 2 (1983), 115–16.
103 John Barnatt, *Stone Circles of the Peak*, Turnstone Books, London, 1978, 176 and fig. 100; see also 54–6. Recent excavations, however, indicate that the stone is not noticeably the main (largest) stone in the ring, as previously thought (John Barnatt, 'Recent research at Peak District stone circles . . .', *Derbys. Arch. J.*, 116 (1996), 27–48, pp. 38–41).
104 Burl 1995, 53 (no. 34).
105 Burl 1976, 205–8.
106 Ritchie 1982, 29–30. See also Ritchie 1985, 122. For a photo see Burl 1976, 181.
107 Ritchie 1976, 12–15. The putative post was not at the centre of the setting.
108 Ritchie 1982, 30. On the other hand, Colin Richards ('Monumental choreography: architecture and spatial representation in Late Neolithic Orkney', in Christopher Tilley (ed.), *Interpretative Archaeology*, Berg, Oxford, 1993, 143–77, p. 174; Richards 1996b, 321) suggests that the central setting was a symbolic hearth.
109 Aubrey Burl, 'Coves: structural enigmas of the Neolithic', *WANHM*, 82 (1988), 1–18.
110 *Ibid.*, 10–11; see also Burl 1979b, 219. Burl has argued more recently that the Avebury cove may have been an 'aggrandised imitation' of the chambers in earlier tombs such as nearby West Kennet (Burl 1995, 84).
111 The presence of a cove here is postulated from excavated features.
112 On Arbor Low and Cairnpapple see Burl 1976, 275–82; on Stanton Drew see *ibid.*, 105; on Avebury see Burl 1979b, 156–8; on Beckhampton see *ibid.*, 44, 52, 191.
113 The azimuths are 147° at Stanton Drew (Burl 1995, 79) and 128° at Beckhampton (Burl 1979b, 255, note 2). On the orientations of the other coves, see Burl 1976, 105.
114 Burl 1979b, 137. Interestingly, the proposed solstitial alignment is actually that of the summit of a hillock (*ibid.*, 255, note 2); all that can be said given the poor present condition of the cove is that it faced approximately in this direction. This once again draws attention to the possibility that structures were not simply built to align upon horizon astronomical events but were carefully located and constructed to correlate with significant topographic features as well.
115 Burl 1979b, 158; 1995, 84. The lunar interpretation bears no relation to the totally insupportable idea of Stukeley's in the 1720s that the north circle at Avebury, within which the cove stands, was dedicated to the moon while the south circle was dedicated to the sun (see Burl 1979b, 49). Note that Malone (1989, 117) wrongly implies that the Avebury cove (rather than the Beckhampton one) is aligned upon the midwinter sun.
116 Burl 1979b, 219.

117 Renfrew, for example, quotes a description of Cherokee festivals at the time of the harvest full moon which 'we can almost imagine . . . as the description of the celebrations at a neolithic henge' (Renfrew 1976, 264–5). See also Knight 1991, 346–7.

118 It is interesting to note that the distribution of henges and RSCs is almost mutually exclusive (Barclay 1997, fig. 8.2).

119 For general descriptions and histories see Patrick J. Ashmore, 'Callanish', in David J. Breeze (ed.), *Studies in Scottish Antiquity Presented to Stewart Cruden*, John Donald, Edinburgh, 1984, 1–31; Burl 1995, 148–51. The title 'Stonehenge of the North' is also popularly applied, for example, to Arbor Low in Derbyshire.

120 Dates have recently been obtained from fifteen pieces of charcoal from Callanish. The stone ring and subsequent cairn date to between 2900 and 2600 BC. The destruction layers contain material of that age and also material dated to between 2000 and 1750 BC, which accords well with the likely age of beaker fragments found in the same layer (Patrick Ashmore, priv. comm., 1997). See also Burl 1995, 150; Ashmore 1996, 73; Patrick J. Ashmore with Ann Macsween, 'Radiocarbon dates for settlements, tombs and ceremonial sites with Grooved Ware in Scotland', in Gibson and Simpson 1998, 139–47.

121 See LH7, 8, 10, 18, 19, 21, 22 and 24 in List 2. For longer descriptions of the stone circles in the Callanish area see Burl 1995, 151–2; Ashmore 1995, 11–18. These other monuments in the vicinity of Callanish have been designated 'Callanish II', 'Callanish III' and so on by various authors (Thom 1967; 1971; Cooke *et al.* 1977; Tait 1978; refs in note 125) and these alternative names are given in List 2, but this nomenclature has also provoked some criticism (e.g. Burl 1995, 148–9). Certainly it would be unwise to presuppose that the different settings of standing stones in this area were necessarily conceived as parts of a single overall scheme rather than possibly reflecting varied activities, evolving and changing over a number of centuries.

122 For the orientations and indicated horizon profiles of each of the rows and of the avenue see Ruggles 1984a, figs 5.5 and 5.6.

123 Amongst them are Lockyer 1909, 342–4; Somerville 1912; and Thom 1967, 122–5. For commentaries see Burl 1983, 13–15—which shows that Callanish provides a good alternative example to Stonehenge of each age interpreting the monument in a way that reflects its own prepossessions—and Burl 1993, 63–5.

124 Gerald S. Hawkins, 'Callanish, a Scottish Stonehenge', *Science* 147, 1965, 127–30, reprinted as Appendix C in Hawkins and White 1970; Thom 1971, 68–9.

125 Cooke *et al.* 1977, 120; Jean-René Roy, 'Comments on the astronomical alignments at Callanish, Lewis', *JRASC*, 74, 1980, 1–9, with comment by Gerald S. Hawkins, *ibid.*, 10–11; Ruggles 1981, 166 (site 9). The Clisham profile is in fact clearly visible from a few metres to the west of the avenue (Ponting and Ponting 1981, 76; Ruggles 1982b, S28). Cnoc an Tursa can be seen in Fig. 8.8e; for other photographs showing it see Ashmore 1995, 34–5.

126 Margaret R. Curtis and G. Ronald Curtis, 'Callanish: maximising the symbolic and dramatic potential of the landscape at the southern extreme moon', in Ruggles 1993a, 309–16, p. 316 and pl. 4.

127 Cf. Ashmore 1995, 40.

128 Ponting and Ponting 1981; Margaret R. Ponting, 'Megalithic Callanish', in Ruggles 1988a, 423–41.

129 Cooke *et al.* 1977.

130 Cleal *et al.* 1995, 57.

131 The NGR and elevation of Robin Hood's Ball are SU 102 460, 132 m; those of Stonehenge SU 1225 4220, 103 m. From this we estimate the azimuth and altitude of Stonehenge from Robin Hood's Ball to be 152° and −0°·4 respectively, yielding a declination of −35°, although Stonehenge would not have been on the skyline. The figures for the reverse direction are 332°, +0°·4, +33°.

Note that the *x*-scales on all area maps in Cleal *et al.* 1995 (figs 33, 35 etc.) are shifted by 1 km to the right, so that 1 km must be added to all eastings: e.g. the position of Stonehenge appears to be SU 1125 4220 but is really SU 1225 4220. Larger-scale maps and plans such as figs 21 and 36 have the scales correctly shown.

132 Cleal *et al.* 1995, 40.

133 *Ibid.*, 476–7.

134 This has emerged from work at English Heritage using GIS with large-scale digital elevation data to explore the visibility potential of different monuments in the Stonehenge landscape (Frances Blore and Miles Hitchen, 'Stonehenge, world heritage site', *English Heritage CAS News*, no. 3 (1995), 4–5; David Batchelor, 'The mapping of the Stonehenge landscape', in Cunliffe and Renfrew 1997, 61–71, pp. 68–70).

135 Burl 1987a, 73.

136 Burl (1987a, 70) argues that observations were made from a central charnel house.

137 This is a crude estimate based upon the period during which the moon's northern limiting declination exceeds +26° (see Astronomy Boxes 4 and 6).

138 Cleal *et al.* 1995, 140–6.

139 The original publication giving Newham's explanation for the entrance postholes was C. A. ['Peter'] Newham, 'Stonehenge: a Neolithic observatory', *Nature*, 211 (1966), 456–8. See also Newham 1993, 20–3 [1972, 15–17].

140 e.g. MacKie 1977, 122–4; Castleden 1993, 56. Burl (1987a, 67–9), however, is more cautious, pointing out that 'perhaps exceptional accuracy was not an important consideration' (*ibid.*, 68).

141 The postholes fall in six rows across the entrance, and Newham proposed that the wooden stakes in a given row marked observations of the full moon rising at midwinter in successive years within one lunar cycle. The main problem with this explanation is that if that was their purpose the posts should be clustered towards the north-west end of each row, closest to the northern major limit, rather than being roughly evenly spaced as they actually are (Heggie 1981a, 202). In addition, it is highly unlikely that they would end up, even approximately, in 'columns' perpendicular to the rows, which mostly seems to be the case (*ibid.*).

An alternative interpretation is that the posts were erected on successive nights or months around a single standstill, but here one runs into the sorts of problems discussed in chapter two, such as the need for extrapolation. Thom, Thom and Thom (1974, 86–7) actually proposed that the postholes *were* an extrapolation device similar to those proposed in Caithness and elsewhere, but see pp. 61–3.

The fact that the grid extends on its south-east side roughly to the position of midsummer sunrise has been offered as supportive evidence for the lunar theory on the grounds that observations of moonrise would have begun around this time (Burl 1987a, 68; Castleden 1993, 56). However, this makes no sense, for around midsummer the moon is new, and hence invisible, when it rises in the north-east.

142 See Burl 1987a, 68–70. If the lunar explanation were correct, then one would expect that two more posts stood in the line to the north-west. A small excavation could reveal whether there was a fifth post, but the sixth, if it existed, would have been destroyed by the later construction of the avenue ditch. Castleden (1993, 58) believes that the discovery of the fifth posthole would vindicate the moonrise interpretation, but while it would support the idea it would not prove it, since a simple line of five equally-spaced posts could have been erected for many reasons and their position in relation to the centre could be entirely coincidental. The lack of a fifth posthole would certainly deal the idea a fatal blow.

143 Cleal *et al.* 1995, 109. While now considered unlikely, the possibility can not be ruled out that the entrance postholes were contemporary with, or even predated, the north-eastern entrance of the ditch and bank, as Atkinson (1979, 26) argued (see also Castleden 1993, 54).

144 Cleal *et al.* 1995, 145.

145 Castleden 1993, 55. Another intriguing possibility is that they represent no more than a planked bridging structure erected rather later to protect the entrance while the sarsens were being dragged through prior to their erection (Julian C. Richards and Mark Whitby, 'The engineering of Stonehenge', in Cunliffe and Renfrew 1997, 231–56, pp. 252–3.

146 Ruggles 1997, 218.

147 Burl 1987a, 103; Castleden 1993, 218–20. But see Ruggles 1997, 218.

148 Ruggles 1997, 218. The postholes were discovered in section C44 (Cleal *et al.* 1995, 94). While appearing to date to Phase 2, they may date to Phase 1 or even before, so that it is still possible that the ring of Aubrey Holes, these three postholes, and all the postholes at the north-eastern entrance predate the enclosure (*ibid.* 107).

149 For example, only this small section of bank was fully excavated and many similar postholes may be waiting to be found elsewhere.

150 Whether some of the many interior postholes represent components of such rings is virtually impossible to determine on current evidence. Many parts of the site are unexcavated. See Cleal *et al.* 1995, 146–50 and fig. 70.

151 Richards 1991, 89–96, 98.

152 The 1995 edition of the English Heritage handbook on *Stonehenge and Neighbouring Monuments* (English Heritage, London) contains the erroneous statement that 'the entrance was . . . reorientated slightly during the lifetime of Stonehenge to compensate for astronomical variation in the midsummer sunrise over many centuries' (p. 20). In fact, the variation in the solstice position is too small (only about 30 arc minutes, or the sun's own diameter, in 4000 years).

153 Ruggles 1997, 219–20.

154 Cleal *et al.* 1995, 268–70; the quote is from p. 270. Castleden (1993, 61, 201–3) argues that the 'Heel Stones' formed a 'ritual doorway' (*ibid.*, 130) through which someone standing at the centre of the monument would have looked. See also Burl 1987a, 77.
155 'In no religion or temple does one enter by a door, walk some way into the building, and then turn round to the entrance to face the chief point of worship' (R. S. Newall, *Stonehenge, Wiltshire*, HMSO, London, 1959, 14, quoting a book on Bronze Age Pottery by John Abercromby published in 1912). See also Ruggles 1997, 219–20.
156 Burl 1997.
157 This has been noted since the 1960s, e.g. Hawkins 1963. It has also been suggested that the latitude of Stonehenge was deliberately selected (to within a few kilometres) in order that solar solstitial alignment to the north-east and a major lunar limit alignment to the south-east would be exactly at right angles to one another. This notion has even entered the mainstream archaeological literature (e.g. Jones 1986, 68). In fact, quite apart from archaeological arguments about the sequence of activities at the site, the alignments concerned are too imprecise for any credence to be given to the idea, and in any case the optimal latitude for this purpose depends upon assumptions about the horizon altitude, refraction, and which part of the sun and moon (centre, upper or lower limb) defines the moment of rising. Assuming a date around 2500 BC, mean refraction, and a horizon altitude of 0°·5, declination values of +24°·0 and −30°·0 (see Table A6.1) are achieved at azimuths 50°·9 and 140°·9 respectively at a latitude of 49°·8, corresponding to a point in the English Channel close to the French coast some 150 km south of Stonehenge. Other assumptions do not greatly alter this result. Cf. Richard J. C. Atkinson, 'The Stonehenge stations', *JHA*, 7 (1976), 142–4; see also Heggie 1981a, 245, note 66.
158 Ruggles 1997, 220.
159 This may even have continued well into the sarsen phase and be reflected in the anomalous radial orientation of bluestone 33e in Phase 3iv (Cleal *et al.* 1995, 231–2 and fig. 117; Ruggles 1997, 221, 224–5).
160 Hawkins (1963) suggested that the WNW–ESE diagonal (between stones 93 and 91) was aligned in both directions upon moonrise and set at the *minor* standstill limit, but the alignment to the ESE actually yields a declination of around −16°, more than 3° off target. Burl (1987a, 146–7; 1997, 4), following Lockyer (1909, 93) and A. R. Thatcher ('The Station Stones at Stonehenge', *Antiquity* 50 (1976), 144–6), interprets the same alignment in the opposite direction (declination +17°) in terms of sunset on May Day, corresponding to the later Celtic festival of Beltane, but see note 60 above.
161 Castleden 1993, 134.
162 It is remarkable that such continuity was achieved in the face of considerable social change. See Cleal *et al.* 1995, 487.
163 Whittle 1997.
164 Darvill 1997.
165 Cleal *et al.* 1995, 284–5.
166 Burl 1994, 90.
167 *Ibid.*, 88. See also Cleal *et al.* 1995, 285.
168 Burl 1994, 93. See also Ruggles 1997, 221.
169 Cf. Cleal *et al.* 1995, 486.
170 Other considerations aside, it seems more likely that the Heel Stone companion was removed by the time that a ditch was dug around the Heel Stone itself, i.e. Phase 3b (Cleal *et al.* 1995, 274), although this is by no means certain (Pitts 1982, 82). By the arguments already given, the erection of the Slaughter Stone is most likely assigned at the earliest to Phase 3ii, when the sarsen circle and trilithons were erected. Taking into account all the available evidence, Cleal *et al.* (1995, 466–70) tentatively place Phase 3ii broadly contemporary with Phase 3b.
171 Cleal *et al.* 1995, 285–7.
172 *Ibid.*, 287.
173 Ruggles 1997, 221. Burl (1994, 90) argues that a fourth stone may have stood to the south-east of the Slaughter Stone, so that the gap forming part of the solar corridor was the middle one of three. This, however, is disputed by Cleal *et al.* (1995, 285). Viewed from the centre, the second 'entrance', between stones *D* and *E*, yields a declination of +27°, close to the northern major standstill limit of the moon (cf. Hawkins and White 1970, 143; Burl 1994, 90). The possible interpretations are almost endless. Castleden (1993, 62), for example, makes the suggestion that stones *D* and *E* came first, on the old lunar axis, the Slaughter Stone being added subsequently when the axis was revised.
174 Atkinson 1979, 73.
175 Cleal *et al.* 1995, 139.
176 Cf. Ruggles 1997, 221.
177 Atkinson 1982, 107.
178 We have seen in chapter three how purely statistical arguments focused on a single monument fail to deal with this problem.
179 Cf. Cleal *et al.* 1995, 145, 483, 485.
180 Bender 1992, 749.
181 On the other hand midwinter sunset, while scarcely less impressive, would have been best viewed from outside the circle to the north-east and hence might have been a rather more public spectacle. People approaching the sarsen circle along the avenue and stopping here rather than proceeding into the interior would have seen the midwinter sun setting into the stones, and finally casting out a beam of light along the 'solar corridor'.
182 It remains questionable whether they were necessarily members of a privileged élite group. Another possibility that arises from recent suggestions that stone monuments may have been associated with ancestors (see chapter nine) is that only occasional visits were made here into the domain of the dead, by small numbers of people, perhaps to communicate with the ancestors at propitious times or as part of a specific initiation ceremony (cf. Parker Pearson and Ramilisonina 1998, 318–19).
183 See Ruggles 1997, 225.
184 It is possible that studies of patterns of movement and approach along the avenue will suggest more. See chapter ten.
185 Cleal *et al.* 1995, 35–7.
186 Ruggles 1997, 221–3 and fig. 8.
187 Burl 1987a, 166. Of eighty-six orientations listed to the nearest one-eighth division of the compass (*ibid.*, table 14), thirty-four (40 per cent) are oriented to the north, and a total of sixty-six (77 per cent) to the N, E, S or W as opposed to NE, SE, SW, or NW.
188 Archie Thom and colleagues suggest that the lozenge functioned as a calendrical device for implementing Alexander Thom's solar calendar (see Astronomy Box 5) (Archibald S. Thom, J. M. D. Ker and T. R. Burrows, 'The Bush Barrow gold lozenge: is it a solar and lunar calendar for Stonehenge?', *Antiquity*, 62 (1988), 492–502; see also Archibald S. Thom, 'The Bush Barrow gold lozenge: a solar and lunar calendar for Stonehenge?', in Ruggles 1993a, 317–23). John North, while criticising Thom *et al.*'s explanation (North 1996, 508–9), offers alternative ideas of his own, involving hundredths of megalithic yards ('HMY's) and relating the corner angles to the horizon rising or setting arcs of the sun and moon (*ibid.*, 511–16).
189 Both Thom *et al.*'s and North's explanations simply fit arbitrary theories to a convenient subset of the available data and are completely untestable. This is amply demonstrated by North's discussion of the other lozenges that exist with different angles, which he suggests could be 'merely tokens, imitating finer and more carefully drafted specimens that were well known for their astronomical purpose' (North 1996, 516).
190 O'Brien 1993, 73 and fig. 7.5.
191 This conclusion is not based on declination data taking account of horizon altitude and should itself be questioned, especially in view of the fact that their westerly orientation pattern invites comparison with the Clava cairns as well as the recumbent and axial stone circles.
192 Hutton 1991, 117.
193 O'Brien 1993, 73.
194 For a distribution map see Burl 1993, fig. 3.
195 Burl 1993, 23.
196 For example Andrew Fleming (*The Dartmoor Reaves*, Batsford, London, 1988, 95) suggests a date around 2500 BC for the Dartmoor rows. But see also Burl 1993, 83.
197 John E. Wood and Alan Penny, 'A megalithic observatory on Dartmoor', *Nature*, 257 (1975), 205–7; Wood 1978, 130–9.
198 Cf. Burl 1993, 86. This interpretation is by no means agreed by all (*ibid.*, and references therein).
199 Burl 1993, 233–4, 236–7.
200 D. D. Emmett, 'Stone rows: the traditional view reconsidered', *Proceedings of the Devon Archaeological Society*, 37 (1979), 94–114, pp. 98–101. This did not stop Lockyer (1909, 145–65, see also 481–4) proposing various stellar alignments, but see Burl 1993, 84–6 and references therein. See also Michael Ferchuck, 'Dartmoor stone rows astronomically considered', *Dartmoor Magazine*, no. 46 (1997), 16–18.
201 See Bradley 1991b and Richard J. Bradley, 'Symbols and signposts—understanding the prehistoric petroglyphs of the British Isles', in Renfrew and Zubrow 1994, 95–106 for case studies in mid-Argyll, Northumberland, and elsewhere. More generally, Bradley argues that rock carvings along the Atlantic seaboard from Scotland to Spain should be interpreted as a series of symbolic messages that are shared between monuments, artefacts and natural places in the landscape (Bradley 1997).

202 Burl 1987a, 186–7. See also Castleden 1993, 216 and fig. 87.
203 Burl 1976, 179 and 1993, 190, and references therein.
204 For example, cup marks in four-posters are predominantly found on easterly stones (Burl 1988b, 34).
205 Cf. Burl 1993, 173–5, 181–3.
206 Ruggles and Burl 1985, S54–6 and table 9 (see also p. 98). Interestingly, this contrasts with the Clava cairns where 'cupmarks were carved almost indiscriminately' (Burl 1976, 178).
207 Powell 1995. The fact that the second is roughly in line with sunrise on May Day raises once again the issue of whether mid-quarter day alignments are preferentially selected by modern investigators because they are viewed as significant 'target dates' in the first place. In fact, the idea is unsupported by any systematic investigations (see note 60).
208 Reading the data rather differently, Powell (1995, S53) points out that all but one of the twenty-three marks fall between declinations of $-23°\cdot2$ and $-24°\cdot4$ as viewed from the assumed centre, suggesting that they correspond quite closely to the horizon position of the rising solstitial sun's orb at around this time; but since the distance from the centre is only some 15 m, quoting declinations to $0°\cdot1$ is dependent on a definition of the geometrical centre precise to within about 3 cm.
209 First, it depends upon the assumption that the geometrical centre, in so far as this can be precisely defined, was meaningful to the constructors. Second, the data are derived from a field plan published in 1940 (H. N. Savory, 'A Middle Bronze Age barrow at Crick, Monmouthshire', *Arch. Camb.*, 95 (1940), 169–91, fig. 8) rather than from first-hand field measurements, and apparently there is some uncertainty over the north point (Powell 1995, S56). Finally, there is no ready explanation for the fact that one stone has cup marks on its outer face while the other has them on its upper face.
210 It is for this reason that the background material presented in the archaeological boxes does not extend beyond 1250 BC.
211 Even when elements of Thom's 'megalithic science' were extrapolated into the Iron Age, as happened for example with the Scottish brochs (e.g. MacKie 1977, 61–9), the same does not seem to have happened with astronomy.
212 For example, all but one of nineteen Celtic shrines in Britain whose orientation is discernible are oriented in the eastern quadrant (G. A. Wait, *Religion and Ritual in Iron Age Britain* (2 vols), BAR (British Series 149), 1985, 172). On the orientations of Celtic cult sites more generally see Jane Webster, 'Sanctuaries and sacred places', in Miranda Green (ed.), *The Celtic World*, Routledge, London, 1995, 445–64, pp. 459–60. Other trends are evident amongst Iron Age domestic structures (see chapter nine).
213 Green 1991; for a summary see pp. 12–13.
214 D. F. Allen (ed. Daphne Nash), *The Coins of the Ancient Celts*, Edinburgh UP, 1980, 149.
215 Alain Bulard, 'Sur deux poignards de la fin de l'époque de La Tène', *Études Celtiques*, 17 (1980), 33–49.
216 Andrew P. Fitzpatrick, 'Night and day: the symbolism of astral signs on Later Iron Age anthropomorphic short swords', *PPS* 62 (1996), 373–98.
217 e.g. MacKie 1988, 225–9.
218 The idea dates back to the nineteenth century (Timothy Champion, 'The European Iron Age: assessing the state of the art', *SAR*, 4 (1987), 98–107, p. 99). While still widely perpetrated, this straightforward picture is questioned by many archaeologists (see, e.g., Andrew P. Fitzpatrick, '"Celtic (Iron Age) religion"—traditional and timeless?', *SAR*, 8 (1991), 123–9 and references therein). On the Iron Age Celts in general see Cunliffe 1997.
219 Hutton 1996, 408–11 and references therein.
220 John King, *The Celtic Druids' Year*, Blandford, London, 1994, 164–6.
221 The Coligny and Villards d'Héria inscriptions date to Gallo-Roman times, probably around AD 200 (Paul-Marie Duval and Georges Pinault, *Recueil des Inscriptions Gauloises, vol. III: Les Calendriers (Coligny, Villards d'Héria)*, Éditions du Centre Nationale de la Recherche Scientifique (XLV[e] Supplément à Gallia), Paris, 1986, 35–7). They represent luni-solar calendars—calendars in which lunar months were reconciled with the solar year through the periodic insertion of intercalary months (Garrett Olmsted, *The Gaulish Calendar*, Habelt, Bonn, 1992; McCluskey 1993b, 102–3; 1998, 54–60).
222 For example, four dates in the Coligny calendar are separated by intervals of between ninety-one and ninety-three days and might correspond to four of the seasonal festivals (Stephen C. McCluskey, 'The solar year in the calendar of Coligny', *Études Celtiques*, 27 (1990), 163–74; see also McCluskey 1993b, 104–5; 1998, 59–60). However, even if these dates do reflect earlier festivals it is arguable that they only became 'fixed' on precise dates in the solar year as a result of contact with the Julian calendar.
223 e.g. Stephen C. McCluskey, 'The mid-quarter days and the historical survival of British folk astronomy', *AA* no. 13 (*JHA* 20) (1989), S1–19.
224 Hutton 1996, 1–4.
225 Hutton (1996, 5) may be right to conclude that 'no overall or enduring pattern of cult may be detected', but it is wrong to base such an assertion merely upon the consideration of alignments upon the rising or setting solstitial and equinoctial sun (cf. *ibid.*, 4–5). Not only does this ignore the sort of orientation evidence presented in this chapter; it also presupposes which types of alignments might be significant. On supposed equinoctial alignments see chapter nine.
226 Francis Pryor, 'Discussion: the Fengate/Northey landscape', *Antiquity*, 66 (1992), 518–31, pp. 528–9; Parker Pearson 1993, 113–17. On Flag Fen in general see the papers in the special section on 'Current research at Flag Fen, Peterborough', *Antiquity*, 66 (1992), 439–531.
227 e.g. Bradley 1991a, 217–18.
228 There is, however, no general agreement on the extent of the social upheaval in Wessex during this time, and some argue for slower histories of change (Whittle 1996; see also Whittle 1997, 147). It has also been suggested that, once transformed into stone, Stonehenge became a place of the ancestors rarely visited by the living (see p. 154).
229 Michael Parker Pearson, review of Armit 1996, *British Archaeology*, no. 19 (Nov. 1996), 12.
230 A typical list is given by Krupp (see note 60). The idea continues to be advocated by MacKie, who in a recent paper (MacKie 1997) proposes that two foresights on the south-westerly horizon at Maes Howe were used to mark the setting position of the sun at intervals of exactly $1/16$ and $1/8$ year before and after the solstice (the latter corresponding to the mid-quarter day).
231 The only possible exception is the group of elliptical, or oval, stone rings in North Wales, the orientations of whose long axes are, according to Burl, clustered around sunset in early May (see note 97).

CHAPTER 9

1 John Henry Newman, *Discourses on the Scope and Nature of University Education*, 1852, reprinted in *Newman: Prose and Poetry*, selected by Geoffrey Tillotson, Rupert Hart-Davis, London, 1957, p. 485.
2 Patrick Moore, foreword in Walker 1996, p. 9.
3 Nigel Barley, *The Innocent Anthropologist: Notes from a Mud Hut*, Penguin Books, Harmondsworth, 1986, p. 9 [Original edn published by British Museum Publications, 1983].
4 North 1996. For the reactions of an archaeologist see Renfrew 1996 and for those of an archaeoastronomer see Aveni 1996.
5 The term 'archaeoastronomy' often seems to be associated exclusively with the ideas of Hawkins and Hoyle on Stonehenge in the 1960s and with the work of Thom through to the 1970s (e.g. Chippindale 1994, 220; Bahn 1996, 287–9), although in the 1960s the term did not even exist. Neither of these sources cites, or even mentions, any archaeoastronomical literature since 1980, although Chippindale (1994, 230) makes the sweeping statements that 'solar alignments are accepted' and 'precise lunar alignments are not'. It is interesting to contrast this with the statement that 'there is no doubt that the movements of the sun and the moon were important to these people, but evidence for plotting the movements of the stars is debatable' (Parker Pearson 1993, 68) and 'it has long been established that many important ritual or religious monuments [all over the world] were carefully aligned upon solar, lunar, or stellar events' (Chris Scarre, in Brian M. Fagan (ed.), *The Oxford Companion to Archaeology*, OUP, 1996, 592).
Similarly, Castleden (1993, 18–27) provides a critique of the old ideas about Stonehenge and the 'astro-archaeology' of the 1960s and 1970s, but despite making much of possible solar and lunar symbolism in his own interpretations makes no reference whatsoever to the more recent literature on solar and lunar symbolism at other Neolithic and Bronze Age monuments in Britain and Ireland.
The polarisation of views about the nature and value of archaeoastronomy is typified in the book by Volker Bialas, *Astronomie und Glaubenvorstellungen in der Megalithkultur. Zur Kritik der Archäoastronomie*, Verlag der Bayerischen Academie der Wissenschaften (Mathematisch-naturwissenschaftliche Klasse, Abhandlungen, Neue Folge, Heft 166), Munich, 1988, and the critique by Hanne Dalgas Christansen, 'A scientific attack on archaeoastronomy?', *AIHS*, 43 (1993), 339–42.
6 According to Renfrew and Bahn (1996, 381), 'Although some of the details of Thom's claims for individual stone circles have been challenged, the cumulative picture argues plausibly for a preoccupation with . . . calendrical events'. Yet the extensive reassessments of Thom's

data described in chapters two and three, which have been available in the archaeoastronomical literature since the early 1980s and widely quoted since, showing that the idea of a solar calendar dividing the year into eight or sixteen equal parts is completely unsupported by the actual alignment data, go unmentioned and uncited.

7 According to Parker Pearson (1993, 65) there are six solar and lunar alignments at Godmanchester and 'it is unlikely that these correspondences are accidental'. But see chapter eight, note 38.

8 Two examples will suffice.

a. 'Orkney is the only place in the British Isles where [the four solstitial directions] are perpendicular to each other' (*ibid.*, 59). In fact, this is impossible to achieve at the latitude of Orkney (59°), and in any case the azimuths of solstitial sunrise and sunset are dependent upon the horizon altitude. The most favourable latitude is around 55°, i.e. the northernmost parts of Ireland and the Scottish/English border, at locations with a horizon altitude of around 0°·5 in each of the intercardinal directions. Locations around latitude 56° would require an altitude of around 1° to the north-east and north-west and 0° to the south-east and south-west; this is reversed at around latitude 54°. Outside these limits the required altitudes in two of the four directions become negative and soon reach values that are unobtainable in practice, even on high ground with a sea horizon. This false statement is already being perpetrated in the literature (e.g. Souden 1997, 122).

b. 'Many astronomical alignments have been claimed among the arrangement of stones at Callanish, but only two stand up to critical examination. The first is southwards to the midsummer moon setting at its southern extreme down the slope of Mount Clisham 26 km away. The second is the Avenue which defines the midwinter sunset' (Darvill 1996, 203). In fact, the orientation of the avenue is far too southerly to have anything to do with midwinter sunset, and Clisham cannot be seen from the stones owing to a local outcrop immediately to the south (see chapter eight).

9 Thus Parker Pearson (1993, 65) speaks of 'solstice or equinoctial positions of the sun and moon'.

10 Examples continue to appear in the wider archaeoastronomical literature. For example, according to Rosa Schipani de Pasquale and Francesco Riccobono ('Originale utilizzo di materiali "da spetramento" in area subetnea', in Romano and Traversari 1991, 118–22), seven enigmatic round towers erected on stepped pyramids found in the foothills of Mount Etna, whose orientations yield declinations of +50°, +47°, +22°, +14°, +8°, 0° and −2°, 'seem to indicate the probable existence of some astronomically interesting alignments' [my translation]. This is special pleading indeed. An extreme example is the work by Eduardo Proverbio and his colleagues on the Sardinian 'Tombe di Giganti' (Eduardo Proverbio, Giuliano Romano and Anthony F. Aveni, 'Astronomical orientations of "Tombe dei Giganti" in Barbagia (Sardinia)', in Romano and Traversari 1991, 52–9; see also Eduardo Proverbio, 'New evidence concerning possible astronomical orientations of "Tombe di Giganti"', in Ruggles 1993a, 324–31), where every measured orientation is interpreted in terms of an astronomical event, be it the rising of the sun at a solstice, the moon at a standstill limit, or a star such as β Crucis or ε Orionis.

11 Cf. Hoskin 1997b, 19. The dangers are even greater when the search for astronomical alignments is itself motivated by another quest entirely, such as the pursuit of evidence of greater cometary or meteoritic activity in ancient skies (see, for example, Mark E. Bailey, 'Recent results in cometary astronomy: implications for the ancient sky', *Vistas in Astronomy*, 39 (1995), 647–71, pp. 663–4).

12 For an example of such a project in the Veneto region of Italy see Franco Posocco and Giuliano Romano, 'Ancient topography and territorial planning in the Veneto region', in Romano and Traversari 1991, 34–7.

13 This raises a number of questions of field procedure some of which are covered in the Appendix. Methods of presenting the results are also an issue.

14 Hoskin, 1997b, 20. Adopting a traditional approach, Hoskin sees data collection and interpretation as a two-stage process, the first of which is 'factual and uncontroversial'. He suggests that the collection of orientation data, as distinct from their interpretation, should be recognised as a field of enquiry in its own right and that a new term such as 'archaeotopography' should be used to describe it (*ibid.*; see also Michael Hoskin, review of Ruggles and Saunders 1993a and Ruggles 1993a, *AA* no. 21 (*JHA* 27) (1996), S85–7).

15 In any case, many archaeologists in fact take the view that all data are 'theory laden', meaning that there is no such thing as 'objective' data free of pre-existing theoretical ideas (see, for example, Stephen J. Shennan, 'Introduction: archaeological approaches to cultural identity', in Shennan 1994, 1–32, p. 2; Alison Wylie, 'Matters of fact and matters of interest', in *ibid.*, 94–109, p. 94; and other papers in the same volume). Thus, for example, although vast numbers of barrows were dug in the last century and early this century, social interpretation was not a concern, and a great deal of evidence relevant to such interpretation has been lost (Bradley 1984, 74).

Certainly it is true that which azimuths and declinations we choose to measure will be influenced by what ideas we already have and what possibilities we consider to be interesting in the first place. Even where we are dealing with groups of similar monuments with a single overriding orientation there is room for different approaches which yield very different results: for example, contrast the work of Lynch (1982) and Ruggles (1994a; 1996) on the Cork and Kerry short stone rows (see chapter six).

16 Cf. chapter four, note 119.

17 See, for example, Lindström 1997.

18 That archaeologists and astronomers should work together is, for example, the main message of the collection of papers in Romano and Traversari 1991, but this is not in itself a prerequisite for good-quality work in archaeoastronomy. See the essay review in Ruggles 1995.

19 Ruggles, *ibid.*, S82; cf. Aveni 1980, 3–8; 1989b, 10–11.

20 Indeed, whether astronomers as such need to be involved at all is open to question. To paraphrase Shennan (1988, 3), who puts a similar argument in the context of archaeological statistics: astronomical specialists may be required to cope with particular problems, but archaeologists must have sufficient awareness to recognise when problems arise which can helpfully be tackled using the methods of archaeoastronomy. Ideally, perhaps, archaeologists should be versed in basic astronomical theory and field techniques, and thus able to include astronomical questions in their broader research design and to attempt to answer them as part of their subsequent investigations. Some of Hoskin's collaborations in the Iberian peninsula (resulting in various papers in *AA*) have begun to achieve this.

21 See p. 161.

22 Kintigh 1992. For a response see Aveni 1992b.

23 'The observation of an alignment, however ingenious, is in and of itself not interesting. However, an argument relating control over privileged knowledge of celestial events to the development of hierarchical power relations at some crucial juncture might have substantial theoretical interest.' (Kintigh 1992, 4.)

24 Marshack 1991, 25.

25 Here we should include ethnoastronomy, or perhaps refer instead to 'cultural astronomy' (see chapter ten, note 103).

26 The principal ones are the Société Européenne pour l'Astronomie dans la Culture (SEAC), founded in 1993, and the International Society for Archaeoastronomy and Astronomy in Culture (ISAAC), founded in 1995.

27 See Ruggles 1993b, 2–4.

28 For a detailed exposition of this view see Ruggles and Saunders 1993b.

29 Aveni (1989b, 9) defines archaeoastronomy as 'the study of the practice and use of astronomy among the ancient cultures of the world based upon all forms of evidence, written and unwritten'. By implication this subsumes, or at least transcends, the longer-established discipline of history of astronomy (cf. Gingerich 1989, 43). On the boundary between history of astronomy and archaeoastronomy see Ruggles 1993b, 6–7.

30 Ruggles and Saunders 1993b, 1. The paper as a whole contains several examples.

31 Nicholas J. Saunders, Review of Aveni 1989a, *New Scientist*, 123 (1989), 57–8, p. 58. See also Ruggles and Saunders 1993b, 4.

32 Ruggles and Saunders 1993b, 9–10 elaborate on this point.

33 Astronomical expertise can yield a degree of apparent control by virtue of the ability to predict certain kinds of phenomena, such as solar and lunar eclipses and heliacal risings and settings. The predictive capacity of astronomy, available to those with even a limited observational expertise, is one of the few universal realities in the study of human societies, and as a result studying cultural astronomy has a unique potential to shed light on the diversity of cultural responses to the 'natural world' (Ruggles and Saunders 1993b, 12).

Comparative studies of cultural astronomy can bring into focus the social and political dynamics of cross-cultural interactions. See, for example, Peter G. Roe, 'The Pleiades in comparative perspective: the Waiwai *Shirkoimo* and the Shipibo *Huishmabo*', in Ruggles and Saunders 1993a, 296–328. For other examples see Ruggles and Saunders 1993b, 10.

34 Stanisław Iwaniszewski, 'Exploring some anthropological theoretical

foundations for archaeoastronomy', in Aveni 1989a, 27–37; 'Some social correlates of directional symbolism', in Ruggles 1993a, 45–56. See also Ruggles and Saunders 1993b, 11–12.

35 According to Anthony F. Aveni (Preface to Aveni 1989a, xi), 'the business of archaeoastronomy [is] to comprehend the meaning of astronomy to ancient peoples . . .'. The whole general question of approaches to understanding systems of meaning is one that has been of great concern recently in theoretical archaeology (see Archaeology Box 8).

36 Clear open skies are not a prerequisite, and this is no less true of people living in a tropical rain forest environment, for example a number of native groups in Colombia (G. Reichel-Dolmatoff, 'Astronomical models of social behaviour among some Indians of Colombia', in Aveni and Urton 1982, 165–81). Perhaps the most conspicuous exceptions are modern city dwellers who rarely if ever see a dark night sky.

37 Ruggles and Saunders 1993b, 1. The paper as a whole contains several examples.

38 The three terms are closely related and in some usages broadly synonymous. Some use the term 'world-view' to refer to an individual's cognitive map, a set of associations developed in the light of personal experience (e.g. Renfrew 1994, 10–11). World-views, in the sense used here, are common features of individual cognitive maps, forming a general framework of understanding, that tend to emerge amongst people living together in a community (see Renfrew and Bahn 1996, 370; Ruggles 1999). The term 'cosmology' is widely used amongst anthropologists, usually to indicate this wider combination of a personal belief system and a social construct system (see Charles Keith Maisels, *The Emergence of Civilization*, Routledge, London, 1990, p. 275), although 'cosmovisión' is preferred by some (cf. Broda 1982; 1993).

39 For an elaboration of these ideas see Ruggles and Saunders 1993b, 6–8 and references therein. On the development of biological classifications within Western science, see Scott Atran, *Cognitive Foundations of Natural History: Towards an Anthropology of Science*, CUP, Cambridge, 1990.

40 Stephen Hugh-Jones, *The Palm and the Pleiades: Initiation and Cosmology in Northwest Amazonia*, CUP, Cambridge, 1979, pp. 138–46.

41 Ruggles and Saunders 1993b, 7. See Stephen Hugh-Jones, 'The Pleiades and Scorpius in Barasana cosmology', in Aveni and Urton 1982, 183–201, p. 191.

42 A number of archaeologists have argued that we would be guilty of cultural imperialism if we expected everyone in the modern world to wish to comprehend other world-views from a Western scientific viewpoint. Nonetheless, from the Western scientific perspective at least, there is a clear distinction between historical facts (including prehistoric people's actions and the thoughts in their minds) and our knowledge or interpretation of them. Thus while the many different philosophical and political positions taken by theoretical archaeologists (see Archaeology Box 8) can engender a variety of ways of interpreting prehistory—on the issues that arise as a result see, for example, the collection of papers in Shennan 1994—it is important to separate such matters from philosophical arguments about whether, for example, objective reality exists independently of the cultural context in which it is observed (see, e.g., Mary Midgely, *Science as Salvation*, Routledge, London, 1992, pp. 47–9). These arise, in effect, from attempts by modern, Western thinkers to question fundamental aspects of the prevailing Western world-view. Cf. Alan Sokal and Jean Bricmont, *Intellectual Impostures: Postmodern Philosophers' Abuse of Science*, Profile Books, London, 1998, esp. 183–5.

43 Knight 1991, 294.

44 The fact that we speak of people's perceptions of (what we see as) their natural environment, using terms such as 'natural world' and 'world-view'—should not be taken to imply necessarily that, as prehistoric people saw things, there was a clear separation between the human perceiver and the external world: in fact, this is unlikely in a non-Western context (cf. Ingold 1993, 153–4).

45 This form of words is preferable to 'the world-view of another culture' as different and incompatible world-views and belief systems may have co-existed side by side; this happens in the Western world today, where one aspect of the 'official' world-view (scientific astronomy) coexists with a widespread folk ideology (astrology). See Ruggles and Saunders 1993b, 15.

46 *Ibid.*, 8, 13.

47 Ruggles 1999. See also D. Miller, 'Artefacts as products of human categorisation processes', in Ian Hodder (ed.), *Symbolic and Structural Archaeology*, CUP, Cambridge, 1982, 17–25, who argues that human categorisation processes are reflected in the organisation of things in the material record. But see Barrett 1994, 77–81, who cautions against assuming that ritual practice expresses a single, shared world-view to which most members of a social group would subscribe.

48 This is a position hinted at by Bradley (1991b, esp. 78–9) in the context of seeking to reconcile the 'scientific' approaches of landscape archaeology and the 'excessive subjectivity' (*ibid.*, 77) of studies of prehistoric art. See also Richard J. Bradley, Tess Durden and Nigel Spencer, 'The creative use of bias in field survey', *Antiquity*, 68 (1994), 343–6 and Bradley 1997, ch. 5, on matching the subjective response to places in the landscape (rock art) to quantification and the use of control samples. The position is also consistent with Hodder's view that 'It is possible to make statements about past meanings which can be strengthened or weakened by consideration of the evidence . . .' (Hodder 1992, 21).

49 Ruggles and Saunders 1993b, 4.

50 On the Borana calendar, see Clive L. N. Ruggles, 'The Borana calendar: some observations', *AA* no. 11 (*JHA* 18) (1987), S35–53; 'Four approaches to the Borana calendar', in Ruggles 1993a, 117–22.

51 See Urton 1981, 109 and references therein.

52 *Ibid*, 5.

53 See note 114 below.

54 Indeed, it is a Western predilection to separate our understanding of the natural world from that of humans and their activities in the first place. As Tilley (1994, 38) says of the landscape as perceived by Aboriginal groups, it 'is not something "natural" and opposed to people, but totally socialized'. While taking this point, we shall continue to talk of the 'natural world' in what follows as an aid to our discourse.

55 Urton 1981, ch. 8.

56 In other words, we should necessarily assume that people have a concept of direction that could be explained in terms of a line joining two points. On general issues relating to human perceptions of proximity and distance, and other concepts of spatial order such as directionality, see Thomas 1996, 85–7.

57 For an overview see M. Jane Young, 'The Southwest connection: similarities between Western Puebloan and Mesoamerican cosmology', in Aveni 1989a, 167–79, esp. p. 170. On Puebloan peoples see McCluskey 1999. Both in Mesoamerica and the US South-West the set of directions may also include the centre, which further illustrates the different ways in which 'directions' are conceived. Sometimes the set of directional schemes include centre, up and down, making seven in all.

As to their interpretation, ethnohistoric and ethnographic evidence attests to the complex symbolism that attaches to these directions, associating them with deities, colours, times of day, sacred mountains, animals and birds, plants, minerals, and so on; and anthropological studies show how this is indicative of a very different conceptual basis for understanding the spatial and temporal organisation of the world and how this needs to be exploited for survival.

58 Thus amongst the villagers of Yalcobá, 'The names of the directions themselves correspond to the *sides* [of a quadrilateral] . . . each of which . . . is therefore not at all a "cardinal point" as has so often been assumed in the Maya and the Mesoamerican literature' (Sosa 1989, 132).

59 Köhler 1991, 134.

60 To take a modern example, Navajo hogans—traditional sacred dwellings round or octagonal in design—have a single door facing 'east' towards the rising sun (Williamson 1984, 151–3, 156–8; Griffin-Pierce 1992, 21). Yet the actual orientations vary over several tens of degrees, because in practice they face a direction such that the light of the rising sun can enter at some time during the year (Stephen C. Jett and Virginia E. Spencer, *Navajo Architecture*, University of Arizona Press, Tucson, 1981, 17–18; cf. Williamson 1984, 332, note 9). The two are not inconsistent.

61 Staying with the example of Navajo hogans, part of their sacred design is that each of their 'sides' is associated with the appropriate one of four sacred mountains (Griffin-Pierce 1992, 95). This is true even though the actual azimuth of each of these mountains clearly varies from place to place.

62 Aveni 1980, 40.

63 The practice, once endemic, of assessing the astronomical potential of architectural alignments in terms of these targets, or of superimposing the targets (or a subset of them) upon plans or graphs of orientations, is still quite widespread. A few diverse examples, taken from both the archaeological and archaeoastronomical literature in more recent years, are Burl 1985, 24; Castleden 1993, fig. 88; Gibson 1994, 187–8; Parker Pearson and Richards 1994, fig. 2.4; North 1996, fig. 118; Papathanassiou and Hoskin 1996, fig. 2; Schlosser and Cierny 1996, fig. 5.4; Oswald 1997, fig. 10.4.

64 Anthony F. Aveni, 'The Thom paradigm in the Americas: the case of the cross-circle designs', in Ruggles 1988a, 442–72, pp. 442–5.

65 But see note 15.

66 An approach where 'statistical rigour precedes interpretative reasoning' has been seen as advantageous by this author in the past (e.g. Ruggles 1988b, 246–50), but a broader perspective suggests that quantitative analysis and cultural interpretation should act as a control on the *sorts of conclusion* reached by the other (Tristan Platt, 'The anthropology of astronomy' (review of Aveni 1989a), *AA* no. 16 (*JHA* 22), S76–83, p. S80).

67 On the 'green' and 'brown' approaches in archaeoastronomy see p. 161.

68 Even then, their main significance may be not as directions-as-points but as the limits of directions-as-ranges—this is the case, for example, amongst the villagers of Yalcobá (Sosa 1989, 132).

69 Thus, for the Chumash of California winter solstice was a time of crisis (Krupp 1997, 157), whereas at summer solstice 'a "big basket" was filled with valuable offerings for the sun . . . with the hope that such respect would inspire a rich harvest in the autumn' (Edwin C. Krupp, 'Summer solstice: a Chumash basket case', in Ruggles 1993a, 251–63, pp. 253–4). Conversely, it has been argued that winter solstice was, and remains, a time of celebration in northern Europe because it marks the onset of lighter, and eventually warmer, days (Christina Hole, *English Custom and Usage*, Batsford, London, 1941, 14; quoted by Green 1991, 17).

70 Relationships are commonly perceived between the sun and cycles of death and regeneration, as well as the dualities of world and underworld, darkness and light. Cf. Hawkes 1962, 48–9; Green 1991, 18, 136–8; Parker Pearson 1993, 59–60; North 1996, 21–2, 530; Krupp 1997, 136, 140.

71 Amongst the Hopi, for example, two of the four places where the sun pauses in his travels along the horizon are true 'houses of the sun': the place of midwinter sunrise in the south-east, where the sun comes out and stands directly above his house eating from a red stone bowl (decorated prayer sticks and other offerings are placed on shrines at the appropriate place on the horizon), and the place of midsummer sunset in the north-west, where the sun stands directly above the house of *Huzruing wuhti*, 'hard-being woman', the deity associated with hard substances such as shells, corals and turquoise, and then descends into it through a hatchway in the roof (McCluskey 1993a, 40).

72 For example see Green 1991 on sun cults in Europe from the Iron Age onwards.

73 On this point see also Ruggles 1997, 206–8; Clive L. N. Ruggles, 'Whose equinox?', *AA* no. 22 (*JHA* 28) (1997), S45–50. *Pace* Darvill (1997, 184, note 8), the point had not to this author's knowledge been made prior to my own presentation at the Royal Society / British Academy meeting on Science and Stonehenge in 1996 (the Cunliffe and Renfrew 1997 volume being the proceedings of this meeting), either amongst British prehistorians or amongst archaeoastronomers at large.

74 This has been suggested, for example, by Hoyle (1966, 271–2) at Stonehenge and by Bernard Yallop at Godmanchester where 'from alignments at the equinoxes they measured East-West to 0·2 of a degree, and they definitely observed the Sun sitting on the horizon' (priv. comm., 1996).

75 The scarcity of equinoctial alignments seems to come as a surprise to some authors: 'The extremes were evidently what counted; the equinoxes were seemingly of lesser importance to early peoples' (North 1994, 2). In an attempt to categorise astronomical terms and terminology amongst Basque people, Endrike Knörr ('Mitología astronómica y lengua vasca', paper presented at SEAC96 conference, Salamanca, 1996) evidently finds it surprising that there is no word for 'equinox' in the Basque language.

76 See, for example, Eogan 1986, 178; Burl 1987a, 44; MacKie 1988; Bradley and Chambers 1988, 286; O'Kelly 1989, 235; Bradley 1993, 62; Parker Pearson 1993, 62; Gibson 1994, 207; Renfrew and Bahn 1996, 382; Darvill 1996, 254; Souden 1997, 125; Waddell 1998, 64. Hutton (1996, 4–5), in attempting to sum up the evidence on astronomy in prehistoric Britain and Ireland, considers only solstitial and supposed equinoctial alignments.

77 See, for example, Castleden 1987, 240–1.

78 This is typified in remarks such as 'Their first attempts to find the exact months when the major and minor standstills occurred would have led to very confusing results' (Wood 1978, 103). The tendency has also been carried well beyond the bounds of prehistoric Britain and Ireland: thus in the US South-west 'it seems clear that Anasazi astronomers were aware of the major and minor standstills of the moon' (J. McKim Malville and Claudia Putnam, *Prehistoric Astronomy in the Southwest*, Johnson Books, Boulder, revised edn 1993, p. 32).

79 Cf. Aveni 1987, 16–18.

80 For a modern example see chapter four, note 92.

81 This point is not appreciated by many otherwise authoritative authors. See, for example, Krupp 1994, xi–xii.

82 For example, the traditional Sun Dance ceremony of North American Plains Indians was held at full moon very near to summer solstice (David Vogt, 'Medicine Wheel astronomy', in Ruggles and Saunders 1993a, 163–201, pp. 168–9).

83 This is because the limiting (northern or southern) monthly declination varies roughly sinusoidally, oscillating between the major and minor standstill limits (see Fig. 1.19b).

84 Thom (1967, 21) noted that this might have been important for a community whose only effective illumination during the long winter nights was the moon. See also Ellegård 1981, 104–5.

85 For examples of ways in which the planets have been perceived and interpreted in a range of world cultures down the ages, see Aveni 1992a. Venus was of particular importance in pre-Columbian Mesoamerica (e.g. Aveni 1980), where, for example, a broad range of evidence suggests that in the prevailing world-view a complex conceptual relationship existed between Venus and rain, maize and fertility (Ivan Šprajc, 'The Venus-rain-maize complex in the Mesoamerican world-view: Part I', *JHA* 24 (1993), 17–70; see also Šprajc 1993).

86 Architectural forms will have different meanings in different contexts, and often in the same context. Thus, just because all recumbent stone circles had the same general form, we cannot assume that they were all put to precisely the same set of uses. When built, architecture acquires meanings not intended by the builders, and it may well continue to acquire new meanings (and uses) over a very long period. Thus the city of Teotihuacan, built and inhabited in the first half of the first millennium AD, was to the Aztecs almost a thousand years later 'the abode of the gods' (cf. Eduardo Matos Moctezuma and Leonardo López Luján, 'Teotihuacan and its Mexica legacy', in Kathleen Berri and Esther Pasztory (eds), *Teotihuacan, Art from the City of the Gods*, Thames and Hudson, London, 157–65, p. 158) and Stonehenge has become in modern times a place of solstitial pilgrimage for New Age travellers, as well as an 'ancient monument', a 'world heritage site', and a major tourist attraction.

87 As already noted (cf. note 15), these data will themselves reflect pre-existing ideas in various ways.

88 See Hodder 1992, 214 on the dangers of implicitly assuming what you then go on to 'prove'.

89 'Research is not a linear process, of course, it is a loop . . .' (Shennan 1988, 6). See also Ruggles 1986, 9. This general observation is common to a wide range of approaches to archaeological explanation. From a 'cognitive processual' viewpoint 'facts modify theory, while theory is used in the determination of facts, and the relationship is a cyclic (but not a circular) one' (Renfrew 1994, 10). Hodder (1991, 151) argues from a post-processual viewpoint that the 'hermeneutic circle' is a question-and-answer procedure rooted in the present, but notes that 'some [answers] can be demonstrated to relate to the evidence better than others'.

It is not necessarily wrong to indulge in unstructured exploration followed by *post hoc* justification, but we must aim to move beyond this at the earliest opportunity, by developing richer sets of ideas as well as by acquiring new evidence. For example, the fact that the Mull stone rows are located on the edge of Ben More viewsheds (chapter seven) may have been discovered and tested in a *post hoc* fashion, but it raises important general questions of how and why relationships between places in the landscape that, for some at least, were imbued with sacred importance might have been deliberately hidden from sight (see chapter ten); questions which can be 'tested' on a broader set of data.

90 Ruggles and Saunders 1993b, 15. On co-existing world-views see note 45.

91 Cf. Orme 1981, 249–50; Parker Pearson and Ramilisonina 1998, 309–11.

92 Attempts to seek such correlations, if taken at too simplistic a level, soon run into problems (see, e.g., Thorpe 1981), but an argument may be made for more subtle correlations or tendencies (see references cited in note 34).

93 An example of such an association would be that between the axial orientation of certain recumbent stone circles, the moon, and prominent hills.

94 We are talking here of evidence that has no direct bearing upon the people whose material culture is being studied. Of course, we may be luckier. Outside prehistory, historical and archaeological evidence may both relate directly to the same societies. In other cases, evidence from historical or modern peoples might arguably have a direct bearing on their forebears in the same geographical area. For example, the fact that

certain astronomically-related beliefs and rites have survived into modern Maya rural communities can be taken to suggest that in pre-Columbian times these were not, as had been proposed before, the exclusive domain of an élite (Šprajc 1993, S45).

There remains, however, the possibility that Neolithic and Bronze Age communities in Britain and Ireland may have been very different from the sorts of indigenous groups that have been and can be studied. Of necessity, the latter tend to be small in size, living in marginal locations, and having at least some interaction with and exposure to modern nation-states, with all that implies in terms of social, political, economic and technological influences—and modes of thought.

95 Bradley 1993, 47.

96 Tilley 1994. On the use of the term 'landscape' in this context see *ibid.*, 24–5.

97 This approach is placed on a more secure anthropological footing through concepts such as 'dwelling' (e.g. Ingold 1993, 152–3; Tim Ingold, 'Building, dwelling, living: how animals and people make themselves at home in the world', in Marilyn Strathern (ed.), *Shifting Contexts: Transformations in Anthropological Knowledge*, Routledge, London, 1995, 57–80; Thomas 1996, 87–9).

98 Bradley 1993, 47.

99 Edwin C. Krupp, message circulated on the HASTRO-L [History of Astronomy] internet mailing list, 8 August 1995. For the description of an expedition to experience minor standstill moonrise at Stonehenge in 1996, see McNally and Ruggles 1997.

100 Tilley 1994, 71–5.

101 Indeed, the significance of particular places will be individually, as well as culturally, defined. See Tilley 1994, 26–8. See also Archaeology Box 7.

102 Clive L. N. Ruggles, 'The North Mull project, 7: Wider problems and perspectives', manuscript in preparation.

103 For many examples see Krupp 1997. See also Hawkins and White 1970, 150; Ruggles and Saunders 1993b, 12; Šprajc 1993, S45.

104 Ruggles and Saunders 1993b, 12.

105 See, for example, Alfred Gell, *The Anthropology of Time: Cultural Constructions of Temporal Maps and Images*, Berg, Oxford, 1992, 306–13, on the Trobriand and others; Janet Hoskins, *The Play of Time: Kodi Perspectives on Calendars, History and Exchange*, University of California Press, Berkeley, 1993, on the Kodi of Indonesia. For a general discussion see *ibid.*, ch. 12.

106 The ritual act of breaking the earth by the Inca himself following the Inti Raymi (winter solstice) festival, which had to take place before ploughing could proceed anywhere else, was a way of asserting his personal mystique and authority over the entire Inca empire (John Hemming, *The Conquest of the Incas* (revised edn), Penguin Books, Harmondsworth, 1983, pp. 172–4). For a fuller account of the relationships between astronomical knowledge and Inca social and political structures see Bauer and Dearborn 1995.

107 See p. 142.

108 Ruggles 1984a, 13.

109 For remarks on the conceptualisation of time in the context of debates about prehistoric astronomy in the early 1980s see Thorpe 1981, 276; 1983, 3–4. For more recent archaeological discussions on time and temporality in the context of British prehistory see, e.g., Bradley 1991a; Gosden 1994, esp. 122–6, 133–5, 162–3; Thomas 1996, chs 2 and 3.

110 For an overview see Ruggles 1999.

111 North (1994, xxiv–xxv) suggests that there is a natural progression of calendrical development relating to astronomical observation, from a simple month-by-month calendar based on the phase cycles of the moon, to one which determines the time in the seasonal year using the rising or setting position of the sun, and finally lunar-solar calendars which combine the two, using solar observations to keep the months in step with the seasons by the judicious, and eventually systematic, addition or subtraction of intercalary months (see Ruggles 1999).

112 For one example, see p. 89.

113 Renfrew and Bahn (1996, 381–2) suggest that the development of units of measurement constitutes a fundamental cognitive step, and that evidence of units of time, length, and weight are particularly susceptible to being recovered archaeologically.

114 Cf. Michael Shanks and Christopher Tilley, *Social Theory and Archaeology*, Polity Press, Cambridge, 1987, ch. 5; McCluskey 1998, 4–5.

115 See, e.g., Whittle 1988, 203; Ruggles 1999. For an example from a classic anthropological account, see Edward E. Evans-Pritchard, *The Nuer*, OUP, Oxford, 1940, 103–8.

116 Broda 1982, 105; Šprajc 1993, S45.

117 At Walpi and other more traditional villages, people observe an elaborate ceremonial calendar, in which the entire year is punctuated with ceremonies, both public and private, which 'assure vital equilibrium . . . and conciliate the supernatural powers in order to obtain rain, good harvests, good health, and peace' (Arlette Frigout, 'Hopi ceremonial organisation', in Ortiz 1979, 564–76.

118 e.g. Renfrew 1976, 262–3; Heggie 1981a, 220; Thorpe 1981, 280–1. On Hopi solar observations, see McCluskey 1977; 1990.

119 McCluskey 1993a, 42.

120 Hieb 1979; McCluskey 1999.

121 McCluskey 1990, S2.

122 For further discussion see Ruggles 1999 and references therein.

123 Tilley 1994, 14–17.

124 Ruggles 1997, 205. See also McCluskey 1998, 6–7.

125 Ronald Goodman, *Lakōta Star Knowledge: Studies in Lakōta Stellar Theology*, Sinte Gleška College, Rosebud, South Dakota, 1990. See also Linea Sundstrom, 'Mirror of heaven: cross-cultural transference of the sacred geography of the Black Hills', *WA*, 28(2) (1996), 177–89.

126 This is evident from the examples in Carmichael *et al.* 1994. See also the review of this book by Richard Bradley, *Antiquaries Journal*, 75 (1995), Book Review Supplement, 6–7.

127 Tilley 1994, 24.

128 Tilley (1996, 169) suggests that there was a desire that prominent tors be visible from stone circles on Bodmin Moor, and that this played a major role in their precise location. A similar idea was first proposed by John Barnatt (*Prehistoric Cornwall: The Ceremonial Monuments*, Turnstone Press, Wellingborough, 1982, 73). In fact, A. L. Lewis suggested over a century ago that the builders of stone rings might have thought hills to be sacred, and hence placed their monuments in sight of conspicuous summits (Lewis 1892, 148). Similar ideas have been proposed for many years by authors such as Burl (see, e.g., Burl 1976, 266; 1979a, 25–6; for other examples see Ruggles and Burl 1985, S47) and have also emerged from the work described in chapters five to seven.

129 Thomas 1991a, 47.

130 Work on the location of cup-and-ring markings suggests they may have been the product of mobile patterns of settlement and land-use, for example the use of summer grazing (Richard J. Bradley, Jan Harding and Margaret Mathews, 'The siting of prehistoric rock art in Galloway, south-west Scotland', *PPS*, 59 (1993), 269–83. See also, for example, Barrett 1994, 140; and Barnatt 1996, 43 with reference to the Peak District.

131 In the Early Neolithic, for example, there is evidence that axe quarries were not worked by specialists but by different groups of people, possibly visiting on a seasonal basis; in addition, they were located, often unnecessarily, in difficult and dangerous mountain or clifftop locations, suggesting perhaps that something—the spectacular nature of a location, maybe—singled it out as of particular (sacred?) importance for this activity (Richard J. Bradley and Mark Edmonds, *Interpreting the Axe Trade: Production and Exchange in Neolithic Britain*, CUP, Cambridge, 1993, 134; Parker Pearson 1993, 36).

132 See, for example, Barrett 1994, 136–46.

133 Tilley 1994, 28.

134 This observation is no less relevant in the Mesolithic, although specific conclusions are limited by the lack of monumental evidence. In fact, it is in the mobile, oral, non-monument building tradition that various kinds of relationship between society and the natural world were almost certainly first forged, creating ways of understanding things—world-views—which may have had much in common with those that developed later, despite elaborations and reorientations of meaning, and the 'concretising' of ideology in the form of monuments. As an example, there are many underlying conceptual similarities between Inca world-view and that of indigenous hunter-gatherer groups elsewhere in South America (see, e.g., Gary Urton, 'The use of native cosmologies in archaeoastronomical studies: the view from South America', in Williamson 1981, 285–304).

135 Tilley 1994, 197. This conclusion was reached as a result of walking along the Dorset cursus and noting the changing relationships with topographic features and contemporary monuments such as long barrows (*ibid.*, 173–96). See also Bradley 1993, 50–3; Barrett 1994, 137–8.

136 e.g. Williamson 1984, 156–62; Griffin-Pierce 1992, 21, 92–6; Trudy Griffin-Pierce, 'The *Hooghan* and the stars', in Ray A. Williamson and Claire R. Farrer (eds), *Earth and Sky: Visions of the Cosmos in Native American Folklore*, University of New Mexico Press, Albuquerque, 1992, 110–30, pp. 113–14.

137 Hieb 1979.

138 Von Del Chamberlain, *When Stars Came Down to Earth: Cosmology of the Skidi Pawnee Indians of North America*, Ballena Press/Center for Archaeoastronomy, Los Altos/College Park, 1982, 155–62, 178–83.

139 Johannes Wilbert, 'Warao cosmology and Yekuana roundhouse symbolism', *Journal of Latin American Lore*, 7(1) (1981), 37–72.

140 The village of Misminay, for example, is laid out in a cross formation that reflects the division of the sky into four by the two axes of the Milky Way when it passes through the zenith. To its inhabitants, the Vilcanota river is regarded as a reflection of the Milky Way in the sky, and both are conceived as part of an integrated system that serves to circulate water through the terrestrial and celestial spheres (Urton 1981, ch. 2).

141 Stephen M. Fabian, 'Ethnoastronomy of the eastern Bororo Indians of Mato Grosso, Brazil', in Aveni and Urton 1982, 283–301.

142 Sosa 1989.

143 Oswald 1997 and references therein.

144 Parker Pearson and Richards 1994, fig. 2.4, which is reproduced from Alastair Oswald, *A Doorway on the Past: Roundhouse Entrance Orientation and its Significance in Iron Age Britain*, BA dissertation, University of Cambridge.

145 J. D. Hill, 'How should we understand Iron age societies and hillforts? A contextual study from southern Britain', in J. D. Hill and C. G. Cumberpatch (eds), *Different Iron Ages: Studies on the Iron Age in Temperate Europe*, Tempus Reparatum (BAR International Series 602), Oxford, 1995, 45–66, p. 53 and fig. 7.

146 J. D. Hill, *Ritual and Rubbish in the Iron Age of Wessex*, Tempus Reparatum (BAR British series 242), Oxford, 1995, esp. ch. 11; Oswald 1997, 94.

147 Fitzpatrick 1997, 77–8.

148 Castleden 1993, 198–201.

149 *Ibid.*, 201. Cf. Parker Pearson and Ramilisonina 1998, 322.

150 *Ibid.* For a development of this theme see 'Part 2: Describing a circle' (chs 6–10) in Richard Bradley, *The Significance of Monuments*, Routledge, London, 1998. See also Gibson 1994, 192. Colin Richards (1996b) also takes this idea further, suggesting that henge monument architecture represents an embodiment of the perceived landscape in which water is a key element.

151 Darvill 1996, 254–5. Unfortunately, the evidence to support the idea seems weak. Unlike the NE–SW axis, the NW–SE one is not marked in any direct way, a disparity that seems strange if the fundamental cosmological principle was a division of the world into four quarters. In addition, the evidence in patterns of distribution of different types of monuments and artefacts around Stonehenge (*ibid.*) is far from conclusive.

152 Tilley 1996, 168–71.

153 In Assyria, for example, temples were oriented south-east and there exists written evidence of the meaning of these orientations (Giovanni Lanfranchi, 'Astronomia e politica in età neo-assira', in Lincei 1995, 131–52, esp. p. 151). There is historical evidence to suggest that certain buildings and roads in ancient Greece were aligned astronomically (Alan C. Bowen and Bernard R. Goldstein, 'Meton of Athens and astronomy in the late fifth century BC', in Erle Leichty, Maria de J. Ellis, and Pamela Gerardi, *A Scientific Humanist: Studies in Memory of Abraham Sachs*, Occasional Publications of the Samuel Noah Kramer Fund 9, Philadelphia, 1988, 39–81, pp. 76–7).

An obvious example is the orientation of Christian churches generally within the arc of the rising sun, which in many cases may be explained by a practice of aligning them towards sunrise on the day when construction began (Henry Chauncy, *The Historical Antiquities of Hertfordshire*, London, 1700 [repr. Kohler and Coombes, Dorking, 1975], vol. 1, 88). There is also evidence that the orientations of medieval churches in many European countries—such as Romanesque churches in Central Europe, and including England before the sixteenth century, when the practice was eliminated by the Puritans—were 'adjusted' towards the rising sun on the patron Saint's Day, perhaps in an attempt to 'preserve' earlier, or pagan, practices in the context of the 'official' Christian religion. For an early work covering this issue see Heinrich Nissen, *Orientation: Studien zur Geschichte der Religion*, Weidmannsche Buchhandlung, Berlin, 1906; see also, for example, Maria G. Firneis and Christian Köberl, 'Further studies on the astronomical orientation of medieval churches in Austria', in Aveni 1989a, 430–5. Writing in 1823, Wordsworth referred to this ancient practice in England: his lines are quoted by William B. Dinsmoor, 'Archæology and astronomy', *Proceedings of the American Philosophical Society*, 80 (1939), 95–173, pp. 101–2 and also by Anthony F. Aveni, 'Evolution of the Maya solar orientation calendar', in Romano and Traversari 1991, 15–22, p. 15. However, see also Richard Morris, *Churches in the Landscape*, Dent, London, 1989, 208–9.

154 Bradley in Barrett *et al.* 1991, 58. The remark refers specifically to the Dorset cursus.

155 Cf. Renfrew 1984, 178–80.

156 Bradley in Barrett *et al.* 1991, 56.

157 Cf. Bradley 1993, 68; Darvill 1996, 177–8. In a similar way, Maori temples were oriented so that the central post would cast a shadow along the ridge pole during March when ceremonial meetings were due to be held (Ettie A. Rout, *Maori Symbolism*, Kegan Paul, London, 1926, 293; see also Thorpe 1981, 284).

158 Paul Gosling, *Archaeological Inventory of County Galway. Volume 1: West Galway (including Connemara and the Aran Islands)*, Stationery Office, Dublin, 1993, no. 57.

159 'Through prehistoric eyes', ch. 13 in Tim Robinson, *Setting Foot on the Shores of Connemara and Other Writings*, Lilliput Press, Dublin, 1996. The chapter derives from an address to the Prehistoric Society in 1995. The quotation is from p. 201.

160 Castleden 1987, 241.

161 Parker Pearson and Ramilisonina 1998. See also Whittle 1997, 151–2.

162 Parker Pearson and Ramilisonina 1998, 317–19, incl. figs 6 and 7.

163 *Ibid.*, 310, 313–14.

164 e.g. Krupp 1997, 2.

165 In the context of Mesoamerica see, for example, Doris Heyden, 'An interpretation of the cave underneath the pyramid of the sun in Teotihuacan, Mexico', *American Antiquity*, 40 (1975), 131–47; 'Caves, gods and myths: world-view and planning in Teotihuacan', in Elizabeth Benson (ed.), *Mesoamerican Sites and World-views*, Dumbarton Oaks, Washington DC, 1981, 1–40. Examples exist from around the world.

166 Cf. Bede, *De Temporum Ratione*, XXVI, ll. 4 and 20 (Bedae, *Opera de Temporibus*, ed. Charles W. Jones, The Mediaeval Academy of America, Cambridge Mass., 1943, 228–9). Likewise the stars may be perceived as close on a clear night, but more remote on a cloudy night or at twilight (Edward G. Ballard, 'The visual perception of distance', in F. J. Smith (ed.), *Phenomenology in Perspective*, Martinus Nijhoff, The Hague, 1970, 187–201, p. 195).

167 In Misminay, for instance, 'One encounters a general disinclination [for people] to point to anything sacred; among other things this includes mountains, stars and rainbows. The ways of getting around the problem of pointing directly at something are ingenious and often frustrating. A sacred mountain, for instance, may be pointed at by holding your hands behind your back and gesturing with one elbow . . . On one occasion, an old man who was very interested in astronomy agreed to discuss the stars with me, but only if I remained seated on the floor inside his hut during the interview.' (Urton 1981, 105–6.)

168 Cf. Chippindale 1994, 227–8. See North 1996 for detailed speculations on Wessex monuments from long barrows to the Stonehenge sarsens.

169 But see McNally and Ruggles 1997.

170 At the base of the northern staircase of this pyramid is the carved head of a serpent. For about an hour before sunset on days close to the equinoxes, the light of the evening sun falling across the stepped corner of the pyramid makes the serpent's body 'appear' (Aveni 1980, 285–6; 1997, 145–6).

171 Anna Sofaer, Volker Zinser and Rolf M. Sinclair, 'A unique solar marking construct', *Science*, 206 (1979), 283–5.

172 For a brief overview and discussion see Ruggles 1999.

173 These are vertical tubes incorporated in specialised ceremonial structures down which the light of the sun shone as it passed close to the zenith at noon on certain days in the year. See Anthony F. Aveni and Horst Hartung, 'The observation of the sun at the time of passage through the zenith in Mesoamerica', *AA* no. 3 (*JHA* 12) (1981), S51–70.

174 An account has recently appeared in the Swedish magazine *Populär Arkeologi*: Curt Roslund and Emília Pásztor, 'Stonehenge ett ljusspel?', *Populär Arkeologi*, 15(4) (1997), 18–22.

175 Krupp 1994, 132.

176 Nicholas J. Saunders, 'Stealers of light, traders in brilliance: Amerindian metaphysics in the mirror of conquest', *Res: Anthropology and Aesthetics*, 33 (1998), 225–52.

177 Cf. Barbara A. Weightman, 'Sacred landscapes and the phenomenon of light', *The Geographical Review*, 86 (1996), 57–71; Richards 1996b.

178 For the Aztecs, mica was *metzcujtatl*, 'the excrement of the moon', associated with cosmic forces, soft, buoyant, and light (Bernardino de Sahagún, *Florentine Codex: General History of the Things of New Spain* (12 books), *Book 11*, transl. and eds Arthur J. O. Anderson and Charles E. Dibble, School of American Research and University of Utah, Santa Fe, vol. 12 (of 13), 1963, 235).

179 Richard J. Bradley, 'Architecture, imagination and the Neolithic world', in Steven Mithen (ed.), *Creativity in Human Evolution and Prehistory*, Routledge, London, 1998, 227–40.

180 See also Ruggles 1999. 'Cosmologies . . . are both drawn from and explain the natural and cultural world' (Richards 1996a, 205).

CHAPTER 10

1 John Ziman, *Prometheus Bound: Science in a Dynamic Steady State*, CUP, Cambridge, 1994, 61.
2 Anthony F. Aveni, 'Archaeoastronomy today', in Williamson 1981, 25–8, p. 25.
3 W. James Judge, 'Archaeology and astronomy: a view from the southwest', in Carlson and Judge 1987, 1–8, p. 5.
4 See Introduction, note 59.
5 North 1996, xxxv.
6 Renfrew 1996.
7 Castleden 1993, 206.
8 Aveni 1997, 91, 85.
9 *Ibid.*, 91.
10 Tilley 1996, 169.
11 This is because most archaeologists feel that by concentrating exclusively on the monuments themselves—considering them purely structurally, or even as points in the physical landscape—we fail to consider the human dimension (see, e.g., Thomas 1988). Their aim, quite rightly, is to provide a broader picture of human activity in the landscape so that ritual activity can be interpreted within that context.
12 For an example of the latter see recent work on the intervisibility of Wessex long barrows (note 16 below).
13 This makes the assumption that the extent to which a given monument has been subject to destruction is independent of any such relationships.
14 Ashbee 1984, appendix 10.
15 See Burl 1987a, 27.
16 There are already some indications that the position of these monuments in the landscape was important, at least with respect to one another. Wheatley (1995; 1996; see also Renfrew and Bahn 1996, 192–3) has concluded that the intervisibility of long barrows on Salisbury Plain is significantly different from the degree to which the barrows are intervisible with random locations. On the other hand, this is not the case in the Avebury region. This seems to indicate that forest cover was not a problem, despite the depiction of long barrows such as Fussell's Lodge (SU 191325) as standing in clearings (Paul Ashbee, 'The Fussell's Lodge long barrow excavations 1957', *Archaeologia*, 100 (1966), 1–80, fig. 9, reproduced, for example, as Burl 1979b, fig. 36; Lynch 1997, fig. 17).
17 At a latitude of about 51°, a 1° uncertainty in altitude for a given azimuth corresponds to an uncertainty in declination of approximately 0°·8 around east and west and 1° around north and south. See also Astronomy Box 2.
18 For example, on passage tombs in the cemeteries of Carrowkeel, Loughcrew, and elsewhere in central and eastern Ireland see Patrick 1975, tables 1 and 2.
19 See Archaeology Box 2, note 4.
20 Cf. Davidson and Henshall 1989, 85.
21 Barclay and Harding 1999.
22 Ruggles 1997, 226–7.
23 There are over a hundred documented stone pairs (see Ruggles 1994a, note 7), though some at least may simply be the remnants of longer rows. The five-stone circles, of which Ó Nualláin 1984a lists over forty, indisputably form a separate morphological category (*ibid.*, 3).
24 For orientation plots of Cork and Kerry stone rows and pairs see Ó Nualláin 1988, fig. 2; for stone rings see Ó Nualláin 1984a, fig. 24; for wedge tombs see de Valera and Ó Nualláin 1982, fig. 36. See also chapter five, notes 100–2.
25 Burl 1987b, 16. For gazetteers see *ibid.*, 17–18 as well as Burl 1993. Rows of many small stones, some attached to stone circles, are also found in some numbers in mid-Ulster. A reconnaissance of the Ulster monuments by this author, including theodolite surveys, took place in 1998 (Ruggles, forthcoming).
26 Burl's (1988a) study of nine prime examples in the Lake District has been described in chapter eight, but over fifty such monuments are known (*ibid.*, 177).
27 There does exist an archaeoastronomical study of Irish henge orientations (Ronald Hicks, 'Irish henge orientation: preliminary results and some problems', in Williamson 1981, 343–50) but the orientation and declination data for each monument are not tabulated, and in any case more reliable surveys are needed (*ibid.*, 347).
28 See Burl 1981a, 257–65.
29 On the Perthshire pairs see Stewart 1966 (which contains a gazetteer in Appendix II) and Burl 1993, 198–201. On the four-posters see Burl 1988b, 144–95. It may be misleading, however, to associate four-posters with stone rows, and especially to view them as 'double pairs'; Burl (1988b, 6) argues that since they rarely form precise rectangles it may be better to view them as a degenerate form of stone circle.
30 Ritchie 1982, 37–9.
31 Ruggles 1985, S126–8.
32 e.g. Andrew Sherratt, review of Burl 1993, *Antiquity*, 68 (1994), 177–9, p. 178.
33 Cf. chapter nine, note 87.
34 Richard Bradley, priv. comm., 1997. This might help to explain the solar orientation of the passage tombs at Balnuaran of Clava in spite of the fact that the orientation pattern of the Clava cairns as a whole seems to be related to the moon (see p. 130).
35 Powell 1994.
36 See, e.g. Thomas 1991a, ch. 4; 1996, 197–205. For a particular example see Joshua Pollard, 'Inscribing space: formal deposition at the Later Neolithic monument of Woodhenge, Wiltshire', *PPS*, 61 (1995), 137–56.
37 See chapter eight, note 147.
38 It is possible that this period of 'original use' was very short, or even that the erection of small stone circles and alignments was the important ceremonial event in itself, and the monuments themselves did not subsequently become a focus for continued ritual activity. 'In Melanesia, for example, megalithic alignments and circles are erected in the context of male initiation ceremonies . . . the initiates are required to sponsor the construction of these monuments and to offer a large feast, but the megaliths themselves do not become a focus for continued ritual activity after the initial ceremony. Each generation of initiates is required either to build a new monument or, in some cases, to extend an existing alignment' (Patton 1993, 120). See in particular John Layard, *Stone Men of Malekula: Vao*, Chatto and Windus, London, 1942.
39 e.g. Thomas 1991a, 29, quoting Bradley; Barrett *et al.* 1991, 1–2. The quote is from Gosden 1994, 127.
40 Bradley in Barrett *et al.* 1991, 106; see also pp. 127–8. The Dorset Cursus, 'even though it went out of direct use fairly quickly . . . influenced the placement of many subsequent monuments, each connected to the rest by their common reference to the Cursus' (Gosden 1994, 98).
41 At Saint-Just, excavations published in the early 1980s revealed a 'ritual landscape' apparently spanning the whole of the Neolithic period and incorporating Early Neolithic long mounds, Middle Neolithic passage tombs, a Late Neolithic lateral entrance grave and a semi-circular setting of stones of uncertain date (Patton 1993, 114–17). The landscape surrounding the more famous Carnac alignments also spans the whole of the Neolithic period (*ibid.*, 117–20). Patton argues (*ibid.*, 120) that older monuments were appropriated by the alignment builders as elements in their own religious system, and the alignments were themselves appropriated in their turn.
42 Patrick and Freeman 1988; Gail Higginbottom and Roger Clay, 'Reassessment of sites in northwest Scotland: a new statistical approach', *AA* no. 24 (*JHA* 30) (1999), in press.
43 The example relates to tombs surrounding the pre-Roman town of Nepi in South Etruria, which are found to cluster in areas away from the settlement, just across a steep ravine (Martin Belcher, Andrew Harrison, and Simon Stoddart, 'Analyzing Rome's hinterland', in Gillings *et al.* 1998, in press, and references therein).
44 Cleal *et al.* 1995, 40.
45 Bender 1992, 745–6.
46 Richards 1990a, 305.
47 Richards 1990a; see also Parker Pearson 1993, 55–9.
48 Richards 1990b; Parker Pearson and Richards 1994, 41–7.
49 Richards 1990b, fig. 5.5; Parker Pearson and Richards 1994, fig. 2.3.
50 It is rather misleading, however, to say that the hearths were 'aligned upon' the four solstitial directions (Parker Pearson 1993, 59), at least if this is taken to imply a precision greater than some tens of degrees; and it is simply incorrect to claim (*ibid.*) that Orkney might have been considered a very special place because the solstitial directions are perpendicular to one another here (see chapter nine, note 8).
51 Richards 1990a, 312–13. One might ask, however, why in this case people did not orient the tomb entrance to the north, where it would never be illuminated by the light of the sun.
52 Fraser 1983, 368–71; Davidson and Henshall 1989, 85.
53 Chapter eight, note 201.
54 Ironically, some of the very sets of monument alignments that led Thom to propose his megalithic calendar might be relevant here. While the possibility of high-precision solar alignments can be effectively ruled out for all the reasons identified in chapter three, it remains possible that patterns of variation in low-precision alignment upon sunrise or sunset at different times of year from one monument to another might give some insight into seasonal patterns of movement.
55 Ruggles 1997, 227.

56 Cf. Bradley 1993, 51–2; Tilley 1994, 173–96.

57 See Glyn T. Goodrick, 'VRML, virtual reality and visualisation: the best tool for the job?', in Vince Gaffney and P. Martijn van Leusen (eds), *CAA97: Archaeology in the Age of the Internet*, Tempus Reparatum (BAR International Series), 1998, in press. At the time of writing, an example under development can be viewed on the Web at http://www.gerty.ncl.ac.uk/jan/interim.htm.

58 On the widespread evidence for correlations between hunting rituals and the moon, and a controversial interpretation, see Knight 1991, ch. 10.

59 However, indirect support comes from elsewhere in prehistory. At the Minoan cemetery at Armenoi in north-west Crete, the spread of orientations of 224 tombs corresponds closely with the range of possible rising positions of the moon (Papathanassiou and Hoskin 1996). Prominent on the skyline to the east of the cemetery is Mount Vrysinas, on whose summit an important Cretan peak sanctuary stood in earlier times which may well have been dedicated to a lunar goddess (Maria Papathanassiou, Michael A. Hoskin and Helen Papadopoulou, 'Orientations of tombs in the Late-Minoan cemetery at Armenoi, Crete', *AA* no. 17 (*JHA* 23) (1992), S43–55, p. S55, note 12). The oldest of the tombs so far excavated is directly oriented upon the peak. The authors suggest that the cemetery was originally located here on a date when the moon was seen rising in line with a mountain sacred to the moon, and that it was important that each subsequent tomb should be oriented within the arc of moonrise.

60 Burl 1980, 195–6; Ruggles 1988b, 245. Cf. p. 154.

61 Cf. Burl 1993, 64: 'The latitude of 58°N is critical. Nowhere farther south in Europe could the major moon between its rising and setting be seen to skim the horizon. It would rise much higher in the sky. Not until 58° south of the equator, around Cape Horn, does the same lunar phenomenon occur.'

62 The moon was apparently an important deity amongst the Scandinavian Lapps (Rafael Karsten, *The Religion of the Samek*, E. J. Brill, Leiden, 1955, 33–5) and Marshack (1972, 336–7, quoting Boris A. Frolov) has claimed that indigenous hunter-gatherers in modern Siberia paid great attention to the moon, but in general there seems little in modern studies of shamanism amongst northern groups to support any prepossession with the moon (see, e.g., Louise Bäckman and Åke Hultkrantz, *Studies in Lap Shamanism*, Almqvist and Wiksell, Stockholm, 1978; Juha Y. Pentikäinen (ed.), *Shamanism and Northern Ecology*, Mouton de Gruyter, Berlin, 1996).

63 Fitzpatrick 1997, 73–5. See also Cunliffe 1997, 188–90.

64 Cunliffe 1997, 190–1.

65 Piggott 1968, 122–5.

66 Gibson 1995.

67 Power 1992, no. 2.

68 Mizen Peak, some 12·5 km distant, is conspicuous between closer, more rounded eminences.

69 The Samhain alignment is shown in a previously unpublished plan by Boyle Somerville, dated 1931 (see William O'Brien, *Megalithic Tombs and Coastal Settlement in Prehistoric South-west Ireland*, Bronze Age Studies 4, Galway UP, Galway, 1998). Calculations from map data confirmed by compass-clinometer measurements in September 1998 give the azimuth and altitude of the summit of Mizen Peak as 242°·8 and 1°·0 respectively, yielding a declination of −16°·0. This corresponds to sunset on around 5 February and 3/4 November.

70 William O'Brien, 'Altar tomb and the prehistory of Mizen', *Mizen Journal*, 1 (1993), 19–26; see also Waddell 1998, 97–8.

71 See p. 147.

72 See, for example, Bell 1994; 'Interpretation and testability in theories about prehistoric thinking', in Renfrew and Zubrow 1994, 15–21.

73 This is in tune with recent developments in archaeological thought, which reject earlier 'hypothesis testing' approaches (see, e.g., Shennan 1988, 5; Hodder 1991, 185–8), shifting the emphasis away from testability as an end in itself towards whether the theories being tested are in themselves interesting, or valuable (Hodder 1992, 2).

74 It is at least clear that 'negative' data—that is, data that do not fit a given theory—should be considered on an equal basis alongside positive data in weighing up the idea, and presented with equal prominence so that others can judge it fairly. To simply say, for example, that 'stone rows can be aligned upon hilltops', even in the context of a broader development of ideas (Bradley 1993, 50), apparently assumes that where this occurs it was deliberate, and begs several questions: at what proportion of rows is this the case, to what precision, are large nearby hills or small distant peaks involved, and so on. We have to be careful not to return to a situation where we are simply drawing attention to the data that fit an idea and ignoring the rest; of using the archaeological record uncritically for *post hoc* justification. See also Ruggles 1994b, 497–9.

An example of more balanced thinking is the following. Despite the solstitial alignments claimed at some other cursus monuments (see chapter eight), the Cleaven Dyke in Perthshire is aligned upon sunrise and sunset at other times in the year. Rather than postulating a deliberate alignment upon the sun at certain mid-year dates, horizon features were examined not just along, but also away from, the actual alignment. This revealed that a solstitial orientation combined with an alignment upon a prominent hill could have been achieved by shifting the orientation by a mere 10°, yet the builders did not do this. Taking this evidence into account, we are led to conclude that solar alignment was probably not a consideration here (Clive Ruggles, 'The possible astronomical alignment of the Cleaven Dyke', in Gordon Barclay (ed.), *The Cleaven Dyke: Linear and Burial Monuments in the Neolithic of Tayside*, Society of Antiquaries of Scotland (Monograph Series), Edinburgh, in press).

75 We have already demonstrated this in the context of statistical reassessments of Thom's work (chapter three). In the context of archaeoastronomy see also Ruggles 1988b, 248–9; Ruggles and Saunders 1993b, 16–17. In the wider archaeological context see Clive R. Orton, *Mathematics in Archaeology*, Collins, London, 1980, 219; James A. Bell, 'On applying quantitative and formal methods in theoretical archaeology', *Sci. Arch.*, 28 (1986), 1–8; Ruggles 1986. Anthropological data too 'often violates all of the assumptions underlying conventional statistical procedures' (quote from Michael Fischer, in Gustaff Houtman, 'Interview with Michael Fischer on computing and anthropology', *Anthropology Today*, 11(2) (1995), 5–8, p. 5).

76 Processual archaeologists placed a good deal of emphasis upon the use of (classical) statistical techniques, but one obvious flaw in doing this in most circumstances is that the archaeologist who forms a hypothesis or model did so in the first place on the basis of a combination of background knowledge, including prior knowledge of relevant material evidence upon which the idea will subsequently be tested. As an example, Bradley (1984) adopts an approach of examining various things in the archaeological record to see how well they accord with particular 'basic models' (*ibid.*, 54). Despite statements like 'According to our model, one would expect . . . In fact, the chronology of the more elaborate graves shows just this evidence' (*ibid.*, 94), it is difficult to believe that the model was formulated in complete ignorance of the evidence cited. (This is not meant as a criticism of Bradley's procedure but as a demonstration that the assumption of independent data is inappropriate.)

Curiously, in an article published in 1984, Christopher Chippindale ('Life around Stonehenge', *New Scientist*, 5 April 1984, 12–17, p. 16) reversed the argument, rejecting Hawkins's and Hoyle's ideas about Stonehenge, and by extrapolation Thom's 'megalithic astronomy', on the grounds that their conclusions are unprovable because we lack an independent data set on which to test them. Such an argument, if extended, would reject virtually every interpretation in the mainstream of prehistoric archaeology.

77 Cf. Shennan 1988, 4.

78 For reviews of methodological developments in archaeoastronomy see, for example, Stanisław Iwaniszewski, 'De la astroarqueología a la astronomía cultural', *Trabajos de Prehistoria*, 51(2) (1994), 5–20; Clive Ruggles, 'The role of excavation in archaeoastronomy', in Emília Pásztor (ed.), *Archaeoastronomy from Scandinavia to Sardinia: Current Problems and Future of Archaeoastronomy, 2*, Roland Eötvös University, Budapest, 1995, 45–54.

79 The 'green' and 'brown' terminology is due to Aveni (Anthony F. Aveni, 'Archaeoastronomy: past, present, and future', *Sky and Telescope*, 72 (1986), 456–60; Aveni 1989b) and reflects the colours of the covers of the respective 'Old World' and 'New World' volumes arising from the first International Symposium on Archaeoastronomy held in Oxford in 1981 (Heggie 1982a; Aveni 1982). On approaches in American archaeoastronomy at this time see, e.g., Aveni 1980; 1982; Aveni and Urton 1982.

While many critiques within 'green' archaeoastronomy around this time addressed issues of data selection and objectivity (e.g. Moir 1981; Heggie 1981a, 1981b; Ruggles 1981, 1982a), similar 'brown' critiques (e.g. Ruggles and Saunders 1984; Köhler 1991) have been comparatively rare.

80 Reaction against this emphasis has even occasionally reached the extreme view that 'the importance of [various celestial phenomena] for the people of a past society must . . . be assessed, before a conscious alignment towards those phenomena can be seriously postulated', in other words, it seems, alignment data should not be considered at all if there is no historical or ethnographic evidence to back them up (Köhler 1991, 131). For a debate on these issues see *A. and E. News* nos 5, 6 (1992), 9 (1993), 17, 18 (1995).

81 See, for example, Anthony F. Aveni, 'On seeing the light: a reply to *Here Comes the Sun* by Dearborn and Schreiber', *AAJ* 10 (1988), 22–4; 'Time among the Inca' [essay review], *AA* no. 17 (*JHA* 23) (1992), S68–73, p. S69; Aveni 1995, S78.

82 Finding satisfactory ways to do this is a non-trivial problem. For example, an idea that three cross-circles, pecked into building floors and rocks in the vicinity of Teotihuacan, could have acted as surveyors' benchmarks in laying out the city (Anthony F. Aveni, Horst Hartung, and Beth Buckingham, 'The pecked cross symbol in ancient Mesoamerica', *Science* 202 (1978), 267–79) was subjected to an archaeological critique by Ruggles and Saunders (1984), most of which was concerned with chronological phasing. The critique began, however, with the observation that there were at least eleven further pecked-cross circles at Teotihuacan which, not being suitably located, clearly could not have been used in this way. The implication is that, by selecting three suitable candidates out of a possible fourteen or more, any apparent support for the original idea is severely diluted: the data have been chosen to fit the theory.

In a counter-critique Aveni (Anthony F. Aveni, 'Comment', *AA* no. 7 (JHA *15*) (1984), S107–10) pointed out that other relevant evidence, together with our wider experience of human behaviour, leads us to believe that the cross-circles had many functions and meanings, and we should not dismiss the interpretation that just three of them were used (amongst other things) as surveyors' benchmarks. While this point is well taken, if we are not careful we end up only ever using carefully selected archaeological evidence for *post hoc* justification. *In extremis*, this would deny the archaeological record any direct role in modifying existing ideas.

83 On the other hand, W. Breen Murray ('The northeast Mexican petroglyphic counting tradition: a methodological summary', in Vesselina Koleva and Dimiter Kolev (eds), *Astronomical Traditions in Past Cultures*, Sofia, 1996, 14–24, pp. 15–16) points out that *post hoc* attempts at a 'comprehensive' interpretation of symbolic artefacts have sometimes achieved great successes, as in the 'decoding' of Maya numerical and calendrical symbols (Michael D. Coe, *Breaking the Maya Code*, Thames and Hudson, London, 1992).

84 This accords with the view of cognitive psychology that 'understanding comes from the interaction between [a person's] prior knowledge and . . . new information, mediated by created human imagination' (Lynton McLain, book review in *The Times Higher Education Supplement*, 18 August 1995, 23).

85 For examples from pre-Columbian America see Ruggles 1986; 1994b.

86 This was how Bayesian methods were presented, rather disparagingly, to archaeologists in an influential book by James E. Doran and Frank R. Hodson (*Mathematics and Computers in Archaeology*, Edinburgh UP, Edinburgh, 1975, 33–6; see also 53–4). For a major attempt to justify Bayesian methods in this context see Merrilee H. Salmon, *Philosophy and Archaeology*, Academic Press, New York, 1982, ch. 3. For commentaries see Ruggles 1986, 11–12; Bell 1994, 160–3.

87 Ruggles 1986; see also Bell 1994, 163–5. Bell clearly and helpfully portrays my own position but confuses my nomenclature: actually, I termed the earlier use of Bayesian inference for inductive assessment 'semi-Bayesian', because I believed it used Bayesian inference in a very restrictive way, as well as misapplying it within archaeology; whereas its proposed use for evaluating data is, I believe, fully in tune with what the Bayesian approach is able to achieve, as well as fully consistent with the needs of archaeologists.

88 Aveni 1995.

89 Similarly, see Richard Reece, 'Are Bayesian statistics useful to archaeological reasoning?', *Antiquity*, 68 (1994), 848–50, and the response by C. E. Buck, J. B. Kenworthy, C. D. Litton and A. F. M. Smith, *ibid.*, 850.

90 People may be able to justify their prior beliefs by pointing to existing evidence from the material record (possibly by showing how certain data combine with earlier priors to produce the proposed distribution as posterior), by arguing from analogy (see Ruggles and Saunders 1993b, 18–22), by arguing from phenomenology and personal experience, or by a combination of these. The justification might have a quantitative element, but could simply rest on personal judgements.

91 Tilley 1994, 11.

92 On avoiding *ad hoc* arguments from a Bayesian standpoint, see Buck *et al.* 1996, 172.

93 Even those who rightly eschew positivism, and hence would certainly not claim any absolute knowledge of the past, can occasionally make statements apparently with total conviction. Thus according to Aveni (Anthony F. Aveni, *Ancient Astronomers*, Smithsonian Books, Washington, 1993, 100–1) 'The Maya . . . built observatories, such as [the Caracol at] Chichén Itzá, to chart the movements of the stars and planets'. In fact, there is no direct evidence that the windows in this ruined circular building 'were used to sight the sun and Venus' (*ibid.*) and although ethnohistory and iconography strongly attest to a Mayan interest in the zenith sun and Venus, the alignment evidence itself from three surviving windows (see, e.g. Aveni 1980, 258–67), is as equivocal as any that has been presented in this book. There can be little objection, however, to the statement of an opinion that 'Astronomical observations seem most successful in accounting for the peculiarities of its structure and orientation' (*ibid.*, 260); only a desire to test it out on further evidence.

94 The notebooks record some eight hundred site visits: some places were visited many times. Thom's field notebooks and most of his original plans are now deposited with The National Monuments Record of Scotland. For a catalogue of the plans see Ferguson 1988. A full list of Thom's publications can be found in Ruggles 1988a, 22–30.

95 Ferguson 1988, 31.

96 Kintigh 1992, 4; see also Aveni 1992b, 1. The same could actually be said, conversely, of those archaeologists who appear content to make sweeping statements with apparently no attempt to understand and critically appraise the relevant archaeoastronomical literature (see chapter nine).

97 The Houston Museum of Natural Science, Texas, for example, provides a thriving local Middle School educational programme which uses archaeoastronomy as a means to increase science awareness and basic technical skills in a wider context through site-based and planetarium-based project work (C. Sumners, A. Knox, and R. Wyatt, 'Archaeoastronomy research and public education in a Digistar computer-generated planetarium', in Fountain and Sinclair 1999, in press).

98 For example, the Griffith Observatory in Los Angeles runs a variety of public lectures and tours and frequently features archaeoastronomical topics in its magazine *The Griffith Observer*.

99 Undergraduate courses have existed, exist currently or are being proposed at several universities in the USA and Europe, many within interdisciplinary programmes. For a more recent list of relevant courses see *A. and E. News*, no. 25 (1997), 1 and 3, and for additions to that list see later issues. See also David Dearborn, 'Teaching archaeoastronomy', *A. and E. News*, no. 8 (1993), 1 and 4; S. M. Nelson and R. E. Stencel, 'Archaeoastronomy in the core curriculum', in Fountain and Sinclair 1999, in press. The 'Oxford 5' conference in 1996 established a Working Group on Education in Cultural Aspects of Astronomy, with the aim of co-ordinating curriculum developments both in the USA and Europe. On multimedia resource bases to support such developments see Clive L. N. Ruggles, 'An image database for learning archaeoastronomy', in Ruggles 1993a, 335–42; Bryan E. Penprase, 'Ancient cosmology resource center', in Fountain and Sinclair 1999, in press.

100 On the one hand, the topic of indigenous perspectives on astronomy (traditional astronomies), by imparting views of the universe that differ substantially from the Western one, has great potential for general training in cultural perspectives and critical thinking. On the other, Native American students (for example) tend to reject science as a Western development unrelated to their societies, and many feel that it cannot be practised without loss of cultural identity. Archaeoastronomy can help by offering a comparative approach in a sympathetic subject area. Projects exploiting both these aspects are under development, for example, in the US south-west (David Dearborn, priv. comm., 1995; Nancy C. Maryboy, Navajo Community College, Tsaile, priv. comm., 1996).

101 Renfrew and Bahn 1996, ch. 10; Hodder 1991, ch. 7. See also Archaeology Box 8.

102 It also has the advantage that its survey-based field techniques are essentially non-destructive, something of particular importance in view of the destructive potential and high costs of excavation itself (Renfrew and Bahn 1996, 67).

103 The term 'cultural astronomy' has been proposed recently (Ruggles and Saunders 1993b; Ruggles 1993b; Stanisław Iwaniszewski, 'Archaeoastronomy and cultural astronomy: methodological issues', in Lincei 1995, 17–26) although the use of the word 'culture' brings its own problems (cf. Archaeology Box 8, note 2) and it is a term that many would prefer to avoid. Perhaps we should move, in name as well as in approach, towards 'archaeocosmology'; indeed, at least one Museum for Ethnocosmology already exists, in Molėtai, Lithuania.

104 Aveni 1994, 34.

105 Anthony F. Aveni, 'The real Venus-Kukulcan in the Maya inscriptions and alignments', in Virginia M. Fields (ed.), *Sixth Palenque Round Table 1986*, University of Oklahoma Press, Norman, 1991, 309–21, p. 309.

106 Ruggles 1997, 205.

107 Examples include Belmonte Avilés 1994; Krupp 1994; Walker 1996; and Aveni 1997. It appears also on books concerning related themes, e.g. Hutton 1996.

108 'The answer to the mysteries of Stonehenge lies in [understanding its role as a symbol of power and a device for legitimising existing power structures], not in the patterns of the heavens' (Geoffrey Wainwright, 'Debating the stones of contention', *The Times*, 20 June 1996, 37).

APPENDIX

1 Cf. p. 144.

2 There have been relatively few attempts to explain the relevant field techniques outside the specialist literature in archaeoastronomy. Exceptions include Anthony F. Aveni, 'Archaeoastronomy', in Michael B. Schiffer (ed.), *Advances in Archaeological Method and Theory*, vol. 4, Academic Press, New York, 1981, 1–77, pp. 25–33; Clive Ruggles, 'Archaeoastronomical field techniques', in Macey 1994, 24–5. See also Schlosser and Cierny 1996, ch. 7; Aveni 1997, appendix B.

3 e.g. Roger Mercer, 'Recording orthostatic settings', in Gibson and Simpson 1998, 209–18.

4 For an example see the inset in fig. 2.16 of Barrett *et al.* 1991, where the sun is shown setting over Gussage Cow Down barrow from a terminal of the Dorset cursus. In contrast, for an example of good practice see Barnatt and Pierpoint 1983, fig. 4.

5 See chapters nine and ten.

6 Cf. Tilley 1994, 74–5.

7 In the literature on archaeoastronomy, declinations are often quoted to a precision of 0°·1 or even greater but doing so is pointless if the questions being asked do not justify anything like this. At Stonehenge, for example, the quality of the material evidence, together with the fact that the horizon around Stonehenge is relatively close (Cleal *et al.* 1995, 37) and devoid of prominent distant features interpretable as precise foresights, do not justify considering declinations to a precision much greater than the nearest degree (Ruggles 1997, 209).

8 Right ascension (RA) can be thought of as longitude on the (rotating) celestial sphere. Just as the Greenwich meridian is assigned longitude 0° on the earth, the line where RA = 0° is taken by convention to be the meridian line through the celestial pole and the point where the sun's path through the stars crosses the celestial equator at the vernal equinox. Declination and RA together determine the position of a star in the sky, from which its azimuth and altitude at any particular time can be deduced. For a formal definition see, e.g., McNally 1974, 34.

9 Actually, a slightly greater precision is generally advisable, since at the latitude of Britain and Ireland the heavenly bodies rise and set at a shallow angle (see Astronomy Box 2). For example, measuring azimuths to the nearest degree and altitudes to the nearest half degree would enable declinations to be quoted to the nearest half degree.

10 The latter can be dispensed with if you are fortunate enough to have the use of gyroscopic attachment for the theodolite that will detect the north direction directly from the earth's rotation and gravity. However such devices are expensive and unlikely to be available to most archaeological surveyors. They also take a good many minutes to set up, and taking readings is not straightforward. Once the gyro has been powered up, it takes about ten to fifteen minutes to determine an azimuth, using an 'approximate' technique reliable to between ±1 and ±3 arc minutes.

11 This is actually the rate of change of the sun's hour angle (HA). (The HA is analogous to longitude on the celestial sphere in a frame of reference that does not rotate with the stars, so that the HA of celestial bodies is constantly changing. It is measured round from the observer's meridian—the circle joining the south and north points on the observer's horizon and the zenith—in a westerly sense. In the case of the sun, its HA is related to the time that has passed since local noon. Cf. Thom 1971, 21–2.) It represents the sun's actual rate of progress across the sky at around the time of the equinoxes, when the sun is moving on a great circle (the celestial equator). Close to the solstices, when the sun is farthest from the celestial equator, the rate is somewhat slower.

12 The highest possible rate of change in the sun's azimuth occurs at around noon at midsummer in southern England. The value can be obtained from Thom 1971, 26, formula (2.4), by setting $\delta = 24°$ and $\phi$ [latitude] = 51°. Closer to the equator, where the sun approaches closer to the zenith at noon in midsummer, the rate of change in the sun's azimuth may be much faster around this time (e.g. Stephen C. McCluskey, 'The probability of noontime shadows at three petroglyph sites on Fajada Butte', *AA* no. 12 (*JHA* 19) (1988), S69–71).

13 Global Positioning System (GPS) receivers obtain the precise time from world-wide broadcasts. They are also excellent for latitude determination and other positioning work where an accuracy no better than, say, 100 m is suitable. They can be used for locating a site using pre-loaded co-ordinates. On their general use in surveying see Frank Prendergast, 'Low cost C/A code GPS receivers', *Civil Engineering Surveyor*, Autumn 1996, GIS/GPS supplement, 14–16; see also Whyte and Paul 1997, 199–203.

14 Electronic Distance Measurement.

15 For this purpose there need in theory to be at least three, and in practice as many as possible, such points. They need to be readily identifiable on photographs, and to have as great a vertical, as well as horizontal, spread as possible. The most common circumstance is for mountain tops to be hidden by low cloud, while points on the lower slopes, and hill junctions, are still visible. Points in the foreground can be used for calibration purposes, but care is needed that if the camera is not placed in exactly the same position as the theodolite, the size of the parallax correction is not significant in relation to the desired precision (see below in the main text for a rule of thumb).

16 If all that is missing is a sun-azimuth determination, an alternative in some circumstances may be to take readings to Ordnance Survey triangulation pillars on the tops of visible hills, or else to prominent landmarks such as church spires. In theory it is possible to determine the north point from three such observations (cf. Ruggles 1984a, 69–70) and computer programs exist to facilitate this (*ibid.*, 71 [TRIGPT]), but in practice errors and misidentifications occur easily, and it may not be possible (even using large-scale maps) to determine their positions sufficiently accurately for the purpose in hand. In practice, four or more such points are generally needed in order to obtain a reliable azimuth calibration.

17 The term 'key position' is used here in preference to 'observing position' (cf. Ruggles 1984a, ch. 4) because it avoids the overtones of 'megalithic science'. Key positions are defined by the archaeological questions, not by the present-day visibility of the relevant horizon profile.

18 The figure is actually about 0°·057, or 3′·4.

19 Under certain conditions, profiles may be partially obscured (for example, by very close bushes or trees) provided that key points on the horizon can be seen and surveyed, and that a photograph of the unobscured horizon can be taken from nearby using the surveyed points for calibration. If the camera cannot be placed sufficiently near to the theodolite for the parallax correction to be negligible, then its position must be measured and noted and a correction applied to the surveyed points before they are used to calibrate the photograph. If at all possible, the camera position should be sufficiently close to the key position that no further correction is required.

20 Larger parallax corrections introduce possible errors, for a number of reasons: (i) perspective effects due to the fact that profiles are not two-dimensional; (ii) junctions between hills at different distances closing or opening up; and (iii) the difficulty of accurately determining the distance of each profile point (Ruggles 1984a, 73). In extreme cases, e.g. where monuments are located within forestry plantations, large parallax corrections may be unavoidable, but the reliability of the results is reduced as a result. In addition, the work involved in establishing the theodolite position in relation to the key position(s) may be considerable.

21 'Left face' readings have the vertical circle to the left of the telescope; 'right face' readings have it to the right. Horizontal and vertical readings on the two faces can be distinguished by using the notation LH, RH, LV, RV.

22 If the instrument is perfectly adjusted then the difference between LH and RH readings on the same object will be exactly 180°. The altitude as measured on either face should be the same, but vertical circles are calibrated in a variety of ways: for example an object at zero altitude may correspond to an LV reading of 90° and an RV reading of 270° with the zenith point yielding 180°; or an LV reading of 0° and an RV reading of 180° with the zenith point yielding 90°; or a number of other possibilities. In the first case, the LV and RV readings of the same object would add up to exactly 360° if the instrument was perfectly adjusted.

23 For example, the tripod being knocked or the horizontal circle clamp being accidentally released, allowing the horizontal circle to move independently of the telescope.

24 These include the (closed loop) traverse, where each point is measured from the previous one and the first from the last (or vice versa), and 'polars', where every other point is measured from a single one. For both techniques azimuth and distance readings are needed in each case. A third technique, the network, involves taking a larger number of readings between pairs of control points, but has the advantage that not every azimuth or distance has to be measured. For details on undertaking traverses the reader is referred to standard surveying texts such as Bannister *et al.* 1992, 161–77; Whyte and Paul 1997, 169–96.

25 This can be done by noting the height of the theodolite when set up at each control point, and measuring the altitude of suitable height gradations of a ranging pole placed vertically on the ground at the remote point.

26 The overall uncertainty in any theodolite scale reading consists of the random (or 'pointing') error associated with any one observation, together with various systematic errors which are associated with a given instrument. The plate-bearing errors are systematic errors of this sort, the relevant ones in the case of the horizontal scale being the *collimation error* and the *centring error*. Similar errors apply to the vertical scale, but a simpler correction can generally be applied since most readings are within a few degrees of altitude 0°. An additional systematic error of possible concern when taking sun-azimuth readings at high altitude is the *transit axis tilt*. It is essential when taking such readings to ensure that the plate bubble is as level as possible, since the resulting error is *not* eliminated by observing on both faces. For details see Bannister *et al.* 1992, 103–6; Whyte and Paul 1997, 129–31.

27 Until recently, most computer programs used for this purpose (including STIMES described in the main text) have deduced the sun's azimuth, as seen from a given location at a given time, from observer-independent parameters (such as the sun's declination and right ascension) which are tabulated at suitable time intervals in astronomical ephemerides and nautical almanacs available in libraries. For the analytical procedure underlying such programs see Thom 1971, 120–1.

Ephemeris information is also now available on-line over the internet, for example (at the time of writing) from the Web site http://ssd.jpl.nasa.gov/horizons.html. Using an on-line ephemeris it is possible to obtain the sun's azimuth directly for a specified location and time.

28 The programs are derived from an earlier 'MEGPAK', described in Ruggles 1984a, 70–3.

29 For an example see Ruggles and Medyckyj-Scott 1996.

30 Since *y* refers to the vertical direction some might prefer the notation (*x*, *z*).

31 Ruggles 1984a, 72 [TRNSFM and PROPIC].

32 Hawkins and White 1970, figs 11 and 14. In fact, information on the differences between the target declinations and the actual indicated declinations is given in an accompanying table (*ibid.*, table I), but the presentation of the data in the figures themselves is very misleading.

33 In fact, the declinations of lunar targets quoted in Hawkins's diagrams and Fig. 1.22a are geocentric (see Astronomy Box 2), and should be reduced by 1° before the comparison is made.

34 But even the path of solstitial sun changes over the centuries, so a particular path is only meaningful if the date is specified and can be justified from independent evidence.

35 This is a problem with many of the profile diagrams in Thom 1971 and Thom and Thom 1978a (cf. Fig. 0.4c), where a picture of the rising or setting moon draws the eye towards a particular notch or other horizon feature and away from others.

36 See, for example, Wilson and Garfitt's diagram published in 1920 where a complete 360° horizon profile is drawn around a site plan. Lines of constant altitude are represented as concentric circles with the 'observing position' at the centre, and the profile at a given azimuth is displayed at that very azimuth from the centre. (J. S. Wilson and G. A. Garfitt, 'Stone circle, Eyam Moor', *Man*, 20 (1920), 34–7, p. 35.)

## ARCHAEOLOGY BOX 1

1 The conventional divisions of prehistory in north-western Europe since the last Ice Age—the Mesolithic (Middle Stone Age), Neolithic (New Stone Age), Bronze Age, and Iron Age—broadly reflect technological change. This general categorisation has clear limitations: many different factors should be taken into account in forming any reasonably wide-ranging picture of life in prehistoric times, which varied from region to region as well as changing through time in various ways. Nonetheless, it continues to be widely used.

2 The archaeological activities of the Office of Public Works have now been transferred to *Dúchas*, the Heritage Service of the Department of Arts, Heritage, Gaeltacht and the Islands.

3 However, there do exist a number of lists of sites and monuments covering smaller areas, in the series *The Archaeological Sites and Monuments of Scotland*, published by RCAHMS from 1978 onwards.

4 For example in Scotland see Diana M. Murray, 'The management of archaeological information—a strategy', in John Wilcock and Kris Lockyear (eds), *CAA93: Computer Applications and Quantitative Methods in Archaeology 1993*, BAR International Series 598, 1995, 83–7.

5 At the time of writing RCHME's heritage database management system, MONARCH, contains over 250,000 detailed records on archaeological sites, shipwrecks and historic buildings. One of the main problems of integrating regional databases is that of inconsistent data models, and for this reason the documentation committee of the International Council of Museums (ICOM), CIDOC, has worked to produce a Core Data Standard for Archaeological Sites and Monuments. See Robin Thornes and John Bold (eds), *Documenting the Cultural Heritage*, Council of Europe/Getty Information Institute, Los Angeles, 1998.

6 Revised and updated versions are being published as lists and maps entitled *Recorded Monuments*.

7 This is part of a wider Arts and Humanities Data Service and is managed by a number of UK academic institutions together with the Council for British Archaeology and RCHME on behalf of all the Royal Commissions. At the time of writing information is available on the Service's World-Wide Web home page http://ads.ahds.ac.uk/ahds/.

8 See the recent publications on north-east and south-east Perth (RCAHMS, *North-East Perth: An Archaeological Landscape*, HMSO, Edinburgh, 1990; RCAHMS 1994), and in particular the preface to each volume.

## ARCHAEOLOGY BOX 2

1 In general discussions such as this, we quote dates accurate to the nearest 250 years. The level of uncertainty in calibrated radiocarbon dates and dates estimated by other means is felt to be such that greater precision is not justified.

2 A total of 1448 Irish tombs were documented in 1989 (for a complete list with maps see Ó Nualláin 1989, 115–44 and figs 84–9). The number of known tombs had increased to 1560 by 1997 (Walsh 1997, 10).

3 On earthen long barrows in Britain see Kinnes 1992. This includes a catalogue (*ibid.*, 7–56). For a general introduction see Lynch 1997.

4 An early synthesis of the chambered tombs of England and Wales was provided by Glyn Daniel, *The Prehistoric Chamber Tombs of England and Wales*, CUP, Cambridge, 1950. On the Cotswold–Severn Tombs see Powell *et al.* 1969, chs 2 and 3 and appendix A. On the megalithic tombs of north Wales see *ibid.*, chs 4 and 5 and appendix B. On those of northern England see T. G. Manby, 'Long barrows of northern England: structural and dating evidence', *Scottish Archaeological Forum* 2 (1970), 1–27.

For an introductory description of Scottish chambered tombs see Ritchie and Ritchie 1981, 19–33. For a detailed inventory, including plans and descriptions of each monument, see Henshall 1963; 1972. On the Clyde tombs in particular see Powell *et al.* 1969, ch. 6 and appendix C. On Orkney see Henshall 1985 and Davidson and Henshall 1989, which includes an updated inventory. On Caithness see Davidson and Henshall 1991, which also includes an updated inventory; likewise for Sutherland Audrey S. Henshall and Graham Ritchie, *The Chambered Cairns of Sutherland: An Inventory of the Structures and their Contents*, Edinburgh UP, Edinburgh, 1996.

For introductions to the Irish tombs see O'Kelly 1989, 85–122; Elizabeth Shee Twohig, *Irish Megalithic Tombs*, Shire Archaeology 63, Shire, Princes Risborough, 1990; Waddell 1998, ch. 3. For detailed inventories see Ruaidhrí de Valera and Seán Ó Nualláin, *Survey of the Megalithic Tombs of Ireland, Vol. I: County Clare*, Stationery Office, Dublin, 1961; *Vol. II: County Mayo*, Stationery Office, Dublin, 1964; *Vol. III: Counties Galway, Roscommon, Leitrim, Longford, Westmeath, Laoghis, Offaly, Kildare, Cavan*, Stationery Office, Dublin, 1972; de Valera and Ó Nualláin 1982; Ó Nualláin 1989; and Stefan Bergh, *Landscape of the Monuments: A Study of the Passage Tombs in the Cúil Irra Region, Co. Sligo, Ireland*, Central Board of National Antiquities, Stockholm, 1995. On court tombs (formerly 'court cairns'), see Ruaidhrí de Valera, 'The court cairns of Ireland', *PRIA* 60C (1960), 9–140 and 'Transeptal court cairns', *JRSAI* 95 (1965), 5–37. On Irish passage tombs, see T. G. E. Powell, 'The passage graves of Ireland', *PPS* 4 (1938), 239–48, and also Seán Ó Nualláin, 'A ruined megalithic cemetery in Co. Donegal and its context in the Irish passage grave series', *JRSAI* 98 (1968), 1–29, which contains a catalogue and map (pp. 20–9).

5 For an early synthesis of the radiocarbon dating evidence from the various Neolithic funerary monuments see I. F. Smith, 'The neolithic', in Colin Renfrew, *British Prehistory*, Duckworth, London, 1974, 100–36, pp. 124–6. For a more recent synthesis of the Scottish evidence see Ashmore 1996. More specifically, on Orcadian tombs see Colin Renfrew and Simon Buteux, 'Radiocarbon dates from Orkney', in Renfrew 1985, 263–9, reproduced in Davidson and Henshall 1989, 95–8; for Caithness see Davidson and Henshall 1991, 83.

6 On Clava cairns, see Henshall 1963.

7 See A. L. Brindley and J. N. Lanting, 'Radiocarbon dates from wedge tombs', *Journal of Irish Archaeology*, 6 (1992), 19–26; O'Brien 1993, esp. pp. 64–6; also 1994, fig. 102.

8 Richard J. Bradley, *priv. comm.*, 1998. The monuments concerned are Balnuaran of Clava north-east, Newton of Petty, Raigmore and (possibly) the central ring cairn at Balnuaran. For details of the Balnuaran dates see Patrick J. Ashmore, 'A list of Historic Scotland archaeological radiocarbon dates', *DES*, 1997, 112–17, p. 114.
9 On causewayed enclosures in general see Rog Palmer, 'Interrupted ditch enclosures in Britain: the use of aerial photography for comparative studies', *PPS*, 42 (1976), 161–86 and Roger J. Mercer, *Causewayed Enclosures*, Shire Archaeology 61, Shire, Princes Risborough, 1990. See also Parker Pearson 1993, 28–38. On Windmill Hill in particular see I. F. Smith, *Windmill Hill and Avebury: Excavations by Alexander Keiller 1925–39*, OUP, Oxford, 1965.
10 The convenient term 'henge' actually covers a range of roughly circular earthworks, varying from huge 'henge enclosures' (or 'great henges') hundreds of metres in diameter, such as Avebury and Durrington Walls, down to tiny 'hengiform monuments' ('mini-henges') only a few metres across. Ironically, it is not really applicable to the earthwork enclosure at Stonehenge itself, which, exceptionally, has the bank inside the ditch.
11 Radiocarbon dating consistently shows the northern examples to be earlier, and the tradition may have originated in the area to the north of the causewayed enclosures. A good introduction to henges is Aubrey Burl, *Prehistoric Henges*, Shire Archaeology 66, Shire, Princes Risborough, 1991 [reprinted with updated bibliography 1997]. For a British corpus see Anthony F. Harding with G. E. Lee, *Henge Monuments and Related Sites of Great Britain: Air Photographic Evidence and Catalogue*, BAR (British Series 175), Oxford, 1987. Irish henges are included in an earlier list by Burl (Aubrey Burl, 'Henges: internal features and regional groups', *Arch. J.*, 126 (1969), 1–28).
12 On timber circles in general see Gibson 1994, 191–213. For corpuses see Barnatt 1989, 512–16; Gibson 1994, 218–23.
13 Gibson 1994, 200–4. There is evidence to suggest that timber circle construction may have continued for considerably longer, perhaps well into the Iron Age, in both Britain and Ireland (Gibson 1995).
14 For an introduction see Burl 1988c. A classic work on stone circles is Burl 1976, which includes a detailed gazetteer. Another comprehensive work including detailed descriptions is Barnatt 1989. For a useful guide book see Burl 1995.
15 Burl 1988c, 10. On dating stone circles, see Burl 1976, 46. On the relationship between henges and early stone circles, see *ibid.*, 24–30.
16 Two good examples of henges that contained rings of timber posts are Balfarg in Fife and North Mains (Strathallan) in Perthshire (Ritchie and Ritchie 1981, 47–51; Ashmore 1996, 67–71). Woodhenge in Wiltshire is another (Richards 1991, 96–8).
17 For a summary and references see Gibson 1994, 205.
18 On the stone circles of Cork and Kerry in particular see Ó Nualláin 1984a, which includes a plan and description of each monument. For a summary of the dating evidence from the south-west Irish stone circles see *ibid.*, 9–10; see also chapter five.
19 For an early discussion see Aubrey Burl, 'Two "Scottish" stone circles in Northumberland', *Archaeologia Aeliana*, 49 (1971), 37–51. On Irish four-posters see Ó Nualláin 1984b, 63–71. For a gazetteer see Burl 1988b.
20 An unpublished gazetteer of cursuses is available in Roy Loveday, 'Cursuses and related monuments of the British Neolithic', Ph.D. thesis, Leicester University, 1985. For a summary of dating evidence for cursus monuments see Bradley in Barrett *et al.* 1991, 51–2. For up-to-date information and interpretation see Barclay and Harding 1999.
21 On stone rows and avenues in general see Burl 1993, which includes a detailed gazetteer. For corpuses of stone rows in south-west England see also Barnatt 1989, appendices 4 and 6. On stone rows in the south of Ireland see Ó Nualláin 1988, which includes a plan and description of each monument. On Avebury and Callanish see chapter eight.
22 The process of transition was by no means straightforward. For example, in some parts of Britain, such as Yorkshire and Perthshire, round mounds and cairns seem to predate the use of long tombs or at least to represent an independent or related tradition (Ian A. Kinnes, *Round Barrows and Ring-Ditches in the British Neolithic*, British Museum (Occasional Paper no. 7), London, 1979, esp. fig. 8.1; cf. Kinnes 1992, 117–19).
23 Cf. Paul Ashbee, *The Bronze Age Round Barrow in Britain*, Phoenix House, London, 1960, ch. 2.
24 Generally, the best gazetteers of such monuments are to be found in regional inventories. On Wessex round barrows in the vicinity of Stonehenge see, e.g., Richards 1991, esp. 106–9. For a variety of examples see Darvill 1996.
25 On the standing stones of Wales and south-west England, for example, and for a list of excavated examples there and in Ireland, see Williams 1988.
26 See, for example, Elizabeth Shee Twohig, *The Megalithic Art of Western Europe*, OUP, Oxford, 1981, 202–27. On Newgrange, see O'Kelly 1982.
27 On Scottish cup-and-ring marks see Burgess 1980, 347–8; Ritchie and Ritchie 1981, 72–5. For regional catalogues see Ronald W. B. Morris and Douglas C. Bailey, 'The cup-and-ring marks and similar sculptures of south-western Scotland: a survey', *PSAS* 98 (1966), 150–72, pp. 158–72, and Ronald W. B. Morris, 'The cup-and-ring marks and similar sculptures of Scotland: a survey of the southern counties, part II', *PSAS* 100 (1968), 47–78, pp. 59–78. See also Ronald W. B. Morris, *The Prehistoric Rock Art of Argyll*, Blandford Press, Poole, 1977; *The Prehistoric Rock Art of Galloway and the Isle of Man*, Blandford Press, Poole, 1979; Morris 1981. For a wider European perspective see Bradley 1997.

ARCHAEOLOGY BOX 3

1 In fact, they are not the earliest constructions in the immediate area: three great timber poles seem to have been erected some four thousand years earlier at the far end of what is now the car-park (see p. 41). These are unrelated to the later developments except by spatial coincidence, unless one is prepared to speculate that the postholes, still visible millennia later and recognisable as having been created by remote ancestors, helped to create a ceremonial focus here in Neolithic times (cf. Parker Pearson and Ramilisonina 1998, 323).
2 On Wessex monuments and landscape in the first half of the fourth millennium BC see Cleal *et al.* 1995, 56–62 and fig. 33; Burl 1987a, ch. 1; Castleden 1993, 35–44.
3 RCHME 1979, 13–15, 19–20; Burl 1987a, ch. 2; Castleden 1993, 44–9.
4 Cleal *et al.* 1995, 60–2 and references therein.
5 *Ibid.*, 63, 113–14. On the entrances, see *ibid.*, 109–11 and on the radiocarbon dating see *ibid.*, 531; Alex Bayliss, Christopher Bronk Ramsey and F. Gerry McCormac, 'Dating Stonehenge', in Cunliffe and Renfrew 1997, 39–60, pp. 41–52.
6 Cleal *et al.* 1995, 94–102.
7 *Ibid.*, 112–13.
8 For the arguments for and against see *ibid.*, 102–7; also 112–13. Atkinson, who excavated two of them, remained convinced that the Aubrey Holes had never held posts.
9 *Ibid.*, 107–8.
10 *Ibid.*, 161–2; see also Bradley 1991a; Castleden 1993, 74–84. For brief notes on the other henges see RCHME 1979, 13, 15–19. On Durrington Walls, see Geoffrey J. Wainwright and I. H. Longworth (eds), *Durrington Walls: Excavations 1966–1968*, Society of Antiquaries, London, 1971.
It has been suggested that there was a period, from about 2500 BC, possibly lasting a number of centuries, during which Stonehenge was abandoned, while the focus of ritual activity moved eastwards towards Durrington (see, e.g., Castleden 1993, 72–4, 92). However, the evidence for this is questionable (Cleal *et al.* 1995, 163).
11 Cleal *et al.* 1995, 115–17.
12 *Ibid.*, 146–50. For the central hut or roundhouse interpretation see Burl 1987a, 55–6; Castleden 1993, 67–71. For the timber circle interpretation see Bradley 1991a.
13 Cleal *et al.* 1995, 150–2.
14 *Ibid.*, 140–6.
15 *Ibid.*, 146.
16 Earlier, it was widely assumed, following Atkinson (1979, 28, 171), that the Aubrey Holes were pits for ritual offerings that were filled in soon after being dug. See also Castleden 1993, 62–7.
17 Cleal *et al.* 1995, 152–5.
18 *Ibid.*, 168–9.
19 Burl 1987a, 123–7; Castleden 1993, 84–8.
20 On possible routes, and a number of related issues, see Castleden 1993, 94–105, 112–23. The question of whether the bluestones were necessarily brought all the way from Wales, or could instead have been local glacial erratics, has been one of considerable debate. For a summary discussion see *ibid.*, 105–12; see also C. P. Green, 'The provenance of rocks used in the construction of Stonehenge', in Cunliffe and Renfrew 1997, 257–70; J. D. Scourse, 'Transport of the Stonehenge bluestones: testing the glacial hypothesis', in Cunliffe and Renfrew 1997, 271–314; O. Williams-Thorpe, C. P. Green and J. D. Scourse, 'The Stonehenge blue-stones: discussion', in Cunliffe and Renfrew 1997, 315–18. For the opposing view, in favour of the glacial erratic explanation, see G. A. Kellaway, 'Glaciation and the stones of Stonehenge', *Nature*, 233 (1971), 30–5; Richard S. Thorpe, Olwen Williams-Thorpe, D. Graham Jenkins and J. S. Watson, 'The geological sources and transport of the bluestones of Stonehenge, Wiltshire, UK', *PPS*, 57(2) (1991), 103–57.
21 The evidence comes from the stoneholes known as the *Q* and *R* holes. Atkinson (1979, 204–6) believed that they formed a double circle that was

never completed, but for a range of possible interpretations see Cleal *et al.* 1995, 185–8.

22 Cleal *et al.* 1995, 187–8; Castleden 1993, 115, 134–6.
23 Cleal *et al.* 1995, 205–6; Castleden 1993, 150–68.
24 Cleal *et al.* 1995, 210–12.
25 On Stonehenge 3iv and 3v see *ibid.*, 229–32. On the Stonehenge 3vi stoneholes, known as the *Y* and *Z* holes, see *ibid.*, 256. See also Castleden 1993, ch. 7.
26 The name 'Heel Stone' comes from a legend about a Friar's Heel and, despite early suggestions, has nothing to do with the Greek *helios*, meaning sun (Atkinson 1979, 204).
27 The existence of this stone, 'stone 97', was discovered during rescue excavations along the side of the A344 road in 1979 (Pitts 1982). See also Michael W. Pitts, 'Stones, pits and Stonehenge', *Nature*, 290 (1981), 46–7; Pitts 1981.
28 Cleal *et al.* 1995, 289. The interpretation is not certain; it is possible, for example, that the companion was erected first, then moved to become the Heel Stone (*ibid.*, 274).
29 Cleal *et al.* 1995, 272. See also RCHME 1979, 8.
30 Cleal *et al.* 1995, 274.
31 *Ibid.* See also RCHME 1979, 8; Castleden 1993, 128–31.
32 Cleal *et al.* 1995, 281–2.
33 Burl 1994. The companions occupied stoneholes *D* and *E*. It is possible that there was a fourth stone in the line to the south-east (*ibid.*, 90) but there is no real evidence for this (Cleal *et al.* 1995, 285).
34 On possible stonehole *B*, some 8m from the entrance, see Cleal *et al.* 1995, 287. On stonehole *C*, whose status as a stonehole is even more doubtful, see *ibid.*, 288 and 290.
35 Cleal *et al.* 1995, fig. 79.
36 Atkinson 1979, 73.
37 Cleal *et al.* 1995, 139–40.
38 *Ibid.*, 140.
39 The closest, Amesbury 11 bell-barrow, is only about 170m east of Stonehenge.
40 Cleal *et al.* 1995, 35–6.

ARCHAEOLOGY BOX 4

1 For good general introductions to British prehistory see Darvill 1987 and Parker Pearson 1993, which covers the Neolithic as well as the Bronze Age, despite its title. On Scotland in particular see Ashmore 1996 and Edwards and Ralston 1997. On Ireland, see O'Kelly 1989; Cooney and Grogan 1994; Waddell 1998.
2 Any attempt to introduce such labels involves gross simplifications, for example in respect of regional variations. Some archaeologists use finer categories, e.g. dividing the Neolithic into Early, Middle, Late, and Final. Dates are only intended to indicate events to the nearest 250 years.
3 Burgess 1980, 33. The heaviest capstone of all, at Browne's Hill portal-tomb in Kernanstown, Co. Carlow (S 754768), is in excess of one hundred tonnes (William C. Borlase, *The Dolmens of Ireland*, vol. II, Chapman and Hall, London, 1897, 433).
4 For an overview see Darvill 1987, ch. 3; see also Parker Pearson 1993, 40–50. For some more detailed arguments for and against ranking, see Bradley 1984, 15–25. On exposure prior to final burial see, e.g., Gordon J. Barclay and Christopher J. Russell-White, 'Excavations in the ceremonial complex of the fourth to second millennium BC at Balfarg/Balbirnie, Glenrothes, Fife', *PSAS*, 123 (1993), 43–210, pp. 178–82.
5 Renfrew and Bahn (1996, 190) summarise the view of causewayed enclosures as focal points in an egalitarian society. For a contrasting interpretation see Bradley 1984, 25–37.
6 In the Boyne Valley, for example, hidden decorated stones in tombs such as Knowth and Newgrange suggest exclusivity and privilege, a distinction 'between insiders and outsiders, between on the one hand those with access to tomb interiors and detailed knowledge of their contents and on the other those with access only to the outside and limited specific knowledge of the contents' (Whittle 1988, 182).
7 Whittle 1988, 181.
8 For an overview and references to work in particular regions see Bradley 1984, 38–46. The quote is from *ibid.*, 41.
9 The dangers of relying upon apparent patterns of distribution of conspicuous monuments are highlighted by the recent discovery of a large timber circle in the vicinity of the cairns and stone monuments at Kilmartin, Argyll (editorial in *Curr. Arch.* 13 [no. 155] (1997), 434; John Terry, *DES*, 1997, 19–21); another at Stanton Drew in Somerset (cf. Parker Pearson and Ramilisonina 1998, 314), the site of three well-known stone circles (Burl 1976, 104–6); and of a major timber monument at Dunragit in Wigtownshire (NMRS nos NX15NE69 and NX15NW76; Patrick Ashmore, priv. comm., 1997).
10 For labour estimates in Orkney see Renfrew 1979, 212–14. See also Anna Ritchie, *Prehistoric Orkney*, Batsford/Historic Scotland, London, 1995, 65, 79, 81.
11 Renfrew 1976, 149–51.
12 For contrasting views on social development in Orkney, see Renfrew 1979, 218–19 and Ritchie 1985, 127–30. On social evolution in the north of Ireland see Timothy C. Darvill, 'Court cairns, passage graves and social change in Ireland', *Man* 14 (1979), 311–27.
13 For descriptions see Malone 1989.
14 On labour organisation and its social implications in Wessex, see Renfrew 1973, reprinted in Renfrew 1984, 225–45. For alternative views see Bill Startin and Richard Bradley, 'Some notes on work organisation and society in prehistoric Wessex', in Ruggles and Whittle 1981, 289–96; Whittle 1981, 320–4.
15 On the social background of Scottish recumbent stone circles see Burl 1976, 167–87; see also Ruggles and Burl 1985, S26–8; Shepherd 1989, 123–9.
16 This occurred at different times in different areas. For instance, there is no direct evidence that it happened anywhere in Scotland before about 2000 BC, and perhaps not until 1750 BC (cf. Ashmore 1996, chs 5–6).
17 Burgess (1980, 79–80) argues that the break in tradition at this time was sudden and dramatic but Bradley (1984, 83–4) points out that the evidence on this point is equivocal. For overviews of the changes see Burgess 1980, 79–159; Bradley 1984, 73–91; Darvill 1987, 103; Parker Pearson 1993, ch. 5. On the Irish evidence see Cooney and Grogan 1994, ch. 5. The quotation is from Burgess 1980, 23.
18 On the social implications see Bradley 1984, 91–5. See also *ibid.*, 80; Castleden 1993, 140–5.
19 Examples include North Mains, Balfarg Riding School, and Meldon Bridge. Cf. Ashmore 1996, 108–9.

ARCHAEOLOGY BOX 5

1 For a comprehensive overview of these and many other techniques of modern archaeology in a world context see Renfrew and Bahn 1996, chs 6–9. Unfortunately, there are few books synthesising the evidence on the prehistoric environment in Britain and/or Ireland as a whole. Earlier works such as Evans 1975 and I. G. Simmons and M. J. Tooley (eds), *The Environment in British Prehistory*, Duckworth, London, 1981 are now out of date in many respects. In Scotland, they have been superseded by Edwards and Ralston 1997, chs 2–6. A good general source on England is Jones 1986, chs 3–5. For a global picture see Bell and Walker 1992, chs 5 and 6. For an overview of the evidence on prehistoric Britain available from aerial photography, illustrated by photographs and discussion of key sites, see Darvill 1996.
2 On Neolithic houses and settlements see Parker Pearson 1993, 50–4; Bewley 1994, ch. 4. On Neolithic buildings in Ireland, in Scotland, and in England and Wales see the papers by Eoin Grogan, Gordon J. Barclay, and Timothy Darvill, chs 4–6 respectively, in Timothy Darvill and Julian Thomas (eds), *Neolithic Houses in NW Europe and Beyond*, Oxbow (Monographs in Archaeology, 57), Oxford, 1996. See also Waddell 1998, 30–42.
3 For overviews see, e.g., Burgess 1980, 28–32; Darvill 1987, 51–4; Cooney and Grogan 1994, 35–52; Parker Pearson 1993, 34–40. For a detailed European perspective see Whittle 1996.
4 But see Kevin J. Edwards, 'Palynological and temporal inference in the context of prehistory, with special reference to the evidence from lake and peat deposits', *J. Arch. Sci.*, 6 (1979), 255–70; Edwards and Whittington 1997.
5 For overviews with different interpretations see Burgess 1980, 43–5; Whittle 1981, 297–9; Bradley 1984, 33–7. See also Alasdair Whittle, 'Resources and population in the British Neolithic', *Antiquity*, 52 (1978), 34–42. On the Céide field system see Seamus Caulfield, *Céide Fields and Belderrig Guide*, Morrigan, Killala, 1988.
6 M. G. L. Baillie, *A Slice Through Time: Dendrochronology and Precision Dating*, Batsford, London, 1995, 77–8, 147.
7 For a discussion see Bradley 1984, 57–67. See also Parker Pearson 1993, 34–8, 76–7.
8 On possible interaction between the Boyne Valley and Orkney see Ritchie and Ritchie 1981, 29; Bradley 1984, 58; Derek D. A. Simpson, 'The stone maceheads of Ireland', *JRSAI* 118 (1988), 27–52, p. 35.
9 There is an extensive literature on beakers, a classic work being David L. Clarke, *Beaker Pottery of Great Britain and Ireland* (2 vols), CUP, Cambridge, 1970. See, e.g., Whittle 1981, 299–320 (which critiques earlier ideas that there was an invasion of 'beaker people' from continental

Europe); Parker Pearson 1993, 84–8; Ashmore 1996, 75–81; Waddell 1998, 114–6
10 Whittle 1981, 324–32; Bradley 1984, 47–57 and 70–3. See also Andrew Sherratt, "Beakers", in Brian M. Fagan (ed.), *The Oxford Companion to Archaeology*, OUP, Oxford, 1996, 88–90.
11 See, e.g., Bradley 1984, 87–91; Darvill 1987, ch. 5.
12 See, e.g., Burgess 1980, 155–7; Bradley 1984, 91–4.
13 On new settlement patterns see Bradley 1984, 106–14; Parker Pearson 1993, ch. 5.

ARCHAEOLOGY BOX 6

1 Some archaeologists contend that the very notion of a site should be abandoned; e.g. Robert C. Dunnell, 'The notion site', in Rossignol and Wandsnider 1992, 21–41.
2 For a general introduction to the nature of the prehistoric legacy in the modern English landscape see R. Muir, *Reading the Landscape* (3rd impression), Michael Joseph / Shell, London, 1993, 28–60. Likewise for Ireland see Frank Mitchell and Michael Ryan, *Reading the Irish Landscape* (revised edn), Town House, Dublin, 1997, ch. 5.
3 The aim of field walking is to examine surface scatters of artefacts, often at very low densities. To be of value while making the best use of limited resources, field walking needs to be undertaken in accordance with strictly applied sampling strategies that ensure fair representation of the overall data. See Renfrew and Bahn 1996, 72–3 and references therein.
4 Renfrew and Bahn 1996, 68–82.
5 Amongst the most destructive human activities are settlement, cultivation, and forestry. On the second of these see George Lambrick, 'The effects of modern cultivation equipment on archaeological sites', in J. Hinchliffe and R. T. Schadla-Hall (eds), *The Past under Plough*, Department of the Environment, London, 1980, 18–21. On general issues relating to the conservation of the cultural landscape see Bell and Walker 1992, ch. 8 and references therein.
6 See Ian Hodder and Clive R. Orton, *Spatial Analysis in Archaeology*, CUP, Cambridge, 1976.
7 On the functionality and the inherent problems in using site catchment analysis and Thiessen polygons, see Amy J. Ruggles and Richard L. Church, 'Spatial allocation in archaeology: an opportunity for reevaluation', in Maschner 1996, 147–73.
8 For a collection of papers following this approach see Rossignol and Wandsnider 1992.
9 For a synthesis of GIS history, principles, tools and methods up to 1990 see Maguire *et al.* 1991. For a computer science perspective see Michael F. Worboys, *GIS: A Computing Perspective*, Taylor and Francis, London, 1995.
10 Some important collections of case studies are K. M. S. Allen, S. W. Green and Ezra B. W. Zubrow (eds), *Interpreting Space: GIS and Archaeology*, Taylor and Francis, London, 1990; Lock and Stančič 1995; Maschner 1996; Mark S. Aldenderfer and Herbert D. G. Maschner (eds), *Anthropology, Space, and Geographic Information Systems*, OUP, Oxford, 1996; Gillings *et al.* 1998.
11 See Renfrew and Bahn 1996, 84.
12 See, e.g., Scott L. H. Madry and Lynn Rakos, 'Line-of-sight and cost-surface techniques for regional research in the Arroux river valley', in Maschner 1996, 104–26.
13 For a commentary see Wheatley 1993.
14 Ruggles and Medyckyj-Scott 1996, 129–30. For a case study see Marcos Llobera, 'Exploring the topography of mind: GIS, social space and archaeology', *Antiquity*, 70 (1996), 612–22.
15 For a case study see Wheatley 1995; 1996.
16 Ruggles, Medyckyj-Scott and Gruffydd 1993; Ruggles and Medyckyj-Scott 1996.
17 See Peter F. Fisher, 'Reconsideration of the viewshed function in terrain modelling', *Geographical Systems*, 3 (1996), 33–58.
18 Wheatley 1993, 135.
19 For example, it was suggested in the midst of the 'megalithic astronomy' debate that the effects of erosion on the cliffs at Hellia on Hoy, Orkney, may have reduced their height by several tens of metres since prehistoric times, significantly changing the profile as seen from the Ring of Brodgar (see pp. 63–6 and Fig. 2.12a).
20 Peter F. Fisher, 'Probable and fuzzy models of the viewshed operation', in Michael F. Worboys (ed.), *Innovations in GIS 1*, Taylor and Francis, London, 1994, 161–75.
21 Although current GIS can generate 3-D–views of topographic data, they only reference data two-dimensionally (i.e. in terms of *x-y* co-ordinates). Digital elevation models (DEMs) actually comprise a set of elevation (*z*) values, one for each *x-y* position. In a truly 3-D GIS, each layer can store information for different positions in three-dimensional space. See, e.g., Jonathan F. Raper and Brian Kelk, 'Three-dimensional GIS', in Maguire *et al.* 1991, 299–317.

ARCHAEOLOGY BOX 7

1 This term is generally preferred to 'ritual landscapes' because the latter is felt by some to imply a clear separation between ritual and mundane activities (e.g. monuments and settlement) that is unproven and (to judge by the evidence from numerous historical and modern indigenous groups) unlikely. See, e.g., Bradley and Chambers 1988, 72–3; Bender 1992.
2 Tilley 1994, 15.
3 e.g. papers in Bender 1993.
4 It has been suggested that ritual practice and monumental architecture were a means by which dominant groups in the community manipulated and recreated the past so as to legitimise their own position (Stephen J. Shennan, 'Ideology, change and the European Early Bronze Age', in Ian R. Hodder (ed.), *Symbolic and Structural Archaeology*, CUP, Cambridge, 1982, 155–61) or to induce social change (Thomas 1988).
5 See Tilley 1994, 37–54 and references therein. 'The aboriginal landscape is one replete with highly elaborate totemic geography linking together place and people' (*ibid.*, 38).
6 Inca sacred places (*huacas*) included rocks, caves, springs, and mountain tops, as well as human constructions such as fountains (Bauer and Dearborn 1995, 15). The fact that such places were situated on lines (*ceques*) radiating out from Cuzco served, through the prevailing ideology, to legitimise the city as the Inca capital (R. T. Zuidema, *The Ceque System of Cuzco: The Social Organisation of the Capital of the Inca*, Brill, Leiden, 1964; see also Aveni 1997, 154–61).
7 For a range of case studies, as well as theoretical ideas concerning human perceptions of landscape, see e.g. Carmichael *et al.* 1994; Peter Ucko and Robert Layton (eds), *The Archaeology and Anthropology of Landscape*, Routledge, London, 1998.
8 This is well illustrated in the Valley of Mexico in Aztec times. See Part II of Davíd Carrasco (ed.), *To Change Place: Aztec Ceremonial Landscapes*, UP of Colorado, Niwot Colorado, 1991; Broda 1993.
9 Many examples are known in pre-Columbian Mesoamerica, from Teotihuacan through to Classic Maya cities. See, e.g., Anthony F. Aveni and Horst Hartung, *Maya City Planning and the Calendar*, American Philosophical Society (Transactions, vol. 76, part 7), Philadelphia, 1986.
10 This is well illustrated by 'folk astronomy' in the Islamic world. See David A. King, 'Folk astronomy in the service of religion: the case of Islam', in Ruggles and Saunders 1993a, 124–38.
11 Nayanjot Lahiri, 'Archaeological landscapes and textual images: a study of the sacred geography of late medieval Ballabgarh', *WA* 28 (1996), 244–64.
12 Richards 1996a.
13 On the reverence and 'use' of earlier monuments see, for example, Patton 1993, 114–24 and 179–96.
14 Bradley 1993, ch. 3.
15 Richards 1996a, 205.
16 Darvill 1997.
17 These include Gabriel Cooney, 'Sacred and secular neolithic landscapes in Ireland', in Carmichael *et al.* 1994, 32–43; B. E. Vyner, 'The territory of ritual: cross-ridge boundaries and the prehistoric landscape of the Cleveland Hills, northeast England', *Antiquity*, 68 (1994), 27–38; Bender 1992 on the Stonehenge landscape; Barnatt 1996 on the Peak District; and Tilley 1996 on Bodmin Moor.
18 Patton 1993.

ARCHAEOLOGY BOX 8

1 See Renfrew and Bahn 1996, 441–2.
2 For a history of traditional methods in archaeology see Bruce G. Trigger, *A History of Archaeological Thought*, CUP, Cambridge, 1989. For a critique see Renfrew and Bahn 1996, 443–8. Use of the word 'culture' still carries the unfortunate overtone of an actual ethnic group; it should really be understood as no more than a classificatory convenience.
3 Renfrew and Bahn 1996, 36–9.
4 The literature on these matters is vast. For overviews see Renfrew and Bahn 1996, ch. 12; Ken R. Dark, *Theoretical Archaeology*, Duckworth, London 1995, ch. 7.
5 Renfrew and Bahn 1996, 442–3; 451–4. For example see Renfrew 1984, 158–9.
6 For many examples see the collection of papers in Renfrew 1984; see also Renfrew and Bahn 1996, ch. 5.
7 For an eminently readable introduction to these and other post processual ideas see Hodder 1991.

8 Hodder 1991, 10.
9 Sosa 1989, 142.
10 Traditionally felt to be one of the most difficult areas in archaeology, cognitive archaeology is a field of enquiry that has only begun to seem productive in the 1990s in the wake of the intensive debates about archaeological theory and method that took place during the two previous decades. It is not just a case of making specific deductions about the mental processes that underlay a particular technological achievement, such as the construction of the sarsen circle and trilithons at Stonehenge, but about developing a broader theoretical framework that suggests more general relationships between human thought and material remains. See, for example, Renfrew and Zubrow 1994; Bell 1994, ch. 9; Renfrew and Bahn 1996, ch. 10.
11 e.g. Michael B. Schiffer, *Behavioral Archeology*, Academic Press, New York, 1976.
12 e.g. Hodder 1991, 2–6. Hodder (*ibid.*) argues that all attempts to 'read off' human patterning from patterning in the material record are doomed to failure.
13 Renfrew and Bahn 1996, 453–4.
14 Hodder 1992, 213–40.
15 Michael Shanks and Christopher Tilley, *Re-Constructing Archaeology: Theory and Practice*, CUP, Cambridge, 1987, chs 2 and 3.
16 See, e.g., Hodder 1992, 1–7; Tilley 1994, 11.
17 Hodder 1991, 4.
18 Aron Crowell, review of Hodder 1991, *Man*, 27 (1992), 882–3.
19 See also Hodder 1992, 170, who seeks to find a middle position between 'objectivism' and relativism.

ASTRONOMY BOX 1

1 Various alternative explanations of the concept of declination, and other relevant concepts from positional astronomy, have appeared in numerous scholarly and popular books and articles on prehistoric astronomy in Britain and on archaeoastronomy in general. These include Thom 1967, ch. 3; Edwin C. Krupp (ed.), *In Search of Ancient Astronomies*, Doubleday, New York, 1977, ch. 1; Wood 1978, ch. 4; Aveni 1980, ch. III; Ellegård 1981; Heggie 1981a, ch. 5; Giuliano Romano, *Archeoastronomia Italiana*, CLEUP, Padova, 1992, 135–71; Belmonte Avilés 1994, ch. 1; and North 1996, appendix 2. For a fuller, more technical, background in this area see for example McNally 1974, chs 1–7; John B. MacKie, *The Elements of Astronomy for Surveyors*, 9th edn, Charles Griffin, High Wycombe, 1985; Robin M. Green, *Spherical Astronomy*, CUP, Cambridge, 1985, chs 1–5.
2 This takes no account of atmospheric refraction close to the horizon (see Astronomy Box 2).
3 Declination is to all intents and purposes equivalent to geographical latitude. The analogy would be exact if the earth were precisely spherical.
4 Note, however, that the term 'celestial latitude' may also be taken, confusingly, to mean the angular distance from the ecliptic (the sun's path through the sky) rather than from the celestial equator (e.g. Walker 1996, 343).
5 Strictly, this applies only to bodies whose motions appear regular and cyclical, such as the sun, moon, planets and bright stars. It does not include one-off events which might have occurred at any place at any time, such as meteors, comets (most of whose periods are so long that we cannot accurately calculate their motions) or novae (stars which, invisible to the naked eye before, suddenly flare up and become visible for a short period).

ASTRONOMY BOX 2

1 On survey procedure, see the Appendix.
2 This can be seen by setting $A = 90°$ and $h = 0°$ in formula (A2.1), whence $\delta = 0°$ for any latitude.
3 Thom (1967, 26), for example, tabulates $\Delta h$ ($= h_0 - h$) values for different $h_0$. According to Thom, when $h_0$ is 0, $\Delta h$ is around 33°, and this is close to the value adopted by the *Astronomical Almanac* (HMSO, London and US Government Printing Office, Washington DC; see p. A12 for any year between 1984 and 1997) in order to tabulate the time of sunrise and sunset. Certain computer programs to calculate declinations, such as GETDEC (see Appendix), also apply a mean refraction correction.
4 Parallax is negligible for other astronomical bodies, which are much more distant than the moon.
5 See, e.g., Cooke *et al.* 1977, 127.

ASTRONOMY BOX 3

1 Further elaboration of the fundamental concepts of positional astronomy is avoided here, since it is not essential to an understanding of the archaeoastronomical issues. For references see Astronomy Box 1, note 1.
2 The factor 0·9856 is equal to 360/365·25, the number of degrees in the circle divided by the number of days in a year (more strictly, in what astronomers know as the 'tropical year').
3 The cruder approximation to $\delta$ given by the sine wave $\delta = \varepsilon \cos(0{\cdot}9856n)$ always yields a value for $\delta$ within 0°·25 of the value given by formula (A3.1).
4 The apparent diameter of the sun varies slightly through the year, but for present purposes can be taken to be 32′.
5 Thom (1969 and subsequent publications) used the symbol $s$ exclusively for the lunar semidiameter. Where minute-of-arc accuracy is not required, it can be used interchangeably for both the solar and lunar semidiameter, since the sun and moon are of a similar apparent size.
6 Alexander Thom, 'An empirical investigation of atmospheric refraction', *Empire Survey Review* 14 (1958), 248–62. See also Thom 1967, 152–4; 1971, ch. 3.
7 e.g. Heggie 1981a, 134–5; Gordon Moir, 'Some archaeological and astronomical objections to scientific astronomy in British prehistory', in Ruggles and Whittle 1981, 221–41, p. 229.
8 Bradley E. Schaefer and William Liller, 'Refraction near the horizon', *PASP* 102 (1990), 796–805. See also Bradley E. Schaefer, 'Basic research in astronomy and its applications to archaeoastronomy', in Ruggles 1993a, 155–77, pp. 162–3.

ASTRONOMY BOX 4

1 The 'month' concerned is the tropical month of 27·3 days, as opposed to the period from one new moon to the next, the synodic month of 29·5 days.
2 This term is used by some authors, e.g. Ellegård 1981, but avoided by others.
3 The cycle arises because the plane of the moon's orbit is inclined to the ecliptic and the two nodes, where they cross, move round once every 18·61 years: hence the name. For references to sources covering the background astronomy see Astronomy Box 1, note 1.
4 $i$ is the inclination of the plane of the lunar orbit to the ecliptic. See previous note.
5 The value of $P$ actually depends upon the observer's latitude and the moon's azimuth and altitude; for declination $\pm(\varepsilon + i)$ it is around $P_M$ = 0°·89 and for $\pm(\varepsilon - i)$ it is around $P_m$ = 0°·83. See Thom 1967, 118. In addition, there is a monthly variation in lunar parallax of amplitude approximately 0°·06 (Derek McNally, priv. comm.; see also Morrison 1980, S73–4).
6 Thom 1971, 18.
7 For example, Thom and Thom (1978a, figs 10.2 and 10.3) show particular alignments captioned 'moon setting [rising] . . . at the minor standstill'; Wood (1978, figs 1.4 and 1.5) shows alignments to 'Moonset (major standstill)' and so on (but see also fig. 1.3); Heggie (1982b, 4) states that 'at a low level of accuracy the extreme positions are simply the "standstill" positions'; and Anna Sofaer, Rolf M. Sinclair and Joey B. Donahue, 'Solar and lunar orientations of the major architecture of the Chaco culture of New Mexico', in Romano and Traversari 1991, 137–50, p. 140, talk of 'orientations within 0°·5 of the lunar major standstill'. Even this author (Ruggles 1984a, appendix I) uses the looser definition and elsewhere (Ruggles 1994a, S15) refers to hill summits falling 'within one degree of the major or minor lunar standstill'.
8 e.g. Burl 1983.
9 e.g. Burl 1979a.
10 e.g. Ellegård 1981.
11 Note that for work at a precision greater than 0°·1, the fact that $P$ varies from line to line must be taken into account, and the geocentric lunar declination must be used (see Astronomy Box 2). Note also that the moon's semidiameter $s$ is generally slightly different from the sun's: like the sun's, it varies somewhat, according to the distance of the moon from the earth.

ASTRONOMY BOX 5

1 Thom 1967, ch. 9.
2 Thom himself (e.g. Thom 1967, 107, 150) just called these 'equinoxes'.
3 Cf. Thom 1967, 23–4 and 109. The factor of 2·07 is $2e \times 180/\pi$, where $e$ = 0·0181 is the (mean) ellipticity of the earth's orbit.
4 Note that the value of the ellipticity also changes slightly with time; its present-day value is nearer 0·017.
5 The minimum declination is the smaller of the sunset and sunrise declinations given by Thom 1967, table 9.1, less the value given in the last column of that table, so as to obtain the smallest value over the leap-year cycle. The maximum declination is derived similarly. Values are quoted here to the nearest 0°·1.

ASTRONOMY BOX 6

1 This sort of movement is called proper motion. The largest known proper motion is that of Barnard's Star, which is approximately 10″ per annum or 0°·3 per century.
2 For a table giving the declinations (and right ascensions—see Appendix, note 8), of the brightest stars (those with apparent magnitudes greater than +3·0) at 100-year intervals from 2500 BC to AD 2500, see Hawkins 1968, 65–88. See also Ruggles 1997, table 4.
3 For example, see Astronomy Box 3 on variable refraction, and Astronomy Box 7 in the case of the moon.
4 Other examples include McNally 1974; Krupp 1994, 10; and Walker 1996, 344.
5 This effect was noted in Astronomy Box 5, but was viewed in the converse way, by expressing the time of year when perihelion occurs as a Gregorian calendar date, or equivalently in terms of the number of days since the June solstice.
6 Cf. André Berger, 'Orbital variations', in S. Schneider (ed.), *Encyclopedia of Climate and Weather*, Robert Ubell Associates/OUP, Oxford, 1996, 557–64, pp. 558–60.
Studies in recent years have suggested that certain long-term periodicities in the earth's orbital elements are correlated with regular fluctuations in the palaeoclimate evident in the geological record. See, e.g., John Imbrie and Katherine P. Imbrie, *Ice Ages: Solving the Mystery*, MacMillan, London, 1979; W. Schwarzacher, *Cyclostratigraphy and the Milankovitch Theory*, Elsevier, Amsterdam, 1993; J. J. Lowe and M. J. C. Walker, *Reconstructing Quaternary Environments* (2nd edn), Longman, Harlow, 1997, 12–14, 361–2. The most important cycles in this respect are a variation in the ellipticity of the earth's orbit, with a period close to 100,000 years; a variation in the magnitude of the obliquity of the ecliptic, which oscillates between about 22° and 25° over a timescale of some 41,000 years (its decrease from about 24° to about 23°·5 over the last 5,000 years merely being part of this longer cycle); and the climatic precession, with a mean period of around 21,000 years. On astronomical and other possible causes of long-term climatic change see Lowe and Walker, *Reconstructing Quaternary Environments*, 361–70.

ASTRONOMY BOX 7

1 For further explanation of the perturbation see, e.g., Heggie 1981a, 95–6; Morrison 1980, S67–9.
2 Thom 1981, 24 and fig. 1.7. Each band corresponds to one of the standstill limits viewed in either the easterly (rising) or westerly (setting) direction.
3 If the lunistice falls mid-way between, say, two visible risings then it will be closer to the intervening setting, but this will occur on a different stretch of horizon where it may not be observable, or may occur in daylight.
4 See Astronomy Box 4, note 5.
5 Heggie 1981a, 98. See Morrison 1980, figs 4 and 5 for graphs showing the monthly extremes of the moon's declination close to particular standstills, together with the phase. See also Ellegård 1981, figs 8 and 10. The discussion of this issue by Wood (1978, 112) is faulty.
6 These include Heggie 1972, McCreery 1980, Ellegård 1981, and Heggie 1981a, 96–8.
7 For a full explanation of all the astronomical effects, together with formulae accurate to 0′·1, see Morrison 1980.
8 On the correlation between lunar 'events' and time of year see Alexander Thom and Archibald S. Thom, 'Observation of the moon in megalithic times', *AA* no. 5 (*JHA* 14) (1983), S57–66, particularly table 2. See also Thom and Thom 1978a, 11–13.

ASTRONOMY BOX 8

1 Burl (1988a, 201) considers this generally improbable, although it is clear (at the very least) that the huge effort involved in preparing and then hauling each of the great sarsens to Stonehenge in around 2400 BC must have involved a degree of pre-planning so as to get no more than the exact number of each size and shape required, and this in turn implies that certain people in Late Neolithic Wessex must have possessed a reasonable set of numerical and related skills.
2 The reason why this interval is not centred upon 0° has to do with the fact that the earth does not travel around the sun at a constant rate (see Astronomy Box 5). For the detailed assumptions made by Thom, see Thom 1967, 109–11.
3 Even in modern Western culture we distinguish between civil, nautical, and astronomical twilight, all defined arbitrarily by the sun being less than a certain altitude (−6°, −12° and −18° respectively). See p. A12 of the *Astronomical Almanac* for any year between 1984 and 1997.

STATISTICS BOX 1

1 See, for example, Barnett 1982, ch. 3. For a full history of the notion of probability see David 1962.
2 Strictly speaking, the converse is not the case. Suppose that a point is to be chosen randomly on a straight line labelled from 0 to 1. First, make a guess at the point to be chosen. The probability is 0 that the random point will coincide with the chosen one. Yet it is clearly not impossible that the random point might land there since it has to land somewhere. This situation arises because of a subtlety to do with countable and uncountable infinities which need not concern us here.
3 In the 'fair die' example, one can arrive at the conclusion that $P(A) = 1/6$ simply by logical argument, since the probability of throwing a number between one and six is 1, and all six outcomes are equally likely, so the probability of any particular outcome must be $1/6$. However, where the different outcomes are not equally likely, the logical argument breaks down and one must resort to the long-run relative frequency definition that we have described. A good example is that of astragali, sheep knuckle bones used for gaming in medieval times. When thrown, they can come to rest in one of four stable configurations, but these are certainly not equally likely. See David 1962, 7–8.

STATISTICS BOX 2

1 Hawkins and White 1970, 171–2.
2 The analogy is not fail-safe. We must consider the possibility that even if factors unrelated to astronomy gave rise to the alignments observed, they might *not* act like the random marksman. Suppose, for example, that the direction of alignments was influenced by the lie of the land, which just happened to run roughly parallel to a direction of astronomical significance. A large number of apparently astronomical alignments might then have arisen for a completely non-astronomical reason. See chapter four, note 119.
3 For further discussion, and useful approximations to this formula in different circumstances, see Heggie 1981a, 242, note 32.
4 Hawkins and White 1970, 172. $p$ was taken to be 0·2 (see below).
5 The errors were pointed out by Atkinson 1966. See also Heggie 1981a, 148–9.
6 Atkinson 1966, 214.
7 The data on errors are presented in a misleading way by Hawkins and White (1970, 143, 169), because their tables present the errors in terms of altitude rather than azimuth. Using appropriate conversion factors (*ibid.*, 142), we find that two lines ($H$ from 93 and 93 from 94) have errors in azimuth of 2°·1, while one (91 from centre) has a massive 5°·1. This even misled Atkinson (1966, 214), who thought that only the latter alignment should have been excluded.
8 The new value of $p$ is $18 \times 5/360 = 0{\cdot}25$. $n$ is still taken as 111; $r$ is now 23.
9 Even these two shots are not actually independent, since they must point in opposite directions; but we suppose for the sake of this argument that they are.

STATISTICS BOX 3

1 This is obtained by taking each possible value of $r$ and multiplying it by the probability of that value occurring, then adding all the results together; i.e. in our example

$$\mu = \sum_{r=0}^{n} r \times P(r)$$

where $P(r)$ is the probability calculated by formula S(2.1).
2 In our example, σ is given by the formula

$$\sigma^2 = \sum_{r=0}^{n} (r - \mu)^2 \times P(r).$$

3 Many good introductory books on statistics describe all this in much greater detail. For an introductory account intended for archaeologists, see Fletcher and Lock 1994, chs 4 and 5.
4 The 'normal distribution' is only a name; other distributions are not in any sense abnormal.
5 See also Shennan 1988, 102–8.
6 Cf. *ibid.*, 101–2.
7 On the principles and problems of radiocarbon dating and calibration see Renfrew and Bahn 1996, 131–6; for a fuller introduction see Aitken 1990, chs 3 and 4. For a classic account of the two 'radiocarbon revolutions' (the introduction of radiocarbon dating in the 1950s and the discovery in the 1960s of their systematic error and the need for calibration) and their impact on European archaeology see Renfrew 1976.
8 For one of the most recent see Stuiver and Reimer 1993.

9 For an example see Cleal *et al.* 1995, appendix 2, which contains many computer-generated plots of such distributions.

STATISTICS BOX 4

1 A number of obvious questions arise. Is it reasonable to assume that the other factors would mimic random processes? If the data are non-random, is the favoured explanation necessarily the only one? See the main text for some discussion of these issues in the context of interpreting sets of indicated declinations.
2 Other problems are discussed in chapter ten.
3 Hence there are any ready-to-hand 'statistical tests' for common problems, such as Kolmogorov–Smirnov tests, which seek to determine whether a sample of observations could have come from a particular distribution, and *t*-tests, which seek to determine whether two samples could have come from the same distribution. Many books on elementary statistics describe these tests and others. For an account intended for archaeologists see Fletcher and Lock 1994, chs 8–11.
4 See p. 77 and also chapter ten.

STATISTICS BOX 5

1 Thom 1955; 1967, chs 5–7.
2 For a discussion see Burl 1976, 41–50.
3 For examples see Ian O. Angell, 'Stone circles: megalithic mathematics or Neolithic nonsense?', *Mathematical Gazette*, 60 (1976), 189–93; 'Are stone circles circles?', *Sci. Arch.* 19 (1977), 16–19.
   Another suggestion is that, although the constructors of the megalithic rings possessed the ability to construct good circles when they so desired and in some cases did so, most shapes should be seen as poor attempts at circles resulting from laying them out by eye. See John Barnatt and Gordon Moir, 'Stone circles and megalithic mathematics', *PPS* 50 (1984), 197–216; John Barnatt and P. Herring, 'Stone circles and megalithic geometry: an experiment to test alternative design practices', *J. Arch. Sci.*, 13 (1986), 431–49.
4 For example, a circle is defined by three parameters, an ellipse by five, and some of Thom's shapes by as many as seven. Given a megalithic ring, it will generally be easier to fit a shape with more parameters, since any or all of them can be varied to suit; but a simpler shape with fewer parameters will generally be more convincing as an explanation of the shape intended. The problem is: how much better a fit should be demanded of, say, a complex shape such as a flattened circle type B (where we can vary seven parameters) before we choose it as a preferable explanation than, say, a circle (only three parameters)?
5 For example, Jon D. Patrick and Chris S. Wallace, 'Stone circle geometries: an information theory approach', in Heggie 1982a, 231–64.
6 Heggie 1981a, 23–31, 60–82.
7 See Burl 1976.
8 The Rollright Stones in Oxfordshire provide an excellent example of the kinds of alteration that can occur (see George Lambrick, *The Rollright Stones*, Historic Buildings and Monuments Commission for England [English Heritage Archaeological Report 6], London, 1988, 41).
9 Examples include Hampton Down in Dorset (Geoffrey J. Wainwright, 'The excavations of Hampton Down Circle, Portesham, Dorset', *PDNHAS* 88 (1966), 122–7); Fowlis Wester West in Perthshire (Young 1943); and Brandsbutt (I. A. G. Shepherd, 'A Grampian stone circle confirmed', *PSAS* 113 (1983), 630–4) and Strichen in Grampian (Hampsher-Monk and Abramson 1982).
10 MacKie 1981, 116–28.
11 Ruggles and Burl 1985, table 8.
12 *Ibid.*; Burl 1979a, 25–32, esp. 30.
13 Thom, Thom and Thom 1974.
14 Burl 1987a, 184.
15 Thom 1955, 280–3. Visual confirmation of Thom's conclusion comes from a curvigram showing the diameters of the forty-six rings (*ibid.*, fig. 7; see also Heggie 1981a, 48; Renfrew and Bahn 1996, 383) which shows clear accumulations close to integer multiples of a unit of approximately 0·83 m.
16 For the data and analysis see Thom 1967, ch. 5. Thom's final value for the MY (which he gives as 2·720 ft ± 0·003 ft) and the quote are from p. 43.
17 For the new data see Thom and Thom 1978a, chs 3, 4, 6, 7, 8 and references therein. They now quote a value for the MY of 2·722 ft ± 0·002 ft and conclude that 'measuring rods of standard length were carried throughout the whole area from Orkney to Carnac' (*ibid.*, 177).
18 See Thom and Thom 1978a, ch. 5.
19 Heggie 1981a, ch. 3.
20 For the classical analysis see David G. Kendall, 'Hunting quanta', *PTRS* A276 (1974), 231–66. For the Bayesian one see Freeman 1976. The statisticians only accepted evidence from circular stone rings, because Thom linked his arguments about geometrical shapes to the proposed unit of measure. Thus geometrical constructions were fitted to monuments by assuming an MY value, while data for the analysis seeking to establish the reality of the MY included the lengths of constructors in these same geometrical models (Ruggles 1990, 137). In order to avoid circular argument (the pun, however, is unavoidable) it seems imperative to view mensuration and geometrical design as formally independent hypotheses, requiring separate testing.

STATISTICS BOX 6

1 The problem is equivalent to setting $n = 6$, $r = 6$, and $p = \frac{1}{3}$ in formula (S2.1).
2 Cf. chapter three, note 74.
3 Or $A_m + \theta - 360$, if $A_m + \theta > 360°$.
4 Actually, the probability is slightly greater than this because the outlier might fall just anticlockwise of the most anticlockwise in the range, $A_m$, forming a new cluster with the most clockwise of the previous range now being the 'outlier'.

STATISTICS BOX 7

1 For a discussion of these differences, see Barnett 1982, ch. 1. For a comprehensive introduction to Bayesian inference using archaeological examples see Buck *et al.* 1996. For more mathematical introductions see Barnett 1982, ch. 6; Peter M. Lee, *Bayesian Statistics: an Introduction*, OUP, Oxford, 1989.
2 This summary is taken, with some alterations, from Clive L. N. Ruggles, 'A statistical examination of the radial line azimuths at Nazca', in Anthony F. Aveni (ed.), *The Lines of Nazca*, American Philosophical Society, Philadelphia, 1989, 245–69, pp. 247–8.
3 See also Buck *et al.* 1996, ch. 4.
4 Strictly, a probability density function is only used for continuous parameters; a probability mass function is used for discrete ones. See Buck *et al.* 1996, 140 and ch. 5.
5 For a slightly less brief outline of the Bayesian approach, and an expression and explanation of Bayes' theorem, intended for an archaeological audience, see C. E. Buck, J. B. Kenworthy, C. D. Litton and A. F. M. Smith, 'Combining archaeological and radiocarbon information: a Bayesian approach to calibration', *Antiquity*, 65 (1991), 808–21, pp. 810–11; see also Ruggles 1994b, 500–2. For a fuller explanation see Buck *et al.* 1996, ch. 7.
   If it is found helpful, the likelihood function can be split into various elements using 'hierarchical models', in an attempt to separate into a series of steps the processes that have led from the thoughts and actions of prehistoric people to the data we observe. For an example see Ruggles 1994b, 502–3.
6 For an early example see Freeman 1976 on the megalithic yard. More recent applications include radiocarbon dating (Buck *et al.* 1996, ch. 9) and other dating methods (*ibid.*, ch. 12); palaeoethnobotany (J. B. Kadane and C. A. Hastorf, 'Bayesian paleoethnobotany', in José M. Bernardo, M. H. DeGroot, David V. Lindley and Adrian F. M. Smith (eds), *Bayesian Statistics 3*, OUP, Oxford, 1988, 243–60); and spatial analysis (Buck *et al.* 1996, ch. 10).
7 This view is shared by Buck *et al.* (1996, 354) who nonetheless give some helpful general guidelines for approaching new problems (*ibid.*, 355–8).
8 Alan E. Gelfand and Adrian F. M. Smith, 'Sampling based approaches to calculating marginal densities', *J. Am. Stat. Ass.*, 85 (1990), 398–409; Alan E. Gelfand, Susan E. Hills, A. Racine-Poon, and Adrian F. M. Smith, 'Illustration of Bayesian inference in normal data models using Gibbs sampling', *J. Am. Stat. Ass.*, 85 (1990), 972–85. For a more general discussion of relevant techniques see Buck *et al.* 1996, ch. 8.

REFERENCE LISTS OF MONUMENTS

1 List 2 includes only those *DES* entries published between 1970 and 1980, because these formed part of the selection criteria determining a site's inclusion in the list (Ruggles 1984a, 25). More recent *DES* entries are included in List 6, but sometimes omitted where all the relevant information is available in the other sources cited.

LIST 2

1 Ruggles 1984a. Further monuments and alleged monuments excluded in the course of the original selection process (*ibid.*, 26 and 44–58) are not listed here.
2 The author's visit was in July 1978, not in 1979 as stated in Ruggles 1984a, 180.

LIST 3

1 Although there is evidence that there was once a circle of seven or eight stones here (Coles 1904, no. 1), there is no direct evidence that it was an RSC.
2 It is possible, though not certain, that there was once a circle here. See Ruggles 1984c, table 1, note *c*.
3 A single 2·7 m-long stone (the Carlin Stone) remains here, which may well be a recumbent stone. According to Coles (1903, no. 32) 'there stood several great stones—none nearly so great as the Carlin—in a circle'.
4 Two large stones stood here in the early years of the twentieth century (Coles 1903, no. 14) but there is no clear evidence that they ever formed part of an RSC (cf. Ruggles 1984c, table 1, note *f*).
5 Only two stones, probably removed from their original positions, remain here, and the evidence that they represent the remains of an RSC is tenuous (see Ruggles 1984c, table 1, note *g*).
6 Although there is evidence that there was once a circle of ten stones here (Fred R. Coles, 'Stone circles, chiefly in Banffshire', *PSAS*, 40 (1906), 164–206, no. 11), there is no direct evidence that it was an RSC.
7 There may be some confusion in the accounts of the destroyed monuments at Clatt (RSC34, 35) and Holywell (RSC36). See Ruggles 1984c, table 1, note *j*.
8 Keiller (1934, 20) mentions a 'now destroyed circle' here, but there is no evidence it was an RSC.
9 It is not certain that there was ever a stone circle at this location, as there may have been confusion between Crookmore (RSC50), Nether Balfour (RSC53) and Druidsfield (RSC52). See Ruggles 1984c, table 1, note *m*.
10 A single stone stood here earlier in the twentieth century but had been moved by 1981. There is no direct evidence that it had once formed part of any sort of stone circle (Ruggles 1984c, table 1, note *n*). Keiller (1934, 12) quotes an azimuth of 242° for a circle at 'Candle Hill, Oyne', although the latitude and longitude (*ibid.*, 23) are identical to that of the 'Carden Stone', i.e. the stone at Mill of Carden. It is possible that this azimuth actually refers to Hatton of Ardoyne (RSC56), which is in the same parish (cf. Barnatt 1989, 463).
11 The diameter quoted by Burl (1976, 350) is that of the ring-cairn.
12 Only a single standing stone remains here. Although there is evidence that there was once a stone circle, the evidence that it was an RSC is weak (see Ruggles 1984c, table 1, note *p*).
13 This monument consists of two erect stones, 1·9 m and 1·8 m high, some 2·3 m apart. They are quite possibly, though not certainly, flankers (see Ruggles 1984c, table 1, note *q*).
14 This monument consists of two erect stones, 3·7 m and 2·2 m high, some 3 m apart. Although the stones are not necessarily flankers (Ruggles 1984c, table 1, note *r*) this seems no less convincing a candidate for an RSC than South Fornet (RSC68).
15 There is no evidence in the informants' accounts quoted by Coles (1900, no. 19) and J. Ritchie ('Notes on some stone circles in the south of Aberdeenshire and north of Kincardine', *PSAS*, 53 (1919), 64–75, p. 64) that the original circle here was an RSC. The remaining three uprights form the arc of a ring at least 40 m in diameter, which would be anomalously large for an RSC.
16 NMRS entry NJ71NW15 appears to be describing the same monument.
17 The figure of *c.* 25 m is deduced from Coles (1900, no. 14), but Burl (1976, 350) quotes 31 m.
18 Despite its original classification as an unlikely RSC (see Ruggles 1984c, table 1, note *t*) this monument, comprising a large recumbent stone adjacent to a cairn, may be comparable to The Camp, Montgoldrum (RSC94) and The Cloch (Supp. List A). See Barclay and Ruggles forthcoming.
19 This is merely a ring cairn, adjacent to a Late Bronze Age / Early Iron Age settlement. Cf. Ruggles 1984c, table 1, note *u*.
20 Only a single 2 m-high standing stone remains here. Although according to Keiller (1934, 20–1) it is 'obviously, to the experienced eye, [a] Pillar Stone', there is no evidence that it was necessarily a flanker. Cf. Ruggles 1984c, table 1, note *v*.
21 A possible standing stone. There is no evidence that it was part of a stone circle.
22 This was almost certainly a robbed cairn or ring-cairn. A ring of stones shown in a plan by Coles (F. R. Coles, 'Notice of the Camp at Montgoldrum and other antiquities in Kincardineshire', *PSAS*, 37 (1903), 193–9, p. 198) were probably the kerbstones. Cf. Ruggles 1984c, table 1, note *w*.
23 This monument comprises a large recumbent stone (now blasted into several pieces) adjacent to a ring cairn.
24 This monument has not in fact been completely destroyed. It comprises a single, large recumbent stone.
25 Although not listed as such in the NMRS, this is definitely an RSC (Barclay and Ruggles, forthcoming).
26 This is probably a ring cairn. There is no recumbent stone and no evidence of any other circle stones. Cf. Ruggles 1984c, table 1, note *x*.
27 This is the southernmost of three adjacent monuments. The other two were originally rectangular settings of eight stones, which have been interpreted as variant four-posters with larger stones at the corners and smaller ones at the midpoints of the sides (Burl 1988b, 166–75). Fortingall South consists of three well-spaced stones, almost in line, which have been interpreted as a variant recumbent stone and flankers (*ibid.*, 175) but their dimensions, and the fact that they are well spaced, argues against this. They might equally well be interpreted as a 5 m-long three-stone row, similar to others in Perthshire (see List 6). A further stonehole to the north-west does not clearly distinguish between these and other possibilities. Cf. Ruggles 1984c, table 1, note *y*.

LIST 4

1 This circle is anomalous, in that it consists (except for the axial stone) of large, rounded boulders with no obvious height gradation. The portal stones are not distinguishable except by their position. Thus the interpretation as an axial stone circle may be open to question.
2 A fourteenth stone, although well set, seems likely to be a secondary insertion (Paul Walsh, priv. comm., 1997).
3 Both Ó Nualláin 1984a and Power 1992 take the SSW stone in the ring to be the axial stone. Naming this stone *a* , and naming the stones anticlockwise *b*, *c* etc. up to missing stone *l*, Ó Nualláin's interpretation is that *a* is the axial stone and *f* and *g* are the portals (see Ó Nualláin 1984a, fig. 9). We feel that a number of factors argue against this. First, *a* is a small, pointed-topped stone characteristic of a circle stone but quite unlike a normal axial stone. Second, height gradations suggest that *e* and *f* are the portals, since they are the tallest stones. This would place the axial between *a* and *k* as missing stone *l*. Third, the prostrate stone *i* which has clearly been moved to its current position, measuring some 1·0 m × 1·0 m × 0·4 m, is of a size and thickness which is consistent with the idea that it is the axial stone, moved from position *l*. Finally, placing the axial stone at *a* yields an axial orientation anomalously far to the south (azimuth approximately 185°); placing it at *l* yields an orientation consistent with the other monuments of this type.
4 There exists an eighteenth-century drawing of this circle: see Roger Stalley (ed.), *Daniel Grose (c. 1766–1838). The Antiquities of Ireland: A Supplement to Francis Grose*, Irish Architectural Archive, Dublin, 1991, 117–19.
5 This information is as given by Ó Nualláin 1984a. The monument was not visited in 1994 because, according to this, the axial stone was missing. However, Power 1992 states that ten stones survive, including the axial stone. At the time of Ó Nualláin's survey a fence crossed the SW end of the circle, which was removed by the time of the Cork survey (Paul Walsh, priv. comm., 1997): this has caused some confusion, and is the reason that Roberts 1996 includes three Knocks circles, nos 56 and 57 in fact being the same monument.

LIST 5

1 Ó Nualláin 1988 lists this monument as a three-stone row but includes the prostrate stone in his plan.
2 This has the appearance of a genuine stone row but there is an ogham inscription on the centre stone, which is why it is not included in Ó Nualláin 1988. See Ó Nualláin 1984b, 75, no. 10, where it is listed as an 'anomalous group of standing stones'.
3 Ó Nualláin 1988 lists this monument as 'three stones with a fourth . . . set roughly in line', the reason being that if it is considered to be a single row, it is unusually long. Burl 1993 follows by listing it as a three-stone row.
4 All but the largest stone shown in Ó Nualláin's plan had been cast down by the time of the author's visit in 1991.
5 The two outer stones shown in Ó Nualláin's plan were cast down in around 1987 and only the central stone now remains standing (Ruggles 1996, table 1, note *h*).
6 The second stone from the SW end had fallen between Ó Nualláin's visit and the author's in 1991, and the third from the SW end, already prostrate, had been shifted (Ruggles 1994a, table 1, note *h*).
7 Burl (1993, 219) separately lists a four-stone row at Carrigaline, giving the NGR as W 69 50, but it is really the same monument (Paul Walsh, priv. comm., 1997).
8 Ó Nualláin lists this as a four-stone row. The fifth stone was recognised as such by a team from the Cork Archaeological Survey in 1983 (Ruggles 1994a, table 1, note *a*).
9 The central stone, which was split in two and half of which is shown fallen on Ó Nualláin's plan, had been reconstructed following Lynch's excava-

tion by the time of the author's visit in 1991 (Ruggles 1994a, table 1, note *p*). Note that the number of stones still standing was shown erroneously as four in the previous publication (*ibid.*).

10 This row straddles the townland boundary.

11 The NE-most stone had fallen between Ó Nualláin's visit and the author's in 1991 (Ruggles 1994a, table 1, note *s*).

12 Although Ó Nualláin and others did not realise it, an ogham inscription survives on the central stone (Richard R. Brash, *The Ogam Inscribed Monuments of the Gaedhil in the British Islands*, George Bell and Sons, London, 1879, p. 158; R. A. S. Macalister, *Corpus Inscriptionum Insularum Celticarum*, vol. i, Stationery Office, Dublin, 1945, 59–61), which casts its status as a Bronze Age row into some doubt. Cf. CKR44a.

13 A plan in the Somerville papers at University College, Cork shows a row of four standing stones oriented ENE–WSW. Somerville recorded that a fifth stone had been removed from the row and broken up and that its remains could be seen in a fence to the north. There is no visible trace of any of these stones today (S. Ó Nualláin, priv. comm., 1991).

14 This is a three-stone row aligned NNE–SSW with a fourth standing 2·5 m to the NW. It was discovered in 1990.

LIST 6

1 Ruggles 1984a, 61–2.

2 Noel Fojut, *A Guide to Prehistoric Shetland*, Shetland Times, Lerwick, 1981, 58.

3 The monument is described here as 'a possible trapezoid setting of prostrate stones'.

4 Only one stone now stands, not three as stated by Burl (1993, 248).

5 Burl (1993, 224, 256) lists this monument twice, once as Ru Ardvule (S Uist) and once as Sligeanach (Skye), where an erroneous grid reference is given.

6 Burl (1993, 254) erroneously gives the total length as 650 m. The standing stone at Ardfernal is 1,000 m to the ENE, not 450 m to the WSW, of the pair 200 m apart at Knockrome. The alignment is ENE–WSW, not NNE–SSW as stated by Burl.

7 Cf. Ruggles 1984a, 170 (IS15).

8 Angus Graham, 'A survey of the ancient monuments of Skipness', *PSAS*, 53 (1919), 76–118, p. 109, no. 4.

TABLE 3.2

1 The coding gives not only the main class to which the indication has been assigned (e.g. Class 3) but also the precise type of indication (e.g. Class 3a, two standing stones, not both slabs oriented along the line joining them). For details see Ruggles 1984a, 62.

2 An archaeological status is also shown: 'A' (reliable) if both sites are 'A', but 'B' (somewhat doubtful) if at least one of the sites is 'B'. This takes no account of whether the sites are in face contemporaneous, since this is generally not known. See Ruggles 1984a, 67.

TABLE 5.1

1 Ruggles (1984c, table 2) quotes Keiller as giving an azimuth of 202° for Candle Hill (RSC39). In fact, Keiller's azimuth refers to 'Old Town, Insch', whose latitude and longitude as tabulated by Keiller (1934, 24) is found to correspond to Stonehead (RSC41).

TABLE 5.2

1 The mean altitude has been estimated by taking the mean of the values $h_E$, $h_C$ and $h_W$ where $h_C$ is the Recumbent Centre altitude; $h_E$ is the East flanker or Recumbent Left altitude where only one of these is quoted or the mean of the two values where they are both quoted; and $h_W$ is the Recumbent Right or West flanker altitude where only one of these is quoted or the mean of the two values where they are both quoted.

TABLE 5.3

1 On how the mean altitude has been estimated, see the note above.

TABLE 6.2

1 The latter is the case at Behagullane (CKR67) (Ruggles 1994a, table 1); in other cases see Table 6.1 to distinguish between the two possibilities.

# Bibliographical abbreviations

*AA* — *Archaeoastronomy* (supplement to *Journal for the History of Astronomy*)
*AAJ* — *Archaeoastronomy* (Bulletin, later Journal, of the Center for Archaeoastronomy, College Park, Maryland)
*A. and E. News* — *Archaeoastronomy and Ethnoastronomy News*
*AIHS* — *Archives Internationales d'Histoire des Sciences*
*ANYAS* — *Annals of the New York Academy of Sciences*
*Arch. Camb.* — *Archaeologia Cambriensis*
*Arch. J.* — *Archaeological Journal*
BAR — British Archaeological Reports
*Camb. Arch. J.* — *Cambridge Archaeological Journal*
CBA — Council for British Archaeology
CUP — Cambridge University Press
*Curr. Anth.* — *Current Anthropology*
*Curr. Arch.* — *Current Archaeology*
*Derbys. Arch. J.* — *Derbyshire Archaeological Journal*
*DES* — *Discovery and Excavation in Scotland*
*Glasgow Arch. J.* — *Glasgow Archaeological Journal*
HMSO — Her Majesty's Stationery Office
*IARF* — *Irish Archaeological Research Forum*
*J. Am. Stat. Ass.* — *Journal of the American Statistical Association*
*J. Arch. Sci.* — *Journal of Archaeological Science*
*JBAA* — *Journal of the British Astronomical Association*
*JCHAS* — *Journal of the Cork Historical and Archaeological Society*
*JHA* — *Journal for the History of Astronomy*
*JKAHS* — *Journal of the Kerry Archaeological and Historical Society*
*JRAI* — *Journal of the Royal Anthropological Institute*
*JRASC* — *Journal of the Royal Astronomical Society of Canada*
*JRSAI* — *Journal of the Royal Society of Antiquaries of Ireland*
*JRSS* — *Journal of the Royal Statistical Society*
NMRS — National Monuments Record of Scotland
*NA* — *Northern Archaeology*
OUP — Oxford University Press
*Oxford J. Arch.* — *Oxford Journal of Archaeology*
*PASP* — *Publications of the Astronomical Society of the Pacific*
*PDNHAS* — *Proceedings of the Dorset Natural History and Archaeological Society*
*PPS* — *Proceedings of the Prehistoric Society*
*PRIA* — *Proceedings of the Royal Irish Academy*
*PSAS* — *Proceedings of the Society of Antiquaries of Scotland*
*PTRS* — *Philosophical Transactions of the Royal Society of London*
*QJRAS* — *Quarterly Journal of the Royal Astronomical Society*
RCAHMS — Royal Commission on the Ancient and Historical Monuments of Scotland
RCHME — Royal Commission on the Historical Monuments of England
*SAR* — *Scottish Archaeological Review*
*Sci. Arch.* — *Science and Archaeology*
SEAC — Société Européenne pour l'Astronomie dans la Culture [European Society for Astronomy in Culture]
*TDGNHAS* — *Transactions of the Dumfriesshire and Galloway Natural History and Antiquarian Society*
*UJ Arch.* — *Ulster Journal of Archaeology*
UP — University Press
UTP — University of Texas Press
*WA* — *World Archaeology*
*WAM* — *Wiltshire Archaeological Magazine*
*WANHM* — *Wiltshire Archaeological and Natural History Magazine*

# Bibliography


Aitken, Martin J. (1990). *Science-based Dating in Archaeology.* London: Longman.

Andresen, Jens, Torsten Madsen and Irwin Scollar, eds (1993). *Computing the Past: Computer Applications and Quantitative Methods in Archaeology, CAA92.* Aarhus: Aarhus UP.

Armit, Ian (1996). *The Archaeology of Skye and the Western Isles.* Edinburgh: Edinburgh UP.

Ashbee, Paul (1984). *The Earthen Long Barrow in Britain* (2nd edn). Norwich: Geo Books.

Ashmore, Patrick J. (1995). *Calanais: The Standing Stones.* Stornoway: Urras nan Tursachan.

——(1996). *Neolithic and Bronze Age Scotland.* London: Batsford/Historic Scotland.

Atkinson, Richard J. C. (1966). 'Moonshine on Stonehenge.' *Antiquity*, 40, 212–16.

——(1975). 'Megalithic astronomy—a prehistorian's comments.' *JHA* 6, 42–52.

——(1979). *Stonehenge* (revised edn). Harmondsworth: Penguin Books.

——(1981). 'Comments on the archaeological status of some of the sites.' Appendix 4.2 to Ruggles 1981. In Ruggles and Whittle 1981, 206–9.

——(1982). 'Aspects of the archaeoastronomy of Stonehenge.' In Heggie 1982a, 107–16.

Aveni, Anthony F. (1980). *Skywatchers of Ancient Mexico.* Austin: UTP.

——, ed. (1982). *Archaeoastronomy in the New World.* Cambridge: CUP.

——(1987). 'Archaeoastronomy in the Southwestern United States: a neighbor's eye view.' In Carlson and Judge 1987, 9–23.

——, ed. (1989a). *World Archaeoastronomy.* Cambridge: CUP.

——(1989b). 'Introduction: whither archaeoastronomy?' In Aveni 1989a, 3–12.

——(1992a). *Conversing with the Planets.* New York: Times Books.

——(1992b). 'Nobody asked, but I couldn't resist: a response to Keith Kintigh on archaeoastronomy and archaeology.' *A. and E. News*, no. 6, 1 and 4.

——(1994). 'Archaeoastronomy.' In Macey 1994, 26–35.

——(1995). 'Frombork 1992: where worlds and disciplines collide' [essay review of Iwaniszewski *et al.* 1994]. *AA* no. 20 (*JHA* 26), S74–9.

——(1996). 'Between a rock and a hard place' (review of North 1996). *Nature* 383, 403–4.

——(1997). *Stairways to the Stars: Skywatching in Three Great Ancient Cultures.* New York: Wiley.

Aveni, Anthony F. and Gary Urton, eds (1982). *Ethnoastronomy and Archaeoastronomy in the American Tropics* (ANYAS 385). New York: New York Academy of Sciences.

Bahn, Paul G., ed. (1996). *The Cambridge Illustrated History of Archaeology.* Cambridge: CUP.

Bailey, Mark E., John A. Cooke, Roger W. Few, Guy [J. G.] Morgan, and Clive L. N. Ruggles (1975). 'Survey of three megalithic sites in Argyllshire.' *Nature* 253, 431–3.

Bannister, A., S. Raymond and R. Baker (1992). *Surveying* (6th edn). Harlow: Longman.

Barber, John W. (1973). 'The orientation of the recumbent-stone circles of the south-west of Ireland.' *JKAHS* 6, 26–39.

Barclay, Alistair and Jan Harding, eds (1999). *Pathways and Ceremonies: The Cursus Monuments of Britain and Ireland.* Oxford: Oxbow Books.

Barclay, Gordon J. (1997). 'The Neolithic.' In Edwards and Ralston 1997, 127–49.

Barclay, Gordon J. and Clive L. N. Ruggles (forthcoming). 'On the frontier: the recumbent stone circles of South Kincardineshire and Angus.'

Barnatt, John (1989). *Stone Circles of Britain. Taxonomic and Distributional Analyses and a Catalogue of Sites in England, Scotland and Wales.* Oxford: BAR (British Series 215). 2 vols.

——(1996). 'Moving beyond the monuments: paths and people in the Neolithic landscapes of the Peak District.' *NA* 13/14, 43–59.

Barnatt, John and Stephen Pierpoint (1983). 'Stone circles: observatories or ceremonial centres.' *SAR* 2, 101–15.

Barnett, Vic (1982). *Comparative Statistical Inference* (2nd edn). Chichester: Wiley.

Barrett, John C. (1994). *Fragments from Antiquity: An Archaeology of Social Life in Britain, 2900–1200 BC.* Oxford: Blackwell.

Barrett, John C., Richard Bradley and Martin Green (1991) (with contributions by Mark Bowden and others). *Landscape, Monuments and Society.* Cambridge: CUP.

Bauer, Brian S. and David S. P. Dearborn (1995). *Astronomy and Empire in the Ancient Andes.* Austin: UTP.

Bell, James A. (1994). *Reconstructing Prehistory: Scientific Method in Archaeology.* Philadelphia: Temple UP.

Bell, Martin and Michael J. C. Walker (1992). *Late Quaternary Environmental Change: Physical and Human Perspectives.* Harlow: Longman and New York: Wiley.

Belmonte Avilés, Juan A., ed. (1994). *Arqueoastronomía Hispana.* Madrid: Equipo Sirius.

Bender, Barbara (1992). 'Theorising landscapes, and the prehistoric landscapes of Stonehenge.' *Man* 27, 735–55.

——, ed. (1993). *Landscape, Politics and Perspectives.* Oxford: Berg.

BEWLEY, ROBERT (1994). *Prehistoric Settlements.* London: Batsford/English Heritage.

BRADLEY, RICHARD J. (1984). *The Social Foundations of Prehistoric Britain.* Harlow: Longman.

——(1991a). 'Ritual, time and history.' *WA* 23, 209–19.

——(1991b). 'Rock art and perception of landscape.' *Camb. Arch. J.* 1(1), 77–101.

——(1993). *Altering the Earth: The Origin of Monuments in Britain and Continental Europe.* Edinburgh: Society of Antiquaries of Scotland (Monograph Series no. 8).

——(1997). *Rock Art and the Prehistory of Atlantic Europe.* London: Routledge.

BRADLEY, RICHARD J. and RICHARD CHAMBERS (1988). 'A new study of the cursus complex at Dorchester on Thames.' *Oxford J. Arch.* 7, 271–89.

BRENNAN, MARTIN (1980). *The Boyne Valley Vision.* Portlaoise: Dolmen Press.

——(1983). *The Stars and the Stones: Ancient Art and Astronomy in Ireland.* London: Thames and Hudson. Reprinted (1994) as *The Stones of Time: Calendars, Sundials and Stone Chambers of Ancient Ireland.* Rochester VT: Inner Traditions International.

BRODA, JOHANNA (1982). 'Astronomy, *cosmovisión*, and ideology in pre-Hispanic Mesoamerica.' In Aveni and Urton 1982, 81–110.

——(1993). 'Astronomical knowledge, calendrics, and sacred geography in ancient Mesoamerica.' In Ruggles and Saunders 1993a, 253–95.

BUCK, CAITLIN E., WILLIAM G. CAVANAGH, and CLIFFORD D. LITTON (1996). *Bayesian Approach to Interpreting Archaeological Data.* Chichester: Wiley.

BURGESS, COLIN (1980). *The Age of Stonehenge.* London: Dent.

BURL, AUBREY [H. A. W.] (1970). 'The recumbent stone circles of north-east Scotland.' *PSAS* 102, 56–81.

——(1976). *The Stone Circles of the British Isles.* New Haven: Yale UP.

——(1979a). *Rings of Stone: The Prehistoric Stone Circles of Britain and Ireland.* London: Frances Lincoln/Weidenfeld and Nicolson.

——(1979b). *Prehistoric Avebury.* New Haven: Yale UP.

——(1980). 'Science or symbolism: problems of archaeo-astronomy.' *Antiquity* 54, 191–200.

——(1981a). '"By the light of the cinerary moon": Chambered tombs and the astronomy of death.' In Ruggles and Whittle 1981, 243–74.

——(1981b). 'Holes in the argument.' *AAJ* 4(4), 19–25.

——(1981c). *Rites of the Gods.* London: Dent.

——(1982). 'Pi in the sky.' In Heggie 1982a, 141–69.

——(1983). *Prehistoric Astronomy and Ritual.* Princes Risborough: Shire (Shire Archaeology, 32) [reprinted with updated bibliography 1997].

——(1985). *Megalithic Brittany: A Guide to over 350 Ancient Sites and Monuments.* London: Thames and Hudson.

——(1987a). *The Stonehenge People.* London: Dent.

——(1987b). 'The sun, the moon, and megaliths: archaeo-astronomy and the standing stones of Northern Ireland.' *UJ Arch.*, 50, 7–21.

——(1988a). '"Without sharp north . . .": Alexander Thom and the great stone circles of Cumbria.' In Ruggles 1988a, 175–205.

——(1988b). *Four-Posters: Bronze Age Stone Circles of Western Europe.* Oxford: BAR (British Series 195).

——(1988c). *Prehistoric Stone Circles* (3rd edn). Princes Risborough: Shire (Shire Archaeology, 9) [reprinted with updated bibliography 1997].

——(1993). *From Carnac to Callanish: The Prehistoric Stone Rows and Avenues of Britain, Ireland, and Brittany.* New Haven: Yale UP.

——(1994). 'Stonehenge: slaughter, sacrifice and sunshine.' *WANHM* 87, 85–95.

——(1995). *A Guide to the Stone Circles of Britain, Ireland and Brittany.* New Haven: Yale UP.

——(1997). 'The sarsen horseshoe inside Stonehenge: a rider.' *WANHM* 90, 1–12.

CAMPBELL, M. and M. SANDEMAN (1961). 'Mid-Argyll: a field survey of the historic and prehistoric monuments.' *PSAS* 95, 1–125.

CARLSON, JOHN B. and W. JAMES JUDGE, eds (1987). *Astronomy and Ceremony in the Prehistoric Southwest.* Albuquerque: Maxwell Museum of Anthropology (Papers, no. 2).

CARMICHAEL, DAVID L., JANE HUBERT, BRIAN REEVES, and AUDHILD SCHANCHE, eds (1994). *Sacred Sites, Sacred Places.* London: Routledge (One World Archaeology, 23).

CASTLEDEN, RODNEY (1987). *The Stonehenge People.* London: Routledge and Kegan Paul.

——(1993). *The Making of Stonehenge.* London: Routledge.

CHILDE, GORDON [V. G.] (1933). 'Trial excavations at the Old Keig stone circle.' *PSAS* 67, 37–53.

——(1940). *Prehistoric Communities of the British Isles.* London: Chambers.

——(1955). Contribution to discussion on Thom 1955. *JRSS* A118, 293–4.

CHIPPINDALE, CHRISTOPHER (1994). *Stonehenge Complete* (revised edn). London: Thames and Hudson.

CIMINO, GUIDO, ed. (1999, in press). *History of Science* (9 vols). Rome: Enciclopedia Italiana.

CLEAL, ROSAMUND M. J., K. E. WALKER, and R. MONTAGUE (1995) (with contributions by Michael J. Allen and others). *Stonehenge in its Landscape: Twentieth-century Excavations.* London: English Heritage (English Heritage Archaeological Report 10).

COLES, FRED R. (1900). 'Stone circles in Kincardineshire (North) and part of Aberdeenshire.' *PSAS* 34, 139–98.

——(1903). 'Stone circles in the N.E. of Scotland. Auchterless and Forgue.' *PSAS* 37, 82–142.

——(1904). 'Stone circles in the N.E. of Scotland. The Buchan district.' *PSAS* 38, 256–305.

COOKE, JOHN A., ROGER W. FEW, GUY [J. G.] MORGAN, and CLIVE L. N. RUGGLES (1977). 'Indicated declinations at the Callanish megalithic sites.' *JHA* 8, 113–33.

COONEY, GABRIEL and EOIN GROGAN (1994). *Irish Prehistory: A Social Perspective.* Dublin: Wordwell.

CORNELL, JAMES (1981). *The First Stargazers.* London: Athlone.

COWIE, TREVOR G. (1980). 'Excavations at Kintraw, Argyll, 1979.' *Glasgow Arch. J.* 7, 27–31.

CUNLIFFE, BARRY W. (1997). *The Ancient Celts.* Oxford: OUP.

CUNLIFFE, BARRY W. and COLIN RENFREW, eds (1997). *Science and Stonehenge* (*Proceedings of the British Academy*, 92). Oxford: OUP.

CUNNINGTON, MAUD E. (1931). 'The "Sanctuary" on Overton Hill, near Avebury.' *WAM* 45, 300–35.

CUPPAGE, JUDITH (1986) (with Isabel Bennett, Claire Cotter, Celie O Rahilly and others). *Archaeological Survey of the Dingle Peninsula.* Ballyferriter: Oidhreacht Chorca Dhuibne.

CURTIS, RONALD [G. R.] (1988). 'The geometry of some megalithic rings.' In Ruggles 1988a, 351–77.

DARVILL, TIMOTHY C. (1987). *Prehistoric Britain.* London: Batsford.

——(1996). *Prehistoric Britain from the Air.* Cambridge: CUP.

——(1997). 'Ever increasing circles: the sacred geographies of Stonehenge and its landscape.' In Cunliffe and Renfrew 1997, 167–202.

DAVID, F. N. (1962). *Games, Gods and Gambling.* London: Griffin.

DAVIDSON, J. L. and AUDREY S. HENSHALL (1989). *The Chambered Cairns of Orkney.* Edinburgh: Edinburgh UP.

——(1991). *The Chambered Cairns of Caithness.* Edinburgh: Edinburgh UP.

DE VALERA, RUAIDHRÍ and SEÁN Ó NUALLÁIN (1982). *Survey of the Megalithic Tombs of Ireland, Vol. IV: Counties Cork, Kerry, Limerick, Tipperary.* Dublin: Stationery Office.

EDWARDS, KEVIN J. and IAN B. M. RALSTON, eds (1997). *Scotland: Environment and Archaeology, 8000 BC–AD 1000.* Chichester: Wiley.

EDWARDS, KEVIN J. and GRAEME WHITTINGTON (1997). 'Vegetation change.' In Edwards and Ralston 1997, 63–82.

ELLEGÅRD, ALVAR (1981). 'Stone age science in Britain?' *Curr. Anth.* 22, 99–125.

EOGAN, GEORGE (1986). *Knowth and the Passage Tombs of Ireland.* London: Thames and Hudson.

EVANS, JOHN G. (1975). *The Environment of Early Man in the British Isles.* London: Paul Elek.

FANE GLADWIN, P. (1985). *The Solar Alignment at Brainport Bay, Minard, Argyll.* Ardrishaig: Natural History and Antiquarian Society of Mid-Argyll.

FENTON, ALEXANDER (1976). *Scottish Country Life.* Edinburgh: John Donald.

FERGUSON, LESLEY J. (1988). 'A catalogue of the Alexander Thom archive held in the National Monuments Record of Scotland.' In Ruggles 1988a, 31–131.

FISHER, PETER F., CHRIS FARRELLY, ADRIAN MADDOCKS, and CLIVE L. N. RUGGLES (1997). 'Spatial analysis of visible areas from the Bronze Age cairns of Mull.' *J. Arch. Sci.* 24, 581–92.

FITZPATRICK, ANDREW (1997). 'Everyday life in Iron Age Wessex.' In Gwilt and Haselgrove 1997, 73–86.

FLEMING, ANDREW (1975). 'Megalithic astronomy: a prehistorian's view.' *Nature* 255, 575.

FLETCHER, MIKE and GARY R. LOCK (1994). *Digging Numbers* (2nd impression). Oxford: Oxford University Committee for Archaeology (Monograph 33).

FOUNTAIN, JOHN and ROLF SINCLAIR, eds (1999). *Archaeoastronomy: Selected Papers from the Fifth Oxford Conference on Archaeoastronomy.* Durham, NC: Carolina Academic Press.

FRANK, ROSLYN M. and JON D. PATRICK (1993). 'The geometry of pastoral stone octagons: the Basque *sarobe*.' In Ruggles 1993a, 77–91.

FRASER, DAVID (1983). *Land and Society in Neolithic Orkney* (2 vols). Oxford: BAR (British Series 117).

FREEMAN, PETER R. (1976). 'A Bayesian analysis of the megalithic yard.' *JRSS* A139, 20–55.

——(1982). 'The statistical approach.' In Heggie 1982a, 45–52.

FREEMAN, PETER R. and W. ELMORE (1979). 'A test for the significance of astronomical alignments.' *AA* no. 1 (*JHA* 10), S86–96.

GIBSON, ALEX (1994) (with contributions by S. H. R. Aldhouse-Green and others). 'Excavations at the Sarn-y-bryn-caled cursus complex, Welshpool, Powys, and the timber circles of Great Britain and Ireland.' *PPS* 60, 143–223.

——(1995). 'The dating of timber circles: new thoughts in the light of recent Irish and British discoveries.' In John Waddell and Elizabeth Shee Twohig (eds.), *Ireland in the Bronze Age*, Stationery Office, Dublin, 87–9.

GIBSON, ALEX and DEREK D. A. SIMPSON, eds (1998). *Prehistoric Ritual and Religion: Essays in Honour of Aubrey Burl.* Stroud: Sutton.

GILLINGS, MARK, DAVID MATTINGLY and JAN VAN DALEN, eds (1998). *Geographical Information Systems and Landscape Archaeology.* Oxford: Oxbow Books (The Archaeology of the Mediterranean Landscape, 3).

GINGERICH, OWEN (1981). Comment on Ellegård 1981. *Curr. Anth.* 22, 117–18.

——(1989). 'Reflections on the role of archaeoastronomy in the history of astronomy.' In Aveni 1989a, 38–44.

GIOT, PIERRE-ROLAND (1988). 'Stones in the landscape of Brittany.' In Ruggles 1988a, 319–24.

GOSDEN, CHRISTOPHER (1994). *Social Being and Time.* Oxford: Blackwell.

GOULD, R. A. (1980). *Living Archaeology.* Cambridge: CUP.

GREEN, MIRANDA (1991). *The Sun-Gods of Ancient Europe.* London: Batsford.

GRIFFIN-PIERCE, TRUDY (1992). *Earth is my Mother, Sky is my Father: Time and Astronomy in Navajo Sandpainting.* Albuquerque: University of New Mexico Press.

GWILT, ADAM and COLIN HASELGROVE, eds (1997). *Reconstructing Iron Age Societies: New Approaches to the British Iron Age.* Oxford: Oxbow Books (Monograph 71).

HADINGHAM, EVAN (1981). 'The lunar observatory hypothesis at Carnac: a reconsideration.' *Antiquity* 45, 35–42.

HAMPSHER-MONK, IAIN and PHILIP ABRAMSON (1982). 'Strichen.' *Curr. Arch.* 8 [no. 84], 16–19.

HAWKES, JACQUETTA (1962). *Man and the Sun.* London: Cresset.

——(1967). 'God in the machine.' *Antiquity* 41, 174–80.

HAWKINS, GERALD S. (1963). 'Stonehenge decoded.' *Nature* 200, 306–8.

——(1964). 'Stonehenge: a Neolithic computer.' *Nature* 202, 1258–61.

——(1968). 'Astro-archaeology.' *Vistas in Astronomy* 10, 45–88.

HAWKINS, GERALD S. and JOHN B. WHITE (1970). *Stonehenge Decoded* (Fontana edn). London: Fontana/Collins.

HAYMAN, RICHARD (1997). *Riddles in Stone.* London: The Hambledon Press.

HEGGIE, DOUGLAS C. (1972). 'Megalithic lunar observatories: an astronomer's view.' *Antiquity* 46, 43–8.

——(1981a). *Megalithic Science: Ancient Mathematics and Astronomy in Northwest Europe.* London: Thames and Hudson.

——(1981b). 'Highlights and problems of megalithic astronomy.' *AA* no. 3 (*JHA* 12), S17–37.

——, ed. (1982a). *Archaeoastronomy in the Old World.* Cambridge: CUP.

——(1982b). 'Megalithic astronomy: highlights and problems.' In Heggie 1982a, 1–24.

HENSHALL, AUDREY S. (1963). *The Chambered Tombs of Scotland, vol. 1.* Edinburgh: Edinburgh UP.

——(1972). *The Chambered Tombs of Scotland, vol. 2.* Edinburgh: Edinburgh UP.

——(1985). 'The chambered cairns.' In Renfrew 1985, 83–117.

HIEB, LOUIS A. (1979). 'Hopi World View.' In Ortiz 1979, 577–80.

HODDER, IAN (1984). 'Burials, houses, women and men in the European Neolithic.' In Daniel Miller and Christopher Tilley (eds), *Ideology, Power and Prehistory*, CUP, Cambridge, 51–68.

——(1991). *Reading the Past* (2nd edn). Cambridge: CUP.

——(1992). *Theory and Practice in Archaeology.* London: Routledge.

HOSKIN, MICHAEL A., ed. (1997a). *The Cambridge Illustrated History of Astronomy.* Cambridge: CUP.

——(1997b). 'Mediterranean tombs and temples and their orientations.' In Jaschek and Atrio 1997, 19–25.

HOYLE, FRED (1966). 'Speculations on Stonehenge.' *Antiquity* 40, 262–76.

——(1972). *From Stonehenge to Modern Cosmology.* San Francisco: Freeman.

——(1977). *On Stonehenge.* London: Heinemann.

HUTTON, RONALD (1991). *The Pagan Religions of the Ancient British Isles.* Oxford: Blackwell.

——(1996). *The Stations of the Sun: The History of the Ritual Year in Britain.* Oxford: OUP.

INGOLD, TIM (1993). 'The temporality of the landscape.' *WA* 25, 152–74.

IWANISZEWSKI, STANISŁAW, ARNOLD LEBEUF, ANDRZEJ WIERCIŃSKI and MARIUSZ ZIÓŁKOWSKI, eds (1994). *Time and Astronomy at the Meeting of Two Worlds.* Warsaw: Centro de Estudios Latinoamericanos [CESLA] (Studies and Materials, 10).

JASCHEK, C. and F. ATRIO BARANDELA, eds (1997). *Actas del IV Congreso de la SEAC 'Astronomía en la Cultura'*. Salamanca: Universidad de Salamanca.

JERMY, A. C. (1978). 'General description and topography.' In Jermy and Crabbe 1978, 3.1–3.11.

JERMY, A. C. and J. A. CRABBE, eds (1978). *The Island Of Mull: A Survey of its Flora and Environment*. London: British Museum (Natural History).

JONES, MARTIN (1986). *England Before Domesday*. London: Batsford.

KEILLER, ALEXANDER (1934). *Megalithic Monuments of North-East Scotland*. London: Morven Institute of Archaeological Research.

KILBRIDE-JONES, H. E. (1935). 'An account of the excavation of the stone circle at Loanhead of Daviot.' *PSAS* 69, 168–222.

KINNES, IAN A. (1992). *Non-Megalithic Long Barrows and Allied Structures in the British Neolithic*. London: British Museum (Occasional Paper, 52).

KINTIGH, KEITH W. (1992). 'I wasn't going to say anything, but since you asked: archaeoastronomy and archaeology.' *A. and E. News*, no. 5, 1 and 4.

KNIGHT, CHRIS (1991). *Blood Relations: Menstruation and the Origins of Culture*. New Haven: Yale UP.

KÖHLER, ULRICH (1991). 'Pitfalls in archaeoastronomy: with examples from Mesoamerica.' In Romano and Traversari 1991, 130–6.

KRUPP, EDWIN C. (1994). *Echoes of the Ancient Skies: The Astronomy of Lost Civilizations*. Oxford: OUP. Originally published (1983) by Harper and Row, New York.

——(1997). *Skywatchers, Shamans and Kings*. New York: Wiley.

LANCASTER BROWN, PETER (1976). *Megaliths, Myths and Men: An Introduction to Astro-archaeology*. Poole: Blandford Press.

LEWIS, A. L. (1892). 'Stone circles in Britain.' *Arch. J.* 49, 136–54.

LINCEI (1995). *Archeologia e Astronomia: Esperienze e Prospettive Future*. Rome: Accademia Nazionale dei Lincei (Atti dei Convegni Lincei, 121).

LINDSTRÖM, JONATHAN (1997). 'The orientation of ancient monuments in Sweden: a critique of archaeoastronomy and an alternative interpretation.' *Current Swedish Archaeology* 5, 111–25.

LOCK, GARY and ZORAN STANČIĆ, eds (1995). *Archaeology and Geographical Information Systems: A European Perspective*. London: Taylor and Francis.

LOCKYER, NORMAN [J. N.] (1909). *Stonehenge and Other British Stone Monuments Astronomically Considered* (2nd edn). London: Macmillan.

LYNCH, ANN (1981). *Man and Environment in South-West Ireland, 4000 BC–AD 800*. Oxford: BAR (British Series 85).

——(1982). 'Astronomy and stone alignments in S.W. Ireland.' In Heggie 1982a, 205–13.

——(1999). 'Excavation of a stone row at Maughanasilly, Co. Cork.' *JCHAS* 104, in press.

LYNCH, FRANCES (1973). 'The use of the passage in certain passage graves as a means of communication rather than access.' In Glyn E. Daniel and Poul Kjærum (eds), *Megalithic Graves and Ritual, Papers Presented at the III Atlantic Colloquium, Moesgård 1969*, Publications XI, Jutland Archaeological Society, Copenhagen, 1973, 147–61.

——(1997). *Megalithic Tombs and Long Barrows in Britain*. Princes Risborough: Shire (Shire Archaeology, 73).

MCCLUSKEY, STEPHEN C. (1977). 'The astronomy of the Hopi Indians.' *JHA* 8, 174–95.

——(1987). 'Science, society, objectivity, and the astronomies of the Southwest.' In Carlson and Judge 1987, 205–17.

——(1990). 'Calendars and symbolism: functions of observation in Hopi astronomy.' *AA* no. 15 (*JHA* 21), S1–16.

——(1993a). 'Space, time and the calendar in the traditional cultures of America.' In Ruggles 1993a, 33–44.

——(1993b). 'Astronomies and rituals at the dawn of the Middle Ages.' In Ruggles and Saunders 1993a, 100–23.

——(1998). *Astronomies and Cultures in Early Medieval Europe*. Cambridge: CUP.

——(1999). 'Puebloan ethnoscience.' In Cimino 1999, in press.

MCCREERY, THOMAS (1979). 'Megalithic lunar observatories—a critique, I.' *Kronos* 5(1), 47–63.

——(1980). 'Megalithic lunar observatories—a critique, II.' *Kronos* 5(2), 6–26.

MCCREERY, THOMAS, A. J. HASTIE, and T. MOULDS (1982). 'Observations at Kintraw.' In Heggie 1982a, 183–90.

MACEY, SAMUEL L., ed. (1994). *Encyclopedia of Time*. New York: Garland.

MACKIE, EUAN W. (1973). 'Duntreath.' *Curr. Arch.* 4 [no. 36], 6–7.

——(1974). 'Archaeological tests on supposed astronomical sites in Scotland.' *PTRS* A276, 169–94.

——(1977). *Science and Society in Prehistoric Britain*. London: Elek.

——(1981). 'Wise men in antiquity?' In Ruggles and Whittle 1981, 111–52.

——(1988). 'Investigating the prehistoric solar calendar.' In Ruggles 1988a, 206–31.

——(1997). 'Maeshowe and the winter solstice: ceremonial aspects of the Orkney Grooved Ware culture.' *Antiquity*, 71, 338–59.

MACKIE, EUAN W., P. FANE GLADWIN, and ARCHIE E. ROY (1985). 'A prehistoric calendrical site in Argyll?' *Nature* 314, 158–61.

MCNALLY, DEREK (1974). *Positional Astronomy*. London: Muller.

MCNALLY, DEREK and CLIVE RUGGLES (1997). 'The minor standstill of the moon and Stonehenge.' *Astronomy and Geophysics* 38, 30–1.

MAGUIRE, DAVID J., MICHAEL F. GOODCHILD and DAVID W. RHIND, eds (1991). *Geographical Information Systems: Principles and Applications* (2 vols). Harlow: Longman.

MALONE, CAROLINE (1989). *Avebury*. London: Batsford/English Heritage.

MARSHACK, ALEXANDER (1972). *The Roots of Civilization*. New York: Weidenfeld and Nicolson.

——(1991). 'The Taï Plaque and calendrical notation in the Upper Palaeolithic.' *Camb. Arch. J.* 1(1), 25–61.

MARTLEW, ROGER D. and CLIVE L. N. RUGGLES (1993). 'The North Mull project (4): Excavations at Ardnacross 1989–91.' *AA* no. 18 (*JHA* 24), S55–64.

——(1996). 'Ritual and landscape on the west coast of Scotland: an investigation of the stone rows of northern Mull.' *PPS* 62, 117–31.

MASCHNER, HERBERT D. G., ed. (1996). *New Methods, Old Problems: Geographic Information Systems in Modern Archaeological Research*. Carbondale, IL: Center for Archaeological Investigations, Southern Illinois University at Carbondale (Occasional Paper no. 23).

MOIR, GORDON (1979). 'Hoyle on Stonehenge.' *Antiquity* 53, 124–8.

——(1981). 'Some archaeological and astronomical objections to scientific astronomy in British prehistory.' In Ruggles and Whittle 1981, 221–41.

MOIR, GORDON, CLIVE L. N. RUGGLES and RAYMOND NORRIS (1980). 'Megalithic science and some Scottish site plans.' *Antiquity* 54, 37–43.

MORRIS, RONALD W. B. (1981). *The Prehistoric Rock Art of Southern Scotland (except Argyll and Galloway)*. Oxford: BAR (British Series 86).

MORRISON, LESLIE V. (1980). 'On the analysis of megalithic lunar sightlines in Scotland.' *AA* no. 2 (*JHA* 11), S65–77.

MOTZ, HANS (1988). 'A personal appreciation of Professor Alexander Thom.' In Ruggles 1988a, 14–20.

MÜLLER, ROLF (1970). *Der Himmel über dem Menschen der Steinzeit*. Berlin: Springer-Verlag.

MYATT, LESLIE J. (1988). 'The stone rows of northern Scotland.' In Ruggles 1988a, 277–318.

NEWHAM, PETER [C. A.] (1993). *The Astronomical Significance of Stonehenge.* Warminster: Coates and Parker. Originally published (1972) by Moon Publications, Shirenewton.

NORTH, JOHN D. (1994). *The Fontana History of Astronomy and Cosmology.* London: Fontana.

——(1996). *Stonehenge: Neolithic Man and the Cosmos.* London: Harper Collins.

O'BRIEN, WILLIAM (1993). 'Aspects of wedge tomb chronology.' In Shee Twohig and Ronayne 1993, 63–74.

——(1994) (with contributions by A. Brindley, F. M. Chambers, R. A. Ixer, J. M. Lanting, S. A. McKeown, T. M. Mighall and M. O'Sullivan). *Mount Gabriel: Bronze Age Mining in Ireland.* Galway: Galway UP.

O'KELLY, CLAIRE (1978). *Illustrated Guide to Newgrange* (3rd edn). Wexford: John English.

O'KELLY, MICHAEL J. (1982). *Newgrange: Archaeology, Art and Legend.* London: Thames and Hudson.

——(1989). *Early Ireland: An Introduction to Irish Prehistory.* Cambridge: CUP.

Ó NUALLÁIN, SEÁN (1984a). 'A survey of stone circles in Cork and Kerry.' *PRIA* 84C, 1–77.

——(1984b). 'Grouped standing stones, radial-stone cairns and enclosures in the south of Ireland.' *JRSAI* 114, 63–79.

——(1988). 'Stone rows in the south of Ireland.' *PRIA* 88C, 179–256.

——(1989). *Survey of the Megalithic Tombs of Ireland, Vol. V: County Sligo.* Dublin: Stationery Office.

ORME, BRYONY (1981). *Anthropology for Archaeologists: An Introduction.* London: Duckworth.

ORTIZ, ALFONSO, ed. (1979). *Handbook of North American Indians, Volume 9: Southwest.* Washington DC: Smithsonian Institution.

O'SULLIVAN, ANN and SHEEHAN, JOHN (1996). *The Iveragh Peninsula: An Archaeological Survey of South Kerry.* Cork: Cork UP.

OSWALD, ALASTAIR (1997). 'A doorway on the past: practical and mystic concerns in the orientation of roundhouse doorways.' In Gwilt and Haselgrove 1997, 87–95.

PAPATHANASSIOU, MARIA and MICHAEL A. HOSKIN (1996). 'The Late-Minoan cemetery at Armenoi: a reappraisal.' *JHA* 27, 53–9.

PARKER PEARSON, MICHAEL (1993). *Bronze Age Britain.* London: Batsford/English Heritage.

PARKER PEARSON, MICHAEL and RAMILISONINA (1998). 'Stonehenge for the ancestors: the stones pass on the message.' *Antiquity* 72, 308–26.

PARKER PEARSON, MICHAEL and COLIN RICHARDS (1994). 'Architecture and order: spatial representation and archaeology.' In Michael Parker Pearson and Colin Richards (eds), *Architecture and Order: Approaches to Social Space*, Routledge, London, 1994, 38–72.

PATRICK, JON D. (1974). 'Midwinter sunrise at Newgrange.' *Nature* 249, 517–19.

——(1975). 'Megalithic exegesis: a comment.' *IARF* 2(2), 9–14.

——(1979). 'A reassessment of the lunar observatory hypothesis for the Kilmartin stones.' *AA* no. 1 (*JHA* 10), S78–85.

——(1981). 'A reassessment of the solstitial observatories at Kintraw and Ballochroy.' In Ruggles and Whittle 1981, 211–19.

PATRICK, JON D. and PETER R. FREEMAN (1988). 'A cluster analysis of astronomical orientations.' In Ruggles 1988a, 251–61.

PATTON, MARK (1993). *Statements in Stone: Monuments and Society in Neolithic Brittany.* London: Routledge.

PENNY, ALAN and JOHN E. WOOD (1973). 'The Dorset Cursus complex—a Neolithic astronomical observatory?' *Arch. J.* 130, 44–76.

PIGGOTT, STUART (1968). *The Druids.* London: Thames and Hudson.

PITTS, MICHAEL W. (1981). 'The discovery of a new stone at Stonehenge.' *AAJ* 4(2), 16–21.

——(1982). 'On the road to Stonehenge: report on investigations beside the A344 in 1968, 1979 and 1980.' *PPS* 48, 75–132.

PONTING, MARGARET R. and GERALD H. PONTING (1981). 'Decoding the Callanish complex—some initial results.' In Ruggles and Whittle 1981, 63–110.

POWELL, ANDREW B. (1994). 'Newgrange—science or symbolizm [sic].' *PPS* 60, 85–96.

POWELL, MARTIN J. (1995). 'Astronomical indications at a bell-barrow in South Wales.' *AA* no. 20 (*JHA* 26), S49–56.

POWELL, T. G. E., J. X. W. P. CORCORAN, FRANCES LYNCH, and J. G. SCOTT (1969). *Megalithic Enquiries in the West of Britain: a Liverpool Symposium.* Liverpool: Liverpool UP.

POWER, DENIS (with Elizabeth Byrne, Ursula Egan, Sheila Lane and Mary Sleeman) (1992). *Archaeological Inventory of County Cork. Volume 1: West Cork.* Dublin: Stationery Office.

——(1994). *Archaeological Inventory of County Cork. Volume 2: East and South Cork.* Dublin: Stationery Office.

——(1997). *Archaeological Inventory of County Cork. Volume 3: Mid Cork.* Dublin: Stationery Office.

RAY, THOMAS P. (1989). 'The winter solstice phenomenon at Newgrange, Ireland: accident or design?' *Nature* 337, 343–5.

RCAHMS (1911). *Third Report and Inventory of Monuments and Constructions in the County of Caithness.* London: HMSO.

——(1912). *Fourth Report and Inventory of Monuments and Constructions in Galloway, vol. 1: County of Wigtown.* London: HMSO.

——(1914). *Fifth Report and Inventory of Monuments and Constructions in Galloway, vol. 2: County of the Stewartry of Kirkcudbright.* Edinburgh: HMSO.

——(1920). *Seventh Report with Inventory of Monuments and Constructions in the County of Dumfries.* Edinburgh: HMSO.

——(1924). *Eighth Report with Inventory of Monuments and Constructions in the County of East Lothian.* Edinburgh: HMSO.

——(1928). *Ninth Report with Inventory of Monuments and Constructions in the Outer Hebrides, Skye, and the Small Isles.* Edinburgh: HMSO.

——(1933). *Eleventh Report with Inventory of Monuments and Constructions in the Counties of Fife, Kinross, and Clackmannan.* Edinburgh: HMSO.

——(1946).[1] *Twelfth Report with an Inventory of the Ancient Monuments of Orkney and Shetland* (3 vols). Edinburgh: HMSO.

——(1963). *Stirlingshire: an Inventory of the Ancient Monuments* (2 vols). Edinburgh: HMSO.

——(1971). *Argyll: an Inventory of the Ancient Monuments, Volume 1: Kintyre.* Edinburgh: HMSO.

——(1975). *Argyll: an Inventory of the Ancient Monuments, Volume 2: Lorn.* Edinburgh: HMSO.

——(1980). *Argyll: an Inventory of the Monuments, Volume 3: Mull, Tiree, Coll and Northern Argyll.* Edinburgh: HMSO.

——(1984). *Argyll: an Inventory of the Monuments, Volume 5: Islay, Jura, Colonsay and Oronsay.* Edinburgh: HMSO.

——(1988). *Argyll: an Inventory of the Monuments, Volume 6: Mid Argyll and Cowal, Prehistoric and Early Historic Monuments.* Edinburgh: HMSO.

——(1994). *South-East Perth: An Archaeological Landscape.* Edinburgh: HMSO.

RCHME (1979). *Stonehenge and its Environs.* Edinburgh: Edinburgh UP.

RENFREW, COLIN [A. C.] (1973). 'Monuments, mobilization and social organization in neolithic Wessex.' In Colin Renfrew (ed.), *The Explanation of Culture Change: Models in Prehistory*, Duckworth, London, 1973, 539–58.

——(1976). *Before Civilization* (Pelican edn). Harmondsworth: Penguin Books.

1 The 'and Historical' had been dropped from the Commission's name when this volume was published.

——(1979). *Investigations in Orkney*. London: Society of Antiquaries (Research Report 38).

——(1984). *Approaches to Social Archaeology*. Edinburgh: Edinburgh UP.

——, ed. (1985). *The Prehistory of Orkney*. Edinburgh: Edinburgh UP.

——(1994). 'Towards a cognitive archaeology.' In Renfrew and Zubrow 1994, 3–12.

——(1996). 'A sceptic's henge' (review of North 1996). *Times Literary Supplement*, 18 October 1996, 10.

RENFREW, COLIN and PAUL G. BAHN (1996). *Archaeology: Theory, Methods and Practice* (2nd edn). London: Thames and Hudson.

RENFREW, COLIN and EZRA B. W. ZUBROW, eds (1994). *The Ancient Mind: Elements of Cognitive Archaeology*. Cambridge: CUP.

RICHARDS, COLIN (1990a). 'Postscript: the late Neolithic settlement complex at Barnhouse Farm, Stenness.' In the 1990 reprint of Renfrew 1985, 305–16.

——(1990b). 'The Late Neolithic house in Orkney.' In Ross Samson (ed.), *The Social Archaeology of Houses*, Edinburgh UP, Edinburgh, 1990, 111–24.

——(1996a). 'Monuments as landscape: creating the centre of the world in late Neolithic Orkney.' *WA* 28, 190–208.

——(1996b). 'Henges and water.' *Journal of Material Culture* 1, 313–36.

RICHARDS, JULIAN C. (1991). *Stonehenge*. London: Batsford/English Heritage.

RITCHIE, GRAHAM [J. N. G.] (1976). 'The stones of Stenness, Orkney.' *PSAS* 107, 1–60.

——(1982). 'Archaeology and astronomy: an archaeological view.' In Heggie 1982a, 25–44.

——(1985). 'Ritual monuments.' In Renfrew 1985, 118–30.

——(1988). 'The Ring of Brodgar, Orkney.' In Ruggles 1988a, 337–50.

——, ed. (1997). *The Archaeology of Argyll*. Edinburgh: Edinburgh UP.

RITCHIE, GRAHAM and ANNA RITCHIE (1981). *Scotland: Archaeology and Early History*. London: Thames and Hudson.

ROBERTS, JACK (1996). *The Stone Circles of Cork and Kerry: an Astronomical Guide*. Skibbereen: Key Books.

ROBERTS, NEIL (1989). *The Holocene: An Environmental History*. Oxford: Blackwell.

ROMANO, GIULIANO and GUSTAVO TRAVERSARI, eds (1991). *Colloquio Internazionale Archeologia e Astronomia*. Rome: Giorgio Bretschneider Editore (Supplementi alla RdA, 9).

ROSSIGNOL, JACQUELINE and LUANN WANDSNIDER, eds (1992). *Space, Time, and Archaeological Landscapes*. New York: Plenum.

ROYAL COMMISSION ON THE ANCIENT AND HISTORICAL MONUMENTS OF SCOTLAND. See RCAHMS.

ROYAL COMMISSION ON THE HISTORICAL MONUMENTS OF ENGLAND. See RCHME.

RUGGLES, CLIVE L. N. (1981). 'A critical examination of the megalithic lunar observatories.' In Ruggles and Whittle 1981, 153–209.

——(1982a). 'Megalithic astronomical sightlines: current reassessment and future directions.' In Heggie 1982a, 83–105.

——(1982b). 'A reassessment of the high precision megalithic lunar sightlines, 1: Backsights, indicators and the archaeological status of the sightlines.' *AA* no. 4 (*JHA* 13), S21–40.

——(1983). 'A reassessment of the high precision megalithic lunar sightlines, 2: Foresights and the problem of selection.' *AA* no. 5 (*JHA* 14), S1–36.

——(1984a) (with contributions by Philip N. Appleton, Stephen F. Burch, John A. Cooke, Roger W. Few, J. Guy Morgan and Raymond P. Norris). *Megalithic Astronomy: A new Archaeological and Statistical Study of 300 Western Scottish Sites*. Oxford: BAR (British Series 123).

——(1984b). 'Megalithic astronomy: the last five years.' *Vistas in Astronomy* 27, 231–89.

——(1984c). 'A new study of the Aberdeenshire Recumbent Stone Circles, 1: Site data.' *AA* no. 6 (*JHA* 15), S55–79.

——(1985). 'The linear settings of Argyll and Mull.' *AA* no. 9 (*JHA* 16), S105–32.

——(1986). '"You can't have one without the other"? I.T. and Bayesian statistics, and their possible impact within archaeology.' *Sci. Arch.* 28, 8–15.

——, ed. (1988a). *Records in Stone: Papers in Memory of Alexander Thom*. Cambridge: CUP.

——(1988b). 'The stone alignments of Argyll and Mull: a perspective on the statistical approach in archaeoastronomy.' In Ruggles 1988a, 232–50.

——(1990). 'Astronomical and geometrical influences on monumental design: clues to changing patterns of social tradition?' In Thomas L. Markey and John A. C. Greppin (eds), *When Worlds Collide: The Indo-Europeans and Pre-Indo-Europeans*, Karoma, Ann Arbor, 115–50.

——, ed. (1993a). *Archaeoastronomy in the 1990s*. Loughborough: Group D Publications.

——(1993b). 'Introduction: archaeoastronomy—the way ahead.' In Ruggles 1993a, 1–12.

——(1994a). 'The stone rows of south-west Ireland: a first reconnaissance.' *AA* no. 19 (*JHA* 25), S1–20.

——(1994b). 'The meeting of the methodological worlds? Towards the integration of different discipline-based approaches to the study of cultural astronomy.' In Iwaniszewski *et al.* 1994, 497–515.

——(1995). 'Co-operation: the ideal and the reality.' *AA* no. 20 (*JHA* 26), S80–6.

——(1996). 'Stone rows of three or more stones in south-west Ireland.' *AA* no. 21 (*JHA* 27), S55–71.

——(1997). 'Astronomy and Stonehenge.' In Cunliffe and Renfrew 1997, 203–29.

——(1999). 'Palaeoscience.' In Cimino 1999, vol. 1, in press.

RUGGLES, CLIVE L. N. and AUBREY BURL (1985). 'A new study of the Aberdeenshire Recumbent Stone Circles, 2: Interpretation.' *AA* no. 8 (*JHA* 16), S25–60.

——(1995). 'Astronomical influences on prehistoric ritual architecture in north-western Europe: the case of the stone rows.' *Vistas in Astronomy*, 39, 517–28.

RUGGLES, CLIVE L. N. and ROGER D. MARTLEW (1989). 'The North Mull project (1): Excavations at Glengorm 1987–88.' *AA* no. 14 (*JHA* 20), S137–49.

——(1992). 'The North Mull project (3): Prominent hill summits and their astronomical potential.' *AA* no. 17 (*JHA* 23), S1–13.

——(1993). 'An integrated approach to the investigation of astronomical evidence in the prehistoric record: the North Mull project.' In Ruggles 1993a, 185–97.

RUGGLES, CLIVE L. N., ROGER D. MARTLEW and PETER HINGE (1991). 'The North Mull project (2): The wider astronomical potential of the sites.' *AA* no. 16 (*JHA* 22), S51–75.

RUGGLES, CLIVE L. N. and DAVID J. MEDYCKYJ-SCOTT (1996). 'Site location, landscape visibility and symbolic astronomy: a Scottish case study.' In Maschner 1996, 127–46.

RUGGLES, CLIVE L. N., DAVID J. MEDYCKYJ-SCOTT, and ALUN GRUFFYDD (1993). 'Multiple viewshed analysis using GIS and its archaeological application: a case study in northern Mull.' In Andresen *et al.* 1993, 125–32.

RUGGLES, CLIVE L. N. and FRANK PRENDERGAST (1996). 'A new archaeoastronomical investigation of the Irish axial-stone circles.' In Wolfhard Schlosser (ed.), *Proceedings of the Second SEAC Conference, Bochum, August 29th–31st, 1994*, Ruhr-Universität, Bochum, 1996, 5–13.

RUGGLES, CLIVE L. N. and NICHOLAS J. SAUNDERS (1984). 'The interpretation of the pecked cross symbols at Teotihuacan: a methodological note.' *AA* no. 7 (*JHA* 15), S101–7.

——, eds (1993a). *Astronomies and Cultures.* Niwot, Colorado: UP of Colorado.

——(1993b). 'The study of cultural astronomy.' In Ruggles and Saunders 1993a, 1–31.

RUGGLES, CLIVE L. N. and ALASDAIR W. R. WHITTLE, eds (1981). *Astronomy and Society in Britain During the Period 4000–1500 BC.* Oxford: BAR (British Series 88).

SADLER, D. H. (1966). 'Prediction of eclipses.' *Nature* 211, 1119–21.

SCHLOSSER, WOLFHARD and JAN CIERNY (1996). *Sterne und Steine: Eine Praktische Astronomie der Vorzeit.* Darmstadt: Wissenschaftliche Buchgesellschaft.

SCOTT, J. G. (1989) (with contributions by Euan N. Campbell and others). 'The stone circles at Temple Wood, Kilmartin, Argyll.' *Glasgow Arch. J.* 15, 53–124.

SHEE TWOHIG, ELIZABETH and MARGARET RONAYNE, eds (1993). *Past Perceptions: The Prehistoric Archaeology of South-West Ireland.* Cork: Cork UP.

SHENNAN, STEPHEN J. (1988). *Quantifying Archaeology.* Edinburgh: Edinburgh UP.

——, ed. (1994). *Archaeological Approaches to Cultural Identity.* London: Routledge (One World Archaeology, 10). Originally published (1989) by Unwin Hyman, London.

SHEPHERD, IAN A. G. (1989). 'The early peoples.' In Donald Omand (ed.), *The Grampian Book,* Northern Times, Golspie, 119–30.

SIMPSON, DEREK D. A. (1967). 'Excavations at Kintraw, Argyll.' *PSAS* 99, 54–9.

SMILES, SAM (1994). *The Image of Antiquity: Ancient Britain and the Romantic Imagination.* New Haven: Yale UP.

SOMERVILLE, BOYLE [H. B. T.] (1912). 'Astronomical indications in the megalithic monument at Callanish.' *JBAA* 23, 83–96.

——(1923). 'Instances of orientation in prehistoric monuments of the British Isles.' *Archaeologia* 73, 193–224.

——(1927). 'Orientation.' *Antiquity* 1, 31–41.

——(1930). 'Five stone circles of West Cork.' *JCHAS* 35, 70–85.

SOSA, JOHN R. (1989). 'Cosmological, symbolic and cultural complexity among the contemporary Maya of Yucatan.' In Aveni 1989a, 130–42.

SOUDEN, DAVID (1997). *Stonehenge: Mysteries of the Stones and Landscape.* London: Collins and Brown/English Heritage.

ŠPRAJC, IVAN (1993). 'The Venus-rain-maize complex in the Mesoamerican world-view: Part II.' *AA* no. 18 (*JHA* 24), S27–53.

STEWART, MARGARET E. C. (1966). 'Excavation of a setting of standing stones at Lundin Farm near Aberfeldy, Perthshire.' *PSAS* 98, 126–49.

STUIVER, MINZE and REIMER, PAULA J. (1993). 'Extended $^{14}$C data base and revised CALIB 3·0 $^{14}$C age calibration program.' *Radiocarbon* 35(1), 215–30.

TAIT, D. (1978). *A Map of the Standing Stones and Circles at Callanish, Isle of Lewis, with a Detailed Plan of Each Site.* Glasgow: University of Glasgow.

THOM, ALEXANDER (1954). 'The solar observatories of megalithic man.' *JBAA* 64, 396–404.

——(1955). 'A statistical examination of the megalithic sites in Britain.' *JRSS* A118, 275–91.

——(1966). 'Megalithic astronomy: indications in standing stones.' *Vistas in Astronomy* 7, 1–57.

——(1967). *Megalithic Sites in Britain.* Oxford: OUP.

——(1969). 'The lunar observatories of Megalithic Man.' *Vistas in Astronomy* 11, 1–29.

——(1971). *Megalithic Lunar Observatories.* Oxford: OUP.

——(1974). 'A megalithic lunar observatory in Islay.' *JHA* 5, 50–1.

THOM, ALEXANDER and ARCHIBALD S. THOM (1971). 'The astronomical significance of the large Carnac menhirs.' *JHA* 2, 147–60.

——(1973). 'A megalithic lunar observatory in Orkney: the Ring of Brogar and its cairns.' *JHA* 4, 111–23.

——(1975). 'Further work on the Brogar lunar observatory.' *JHA* 6, 100–14.

——(1977). 'A fourth lunar foresight for the Brogar Ring.' *JHA* 8, 54–6.

——(1978a). *Megalithic Remains in Britain and Brittany.* Oxford: OUP.

——(1978b). 'A reconsideration of the lunar sites in Britain.' *JHA* 9, 170–9.

——(1980a). 'A new study of all megalithic lunar lines.' *AA* no. 2 (*JHA* 11), S78–89.

——(1980b). 'Astronomical foresights used by Megalithic Man.' *AA* no. 2 (*JHA* 11), S90–4.

THOM, ALEXANDER, ARCHIBALD S. THOM, and AUBREY BURL (1980). *Megalithic Rings.* Oxford: BAR (British Series 81).

——(1990). *Stone Rows and Standing Stones* (2 vols). Oxford: BAR (International Series 560).

THOM, ALEXANDER, ARCHIBALD S. THOM, and J. M. GORRIE (1976). 'The two megalithic lunar observatories at Carnac.' *JHA* 7, 11–26.

THOM, ALEXANDER, ARCHIBALD S. THOM, and ALEXANDER S. THOM (1974). 'Stonehenge.' *JHA* 5, 71–90.

——(1975). 'Stonehenge as a possible lunar observatory.' *JHA* 6, 19–30.

THOM, ARCHIBALD S. (1981). 'Megalithic lunar observatories: an assessment of 42 lunar alignments.' In Ruggles and Whittle 1981, 13–61.

——(1988). 'A personal note about my late father, Alexander Thom.' In Ruggles 1988a, 3–13.

THOMAS, JULIAN (1988). 'The social significance of Cotswold–Severn burial practices.' *Man* 23, 540–59.

——(1991a). *Rethinking the Neolithic.* Cambridge: CUP.

——(1991b). 'Reading the Neolithic.' *Anthropology Today* 7(3), 9–11.

——(1996). *Time, Culture and Identity: An Interpretive Archaeology.* London: Routledge.

THORPE, NICK [I. J.] (1981). 'Ethnoastronomy: its patterns and archaeological implications.' In Ruggles and Whittle 1981, 275–88.

——(1983). 'Prehistoric British astronomy—towards a social context.' *SAR* 2, 2–16.

TILLEY, CHRISTOPHER (1994). *A Phenomenology of Landscape.* Oxford: Berg.

——(1996). 'The powers of rocks: topography and monument construction on Bodmin Moor.' *WA* 28, 161–76.

TURTON, DAVID A. and CLIVE L. N. RUGGLES (1978). 'Agreeing to disagree: the measurement of duration in a southwestern Ethiopian community.' *Curr. Anth.* 19, 585–600.

URTON, GARY (1981). *At the Crossroads of the Earth and Sky: an Andean Cosmology.* Austin: UTP.

WADDELL, JOHN (1998). *The Prehistoric Archaeology of Ireland.* Galway: Galway UP.

WALKER, CHRISTOPHER, ed. (1996). *Astronomy Before the Telescope.* London: British Museum Press.

WALSH, PAUL (1997). 'In praise of field-workers: some recent "megalithic" discoveries in Cork and Kerry.' *Archaeology Ireland,* 11(3), 8–12.

WHEATLEY, DAVID (1993). 'Going over old ground: GIS, archaeological theory and the act of perception.' In Andresen *et al.* 1993, 133–8.

——(1995). 'Cumulative viewshed analysis: a GIS-based method for investigating intervisibility, and its archaeological application.' In Lock and Stančič 1995, 171–85.

——(1996). 'The use of GIS to understand regional variation in Earlier Neolithic Wessex.' In Maschner 1996, 75–103.

Whittle, Alasdair W. R. (1981). 'Later Neolithic society in Britain: a realignment.' In Ruggles and Whittle 1981, 297–342.

——(1988). *Problems in Neolithic Archaeology*. Cambridge: CUP.

——(1996). *Europe in the Neolithic*. Cambridge: CUP.

——(1997). 'Remembered and imagined belongings: Stonehenge in its traditions and structures of meaning.' In Cunliffe and Renfrew 1997, 145–66.

Whyte, W. S. and R. E. Paul (1997). *Basic Surveying* (4th edn). Oxford: Laxton's.

Wickham-Jones, Caroline R. (1994). *Scotland's First Settlers*. London: Batsford/Historic Scotland.

Williams, George (1988). *The Standing Stones of Wales and South-West England*. Oxford: BAR (British Series, 197).

Williamson, Ray A., ed. (1981). *Archaeoastronomy in the Americas*. Los Altos/College Park: Ballena Press/Center for Archaeoastronomy.

——(1984). *Living the Sky: The Cosmos of the American Indian*. Boston: Houghton Mifflin.

Wood, John E. (1978). *Sun, Moon and Standing Stones*. Oxford: OUP.

Young, Alison (1943). 'Report on standing stones and other remains near Fowlis Wester, Perthshire.' *PSAS* 77, 174–82.

# Index

References to illustrations are in italics. For references in the notes the note number is shown following the page number, with the section (e.g. Arch Box 1) also given where there is any ambiguity. Where a topic is covered in the main text and also in one or more notes cited therein, only the page number in the main text is given. Particular monuments or people are only included where they receive a substantial mention in the narrative. Entries in the Reference Lists are not indexed.


9′ wobble *See*: lunar perturbation 'Δ'
18·6-year cycle *See*: lunar node cycle

Abney Level 165
aboriginals, Australian 120, 236 n. 61
abstract concepts of space and time 147–8, 149
access to monuments x
accuracy ix
adjacent azimuth range (AAR) 70, 171
afforestation 85, 88, 112, 125, 164 *See also*: forest clearance
aligned pairs of standing stones 108
  in western Scotland 198
alignment, concept of 148
'alignment studies' 78, 144
alignments
  astronomical *See*: astronomical alignments
  between stone circles and outliers 53, 58, 233 n. 22, 246 n. 99
  'inter-site' 70, 158, 233 n. 48
  upon astronomical events above the horizon 154
  upon sacred places, possible avoidance of 154
Altar wedge tomb 159
alternative archaeology 3
altitude ix, 22
  correction of, for lunar parallax 23
  correction of, for refraction 23
  dependence of declination on *23*
  measurement of, using a clinometer 165
Amazonia *See*: Barasana; Yanomamö
analogies, cultural 151, 237 n. 90
ancestors
  and astronomy 129–30
  orientations in direction whence they came 90
  power of 128
  veneration for conspicuous works of 85
ancestral monuments 87, 128
ancient Greece, astronomical alignment of buildings and roads in 254 n. 153
Andean perceptions of space and time 148
*Antiquity* 7, 8, 91–2
Apollo 88
Arbor Low 14
  cove at 133
archaeoastronomical
  data, in archaeological inventories 144
  evidence, nature of 41–7
  field techniques 164–71
  'toolkit' 100, 149 *See also*: 'Thom paradigm'
archaeoastronomy 9, 144, 145, 162
  and history of astronomy 250 n. 29
  and mainstream archaeology 78, 144, 162–3
  'brown' 161
  definitions of 250 n. 29
  'green' 148, 161
  in the Americas 9, 10, 148
  journals 226 n. 94
  methodological debates within 161
  misuse of the term 249 n. 5
  'Oxford' international symposia on 10, 80
  use of statistics in 148
archaeological site, concept of 116
archaeological theory 146 *See also*: processual archaeology; post-processual archaeology
Archaeology Data Service 4
archaeotopography 250 n. 14
architecture reflecting cosmic order 153
ard-marks 114, *115*, 123 *See also*: cultivation
Ardfernal 232 nn. 87, 5
Ardnacross 37, 110, 113–15, 121–2, 123
  dating evidence 114
  excavations at 113–15
Argyll, inventories of monuments in 4
Arran, Isle of 47–8, 133
artefact deposition *See*: deposition
Ashanti 238 n. 118
Assyria, temple orientations in 254 n. 153
astro-archaeology 226 n. 93
astrology 251 n. 45
astronomer-priests 81, 88
astronomical alignments
  as basic data 144
  as reflection of cosmic order 154
  as symbolic expressions of correspondences in non-Western world-views 155
  development of interest in 6
  emphasis on, in 'green' archaeoastronomy 161
  empowering a monument at certain times 154
  expressing links with ancestors 154
  guidelines for marking on site plans 170
  in domestic architecture 87
  interpretation as calendrical markers 83
  of buildings and roads in ancient Greece 254 n. 153
  of high precision 75 *See also*: high-precision lunar foresights; high-precision lunar observations; high-precision solar foresights
  probability of fortuitous occurrence 17–19, 23–5, 32, 33–4, 35, 38–40, 41, 42–3, 51, 73, 129–30, 228 n. 111, 244–5 n. 38
  serving to constrain movement 127
  symbolic function of 87
  use of, for social manipulation 154
  *See also*: cosmological symbolism; lunar alignments; solar alignments; stellar alignments
astronomical dating
  dangers of 227 n. 36, 230 n. 20
  use of, to disprove astronomical hypotheses 66
astronomical knowledge, as political resource 152
astronomical observations
  effect of weather upon 150
  from recumbent stone circles 98
astronomical symbolism
  and proto-science 155
  and social change 152
  importance of in interpreting perceptions of cosmic order 155
  seeking in material record and interpreting 149
astronomical targets *See*: archaeoastronomical 'toolkit'
astronomy
  amongst hunter-gatherers 83, 159, 253 n. 134 *See also*: Barasana
  ancient Greek 148
  and ancestors 129–30
  and conceptualisation of space and time 152
  and developments in human thought 152–3, 155
  and funerary ritual 19, 25
  and ideology 145
  and landscape in the Late Neolithic 130–6
  and other possible functions of stone monuments 89
  as an influence on orientation 76
  as integral part of cosmology 163
  as 'oldest science' 155
  cultural correlates for 145, 151
  importance of, in non-Western world-views 145
  meaning of the term 81
  meaning of, to ancient peoples 145
  regulating seasonal movement through landscape 153
  relationship to geometry 131–3
  universality of 83

use of, for social manipulation 145, 152
*See also*: prehistoric astronomy
astronomy (modern Western), origins of 235 n. 29
Atkinson, Richard 7, 8, 9, 35, 40, 41, 79
atmospheric extinction 52
atmospheric refraction *See*: refraction, atmospheric
Auldgirth 53
Avebury 14, 16, 81, 83, 85, 131, 225 n. 42
avenue 158
cove 133
landscape, as 'domain of the dead' 154
Aveni, Anthony 148, 156
axe quarries in the Early Neolithic, location of 253 n. 131
axial stone circles 9, 99–101, 140, 158, 189–91
comparison with recumbent stone circles 99–100
consistency in orientation of 90, 100
dates of 100, 134
morphological variations of 239 n. 77
orientation of visibility at 100
possible connections with recumbent stone circles 100–1
*See also*: Bohonagh; Derreenataggart West; Drombeg; five-stone circles; Gortanimill; Reanascreena South
azimuth 22
determination of, from theodolite plate-bearing 168–9
measurement of, using a magnetic compass 165
azimuthal distribution of horizon distance *See*: orientation of visibility

Babylonia 145
backsights of doubtful authenticity 58
Baity, Elizabeth Chesley 9
Balfarg 14
Balliscate 113, 121
Ballochroy 19–25, *20*, *26*, 41, 76
as high-precision 'observatory' 21–5
solstitial foresights at 23–5, *26*
Ballymeanoch 53, 109
Balnuaran of Clava *131*, 155, 255 n. 34
as exceptions amongst Clava cairns 246 n. 78
solstitial orientation of passage tombs at 130
Balquhain 98
bank barrows 16
Banks, Joseph 226 n. 48
Barasana 145, 155
Barber, John 9
Barbreck 76, *77*
Barnard's Star 264 n. (Ast Box 6) 1
Barnatt, John 133
Barnhouse 158, 245 n. 52
barrows
long *See*: long barrows
round *See*: round barrows
Basque
country 89
language 89
stone octagons 235 n. 20
Bayes' theorem 160
Bayesian hierarchical models 266 n. (Stat Box 7) 5
Bayesian inference 160
applicability to archaeoastronomical investigations 161–2
archaeological applications of 266 n. (Stat Box 7) 6
Bayesian posterior 160–1
Bayesian prior 160, 161–2
Beacharr *8*
'beaker people' 262 n. (Arch Box 5) 10
Beaker pottery 85
Beckhampton cove 133
Beinn Talaidh (Mull) 122
bell-barrows 16, 141
Beltane, alleged alignments upon sunrise or sunset at 142, 237 n. 86, 246 n. 97, 248 n. 160
Beltany stone circle 237 n. 86
Ben More (Mull) 115–18, 122, 124, 242 n. 36
alignment of Glengorm stone row upon 115, *118*
association with midsummer full moon at North Mull rows 115, 123
broad alignment of north Mull stone rows upon 117
location of stone rows on limit of visibility for 115, *119*, 123, 158, 252 n. 89
Bernoulli's formula 42
Berrybrae 83, 91, 97, 239 n. 67
Black Hills of Dakota 153
Blashaval stone row 75
'blind spots' in the landscape 158
Bodmin Moor 156, 253 n. 128
Boece, Hector 238 n. 5
Bohonagh 100, *101*
Borana 147
Boyne Valley tombs 12, *13*, 14, 84–5, 140–1, 157 *See also*: Dowth; Knowth; Newgrange
hidden decorated stones in 261 n. 6
solar alignment of 129
Bradley, Richard 11, 120, 140, 155
Brainport Bay 29–34, *30–1*
midsummer sunrise at 29–33, *31*
Oak Bank stone 33, *33*
Brats Hill 231 n. 66
Brenanstown 84
brilliance 155
British Academy 9
Brodgar, Ring of 14, *64*, 81, 85, 120, 225 n. 42
as high-precision lunar observatory 63–6, *64–5*
Comet Stone 63
Hellia, cliffs of, as lunar foresight from *65*, 66, 262 n. 19
'brown archaeoastronomy' 161
Browne's Hill portal-tomb 261 n. 3
Bryn Celli Ddu 129
burials 16, 84–5, 139
Burl, Aubrey 7, 10, 40, 69, 76, 79, 91–2, 102, 104, 126, 130, 131, 138, 140
Bush Barrow gold lozenge 139, *140*

Cairnpapple, cove at 133
cairns, as foresights or backsights 231 n. 62
Calanais *See*: Callanish
'calendar stones' at Knowth 129, *129*
calendars *See also*: Celtic calendar; Hopi horizon calendar; Thom's solar calendar
diversity of possible 152
early development of 152
possible continuity of tradition since prehistory 88–9, 142, 245 n. 60
calendrical
development 253 n. 111
festivals 149, 151
knowledge, as political resource 152
Callanish 16, 66, 75, 88, 108, *134–5*, 134–6, 225 n. 42, 250 n. 86
alleged stellar alignments at 136
dating and chronology of 135
possible lunar significance of 136
Callanish area *134*, 232 n. 7
as 'lunar complex' 136
stone monuments in 135
Camster tombs 130
Cara Island, as foresight from Ballochroy 23, 25, *26*
Caracol at Chichen Itza 257 n. 93
cardinal directions
non-universality of concept of 148
orientation upon 94, 131
Carn Ban 47–8
Carnac 8
sacred geography of *120*, 255 n. 41
stone rows 34, 158, 225 n. 46
Carrowkeel 14, 130
Carrowmore 14
Cashelkeelty 101, 240 n. 15
Castleden, Rodney 41, 156
Castlerigg 14, 53, *56*, 131, 231 n. 66
categorisation of the world *See*: cosmology
Caterpillar Jaguar constellation 145, 155
causewayed camps *See*: causewayed enclosures
causewayed enclosures 14, 44, 81 *See also*: Robin Hood's Ball
possible functions of 84
caves, as points of contact with the underworld 154
Céide fields 85
celestial
equator 18, 148, 150
latitude 263 n. (Ast Box 1) 4
poles 18
sphere 18, *18*, *22*, 57, 258 nn. 8, 11
Celtiberians 89
Celtic calendar 88, 128–9, 131, 141–2, 159 *See also*: Beltane
central places 120
*ceque* lines 262
Chaco Canyon 88, 155
chambered tombs *5*, 14, 89, 157
as foresights or backsights 231 n. 62
consistency of orientation amongst regional groups of 90, 130
dating evidence 259–60 n. (Arch Box 2) 5
solar alignment of passages in 129–30
*See also*: Boyne Valley tombs; Orkney chambered tombs; Clava cairns; wedge tombs
Cherokee festivals 247 n. 117
Chichen Itza 155, 257 n. 93
chiefdoms 85
Childe, Gordon 7, 98
Chippindale, Christopher 10, 41
Chumash 155, 252 n. 69
church orientations and their interpretation 254 n. 153
circular enclosures 40, 157 *See also*: 'Stonehenge 1'
circular monuments
significance of 153–4
*See also*: circular enclosures; henges; timber circles; stone circles
circularity of argument, avoidance of 151
cists 16
Clach an't Sagairt 53
Clark, Grahame 226 n. 56
classical statistics 160
and processual archaeology 256 n. 76
general inapplicability of, for testing archaeological ideas 161
underlying paradigm 77
Classical writers 88–9
Clava cairns 14, 91, 130, 142, 157, 255 n. 34 *See also*: Balnuaran of Clava
apparent lunar significance of 130, *131*
consistency in orientation of 90, 130, *131*
dates of 14
Cleaven Dyke 256 n. 74
climate
long-term variations in 57
possible deterioration since the Early Bronze Age 87, 236 n. 69
possible effect of volcanic eruptions on 85

*See also*: weather
clinometers 165
cluster analysis 233 n. 44
Clyde tombs 14, 47–8, 130
cognition 116
cognitive archaeology 146
cognitive maps 251 n. 38
Coligny calendar 142, 159
comets 155, 263 n. (Ast Box 1) 5
compass and clinometer, use for field survey 165
computer
programs for field survey data reduction 169
simulations of light-and-shadow effects 155
virtual reality modelling 159
Comrie (Tullybannocher) 53
Coneybury 44, 138
confidence interval 73
constellations, association of with landmarks 153
control data in North Mull project 115
control points
during field survey 167
in North Mull project 118, 119, 121, *121*, 122, *122*
Cooke, John 68
Cork–Kerry five-stone circles 157
Cork–Kerry rows 101, 103–7, 192–5
apparent direction of indication at 104
astronomical alignments claimed prior to 1991 105
consistency of orientation in 104
correlation between apparent direction of indication and conspicuous hill summits 106–7
general pattern of lunar alignment at 106
geographical distribution of *104*
orientation of visibility at 104–5, *105*
quartz scatters at 123–4
*See also*: Gortnagulla
Corogle Burn 231 n. 61
Corra Bheinn, as foresight from Ballochroy *20*, 23, 25, *26*
cosmic order
importance of astronomical symbolism in revealing perceptions of 155
reflected in astronomical alignments 154
reflected in layout of traditional buildings 153
reflected in layout of traditional villages 153
reflected in orientations of monuments 154
reflected in the play of solstitial sunlight upon rock art 155
reflected in spatial patterns of activity 153
cosmological metaphors in material record 147
cosmological symbolism
in city plans 120
in domestic architecture 87, 158, 159
in henges 120–1
in houses in Neolithic Orkney 158
in Iron Age houses 159
in sacred architecture 120
cosmologies
indigenous 120, 148
quadripartite 148, 152, 153–4, 158
cosmology 81, 145, 251 n. 38 *See also*: world-view
'cosmovisión' 145 *See also*: cosmology; world-view
Cothiemuir Wood *94*
Cotswold–Severn tombs 14
coves 133
Cowan, Thaddeus 235 n. 21
Crab supernova, alleged depictions of 88
Cranborne Chase 237 n. 103, 238 n. 119
Crete 256 n. 59
Crick bell-barrow 141, *141*
Croagh Patrick 76
cross-quarter days *See*: mid-quarter days
cultivation
at Ardnacross 114
traditional methods in Scotland 242 n. 29
Cultoon 82, 133
cultural analogies 151, 237 n. 90
'cultural astronomy' 226 n. 93, 257 n. 103
cultural imperialism 146, 251 n. 42
cultures 146
Cumbrian stone circles 131, *132*, 157, 223 *See also*: Castlerigg; Long Meg and her Daughters; Swinside
cumulative probability histograms 51–2, *52*, *56*, *59*, *72*, 161
cup marks 16, *98*, 140
lunar alignments involving 88, 97, 98, *98*, *99*
solar alignments involving 33–4, 141, *141*
cup-and-groove marks 33, *33*
cup-and-ring marks 16, 83, 88, 140
and mobile patterns of settlement 253 n. 130
*Current Anthropology* 9
cursus monuments 16, 81, 84, 127, 129, 157, 256 n. 74
associations with long barrows 127, 128
*See also* Dorchester Cursus; Dorset Cursus; Godmanchester; Stonehenge Cursus
'curvigrams' *See*: cumulative probability histograms
Cuzco, sacred geography of 262 n. (Arch Box 7) 6

Daniel, Glyn 7, 8, 9
data recording during field survey 168, *169*
data selection 43, 56–8, 129–30, 156
in western Scottish project 69–70
methodologies 34–5, 41, 52–3, 58, 61, 67–70, 77, 92, 107–8
dating *See*: astronomical dating; radiocarbon dating; tree-ring dating
decay and destruction of monuments and other material remains 41, 82, 116
effect on our knowledge of the spatial distribution of monuments 156
decision making, statistical 160
declination 18, *18*, 19
calculation of, from field survey data 168–70
calculation of, from map data 169
contour maps *114*, 115–17, 124
daily rate of change of, for sun near equinoxes 24, 150
dependence on altitude *23*
distributions: interpretation of 144
formulae for determining 22–3
geocentric lunar 23, 58, 60–1
of certain stars 18
of moon at standstill limits 57
of moon, variation over a lunar node cycle 34, 36, *36*
of moon, variation over a month 36, *36*
of prominent hill summits from North Mull rows 121
of stars, long-term variation 57
of sun, at epochs in Thom's solar calendar *54*, 54–5
of sun at solstices 18
of sun at solstices, long-term variation 57
of sun, close to a solstice *24*, 24–5
of sun, variation over year 18, 19, *24*, 24–5, 54–5
of the lunar limbs 37, 57, 60
of the solar limbs 24, 57
use of, to avoid prejudging significant astronomical targets 148
declination (magnetic) *See*: magnetic correction
declinations
indicated by Cork–Kerry rows 106, *106*, 218–19
indicated by the axes of axial stone circles *96*, 100, 217
indicated by the axes of recumbent stone circles 96, *96*, 214–15
indicated by western Scottish rows *106*, 109, 221
of Ben More from the North Mull rows 222
of conspicuous hilltops indicated by western Scottish rows *108*, 109
of horizon above cup marks at recumbent stone circles 98, *99*, 216
of hill summits above the axial stone at axial stone circles *99*, 100, 218
of hill summits above the recumbent stone at recumbent stone circles 98, *99*, 216
of hill summits indicated by Cork–Kerry rows 107, *108*, 220
of hill summits indicated by north Mull rows *123*
of hill summits indicated by western Scottish rows 222
of outliers, entrances and tallest stones at Cumbrian stone circles 131
deconstruction of Western astronomical concepts 147–9
dendrochronology 51
Department of the Environment for Northern Ireland 4
deposition, spatial patterning of 87–8, 112, 138, 153, 158
Derreenataggart West *101*
Dervaig N *103*
Dervaig S 242 n. 16
Devil's Quoit (Stackpole Warren) 242 n. 3
difference between magnetic north and true north 165
Diodorus of Sicily 88–9
direction, concepts of 148
Dirlot *63*, 232 n. 92
Dorchester Cursus 128
solstitial alignment of 127
Dorset Cursus 16, 127, *127*, 128, 143, 153, 157, 158, 159
solstitial alignment of 127, *127*
Dowth 12, 81
solar alignment of 129
Drombeg 100, *101*, 140
Druids 1, 6, 88, 159
Druid's Circle 133
du Cleuziou, H. 6
Duachy *5*, 53, 76
*Dúchas* the Heritage Service 4
Dunragit 261 n. 9
Dunskeig 66, *66*
Durrington Walls 44, 81, 85, 132, 138
dwelling, concept of 253 n. 97

Early Bronze Age period 83, 85, 86–7, 139–41
Early Neolithic period 81, 84, 125–9
methods of subsistence in 85
earthen long barrows 81, *126*, 157, 237 n. 103
Earthwatch 152, 242 n. 11
East Cult stone row 124
Easter Aquorthies *92*, *94*
eclipse seasons 229 n. 132
eclipses, association with catastrophes 155
ecliptic, obliquity of the 24
Eigg, Isle of 224 n. 11
elevation ix
élite groups 84–5
elliptical stone rings 82–3
orientation of longer axis 133
English Heritage 4, 10, 128, 157
environment
and astronomy 88
sources of evidence on 85
*See also*: afforestation; climate
epochs (in Thom's solar calendar) 54–5

equal day and night, day of
arbitrariness in defining 150
non-equivalence to equinox 150–1
equinoctial alignments 149
Brainport Bay 33, 34
Knowth 129
Stonehenge cursus 127
equinox
as questionable concept outside Western context 148–9, 150–1
hierophany at Chichen Itza 155
non-equivalence to mid-point between solstices 149, 150, *150*
equinoxes 54, 88, 142
'megalithic' 54 *See also*: Thom's solar calendar
precession of the 18, 52, 57
equipment inventory for field survey 166
Er Grah *See*: Grand Menhir Brisé
Ethiopia *See*: Borana; Mursi
ethnoastronomy 226 n. 93
ethnocentrism 8, 11, 80
ethnographic evidence, use of 89, 237 n. 90
ethnographic parallels 151, 237 n. 90
Europe, medieval, concepts of direction in 148
excavation 41, 112
excavations
at Ardnacross 113–15
at Glengorm 113, 242 n. 27
exploration loop 162 *See also*: hermeneutic circle
extinction, atmospheric 52
extrapolation between nightly risings and settings of moon 61 *See also*: lunar extrapolation devices; lunar extrapolation length

Fajada Butte 155
Fane Gladwin, Peter 29
farming in the Early Neolithic 81
Few, Roger 68
field equipment inventory 166
field survey
choice of instrumentation 166
data recording during 168, *169*
data reduction 168–71, *170*
goals 164
procedure 166–8
strategy 164
using a theodolite 165–8
using compass and clinometer 165
weather conditions during 164, 165, 166–7
field systems, Neolithic 85
field techniques, archaeoastronomical 164–71
field walking 262 n. (Arch Box 6) 3
fieldwork methodology
at Cork–Kerry rows 103, 105–6
in North Mull project 115
five-stone circles in Cork and Kerry 157
Flag Fen 142
flankers *See*: recumbent stone circles
Fleming, Andrew 8–9
flint mining 85
'folk astronomy', Islamic 262 n. (Arch Box 7) 10
folk traditions, modern 1, 89
foresights *See*: high-precision lunar foresights; high-precision solar foresights; notches as foresights
forest clearance 44, 85
form and meaning 157, 252 n. 86
four-part division of world *See*: quadripartition of cosmos
four-posters *5*, 14–16, 110, 157 *See also*: Corogle Burn
frameworks of understanding *See*: cosmologies; world-views
Fraser, David 245 n. 66
free festivals at Stonehenge 1

full moon
nearest solstice 149
rituals at 89

Galileo 80
Garrough 106–7
gaussian humps 50, 51, 70 *See also*: normal distributions
generalisation, levels of, in archaeological explanation 146–7, 151, 152
geocentric lunar declination 23, 58, 60–1
Geographical Information Systems *See*: GIS
geographies, sacred *See*: sacred geographies
geometrical shapes, number of parameters required to define 265 n. (Stat Box 5) 4
geometry of stone rings 7, 9, 80, *82*, 82–3, 131–2, 233 n. 22, 235 n. 21
different perceptions of 235 n. 20
relationship to astronomy 131–3, 233 n. 22
Gibbs sampling 161
GIS 116–17, 169
applications of in archaeology 117
as alternative to field survey 165
use of, for studying properties of place 124
viewshed function *116*, 117, *119*
Glengorm 113, *118*, 122, 152
excavations at 113, *114*, 242 n. 27
orientation of 115
orientation of visibility at 121
possible timber reference marker at 113
Gleninagh stone row 154
Global Positioning System (GPS) receivers 166
Godmanchester 128, *128*, 252 n. 74
Gortanimill 100, *100*, 158
Gortnagulla *107*
Grand Menhir Brisé 34–5, *35*, 76
as universal lunar foresight 34–5, *35*
Grange [The Lios] stone circle 235 n. 21
graze effect 67
great henges *See*: henge enclosures
Greece, ancient 145, 254 n. 153
'green archaeoastronomy' 148, 161
Green, Miranda 141
grid references x
Grooved Ware pottery 16, 81, 85
ground-plan generation, during field survey 168
ground-truthing 117, 165
groups of monuments, importance of evidence from 41–8, 76, 90, 142–3, 157
gyroscopic attachments for theodolites 258 n. 10

half-quarter days *See*: mid-quarter days
Hambledon Hill 14
Hawkes, Jacquetta 6, 7, 156
Hawkins, Gerald 3–6, 9, 38–40, 41, 43, 79, 136, 234 n. 7
Headrick, James 226 n. 48
hearths in Orcadian houses, orientations of 158
heavenly bodies, apparent motions of 22
Hecateus of Abdera 88
heel-shaped cairns 130
Heggie, Douglas 8, 9, 10, 77, 78
height ix
heliacal rising and setting 83, 98, 235 n. 25
Hellia, cliffs of, as lunar foresight from Brodgar 66, 262 n. 19
henge enclosures 131, 260 n. (Arch Box 2) 10 *See also*: Avebury; Durrington Walls
henge monuments *See*: henges
henges 14, 40, 85, 91, *93*, 120–1, 131, 138, 157 *See also*: Arbor Low; Brodgar, Ring of; henge enclosures; 'hengiform monuments'; Stenness, Stones of; Wessex henges
'hengiform monuments' *93*, 260 n. (Arch Box 2) 10
hermeneutic circle 146, 252 n. 89

Hesiod 80
hidden places 158
high-precision
alignments, lack of evidence for 75
lunar observations, practical difficulties with 60–1, 63
high-precision lunar foresights *8*, 49–51, 58, 59–61, 63, *63*, *66*
high-precision solar foresights
methods of setting up 23, 29
methods of use *21*, 21–3
hill summits
and the location of North Mull rows 121
general relationship of recumbent stone circles and short stone rows to 140
(lack of) preferential orientation of axial stone circles upon 100
orientation of Cork–Kerry rows upon 106–7
preferential orientation of recumbent stone circles upon 98
hillfort entrances, orientations of 153
historical evidence 159, 252–3 n. 94, 254 n. 153
hogans 153, 251 n. 60
Hopi 83, 153, 252 n. 71
horizon calendar 152
kivas 153
horizon
astronomical events, unindicated 113, 133
astronomical targets, selection of 42–3
distance, azimuthal distribution of *See*: orientation of visibility
foresights *See*: high-precision lunar foresights; high-precision solar foresights; notches as foresights
'scans' 94
significance of, as boundary between earth and heavens 154
horizon position of solstitial sunrise and sunset
general qualities of 148
importance for Hopi 252 n. 71
horizon profile diagrams 70
generation of, from survey data and photographs 170
horizon profiles
photographing 167–8
surveying 168
hour angle 258 n. 11
Hoyle, Sir Fred 7, 41, 227 n. 66, 234 n. 9
hunter-gatherers 83, 253 n. 134 *See also*: Barasana
interest in moon amongst 159
Hyperboreans, island of the 88
hypothesis testing 56, 72, 73, 146, 160
archaeological tests of astronomical hypotheses 9, 25–9, 34
fundamental difficulties 77

Iberian peninsula 89
ideas, sources of 149–52
ideologies, co-existing 151
Inca empire 152
Inca sacred places 262 n. (Arch Box 7) 6
independence of data 43
indicated azimuth range (IAR) 70, 170–1 *See also*: indicated horizon range
indicated foresights 53
indicated horizon range 106 *See also*: indicated azimuth range
indications 49–51 *See also*: indicated foresights
displaying cumulative data from *See*: cumulative probability histograms
selection criteria for 53
types considered by Thom 53, 58
Indonesia, funerary ceremonies in 89
inductive paradigm 73
Indus valley 234–5 n. 18

inference, statistical 160
inhumations, orientation of 89
intellect, level of 79
International Astronomical Union 9
International Society for Archaeoastronomy and Astronomy in Culture (ISAAC) 250 n. 26
'inter-site' alignments 70, 158, 233 n. 48
intervisibility
of long barrows on Salisbury Plain and Avebury area 255 n. 16
of monuments 70, 156
Iron Age 141–2, 159
hillforts, entrance orientations 153
roundhouses, consistency of orientation amongst 153
Islamic 'folk astronomy' 262 n. (Arch Box 7) 10

Jones, Inigo 6
*Journal for the History of Astronomy* 8
Jura, Paps of *See*: Paps of Jura

Kendall, David 8
kerb-cairns 157
at Ardnacross 114
Kermario stone rows 34
Kern DKM1 theodolite 166
key positions, during field survey 167
Killichronan 117
'Kilmartin stones' *See*: Nether Largie stones; Temple Wood
Kilmartin Valley area 109
Kilmartin valley monuments 261 n. (Arch Box 4) 9 *See also*: Temple Wood
Kilmartin valley rows 197–8
general pattern of lunar alignment at *109*, 109–10
orientation of visibility at 119–21
*See also*: Ballymeanoch; Nether Largie stones
Kintraw 25–9, *27*
'early warning' stone at 26, 29
lunar horizon profile at *63*
observing platform at 26–9, *28*
Kippagh 106
Knocknakilla 124
Knockrome 232 nn. 87, 5
Knowth 12, 81
'calendar stones' at 129, *129*
equinoctial alignment of passages at 129
Kolmogorov–Smirnov tests 231 n. 57, 243 n. 60, 265 n. (Stat Box 4) 3
Krupp, Edwin 9

labour estimates for monument construction 85
labour organisation and social implications 84–5
Lakota 153
landmarks, association of with constellations 153
landscape
and astronomy in the Late Neolithic 130–6
archaeology 112–13, 116
astronomical potential of different points in 133
'blind spots' in 158
features, as symbolic resource 153
perception of 151
seasonal movement through 153
transformation of, through decay and destruction 156
landscapes, ritual *See*: sacred geographies
Lapps 256 n. 62
largest monuments of their classes 157
Late Neolithic period 81, 84–5
farming developments in 85
latitude
optimal for perpendicularity of midsummer sunrise and moonrise at southern major standstill limit 248 n. 157
optimal for perpendicularity of the four solstitial directions 250 n. 8a
lazy beds 242 n. 34
Le Manio 234 n. 64
Lewis, A. L. 6, 253 n. 128
ley lines 3, 7
Lhuyd, Edward 25
light, meanings associated with 155
light-and-shadow effects 141, 154, 155, 245 nn. 47, 49
likelihood function 160
linear monuments and patterns of movement through the landscape 153, 154
line-of-sight (GIS) function 117
Linnaean classification 145
literacy 235 n. 32
Loanhead of Daviot *94*, 98
location of later monuments with respect to earlier ones 158
Lochbuie 232 n. 5
Lockyer, Sir Norman 6–7, 17, 41
long barrows 14, 44, 125–7, 128 *See also*: earthen long barrows
astronomical interpretations of orientations 126–7, 244 n. 18
consistency of orientation amongst regional groups of 90, 125–6, *126*
in Wessex 157
intervisibility of 255 n. 16
on Salisbury Plain 126, *126*
Long Meg and her Daughters 131, *132*
long tombs 14, 81, 84, *126*
Lord Renfrew of Kaimsthorn *See*: Renfrew, Colin
Loughcrew 14
lunar alignment
of Callanish avenue 136
of cup marks at recumbent stone circles 98
of NE entrance at Stonehenge 136
of tombs at Armenoi, Crete 256 n. 59
lunar alignments 49–51, 52, 55–6, 58–67
at Clava cairns 130
at Cork–Kerry rows 106
at Godmanchester 128–9
at recumbent stone circles 92, 97
high-precision 9, 49, 51, 58–67, 88
low-precision 74–5, 76
preplanning of 109
'primary–secondary' pattern in Kilmartin valley and north Mull rows *109*, 109–10
revealed by western Scottish project 74–6
Thom's data and their reassessment 200–5
'lunar bands' 60–1, *61*, 232 n. 82
lunar extrapolation devices 61–3, 235–6 n. 37
lunar extrapolation length 61–3
lunar limbs, practical difficulties in observing 61
lunar node cycle 49, 61, 88, 263 n. (Ast Box 4) 3
awareness of, in modern Ireland 89
possible perceptions of 149
'lunar observatories' 8, 58–67
lunar parallax *See*: parallax, lunar
lunar perturbation 'Δ' 59, 60, 63
lunar phase cycle
alleged representation at Knowth 129
indigenous observations of 83
*See also*: synodic month
lunar standstill limits 60 *See also*: major standstill limits, minor standstill limits
alignments upon 55–6, 58–67, 75
and Ben More, as viewed from monuments in northern Mull 117
as possible horizon targets 149
declination of moon at 57, 60
inherent qualities of 149
lunar symbolism to solar symbolism, shift in the Neolithic 128, 142
lunistices 36, 60 *See also*: tropical month
Lynch, Ann 105, 106

Machrie Moor 133
MacKie, Euan 9, 10, 28–9, 32, 81, 88, 133
Madagascar 88, 154
Maes Howe 81, 129, 130, 158
date of 245 n. 52
solar alignment of 129
'magic circles' 153
magnetic compass, use of for measuring azimuths 165
magnetic correction 165
major limits *See*: major standstill limits
major standstill limit, moon close to horizon at 154, 159
major standstill limits 34, 36–7
major standstills 36
Maol Mor 113
Maori temples, orientation of 254 n. 157
marginality, perceived 158
Marshack, Alexander 125, 144
Martin Martin 6
material record
as source of ideas 149–51
cosmological metaphors in 147
direct evidence for astronomical observation in 83
limitations of 151
symbolic expression in 146–7
use of, to test archaeological ideas 147
Maughanasilly 244 n. 89
Maya
ancient 81
glyphs, decoding of, as *post hoc* interpretation 257 n. 83
modern 148, 153, 252–3 n. 94
mean
of a normal distribution 50, *50*
of any probability distribution 50, *50*
meaning
of material objects, specificity of 146
of monuments, not necessarily correlated with architectural form 252 n. 86
megalithic art *See*: rock art
'megalithic astronomy' 9–10, 19, 26, 68–78, 80
Level 1 analysis 49, 51–2, *52*
Level 1 reassessment 52–5, 68–75, *72*, *74*
Level 2 analysis 49, 55–6, *56*
Level 2 reassessment *56*, 56–8, 202–3
Level 3 analysis 49, 58–9, *59*
Level 3 reassessment *59*, 59–63, 204–5
Level 4 analysis 49–51, 63–6
Level 4 reassessment 66–7
*See also*: prehistoric astronomy
'megalithic calendar' *See*: Thom's solar calendar
'megalithic geometry' *See*: geometry of stone rings
'megalithic inch' 83
'megalithic lunar observatories' 8, 58–67
'megalithic man' 7, 8, 79
megalithic monuments 4
and social organisation 84–5
of Brittany 228 n. 105
*See also*: chambered tombs; stone circles; stone rows; standing stones
'megalithic observatories' 80 *See also*: 'megalithic lunar observatories'
megalithic rings *See*: stone circles; elliptical stone rings; oval stone rings
'megalithic science' 79–81 *See also*: 'science or symbolism' debate
'megalithic yard' 83
and European units of measurement 236 n. 40
statistical critiques of 83
*See also*: mensuration
megaliths *See*: standing stones

Melanesia 255 n. 38
Men-an-Tol 224 n. 11
mensuration 7, 9, 80, 81, 83 *See also*: 'megalithic yard'
Merrivale *5*, 140
Mesolithic period 79, 125, 253 n. 134
metaphors 147
meteors 155, 263 n. (Ast Box 1) 5
methodological issues 78, 159
  in studying properties of place 124
methodology 40–1, 113
  basic procedural principles 76
  for studying orientations 90
  of data selection *See*: data selection methodologies
  *See also*: fieldwork methodology
Metonic cycle 88–9
mica, association with moonlight 254 n. 178
mid-Argyll Archaeological Society 29
mid-point between solstices
  different possible definitions of 149, 150
  non-equivalence to modern 'equinox' 149, 150
mid-quarter days 88, 142
Midmar Kirk *94*, 97, *97*
midsummer ix
midsummer full moon
  and Callanish 136
  and North Mull rows 121–3
  and recumbent stone circles 97, 98
  and western Scottish rows 109
midsummer sun, and North Mull rows 121
midsummer sunrise and sunset *See*: solstices; solstitial alignments; solstitial orientation; solstitial sunrise and sunset
mid-Ulster monuments 110, 157
midwinter ix
midwinter sunrise and sunset *See*: solstices; solstitial alignments; solstitial orientation; solstitial sunrise and sunset
Milankovitch cycles 264 n. (Ast Box 6) 6
Milky Way, in Andean cosmologies 147
Minard *See*: Brainport Bay
Minoan Crete, tomb orientations in 256 n. 59
minor limits *See*: minor standstill limits
minor standstill limits 34, 36–7
  alignments upon 129
minor standstills 36
Misminay 148, 254 n. 140
Mither Tap 238 n. 32
Moir, Gordon 227 n. 101
Monte Carlo testing 72, 73
  of western Scottish alignment data 72–4, *74*
month
  synodic 263 n. (Ast Box 4) 1
  tropical 263 n. (Ast Box 4) 1
monument construction and cosmological principles 155
monuments
  access to x
  ancestral 87, 128
  as central places 120
  as conspicuous places in later landscapes 158
  as points of reference in the landscape 153, 158
  changing use over time 77, 112, 114, 157, 234 n. 68
  decay and destruction since prehistory 157, 234 n. 67
  inventories of 4
  orientation of, reflecting cosmic order 154 *See also*: orientation; orientations
  possible associations with prominent landmarks 153
  regional groups of, importance of 139–40, 156
  siting of, in relation to existing monuments 127–8
  spatial distribution of 156
  *See also*: causewayed enclosures; chambered tombs; cursus monuments; henges; long barrows; round barrows; standing stones; stone circles; Stonehenge; stone rows; timber circles
moon
  apparent diameter 264 n. 11
  association of short stone rows with 76, 106, 109–10, 140, 142
  association of stone circles with 91, 92, 97, 133, 136, 140, 142
  blood of 236 n. 50
  declination of *See*: declination of moon
  interest in, amongst hunter-gatherer groups 159, 256 n. 62
  lack of horizon observations of, in anthropological record 159
  light of, symbolic associations with quartz or mica 155
  perceived as nearer when close to horizon 151, 154
  phase of, in relation to high-precision horizon observations 61
  proximity to horizon near major standstill limit 154, 159
  representations of, in Iron Age 141
moonlight, duration of, in wintertime 149
Morgan, Guy 68
movement through landscape, seasonal patterns of 120, 153, 158–9
Mull, Isle of
  prehistoric activity in 113
  prehistoric subsistence on 158
  sacred geography of, in Bronze Age 123–4
  suitability of, for prehistoric farming 113
  topography of 242 n. 20
  *See also*: Ben More (Mull); North Mull project; north Mull rows
multiple-stone circles *See*: axial stone circles
multiple stone rows 83
multiple viewsheds 124
Mursi 80, 89, 236 n. 53

names of sites ix–x
National Grids x
National Monuments Record of Scotland (NMRS) 69, 233 n. 12
natural cycles, harmonising human activity with 152
natural foresights *See*: high-precision lunar foresights; high-precision solar foresights; notches as foresights
*Nature* 6, 8, 40
Navajo hogans 153, 251 n. 60
navigation 79
'negative' data 256 n. 74
Nether Largie stones 59, *62*, 76, *77*, 108, 109, 110, 231 n. 79, 232 n. 88, 233 n. 49
'New Archaeology' 146
Newgrange 12–19, *13*, *17*, 81, 87
  architecture of 12–17
  blocking of entrance 19
  Great Circle 12–17
  midwinter sunrise at 1, *2*, 9, *17*, 17–19
  roof-box 1, 17–19, 80, 129
  sunlight-and-shadow phenomena at 245 n. 47
Newham, Peter 41, 47, 136
Nine Stone Close 133
non-circular rings, as poor attempts as circles 265 n. (Stat Box 5) 3
non-local stone, use of 45, 97
non-Western world-views 80, 81, 152–3, 154, 155
  comprehension of, from a Western viewpoint 251 n. 42
normal distributions *50*, 50–1, 73
North, John 156
North Mull project 110, 113, 121, 123–4
  control data in 115
  control points in 118, 119, 121, 122
north Mull rows 113–24, 197
  as sacred boundary markers 123
  general pattern of lunar alignment at 76, *109*, 109–10
  geographical distribution of *114*
  location of, with respect to prominent hills and their astronomical potential 121–3
  orientation of visibility at 119
  possible factors influencing the location of 119, 123
  *See also*: Ardnacross; Balliscate; Dervaig N; Dervaig S; Glengorm; Maol Mor; Quinish
notches
  as foresights 21, *21*, 26, 33–4, 58
  use of to determine solstice by 'halving the difference' 29–32
novae 263 n. (Ast Box 1) 5
null hypothesis 73, 160

O'Kelly, Claire 19
O'Kelly, Michael 9, 12, 17, 19
Ó Nualláin, Seán 103
objectivity 77, 162
obliquity of the ecliptic 24, 57
  variation with time of 57, 61, 66, 264 n. (Ast Box 6) 6
observing platform at Kintraw 26–9
odds 39, 160
Office of Public Works 4
Ordnance Survey (Britain) 4
Ordnance Survey Ireland 4
orientation
  importance of analyses of 156
  of Assyrian temples 254 n. 153
  of Christian churches 254 n. 153
  of enclosure entrances at Stonehenge 1 136
  of longer axis at elliptical and oval stone rings 132–3
  of Maori temples 254 n. 157
  of monuments, reflecting cosmic order 154
  of Navajo hogans 251 n. 60
  of Neolithic houses in Orkney 158
  of Station Stone rectangle at Stonehenge 138
  of stone rows on Dartmoor 140
  of tombs, reflecting that of domestic structures 125
  patterns of, and their interpretation 90
  possible influences upon 89
  trends, statistical testing of 90
  upon cardinal directions 94
  upon prominent landmarks 98, 130
  upon the sun low in the sky 95
orientation, consistency in
  amongst Iron Age roundhouses 153
  amongst non-circular stone rings in north and central Wales 246 n. 97
  amongst regional groups of chambered tombs 90, 130
  amongst regional groups of long barrows 90, 125–6, *126*
  amongst various south-west Irish monuments 101
  between recumbent stone circles and axial stone circles 100, 134
  between recumbent stone circles and Clava cairns 130, 133–4
  of axial stone circles 90, 100
  of Clava cairns 90, 130
  of Cork–Kerry rows 104
  of recumbent stone circles 98
  of wedge tombs 90, 101, 240 n. 102

of western Scottish rows *75*, 108, 110
orientation of visibility
at axial stone circles 100
at Cork–Kerry rows 104–5, *105*
at Kilmartin valley rows and control points 119–21
at North Mull rows and control points 119, 121, *122*
at Orkney chambered tombs 245 n. 66
at recumbent stone circles 94, *96*
*See also*: 'horizon scans'
orientations
relating to the moon *See*: lunar alignments
relating to the stars *See*: stellar alignments
relating to the sun *See*: solar alignments
sets of, between the solstitial limits, and their interpretation 245 n.67
Orkney chambered tombs 84–5, 237
orientation of visibility at 245 n. 66
*See also*: Maes Howe
orthostatic stone settings *See*: standing stones, settings of
outliers (standing stones) 53, 58, 233 n. 22, 246 n. 99 *See also*: stone circles with outliers
outliers (statistical) 95
oval stone rings: orientation of longer axis 132
'Oxford' international symposia on archaeoastronomy 10, 80

Paps of Jura *8*
as foresight from Ballochroy 19, *20*, 23, 25, *26*
as foresight from Kintraw 26, *28*
parallax corrections to theodolite readings 167
parallax, lunar 23, 36, 58, 60–1, 63
variation in, with time 61
Parc-y-Meirw 232 n. 89
passage graves *See*: passage tombs
passage tombs 14 *See also*: Boyne Valley tombs; Carrowkeel; Clava cairns; wedge tombs
entrance blocking devices at 129
sunlight-and-shadow phenomena at 245 nn. 47, 49
Patrick, Jon 9, 19
pecked cross-circles 257 n. 82
Pedersen, Olaf 237 n. 89
perihelion 54, 57
Perthshire
monuments 157
short stone rows 110
*See also*: Cleaven Dyke
Peterborough Ware pottery 85
petrofabric analysis 28
petroglyphs *See*: rock art
phenomenology 151–2, 156, 159, 165
photography
as alternative to survey 165
of horizon profiles 167–8
Pierpoint, Stephen 133
Piggott, Stuart 7, 10
pit circles 40
place, concept of 153
place-names and possible continuities of astronomical tradition 89
Plains Indians of North America 252 n. 82
planets
difficulty of recognising alignments upon 149
perceptions of 252 n. 85
*See also*: Venus
plate bearing errors 259 n. 26
plate bearings 168
pollen evidence 88
popular perceptions of prehistoric astronomy 1, 2
portals *See*: axial stone circles
post-processual archaeology 146–7
precession of the equinoxes 18, 52, 57
precision ix
of field measurements 165
of observing position 229 n. 123, 231 n. 73
of quoted declinations 227 n. 34
predictive power of astronomical correspondences 155
prehistoric astronomy 81–7
and alternative archaeology 3, 7, 9–10
and mainstream archaeology vii, 6–7, 8–9, 10
and mainstream astronomy 9
antiquarian interest in 6
localised nature of 142
patterns of continuity and change 142–3
perceptions of amongst archaeologists 1–2
popular perceptions of 1, 2
prestige goods 85, 87
prevailing wind 89, 90, 94
probability 39, 160
analysis of alignment data from Cork–Kerry rows 105
distributions *50*, 50–1 *See also*: normal distributions
of chance alignments upon astronomical targets *See*: astronomical alignments, probability of fortuitous occurrence
that *n* random azimuths will fall within θ degrees 93, 95
processual archaeology 116, 146
and classical statistics 256 n. 76
prominence of topographic features
measures of 121
personal assessments of 152
proper motion 264 n. (Ast Box 6) 1
proto-science 155
Pueblo Indians 236 n. 52

quadripartition of cosmos 148, 152
as basis for interpreting Neolithic Orkney 158
as basis for interpreting Stonehenge 3 and its landscape 153–4
quartz 12, 19, 29, 32, 123–4
assemblages, spatial distribution of 88
found in a variety of archaeological contexts 243 n. 86
possible association with the moon 88, 98, 124
symbolic association with moonlight 155
quartz scatters 158
at Cork–Kerry rows 123–4
at Glengorm and Ardnacross 123
at recumbent stone circles 97
Quinish 113

radiocarbon dating ix, 51
random processes 73
random variables 73
Ray, Tom 19
RCAHMS 4
inventories 69, 232–3 n. 11
RCAHMW 4
RCHME 4
Reanascreena South *101*
recumbent stone circles 53, 85, 88, 91–9, 100, 133–4, 140, 142, 155, 159, 161, 185–8
and the cosmos 98
apparent lunar significance of 91, 92, 97
careful placement of recumbent stone at 97, *97*
'Perpendicular Line' axis 93, 94, *94*, 212–13
consistency in orientation of 90, 91, 98
cup marks at 98, *98*
dates of 91, 134, 239 n. 67
design of 91
distinctiveness of orientation tradition at 133–4
geographical distribution of *93*
location of 97–8
morphological variations of *94*, 98–9
orientation of visibility at 94, *96*
orientation of, upon prominent landmarks 98
'Centre Line' axis 93, 94, *94*, 212–13
possible connections with axial stone circles 100–1
possible variants with just a recumbent stone on the edge of a cairn 188, 239 n. 73
possible variants with recumbent and flankers only 99
spatial distribution of quartz scatters at 97, 158
type of stone used in 97
*See also*: Berrybrae; Cothiemuir Wood; Easter Aquorthies; Loanhead of Daviot; Midmar Kirk; Strichen; Sunhoney
recumbent stone circles (Irish) *See*: axial stone circles
reference object 167
reflectivity, meanings associated with 155
refraction, atmospheric 23, 25, 29, 61, 63, 227 n. 35, 228 n. 76
refraction, terrestrial *See*: graze effect
Renfrew, Colin 9, 81, 156, 234–5 n. 18
right ascension 258 n. 8
rings of pits *See*: pit circles
Rinyo 158
Ritchie, Graham 80
ritual
and mundane activity 87
landscapes *See*: sacred geographies
performances, role of in reaffirming cosmic order 152
public, slow timescale for changes in 142
ritual tradition
discontinuities of, and social change 152
possible continuity of, into Iron Age 88
possible continuity of, into modern times 89
rituals relating to the moon at recumbent stone circles 88
Robin Hood's Ball 44, 136
Robinson, Tim 154
rock art 16, 87, 140–1
as literal representation of astronomical bodies or events 88, 129, 140
light-and-shadow effects upon 141, 155, 245 n. 47
marking points of reference or sacred places in the landscape 140, 153, 158, 243 n. 84
*See also*: cup marks; cup-and-ring marks
Rollright Stones 237 n. 87, 265 n. (Stat Box 5) 8
round barrows 16
in Wessex 16, 47, 83, 85
visibility of, from Stonehenge 139
*See also*: Bush Barrow gold lozenge; Crick bell-barrow
roundhouses
Amazonian 153
Iron Age 153
Royal Astronomical Society 10
Royal Commission on the Ancient and Historical Monuments of Scotland *See*: RCAHMS
Royal Society 9
Royal Statistical Society 7

sacred geographies 120–1, 153
in Brittany 255 n. 41
in late medieval India 120
in the Late Neolithic 130–6
of Bronze Age Mull 123–4
sacred places 120, 153
avoidance of alignments upon 154
cosmological significance of 120
*See also*: hidden places
Saint-Just, sacred geography of *120*, 255 n. 41
Salisbury Plain, long barrows on 126
Samhain, alleged alignments upon sunrise or sunset at 131, 159

Sanctuary, the 131
*saroeak See*: Basque stone octagons
Schaefer, Bradley 52
'science or symbolism' debate 80–1, 155
Scilly Isles, chambered tombs in 237 n. 104
Scott, Jack 234 n. 68
seasonal
festivals, in Celtic calendar *See*: Celtic calendar
indicators, in indigenous calendars 89
patterns of movement through the landscape 120, 158–9
segmentary societies 84
selection criteria *See*: data selection methodologies
settlements and settlement patterns
Early Bronze Age 87
Neolithic 85, 87, 153
Seven Stones of Hordron 133
Shepherd, Ian 91
shift from lunar to solar symbolism in the Neolithic 128, 142
shiny materials, meanings associated with 155
short stone rows *5*, 16, 102, 124, 140, 142
as boundary markers 154
built of quartz 124
dates of 102
in Perthshire 110
possible chronological development of 102
possible functions of 102–3
relationships between local groups of 110
*See also*: Cork–Kerry rows; western Scottish rows
significance levels 73, 77
Silbury Hill 81, 85, 131, 158
simulations, Monte Carlo 73
single standing stones *5*, 89
consistency in orientation of, in Cork and Kerry 101
site
catchment analysis 116, 117
concept of 116
definition of, as part of data selection methodology 68, 69
names ix–x
plans, marking astronomical alignments on 170
Sites and Monuments Records (SMRs) 4
Skara Brae 81, 158
Skidi Pawnee earth lodges 153
Skipness, Kintyre 66, *66*
sky
as metaphorical resource 145
as part of 'natural world' 145
cultural specificity of perceptions of 147
immutability of 145
visibility of, in forest 251 n. 36
slab, definition of, as part of data selection methodology 241 n. 57
Smith, John 6
social archaeology 146
social ranking 84–5
Société Européenne pour l'Astronomie dans la Culture (SEAC) 250 n. 26
solar alignment
of chambered tomb passages 129–30
of longer axis at elliptical and oval stone rings 133
of Maes Howe 129
solar alignments 49
at Godmanchester 128–9
*See also*: Ballochroy; Brainport Bay; equinoctial alignments; Kintraw; solstitial orientation; solar symbolism; Thom's solar calendar
solar horizon calendars 83
Thom's *See*: Thom's solar calendar
solar symbolism, alleged shift to, from lunar symbolism in the Neolithic 128, 142
solstices
cultural significance of 148
declination of sun at 18
declination of sun close to 24–5
determination of, by 'halving the difference' 29–32, 38, 83
determination of, by shadow casting at noon 236 n. 55
festivals 88
full moon nearest to 149
solstitial alignments, evidence for in Thom's work 52, 56, 231 n. 51
solstitial orientation
of Ballochroy 23–5
of Balnuaran of Clava passage tombs 130
of Brainport Bay 29–33
of Cultoon 133
of Cumbrian stone circles 131
of Dorchester Cursus 127
of Dorset Cursus 127
of Drombeg 100, 140
of Gleninagh stone row 154
of Newgrange 17–19
of Stonehenge 7, 138–9
of Wessex henges 138
solstitial sunrise and sunset
cultural significance of 148
directions of 37, *37*
Somerville, Boyle 7
space
conceptualisation of 152–3
indigenous notions of 147–8
spatial analysis 116
spatial patterning of deposition 87–8, 112, 138, 153, 158
Speinne Mór (Mull) 121–2
Stackpole Warren 242 n. 3
standard deviation
of a normal distribution 50, *50*
of any probability distribution 50, *50*
standard units of length *See*: 'megalithic inch'; 'megalithic yard'
standing stones 16
aligned pairs of 108
as outliers to stone circles *See*: outliers (standing stones)
erected by modern indigenous groups 91, 255 n. 38
possible uses of 89
settings of 68, 89 *See also*: Nether Largie stones; stone circles; stone rows
single 89, 101 *See also*: Grand Menhir Brisé; Kintraw
standstill limits *See*: lunar standstill limits
Stanton Drew 261 n. 9
cove at 133
statistical approaches to the analysis of alignment data 76–7
statistical inference 160
statistical tests, classical 73
statistics
Bayesian *See*: Bayesian inference
classical *See*: classical statistics
use of, in archaeoastronomy 148
stellar alignments 52
at Callanish 136
use of, for dating monuments 230 n. 20
Stenness, Stones of 14, 66, 81, 85, 120, 224 n. 11, 225 n. 42
stone setting at the centre of 133
stone
metaphorical association with the dead 154
type of, used at Stonehenge 45
type of, used in recumbent stone circles 97
Stone Age 259 n. (Arch Box 1) 1 *See also*: Mesolithic period; Early Neolithic period; Late Neolithic Period
'stone alignments' 230 n. 31 *See*: stone rows
stone avenues 16
stone axe production 85
stone circles *5*, 14–16, 85, 87, 131, 136
centre stones 133, 230 n. 35
changing use in prehistory 82
decay and destruction 82
geographical distribution of *15*
geometrical centre of 133
geometrical construction of *See*: geometry of stone rings *See also*: elliptical stone rings
in Cork and Kerry *See*: axial stone circles
in north-east Scotland *See*: recumbent stone circles
on Bodmin Moor 156, 253 n. 128
on Machrie Moor 133
position of observers within 133
relationship to prominent landmarks 253 n. 128
with outliers 53, 230 n. 35, 232 n. 5
*See also*: Brodgar, Ring of; Cumbrian stone circles; Stenness, Stones of; Stonehenge 3; Temple Wood
stone height gradation
at axial stone circles 100
at circles surrounding Clava cairns 130
at Cork–Kerry rows 104, 105, 107
at recumbent stone circles 97
in Balnuaran of Clava kerbstones 155
stone monuments
seen as inhabited by ancestors 154
stone pairs
aligned *See*: aligned pairs of standing stones
in Perthshire 110
stone pairs in Cork and Kerry 157
consistency in orientation of 101
stone rings *See*: stone circles
stone rows *5*, 16, 87, 140, 232 n. 89
attached to stone circles 16
more than 25 m long in western Scotland 108
multiple 83
of mid-Ulster 110
*See also*: short stone rows; stone pairs
Stonehenge 35–41, *38*, 44–7, 79, 81, 88, 136–9, 144, 156, 163
'A' holes 136
Altar Stone 45, 138
as cosmic temple 7, 139, 156
as eclipse predictor 40, *40*, 41
as focus for new-age travellers 154
as universal lunar backsight 40
astronomy, attitudes towards 10–11
Atkinson's phasing of 35, 37
Aubrey Holes 37, 40, 41, 44, 47, 76, 79, 156
auspicious times of approach along avenue 159
avenue 38, 47
axis of, shift in 47, 138
bluestones 37, 38, 45, 85, 138
car park postholes 41, 260 n. 1
ditch-and-bank enclosure *See*: Stonehenge 1
entrance postholes 44, 136–8
Hawkins's interpretation of 38–40, *40*, 43, 50
Heel Stone 37, 38, 47, 138–9
Heel Stone companion 37, 41, 47, 138–9
latitude of 248 n. 157
location of, choice of 136
meanings of, in modern times 252 n. 86
midsummer sunrise at 1, *2*, 138–9, 225 n. 43
orientations and astronomical potential 136–9, *137*
patterns of deposition at 44, 138
phases in the development of *46*
sarsen stones 38, 45, 83, 85, 153–4 *See also*: Stonehenge 3

Slaughter Stone 47, 138
solstice free festivals 1
solstitial orientation of 7, 138–9
Station Stones 38, 47, 138
symbolic power of 138, 139
timber structures in the interior of 44
visibility of 136
visibility of Bronze Age round barrows from 139
'Stonehenge 1' 37, 40, 44, *46*, 136–8, 139
possible lunar orientation of NE entrance 136
'Stonehenge 2' 37, 40, 44, *46*, 136–8, 139
'Stonehenge 3' 37–8, 44–7, *46*, 138–9
numerical skills implied by pre-planning of 264 n. (Ast Box 8) 1
solstitial orientation of axis of 138–9
Stonehenge and its landscape *45*
as domain of the dead 154, 248 n. 182
sacred geographies of 121
Stonehenge Avenue, 'elbow' in 158
Stonehenge Cursus 44, 127, 136
equinoctial alignment of 127
*Stonehenge Decoded* 3–6, 7
'Stonehenge I' etc *See*: Stonehenge, Atkinson's phasing of
Strabo 89
Strichen 82–3, 238 n. 5
Strontoiller *5*, 232 n. 5
Stukeley, William 6
sun
alleged symbolic representation of, on roof-box corbel at Newgrange 129, 140
annual motion along the horizon 21
apparent diameter 263 n. (Ast Box 3) 4
cult of the 141
declination of *See*: declination of sun
motion close to a solstice 21, 23, *24*, 24–5
upper limb of 24, *24*
symbols, Iron Age 141
sun-azimuth readings 167
'sun dagger' at Fajada Butte 155
Sun Dance ceremony 252 n. 82
Sunhoney *94*, *98*
sunlight-and-shadow effects 141, 154, 155, 245 nn. 47, 49
sunrise opposite sunset, day of, non-equivalence to equinox 151
sunrise position, as merely one seasonal indicator in a lunar-regulated calendar 89
superhenges *See*: henge enclosures
survey data reduction 168–70
Swinside 131, *132*
symbolic expression in the material record 146–7
synodic month 263 n. (Ast Box 4) 1 *See also*: lunar phase cycle

*t*-tests 265 n. (Stat Box 4) 3
tally marks 87, 125
Temple Wood 231 n. 79, 234 n. 68
Tenga 117
Teotihuacan 252 n. 86, 257 n. 82
terrestrial refraction *See*: graze effect
theodolite
accessories 166
determination of position during survey 167
gyroscopic attachments 258 n. 10
plate bearing errors 259 n. 26
plate bearings 168
types of 166
use of 165–8
theory in archaeology *See*: archaeological theory
Thiessen polygons 116
Thom, Alexander 7–9, 10, 19, 21, 23, 26–8, 29, 34–5, 40, 41, 49, 54–5, 58, 59, 61–6, 67, 69, 79, 80, 81–3, 88, 89, 90, 131, 136, 162, 170, 171, 233 n. 52
Thom, Archibald 34, 40, 63–6, 67
'Thom paradigm' 148, 149 *See also*: archaeoastronomical 'toolkit'
Thom's solar calendar 49, 52, 53, 54–5, 81, 88, 128–9, 142, 150, 231 n. 56, 235 n. 23, 255 n. 54
Bush Barrow gold lozenge as device for implementing 248 n. 188
lack of evidence for 75
Thomas, Julian 127
timber circles 14, 131, 154, 159, 261 n. 9
reconstruction of, in stone 14
time
as abstraction 148–9
conceptualisation of 152
determination of the, during theodolite survey 166
indigenous notions of 147–8
timing devices, calibration of 166
Tinkinswood 84
tomb orientations in Minoan Crete, possible lunar significance of 256 n. 59
Tombe di Giganti 250 n. 10
tombs
factors influencing orientation of 89
in Etruria 255 n. 43
*See also*: chambered tombs; long barrows; passage tombs
Tostarie 117
trade and exchange networks 85, 87
traverses 258 n. 24
tree cover *See*: afforestation
tree-ring dating 51
Triangulum 147
tropical month 263 n. (Ast Box 4) 1 *See also*: lunistices
Twelve Apostles stone circle 133
twilight, definitions of 264 n. (Ast Box 8) 3

Uluvalt 117
underworld, caves as points of contact with 154
units of measurement
of length *See*: 'megalithic inch'; 'megalithic yard'
of time 152
of weight 234–5 n. 18
Unival chambered tomb *5*, *63*, 231 n. 50
Upper Fernoch (Tayvallich) 53
Upper Palaeolithic engravings, as possible tally marks related to the phases of the moon 125, 144–5

vegetation
effect of, on astronomical observations and sightlines 61, 70, 88, 112, 117
in prehistoric Mull 113
Venus
claimed alignments upon, at axial stone circles 100
difficulty of recognising alignments upon 149
importance of, in pre-Columbian Mesoamerica 252 n. 85
viewshed (GIS) function 117
Villards d'Héria inscription 142
virtual reality modelling 159
visibility
combined with inaccessibility 158
from Glengorm 121
of Ben More from north Mull rows 115, 123, 158
use of GIS to assess 117
volcanic eruptions, possible effect of on climate 85

Walpi 152
water, possible symbolic significance as shiny material in landscape 155
weather
conditions during field survey 164, 165, 166–7
effect of, on astronomical observations 61, 88, 150
*See also*: climate
wedge tombs 14, 130, 140, 157
consistency in orientation of 90, 101, 240 n. 102
dates of 14
weights and measures 234–5 n. 18
Wessex
forest clearance in 85
henges 138 *See also*: Coneybury; Durrington Walls; Woodhenge
long barrows *126*, 157
monuments astronomically interpreted 144
Western
abstractions 147
astronomical concepts, deconstruction of 147–9
scientific tradition 80, 145, 148
viewpoint, interpretation of other world-views from 251 n. 42
western Scottish project 68–78, *69*, *71*, *72*, 170
summary of the data obtained 206–12
western Scottish rows 75–6, 102, 107, 110, 159, 196–8
consistency in orientation of *75*, 108, 110
difficulties with solar explanation of orientations 109
orientation upon conspicuous hilltops 108
relative horizon distance in the two directions along the alignment 108
*See also*: Ballochroy; Duachy; Kilmartin valley rows; North Mull rows
wind, prevailing 89, 90, 94
Windmill Hill 14
winter solstice
as time of celebration 252 n. 69
as time of crisis 252 n. 69
Wood, John 6
Woodhenge 44, 132
solstitial orientation of 7, 138
spatial patterning of deposition at 255 n. 36
world-view 78, 145, 251 n. 38
and subsistence strategy 145
Andean 147
Hopi 152
reflected in symbolic relationships in material record 156
*See also*: cosmology; Western scientific tradition
world-views
amongst hunter-gatherers 253 n. 134
co-existing 251 n. 45
indigenous American 155
non-Western *See*: non-Western world views
similarities between, in South America 253 n. 134

Yalcobá 153, 251 n. 58
Yallop, Bernard 252 n. 74
Yanomamö 236 n. 50

Zafimaniry 88
zenith passage of sun 148
'zenith tubes' 155

# Figure Acknowledgements

## Photographs

Copyright owners of photographs in this book are as follows, and permission to reproduce them is gratefully acknowledged. *Dúchas* The Heritage Service: 0.1, 1.1d, 8.5a; English Heritage: 1.21; Archaeological Branch, Ordnance Survey Ireland: 5.9; RCAHMS [Crown copyright]: 8.8e; Alex Gibson: 1.18a; Chris Jennings: 0.3a–g, 0.4a,b, 1.5d, 1.14b, 2.5, 2.11d, 2.13b, 5.1, 5.6, 8.7c; Euan MacKie: 1.14a, 1.16c; and Roger Martlew: 7.3. All other photographs are by the author.

## Line drawings

Permission has been obtained, or has been sought, from the following to reproduce figures already published elsewhere by this author and others (as cited in the captions): The British Academy: 1.20, 8.2, 8.9, 9.1; Aubrey Burl: 8.6b; The Prehistoric Society: 7.2; Science History Publications Ltd: 5.3, 5.4, 6.2, 7.1b, 7.8. In addition, permission has been obtained, or has been sought, from the following to make and use redrawings based upon published figures: Patrick Ashmore and Urras nan Tursachan Ltd: 8.8c,d; Antiquity Publications Ltd: 8.10; Martin Brennan: 8.5b; The British Academy: 1.19b; Aubrey Burl: 1.5c, 8.2, 8.7a; Cambridge University Press: 8.1, 8.3a,b, 8.7a; Peter Dunham: 1.8; Elsevier Science Ltd: 8.6a; English Heritage: 1.23b–d, 1.24, 8.9; P. Fane Gladwin: 1.15c,e, 1.16d; Gerald Hawkins; 1.22a; Heinemann Educational Publishers (a division of Educational and Professional Publishing Ltd) and Sir Fred Hoyle: 1.22b; Macmillan Magazines Ltd: 1.3c, 1.12; Claire O'Kelly: 1.1c, 1.3a,c; Oxford University Press: 0.4c, 1.19, 2.9c, 2.10a,b, 2.12a; New Scientist: 1.6; Jon Patrick: 1.3c; Mark Patton: 7.7; Pearson Television: 8.3c; RCAHMS [Crown copyright]: 1.13c, 2.8 [Thom collection], 2.9c; The Royal Society and Euan MacKie: 1.11, 1.13c, 1.14c; Science History Publications Ltd: 1.18b, 2.10c, 2.11c, 2.12b–d, 3.6, 5.5a,b, 6.3, 7.2, 7.9, 7.10, 8.11; Board of Trustees, Southern Illinois University: 7.6; Thames and Hudson Ltd: 1.1c, 1.3a,c, 7.4.